# Geomorphology

Panorama of the Grand Canyon, Arizona, from Point Sublime, looking east. *Drawn by William Henry Holmes.*

# Geomorphology

**RICHARD J. CHORLEY, STANLEY A. SCHUMM, DAVID E. SUGDEN**

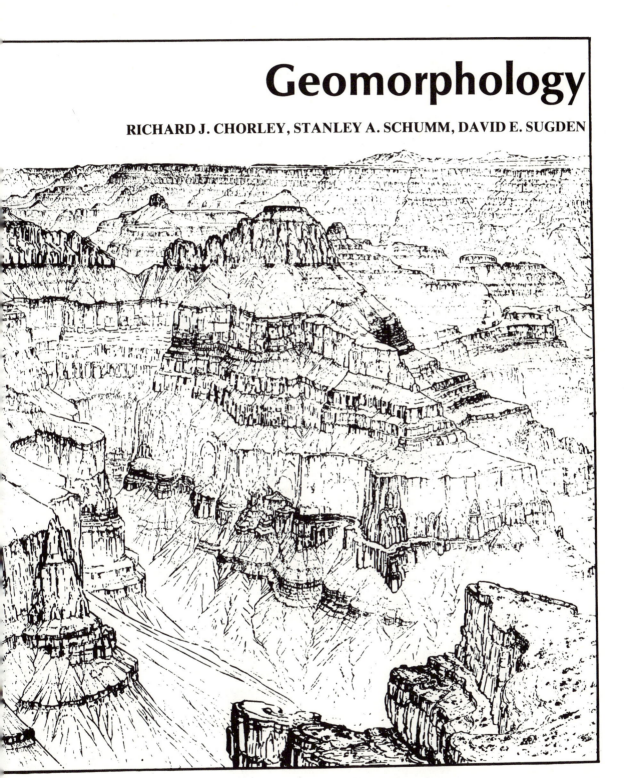

**METHUEN** London and New York

First published in 1984 by
Methuen & Co. Ltd
11 New Fetter Lane, London EC4P 4EE

First published in the USA in 1985 by
Methuen & Co.
in association with Methuen, Inc.
733 Third Avenue, New York, NY 10017

Typeset by Colset Pte Ltd, Singapore
Printed in Great Britain at the
University Press, Cambridge

*British Library Cataloguing in Publication Data*

Chorley, Richard J.
  Geomorphology.
  1. Geomorphology
  I. Title      II. Schumm, Stanley A.
  III. Sugden, David E.
  551.4      GB401.5
  ISBN 0-416-32590-4

*Library of Congress Cataloging in Publication Data*

Chorley, Richard J.
  Geomorphology.
  Bibliography: p.
  Includes index.
  1. Geomorphology.    I. Schumm, Stanley Alfred,
1927–  .   II. Sugden, David E.    III. Title.
GB401.5.C5    1984        551.4    84-14709
  ISBN 0-416-32590-4

# Outline contents

Contents                                                                *page*   vi
Acknowledgements                                                                 xv
Preface                                                                          xxi

**Part One**        *Introduction*                                               1

One            Approaches to geomorphology                                       3
Two            Morphologic evolutionary systems                                  17
Three          Cascading process systems                                        43

**Part Two**        *Geological geomorphology*                                   77

Four           Minerals, rocks and sediments                                    79
Five           Diastrophism                                                      98
Six            Igneous activity and landforms                                    123
Seven          Structure and landforms                                          150
Eight          Lithology and landforms                                          177

**Part Three**      *Geomorphic processes and landforms*                         201

Nine           Weathering                                                        203
Ten            Mass movement                                                     230
Eleven         Hillslopes                                                        255
Twelve         Rivers                                                            278
Thirteen       Drainage basins                                                   316
Fourteen       Fluvial depositional landforms                                   341
Fifteen        Coastal geomorphology                                            371
Sixteen        Aeolian processes and landforms                                  410
Seventeen      The glacier sedimentary system                                   431

**Part Four**       *Climatic geomorphology*                                     465

Eighteen       Morphogenetic landforms                                          467
Nineteen       Geomorphological effects of former glacier expansion             508
Twenty         Climatic change and polygenetic landforms                        534

**Appendix**        *Applied geomorphology*                                      571
**Plates**                                                                       591
**Index**                                                                        593
**Endpiece**        *The geological time scale*                                  607

# Contents

**Acknowledgements** *page* xv

**Preface** xxi

**Part One**  *Introduction* 1

**One**  *Approaches to geomorphology* 3

    1.1   Concepts 3
    1.2   The geomorphic system 5
        1.2.1   Systems structure 5
        1.2.2   Complex response and thresholds 11
    1.3   Geomorphic scale 12
        1.3.1   Timescales 12
        1.3.2   Spatial scales 14

**Two**  *Morphologic evolutionary systems* 17

    2.1   The cycle of erosion 17
    2.2   Interruptions of the cycle of erosion 20
    2.3   Denudation chronology 22
    2.4   Criticisms of the cycle and alternative models 26
        2.4.1   The Penck model 28
        2.4.2   The L. C. King model 30
    2.5   Strategies for inferring landform evolution 31
        2.5.1   Ergodic assumptions 32
        2.5.2   Direct observation and measurement 33
        2.5.3   Simulation modelling 35
    2.6   Equilibrium landforms 37
    2.7   New evolutionary concepts 39

| | | |
|---|---|---|
| **Three** | *Cascading process systems* | 43 |
| | 3.1 The solar energy cascade | 43 |
| | 3.2 The hydrological cycle | 47 |
| | 3.3 Denudation: the sediment cascade | 50 |
| | 3.3.1 Transported loads in rivers | 50 |
| | 3.3.2 Erosion rates over space and time | 53 |
| | 3.3.3 Regional denudation | 60 |
| | 3.4 Diastrophism: the geophysical cascade | 64 |
| | 3.4.1 Orogenic movements | 65 |
| | 3.4.2 Isostatic and epeirogenic movements | 68 |
| | 3.5 Diastrophism and erosion | 70 |
| **Part Two** | *Geological geomorphology* | 77 |
| **Four** | *Minerals, rocks and sediments* | 79 |
| | 4.1 Igneous minerals and rocks | 79 |
| | 4.1.1 Minerals | 79 |
| | 4.1.2 Igneous rocks | 80 |
| | 4.2 Sediments | 83 |
| | 4.2.1 Provenance | 83 |
| | 4.2.2 Particle size and sorting | 84 |
| | 4.2.3 Particle shape | 88 |
| | 4.2.4 Sedimentary fabric | 89 |
| | 4.3 Sedimentary rocks | 89 |
| | 4.3.1 Bedding | 90 |
| | 4.3.2 Diagenesis | 91 |
| | 4.3.3 Classification | 91 |
| | 4.3.3.1 Clastic rocks | 91 |
| | 4.3.3.2 Chemically precipitated rocks | 93 |
| | 4.3.4 Facies | 94 |
| | 4.4 Metamorphic rocks | 94 |

**Five**          *Diastrophism*                                                           98

          5.1    Earth structure                                                   98
                 5.1.1    Global topography                                        98
                 5.1.2    Geophysical evidence                                    102
                 5.1.3    Interpretations of the evidence                         108
          5.2    Global tectonics                                                 112
                 5.2.1    Continental drift and plate tectonics                   112
                 5.2.2    Zones of spreading                                      115
                 5.2.3    Subduction zones and orogeny                            118

**Six**           *Igneous activity and landforms*                                        123

          6.1    Igneous activity in space and time                               123
          6.2    Intrusive constructional forms                                   129
                 6.2.1    Plutons                                                 129
                 6.2.2    Smaller intrusions                                      130
          6.3    Extrusive constructional forms                                   133
                 6.3.1    Types of eruption                                       137
                 6.3.2    Basaltic magmas                                         139
                 6.3.3    Acidic magmas                                           140
          6.4    Igneous tectonism                                                143

**Seven**         *Structure and landforms*                                               150

          7.1    Horizontal and domed structures                                 150
          7.2    Homoclinal structures                                           152
          7.3    Folded structures                                               160
                 7.3.1    Simple folding                                          160
                 7.3.2    Complex folding                                         165
          7.4    Faulted structures                                              165
                 7.4.1    Faulting                                                165
                 7.4.2    Faulted landforms                                       172

**Eight**          *Lithology and landforms*                          177

    8.1   Arenaceous landforms                              177
    8.2   Argillaceous landforms                           180
    8.3   Calcareous landforms                             181
    8.4   Igneous destructional landforms                  190
    8.5   Metamorphic landforms                            194
    8.6   Rock strength                                    196


**Part Three**    *Geomorphic processes and landforms*                 201

**Nine**          *Weathering*                                         203

    9.1   The earth–atmosphere interface                   203
    9.2   Processes of weathering                          205
          9.2.1   Physical weathering                205
          9.2.2   Chemical weathering                207
          9.2.3   Biochemical weathering             219
    9.3   Rates of weathering                              220
    9.4   The weathered mantle                             224


**Ten**           *Mass movement*                                      230

    10.1   Significance                                    230
    10.2   Gravity tectonics                               232
    10.3   Classification                                  234
    10.4   Location of mass movement                       244
    10.5   Causes of mass movement                         247
    10.6   Mass movement and landform evolution            252


**Eleven**        *Hillslopes*                                         255

    11.1   Introduction                                    255
    11.2   Characteristic slopes                           255

| | | |
|---|---|---|
| 11.3 | Classification of hillslopes | 256 |
| 11.4 | Origin of hillslopes | 257 |
| 11.5 | Hillslope erosion | 258 |
| | 11.5.1   Creep | 258 |
| | 11.5.2   Overland flow | 260 |
| | 11.5.3   Rainsplash | 260 |
| | 11.5.4   Slope erosion | 261 |
| | 11.5.5   Rills | 264 |
| | 11.5.6   Throughflow | 265 |
| 11.6 | The evolution of hillslopes | 266 |
| 11.7 | Summary | 275 |

**Twelve**      *Rivers*                                             278

| | | |
|---|---|---|
| 12.1 | Significance | 278 |
| 12.2 | Open-channel hydraulics | 279 |
| 12.3 | Sediment transport | 283 |
| 12.4 | Hydrology | 289 |
| 12.5 | River morphology | 290 |
| | 12.5.1   Hydraulic geometry | 290 |
| | 12.5.2   Channel patterns | 295 |
| 12.6 | Channel stability | 302 |
| | 12.6.1   Stable channels | 303 |
| | 12.6.2   Unstable channels | 305 |
| 12.7 | Examples of river metamorphosis | 306 |
| | 12.7.1   Historical river metamorphosis | 306 |
| | 12.7.2   Geological river metamorphosis | 308 |
| 12.8 | Rivers and valley morphology | 309 |

**Thirteen**    *Drainage basins*                                     316

| | | |
|---|---|---|
| 13.1 | The basin geomorphic unit | 316 |
| 13.2 | Morphometric analysis | 317 |
| 13.3 | Morphometric controls | 324 |

| | | |
|---|---|---|
| 13.4 | Drainage basin evolution | 328 |
| | 13.4.1 The ergodic method | 328 |
| | 13.4.2 Physical simulation | 330 |
| 13.5 | Drainage basin response | 333 |

**Fourteen**     *Fluvial depositional landforms*     341

| | | |
|---|---|---|
| 14.1 | Alluvial fans | 341 |
| | 14.1.1 Fan structure | 342 |
| | 14.1.2 Dry fans | 344 |
| | 14.1.3 Wet fans | 347 |
| | 14.1.4 Depositional belts | 347 |
| 14.2 | Valley fills | 349 |
| | 14.2.1 Floodplains | 351 |
| | 14.2.2 River terraces | 355 |
| 14.3 | Deltas | 359 |
| | 14.3.1 Delta morphology | 361 |
| | 14.3.2 Experimental study of delta morphology | 364 |
| | 14.3.3 Avulsion | 366 |

**Fifteen**     *Coastal geomorphology*     371

| | | |
|---|---|---|
| 15.1 | Sea level, waves and currents | 371 |
| 15.2 | Beach processes and profiles | 376 |
| 15.3 | Shoreline processes and depositional forms | 379 |
| | 15.3.1 Longshore movement | 379 |
| | 15.3.2 Shoreline configuration | 384 |
| | 15.3.3 Barrier islands | 387 |
| 15.4 | Erosional coasts | 388 |
| | 15.4.1 Processes of coastal bedrock erosion | 388 |
| | 15.4.2 Cliffs | 390 |
| | 15.4.3 Shore platforms | 395 |
| 15.5 | Sea-level variations | 398 |
| | 15.5.1 Eustatic changes | 398 |

|  |  |  |
|---|---|---:|
| | 15.5.2 Submergence features | 399 |
| | 15.5.3 Emergence features | 401 |
| | 15.5.4 Coastal classification | 402 |
| 15.6 | Organic coasts | 404 |
| | 15.6.1 Coral reefs | 404 |
| | 15.6.2 Salt marshes and mangrove coasts | 406 |
| 15.7 | Coastal management | 407 |

**Sixteen**   *Aeolian processes and landforms*   410

| 16.1 | Aeolian environments | 410 |
| 16.2 | Aeolian sand movement | 410 |
| 16.3 | Wind abrasion | 413 |
| 16.4 | Aeolian bedforms | 415 |
| 16.5 | Coastal sand dunes | 425 |
| 16.6 | Loess | 427 |
| 16.7 | Snow drifting | 428 |

**Seventeen**   *The glacier sedimentary system*   431

| 17.1 | Glaciers | 431 |
| 17.2 | Glacier ice | 434 |
| 17.3 | Glacier flow | 437 |
| | 17.3.1 Internal deformation | 437 |
| | 17.3.2 Sliding and bed deformation | 438 |
| | 17.3.3 Velocities | 440 |
| | 17.3.4 Ice surface features | 441 |
| 17.4 | Rock debris in glaciers | 442 |
| 17.5 | Processes affecting debris at the glacier sole | 442 |
| | 17.5.1 Entrainment | 443 |
| | 17.5.2 Rock particles in traction at the bed | 444 |
| 17.6 | Erosion by glaciers | 445 |
| 17.7 | Deposition by glaciers | 448 |
| 17.8 | Landforms of glacial deposition | 452 |
| 17.9 | The glacier meltwater subsystem | 457 |

**Part Four**      *Climatic geomorphology*                                      465

**Eighteen**       *Morphogenetic landforms*                                     467

           18.1   Morphogenetic regions                                 468
           18.2   Humid tropical landforms                              472
           18.3   Tropical wet–dry landforms                            477
           18.4   Arid and semi-arid landforms                          485
                  18.4.1   Pediments                                    487
                  18.4.2   Inselbergs                                   491
                  18.4.3   Desert surfaces                              493
           18.5   Cold region landforms                                 494
                  18.5.1   The periglacial system                      495
                  18.5.2   The glacial system                          502

**Nineteen**       *Geomorphological effects of former glacier expansion*        508

           19.1   Introduction                                         508
           19.2   Direct erosional effects                             511
                  19.2.1   Alpine glaciation                           511
                  19.2.2   Ice sheet glaciation                        513
           19.3   Direct depositional effects                          518
                  19.3.1   Glacial deposition                          518
                  19.3.2   Models of ice retreat                       522
           19.4   Indirect effects: glacioisostasy and glacioeustasy   529

**Twenty**         *Climatic change and polygenetic landforms*                   534

           20.1   Climatic change                                      534
                  20.1.1   Tertiary climatic changes                   536
                  20.1.2   Pleistocene climatic changes                536
                  20.1.3   Holocene climatic changes                   542
                  20.1.4   Historical climatic changes                 542
           20.2   The geomorphic effects of climatic change            547

|  |  |  |
|---|---|---|
| 20.2.1 | Precipitation changes dominant | 549 |
| 20.2.2 | Temperature changes dominant | 560 |
| 20.3 | Conclusion | 567 |
| **Appendix** | *Applied geomorphology* | 571 |
| A1.1 | Nature of applied geomorphology | 573 |
| A1.2 | Objectives of applied geomorphology | 573 |
| A1.3 | Geomorphic hazards | 574 |
| A1.4 | Fluvial hazard evaluation | 578 |
| A1.5 | Case studies | 582 |
| **Plates** |  | 591 |
| **Index** |  | 593 |
| **Endpiece** | *The geological time scale* | 607 |

# Acknowledgements

The authors would like to thank the following editors, publishers, organizations and individuals for permission to reproduce figures, tables and other material. All primary sources shown in the figure captions are copyright © except for US government publications.

## Editors

*Advances in Agronomy* for Table 18.4.

*American Journal of Science* for Figures 1.9, 2.15, 3.10A, 3.13B, 3.21, 3.28, 3.29, 4.6, 10.1, 10.8, 11.1, 11.6, 11.20, 11.24, 11.27A, 11.31, 13.19, and for Tables 1.1, 1.2, 3.12, and some data.

*American Scientist* for Figure 18.24.

*Annales de Géographie* for Table 3.13.

*Annales Société Géologique de Belgique* for Figure 20.33.

*Annals of the Association of American Geographers* for Figures 18.2, 20.24 and 20.25.

*Annals of Glaciology* for Figures 17.2, 17.17, 17.19, 19.15B, 19.18 and 19.19.

*Annual Review of Ecology and Systematics* for Table 9.2.

*Arctic and Alpine Research* for Figures 17.5, 18.32 and 18.38.

*Area* for Figure 9.13.

*The Australian Geographer* for Figure 13.21.

*Australian Geographical Studies* for Figures 20.15 A–C, 20.17, and Table 18.5.

*Biological Reviews* for Figure 15.38.

*Boreas* for Figure 17.14.

*Bulletin of the American Association of Petroleum Geologists* for Figures 7.14A and 13.3.

*Bulletin Géodesique* for Figure 3.27A (Crown Copyright Reserved).

*Bulletin of the Geological Society of America* for Figures 2.12, 3.8, 3.24, 8.18, 10.20, 11.26, 11.33, 12.11, 12.26, 12.28, 12.30, 12.38, 13.8, 13.9, 13.22, 13.30, 14.34, 15.22, 15.31, 16.4, 16.6, 16.13, 18.7, 18.18, 20.19, 20.26, and Tables 13.2 and 18.3.

*Bulletin of the International Association of Scientific Hydrology* for Table 9.11, and some data.

*Bulletin of the Seismological Society of America* for Figure 5.4.

*Canadian Geographer* for Table 3.10.

*Canadian Journal of Earth Science* for Figures 14.6 and 19.24B.

*Deep Sea Research* for Figure 5.7.

*Earth Science Reviews* for Figure 12.9B (© Elsevier, Amsterdam).

*Earth Surface Processes* for Figures 3.17, 12.7, 13.31, and Table 3.9 (© John Wiley & Sons).

*The Ecologist* for Figure 20.12B.

*Engineering Geology* for Figure 10.16 (© Elsevier, Amsterdam).

*Estratto da Geologia Applicata e Idrogeologia* for Figure 10.24.

*Fennia* for Figure 17.25.

*Geo Books Ltd* for Figures 11.27B, 11.29, and some formulae.

*Geografiska Annaler* for Figures 10.6, 10.25, 12.4, 12.14, 17.18, 17.31, 18.13, 18.14, 18.15B, 18.16, 19.6 and 19.7.

*Geographical Journal* for Figures 9.18, 15.19 and 18.23.

*Geographical Magazine* for Figures 15.37 and 19.14.

*Geographical Review* for Figures 13.18 and 16.20.

*Geological Magazine* for Figures 7.10, 11.25 and 18.20.

*Geologie en Mijnbouw* for Figure 14.20.

*Geoscience and Man* for Figures 14.33 and 20.3.

*Journal of Geological Education* for Figures 5.8 and 14.3.

*Journal of the Geological Society of India* for Figure 1.8.

*Journal of the Geological Society of London* for Figures 6.1 and 17.27.

*Journal of Geology* for Figures 3.16, 5.18, 8.1, 8.3, 9.4, 13.2A, 13.4, 13.23, 17.21, 17.22 and Tables 8.6 and 17.5.

*Journal of Geophysical Research* for Figures 5.12, 5.15, 5.17, 5.19 and 5.20.

*Journal of Glaciology* for Figures 17.9, 17.12, 17.29, 18.37 and 19.5.

*Journal of Hydrology* for Figures 12.25, 18.8, 18.12, and Table 3.14 (© Elsevier, Amsterdam).

*Journal of Sedimentary Petrology* for Figures 16.1, 16.9, and Table 16.2.

*Journal of Soil Science* for Figure 9.6.

*Marine Geology* for Figure 15.30 (© Elsevier, Amsterdam).

*Mitteilungen der Versuchsaustalt für Wasserbau, Hydrologie und Glaziologie* for Figure 17.10.

*Nature, Physical Science* for Figures 9.8 and 12.29.

*Norges Geologiske Undersøkelse* for Figures 19.10 and 19.21.

*Norsk Geografisk Tidskrift* for Figure 19.13.

*Photogrammetric Engineering* for Figures 13.12 and 13.13.

*Photogrammetric Journal of Finland* for Figure 19.16.

*Physical Geography* for Figure 12.33.

*Physics and Chemistry of the Earth* for Figure 6.26.

*Polarforschung* for Table 17.1.

*Proceedings of the Geologists' Association* for Figure 3.27B.

*Proceedings of the Indiana Academy of Science* for Figure 2.9.

*Progress in Geography* for Figure 18.3.

*Progress in Physical Geography* for Figures 14.1, 15.27, 16.15D, 20.4C and 20.6B (© Edward Arnold).

*Quarterly Journal of Engineering Geology* for Table 10.6.

*Quaternaria* for Figure 15.32.

*Quaternary Research* for Figures 18.32, 19.24A, 19.25, 19.26, 19.27 and 20.9.

*Questiones Geographicae* for Figure 20.29.

*Revue de Géographie physique et de Géologie dynamique* for Figure 18.2.

*Revue de Géomorphologie dynamique* for Figure 18.9.

*Royal Astronomical Society, Geophysical Journal* for Figure 6.2.

*Science* for Figures 11.7, 18.4, 20.7, 20.11, and Tables 18.2 and 20.2.

*Scientific American* for Figures 5.6, 5.14 and 6.4 (© Scientific American).

*Scientific Monthly* for Figure 7.27B.

*Sedimentology* for Figures 12.13 and 14.25 (© Blackwell Scientific Publications Ltd).

*Southeastern Geologist* for Figure 18.2.

*Soviet Hydrology* for Figures 14.16 and A1.1.

*Tellus* for Figure 15.33.

*Transactions of the American Geophysical Union* for Figures 11.12, 13.1 and 15.11.

*Transactions of the American Society of Civil Engineers* for Figure 10.7.

*Transactions of the Institute of British Geographers* for Figures 2.6, 2.7, 2.11, 7.7, 11.21, 11.28, 15.18, 17.15, and Tables 3.8, 18.1, and one quotation.

*Transactions of the International Congress of Soil Science* for Figure 11.4.

*Water Resources Research* for Figures 13.20, 14.13, and Table 3.11.

*Zeitschrift für Geomorphologie* for Figures 3.2C, 3.2D, 8.20, 9.11, 9.15, 11.11, 11.23, 11.30, 15.28, 18.6A, 18.15A, 18.21, 18.22, 18.23, 20.35, and Tables 8.1 and 8.7.

## Publishers

Academic Press, London, for Figures 14.9, 14.10, 20.10, and chemical equations.

Adam Hilger, Bristol, for Figures 8.4, 8.9B, and Table 8.5.

Australian National University Press, Canberra, for Figures 6.7, 8.13 and 8.15.

Balkema, Rotterdam, for Figure 19.12.

Batsford, London, for Figures 16.1, 16.7E, 16.16, and Table 16.3.

Blackwell, Oxford, for Figures 12.13, 14.25, 18.26, 18.27, 18.28, 18.29, 18.31, 19.1, 19.2 and 19.9.

Cambridge University Press for Figures 2.13, 10.3, 10.4, 11.9, 11.12, 11.14, 11.22, 14.22, endplate, Table 9.8, and some formulae and data.

Chapman & Hall, London, for Figures 6.6, 7.6, 7.14B and 7.21.

Collins, Glasgow, for Figure 7.8.

Columbia University Press, New York, for Figures 2.4, 2.5, 3.5, 3.7 and 10.25.

Constable, London, for Figure 20.12A.

Dowden, Hutchison & Ross, Inc., Stroudsburg, Pa., for Figures 15.22B and 15.22C.

Edward Arnold, London, for Figures 3.9, 7.15, 7.19A, 15.9, 15.10, 15.11B, 16.2B, 16.5, 16.10, 16.11, 16.14, 17.1, 17.15, 17.24, 17.26, 17.30, 18.30, 18.32, 18.36, 19.20, 20.32, 20.35, and Tables 3.4, 3.7, 3.10B, 3.11, 9.9, 9.10, and some data.

Elsevier, Amsterdam, for Figures 3.24, 15.1, 15.13, 15.14, 15.16 and 20.2B.

Elsevier, New York, for Figures 9.2, 9.4A, 9.7, 9.10, 14.12, 14.15, 14.24, 14.26, and Table 9.5.

Ferdinand Enke Verlag, Stuttgart, for Figures 6.6 and 6.21.

Freeman, San Francisco, for Figures 3.11, 3.15, 4.5, 4.7,

4.8, 4.10B, 7.12A and B, 20.13B, and Tables 4.2 and 9.6.

George Allen & Unwin, London, for Figure 20.18.

Harper & Row, New York, for Figures 4.3, 4.4, 4.7, 4.8, 4.12, 4.17, 5.18A, 6.25, 7.13, 15.5B, 15.34 and 15.36.

Houghton Mifflin, Boston, for Figures 11.3, 13.9 and 16.22.

International Book Production, Stockholm, for Figure 20.2.

Isdatl'stvo Nauka, Moscow, for Table 9.1.

Longman, London, for Figures 2.2B, 2.10, 4.2, 6.7, 6.9, 8.5A, 8.17, 11.34, 12.39A and B, 14.18, 15.24, 18.3, 20.31, 20.34 and 20.37.

McGraw-Hill, New York, for Figures 6.3, 6.5, 6.16, 7.11A, 7.12, 7.20A, 7.22, 7.23, 7.24, 8.7B, 8.9A, 9.14, 10.17, 13.6, 13.27, and Table 9.6.

Macmillan, London and Basingstoke, for Figures 2.9, 8.7A, 8.11, 18.10, 18.11, 18.20, 18.21, 18.22, 18.23, 18.25, 20.4D, and Tables 8.3 and 18.4.

The Macmillan Co., New York, for Figures 7.9, 7.16, and Table 10.5.

Masson, Paris, for Figure 18.11.

Methuen, London, for Figures 2.8, 3.2B, 3.3B, 3.5, 3.20, 4.1, 8.8A and B, 8.9A, 9.12, 12.3, 12.17, 12.24, 15.23, 16.2A, 16.3, 18.3, 20.8, 20.13 and 20.30.

MIT Press, Cambridge, Mass., for Figures 6.7, 8.5B, 8.6, 8.8C, 8.13, 8.15 and 18.17.

Nelson, London, for Figures 6.8 and 7.19B.

Oliver & Boyd, Edinburgh, for Figure 9.16.

Oxford University Press for Figures 10.5, 14.19, 15.29, 20.2A, 20.4A and B, 20.5, 20.36, and Table 20.1.

Pergamon Press, New York, for Figure 9.9.

Pergamon Press, Oxford, for Figures 5.10, 5.16, 5.17, 5.18A, 17.11 and 19.22.

Prentice-Hall, New Jersey, for Figures 6.15, 6.17, 6.19, 6.20, 6.23, 15.1, 15.3, 15.4, 15.12, 15.20, 15.22A, 15.33, and Tables 6.2, 15.2, and some formulae.

Presses Universitaires de France, Paris, for Figures 3.18 and 3.20.

Princeton University Press for Figures 3.21, 15.33, 20.22 and 20.23.

Reinhold, New York, for Figure 15.32, and Table 15.1.

Ferdinand Schöningh, Paderborn, for Figures 18.33, 18.34 and 18.35.

Shaanxi Peoples' Art Publishing House for Figure 16.21.

Springer-Verlag, New York, for Figures 4.3, 4.11, 4.13, 4.14, 4.15, 15.8 and 15.14.

Texas University Press, Austin, for Figure 6.24.

Valgus, Tallinin, for Figures 3.22, 3.25 and 3.26.

Van Nostrand Reinhold (UK) Co. Ltd for Figures 6.8, 6.15, 7.19B and 15.32.

Victor Gollancz, London, for Figures 15.25A and B.

Water Resources Publications, Littleton, Colorado, for Figure 12.35.

Whitcombe & Tombs, New Zealand, for Figures 7.2, 7.27A, 8.10, 8.12, 8.14 and 8.16.

Wiley, Chichester, for Figures 2.14B, 3.4, 9.3, 10.19, 10.22B, 11.8, 11.19, 15.26, 18.5, 20.27, 20.38, and some formulae.

Wiley, New York, for Figures 1.2, 2.1, 2.2A, 2.14A, 2.19, 5.1, 5.2, 5.5A, 5.9, 5.11, 6.6, 6.14, 6.21, 7.11B, 7.18, 7.20B, 7.26, 10.2, 12.7, 12.9B, 12.39C, 12.40, 13.5, 13.17, 13.31, 14.8, 14.17, 14.18, 14.27, 14.28, 14.29, 14.30, 15.15, 15.17, 16.1, 19.4, A1.8, and Plates 2, 6 and 7.

## Organizations

Aberdeen University, Department of Geography, for Figure 16.23.

American Society of Civil Engineers for Figures 10.7, 10.9, 10.10, 12.6, 12.10, 14.8, 15.6, A1.7, and Table 14.1.

Arizona Department of Transportation for Figures A1.5 and A1.6.

California Division of Mines for Figures 2.3, 7.25, and Table 9.3.

Cambridge University Collection for Plates 3, 19, 23 and 31.

Carnegie Institution of Washington for Figure 6.23.

Chuo University, Faculty of Engineering, for Figure 12.15.

Colorado State University, Department of Earth Resources, for Figures 11.16, 11.17, 11.18, 12.27, 13.15, 13.24, 13.25, 13.26, 13.28, 13.29, 14.23, 18.19, and Table 9.7.

Conference on Coastal Engineering for Figure 15.7.

Copyright Agency of the USSR for Figure 3.3A.

CSIRO for Plate 25.

Federal Highway Administration, Washington DC, for Figures 12.31, A1.2, A1.3, and Table A1.3.

Geodetic Institute, Copenhagen, for Plates 22 and 29.

Geological Society of America for Figures 1.7, 6.16, 6.22, 7.5, 10.14, 15.22A, 16.22, 19.15A, and Tables 10.1 and 10.2.

Geological Society of Canada for Figure 19.17.

Geological Survey for Figure 6.13 (Crown Copyright Reserved).

Geological Survey of Canada for Figure 19.3.

Geologists' Association, London, for Figures 3.27B and 15.25C.

Houston Geological Society for Figure 4.31.

Illinois State Geological Survey, Urbana, for Figure 14.11.

Illinois University, Water Resources Center, for Figures 12.37 and 13.10.

Instytut Geologiczny, Warsaw, for Figure 20.34.

International Association of Scientific Hydrology for Figure 17.28.

International Atomic Energy Symposium for Table 9.8.

International Clay Conference for Figure 9.5.

International Society of Soil Science for Figure 11.4.

International Union of Geodesy and Geophysics for Figures 3.22, 3.24, 3.25 and 3.26.

Mississippi River Commission, Vicksburg, for Figure 12.34.

Museum of New Mexico, Santa Fé, for Plate 10A.

NASA for Plate 9.

National Academy of Sciences, Washington DC, for Figure 10.5, and Table 10.4.

National Oceanographic and Atmospheric Administration, Miami, for Figure 5.13, and Table 5.4.

Netherlands Government Printing Office for Figure 6.17.

New South Wales Department of Lands for Plate 32.

Norsk Polarinstitut, Copenhagen, for Plate 28.

Nottingham University, Department of Geography, for Figure 20.15D.

St Andrews University, Department of Geography, for Figure 19.11.

Society of Economic Paleontologists and Mineralogists for Figures 1.3 and 12.32.

State University of New York at Binghamton for Figures 2.18 and 11.32.

Texas Gulf Coast Association of Geological Societies for Figure 14.32.

Tokyo University, Department of Geography, for Figure 3.23.

US Army for Figures 12.23, 14.4, 14.31 and 15.16.

US Corps of Engineers for Figure 2.14A.

US Department of Agriculture for Figures 11.9, 11.13, and 12.12.

US Forest Experiment Station, Berkeley, for Figure 10.11.

US Geological Survey for Figures 3.12, 3.13A, 3.30, 5.3, 6.10, 6.11, 6.18, 7.1, 7.3, 7.4, 7.5, 8.2, 8.19, 10.12, 10.13, 11.10, 12.1, 12.9A, 12.16, 12.18, 12.19, 12.20, 12.21, 12.22, 12.36, 13.11, 13.14, 13.16, 14.2, 14.5, 14.7, 14.17, 14.21, 16.7, 16.8, 16.12, 16.15, 16.17, 16.18, 16.19, 20.6A, 20.20, 20.21, 20.28, A 1.7, and Tables 14.2, 16.1, 18.6, Frontispiece, and Plates 1, 11, 15, 16, 20 and 21.

US National Park Service for Plate 10B.

US Navy for Figures 8.1, 13.1, 13.7 and 18.1.

University of Guelph, Department of Geography, for Figures 11.27B and 11.29.

University of Texas, Bureau of Economic Geology, for Figures 14.15 and 14.26.

University of Wisconsin-Madison, Institute of Environmental Studies, for Figure 20.13A.

Utah Geological and Mineralogical Survey for Figure 10.18.

Virginia Geological Survey for Figure 7.14C.

Virginia University, Department of Environmental Sciences, for Figure 15.35.

**Individuals**

Dr G. A. Brook for Plate 27.

T. E. Carter, Colorado State University, for Figure 18.19.

Dr C. Clapperton for Plate 5.

Dr H. Cloos for Figures 6.14, 7.20B and 7.26.

Dr D. O. Doehring for Figure 11.32.

Dr Hai-Yang Chang for Figures 14.27, 14.28, 14.29 and 14.30.

Dr A. M. Hall, St Andrews University, for Figure 19.11.

Dr B. A. Kennedy, Oxford University, for Figures 1.9, 2.16, 3.14A and B, 4.9, 13.32 and 20.14.

Dr H. R. Khan, Colorado State University, for Figure 12.27.

P. D. Komar, Oregon, for Figures 2.17, 15.1, 15.4, 15.20, 15.22A, 15.33, and Table 15.2.

Dr L. B. Leopold for Plate 17.

C. F. McLane, Colorado State University, for Figure 13.29.

Dr C. M. Madduma Bandara, Sri Lanka, for Figure 18.6B.

Dr H. W. Menard for Figure 6.3.

Dr M. P. Mosley, Colorado State University, for Figures 11.17 and 11.18.

Dr P. A. Mour for Figure 7.23.

Dr R. S. Parker, Colorado State University, for Figures 13.24, 13.25, 13.26 and 13.28.

Dr J. S. Shelton for Plates 2, 6 and 7.

Dr D. R. Stoddart, Cambridge University, for Figure 5.13.

Dr A. N. Strahler for Figures 2.14A, 3.1, 3.6, 10.25, 13.6 and 15.15.

J. Sugden for Plate 18.

M. M. Sweeting for Figures 8.7 and 8.11.

Dr L. R. Sykes for Figure 5.5B.

Dr G. Thom for Figure 18.31.

Dr R. G. Ward, Aberdeen University, for Figure 16.23.

N. A. Wildman, Colorado State University, for Figure 14.23.

Dr G. L. Zimpfer, Colorado State University, for Figure 13.15.

Dr T. Zingg for Figure 4.10.

Thanks are also due to the following scholars for further permission to use illustrative material:

F. Ahnert, T. C. Atkinson, V. R. Baker, W. C. Bradley, D. Brunsden, F. M. Bullard, N. Caine, M. A. Carson, J. A. Catt, M. W. Clark, E. T. Cleaves, D. R. Coates, J. D. Collinson, K. C. Condie, R. U. Cooke, D. R. Crandell, L. U. De Sitter, E. G. Ehlers, T. Elliott, P. Elter, R. A. Freeze, H. M. French, H. C. Fritts, J. C. Frye, R. J. Gibbs, A. Goudie, K. J. Gregory, J. Hooke, W. T. Horsfield, J. H. Hoyt, D. W. Hyndman, R. G. Jackson II, L. Jakucs, R. J. Janda, C. A. M. King, P. D. Komar, L. B. Leopold, J. Lewin, F. C. Loughnan, J. A. Mabbutt, J. F. McCauley, G. A. Macdonald, G. T. Moore, T. R. Oke, C. D. Ollier, F. J. Pettijohn, K. S. Richards, R. V. Rume, M. J. Selby, R. C. Selley, R. P. Sharp, R. L. Shreve, B. W. Sparks, L. Starkel, A. N. Strahler, M. F. Thomas, J. B. Thornes, T. J. Toy, J. M. Trefethen, J. Tricart, S. T. Trudgill, R. E. Wallace, A. Warren, A. L. Washburn, R. G. West, E. H. T. Whitten, P. J. Wyllie, A. Young.

Thanks are also due in respect of:

*editorial work* – Dr R. J. M. More
*drawing* – I. Gulley, A. Shelley and M. Young
*photography* – D. A. Blackburn and R. Coe
*helpful general discussions* – Professor D. Brunsden.

The preparation of this extensive work has naturally involved the use of a great deal of source material, both primary and secondary. This book has been designed as a basic textbook and, as such, it has not been thought necessary to overload the text with source references. All important references are clearly associated with their derived material by the complete chapter referencing system, detailed references being given for each figure.

The authors and publisher have made every effort to obtain permission to reproduce copyright material throughout this book. If any proper acknowledgement has not been made, or permission not received, we would invite any copyright holder to inform us of this oversight.

# Preface

In writing *Geomorphology*, we have been concerned to meet a number of very important challenges in this rapidly evolving discipline. We have been at pains to do justice to the physical bases of modern 'process geomorphology' in the fluvial, marine, aeolian and glacial fields; as well as to integrate work on glacial and periglacial geomorphology with that on climatic change. At the same time a balance has been achieved between the equilibrium models relating landforms to inferred processes and the longer-term historical evolution which has involved most landforms throughout many different process regimes. Developing from the foregoing, a unified scientific philosophy of the discipline has been outlined which encompasses systems notions such as equilibrium, feedback, process–response and thresholds, which are so obviously adapted to the functional study of rapidly changing process–response systems, as well as historical and evolutionary ideas involving progressive change through long time periods. Here the important linking notions of complex response, landscape sensitivity and response time have been highlighted to suggest a coherent theoretical approach to geomorphology. We have further addressed ourselves to the fundamental question as to the *subject-matter* of geomorphology, and particularly to whether it should deal with *landforms* as the central objects of study or whether emphasis should be placed on the study of observable *processes*, which seems to be so much in harmony with current preoccupations with scientific observation and measurement, environmental management and applied geomorphology. From the above considerations has emerged a balanced treatment of the philosophy of geomorphology, the influences of geology, geomorphic processes, the detailed geometry of landforms, the effects of climate and practical applications of geomorphology.

The historical v. the functional controversy, which has recently swung so much in favour of the latter, forms a major balanced theme in this book. It is recognized to have generated a fundamental scale problem both in time and space which is treated in some depth. Classical geomorphologists were in the habit of viewing landform changes in terms of 'cycles' tens of millions of years in length, or at least of parts of cycles, whereas process geomorphologists often base their work on observations lasting a decade or less. This temporal difference has led to a spatial shift from the study of middle-scale landform assemblages hundreds or thousands of square kilometres in extent to geomorphic units hundreds or thousands of square metres in extent. We have met the scale challenge of modern geomorphology by treating and integrating landforms on a wide variety of scales from the forms and movement of stream bed ripples to those of continents. It has also been necessary to recognize and counter the tendencies of national parochialism in geomorphology, such as a British preoccupation with chalk landforms and a North American scant regard for the tropical wet–dry and equatorial environments. Geomorphic examples, therefore, have been widely chosen and global comparisons and generalizations made.

Enough basic geology has been included in Chapters 4–8 to serve the requirements of students approaching geomorphology from other fields, but not so much as to prove unduly repetitive for those specializing in geology and the earth sciences. Indeed, the students in many geomorphology courses may well include hydrologists, civil and agricultural engineers, glaciologists and environmental decision-makers, as well as geographers and geologists.

It is clearly no longer possible to consider adequately all the significant current advances in geomorphology and its related fields in a single volume, and the extensive modern geomorphic literature has been increasingly characterized by the publication of books dealing exhaustively with specialized aspects of the discipline. It has been necessary, therefore, for us to be selective in our presentation, recognizing that some outstanding research has only received cursory consideration. However, we hope that this book will lead the student on to ask further questions regarding landforms. The book is subdivided into four parts and an integral appendix.

In Part I the background material needed to appreciate that discussed in subsequent chapters is presented. Three chapters consider the approaches to geomorphic problems, together with concepts that are basic to geomorphology and, indeed, to much of earth science (Chapter 1), as well as the two interlocking aspects of the total geomorphic system, morphologic and cascading systems (Chapters 2 and 3). The fact that there is an historical development of landforms requires that, for a true understanding of the modern landscape, this evolution must be clearly recognized. This is necessary for the prediction of future landform change, both natural and as a result of man's activities. In addition, an appreciation of the flow of materials and energy within a landscape is basic to an understanding of the erosional and depositional processes acting upon landforms and the prediction of future changes. Geomorphology in its fullest development, therefore, is dependent upon both historical and functional approaches to landscape study.

Part II is devoted to the geological basis of geomorphology. Primary landforms are those formed by endogenic processes, tectonics and volcanism. The exposed rocks (Chapters 4 and 8) and structures (Chapter 7) are immediately acted upon by exogenic weathering and erosional processes to form secondary landforms, which reflect geologic controls at a wide range of spatial scales. Rates of diastrophism, already considered in

Chapter 3, are supplemented by consideration of global structures and endogenic processes (Chapter 5) in order to understand the first- and second-order landforms of the continents and ocean basins. Volcanoes are, of course, smaller primary features that are formed by endogenic constructional processes. However, vast areas of the earth, moon and the planets are volcanic in origin, and both intrusive and extrusive igneous bodies are discussed in Chapter 6. Secondary landforms are modified by erosional or depositional processes. A major factor determining the character of secondary landforms is the geologic character of a landscape both structural and lithologic (Chapters 7 and 8). Indeed, an important practical application of geomorphology has been the interpretation of structure and lithology based on landform characteristics depicted on aerial photographs.

Acting upon the earth's surface are a variety of erosional and depositional processes, which both modify existing landforms and produce new landforms. In Part III the major process–response systems are identified and described, the resulting landforms and their evolution with respect to these processes are considered, and the dual theme of the historical and functional approaches to landform study continues to be explored. The major systems considered are weathering (Chapter 9), mass movement (Chapter 10), hillslopes (Chapter 11), fluvial systems (in respect of rivers and valleys) (Chapter 12), drainage basins (Chapter 13), fluvial depositional features (Chapter 14), coastal systems (Chapter 15), aeolian systems (Chapter 16) and those associated with the production of glacial landforms (Chapter 17).

It is clear that the present landforms of large areas of the world have evolved under previously more extensive tropical or polar climates, and that the mid-latitudes especially show evidence of the operation of such processes in earlier times. It is, therefore, the purpose of Part

IV to examine the extent to which different climatic regimes are potentially capable of exerting direct and indirect influences on geomorphic processes, and thereby generating different 'morphogenetic' landform assemblages (Chapter 18). Chapter 19 discusses the important morphogenetic class of glacial landforms which, for the most part, are the result of past climatic conditions in that they are only fully revealed following deglaciation. Chapter 20 is concerned with 'polygenetic' landforms which exhibit the morphological effects of multiple climates, particularly involving relatively recent changes in precipitation and the lowering of temperature under periglacial conditions.

The important subject of applied geomorphology has been treated in the Appendix, rather than as part of the unfolding theme of the book, because in a very real sense it flows from and relates to almost every part of the work. During the past two decades applied geomorphology has widened and become a highly professional discipline as its possibilities and techniques have been explored. In a work such as this it has only been possible to introduce something of the range of this subject and to provide some examples of work with which one of the authors has been personally involved. However, it is hoped that thereby the student will realize the increasing importance of applied geomorphology and its integral relationship with much of the modern interpretation of the discipline of geomorphology as a whole.

# Part One  *Introduction*

The three chapters of Part I present the basic theoretical background to geomorphic concepts and problems (Chapter 1), an historical perspective of models of landform evolution (Chapter 2) and an introduction to the external and internal processes affecting landform development and erosional evolution (Chapter 3).

# One *Approaches to geomorphology*

Geomorphology (Greek: *ge* – 'earth', *morphe* – 'form', *logos* – 'a discourse') is the scientific study of the geometric features of the earth's surface. Although the term is commonly restricted to those landforms that have developed at or above sea level, geomorphology includes all aspects of the interface between the solid earth, the hydrosphere and the atmosphere. Therefore, not only are the landforms of the continents and their margins of concern, but also the morphology of the sea floor. In addition, the close look at the moon, Mars and other planets provided by spacecraft has created an extraterrestrial aspect to geomorphology.

## 1.1 Concepts

Geomorphic studies comprise a spectrum of approaches between two major, interrelated conceptual bases:

(1) Historical studies which attempt to deduce from the erosional and depositional features of the landscape evidence relating to the sequence of historical events (e.g. tectonic, sea level, climatic) through which it has passed.

(2) Functional studies of reasonably contemporary processes and the behaviour of earth materials which can be directly observed and which help the geomorphologist to understand the maintenance and change of landforms.

Functional studies explain the existence of a landform in terms of the circumstances which surround it and allow it to be produced, sustained, or transformed such that the landform functions in a manner which reflects these circumstances. Historical studies explain the existing landform assemblage as a mixture of effects resulting from the vicissitudes through which it has passed. Thus functional explanation is most applicable to those land-

forms which most clearly manifest the effects of recent processes to which they have readily responded, whereas historical explanation is reserved most obviously for landforms whose features have evolved slowly and which bear witness to the superimposed effects of climatic and tectonic vicissitudes, i.e. they are a *palimpsest* (like a surface which has been written on many times after previous inscriptions have been only partially erased; Greek: *palin* – 'again', *psegma* – 'rubbed off'). It is clear that most objects of geomorphic interest show evidence of both functional and historical influences and this is one of the reasons why so many geomorphic problems are open to widely differing approaches. Most functional explanation is directed towards *prediction*, the deducing of effects produced by causative factors (i.e. independent variables); whereas historical explanation rests on *retrodiction*, the derivation of a chronology of a sequence of past landscape-forming events. Both functional and historical studies require a *description* of the landform or landscape, either quantitatively or qualitatively. This is the groundwork of research, but description itself rests on one of the conceptual bases which define the rules by which description is carried out.

An understanding of the erosional and depositional processes that fashion the landform, their mechanics and their rates of operation must also be obtained in order that the past evolution can be explained and the future evolution predicted. This 'process geomorphology' has a strong utilitarian aspect. The great complexity and diversity of landscape features has led to different approaches to the study of landforms. The engineer is interested in a description of the landform and an assessment of its stability and short-term rate of change, which is of great practical concern. The geologist wants to know how various lithologic units affect the landscape, so that this understanding of geologic control of landforms can be used to map the rocks and structures from aerial photographs or satellite images. Geomorphologists use different approaches and techniques of study depending on their goals which may be description, retrodiction, prediction, or all three.

It is commonly believed that scientific investigation proceeds by the method of *multiple working hypotheses*. Stated in a highly oversimplistic manner, this method is thought to involve the collection of a body of observations, and the formulation of a number of distinct hypotheses which might explain the observations. The next step is the deduction of further possible 'facts' which would logically be expected to result from the reality of each hypothesis. Finally, there is the testing of the hypotheses by trying to verify the deduced 'facts' by further observation, and the modification and combining of hypotheses to produce the most probable one, which can then be elevated to the rank of theory. Of course, the strict application of this method is not possible because all scientists approach the problems of the real world from conceptual bases the sources of which are difficult to determine, and the effects of which are to set in train complex loops of description, data generation, hypothesis-building and hypothesis-testing (Figure 1.1). Each of the two most important conceptual bases for theory-building in geomorphology, the historical and the functional, prompts a particular type of investigation. It is clear, however, that certain explanations of complex landforms must involve elements of both.

Although there are different approaches to geomorphology and investigations may have very different objectives, nevertheless, there are several concepts that are basic to landform studies. Expressed in four words, these are uniformity, evolution, complexity and systems.

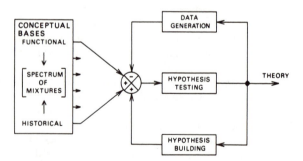

**Figure 1.1**   The relations between conceptual bases, data generation, hypothesis-building, hypothesis-testing and theory construction. Conceptual bases and hypothesis-building create more ( + ) hypotheses for testing, whereas data generation by field observations may decrease ( – ) the number of viable hypotheses for testing; ( + ) and ( – ) represent positive and negative feedback, respectively.

When present landform evolution and the operation of erosional and depositional processes are understood, the principle of *uniformity* (i.e. that the present is the key to the past) can be invoked to extend the results into the future (prediction) or into the past (retrodiction). Great care must be exercised in the application of this principle, mainly because of the complexity of landform evolution and the interruption of the evolution by other factors (tectonic, climatic and human). Nevertheless, in its simplest form, uniformity means that basic physical and chemical relationships apply equally to the present, the future and the past. For example, although rates of erosion may be very different, the behaviour of fluids on a slope or in a channel is known and the hydraulic relations can be extrapolated forward and backward in time.

The earth's surface is dynamic and landforms change through time. This *evolution* occurs in different ways, at different rates and during variable periods of time. Between 1884 and 1899, in the wake of Charles Darwin's *Origin of Species* (1859), the American geomorphologist William Morris Davis (1850–1934) developed his *cycle of erosion* theory for landforms based on the view that, for them, evolution implies an inevitable, continuous and irreversible process of change producing an orderly sequence of transformation stages of landform assemblages from youth through maturity to old age. In this way Davis was using a *paradigm* (Greek: *para* – 'beside', *deiknoomi* – 'to show'); in other words, a model developed in another discipline which appears to possess such general power, pervasion and applicability that it can be usefully employed in geomorphology. Davis's cycle of erosion concept represented an application of Darwin's biological paradigm and, in much the same way, the systems approach to geomorphology finds its roots in the thermodynamic paradigm which emerged in the latter half of the last century. Davis thus provided a theoretical model for the cyclic landforms which he would expect to evolve in the period between the initial uplift of a land surface and its subsequent reduction to a surface of low relief (i.e. a *peneplain*). However, as will be seen later in this chapter, there may be no evidence for such an orderly evolution in the sequence of changes undergone by an assemblage of landforms.

Although during very long spans of time one can conceive a slow progressive evolution towards an increasing uniformity of landforms (i.e. increasing *entropy*), the details of this change are usually *complex* with periods of erosion or incision being followed by periods of deposition, as landforms respond to changed conditions (e.g. climate, baselevel, land use). This complexity has been especially marked during the last few million years of earth history which have been characterized by climatic and tectonic change and by the increasing impact of man's activities. Therefore, landscape evolution may be expected to involve the development of complex landform assemblages (e.g. those developed over long timespans, covering large areas, influenced by many factors, subjected to many threshold effects etc.), the explanation of which employs elements of a wide spectrum of conceptual bases. Thus the dimensions of a river may depend partly on the effects of the mechanics of water and sediment movement through the channel (i.e. functional), and partly on the dimensions and slope of the whole valley form which may have been produced by a sequence of geological events (i.e. historical). Once again, the complexity of geomorphic problems is obvious, but if the student of landforms is aware of, and uses, the concepts presented in this chapter, much order can be introduced into attempts to come to grips with their formation, maintenance and evolution.

A landform is a part of a larger *system*. This system, which consists of both the landforms (*morphologic systems*) and the mass (sediment) and energy flows through the landscape (*cascading systems*), cannot be understood without an understanding of individual landforms. In turn, the individual landforms cannot be understood except as part of the larger system which is tending towards an equilibrium state.

Two important concepts which are basic to the organization of this book are, first, the systems approach to landforms and, second, the view that landforms can be considered in different ways depending upon the temporal and spatial perspective of the investigator. The remainder of this chapter will consider these basic concepts.

## 1.2 The geomorphic system

A complete explanation of a landform must involve a description of the feature and an understanding of the processes involved in its formation, as well as its development through time. It is in this way that the morphologic components of the system can be related to the cascading (energy and material flow) components of a geomorphic process–response system.

### 1.2.1 Systems structure

A geomorphic system is a structure of interacting processes and landforms that function individually and jointly to form a landscape complex. The easiest landscape complex to visualize is that of a drainage basin with its interrelated summits (divides), hillslopes, drainage network and major alluvial channels. The maintenance of such a system is dependent upon *inputs, throughputs*

and *outputs* of mass and energy. A reach of a river through which water and debris are passing is easy to conceive of as a subsystem. However, the most important aspect of this concept for earth scientists is that it reminds us that this river reach is part of a larger fluvial system and that the behaviour of the reach is strongly influenced by both upstream and downstream variables.

For reasons of simplicity and convenience of discussion, the fluvial system may be divided into three parts that are referred to as zones 1, 2 and 3 (Figure 1.2). The uppermost zone is the drainage basin or watershed from which water and sediment are derived. It is primarily the zone of sediment production, although sediment storage occurs there in important ways. Zone 2 is the transfer or transportation zone, where major streams move water and sediment from zone 1 to zone 3, which is the sediment 'sink' or zone of deposition. Although sediments are stored, eroded and transported in all of these zones, within each a single process is usually dominant, and it is convenient to discuss the fluvial system in this manner. Figure 1.2 indicates in a very simple way the characteristics of the three zones. Each zone can be considered to be composed of two basic parts, the morphologic system (the landforms that make up each zone) and the cascading system (the energy and materials flowing through that zone).

There is little value in describing the morphologic system without a consideration of the associated cascading system, and Table 1.1 relates the morphology and hydrology of the fluvial system to the controlling *variables*, which produce the morphologic and cascading characteristics of zone 1 and which, in turn, significantly influence zones 2 and 3. The variables of Table 1.1 are arranged in a sequence that reflects increasing degrees of dependence, in so far as this can be done for the fluvial system. A variable is any property (i.e. of form, rate of

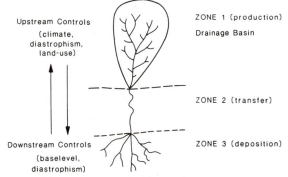

**Figure 1.2** The idealized fluvial system.
*Source:* S. A. Schumm, *The Fluvial System*, 1977, figure 1-1, p. 3, copyright © John Wiley and Sons, by permission.

change, etc.) which may assume one of a specified set of values, a *dependent* variable being an effect whose variation is determined or constrained by values assumed by other *independent*, or causal, variables. Time, initial relief, geology and climate (variables 1–4) are the dominant independent variables that influence the progress of the erosional evolution of a landscape and its hydrology. Vegetation type and density (variable 5) depend on lithology (soil) and climate (variables 3 and 4). As time passes the relief, or the volume of the drainage system remaining above baselevel (variable 6), is determined by the factors above it in the table and relief, in turn, significantly influences runoff and sediment yield per unit area within the drainage basin (variable 7). Runoff acting on the soil and geologic materials produces a characteristic drainage network morphology (variable 8: drainage density, channel shape, gradient and pattern) and hillslope morphology (variable 9: slope angle, length and profile form). These morphological variables, in turn, strongly influence the cascading system, the volumes of runoff and sediment that are eventually discharged from zone 1 (variable 10). It is the volume and type of sediment and the discharge and flow character of water that, to a major extent, determine channel morphology and the nature of the fluvial deposits that form in zones 2 and 3 (variables 11 and 12).

**Table 1.1** Fluvial system variables

*Drainage system variables*

1  Time
2  Initial relief
3  Geology (lithology, structure)
4  Climate
5  Vegetation (type and density)
6  Relief or volume of system above baselevel
7  Hydrology (runoff and sediment yield per unit area within zone 1)
8  Drainage network morphology
9  Hillslope morphology
10 Hydrology (discharge of water and sediment to zones 2 and 3)
11 Channel and valley morphology and sediment characteristics (zone 2)
12 Depositional system morphology and sediment characteristics (zone 3)

*Source:* Schumm and Lichty, 1965, table 1, p. 112.

In Table 1.1 only upstream controls are listed, but the fluvial system can also be significantly influenced by downstream baselevel variations (Figure 1.2). Lowering of baselevel will rejuvenate the drainage system, and the effect on zones 1 and 2 will be significant with a feedback

to zone 3 of greatly increased sediment production and a change of sediment characteristics. The complex zone 1 landscape is composed of a number of landform elements, drainage divides, hillslopes, floodplains and channels, and the response of this complex landscape to change will not be simple.

The morphology of zone 1 can be as different as the range of variables acting upon it. Consider, for example, the changes of drainage density (length of channels per unit area of drainage basin), as geologic materials vary from highly erodible shales and siltstones to resistant crystalline rocks, or as climate varies from semi-arid to humid, or as relief varies from high to low. In each case drainage density and the number and length of channels will decrease. Figure 1.3 summarizes the variability of a fluvial system under the influence of only three controls: stage or time, relief and climate. Only two examples are shown in Figure 1.3 which, nevertheless, clearly demonstrates the power of the systems approach to the study of landforms. Example 1 is a young, high-relief, dry-climate drainage basin, and example 2 is an old, low-relief, humid-climate drainage basin. In each of these examples the three controls act to produce a general overall effect, so they can be lumped together. A discussion of each of these variables is presented later.

There is a continuum of drainage basin, channel and depositional environments depending upon the variables listed in Table 1.1, but in Figure 1.3 only the end-members of this continuum are shown. Figure 1.3 reviews both the morphologic and the cascading system characteristics for each of the three zones. For the youthful, high-relief, sparsely vegetated drainage basin, the drainage density $D$ will be high, and both hillslope inclination and stream gradient $S$ will be steep. A well-developed drainage network will result in high water discharge $Q$ per unit of precipitation, high peak discharge $Q_p$, relatively low base (groundwater) flow $b$ and high sediment production $L$. The well-developed, efficient, fine-textured drainage network will move water and sediment rapidly from the basin and deliver it to zone 2. In zone 2 the highest sediment load, the high bedload and the flashy nature of the discharge will produce a bedload (sand and gravel) channel of steep gradient, large width–depth $w/d$ ratio, low sinuosity $P$ and braided pattern. Downstream (zone 3) the large quantity of coarse sediment may rapidly form an alluvial fan, bajada, or fan delta, with sedimentary discontinuities (Allen, 1978).

At the other extreme is example 2, an old, low-relief, humid, well-vegetated drainage basin that has a low drainage density $D$, low slopes and low discharge $Q$ per unit of precipitation. A high percentage of precipitation infiltrates, or is lost to evapotranspiration. Peak discharge $Q_p$ will be relatively low, and groundwater will be abundant, leading to high base flow $b$. Sediment loads will be low. In zone 2 the result is a suspended-load (silt–clay) channel, which transports relatively fine sediments at a low slope in a low width–depth ratio $w/d$ channel with high sinuosity. Discharge will be relatively steady, although during major precipitation events large floods will move through the valley. In zone 3 the fine sediment and the steady nature of the flow will cause slower rates of deposition and an alluvial plain or delta will form.

A change of climate can transform example 1 to example 2 or vice versa or to some intermediate stage. The result of this is that the character of the sediments delivered to zones 2 and 3 will change and significant channel adjustments result (Schumm, 1977).

Without tectonic interruptions through geologic time, the erosional evolution of a landscape should result in a transition from example 1 drainage basins and channels to those of example 2 (Figure 1.3). As the relief of the drainage basin is reduced during the erosional evolution,

drainage density will decrease, slopes will decline, sediment load and sediment size will decrease. The result will be a transition from a braided to a meandering channel in zone 2 and to finer-grained, more uniform deposits in zone 3.

In this way changes of inputs (i.e. process magnitudes) into systems produce changes in (sediment) output and changes in the forms or structures of the internal components of the system (i.e. *subsystems*). The speed with which changes in input are reflected in these resulting form changes is expressed by the so-called *relaxation time* of the system (Melton, 1958). An abrupt change in the energy of the processes will result in a progressive adjustment of form (Figure 1.4A). For example an increase of runoff will very quickly be expressed by a change in the width and depth of alluvial channels but more slowly by the development of fingertip channels which increases drainage density (Figure 1.4B). Thus landform response can be measured in terms of *sensitivity* (i.e. recurrence interval/relaxation time) (see Brunsden and Thornes, 1979) (Figure 1.4C), where

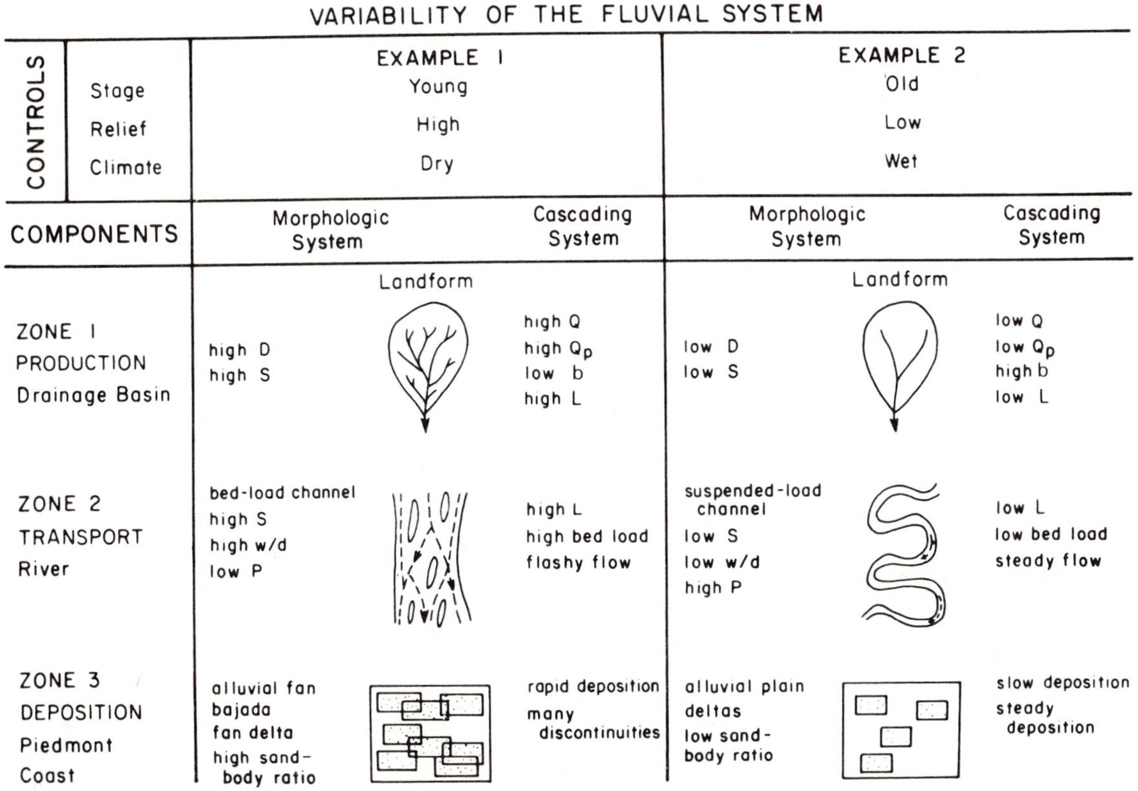

## VARIABILITY OF THE FLUVIAL SYSTEM

| CONTROLS | | EXAMPLE 1 | | EXAMPLE 2 | |
|---|---|---|---|---|---|
| Stage | | Young | | Old | |
| Relief | | High | | Low | |
| Climate | | Dry | | Wet | |
| **COMPONENTS** | | Morphologic System | Cascading System | Morphologic System | Cascading System |
| | | Landform | | Landform | |
| ZONE 1 PRODUCTION Drainage Basin | | high D<br>high S | high Q<br>high $Q_p$<br>low b<br>high L | low D<br>low S | low Q<br>low $Q_p$<br>high b<br>low L |
| ZONE 2 TRANSPORT River | | bed-load channel<br>high S<br>high w/d<br>low P | high L<br>high bed load<br>flashy flow | suspended-load channel<br>low S<br>low w/d<br>high P | low L<br>low bed load<br>steady flow |
| ZONE 3 DEPOSITION Piedmont Coast | | alluvial fan<br>bajada<br>fan delta<br>high sand-body ratio | rapid deposition<br>many discontinuities | alluvial plain<br>deltas<br>low sand-body ratio | slow deposition<br>steady deposition |

**Figure 1.3** The variability of the fluvial system under the influence of stage, relief and climate for two extreme examples: see text, for explanation.
*Source:* Schumm, 1981, figure 2, p. 20.

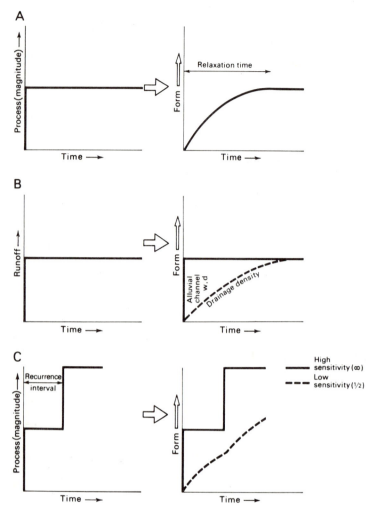

**Figure 1.4** The manner in which sensitivity of landforms is defined in terms of ratio between recurrence interval and relaxation time in response to a step input change of process magnitude; see text, for description. The values of the sensitivity (i.e. ∞ and ½) are obtained by dividing the mean recurrence interval *C* by relaxation time *A*.

*recurrence interval* is the average length of time separating events of a relevant magnitude. Systems with low sensitivity possess long relaxation times, especially when the input changes are small.

Thus large-scale landforms developed on resistant rocks over long periods of time, like the quartzite ridges of the Appalachian Mountains, may be expected to reflect conditions of the distant past rather than those of present weathering and erosional rates. On the other hand, systems composed of weak materials (e.g. alluvial channels, beaches, etc.) will exhibit short relaxation times such that even short-term or small changes of input may be expected to be manifested not only in changes of sediment output, but also in changes of form and structure within the system itself. For example, an increase in average wave size during a storm will not only change the rate at which sand and cobbles are being

moved along the beach, but also the geometry of the beach itself. The extent to which a geomorphologist might be committed to a study of process in order to understand the origin of landforms is largely dictated by the relaxation times of the geomorphic system concerned. It is no accident that historical types of explanation are more usually applied to systems which have long relaxation times, whereas functional explanations are reserved for those with short relaxation times.

Another important concept is that of landform *recovery* (i.e. recurrence interval/recovery time) (see Wolman and Gerson, 1978), where recovery time is that required for an erosional feature (e.g. a channel widening, a gully on a hillslope, etc ) which has been produced by an isolated high-magnitude process pulse to be obliterated or returned to its original form (Figure 1.5A). Alluvial channels which have been enlarged by

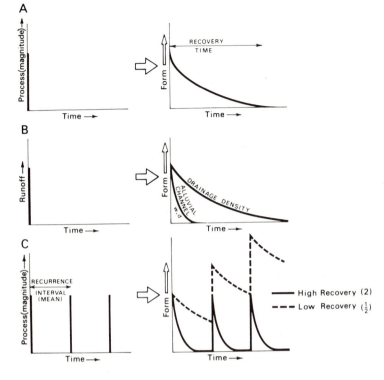

**Figure 1.5** The manner in which recovery of landforms can be defined in terms of ratio between recurrence interval and recovery time in response to unit pulse inputs of processes of given magnitude; see text, for description. The values of recovery (i.e. 2 and $\frac{1}{2}$) are obtained by dividing mean recurrence interval *C* by recovery time *A*.

hurricane floods generally recover their pre-existing geometry in a mere year or two, whereas associated gullies may persist (in the absence of renewed high-magnitude events) for 10–100 times as long (Figure 1.5B). Landforms with high recovery rates may be expected to exhibit considerable temporal adjustment to the general magnitude of frequent processes, whereas those of low recovery usually show the effects of infrequent high-intensity events (Figure 1.5C). The palimpsestic view of the landform system is that it is composed of a nested hierarchy of subsystems each having different levels of sensitivity and recovery, the whole being subject to a temporal stream of input (i.e. process) changes (Figure 1.6). The result of this is that in any given time *t* each part of the landscape may exhibit varying degrees of adjustment to present processes.

The degree to which the internal state or the output of a system is adjusted to its input is a measure of system *equilibrium*. In very simple systems equilibrium is attained by the output responding in a purely passive manner (i.e. the discharge of water from a rock-floored lake varying with flow into the lake), but more often changes in the throughput have an effect upon the input by a process known as *feedback*. In 1884, the same year as the American geomorphologist W. M. Davis published his first tentative statement of the cycle of erosion, the French scientist Henri-Louis Le Châtelier stated the

principle that a change undergone by any of the factors governing the equilibrium of a chemical thermodynamic system will result in a compensating change in that factor in the opposite sense to the original change, such that the effect of the change will be halted and absorbed. This principle, which we now term homeostasis, self-regulation, or *negative feedback* ( ), has long been known in a general sense in the physical and engineering sciences; indeed, James Watt used it in the construction of a centrifugal governor on his steam engine in the last decade of the eighteenth century. What is more important from our point of view is that, seven years before Le Châtelier, the American geomorphologist G. K. Gilbert had used the concept, termed by him *grade*, in the context of landform development (Plate 1).

Gilbert applied the concept of negative feedback to explain the graded condition in which a stream is provided with as much load as it is capable of transporting, and unless external supplies of discharge or debris are changed, the stream will neither erode nor deposit in the short term. If a graded stream encounters a steeper reach, the accompanying increase in velocity will lead to entrainment of more debris, erosion and the production of a lower bed slope; the interruption of flow by a less steep section will have the opposite effect of increasing

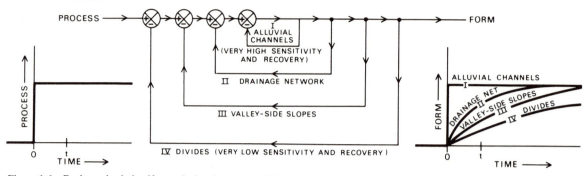

**Figure 1.6**   Drainage basin landforms depicted as a nested hierarchy of differing sensitivity and recovery, so that at any time *t* a palimpsest of differentially adjusted landforms (I–IV) may result from a given initial change in process. The nested negative feedback loops (described in text) allow the process of mutual adjustment of forms to operate both from landforms I–IV and from IV–I.

bed slope by deposition. In this way a graded stream tends to equalize its work of erosion, transportation, deposition and corrasion in all parts of its course to produce a smooth, longitudinal *profile of equilibrium* which, to Gilbert, was the visible manifestation of the graded state. Every reach of a graded stream is part of a series receiving and discharging water and debris, and any local interruption of this condition will produce yoked waves of erosion and deposition which will move headward and mouthward, respectively, until new equilibrium forms have been produced. Thus the graded condition implies interdependence throughout the fluvial system. As will be seen in a later chapter, this simple equation of stream-bed slope with fluvial activity has been much criticized of recent years (Chorley and Beckinsale, 1980).

In most geomorphic systems for the short term, and in all for the long term, such *negative feedback* processes operate and have the effect of counteracting and stabilizing system change by means of adjustments within the system. Thus when coarse debris is brought into a stream reach, much of it is deposited thereby steepening the gradient of the channel and increasing the velocity of waterflow until a steady throughput of the new debris is brought about. It is this opposing tendency of input change and negative feedback which results in much of the observed oscillation of output (e.g. sediment) and of the internal forms of geomorphic systems.

*Positive feedback* ( ⊗ ) occurs when a change in input is magnified by the system operation such that its effect is enhanced or continued. Obviously, such a state of affairs can only exist within limited scales of space and time, otherwise the whole terrestrial system would be caught up in accelerating change; but an example of positive feedback is when an increase in precipitation

causes an increase in overland flow, which causes an increased rate of removal of open-textured surface soil which, in turn, exposes underlying less permeable material at the surface allowing less rainfall to infiltrate and leading to more surface runoff and erosion, and so on. Such a positive feedback loop can have one of two outcomes: accelerating soil erosion will continue until either more resistant bedrock is exposed and the local soil subsystem is destroyed, or until a certain soil level is reached at which permeability (i.e. its ability to transmit water) does not decrease with depth and the positive feedback loop is broken. It would be wrong, however, to view positive feedback merely as affecting geomorphic subsystems of limited temporal and spatial scales. On a much grander level the progressive, irrecoverable historical changes of landforms are also examples of the operation of positive feedback loops. This is illustrated by the non-stationary means shown in Figure 1.7A, C and D. Thus study of the nature of these large-scale positive feedback loops may prove a means of linking the historical and the functional approaches to landform modelling.

The output of a geomorphic system can be expressed in two ways: first, by the rate at which mass (e.g. sediment) is evacuated from it, and second, the energy which has been expended in sustaining or transforming it. This leads us to the important concept of system state termed *equilibrium*. System equilibrium manifests itself both in terms of rate of mass transfer and the maintenance of forms expressive of given energy levels. Paradoxically equilibrium can only be expressed with reference to directions of change; Figure 1.7 illustrates four types of equilibrium:

(1) A condition when the rate of change of form declines (*decays*) through time to a state of relatively slow change. The late-stage surface of low relief (pene-

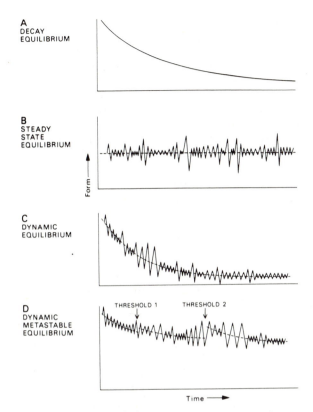

A
DECAY
EQUILIBRIUM

B
STEADY
STATE
EQUILIBRIUM

Form

C
DYNAMIC
EQUILIBRIUM

D
DYNAMIC
METASTABLE
EQUILIBRIUM

THRESHOLD 1    THRESHOLD 2

Time ⟶

**Figure 1.7** Types of equilibria; see text, for explanation. *Source:* Chorley and Beckinsale, 1980, figure 1, p. 130.

plain) of the cycle of erosion is such a quasi-equilibrium feature (Figure 1.7A).

(2) A condition wherein form oscillates around a stable average value due to the operation of interacting feedback loops in the system or to the 'hunting' of a complex system, which will be referred to shortly. This type of equilibrium is termed a *steady state* (Figure 1.7B).

(3) A condition of oscillation about a mean value which is itself trending continuously through time. This trend might be due to externally controlled changes of input (e.g. climatic change involving increases in slope runoff, stream discharge, wind velocity, or waveheight). This equilibrium about a moving average, termed *dynamic equilibrium* (Figure 1.7C), could be applied to a river whose general downcutting is effected by means of alternating cut and fill which can be identified stratigraphically.

(4) A condition of oscillation about a mean value of form which is trending through time and, at the same time, is subjected to steplike discontinuities as a *threshold effect* (see p. 39) operates to promote a

sudden change of form. A river subject to net downcutting by a process of cut and fill upon which is superimposed discontinuous uplift exhibits this type of *dynamic metastable equilibrium* (Figure 1.7D).

It is thus clear that 'true' equilibrium is a theoretical state towards which the system behaviour is tending with greater or lesser rapidity, by attempting to absorb the successive effects of a sequence of process inputs of differing magnitude and frequency.

### 1.2.2 Complex response and thresholds

Two additional geomorphic concepts that have potential for aiding in the development of an understanding of the nature of landscape evolution are those of *complex response* and geomorphic *thresholds*.

The intricate connections between the various parts of a geomorphic system (e.g. between alluvial channels, the tributary stream network, foot slopes, major valley-side slope elements, valley heads and divides) implies that any externally imposed change may diffuse through the system in a complex manner, with many lags, such that any one part may suffer a complex of change through time while a new equilibrium is being achieved. This means that the behaviour of such a system is indeterminate in so far as it is not always possible to predict a unique, immediate pattern of outcomes resulting from a given stimulus for change. For example, when a small experimental drainage basin is rejuvenated, the system responds not simply by incising, but by hunting for a new equilibrium by incision, aggradation and renewed incision. Such behaviour is an example of the complex response of the system. Complexity of response is also compounded by the existence of thresholds, which may be due to *extrinsic* or *intrinsic* causes. A threshold may be externally triggered if, for instance, a drastic change in the climatic regime destroys vegetation and sets in train an intensity of erosion and terrain modification quite distinct from that which preceded it and from which the terrain is incapable of recovery, even after a return to initial climatic conditions. An intrinsic threshold would operate if, for example, sediments stored within a fluvial system (e.g. on the floodplain or on slopes) became unstable at critical threshold slopes, leading to accelerated erosional events.

When the influence of external variables such as isostatic uplift is combined with the effects of complex response and geomorphic thresholds, it is clear that denudation – at least during the early stage of the geomorphic cycle – cannot be a simple progressive process as assumed in the Davis cycle. Rather, it is comprised of episodes of erosion separated by periods of relative stability, a complicated sequence of events. Much of this complexity is the result of a delayed transmission of effects through the landscape. For example, channel changes that take place near the mouth of a drainage basin following incision are responding to the conditions at that time and location, but the channel may not be prepared for the changes that this incision induces within the river system upstream. Hence downcutting may be followed by deposition when an upstream response occurs.

## 1.3 Geomorphic scale

A scale problem arises from the understanding that space and time are not passive frameworks within which events occur, but that they both reflect, and impose, choices to do with cause v. effect and change v. equilibrium (Figure

1.8). At different scale resolution levels, which are mapped out according to our aims and abilities, different problems are identified; different types of explanation are relevant; different levels of generalization are appropriate; different variables are dominant; and different roles of cause and effect are assigned. It is clear, for example, that for studies of large-scale landforms developed over long timespans, much larger degrees of sweeping historical explanations must be acceptable than in respect of small-scale landforms changing within a short timespan. It is also apparent that conclusions derived from studies made at one scale need not necessarily be expected to apply to another.

### 1.3.1 Timescales

Three classes of timespan are significant for geomorphic interpretation (see Figure 1.9).

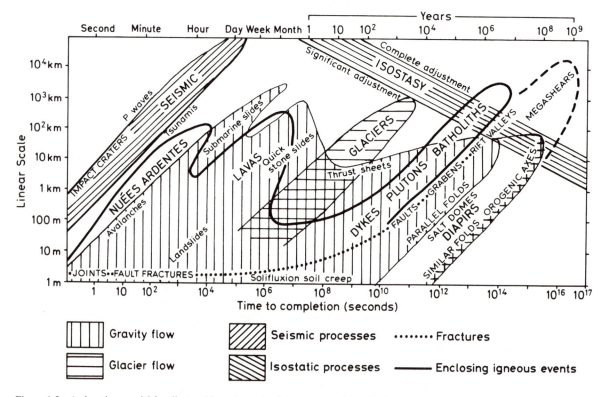

**Figure 1.8** A size–time model for diastrophic and geomorphic processes. Most displacements are completed in times proportional to their sizes and form an *en échelon* sequence sloping up to the right at 45°, indicating that larger-scale movements are accomplished in generally longer time periods. The sequence of such movements is arranged from left to right with lower viscosity materials on the left. These sloping units do not take account of the episodic nature of many movements – e.g. a significant proportion of glacier movements takes place by surges having a periodicity of the order of 10–100 years. Contrasting with the above movements are isostatic recoveries (sloping down to the right at 30°) with large areas achieving general recovery more rapidly than smaller locations.
*Source:* Carey, 1962, figure 1, p. 98.

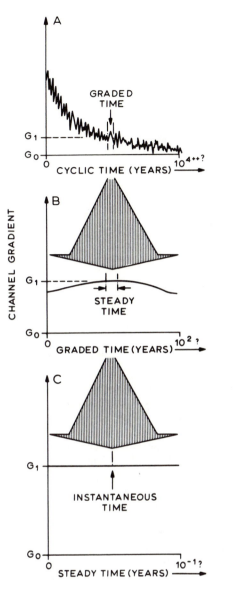

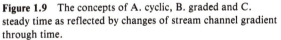

**Figure 1.9** The concepts of A. cyclic, B. graded and C. steady time as reflected by changes of stream channel gradient through time.
*Sources:* Schumm and Lichty, 1965; Chorley and Kennedy, 1971, figure 7.2, p. 254.

(1) *Cyclic time.* This includes long geologic-type timespans measured in millions of years and is suitable for viewing larger areas and changes which are slow and cumulative. Table 1.2 shows that the major independent variables controlling cyclic landforms are: time; initial relief, or the general elevation differences between high and low areas created by earth movements, volcanism, or

sea-level changes, and influencing the land slope and consequently rate of erosional downcutting; geology, or rock type and structure; and climate, as affecting the amount of surface runoff, vegetational cover, etc. Conversely, vegetation type and density, the actual relief at a given time, river and slope runoff and sediment discharge per unit area, the form of the drainage network (e.g. the density of drainage lines in kilometres per square kilometre), the form of erosional hillslopes, and the total discharge of water and sediment from a wide area are clearly dependent effects. Only equilibrium concepts of the time-trending type are relevant here (see Figure 1.7A, C and D).

(2) *Graded time.* This refers to a shorter span of time during which smaller areas such as individual hillslopes and stream reaches may achieve a steady-state equilibrium (see Figure 1.7B), with negative feedback causing fluctuations about a mean condition of form or rate of transfer which is related to the motivating process concerned. And this equilibrium state makes time and initial relief irrelevant as controlling variables; geology and climate are joined by vegetation, actual relief and by water and sediment discharge per unit area as independent variables; and form of the drainage network, hillslope form and total discharge of water and sediment remain dependent variables. That this change in operation of the variables must be so is implied by the short timespan during which, for example, vegetation, actual relief, and water and sediment discharge averaged per unit area may be expected to remain constant. During this time they impart their influence to the existing form of drainage networks, hillslopes, and the total regional discharge of water and sediment whose variations are those which oscillate about stationary mean values (Table 1.2, Figure 1.7B).

(3) *Steady time.* This is short and viewed over a small area (e.g. a short stream reach or a slope segment). Only the total discharge of water and sediment is clearly a dependent variable under these circumstances and all other variables may be looked upon as independent.

In respect of these three timescales any given landscape may be conceived as a nested hierarchy of spatial units. The smaller of these are best analysed in steady or graded time over which their shorter relaxation times may find expression in functional/equilibrium terms; the larger prompt an historical/evolutionary approach with cyclic timespans. To the geologist, concerned with long-term earth history, the form of a drainage network is a dependent variable controlled by time, climate, geology, vegetation, relief, and runoff of water and sediment; whereas to the hydrologist, interested in day-to-day

**Table 1.2**   Status of drainage basin variables during timespans of decreasing duration

| Drainage basin variables | Status of variables during designated timespans | | |
|---|---|---|---|
| | Cyclic | Graded | Steady |
| 1 Time | Independent | Not relevant | Not relevant |
| 2 Initial relief | Independent | Not relevant | Not relevant |
| 3 Geology (lithology, structure) | Independent | Independent | Independent |
| 4 Climate | Independent | Independent | Independent |
| 5 Vegetation (type and density) | Dependent | Independent | Independent |
| 6 Relief or volume of system above baselevel | Dependent | Independent | Independent |
| 7 Hydrology (runoff and sediment yield per unit area within system) | Dependent | Independent | Independent |
| 8 Drainage network morphology | Dependent | Dependent | Independent |
| 9 Hillslope morphology | Dependent | Dependent | Independent |
| 10 Hydrology (discharge of water and sediment from system) | Dependent | Dependent | Dependent |

*Source:* Schumm and Lichty, 1965, table 1, p. 112.

water supply, drainage network form is an independent variable exercising a control over the temporal pattern of runoff for the area (Table 1.2).

**1.3.2 Spatial scales**

As with time, spatial scales are not passive in that they do not merely distinguish between investigations of differing extent which are similar in other respects. Spatial scale itself implies intrinsic characteristics of structure and process which give a particular flavour to geomorphic work. At different scales of space different variables become dominant, different levels of generalization may be employed and even different problems identified.

Although geomorphologists have conducted investigations on a wide variety of spatial scales, it is only recently that the dynamism of scale has become apparent. In most classical work spatial scale is implicit, rather than explicit. Fenneman's (1914) *physiographic regions* were largely based on considerations of structural geology (e.g. the Ridge and Valley Province of the folded Appalachians) and were considered to be at the largest scale at which coherent geomorphic unity could be observed. These units could be broken down into smaller areas by a process of logical subdivision. A contrasting 'accretion' approach to the problem of spatial scale was exemplified by Wooldridge (1932), who attempted to identify 'the

physiographic atoms out of which the matter of regions is built'. These atoms were the facets of 'flats' and 'slopes' composing a polycyclic landscape (see Chapter 2). Another important geomorphic spatial scale unit has been the erosional drainage basin with its implications of form and process interaction (see Chapter 13).

Spatial constraints in geomorphology are inherent in the temporal ones. Processes are not homogeneous in operation over a wide variety of terrestrial scales. An example of this is that because the heaviest rain is usually produced by thunder cells which have a restricted size, smaller basins which on occasion may be completely blanketed by such rainstorms exhibit higher maximum runoff and erosion rates per unit area than larger areas which are only partly influenced by high-intensity rainfall. A further instance of this interlinkage of process with spatial scale is given by stream-bed forms. On sand beds a variety of microforms (e.g. ripples and dunes) are associated with various ranges of flow, but these microforms which have wavelengths measurable in centimetres are superimposed on larger-scale riffles and pools and bars which give a series of shallows and deeps in a river reach with wavelengths of tens of metres. At a larger scale the river may meander with wavelengths measured in kilometres. Clearly, the geomorphologist must be prepared to deal with the morphologic system at all temporal and spatial scales.

**Table 1.3** Hierarchical spatial ordering of landforms

| Order | Examples |
| --- | --- |
| 1st | Continents, oceans, plates, convergence zones, divergence zones |
| 2nd | Physiographic provinces, mountain ranges, massifs, plateaus, lowlands, accumulative plains, tectonic depressions (termed 'morphostructural' units by the Russians) |
| 3rd | Medium-scale geological units, such as folded sequences, fault blocks, domes and volcanoes |
| 4th | Large-scale erosional/depositional units, such as large valleys, deltas and long continuous beaches |
| 5th | Medium-scale erosional/depositional units; smaller valleys, floodplains, alluvial fans, cirques, moraines |
| 6th | Small-scale erosional/depositional units; small valleys, offshore bars, sand dunes |
| 7th | Hillslopes, stretches of stream channel |
| 8th | Slope and flat facets, pools, riffles |
| 9th | Stream bed and aeolian sand ripples, slope terracettes |
| 10th | Microroughness represented by the diameter of individual pebbles or sand grains |

It may be helpful to suggest a spatial-ordering classification which accommodates the landform features of concern to the geomorphologist (Table 1.3). This *taxonomic* approach (Greek: *tassein* – 'to arrange') encourages one to place geomorphic phenomena into classes in the hope of thereby generating more information about them. However, this classification is ambiguous to the extent that it combines together features which are defined both *genetically* (i.e. based on assumptions of origin) and *generically* (i.e. based on appearance). An alluvial fan and a cirque are examples of genetic classification, whereas plateaus and slope facets are defined generically on the basis of their form. This classification according to scale is useful because it enables us to see how the differing conceptual bases for geomorphological theory-building rely on differing scale emphases. Some geologists deal only with first-order features, whereas sedimentologists will often be concerned with tenth-order features. The geomorphologist is concerned with all, but his emphasis is generally on the third- to ninth-order features.

# References

Allen, J. R. L. (1978) 'Studies in fluviatile sedimentation: an exploratory quantitative model for the architecture of avulsion-controlled alluvial suites', *Sedimentary Geology*, vol. 21, 129–47.

Beckinsale, R. P. and Chorley, R. J. (1968) 'History of geomorphology', in R. W. Fairbridge (ed.), *The Encyclopedia of Geomorphology*, New York, Reinhold, 410–16.

Bennett, R. J. and Chorley, R. J. (1978) *Environmental Systems: Philosophy, Analysis and Control*, London, Methuen.

Brunsden, D. and Thornes, J. (1979) 'Landscape sensitivity and change', *Transactions of the Institute of British Geographers*, n.s., vol. 4, 463–84.

Carey, S. W. (1962) 'Scale of geotechnic phenomena', *Journal of the Geological Society of India*, vol. 3, 97–105.

Chorley, R. J. (1962) 'Geomorphology and general systems theory', *US Geological Survey Professional Paper 500-B*.

Chorley, R. J. (1965) 'A re-evaluation of the geomorphic system of W. M. Davis', in R. J. Chorley and P. Haggett (eds), *Frontiers in Geographical Teaching*, London, Methuen, 21–38.

Chorley, R. J. (ed.) (1972) *Spatial Analysis in Geomorphology*, London, Methuen.

Chorley, R. J. (1978) 'Bases for theory in geomorphology', in C. Embleton, D. Brunsden and D. K. C. Jones (eds), *Geomorphology: Present Problems and Future Prospects*, Oxford University Press, chapter 1, 1–13.

Chorley, R. J. and Beckinsale, R. P. (1980) 'G. K. Gilbert's geomorphology', *Geological Society of America Special Paper* 183, 129–42.

Chorley, R. J., Beckinsale, R. P. and Dunn, A. J. (1973) *The History of the Study of Landforms*, Vol. 2, *The Life and Work of William Morris Davis*, London, Methuen.

Chorley, R. J., Dunn, A. J. and Beckinsale, R. P. (1964) *The History of the Study of Landforms*, Vol. 1, *Geomorphology before Davis*, London, Methuen.

Chorley, R. J. and Kennedy, B.A. (1971) *Physical Geography: A Systems Approach*, London, Prentice-Hall International.

Coates, D. R. and Vitek, J. D. (eds) (1980) *Thresholds in Geomorphology*, London, Allen & Unwin.

Cullingford, R. A., Davidson, D. A. and Lewin, J. (eds) (1980) *Timescales in Geomorphology*, Chichester, Wiley.

Davis, W. M. (1889) 'The rivers and valleys of Pennsylvania', *National Geographic Magazine*, vol. 1, 183–253.

Davis, W. M. (1899) 'The geographical cycle', *Geographical Journal*, vol. 14, 481–504.

Derbyshire, E. (ed.) (1976) *Geomorphology and Climate*, London, Wiley.

Derbyshire, E., Gregory, K. J. and Hails, J. R. (1979) *Geomorphological Processes*, Folkestone, Dawson.

Dury, G. H. (1972) 'Some current trends in geomorphology', *Earth Science Reviews*, vol. 8, 45–72.

Embleton, C. and Thornes, J. (eds) (1979) *Process in Geomorphology*, London, Arnold.

Fenneman, N. M. (1914) 'Physiographic boundaries within the United States', *Annals of the Association of American Geographers*, vol. 4, 84–134.

Flemal, R. C. (1971) 'The attack on the Davisian system of geomorphology: a synthesis', *Journal of Geological Education*, vol. 19, 3–13.

Gilbert, G. K. (1877) *Report on the Geology of the Henry Mountains*, Washington, DC, US Department of the Interior, chapter 5, 'Land sculpture', 93–144.

Gilbert, G. K. (1886) 'The inculcation of scientific method by example, with an illustration drawn from the Quaternary geology of Utah', *American Journal of Science*, 3rd ser., vol. 31, 284–99.

Gilbert, G. K. (1914) 'The transportation of debris by running water', *US Geological Survey Professional Paper* 86.

Haggett, P., Chorley, R. J. and Stoddart, D. R. (1965) 'Scale standards in geographical research: a new measure of areal magnitude', *Nature*, vol. 205, 844–7.

Horton, R. E. (1945) 'Erosional development of streams and their drainage basins: hydrophysical approach to quantitative morphology', *Bulletin of the Geological Society of America*, vol. 56, 275–370.

Howard, A. D. (1965) 'Geomorphological systems: equilibrium and dynamics', *American Journal of Science*, vol. 263, 302–12.

Isachenko, A. G. (1973) *Principles of Landscape Science and Physical-Geographic Regionalization*, trans. R. J. Zatorski, Melbourne University Press.

Langbein, W. B. and Leopold, L. B. (1964) 'Quasi-equilibrium states in channel morphology', *American Journal of Science*, vol. 262, 782–94.

Leopold, L. B., Wolman, M. G. and Miller, J. P. (1964) *Fluvial Processes in Geomorphology*, San Francisco, Calif., Freeman.

Mackin, J. H. (1948) 'Concept of the graded river', *Bulletin of the Geological Society of America*, vol. 59, 463–512.

Melton, M. A. (1958) 'Correlation structure of morphometric properties of drainage systems and their controlling agents', *Journal of Geology*, vol. 66, 442–60.

Mitchell, C. (1973) *Terrain Evaluation*, London, Longman.

Mosley, M. P. and Zimpfer, G. L. (1976) 'Explanation in geomorphology', *Zeitschrift für Geomorphologie*, vol. 20, 381–90.

Schumm, S. A. (1973) 'Geomorphic thresholds and complex response of drainage systems', in M. Morisawa (ed.), *Fluvial Geomorphology: Proceedings of the Fourth Annual Geomorphology Symposium*, Binghamton, NY, State University of New York, 299–310.

Schumm, S. A. (1975) 'Episodic erosion: a modification of the geomorphic cycle', in W. L. Melhorn and R. C. Flemal (eds), *Theories of Landform Development*, Binghamton, NY, State University of New York, 70–85.

Schumm, S. A. (1977) *The Fluvial System*, New York, Wiley.

Schumm, S. A. (1981) 'Evolution and response of the fluvial system, sedimentologic implications', *Society of Economic Paleontologists and Mineralogists Special Publication* 31, 19–29.

Schumm, S. A. and Lichty, R. W. (1965) 'Time, space, and causality in geomorphology', *American Journal of Science*, vol. 263, 110–19.

Stoddart, D. R. (1969) 'Climatic geomorphology: review and assessment', *Progress in Geography*, vol. 1, 159–222.

Strahler, A. N. (1952) 'Dynamic basis of geomorphology', *Bulletin of the Geological Society of America*, vol. 63, 923–38.

Thorn, C. E. (ed.) (1982) *Space and Time in Geomorphology*, London, Allen & Unwin.

Thornes, J. and Brunsden, D. (1977) *Geomorphology and Time*, London, Methuen.

Wolman, M. G. and Gerson, R. (1978) 'Relative scales of time and effectiveness of climate in watershed geomorphology', *Earth Surface Processes*, vol. 3, 189–208.

Wooldridge, S. W. (1932) 'The cycle of erosion and the representation of relief', *Scottish Geographical Magazine*, vol. 48, 30–6.

# Two  *Morphologic evolutionary systems*

The morphologic system consists of the interrelated geometrical aspects of landforms which together make up the previously defined geomorphic system. As we have seen, landforms can be considered both from their historical and functional perspectives with reference to different scales of time and space. The classic models of landform evolution (e.g. Davis, Penck and King) relate to long periods of cyclic time, whereas the short-term functional models of landform response to change are of especial practical significance to engineers and conservationists. Much of Part III of this book relates to the operation of functional models.

In Chapter 1 we introduced two other important geomorphic ideas, those of *evolution* and *equilibrium*. The aim of the present chapter is to explore these further, to examine the linkages between them and identify their outward and visible manifestation in terms of morphologic landscape systems. It may seem strange to link evolution and equilibrium in that the former is commonly associated with a time sequence of changes along an irreversible (i.e. timebound) path, whereas the latter is associated with landforms which tend to persist in some manner throughout the passage of time (i.e. timeless). However, this linkage becomes more reasonable when it is realized that the speed, direction and progress of change has meaning only in the context of some equilibrium state towards (or away from) which the system is moving. For the geomorphologist, the evidence for change and equilibrium comes from the form and structure of landforms (i.e. their morphology). Thus the morphology of the landscape system potentially embodies information regarding both present and past equilibrating tendencies; information regarding the timing, rapidity and nature of past changes; and information regarding the manner of behaviour of residual parts of the landscape at various times in the past. It must be

clear, however, that this information furnished by landform morphologic systems is usually partial, blurred and difficult to date or interpret unambiguously. For this reason, the scientific study of landforms has been the subject of considerable dispute and a forum for rival models. Therefore, an objective of this chapter is to review three major models of landform evolution – those of Davis, Penck and King – and to show that, under certain circumstances, each can be useful in interpreting real landform assemblages, but that more recent work has demonstrated that all three are too general.

## 2.1 The cycle of erosion

The first – and still most important – geomorphic model was developed by the American geographer William Morris Davis (1850–1934) between 1884 and 1899 under the title of 'the cycle of erosion'. During this period the evolutionary concepts of Charles Darwin were being incorporated into the philosophy of a wide variety of scientific and social fields under the leadership of T. H. Huxley and Herbert Spencer, and it was natural that Davis should have developed a model of sequential landform changes through time, so that, as he put it, 'land forms, like organic forms shall be studied in view of their evolution'. For Davis, the concept of evolution implied an inevitable, continuous and broadly irreversible process of change producing an orderly sequence of landform transformation, wherein earlier forms could be considered as *stages* in a progression leading to later forms (Figure 2.1.). By this model, time became not a temporal framework within which events could occur, but *a process itself* leading to an inevitable progression of change (see Table 1.1). Thus the passage of time, according to Davis, imprinted on landforms progressive changes in their geometry from which the passage of time

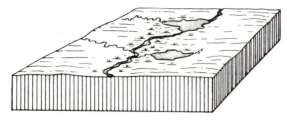

**A** In the initial stage, relief is slight, drainage poor.

**B** In early youth, stream valleys are narrow, uplands broad and flat.

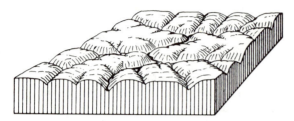

**C** In late youth, valley slopes predominate but some interstream uplands remain.

**D** In maturity, the region consists of valley slopes and narrow divides.

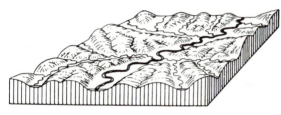

**E** In late maturity, relief is subdued, valley floors broad.

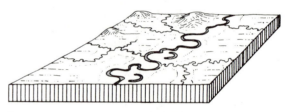

**F** In old age, a peneplain with monadnocks is formed.

**G** Uplift of the region brings on a rejuvenation, or second cycle of denudation, shown here to have reached early maturity.

**Figure 2.1** The cycle of erosion, proposed by W. M. Davis, drawn by E. Raisz.
*Source:* A. N. Strahler, *Introduction to Physical Geography*, 1965, figure 21.1, p. 304, copyright © John Wiley and Sons, by permission.

could be inferred as surely as from the changing geometrical relations of the hands of a clock.

The second paradigm which Davis built into his cycle of erosion, although much more subtly than his rather crude evolution, involved the principles of physical and chemical thermodynamics the development of which was the most important scientific advance of the second half of the nineteenth century. The Second Law of Thermodynamics, formulated just after Davis's birth, stated that in a closed system (i.e. one incapable of receiving or giving off energy) *entropy* must increase irreversibly. Entropy (Greek: *en* – 'inside'; *trope* – 'transformation') is a measure of the extent to which energy within the system has ceased to be free – in the sense of being able to perform work on the system. A low-entropy closed system is one possessing within it differentials in the distribution of energy such that the flows from locations of high to low energy are capable of performing work. As time passes and the energy within the system becomes more equally distributed the entropy increases until, at the state of maximum entropy, all parts of the closed system have the same energy levels, no flows of energy are taking place and no work is being done in the system.

Of course, landform assemblages do not constitute closed systems. They are open to inputs of potential energy associated with their uplift, of kinetic energy from precipitation, of thermal energy from the sun and chemical energy released as the result of the breakdown of rocks at or near the earth's surface. They are open to the export of thermal energy as the result of stream discharge and air movement, of kinetic energy by stream-flow and sediment load and chemical energy by the dissolved stream load. However, one of the great simplifying assumptions underlying the Davisian cycle of erosion is that the potential energy of landform initial uplift is the dominant source of energy input and that, thereafter, there is an irreversible equalization of energy levels throughout the landform assemblage, leading ultimately to a spatially uniform terrain (the *peneplain*, or *peneplane*). Thus, for Davis, each stage of his cycle was associated with declining potential energy as the relief was worn down, and each stage was characterized by an assemblage of landforms (i.e. valley-side slopes, drainage patterns, etc.) having geometries appropriate to the local potential energy expressed by the difference in level between the land surface and some lower elevation (*baselevel*) towards which degradation was directed (see Figure 2.8A). Downwearing may be interrupted by renewed uplift which initiates a set of higher-energy landforms (i.e. steeper slopes, more dense drainage networks, etc.), which coexist with the 'older' lower-energy forms on which they were superimposed (i.e. polycyclic or multicycle landforms) until the latter are completely consumed by, or integrated into, the former set of landforms as they themselves evolve into older forms.

Between 1884 and 1899 Davis developed his model of the cycle of erosion in which assemblages of landforms passed through an anthropomorphic series of irreversible changes associated with the so-called youth, maturity and old-age sequence of landforms. Davis incorporated into his cycle Gilbert's concept of the *graded condition*, as manifested by the development of smooth curves of longitudinal stream and valley-side slope profiles, which are first achieved near the mouth of a river and then spread headwards throughout the drainage lines and up the associated valley-side slopes towards the divides. He

proposed that each type of regional geological *structure* (see Chapter 7) could be viewed as evolving under the protracted operation of a given *process* (e.g. temperate fluvial, hot arid, mountain glacial, etc.) to produce a sequence of *stages* of landform assemblages. Each stage would possess a suite of landforms appropriate to and characteristic of it. For simplicity, Davis gave the most complete elaboration of his ideal *cycle of erosion* on the assumption of:

(1) uniform lithology:

(2) rapid uplift, accompanied by little erosion, to give a considerable *available relief* (i.e. the height difference between the uplifted surface and the general *baselevel*, or the inland projection of sea level which forms the general lower limit of fluvial degradation). He allowed that other types of uplift were possible, for example, a slow uplift of small relative relief producing a landscape of low convex divides from the start, not allowing high relief to develop; but he considered these types to be much less common.

Davis named his cyclic states after the periods of human life, so emphasizing the biological origins of his theory.

*Youth.* The uplift of a subaerial or submarine surface of low relief (Figure 2.1A) produces a region of broad, poorly defined stream divides separated by a number of trunk rivers and few large tributaries, but with numerous short tributaries engaged in aggressive headward cutting (Figure 2.1.B). This headward cutting, together with vertical incision of the whole drainage network, rapidly increases the relief during the whole of youth to produce steep, V-shaped valleys flanking stream courses, which are initially irregular with falls and rapids produced by minor structural controls. By the end of early youth any initial lakes have been drained and removed by stream incision, and by the end of middle youth the major falls and rapids have been removed by erosion in the trunk rivers. In late youth the original flat summit areas are restricted and much of the landscape is dominated by steep valley sides close to the angle of repose of the thickening layer of surface-weathered debris (Figure 2.1C). At the end of late youth the major rivers are graded, as evidenced by their smooth, unbroken curves, and lateral river-cutting to produce small floodplains is beginning in the lower parts of the main rivers.

*Maturity.* At the beginning of early maturity the region achieves its maximum relief allowed by the available initial relief, the drainage network ceases to increase and the area is covered with a well-integrated drainage network, which shows a maximum relationship to bedrock geological structures. The graded condition spreads headward up the tributaries, as well as the valley-side slopes, which progressively decline in angle after their early mature maximum. At middle maturity (Figure 2.1.D) the divides are narrow, meandering is the rule along the major rivers and the lateral cutting produces significant floodplains which, however, do not greatly exceed the width of the meander belts. Valley floors are lowered exceedingly slowly by the graded rivers, whereas the more rapid lowering of the summits and ridgetops brings about a progressive decrease in relief. In late maturity (Figure 2.1E) the topography is composed almost entirely of valley-side slopes and, to a lesser extent, floodplains.

*Old age.* By the start of old age all streams, valley-side slopes and divide crests are graded and the landscape is composed of broad, open and gently sloping valleys containing broad floodplains, many times the width of the associated meander belts, surmounted by the very slowly lowering rounded divides. The whole surface approaches closer and closer to baselevel under a thickening cover of creeping soil, and the relations between terrain and geology, so marked in maturity, become progressively more masked. Exceptions to this are possible topographic residuals (i.e. *monadnocks*), supported by resistant outcrops, which may persist above the general level of the peneplain of later old age (Figure 2.1F).

Renewed uplift can cause the peneplain to be redissected and to exhibit a composite set of landforms in which different cyclic stages are superimposed upon one another (Figure 2.1G).

## 2.2 Interruptions of the cycle of erosion

The general, spatially integrated degradation of the landscape envisaged in the cyclic model may be subjected to *interruptions*; these are of two kinds:

(1) Climatic (*accidents*). It is recognized that climatically induced alterations of the relationship between variables, such as surface runoff, stream discharge, vegetation cover, erosion rates, debris supply, debris calibre, etc., may lead to the local initiation of accelerated erosion or deposition the effects of which might progressively spread through the landscape as new graded forms of channels, slopes and divides evolve.

(2) Baselevel changes. Changes of baselevel – positive for a rise, negative for a fall – would clearly introduce tendencies for accelerated alluviation and renewed erosion, respectively. As peneplanation may require anything from 10 to 50 million years to accomplish, depending on the importance of the resulting isostatic compensatory uplift (see Section

5.1), it is unlikely that any one erosional cycle will run its course uninterrupted by baselevel changes induced by either tectonic or eustatic causes.

Positive movements of baselevel result in the drowning of lower parts of valleys and associated aggradation in the form of deltaic outbuilding and the build-up of floodplains by backfilling upvalley, which has the effect of reducing stream gradients and gives the appearance of old age in valleys such as that of the Mississippi River. However, there are several other possible causes of valley aggradation, including uplift of the source area, climatic change, increased debris supply from certain tributaries and the flush of sediment movement into a reach due to the complex response of the basin to a wide range of changes (see Sections 1.2.2, 2.7 and 13.5).

Negative movements of baselevel produce *rejuvenation* of landforms with youthful zones of steeper slopes spreading headward as waves of incision and planation, first affecting river profiles and the outcrops of softer rocks, and later the interfluves and harder outcrops. The leading-edge of this rejuvenation wave (*epicycle*) is marked by steeper breaks of slope, terrace edges and nickpoints on river courses producing *polycyclic* (*multicyclic*) features of valley-in-valley forms (Figure 2.2.)

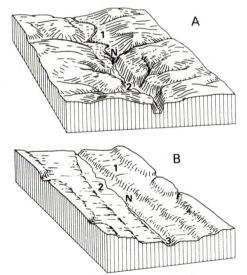

**Figure 2.2** Valley-in-valley features produced by discontinuous incision and limited planation: A. a two-cycle valley (1 and 2) separated by retreating nickpoint N; B. a three-cycle valley (1, 2 and 3) exhibiting substantial rock-cut terraces (drawn by W. J. Wayne).
*Sources:* W. D. Thornbury, *Principles of Geomorphology*, 1955, figure 6.1, p. 145, copyright © John Wiley and Sons, by permission; B. W. Sparks, *Geomorphology*, 1971, figure 9.4, p. 296, copyright © Longman.

(see Section 14.2.2). Where negative movements of base-level lead to the dissection of a peneplain, a summit peneplain remnant may result, usually preserved on the resistant rock outcrops of the major interfluves, with the lower terrain exhibiting a succession of partial peneplains (straths) on successively less-resistant outcrops, valley-side benches and river terraces. Peneplain remnants may be confused with those of marine erosion surfaces (see Section 15.4.3), stripped plains (see Section 7.1), pediplains (see Section 2.4.2), or excavated unconformities (i.e. exhumed peneplains). An example of an exhumed peneplain is thought to be the Fall Zone in the Coastal Plain Province (see Section 2.3). The absence of over-lying datable deposits on uplifted peneplain remnants has often led to relative dating and correlation of surfaces on the mere basis of their elevation, but this is very questionable because of possible distortion of the older surfaces.

Uplift of an erosion surface and the resulting stream incision may produce a drainage network that tends to transgress structural and lithological controls. Similar, but more striking, results may arise from other causes. One may involve a structural discordance of drainage patterns due either to *superimposition* of a drainage network from an unconformable cover of rock or marine or floodplain deposits on to a set of buried structures, or to *antecedence* where rivers cut across young mountains or uplifted structures of recent origin. It was commonly assumed that the latter had formed across the course of an existing river and that erosion had been able to keep pace with the uplift. In practice it is often difficult to assign a cause to such discordance; indeed, the type example of supposed antecedence, that of the Green River cutting across the Uinta Mountains in northern Utah, is now considered to be due, in part at least, to superimposition. However, the cutting of the Dolores River across the Paradox Valley anticline in western Colorado (see Figure 6.25), although less striking, may be more truly antecedent, as is that of the Columbia River across the Cascade Mountains. A spectacular example of probable antecedence is that of the gorge of the River Arun across the axis of the Himalayas between Mount Everest and Kanchenjunga. Two supposed instances of superimposition at least partly due to negative baselevel movements have been, first, certain features of the drainage of the folded Appalachians, in particular the watergaps, and, second, the drainage on the flanks of the Wealden dome and in the associated area of Wessex in south-central England (see Section 2.3). The Wessex drainage may have been superimposed from a gently dipping, thin cover of marine sediments associated with an Early Pleistocene (Calabrian) sea-

level transgression which reached up to some 200 m (c.650 ft) above present sea level (see Figure 2.7). The subsequent negative baselevel movements may have caused this southward-flowing drainage to be super-imposed across a series of truncated folds formed during the Alpine orogeny, but recent work has questioned this interpretation (see Section 2.3).

A drainage anomaly resulting from polycyclic development is *river capture*, although it may occur as a natural development within one cycle, as with the head-ward growth of subsequent rivers on homoclinal structures (see Section 7.2). Capture may take place when the accelerated downcutting of one river, due to its steeper gradient or softer outcrop location, gives it an erosional advantage over an adjacent river at the same elevation resulting in headward cutting and diversion. Figure 2.3 shows some of the features associated with capture,

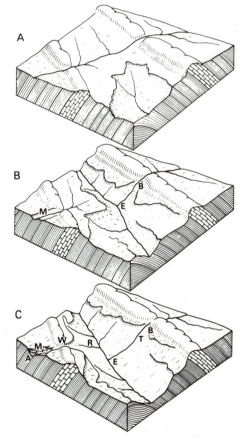

**Figure 2.3** Three successive stages of stream piracy resulting from a stream at a lower elevation cutting back and beheading a stream flowing at a higher level; see text, for description. *Source:* Jenkins, California Division of Mines Bull. 135, 1964, p. 135.

which include an elbow of capture (E), renewed trenching (T) by both the capturing stream and the captured segment headed by a nickpoint (B), a wind gap (W), aggradation (A) by the 'misfit' beheaded stream (M) and the reversal of drainage (R) near the elbow of capture which becomes more pronounced as time passes. Multiple captures obviously can produce very complex drainage patterns.

## 2.3 Denudation chronology

Denudation chronology is concerned with the reconstruction of the evolution of landforms for a given area through time by means of the absolute or relative dating of erosional and depositional events occurring under the influence of tectonic, eustatic, climatic, or other variations. The most simple reconstructions of denudation chronology rest on the assumption that landform assemblages are a palimpsest of superimposed parts of different erosional cycles each initiated by a change of baselevel of tectonic or eustatic origin. An obvious difficulty is that, as we have seen, the possible geomorphic results of negative movements of baselevel are more widespread and striking than those of positive movements, but because these results are predominantly erosional, they are more difficult to date. Unlike sedimentation, upon which most of the study of historical geology is based, erosional processes are historically defective in that they tend to destroy the signs of their own agency. It is for these reasons that large-scale cyclic studies of denudation chronology tend to identify the results of a supposed sequence of sporadic negative movements of baselevel, and they commonly apply to a long time period with few internal time constraints. The most ambiguous denudation chronologies employ widespread morphological evidence of accordances, flats and breaks of slope of supposed erosional origin to postulate a widespread series of complete or partial peneplains, or of planes of marine denudation formed close to pre-existing baselevels, and which have been relatively or absolutely uplifted by negative baselevel movements. By contrast, the most unambiguous denudation chronologies are those which are spatially limited to the immediate vicinity of rivers within which a datable sedimentary record may give a much clearer picture of erosional and depositional events over a short historical timespan.

A classic example of denudation chronology is that developed for the Appalachian Mountains in Pennsylvania and adjacent areas. Athough in many respects there is a striking relation between geology and terrain in the region, a multicyclic history has been proposed because here, as in most other older localities, there are significant departures of landforms from those which could be predicted on the basis of tectonics and geological structure alone. These departures include:

(1) Possible accordant summit levels, notably at about 2200–2400 ft, existing in the distinct structural provinces of the Ridge and Valley Province of the folded Appalachians, the Appalachian plateaus and the northern part of the crystalline Blue Ridge Province (see Figure 7.13); these suggest peneplanation.

(2) The transverse river courses and aligned watergaps of the Ridge and Valley and the northern Blue Ridge provinces; these suggest antecedence, superimposition, or piecemeal headward cutting of drainage lines along weak structural trends.

(3) The asymmetry of the whole drainage, whereby much drainage originating on the west side of the axis of tectonic symmetry (see Figure 7.14A) flows across the range to the east; this suggests a mass reversal of drainage at some stage in the denudation history.

A major difficulty in explaining these features is that, with the exception of the formation of the fault troughs in the Triassic (see Figure 7.24), there are no datable events in the region between the folding of the lowest Permian rocks and the deposition of Early Pleistocene high river terraces – a period of some 250 million years! On the other hand, this temporal vacuum permits great latitude for creative flights of fancy on the part of the denudation chronologer. Most theories of Appalachian topographic evolution have concentrated on the drainage considerations in points 2 and 3, above. These postulated, variously, the existence of original master antecedent rivers, for which there is no evidence; that the original consequent drainage divide was located in the Ridge and Valley Province, or relocated there by renewed

---

**Figure 2.4**   Johnson's theory of the development of Appalachian landforms:
1, rejuvenated Appalachians in Post-Triassic time; 2, Fall Zone Peneplain; 3, marine overlap and deposition of Cretaceous beds; 4, arching of Fall Zone Peneplain and superimposition of south-east-flowing rivers; 5, Schooley Peneplain; 6, arching of the Schooley Peneplain; 7, dissection of Schooley Peneplain and erosion of the (partial) Harrisburg Peneplain on less-resistant limestones and shales; 8, further uplift and erosion of the (partial) Somerville Peneplain on least-resistant limestones; 9, latest uplift and trenching of the present valleys.
*Source:* Johnson, 1931, figures 1–9, pp. 15, 17, 19, copyright © Columbia University Press, reprinted by permission.

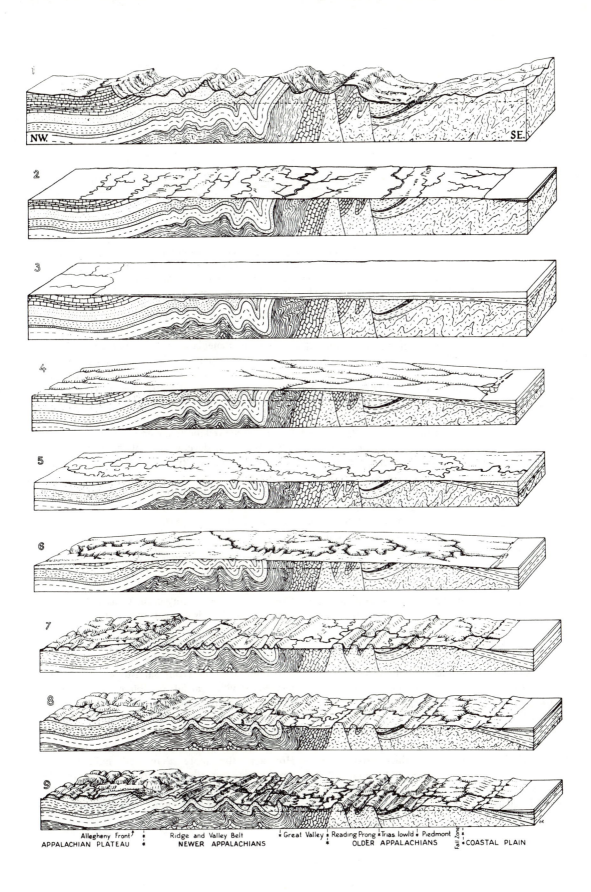

NW.      SE.

Allegheny Front  |  Ridge and Valley Belt  |  Great Valley  |  Reading Prong  |  Trias lowl'd  |  Piedmont  |  Fall Zone

APPALACHIAN PLATEAU     NEWER APPALACHIANS     OLDER APPALACHIANS     COASTAL PLAIN

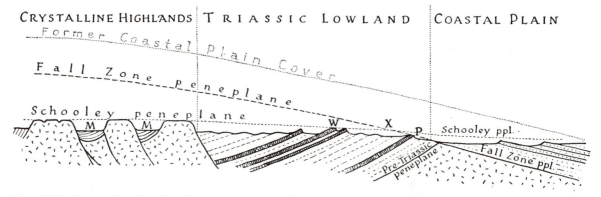

**Figure 2.5**  Cross section of northern New Jersey showing the geomorphic evolution, according to the theory of regional superimposition in Pre-Schooley time: M, M = Musconetcong and similar adjusted streams of the highland area; W = Watchung Mountains diabase ridges; P = Palisades diabase ridge; X = intersection of Schooley and Fall Zone peneplains. *Source:* Johnson, 1931, figure 15, p. 77, copyright © Columbia University Press, reprinted by permission.

buckling, for which there is no evidence; the reversal of drainage by headward extension of drainage westward involving capture and ridge breaching along aligned zones of weakness due to faulting or stratigraphic thinning of the resistant beds, for which there is no structural evidence; and a piecemeal drainage reversal by Triassic faulting, progressive capture, local tilting of peneplains and local superimposition. The latter formed part of Davis's classic denudation chronology in his *Rivers and Valleys of Pennsylvania* (1889), which was replaced in 1931 by the scheme of regional superimposition suggested by his student Douglas Johnson, even though it was not based on any substantially new or more secure evidence. This replacement is a good example of the philosophical principle of Occam's Razor (entities are not to be multiplied without necessity), that the most logical theory is the one based on the fewest assumptions. Johnson's thesis, strikingly supported by block diagrams (Figure 2.4), involved:

(1)  High relief (rejuvenated?) in the Triassic, with the major divide in the vicinity of the older Appalachians and arid fan deposits being laid in the New Jersey faulted trough (1 in Figure 2.4).

(2)  The formation of the Fall Zone Peneplain, Post-Triassic and Pre-Cretaceous in age (2 in Figure 2.4).

(3)  An extensive Cretaceous marine overlap extending to the west of the present Allegheny Front (3 in Figure 2.4).

(4)  Broad uparching of the Fall Zone Peneplain and its cover, with the crest over the western part of the Ridge and Valley Province, from which east- and west-flowing drainage was superimposed (4 in Figure 2.4).

(5)  Development of the widespread Schooley Peneplain (5 in Figure 2.4).

(6)  Uparching of the Schooley Peneplain (6 in Figure 2.4). The relation between the Fall Zone and Schooley Peneplains is shown in Figure 2.5.

(7)  Development of the Harrisburg partial peneplain on the less-resistant shale and limestone outcrops of the Ridge and Valley Province, leaving remnants of the Schooley surface bevelling the quartzite ridge crests and the Reading Prong (7 in Figure 2.4) (Plate 2).

(8)  Renewed moderate uplift of the region allowing the restricted development of the Somerville partial peneplain on the Ridge and Valley limestones (8 in Figure 2.4).

(9)  Slight rejuvenation uplift causing the present rivers to be trenched below the Somerville level (9 in Figure 2.4).

This is an elegant thesis, but certain weaknesses must be noted. Rejuvenation and relative uplift are considered to be entirely a tectonic matter; there is no evidence for the Fall Zone Peneplain except that related to the Sub-Cretaceous unconformity in the Coastal Plain Province (Figure 2.5); there is no evidence for the Schooley Peneplain except for quite a wide vertical range of ridgetop elevations, many of which in the Appalachian Plateau province may be structurally controlled by the near-horizontal Carboniferous sandstones; the Harrisburg and Somerville partial peneplains are restricted to separate outcrops of differing resistance and may be the result of differing rates of erosion rather than of different cycles of erosion (Hack, 1960); and there is no evidence for a Cretaceous cover inland of the Fall Zone. However, the more than 250 million years of Appalachian erosional history, bereft of datable deposits, means that theories of denudation chronology alternative to that of 'regional superimposition' are also difficult to substantiate. These include the 'consequent theory' which assumes an initial tectonic drainage divide north-west of the folded Appalachians *ab initio*, and that of westward drainage divide migration by 'progressive piracy' (Von Engeln, 1948; Strahler, 1945).

Another important cyclic interpretation of denudation chronology is that of south-east England by Wooldridge and Linton (1939) (see analysis by Sparks, 1972; Small, 1978). It relates to an area smaller than the central Appalachians, is more detailed and draws on a much wider range of evidence, including that of structure, lithology, topography, drainage anomalies and, most important, datable deposits. The major observations and inferences can be listed as follows:

(1) There is a Sub-Eocene unconformity (UCF) suggesting uplift and planation of the Chalk, which finds very limited surface expression as a resurrected fossil feature on the dip slopes of the Chalk where the overlying Eocene rocks have been stripped back (Figure 2.6). (For Tertiary chronology, see Figure 20.1).

(2) After further marine sedimentation in the Early Tertiary, the Weald was domed up during the Alpine orogeny (Late Oligocene–Early Miocene) in association with the production of a series of minor anticlinal axes (see Figure 7.7).

(3) There is a rough accordance of summit levels in south-east England (at 215–275 m) (700–900 ft) on the higher parts of the Chalk cuestas, and Lower Greensand and the central Weald. The irregular form and weathered character of this surface implies that this is a dissected Mio-Pliocene peneplain which is younger than the Early Miocene (as it truncates the assumed Alpine folds).

(4) The Chalk cuestas of the North and South Downs appear to be bevelled by a level marine surface, at two localities of which (Netley Heath at 200 m and Lenham at 180 m: Figure 2.7) are located marine deposits of Late Pliocene–Early Pleistocene (Calabrian) age (Plate 3). Where bevelling is not complete, the inner edge of this surface is marked by a steepening of slope inferred to be a possible degraded marine cliff (Figure 2.6).

(5) On either side of this assumed Calabrian shoreline the adjustment of drainage to geological structure is strikingly different (Figure 2.7). 'Landward' (i.e. on the Wealden 'island' and north-west of the Chilterns) the drainage is well adjusted to structure and there is a general absence of surficial flint debris, suggesting that the Chalk was stripped off during an earlier cycle of erosion. 'Seaward' of the Calabrian shoreline the drainage is markedly less well adjusted to structure suggesting that it may have been superimposed – possibly from a Late Pliocene–Early Pleistocene marine cover.

(6) Below about 200 m (650 ft) there are graded reaches on certain rivers and a possible flight of marine benches on the dip slope of the South Downs suggesting a discontinuous drop in sea level since the Early Pleistocene, with an approximate 150-m (475-ft) level being especially marked in the Hampshire Basin.

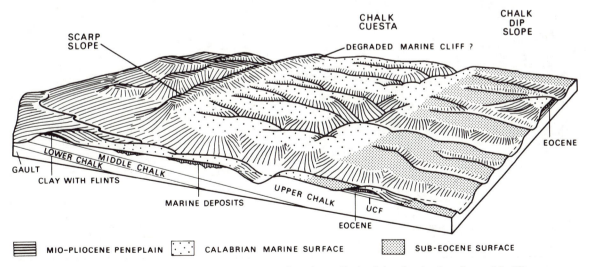

**Figure 2.6** Generalized features of the Upper Chalk cuestas of south-east England showing the three facets: Mio-Pliocene Peneplain; Calabrian (Plio-Pleistocene) marine surface and Sub-Eocene surface.
*Source:* Wooldridge and Linton, 1939, figure 16, p. 60.

Putting this supposed evidence together, a complex denudation chronology was proposed by Wooldridge and Linton involving the production of an extensive Sub-Eocene marine (subaerial?) erosion surface, subsequently deformed by earth movements; the production of a subaerial peneplain during the Mio-Pliocene, the margins of which were eroded into marine benches by the encroaching Calabrian sea; the downwarping of the eastern part of the Weald by at least 100 ft; the overall intermittent relative uplift of the region of the order of more than 650 ft during Pleistocene times mainly by negative eustatic movements during which there was extensive superimposition of drainage, together with the formation of marine and river terraces; and finally, a post-glacial marine transgression during the last 18,000 years.

Unlike Johnson's more speculative Appalachian denudation chronology, Wooldridge and Linton's imaginative scheme has been the object of detailed criticism in recent years (Jones, 1981). Early Tertiary tectonic instability is now held to have been prolonged

(i.e. Paleocene to Oligocene) and complicated during which considerable denudation produced an extensive polycyclic erosion surface on the Chalk; similarly, the time between the Mid-Oligocene and Upper Pliocene had a much more complicated tectonic history than is suggested by a simple Alpine (i.e. Late Oligocene to Early Miocene) deformation of the Sub-Eocene surface (see Figure 7.7); the origin of the proposed Mio-Pliocene Peneplain was probably complicated and prolonged, perhaps involving exhumation; the Calabrian transgression was not capable of achieving the drainage modifications attributed to it, which may have been partly due to antecedence; the Lenham sea was probably restricted to the eastern Weald; and Pleistocene chronology, particularly of the Thames Valley, has been shown by stratigraphic studies to have been much more complex than was hitherto believed.

## 2.4 Criticisms of the cycle and alternative models

The concept of the cycle of erosion has faced increasing

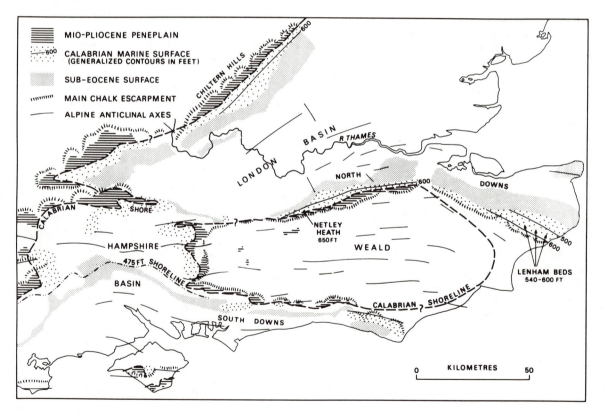

**Figure 2.7**   The present-day distribution of principal features of the denudation chronology of south-east England. The Calabrian shoreline (approx. 200 m; 650 ft) is shown by broken line.
*Source:* Wooldridge and Linton, 1939, figures 15 and 17, pp. 58, 62.

criticism. It has been held that the cyclic concept is too theoretical, that landscape systems behave in a much more complex manner than Davis allowed, that notions of equilibrium and non-cyclic erosion do not find a sufficient place in the thesis and, finally, that many of the premises of denudation chronology with which the concept of the cycle became almost entirely associated are questionable and highly subjective.

The theoretical emphasis of Davis's cyclic concept was criticized from its inception; indeed, Davis, an unashamed model-builder, added fuel to these criticisms by declaring that 'the scheme of the cycle is not meant to include any actual examples at all, because it is by intention a scheme of the imagination and not a matter for observation'. However, one of the most attractive features of Davis's cycle is its simplicity, and this was achieved by playing down possible complications, by the pruning of supposedly 'irrelevant' information and the adoption of restrictive assumptions. The net effect of these was to make the cyclic landform system dominantly a closed one derived largely through initial inputs of potential energy from uplift and operating under the influence of declining energy gradients against a background of static tectonics, baselevel, climate, hydrometeorology and vegetation. Of course, Davis claimed that such assumptions were only made to facilitate his *explanation* of a more complex reality, but he was astute enough to recognize that the very existence of his model was impossible without them. The tectonic assumptions have been questioned by German workers and more flexible models proposed (see Section 2.4.1); eustatic, climatic, hydrologic and vegetational changes are now believed to have been virtually continuous, strongly episodic and to have exerted profound controls over denudation rates; relief and slope are not now believed to be the sole determinants of intensity of erosion and rates of change in landforms; and the recent concepts of thresholds and the complex response of geomorphic systems (see Sections 1.2.2, 2.7 and 13.5) are a significant departure from cyclic thinking.

Apart from a largely outmoded classificatory model based on questionable assumptions, what remains of the cycle of erosion is largely embedded in 'classical' pre-Second World War denudation chronology. Even here, as we have seen, simplistic cyclic interpretations have been questioned. In particular, elevated flat areas or accordant summit planes have been variously ascribed to structural controls; to the geometrical result of equal stream spacing flanked by uniform slopes; to the increase of erosional intensity with elevation; to the achievement of some balance between uplift and erosion (see Section 2.4.1); to the maintenance of balance between erosion

and resistance (see Section 2.6); to the uplift of other types of erosion surface (e.g. surfaces of marine planation); and to the formation of erosion surfaces at high levels unrelated to grand baselevel (e.g. arid pediplains). An even greater difficulty arose from the practice of treating all topographic benches, straths and terraces as embryonic or partially developed peneplains. As we have seen, studies of the Appalachians exemplified these difficulties and, indeed, in the 1930s remnants of some seventeen partial peneplains were supposedly identified – most of them restricted to a single major rock type! It is now recognized that a single 'cycle' may result in a number of accordant surfaces at different levels on different geological formations, and that supposedly polycyclic features like river terraces may result just as easily from variations in runoff, vegetation and sediment supply as from baselevel changes. In addition, the common lack of datable deposits overlying most of these assumed surfaces, or indeed of any deposits at all, gave an ambiguity and elbow-room for the imagination which was exploited to the full by historians of landscape. In short, denudation chronology in its cyclic guise developed into a highly stylized game based on such initial steps of faith as 'topographic flat equals baselevel still-stand', 'higher equals older' and 'uplift has been infrequent and discontinuous'. Historical ambiguity resulted both from Occam's Razor being used with abandon and in the cavalier employment of *ad hoc* postulates (e.g. the proposed superimposition of drainage from sedimentary covers of which there is no evidence or probable means of verification).

The preceding schemes of denudation chronology have the common underlying cyclic assumption that all parts of a given erosion surface have been developed with respect to a distinct and identifiable baselevel. This important principle is most clearly manifested by a sub-aerial surface which is being reduced according to the Davisian cyclic model for, after a short youthful period, all parts of the surface – hilltops, slopes and valley bottoms – are continually reduced (although at very different rates), so that the landscape can be regarded as a response surface all parts of which represent changing forms which are *synchronous*. However, old surfaces of low relief, even if uplifted, change at very slow rates under Davisian cyclic assumptions until they are caught up in a new erosion cycle wherein trenching streams extend their influence with respect to a new lower baselevel. In contrast, some geomorphologists believe that, to a greater or lesser degree and under certain ranges of climates, active slope retreat can take place by means of unconcentrated slope processes which are unrelated to the location of stream channels in space or to the elevation of baselevel, and that the rate of slope retreat is directly related to the steepness of the slope. Thus steep slope elements are considered to be able to migrate across country high above general baselevel, maintaining their steepness and suffering only small incremental reductions in their height as they leave in their wake a gently sloping surface of low relief. The two most important of these models are those of W. Penck and L. C. King (Figure 2.8).

### 2.4.1 The Penck model

Even during the heyday of Davis's propagation of his cycle of erosion many Central and East European geomorphologists resisted his simplifying assumptions, particularly those relating to crustal behaviour. Of these opponents, the only one to develop a coherent alternative to the Davis cycle was the German Walther Penck. Penck's views have never been popular in the English-speaking world, partly because his early death in 1923 occurred when his major work, *Die Morphologische Analyse* (1924), was incomplete, and because of his obscure composition and ill-defined terminology. It was also unpopular because a critical and misleading review by Davis which appeared about a decade later provided, for many years, the only access of Penck's views to non-German speakers (see Figure 2.8B and C), and because certain of Penck's ideas seem contradictory and impossible to support. Nevertheless, Penck's system threw an interesting and novel emphasis on the possible geomorphic effects of diastrophic causes.

Penck believed that landforms should be interpreted by means of the ratios which might be expected to occur between erosional (exogenetic) processes and diastrophic (endogenetic) processes. Erosional processes were held to operate according to the following worldwide laws, differing only in rate between different climates:

(1) Local intensity of erosion is directly related to the steepness of the slope segment.

(2) The inclination of each segment of an erosional slope is determined by the sizes of the mobile debris.

(3) The largest debris size which is mobile on a slope segment varies with the inclination of the latter; the greater the inclination, the greater is the largest size which is mobile.

(4) If the production of debris by weathering is uniform on a slope segment, erosion will cause the segment to retreat parallel to itself.

(5) If some eroded material is allowed to collect at the base of a retreating slope segment, a new segment of lower inclination will develop.

Most important, Penck believed (as a result of his studies of sedimentary facies flanking Alpine ranges) that most tectonic movements began and ended slowly, and that the common pattern of such movements involved a slow initial uplift, an accelerated uplift, a deceleration in uplift and, finally, quiescence. Regional updoming began with a major phase of waxing development in which the accelerating uplift rates were generally in excess of stream degradation, and the resulting landforms were dominated by the crustal instability (see Figure 2.8C). The term 'waxing' referred to a progressive increase in the erosive rate of the rivers which could, of course, be achieved by means other than accelerating uplift (e.g. increased discharge). Increase of uplift rate during waxing development resulted in a progressive steepening of the slope segments (four of which are shown in Figure 2.9B) producing a general convexity of slope forms on the initial surface or primary peneplain (*Primärrumpf*) (see Figure 2.8B). With the passage of time, the more rapid parallel retreat of the steeper slope segments would tend to reduce the radius of convexity. It should be noted that the Davisian interpretation of such upper slope convexity was to attribute it to the action of soil creep. As the uplift accelerated, the primary peneplain would be surrounded by a series of benches (*Piedmonttreppen*), each of which had originated as a piedmont flat (*Piedmontfläche*) on the slowly rising dome margin. Penck believed that convex breaks of slope (*Knickpunkte*) form along the radially draining river courses during accelerating uplift, leaving one convex nick after the other, below each of which there begins a narrow, steep course flanked by convex valley slopes, while above each there is a broader reach flanked with concave valley-side slopes. The concave stream-reaches between the convex nicks are formed in association with the Piedmonttreppen and each tends to act as an independent local baselevel for the subsequent valley widening on either side of the stream course. Penck made no clear distinction between continuous acceleration of uplift and continuous but intermittently accelerated uplift; the mechanisms that he evoked for the production of Piedmonttreppen and Knickpunkte also lacked

**Figure 2.8** Cyclic models A.–D. of landscape evolution showing the relationships between elevation and time (with schematic slope profiles), assuming a stable baselevel: A. Davis, 1909; B. Von Engeln's (1942, figure 138, p. 259) interpretation of W. Penck; C. a more faithful interpretation of the Penck model; D. model of L. C. King.
*Source:* Thornes and Brunsden, 1977, figure 6.2, p. 122.

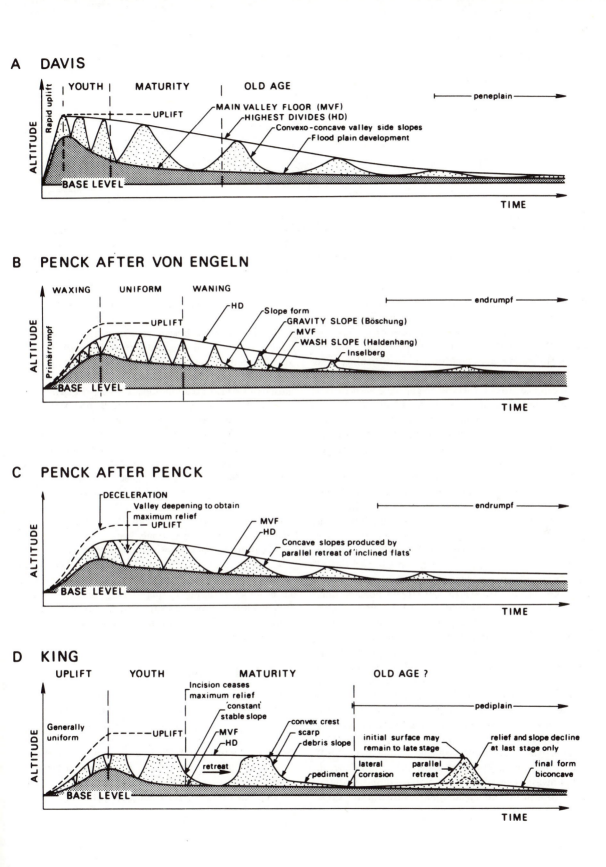

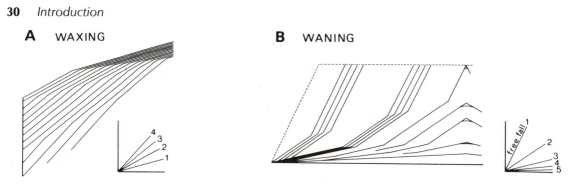

**Figure 2.9** Penck's models for slope development: A. waxing development leading to steepening slope elements, four of which are shown, which tend to retreat parallel to themselves at rates proportional to their inclination; B. waning development showing the free fall (boschungen) slope grading into the haldenhang (segment 2) and the latter progressively replaced by the development and retreat of other lower segments (3–5). The whole process of slope retreat by mutual consumption producing concave slopes. *Sources:* Penck, 1953, figures 3 and 7, pp. 137, 158; Tuan, 1958, figure 2, p. 213.

clarity. Davis, who explained such breaks of slope by intermittent negative movements of baselevel, made much of these objections but the application of Davisian ideas of the development of polycyclic landforms to young mountain ranges runs into severe difficulties in so far as there is a marked lack of time during which flights of widespread erosion surfaces might have been *successively* developed, at least by conventional estimates of erosion rates. The piedmont benchland mechanism of *synchronous* erosion appeared to circumvent this difficulty and had the merit of introducing the concept of episodic behaviour of landforms.

The period of waxing development was believed to be followed by a general decline in the rate of uplift during which a short period of uniform development in which the rate of erosion by streams overtook that of uplift was succeeded by a dominantly waning phase during which the rate of uplift decreased and the landscape became progressively dominated by the erosional processes of valley widening. Waning development is associated with a marked decrease in the rate of stream downcutting, either by deceleration of uplift or by a change in stream activity itself (e.g. a decreasing discharge) – the two mechanisms producing indistinguishable geomorphic effects. Waning slope development is concerned with the parallel retreat of slope segments and, in particular, of the steepest element (the *boschungen*), an angle of repose or free fall slope (segment 1 in Figure 2.9B). The rapid parallel retreat of the boschungen soon consumes much of the convex waxing slope and leaves behind at its base a slope element of lower inclination composed of talus material (the *haldenhang*: segment 2 in Figure 2.9B) susceptible to parallel retreat under the action of creep and rainwash. The collection of mobilized finer material at the foot of the haldenhang may cause a lower slope

segment to develop, and so on (segments 3–5 in Figure 2.9B).

The retreat of the boschungen gives rise at a later time to the production of steep-sided residual *inselbergs* at the divides, and when these are consumed, the whole landscape is made up of concave slopes of low angle and composed of slowly retreating slope segments. Such a surface is termed an *endrumpf* (see Figure 2.8B and C). It is clear that the inclinations of the retreating slope segments, including the boschungen, depend upon the dominant sizes of weathered material successively produced by weathering. At one extreme, if the initial production of coarse weathered blocks was succeeded by their complete collapse into very fine material, the landscape is dominated by only the boschungen and the haldenhang, intersecting at a sharp angle, with the former retreating actively and the latter virtually static (see Figure 2.8B). At the other extreme, debris susceptible to a continuous decrease in size during weathering generates an infinite number of slope segments, producing continuous concave slopes reclining during their retreat (see Figure 2.8C). Little support exists for Penck's tectonic speculations (see Section 3.5) but his views on weathering and slope retreat are important.

### 2.4.2 The L. C. King model

The concept of large-scale pediplanation was developed by L .C. King more than thirty years ago to account for extensive surfaces of low relief in Africa and the tropics, as well as for uplifted surface remnants in higher latitudes (Figure 2.8D). Pediplanation was conceived as dominated by slow parallel scarp retreat under arid, semi-arid and savanna (seasonal wet–dry) climates leaving in its wake broadly concave surfaces ($<6°–7°$), often studded with steep-sided residual hills (inselbergs), varying in size depending on their degree of erosive consumption (e.g. mesas and buttes) and in shape depending on underlying rock structure. Thus angular slope profiles and residuals are associated with horizontal sedimentary rocks or with surfaces capped by material cemented by some processes of chemical weathering (i.e. duricrusts, including silcretes and calcretes (caliche)), whereas rounded profiles and residuals (bornhardts) are supported by curved weathering or pressure-release jointing, notably in granites (Plate 27). Once formed, the low-angle pediplain surfaces are subject to only minute erosion and are capable of persisting for very long periods until they are themselves consumed by a new sequence of steep slope (15°–30°) retreat initiated by isostatic uplift on a regional scale, perhaps in response to metamorphic changes of geochemical origin in the subcrust. In this way one phase of pediplanation may be succeeded by uplift, marginal canyon-cutting and extension, leading to a wave of scarp retreat sweeping across country for long distances, and leaving inselbergs of up to 300 m (1000 ft) in its wake, until their reduction blurs them into the generally concave pediplain surface. Particularly where formed in resistant rocks, pediplains and pediplain remnants are believed to achieve great antiquity, so much so that the highest pediplain remnants are believed by King to have formed before the break-up of the southern hemisphere continental plates in the Jurassic (see Chapter 5). King correlates the highest Gondwana pediplain (Jurassic) in Southern Africa (>4000 ft: 1300 m) with a surface of similar age in Brazil (700–1000 m: 2300–3300 ft). Break-up of the southern continents is believed to have led to further piecemeal continental uplift in the Early Tertiary to produce the African pediplain (650–800 m: 2000–2500 ft near the coast and 1000–1600 m: 3000–5000 ft in the interior of Southern Africa) and in the Cretaceous to produce the Australian pediplain (400–500 m: 1200–1500 ft). It has been suggested that ancient drainage lines on this surface in Western Australia were formed before the separation of the Australian and Antarctic plates (Figure 2.10). Late Tertiary uplift initiated a new phase of pediplanation in Southern Africa, leading to a model of its continental surface as a giant staircase of mutually consuming steps. Despite the existence of these extensive surfaces of low relief separated by clifflike escarpments in the tropics, the concept of antique pediplanations must remain questionable, if only because of the vast periods of time involved and our lack of knowledge regarding the nature and rapidity of erosional processes in subhumid environments.

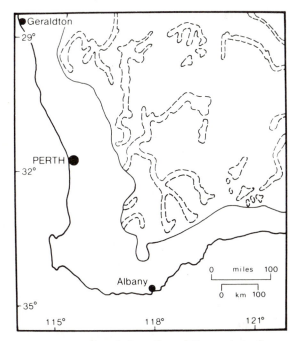

**Figure 2.10** Ancient drainage lines of Western Australia on an old, uplifted drainage surface bounded by the Meckering Line to the west and the Southern Ocean Watershed to the south. The headwaters of the broad drainage lines appear to have originated to the south before plate movements brought about separation of Australia and Antarctica.
*Source:* Ollier, 1981, figure 12.16, p. 175, copyright © Longman, London.

## 2.5 Strategies for inferring landform evolution

When one considers the range of strategies available to students of landform evolution (see especially Thornes and Brunsden, 1977), it is interesting to note the restricted bases employed in the construction of the classic geomorphic models of Davis, Penck and King. Owing to the difficulty in dating the upland areas where erosion is dominant, the models were based on the ergodic assumption which is considered below. Other strategies such as direct observation and measurement or simulation-modelling have also been employed. More recently geochemical dating methods (see Section 5.1.2), such as the use of $^{14}C$, and botanical methods, such as pollen analysis, are proving increasingly useful in dating geomorphological events, particularly more recent ones. For example, radiocarbon dates have been used to estimate the rate of Holocene alluviation in southern England. Such geological techniques will doubtless become more and more important to geomorphological chronology.

### 2.5.1 Ergodic assumptions

Although strict 'ergodicity' (Greek: *ergon* – 'energy', 'work', *hodos* – 'way') requires rigorous statistical assumptions, the *ergodic hypothesis* suggests that under certain circumstances sampling in space can be equivalent to sampling through time; and that space–time transformations are permissible as a working tool. This is based on the assumption that, when individual members of a population of landforms are changing regularly through time, the spatial frequency of occurrence of given types of landforms is inversely proportional to their rate of change (i.e. there are fewer forms representing stages of rapid change). Clearly, not all morphometric features exhibit ergodicity; however, interesting results have been obtained by placing such

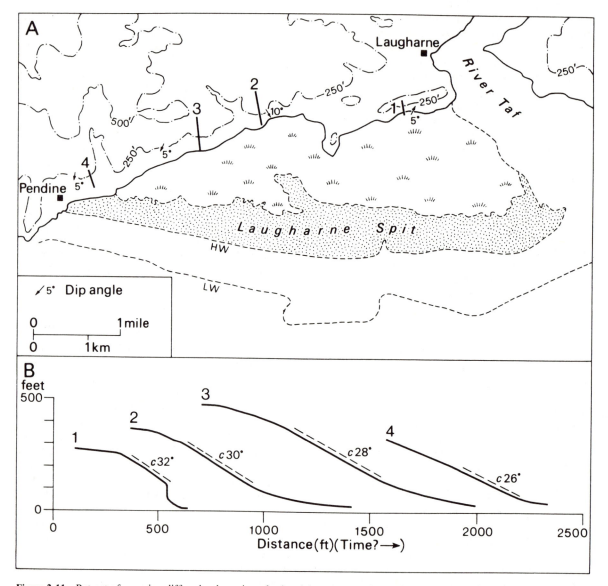

**Figure 2.11** Retreat of a marine cliff under the action of subaerial processes as it was progressively protected from wave attack by the eastward-growing Laugharne Burrows split. A. Cliffed coast between Pendine and Laugharne, South Wales; locations of the cliff profiles depicted in B. are shown; B. Four profiles of cliff sections arranged, so that the subaerially youngest is on left. *Source:* Savigear, 1952, figures 1 and 2, pp. 32, 36.

forms as regional valley-side slope profiles and drainage networks in assumed time sequences. The ergodic method, as first applied in thermodynamics, is based on a prior assumption of the energy path of a system and is not a means of proof in itself. In the absence of absolute dating methods it can sometimes be hypothesized that spatial morphologic assemblages may be genuinely representative of time sequences of individual landforms, and this hypothesis may be of assistance in the development of geomorphic ideas. Indeed, the concept of the cycle of erosion is based to a large extent on ergodic assumptions. Section 13.4.1 shows how the ergodic hypothesis has been used to infer drainage basin evolution, and another simpler example – of sequential slope development – has been provided by a study of the extent of subaerial erosion of a sea cliff in South Wales. The cliff has been progressively protected from wave attack as the Laugharne Burrows marine spit (Figure 2.11A) has extended eastwards and, as a consequence, the cliff profiles can be placed in a simple locational time sequence with respect to their relative protection from all but subaerial processes (Figure 2.11B).

Obviously, ergodic assumptions are dangerous in that landforms may be assembled into assumed time sequences simply to fit preconceived theories of denudation. Additionally, the investigator always runs the risk of arguing in circles, by assuming that a temporal sequence exists because there are spatial variations when, in reality, the latter merely represent chance fluctuations around an equilibrium state (see Section 2.6). A much deeper cause for concern is that only one variable is commonly assumed to be an ergodic indicator. In this connection it is well to note that in Figure 2.11 cliff form may also be influenced by cliff height and direction and dip of bedrock, each of which vary with the locations of individual cliff profiles. Similarly, in Chapter 13 the basins whose hypsometric integrals and drainage densities will be compared (see Figure 13.22) were of different absolute magnitudes of area and height.

### 2.5.2 Direct observation and measurement

The traditional geomorphic view of change through time is that it is generally slow compared with periods of human observation such that, for example, Davis never thought it worth while to make field measurements of rates of change. Of course, short-term changes of landforms composed of weak, depositional materials (e.g. beaches, floodplains, channel reaches, sand dunes, etc.) have long been recognized to take place, but it appeared difficult to rationalize such apparently capricious changes in terms of protracted landform evolution. Modern work is showing, first, that a larger group of landforms than was hitherto thought is susceptible to measurement, demonstrating significant changes through time, and second, that concepts of episodic erosion (see p. 40) and of the magnitude and frequency of geomorphically significant processes help to make long-term evolutionary sense out of short-term measurements of landform change. In particular, it is now clear that erosional and depositional landforms in semi-arid regions are susceptible to quite rapid and measurable systematic changes, especially those developed on shale outcrops. Examples will be given here of landform changes deduced from *point measurements* (i.e. those restricted to the small spatial scales of individual landforms and to limited time periods of experimental field observation), and these should be distinguished from those changes which are deduced from lumped measurements (i.e. those relating to erosional and depositional changes deduced from longer-term estimates relating to extensive areas – see Chapter 3).

An example of point measurement is given in Figure 2.12 which shows observed erosion on four clay badland slopes at Perth Amboy, New Jersey, measured during a ten-week period in the summer of 1952. These results have been plotted together, using a dimensionless percentage distance downslope from the top of the straight segment. Two features appear from this latter plot: the variation of the observed amount of erosion at a given distance downslope is quite large, suggesting a departure from a simple spatial pattern; but nevertheless, the average depth of erosion (about 0.9 in) is more or less uniform at all parts of the slopes, suggesting that the straight segments are retreating parallel to themselves. Successive surveys can give some timespan to estimates of rates of landform translation, as with a sequence of maps of the western end of Scolt Head Island, Norfolk, England (Figure 2.13) and with superimposed maps showing channel changes in part of the Mississippi River and the River Sid in Devon, England (Figure 2.14A and B). The former show little change in the Mississippi River meander magnitude following the catastrophic cutoff of Moss Island in 1821, whereas the latter suggest an increasing sinuosity of the River Sid during the last 100 years, possibly due to human agricultural activities. On occasion vegetation dating can suggest temporal and spatial patterns of landform evolution, as with tree-ring dating of cottonwoods on a section of floodplain of the Little Missouri River (Figure 2.15). This study indicated that during the past 200 years the elevation of the river has remained unchanged and that in this period the river has redistributed by bank erosion and point-bar deposi-

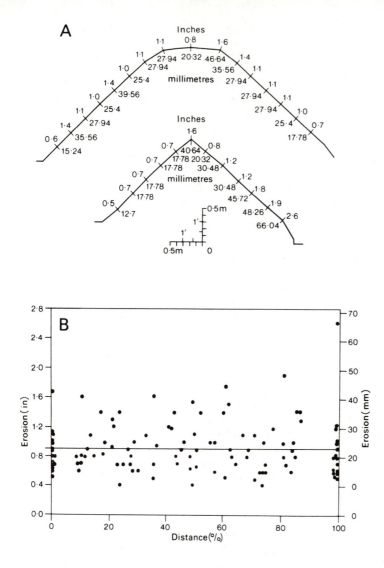

A

Inches
0·8
1·1        1·6
1·1    27·94  20·32  46·64    1·4
     27·94         35·56    1·1
1·0    25·4   millimetres   27·94    1·1
1·4    39·56                     1·1
1·0    25·4                  27·94   1·1
1·1    27·94                 27·94   1·0
1·4    27·94                      25·4   0·7
0·6    35·56                          17·78
15·24

Inches
1·6
0·7    40·64   0·8
0·7  17·78  20·32   1·2
0·7  17·78  30·48   1·2
0·7    millimetres   30·48   1·8
0·7  17·78         45·72   1·9
0·7  17·78              48·26   2·6
0·5  17·78              66·04
0·5    17·78
12·7

0·5m
1'
1'
0·5m    0

**Figure 2.12** Slope profiles at Perth Amboy, New Jersey, showing A. depth of erosion (in inches and millimetres) during the period June–September 1952; B. a regression fitted to scatter diagram of depth of erosion at all measured points at Perth Amboy during the period June–September 1952, expressed in terms of percentage distance from the top of respective straight slope segment.
*Source:* Schumm, 1956, figures 26 and 27, pp. 623–4.

B

[scatter plot: Erosion (in) and Erosion (mm) vs Distance (%)]

---

1927                      1932                      1937

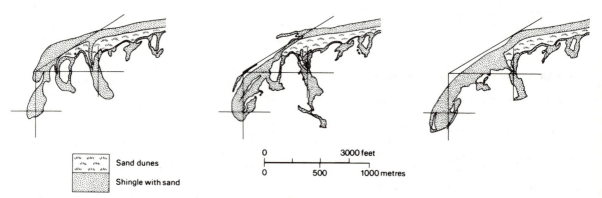

 Sand dunes

Shingle with sand

0          3000 feet
0    500   1000 metres

**Figure 2.13** Development of Tenery and Far Point, Scolt Head Island, Norfolk, England, in 1927–37.
*Source:* Steers, 1964, figure 80, p. 365.

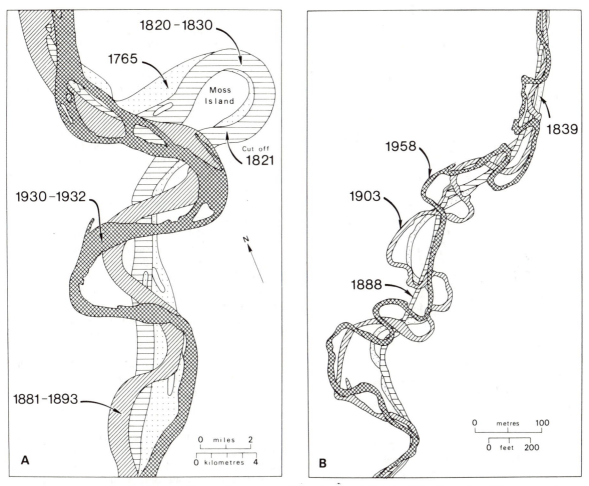

**Figure 2.14** Surveys showing changes of course for two meandering rivers: A. Mississippi River in northern Tennessee during the period 1765–1932; B. River Sid in east Devon during the period 1839–1958.
*Sources:* US Army Corps of Engineers, from Strahler, 1965, figure 20.13, p. 295; Hooke, in K. J. Gregory (ed.), *River Channel Changes*, 1977, figure 17.8A, p. 276, copyright © John Wiley and Sons, by permission.

tion an amount of sediment equal to all that presently available in the floodplain.

Point measurements of landform change still present problems in extrapolation, especially as erosion and deposition are irregular and episodic both in space and time. Questions of spatial generalization from point measurements and of their extrapolation through time in terms of total geomorphic work accomplished by intense, infrequent processes, as distinct from average and frequent ones, will be treated elsewhere. Questions of climatic change, the existence of erosional thresholds, the effect of vegetation (which may affect local erosional rates by up to two orders of magnitude) and human influence present problems of temporal extrapolation, even if we assume that measured changes are accurate.

### 2.5.3 Simulation modelling

An increasingly popular way of investigating landform evolution through time is by the use of simulation models which enable the effects of geomorphic processes, or more commonly surrogates of them, to be speeded up under controlled conditions. There are three main classes of such simulation models: hardware scale models, analogue models and mathematical models (Chorley and Kennedy, 1971).

Hardware scale models are closely imitative of a segment of the real world, which they resemble in some very obvious respects (i.e. being composed mostly of the same types of materials), and the resemblance may sometimes be so close that the scale model becomes merely a

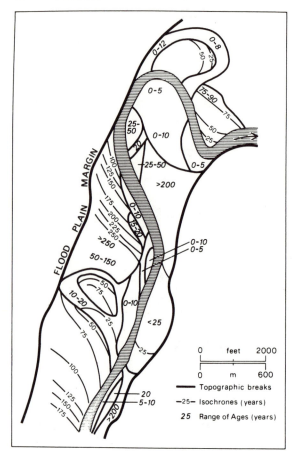

**Figure 2.15** Age distribution of cottonwood trees on floodplain of the Little Missouri River, near Watford City, North Dakota.
*Source:* Everitt, 1968, figure 6, p. 431.

suitably controlled portion of the real world. Two obvious instances of this are wind-tunnel observations of sand movement and investigations of the erosional transformations of small-scale controlled rainfall-erosion facilities. This, of course, gives the clue to the basic difficulty of the construction of hardware models in geomorphology; that nature's complexity imposes scaling or changes of media requirements of a very high order of sophistication. Of these problems, the most difficult is that changes of scale affect the relationships between certain properties of the model and the real world (e.g. scale ratios) in different ways, such that, for example, the kinematic scale ratios (i.e. those involving velocities and accelerations) behave differently from linear scale ratios (i.e. those involving lengths and shapes). Similar difficulties are involved with attempts to produce meaningful dynamic scale ratios (i.e. those

involving gravity forces, such as mass and inertia). Such discrepancies can be commonly circumvented either by distorting one variable to represent another faithfully (e.g. a distortion of the vertical linear scale of river models enables the effects due to turbulence – the kinematic ratio – to be more faithfully reproduced) or by the dimensionless combination of attributes (as in the Reynolds and Froude numbers), which allow individual variables to be reproduced (in this case viscous and gravity effects, respectively). Difficulties of scaling natural geomorphic phenomena explain why attempts to reproduce whole fluvial landform associations and their transformation have been less successful than those reproducing more limited features like river meander reaches (see Chapter 12) and beach segments.

Analogue models involve radical changes in the media of which the model is constructed. They have much more limited aims than scale models in that they are intended to reproduce only some limited aspects of reality. Such transformations are obviously rather difficult in that large and often questionable assumptions must be made regarding the appropriateness of the changes of media involved. One example of such an analogue hardware model is the use of a kaolin mixture to simulate some features of the deformation and crevassing of a valley glacier. Of recent years analogue models have been almost totally replaced by computer-based mathematical models.

A mathematical model is an abstraction in that it replaces objects, forces, events, etc. by expressions containing mathematical variables, parameters and con-stants, involving the adoption of a number of idealiza-tions of the various phenomena studied and ascribing to the various entities involved some strictly defined prop-erties. The essential features of the phenomena are then analogous to the relationship between certain abstract symbols and resemble closely a very simple version of the real world, so that the equations are a kind of working model from which one may predict evolutionary behaviour. The common type of geomorphic math-ematical model is concerned with some simplified state-ment of certain important features of the real world (usually geometric ones) which can be transformed according to mathematical assumptions regarding the basic operation of the system (usually related to changes through time) to yield a succession of geomorphic changes through time. These mathematical conclusions are then checked against the real world, and the correspondence between the real world and the effects predicted by the model indicate the success which has been had in its construction – the differences revealed may then lead to improvements in the model such that observed facts may become better understood. Mathematical models are commonly divided into deterministic and stochastic.

Deterministic mathematical models are based on classic mathematical notions of exactly predictable relationships between independent and dependent variables (i.e. between cause and effect) and consist of sets of exactly specified mathematical assertions (derived from experience or intuition) from which unique consequences can be derived by logical mathematical argument. The most common type of deterministic geomorphic model involves the transformation of slope profiles under various assumptions regarding the original slope geometry and the manner of its transformation. Deterministic mathematical models have been extensively employed in research on slope development (see Chapter 11).

Stochastic mathematical models consist of sets of equations involving mathematical variables, parameters

and constants together with one or more random components, the latter arising from random effects which may be extremely important in influencing natural processes and in producing unpredictable fluctuations in observational or experimental data. A simple type of stochastic mathematical model is the random walk process used to generate simulated branching stream networks by allowing drainage from each of a grid of squares with differing probability, depending on direction. Figure 2.16A presents a network resulting from flow which is equally probable in all four cardinal directions (i.e. $P{\downarrow} = P{\uparrow} = P{\leftarrow} = P{\rightarrow} = 0.25$) and Figure 2.16B shows a network developed from a directionally biased model (i.e. $P{\downarrow} = 0.4$; $P{\uparrow} = 0.1$; $P{\leftarrow} = P{\rightarrow} = 0.25$). This type of model illustrates one of the major difficulties in employing mathematical simulation to explore system changes through time, namely that the model may have to be so simplified and its development proceed through such unreal and artificial steps that its only value lies not in its evolutionary representations, but in its ability to approach and exemplify some equilibrium state.

Despite the attractions of mathematical modelling, it presents the major problem that very different models are often found to generate rather similar results, and that the differences between the various simulated landforms and real landforms are not easy to test as to their significance.

## 2.6 Equilibrium landforms

Two themes have been interwoven in twentieth-century geomorphology; first, the idea of progressive, irreversible change, and, second, that of compensatory change about some characteristic condition. The former stems from ideas of organic evolution, entropy maximization and positive feedback; the latter from the notions that form and process are related by negative feedback mechanisms and that processes are subject to oscillations in intensity, causing the associated landforms to respond sympathetically in their form. Progressive change and oscillatory change (or, put more crudely, evolution and equilibrium) cannot be considered independently, for change can only be measured with reference to some equilibrium datum and oscillations are seldom compensatory over the long term. These notions have been introduced in Figure 1.7, suggesting that equilibrium can be variously viewed as a condition towards which landforms evolve or decay in the longer term (Figure 1.7A), as Davis proposed, or as a condition about which they fluctuate in a steady state in the shorter term, as defined by Gilbert (Figure 1.7B). As will be seen later, modern work

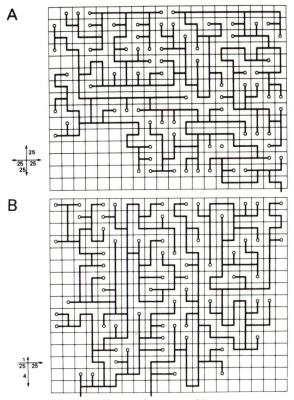

**Figure 2.16** Channel networks generated by computer simulation: A. random model in which the probability of 'steps' in all four major directions are equal: B. directed model in which the probability of steps towards the bottom of the figure is four times that towards the top.
*Source:* Chorley and Kennedy, 1971, figure 7.28, p. 284, after Smart, Surkan and Considine.

in geomorphology has attempted to unite the two approaches both in terms of the idea of change within a framework of dynamic equilibrium (Figure 1.7C) and by the inclusion of discontinuous thresholds of more catastrophic change characteristic of dynamic metastable conditions (Figure 1.7D) (Chorley and Kennedy, 1971).

Concepts relating to the equilibrium adjustments of landforms to their formative processes are found throughout this book, but it is instructive here to examine some of the more important geomorphic manifestations of equilibrium. In doing so it soon becomes obvious that all of these are closely interrelated. Four such manifestations are:

(1) Statistical stability of form. This appears as the statistical clustering of certain morphometric variables around values characteristic of given lithology, soils, vegetation, climate and stage of dissection, imparting to a geomorphic region its aspects of uniformity.

(2) Correlation of forms and processes, such as between stream channel and associated valley-side slope angles in given regions (see Section 13.5) and between stream discharge and alluvial channel cross-sectional form (see Section 12.5.1).

(3) Balance over time. The most common temporal geomorphic balances occur when oscillations of process (i.e. flood discharges or seasonal variations of process intensity) affect deposits of erodible sediments. Seasonal cut and fill is well exemplified by the summer advance and the winter retreat of a beach at Carmel, California, during the period April 1946 to February 1947 (Figure 2.17) under the action of swell

and storm waves, respectively. Here the volume of sand remains relatively constant but assumes an overall steeper profile during the constructional wave action of summer than during the destructional winter wave attack.

(4) Balance over space. This implies some uniformity of energy expenditure or work done over space which is mirrored in the spatial association of landform assemblages, as was implied by the notion of statistical stability of form. For example, it has been suggested that, in the folded Appalachians, the local relief and slope angles have been so adjusted that each major geological outcrop yields an equal sediment load per unit area (i.e. hard rocks – high, rugged and steep; soft rocks – low, gently rolling and with low slopes) (Hack, 1960). Although this is an attractive alternative explanation for geologically limited 'cyclic' surfaces, it is difficult to support. Classical ideas of grade contain elements of spatial equality and continuity of work, with the longitudinal profile consisting of a set of reaches each having a gradient necessary to provide the flow velocity required to transport the load supplied from upstream. In this respect the common concave-up longitudinal stream profile is viewed as the result of systematic downstream changes of discharge, load–discharge ratio and debris calibre.

The above consideration of equilibrium landforms leads inevitably to the problem of how landforms exhibiting equilibriating tendencies in the shorter term can *evolve* in the longer term. In Chapter 1 it was suggested that the

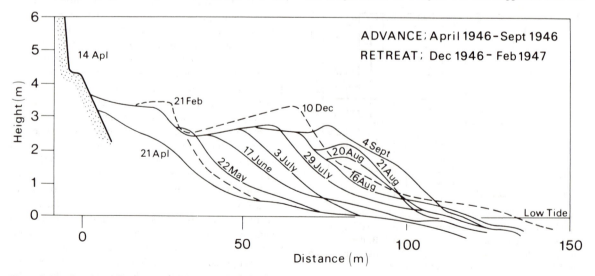

**Figure 2.17**   Beach profile changes at Carmel, California, between April 1946 and February 1947.
*Source:* Surveyed by Bascom, from Komar, 1976, figure 11-3, p. 291.

time required for the denudation of a landscape can be subdivided into cyclic, graded and steady time periods (see Figure 1.9). Under the category of cyclic time are timespans of geologic duration, that is the period of time required for the denudational evolution of a landscape. For example, during this period, one expects an essentially exponential decrease of stream gradients, which is a landscape component that reflects broad changes in the fluvial system. However, cyclic time can itself be subdivided into graded time and steady time periods. During graded time average grade will remain relatively constant, but there will be, through time, fluctuations about this mean. During the very short period of steady time there is no change. When considering a landscape or its components, it is helpful to think in terms of the timespans and how a landscape is altered during the timespan under consideration.

Of course, it is unlikely that denudation will continue for any appreciable period of time without external interruption by climate change, high-intensity catastrophic events, tectonic movements, or by isostatic adjustment. So the smooth curves presented in Figure 1.9 can be expected to be complicated when external influences act on the system. In addition, it is likely that the internal workings of the fluvial system itself will prevent progressive reduction of a valley floor or stream gradient. A change of an external variable will interrupt the progress of the geomorphic cycle and changes of stream profile and variations of gradient during graded time are readily understood as reflecting variations of discharge and sediment load. Nevertheless, Figure 1.9 poses a problem, for it is difficult to image how the graded time curve can be compatible with the cyclic time curve. This line of reasoning apparently requires the elimination of either the concept of progressive erosion or grade. However, there is an alternative solution. If valley floors and stream gradients do not evolve progressively, but rather change rapidly during brief periods of instability that separate longer periods of grade, then a model incorporating both progressive change and grade can be proposed (i.e. episodic behaviour, Figure 2.18).

As a landscape changes components of the landscape (e.g. divides, hillslopes, tributaries, and alluvial channels) will not necessarily be adjusted to one another or be graded. That is, a channel adjusting to uplift may not be ready to cope with the effects of the rejuvenation of the watershed upstream. Hence, an actively eroding system will be continually hunting for a stability that cannot be maintained. In this way an uplifted drainage system will have difficulty in immediately disposing of all the sediment delivered to its major channels from minor tributaries and interfluve areas, and sediment will be temporarily stored within the system.

## 2.7 New evolutionary concepts

Two geomorphic concepts that have potential for aiding in the development of an understanding of this complexity of landscape evolution were introduced in Chapter 1.2.2, namely geomorphic thresholds and complex response. When a small experimental drainage basin is rejuvenated, the system responds not simply by incising, but by hunting for a new equilibrium by incision, aggradation and renewed incision. This can be referred to as the *complex response* of the system. The concept of *geomorphic thresholds* suggests that there can be changes *within* the fluvial system itself that are not due to external influences, but rather to geomorphic controls inherent in the eroding system. Field and experimental studies demonstrate that when sediments are stored within a fluvial system, they become unstable at critical threshold slopes, leading to accelerated erosional events. This, for example, seems to be a reasonable explanation for the formation and distribution of some arroyos and

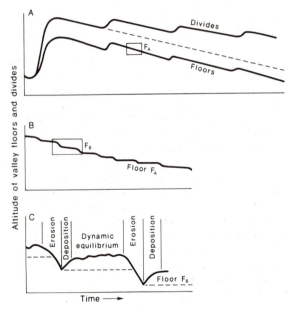

**Figure 2.18** Modified concept of geomorphic cycle:
A. erosion cycle following uplift, as envisioned by Davis (broken line), and as affected by isostatic adjustment to denudation (full lines), B. portion of valley floor $F_A$ showing episodic nature of decrease of valley floor attitude; C. portion of valley floor $F_B$ showing periods of instability separated by longer periods of dynamic equilibrium.
*Source:* Schumm, 1975, figure 5, p. 77.

gullies in the American West. When the influence of external variables such as isostatic uplift is combined with the effects of complex response and geomorphic thresholds, it is clear that denudation, at least during the early stage of the geomorphic cycle, cannot be a progressive process as assumed in the Davis cycle. Rather, it is comprised of episodes of erosion separated by periods of relative stability, a complicated sequence of events. Much of this complexity is the result of a delayed transmission of effects through the landscape. For example, channel changes that take place near the mouth of a drainage basin following incision are responding to the conditions at that time and location, but the channel may not be prepared for the changes that this incision induces within the river system upstream. Hence downcutting may be followed by deposition when an upstream response occurs (Schumm, 1973, 1975).

When the concepts of geomorphic thresholds and complex response are applied to landscape evolution, the model becomes as summarized in Figure 2.18, as the progress of denudation is here interrupted by periods of isostatic adjustment. The upper line of Figure 2.18A represents changes of divide elevations and the lower line those of valley-floor elevations. Only major external

influences affect the divides which are subjected to a relatively uniform downwearing. However, if the valley floor is considered in greater detail over a shorter span of time (Figure 2.18B), a stepped pattern of valley-floor reduction emerges as a result of storage and flushing of sediment from the valleys. This model ignores variations due to external influences, and it shows a system that is in dynamic metastable equilibrium (Figure 2.19). As we have seen, a steady state equilibrium involves fluctuations about an average, but a metastable equilibrium occurs when an external influence carries the system over some threshold into a new equilibrium regime. The effects of external variables on equilibrium systems are expected, but in the case of landscape denudation the dynamic metastable equilibrium may reflect the response of the system to *inherent* geomorphic thresholds, for example, the accumulation of sediment to an unstable condition. When a geomorphic threshold is crossed, the drainage system will be rejuvenated and the complex response will come into play (Figure 2.18C). Figure 2.18C shows periods of instability separated by longer periods of dynamic equilibrium or grade. Because periods of erosion are followed by periods of deposition, the bedrock floor of the valley will be reduced in a discontinuous manner through time as shown by the broken lines. Hence separating the periods of erosion will be periods of deposition and storage of alluvium. In addition, during these periods of relative stability, channel pattern may change from straight to increasingly sinuous, as the nature of the sediment moved through the channel changes. Hence sinuosity may also increase to a condition of incipient instability at which time a large flood can cause an abrupt shortening of the river course and the establishment of a period of valley erosion.

This dynamic metastable equilibrium model of episodic erosion shows, in addition, that many of the details of the landscape (e.g. small terraces and recent alluvial fills) do not need to be explained by the influence of external variables because they develop as an integral part of system evolution. More than this, modern studies of thresholds and complex response have suggested how the Davisian cyclic decay model and the steady state model of Gilbert may be effectively combined into a unified vision of landform evolution.

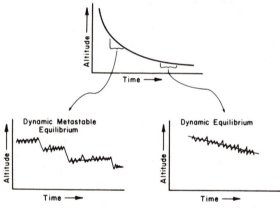

**Figure 2.19** Erosional evolution of valley floors under dynamic metastable equilibrium conditions and those of dynamic equilibrium, depending on time (stage) of the cycle of erosion.

*Source:* S. A. Schumm, *The Fluvial System*, 1977, figure 4-18, p. 90, copyright © John Wiley and Sons, by permission.

# References

Brown, E. H. (1960) *The Relief and Drainage of Wales*, Cardiff, University of Wales Press.

Chorley, R. J. (1963) 'Diastrophic background to twentieth-century geomorphological thought', *Bulletin of the Geological Society of America*, vol. 74, 953–70.

Chorley, R. J. (1965) 'A re-evaluation of the geomorphic system of W. M. Davis', in R. J. Chorley and P. Haggett (eds), *Frontiers in Geographical Teaching*, London, Methuen, 21–38.

Chorley, R. J. (1967) 'Models in geomorphology', in R. J. Chorley and P. Haggett, *Models in Geography*, London, Methuen, 59–96.

Chorley, R. J. (1974) 'Walther Penck', in *Dictionary of Scientific Biography*, New York, Charles Scribner's Sons, Vol. 10, 506–9.

Chorley, R. J., Beckinsale, R. P. and Dunn, A. J. (1973) *The History of the Study of Landforms*, Vol. 2, *The Life and Work of William Morris Davis*, London, Methuen.

Chorley, R. J. and Kennedy, B. A. (1971) *Physical Geography: A Systems Approach*, London, Prentice-Hall.

Coates, D. R. and Vitek, J. D. (eds) (1980) *Thresholds in Geomorphology*, London, Allen & Unwin.

Cullingford, R. A., Davidson, D. A. and Lewin, J. (eds) (1980) *Timescales in Geomorphology*, Chichester, Wiley.

Davis, W. M. (1909) *Geographical Essays*, Boston, Ginn.

Davis, W. M. (1932) 'Piedmont benchlands and the primärrümpfe', *Bulletin of the Geological Society of America*, vol. 43, 399–440.

Dury, G. H. (1975) 'Neocatastrophism?', *Annals of the Brazilian Academy of Sciences*, supplement, 135–51.

Everitt, B. L. (1968) 'Use of cottonwood in an investigation of the recent history of a flood plain', *American Journal of Science*, vol. 266, 417–39.

Fenneman, N. M. (1936) 'Cyclic and non-cyclic aspects of erosion', *Science*, vol. 83, 87–94.

Flemel, R. C. (1971) 'The attack on the Davisian system of geomorphology', *Journal of Geological Education*, vol. 19, 3–13.

Gilbert, G. K. (1877) *Report on the Geology of the Henry Mountains*, Washington, DC, US Department of the Interior.

Green, J. F. N., Bull, A. J., Gossling, F., Hayward, H. A. and Wooldridge, S. W. (1934) 'The River Mole: Its physiography and surficial deposits', *Proceedings of the Geologists Association*, vol. 45, 35–67.

Gregory, K. J. (ed.) (1977) *River Channel Changes*, Chichester, Wiley.

Hack, J. T. (1960) 'Interpretation of erosional topography in humid temperate regions', *American Journal of Science*, vol. 258A, 80–97.

Harbaugh, J. W. and Merriam, D. F. (1968) *Computer Applications in Stratigraphic Analysis*, New York, Wiley.

Hooke, J. and Kain, R. (1982) *Historical Change in the Physical Environment*, London, Butterworth.

Johnson, D. W. (1931) *Stream Sculpture on the Atlantic Slope*, New York, Columbia University Press.

Jones, D. K. C. (1981) *Southeast and Southern England*, London, Methuen.

Kesseli, J. E. (1940) 'The development of slopes', Berkeley, Calif., Department of Geography, University of California, Berkeley, mimeo.

King, L. C. (1950) 'The study of the world's plainlands', *Quarterly Journal of the Geological Society of London*, vol. 106, 101–31.

King, L. C. (1967) *Morphology of the Earth*, 2nd edn, Edinburgh, Oliver & Boyd.

Komar, P. D. (1976) *Beach Processes and Sedimentation*, Englewood Cliffs, NJ, Prentice-Hall.

Lobeck, A. K. (1939) *Geomorphology*, New York, McGraw-Hill.

Ollier, C. D. (1981) *Tectonics and Landforms*, London, Longman.

Penck, W. (1953) *Morphological Analysis of Landforms*, trans. H. Czech and K. C. Boswell of *Die morphologische Analyse* (1924), London, Macmillan.

Savigear, R. A. G. (1952) 'Some observations on slope development in South Wales', *Transactions of the Institute of British Geographers*, vol. 18, 31–51.

Schumm, S. A. (1956) 'Evolution of drainage systems and slopes in badlands at Perth Amboy, New Jersey', *Bulletin of the Geological Society of America*, vol. 67, 597–646.

Schumm, S. A. and Lichty, R. W. (1965) 'Time, space and causality in geomorphology', *American Journal of Science*, vol. 263, 110–19.

Schumm, S. A. (1973) 'Geomorphic thresholds and complex response of drainage systems', *Proceedings of the 4th Annual Geomorphology Symposium, Binghamton*, 299–310.

Schumm, S. A. (1975) 'Episodic erosion: a modification of the geomorphic cycle', *Proceedings of the 6th Annual Geomorphology Symposium, Binghamton*, 70–85.

Schumm, S. A. (1977) *The Fluvial System*, New York, Wiley.

Simons, M. (1962) 'The morphological analysis of landforms: a new review of the work of Walther Penck (1888–1923)', *Transactions of the Institute of British Geographers*, no. 31, 1–14.

Small, R. J. (1978) *The Study of Landforms*, 2nd edn, Cambridge University Press.

Sparks, B. W. (1972) *Geomorphology*, 2nd edn, London, Longman.

Steers, J. A. (1964) *The Coastline of England and Wales*, 2nd edn, Cambridge University Press.

Strahler, A. N. (1945) 'Hypotheses of stream development in the folded Appalachians of Pennsylvania', *Bulletin of the Geological Society of America*, vol. 56, 45–88.

Strahler, A. N. (1965) *Introduction to Physical Geography*, New York, Wiley.

Thornbury, W. D. (1969) *Principles of Geomorphology*, 2nd edn, New York, Wiley.

Thornes, J. B. and Brunsden, D. (1977) *Geomorphology and Time*, London, Methuen.

Tuan, Y. -F. (1958) 'The misleading antithesis of Penckian and Davisian concepts of slope retreat in waning development', *Proceedings of the Indiana Academy of Science*, vol. 67, 212–14.

Von Engeln, O. D. (1948) *Geomorphology*, New York, Macmillan.

Wooldridge, S. W. and Linton, D. L. (1939) 'Structure, surface and drainage in south-east England', *Institute of British Geographers Special Publication*; reprinted with additions, London, George Philip, 1955.

# **Three**  *Cascading process systems*

The concept of the cascading system was briefly touched upon in Chapter 1 and it is now necessary to elaborate this with particular reference to the solar energy, hydrological and sedimentary cascades of the *exogenetic* (Greek: *exo* – 'outside', *gegnesthae* – 'to beget') processes at or near the earth's surface (Sections 3.1–3), as well as to the less obvious diastrophic cascades of the *endogenetic* (Greek: *endo* – 'inside') processes of the earth's interior (Section 3.4). Finally, the geomorphic effects of the interaction of these two sets of processes will be considered (Section 3.5). This is predominantly a functional approach to geomorphology.

Cascading systems are composed of connected chains of subsystems, through and between which may flow a cascade of mass or energy. Thus the output from one subsystem may, in whole or part, become the input for another, perhaps triggering off threshold reactions or complex responses in the latter. The behaviour of cascading systems may be analysed in terms of three types of model which form a continuum:

(1) White box models which represent attempts to include all known details of flows, storages, subsystem states, detailed responses to inputs, etc., in the hope of predicting the details of system behaviour.

(2) Black box models which treat the whole cascading system as a unit, without concern for details, directing attention solely to the outputs and changes of gross system state resulting from given inputs.

(3) Grey box models which involve consideration of only selected details of system operation and thus lie intermediate between the white and black box models.

In an attempt to to give an introduction to the detailed exogenetic processes treated in Part III we shall, where possible, in this chapter concentrate on the white box approach. Although fluvial sediment cascades are high-lighted in this chapter, it will become clear in Part III that glacial, coastal and aeolian processes are also examples of cascading systems.

## 3.1 The solar energy cascade

Solar energy flows, or flux densities, are commonly expressed in watts per square metre (W/m$^2$ = joules per second per square metre) or in langleys (ly = 1 calorie per square centimetre), where 1 ly/min = 697·3 W/m$^2$. The flows involved in local diurnal energy balances are shown in Figure 3.1A as:

$$R_n = H + LE + G$$

where:  $R_n$ = net all-wave radiation;
  $H$ = turbulent sensible heat flux density;
  $LE$ = turbulent latent heat flux density (associated with evaporation or evapotranspiration);
  $G$ = subsurface heat flux density.

All these are positive (i.e. $H$, $LE$ and $G$ directed away from the surface) during the day and negative at night, when the surface radiation loss (i.e. $R_n$, negative) is made up partly of low-level atmospheric heat conduction (i.e. $H$, negative) but mostly of subsurface heat flux (i.e. $G$, negative). When, as is usual, net daily values of $R_n$ are not zero, this implies a net gain or loss in the energy $\triangle S$ stored in the near-surface soil or vegetation layers: where $\triangle S$ = the net energy storage (i.e. rate per unit volume, per unit horizontal area) made up of the residual between positive (mostly diurnal) and negative (mostly nocturnal) values of $G$. The biochemical utilization of heat by plant growth (in photosynthesis) is normally very small (maximum 0·009 – 0·023 ly/min = 6 – 16 W/m$^2$). The above equation can, therefore, be rewritten:

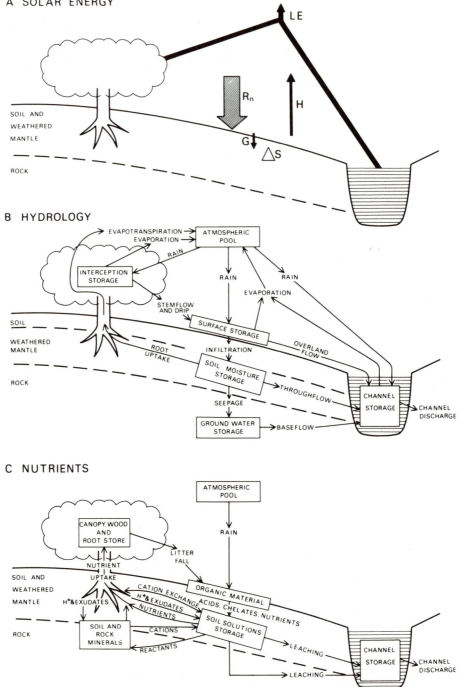

**Figure 3.1** Three important geomorphic cascades: A. solar energy cascade: B. hillslope hydrological cascade; C. nutrient cycle cascade.

$$R_n = H + LE + \Delta S.$$

The energy exchanges together with the associated thermal levels, involving non-vegetated bedrock and soil surfaces, have been studied especially at high elevations and in hot deserts. Figure 3.2A shows the energy exchanges at noon and at 1800 hours on 18 July 1963, at an elevation of 1610 m (5280 ft) in the Sierra Nevada, California, producing temperatures at the granite surface of 52°C (125·6°F) and 22°C (71·6°F), respectively. More detailed temperature readings are available for bare dark-coloured basalt and light-coloured

sandstone in the Tibesti region of the central Sahara for a mid-August day in 1961 (Figure 3.2C and D). Although the maximum surface temperatures of both rocks reached almost 79°C (174°F), the greater conductivity of the basalt ($7·4 \times 10^{-3}$ v. $5·7 \times 10^{-3}$ cal cm$^{-1}$ s$^{-1}$ °C$^{-1}$) gave it a lower night minimum temperature, a greater diurnal temperature range (43°C v. 37°C) and a greater depth of penetration of the diurnal temperature wave. This wave penetrated to a depth of about 1 m into the basalt but much less into the sandstone, giving a steeper temperature gradient near the surface of the latter.

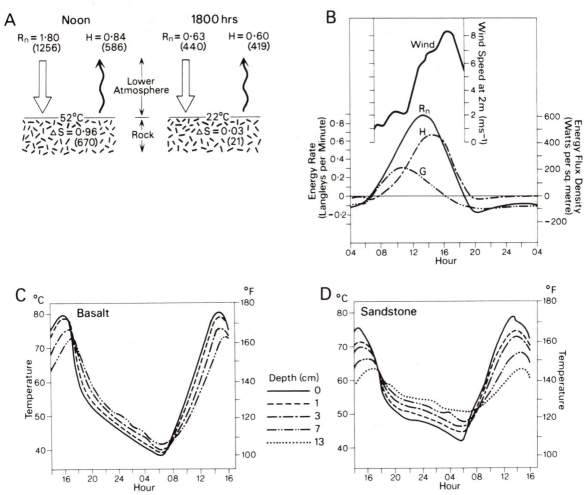

**Figure 3.2** A. and B. Solar energy flows involved in the energy balance of some non-vegetated natural surfaces., A. Bare granite in Blue Canyon, California, at noon and 1800 hours on 18 July 1963, when the sun's altitude was 70° and 10°, respectively; values are given in ly/min and in W/m² (in parenthesis). B. Dry playa lake surface at El Mirage, California (38°N), on 10–11 June 1950; wind speed due to surface turbulence was measured at a height of 2 m. C. and D. Diurnal temperatures near, at and below the surface in the Tibesti region, central Sahara, in mid-August 1961: basalt, at the surface and at 1 cm, 3 cm and 7 cm below it, and light-coloured sandstone, at the surface and at 1 cm, 3 cm, 7 cm and 13 cm below it, respectively.
*Sources:* Miller, 1965; Oke, 1978, figure 3.1, p. 65; Peel, 1974, figures 1 and 2, pp. 24–5.

**Figure 3.3** A. and B. Solar energy flows involved in the energy balance of forested surfaces.
A. A thirty-year-old oak stand in Voronezh Province in the Soviet Union on an average summer day (June–August); values are given in ly/day and in W/m² (in parenthesis). B. A stand of Scots and Corsican pine at Thetford, eastern England, on 7 July 1971; there was cloud cover during the period midnight to 0500 hours. C. Diurnal temperatures in the treetops of the highest canopy (24 m) and in the undergrowth (0·7 m) of a tropical rainforest in Nigeria on 10–11 May 1936, during the wet season.
*Sources:* Sukachev and Dylis, 1968, figure 4, p. 69; Oke, 1978, figure 4.21a, p. 129; Richards.

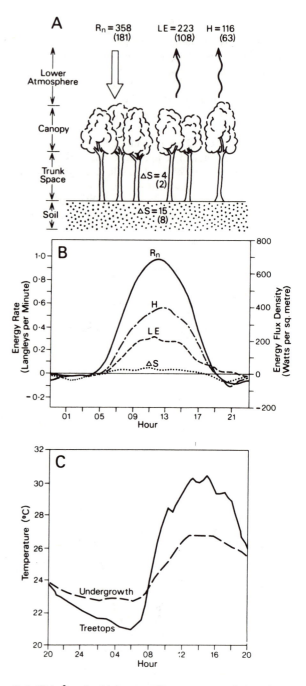

The features of the energy flows in respect of clastic desert surfaces are exemplified by a dry lake bed at El Mirage, California, during 10–11 June 1950, when the air temperature varied between 17°C (62·6°F) and 29°C (84·2°F) (Figure 3.2B). At noon the air temperature at a height of 2 m was some 28°C less than that at the surface. In places the temperature of desert surfaces may reach as much as 70°C (158°F) and, for example, even at a height of 1·5 m the diurnal temperature range in the air at Tucson, Arizona, may reach 56°C in summer. $R_n$ varies systematically at desert surfaces, reaching a maximum at about 1300 hours and falling to small negative values (i.e. net radiation) between an hour after sunset and an hour after sunrise. For much of the morning, ground heating $G$ exceeds the transfer of sensible heat to the atmosphere $H$, but after noon $H$ greatly exceeds $G$ due to the surface heat losses being increased by the convective turbulence of the lowest atmosphere. During a 24-hour period in summer $R_n$ is roughly apportioned 90 per cent to $H$ and 10 per cent to $G$.

Vegetated surfaces exhibit the above effects in a more complicated way. Figure 3.3A shows solar energy conditions for a Russian oak forest on an average summer day when most of the radiation is being processed and dissipated in the crown canopy and only small proportions are heating the trunk space and soil, which suffer small diurnal temperature changes. In temperate deciduous forests mean monthly trunk space temperatures may be more than 5°C (9°F) less than in the surrounding open country. Forest humidity is naturally higher and less variable than in surrounding localities. Figure 3.3B gives the diurnal energy flows for a pine forest in eastern England on 7 July 1971, showing similarly low values for $\triangle S$ but significantly lower values of $LE$ compared with the oak which transpires with greater facility than the pine. Like short green crops, only a very small proportion of $R_n$ is used ultimately for tree growth, an average figure being about 1000 ly/year

(1·3 W/m²), of which some 60 per cent produces wood tissue and 40 per cent forest litter. The effect by trees of creating a stable ground-level microclimate is even more marked in tropical rainforests which have stratified trees up to 46–55 m (150–180 ft) high and a great variety of different species which produce three or more layered

**Table 3.1** Interception differences between a tropical forest and temperature woodlands

|  | Gross interception (%) | Stemflow and drip (%) | Net interception (%) |
|---|---|---|---|
| Tropical rainforest (Brazil) | 67 | 27 | 40 |
| White pine | 30 | 4 | 26 |
| Aspen/birch | 15 | 5 | 10 |

canopies leading to very constant temperatures (Figure 3.3C), high humidities and low light intensities at the forest floor.

## 3.2 The hydrological cycle

Figure 3.1B gives details of the near-surface circulation of water which results from the solar energy budget and which provides the primary vehicle for the cycling of plant nutrients. This water is shown as circulating through a number of stores and becoming evaporation, evapotranspiration, or channel discharge (Figure 3.1B; see also Section 12.4). A *storm hydrograph* is the plot of stream discharge against time for a basin subjected to a given rainstorm event. It is usually a right-skewed, curve with a short-lived rising limb, a peak and a prolonged recession limb (see Figures 13.15 and 20.25). The form of the hydrograph is controlled by the nature of the storm event, the hydrological and geomorphic state of the basin, and the contribution of the different basin storages to the hydrological cascade through time.

Rainfall, depending on its location, becomes temporarily either channel storage, surface storage, or interception storage. Interception storage depends very much on the density and structure of vegetation. For heavy-crowned, open-grown temperate deciduous forests, gross interception commonly consists of the first millimetre of any rainstorm plus 20 per cent of the rest of the rainfall. Of this remainder, 15–19 per cent (termed 'net interception') subsequently evaporates and 1–5 per cent reaches surface storage as stemflow and drip. In Table 3.1 the annual figures show interception differences between tall, layered tropical forests and less-dense temperate woodlands. Surface storage is made up of moisture retained within the litter layer, depression storage and water in transit over the surface. In temperate woodlands litter storage may reach 5 per cent of the total precipitation and on slopes of moderate angle microdepressions may be expected to retain 7–13 mm of rainfall as depression storage. The amount of overland flow is commonly determined as the excess of surface

water remaining after infiltration into the soil has occurred. Infiltration capacity is the maximum rate (e.g. millimetres per hour) at which a given soil in a specified condition can absorb rainfall. Infiltration capacity varies with vegetation and litter cover, soil permeability and existing soil moisture content, variations in the last causing a decrease of infiltration capacity during individual storms to a limiting value. The general effects of surface cover on average infiltration capacities are shown in Table 3.2.

**Table 3.2** Effects of surface cover on infiltration capacity

|  | Infiltration capacity (mm/h) |
|---|---|
| Oak/hickory forest | 76 |
| Mature pine forest | 63 |
| Forest with little humus or litter | 49 |
| Old pasture | 43 |
| Bare loamy sand | >25 |
| Bare loam | 12·5–25 |
| Bare silt and loam | 7·5–12·5 |
| Bare clay and loam | 2·5–5·0 |

Overland flow is, therefore, greater on surfaces with lower infiltration capacities $f$ and may reach velocities of 180–270 m/h (i.e. 0·05–0·075 m/s, compared with average streamflow velocities of 0·45 m/s). Overland flow may be either unsaturated (i.e. Hortonian; $q_h$) where rainfall intensity exceeds the surface infiltration capacity plus evaporation and depression storage, or saturated ($q_s$) when the underlying soil layers are saturated as is common in concavities, near channels and at streamheads after prolonged rain. These saturated surfaces (or 'partial areas') vary in extent during and between storms; for example, covering 5 per cent of Appalachian forested drainage basins during periods of moderate rainfall and 20–40 per cent during wetter periods. This 'dynamic basin model' has an important bearing on theories relating to the development of new drainage lines (see

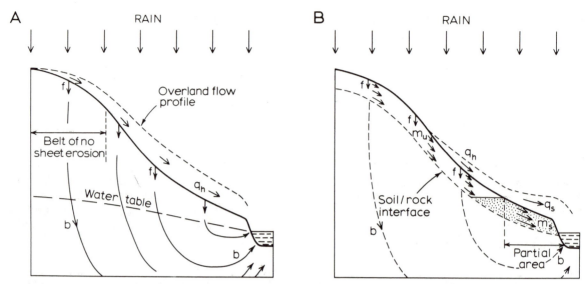

**Figure 3.4** Schematic representation of the A. Hortonian and B. dynamic partial area models of hillslope hydrology; see text for description.

*Source:* Chorley in M. J. Kirkby (ed.), *Hillslope Hydrology*, 1978, figure 1.7, p. 26, copyright © John Wiley and Sons, by permission.

Chapter 13) in so far as they are most likely to be excavated during extremely high runoff conditions and in the saturated areas. Over the past twenty years the simple Horton overland flow model of hillslope hydrology (Figure 3.4A) has been shown to be of restricted application to slopes with sparse vegetation and thin soils (e.g. badlands), whereas the dynamic, partial area model (Figure 3.4B) is probably of much wider application in vegetated, soil-covered areas (see Section 11.5).

Water exists in soil storage in intergranular films (Figure 3.5) through which it flows partly vertically to the bedrock and thence to baseflow (*b*), and partly as lateral throughflow, either under unsaturated ($m_u$) or saturated ($m_s$) conditions. Velocities of the lateral flows may be many orders of magnitude greater than the vertical ones where a permeable soil lies on a less-permeable bedrock, as is commonly the case.

Figure 3.5 suggests that water which has been delivered to the surface of a porous soil or rock tends (under the conflicting forces of surface tension and gravity) to dispose itself into four zones. Three of these are above the water table and have a positive capillary tension ($P_c$) in the interstitial water films, with $P_c$ increasing in zones 1 and 2 inversely with the size of the films (i.e. as $r_1$ and $r_2$ decrease; see Figure 3.5) above the capillary rise, so that water tends to be drawn up towards the capillary fringe if evapotranspiration decreases the film sizes there. Below the water table there is complete saturation of the pore spaces and porewater *pressure* occurs (i.e. $P_c < 0$).

Below the water table (i.e. in the zone of true ground-water flow in porous rocks – *b*, or in that of saturated soil flow – $m_s$) flow is largely governed by Darcy's Law, which states that the rate of waterflow through a permeable medium is directly proportional to the head loss and inversely to the length of the flow path:

$$V = K \frac{H}{L}$$

where: $V$ = velocity of flow (cm/day);
$K$ = coefficient of permeability (see later);
$L$ = flow length (cm);
$H$ = hydraulic head (or piezometric head loss in distance $L$).

This assumes the following to be constant:

$Q$ = discharge into and out of the system (cm³/s);
$A$ = cross-sectional area of flow (cm²);
$\mu$ = viscosity of water (in centipoises: 1 centipoise = 0·01 dyne s/cm²: water at 21°C has $\mu$ = 1 centipoise);
$t$ = temperature (°C).

See Figure 3.6 for a simple diagrammatic explanation, where $\theta$ is the slope of the surface of piezometric head loss due to friction. Darcy's Law only holds for laminar flow, as is shown by the variable hydraulic gradient experiments on loosely and densely packed sand (Figure

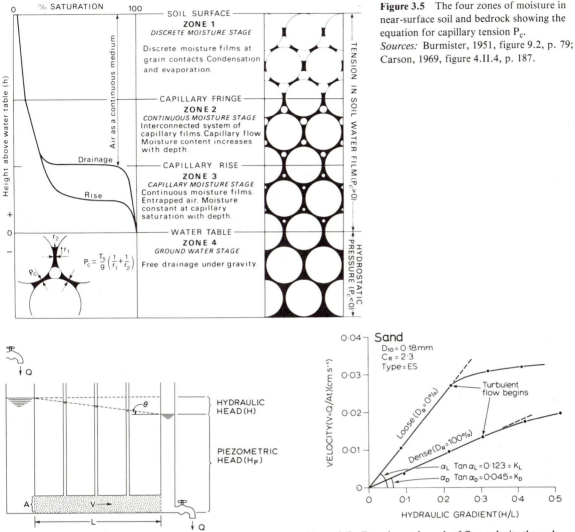

**Figure 3.5** The four zones of moisture in near-surface soil and bedrock showing the equation for capillary tension $P_c$.
*Sources:* Burmister, 1951, figure 9.2, p. 79; Carson, 1969, figure 4.II.4, p. 187.

**Figure 3.6** A simple hydraulic analogue model illustrating Darcy's Law in respect of water moving through a granular medium (stippled) where the piezometric head is analogous to the water table.

**Figure 3.7** Experimental graph of flow velocity through loose and dense sand achieved under varying hydraulic gradients. The slopes of the linear graphs express coefficients of permeability and the non-linear graphs the flow regimes wherein Darcy's Law does not apply.
*Source:* Burmister, 1951, figure 13.2, p. 116.

3.7) which yield two different values for $K$ (0·123 and 0·045). $K$ is defined as follows:

$1K \simeq$ discharge of 4 cm³ water per day, at 15·6°C, through a cross section of 1 cm² under a hydraulic head of 100%.

It is modern practice to express sediment permeability in pore area, as 'specific permeability' $k$, expressed in square centimetres of bulk volume. The unit for expressing $k$ is the darcy, where:

1 darcy = discharge of 1 cm³ per second, at a viscosity of 1 centipoise, through a cross section of 1 cm² under a pressure gradient of 1 atm per centimetre length;

1 darcy = $0.987 \times 10^{-8}$ cm² (for water at 15·6°C, 1 darcy = $18.2K$).

Table 3.3 gives some comparative values of $k$ and $V$.

**Table 3.3** Comparative values of $k$ and $V$

| Material | $k$(darcy) | | $V(m/day)$ ( $\frac{H}{L} = 0.1\%$) |
|---|---|---|---|
| Gravel (well sorted) | $4.3 \times 10^4$ | | $>9$ |
| Coarse sand (well sorted) | $3.1 \times 10^3$ | | $1.96$ |
| Medium sand (well sorted) | $2.6 \times 10^2$ | | $0.36$ |
| Very fine sand (well sorted) | $9.9 \times 10^0$ | Good | |
| Sandstone (29% porosity) | $2.4 \times 10^0$ | aquifers | |
| Sandstone, coarse (12% porosity) | $1.1 \times 10^0$ | | |
| | | $10^0$ | |
| Limestone (16% porosity) | $1.4 \times 10^{-1}$ | | |
| Sandstone, silty (12% porosity) | $2.6 \times 10^{-3}$ | | |
| Silt | | Poor | $0.02$ |
| Fine sands, silts, glacial till | $10^0 - 10^{-4}$ | aquifers | |
| Kaolinite clay | $10^{-3}$ | | |
| | | $10^{-4}$ | $<0.001$ |
| Montmorillonite clay | $10^{-5}$ | Impervious | |

## 3.3 Denudation: the sediment cascade

Central to any understanding of landform change is some conception of the rates at which sediment is removed from one area and cascades through a series of others, so changing the morphology of all of them. Estimates of rates of denudation are important both in developing historical theories of landform evolution through cyclic time and for the more utilitarian and functional purposes of shorter timespans. Not least it is important for the geomorphologist to be able to make some comparison, however rough, between the rates of tectonic movements and of erosion and deposition, the balance of which determines the large-scale changes of landforms.

### 3.3.1 Transported loads in rivers

Frequently in order to estimate rates of landform change the only data available are sediment loads transported by rivers.

There are three components of the total sediment load of a river: the dissolved load or materials in solution; the suspended load or the sediment held in the water by its turbulence; and bedload or the sediment moving on or near the bed. The size of the particles moved in suspension changes with flow velocity and turbulence, but the larger particles quickly return to the bed. Further discussion of the components of total sediment load will be presented in Chapter 12.

The most simple and direct measurements involve the *solution* or *dissolved load*. It is often assumed that, unlike the suspended sediment load, solute concentrations are fairly evenly distributed across the channel cross section such that a single sample taken with a clean bottle is representative. Such a sample when evaporated will yield total dissolved solids (mg/l), which when multiplied by discharge will give a dissolved load discharge (kg/s) that can be readily converted into volume of rock removal ($m^3$ $km^{-2}$ $year^{-1}$) or an average rate of denudation (mm/1000 year). However, there are two major difficulties in interpreting erosion rates from measurements of total dissolved solids. First, unlike suspended load, dissolved load reaches a *concentration* peak at relatively low discharges, and thereafter concentration decreases with increasing discharge. As low concentrations are difficult to measure accurately, the estimates of high *total* rates of dissolved load transport associated with high discharges may be much in error. The second problem is the identification of sources of solids dissolved in river water – all of which do not come from sources that contribute to denudation. Figure 3.8 shows 'natural' denudation and non-denudation components of the dissolved load of rivers, to which must be added the considerable effects of human pollution. Although difficult to calculate, it has been suggested that inclusion of non-denudation natural components plus those resulting from human agency has led to natural denudation rates by solution having been overestimated by 1.4–2.4

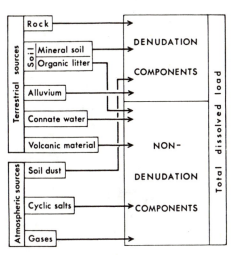

**Figure 3.8** Graphic representation of the principal natural sources of dissolved load in rivers.
*Source:* Janda, 1971, figure 1, p. 68.

times. For example, it is possible that 25–55 per cent of the total weight of dissolved solids from crystalline silicate rocks consist of anions derived from the atmosphere or from the interaction of atmosphere and biosphere. Twenty-five per cent of the dissolved material carried by the Atlantic rivers of the eastern United States may have derived from the atmosphere, pollutants and recycled sea salts, and a further 10 per cent from human industrial and agricultural waste material (Janda, 1971).

Although measurements of *suspended sediment load* in rivers have formed the bases for most of the estimates of land surface degradation rates during the past century or more, they too are not without substantial sources of error. The sample of suspended load is filtered, multiplied by a factor to relate its associated water to mean annual river discharge, converted to the equivalent of solid rock of specific gravity 2·5 (although most material is derived from soil of SG < 1·0) and then is assumed to have been derived uniformly from the whole drainage basin. The main difficulties are that through time suspended sediment transport bears a non-linear relationship to discharge (when collected in a rising-stage sampler: Figure 3.9A) during the passage of a single floodwave, and at any given time sediment concentration varies greatly within the river cross section. Suspended sediment concentration shows a striking exponential decrease from the bed of the river to its surface, whereas the velocity of normal turbulent flow shows a significant increase from just above the bed to the surface. A single sample of suspended load, such as that taken by the

single 'gulp' of an instantaneous horizontal sampler (Figure 3.9B) is, therefore, difficult to extrapolate, even if it is taken at the usual point of mean stream velocity, at 0·6 depth below the river surface.

A sample much more representative of the whole cross section is provided by a depth-integrating sampler (Figure 3.9C) which is raised from the river bottom to the surface at a constant rate, collecting sediment at a rate proportional to the flow velocity at each level. Even with the most sophisticated modern suspended-sediment samplers (e.g. pumping samplers and turbidity metres), direct measurements tend to underestimate rates of transport of suspended sediments, and reservoirs have been shown to capture some 1·8 times the suspended sediment predicted from streamflow sample measurements. In addition, attempts to predict suspended-sediment production from drainage basin morphology (e.g. area, channel length and mean slope) are less than half as accurate as similar attempts to predict dissolved load production.

Movement of *bedload* by rivers is especially difficult to measure. Direct measurement by means of basket or Polyakoff-type tray samplers (Figure 3.9D and E) is very inaccurate because these devices are on the one hand too small and fragile to trap large boulders moving in high discharges, and on the other cumbersome enough seriously to impede the free movement of smaller debris at lower discharges to which they are best adapted. A further serious problem is that, as the level of the effective bed of the river may vary greatly with discharge, it is very difficult to keep the trap correctly positioned. In the absence of precise measurements estimates of the 'bedload function' (i.e. bedload transported through a given river cross section) are made using calculated bed velocities and sizes of bed material. It is commonly assumed to comprise less than about 10 per cent of the total load, increasing to as much as 55 per cent in favourable circumstances. The amount of material moving as bedload varies greatly both in space and time, most being in steep mountain headwaters and at infrequent times of high discharge. In the Appalachians the ratio of bedload to suspended load has been estimated to vary between 1:50 and 1:3. As for the immediate sources of the solid load transported by rivers little is known, but in the central and southern Appalachians it has been suggested that 92 per cent has been supplied by sheet and gully erosion, soil creep and rapid slope failures, as against 8 per cent of solid material derived from channel erosion.

An interesting example of the application of geomorphic processes was provided more than a century ago when Archibald Geikie (1868) examined the conflicting claims regarding the relative efficacy of marine v. sub-

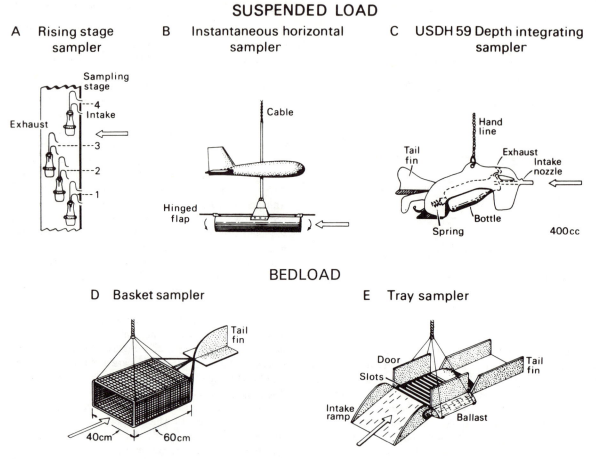

# SUSPENDED LOAD

**A  Rising stage sampler**

**B  Instantaneous horizontal sampler**

**C  USDH 59 Depth integrating sampler**

# BEDLOAD

**D  Basket sampler**

**E  Tray sampler**

**Figure 3.9**  Some equipment for sampling the suspended load and bedload of rivers.
*Source:* Gregory and Walling, 1973, figures 3.21 and 3.22, pp. 152, 160.

**Table 3.4**  Proportions of suspended load, bedload and dissolved load transported by certain rivers

| Location | Solid load | | Dissolved load |
| --- | --- | --- | --- |
| | Suspended Load (%) | Bedload (%) | (%) |
| *PERCENTAGE OF TOTAL LOAD* | | | |
| East Devon rivers, England | 38·5 | 1·5 | 60 |
| River Tyne, England | 56·5 | 8·5 | 35 |
| River Volga, USSR | 35·6 | 0·4 | 64 |
| Colorado River, Grand Canyon | 94 | | 6 |
| Wind River, Wyoming | 73 | | 27 |
| Iowa River, Iowa | 83 | | 17 |
| *PERCENTAGE OF SOLID LOAD* | | | |
| Upper Niger | 93·5 | 6·5 | |
| Lower Niger | 95·0 | 5·0 | |
| Mississippi River | 94·9 | 5·1 | |
| Alpine mountain rivers | 30·0 | 70·0 | |

*Sources:* Various, assembled by Gregory and Walling, 1973, table 4.4, pp. 202–3.

aerial erosion (the sea v. 'rain and rivers'). Relating suspended sediment measures to crude estimates of drainage basin vertical denudation, he arrived at an average figure for the temperate latitudes of about 50 mm/1000 year (valleys 250 mm/1000 year; divides 30 mm/1000 year), which he compared with maximum average rates of horizontal marine erosion of about 3 m/1000 year. From this, he deduced that Europe might be peneplained in little more than 4 million years, during which time the coast would have been cut back only 120 km. His conclusion that 'before the sea could pare off more than a mere marginal strip of land between 70 and 80 miles in breadth, the whole land would be washed into the ocean by atmospheric denudation', despite its very questionable premisses, has exercised a lasting influence on geomorphic thinking.

### 3.3.2 Erosion rates over space and time

Short-term measurements of point rates of erosion and of rates of sediment transport across given cross sections are difficult to extrapolate over larger areas, and through longer time periods, because neither rates are scale-free.

It has been commonly observed that sediment yields per square kilometre are greater for small drainage basins than for larger ones – at least up to basins about 2000 km² in area. For otherwise similar small basins, a halving of the area may increase sediment yield by ten times, while it has been estimated for the Upper Mississippi basin that a 259 km² (100 sq. mile) basin yields 1·4–2 times the sediment per unit area of a 2590 km² (1000 sq. mile) basin. For basins in the western United States, the following relationship has been proposed (summarized in Schumm, 1963):

$$S \propto A^{-0·15}$$

where: $S$ = annual sediment yield (acre-feet/sq. mile); $A$ = basin area (sq. mile).*

This is exemplified in Table 3.5 for basins in the United States.

This inverse relationship is due to the following properties of smaller basins in relation to larger ones:

(1) They commonly have steeper valley-sides and stream-channel gradients encouraging more rapid rates of erosion.

* 1 acre = 4046.86 m²; 1 foot = 0.3048 m; 1 mile = 1.60934 km; 1 sq. mile = 2.59 km².

**Table 3.5** Mean denudation rate and basin area

| Basin area (km²) | Mean denudation rate (mm/1000 year) |
|---|---|
| 0·3 | 12,600 |
| 3·0 | 2,550 |
| 80·0 | 220–60 |
| 3,900 | 100–30 |
| 37,000–3,280,000 | 60–30 |

(2) Lack of floodplains gives less opportunity for sediment storage within the basin after weathering and removal from the slopes. This gives a shorter sediment residence time in the basin and a more efficient sediment transport, as expressed by a higher sediment delivery ratio (i.e. rate of sediment evacuation/total rate of sediment production, expressed as a percentage) (Figure 3.10A).

(3) Small basins may be totally blanketed by high-intensity storm events, giving high maximum erosion

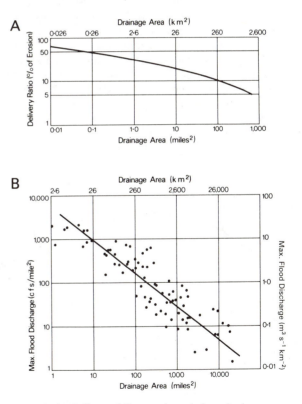

**Figure 3.10** Sediment delivery ratio v. drainage basin area for the United States, and maximum flood discharge per unit area for drainage basins in Colorado.
*Sources:* Trimble, 1977, figure 2, p. 881; Follansbee and Sawyer.

rates per unit area. This effect is well illustrated by the maximum flood discharges per unit area for drainage basins in Colorado (Figure 3.10B).

This relationship may be mitigated to a limited extent because some solid load in smaller tributary basins may become dissolved load in the larger channels and may escape measurement, but this inverse relationship with area is nevertheless very pronounced. A further spatial scale feature is that for small basins, a larger fraction of the total solid load is transported during less frequent, higher-intensity runoff events. In 192 basins of less than 20 km² in the southern California Coast Ranges about 90 per cent of the debris movement is accomplished by storms of recurrence intervals of more than five years and 50 per cent by those of greater than twenty-five-year recurrence intervals (see p. 57). All the above observations make it clear that the simple extrapolation of erosion rates from smaller areas to larger ones is very questionable.

Clearly, terrain morphometry exercises a profound control over local rates of erosion in terms of elevation, slope and relief. Elevation can often be used as a surrogate for climatic and vegetative controls. Erosion of a 650,000-year-old andesite stratovolcano in the Hydrographers' Range, north-east Papua, has been estimated from maps and air photographs at 80 mm/1000 year at an elevation of 60 m, and 800 mm/1000 year at 760 m – clearly, a function of the increase of rainfall amount and intensity with elevation in the tropics (Ruxton and McDougall, 1967). It is interesting that a pumice and ash

volcano formed in New Guinea, in 1947, exhibits an average erosion rate of as much as 19000 mm/1000 year, indicating both the initial rapidity of erosion on steep, unvegetated surfaces and the erodibility of unconsolidated fine-grained material (Ollier and Brown, 1971). An inverse effect of elevation on denudation rates has been estimated from the suspended sediment transported by the Colorado River and its tributaries, the Little Colorado, the Green and the Yampa (Figure 3.11). These basins are generally large enough for spatial scale effects to be less important, and the altitudinal pattern of denudation suggests that the maximum rates occur towards lower elevations where there is less vegetational protection of the surface (see Section 3.3.3). The control effected by slope, expressed by basin relief ratio ($R_h$ = basin relief/basin length; see Chapter 13) or by mean quadrat relief ($H'$), over mean denudation rates is very significant but, as Table 3.6 shows, estimates between authors vary by a factor of four to five times.

**Table 3.6**   Control effected by slope on denudation rates

|  | Mean denudation rate (mm/1000year) | | |
|---|---|---|---|
|  | *Corbel* | *Young* | *Schumm* |
| 'Normal' relief | 22 | 46 | 72 |
| Steep relief | 206 | 500 | 915 |

Figure 3.12 shows the exponential increase of annual sediment yields with relief ratio for basins of about 1 sq.

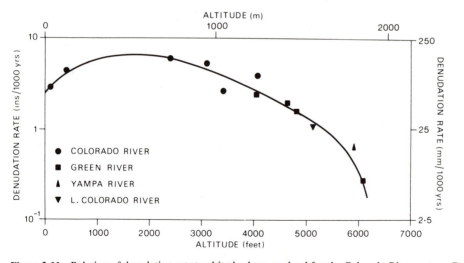

**Figure 3.11**   Relation of denudation rate to altitude above sea level for the Colorado River system. Denudation rate is derived from suspended load only, expressed as inches or mm derived from the drainage basin per unit of time.
*Source:* L. B. Leopold *et al.*, *Fluvial Processes in Geomorphology*, 1964, figure 3-22, p. 79, copyright © W. H. Freeman and Co., San Francisco, by permission.

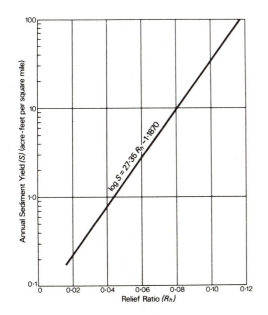

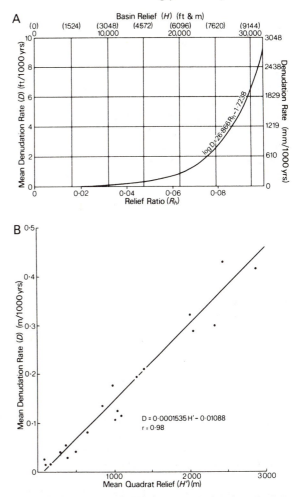

**Figure 3.12** Relation between annual sediment yield per unit area and relief ratio (i.e. maximum basin relief divided by length) in respect of drainage basins of approx. 1 sq. mile in area underlain by sandstone or shale in the semi-arid western United States.
*Source:* Schumm, 1963, figure 1, p. 5.

mile (2·59 km²) on sandstone and shale in the semi-arid south-western United States. The values of sediment yields have been converted to denudation rates for basins of 1500 sq. mile (3885 km²) and plotted as an exponential curve in Figure 3.13A. This relationship contrasts with the arithmetic one obtained between mean denudation rate and mean relief for twenty basins in the United States and Western Europe (size varying from several hundred to over 100,000 km²; mean annual precipitation, 250–2500 mm); relief being calculated from sample quadrat areas of 400 km² (Figure 3.13B).

Local rock type is also a very important control over denudation rates. For example, denudation rates calculated from sedimentation rates in small stock reservoirs in the northern Great Plains are 61 mm/1000 year for sandstones and 853 mm/1000 year for shales. Similar observations of suspended load trapped in small reservoirs in Utah, New Mexico and Arizona give estimates of sediment loss (m³ km⁻² year⁻¹) at 95–143 for resistant conglomerates, limestones and sandstones, 523–571 for medium-resistant friable sandstones and 761–1237 for soft shales and gypsum. Table 3.7 shows the large differences in dissolved load carried by streams draining different rock types in Bohemia.

**Figure 3.13** A. Mean denudation rate as a function of relief ratio adjusted to drainage basins of 1500 sq. miles, based on the relation shown in Figure 3.12. B. Mean denudation rate as a function of mean relief of 400 km² quadrats for twenty basins in the United States and Western Europe.
*Sources:* Schumm, 1963, figure 2, p. 6; Ahnert, 1970, figure 3, p. 251.

**Table 3.7** Dissolved load carried by streams in Bohemia

| Rock Type | Total ion concentration (mg/l) |
| --- | --- |
| Phyllite | 59 |
| Granite | 74 |
| Mica schist | 87 |
| Basalt | 480 |
| Cretaceous sedimentary rocks | 792 |

*Source:* Data from F. W. Clarke and E. Gorham in Gregory and Walling, 1973, table 6.7, p. 321.

The effect of vegetation and land use differences on sediment yields and denudation rates is even more pronounced than those of terrain and rock type, and will be considered further in Section 3.3.3. In the central and southern United States average sediment yields from cultivated lands are 10 to 100 times those from forested areas, and yields from especially clean tillage 200 to 1800 times! It has further been estimated that the present sediment loads of the Atlantic-draining rivers of the United States are four to five times greater than before European settlement. Sediment yields ($m^3$ $km^{-2}$ $year^{-1}$) from humid tropical catchments show similar results (Table 3.8).

**Table 3.8** Sediment yields from humid tropical catchments

|  | Sediment yield ($m^3$ $km^{-2}$ $year^{-1}$) | |
| --- | --- | --- |
| *Region* | *Forested* | *Cultivated* |
| Northern Range, Trinidad | 1·8 | 16·0 |
| Barron, Queensland | 5·7 | 13·6 |
| Mbeya Range, Tanzania | 6·9 | 29·5 |
| Cameron Hills, Malaysia | 21·1 | 103·1 |
| Apiodoume, Ivory Coast | 97·0 | 1700·0 |
| Tjiloetoeng, Java | 900·0 | 1900·0 |

*Source:* Data from Douglas, 1969.

Recent research on an interfluve in the Colorado Front Range (elevation 3300–3700 m; slopes 5°–30°), while giving an average present erosion rate of 0·1 mm/year mostly from rainsplash and snowmelt during June–September, has emphasized that erosion rates may differ by whole orders of magnitude depending on the character of the surface, with nivation being particularly important (Table 3.9).

From the foregoing it is apparent that topographic, lithological and vegetational spatial variations may each be responsible for varying sediment yields by ten to more than a hundred times. It is also the rule that different parts of drainage basins of all scales contribute sediment at very different rates. Ninety-five per cent of the load of

the Mississippi River basin is derived from 5 per cent of its headwater area, and 82 per cent of the load of the Amazon River is provided by the 12 per cent of the basin area occupied by the Andes. Even for small basins, most of the surface sediment erosion is derived from the 'partial areas' of low infiltration capacity (see Section 3.2), which for areas in the north-eastern United States occupy an average of 15 per cent of the basins in the summer and 50 per cent in the spring.

Just as it is difficult to generalize over space regarding rates of denudation, generalizing with regard to rates at any one location through time is similarly so. Erosional processes may vary greatly in intensity, both in the long term and short term, with high-intensity processes occurring during relatively short periods and medium-intensity ones operating for most of the time. During much of the Mesozoic and Cenozoic denudation for the Appalachian region averaged about 3 mm/1000 year, with oscillations particularly during the Cenozoic within about one order of magnitude. The present assumed rate of 22 mm/1000 year (dissolved load, 16; suspended load, 5; bedload, 1) may involve an overestimation of 'natural' erosion rates by a factor of more than 2 due to human activity.

Turning to the shortest timescales, over a period of years the annual sediment yield of a given river can vary about an order of magnitude and, within a few days, by several orders of magnitude under flood conditions. It is a feature of the frequency distributions of such events as the magnitudes of river discharges, rainfall intensities, waveheights, temperatures, and so on, that they are characteristically right-skewed (i.e. are most frequent in the lower- to middle-magnitude ranges, with a small proportion of high-magnitude events). If extreme values of such events are considered (e.g. maximum flood discharges, maximum daily rainfalls, maximum waveheights, maximum temperatures per year), then the resulting frequency distributions are even more right-skewed, and have been the subject of much research by E. J. Gumbel who developed a method of relating frequency and magnitude of events in an extreme probability form (Figure 3.14). A feature of this is the

**Table 3.9** Erosion rates in the Colorado Front Range

| *Surface character* | *% total area* | *% total eroded material* | *Mean erosion rate (mm/year)* |
| --- | --- | --- | --- |
| Tundra meadow | 50 | 5 | 0·01 |
| Dry tundra | 35 | <50 | 0·1 |
| Nivation hollows with late-lying snow | 3 | 50 | 1·0 |

*Source:* Bovis and Thorn, 1981, p. 151.

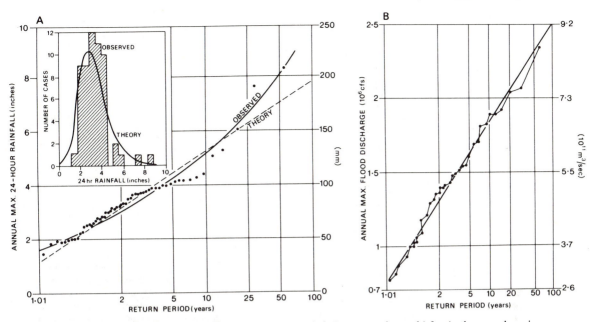

**Figure 3.14** Extremal probability plots showing average return periods (recurrence intervals) for A. the annual maximum 24-hour precipitation for Baltimore, Maryland, (1900–56); and B. the annual peak flood discharges for the Mississippi River at Vicksburg, Mississippi (1898–1949).
*Sources:* Chorley and Kennedy, 1971, figure 5.15, p. 176, data by Hershfield and Kohler; ibid., figure 5.18A, p. 180, data by Gumbel.

employment of the *return period* (or *recurrence interval*) – the average length of time separating events of similar extreme magnitude. These distributions of event magnitudes make it difficult to make long-term estimates of erosion from short-term observations – the more so because sediment movement events (i.e. river sediment loads, material moved by mass movements, longshore drift of debris, etc.) appear to be more highly right-skewed than the rainfall, river discharge and waveheights which generate such movements. Thus the fifty-year river sediment load is closer to the mean annual load than is the fifty-year river discharge to the mean annual discharge.

The relative effects of event classes of different magnitude on the movement of stream load is illustrated by records and calculations extending over some five and a half years for the Bighorn River at Thermopolis, Wyoming (Figure 3.15). The amount of material moved by events in each magnitude class is multiplied by its relative frequency of occurrence and all classes are summed to get the total moved. In this case discharges between 1000–2000 cfs (28–57 m³/s) are estimated to transport more than 25 per cent of the total suspended load, and those exceeding 10,000 cfs (280 m³/s) only about 9 per cent. The percentage of suspended load transported by given magnitudes of discharge is shown

for the Bighorn and five other American rivers in Figure 3.16. These rivers fall into two groups; the 'higher-magnitude' Rio Puerco and Cheyenne River (only 30 per cent of load carried by discharges more frequent than ten days per year, 2·7 per cent of the time, and 50 per cent of load transported on an average of four days per year) and the more normal transporting rivers (70–82 per cent by discharges more frequent than ten days per year, and 50 per cent of load transported on more than thirty days per year). Rivers which are subject to considerable variations in discharge, like some semi-arid ones and very small humid headwater tributaries, carry much of their solid load during comparatively short periods. Small catchments in east Devon, England, transport 78 per cent of their total suspended load on an average of only four days per year. Table 3.10 gives details of percentages of suspended and dissolved load transported by different magnitudes of runoff event; the table shows that

(1) the smaller the basin area, the greater the percentage of suspended load moved by infrequent, high-magnitude events;

(2) for all basins, the majority of suspended load is carried by moderate- and small-magnitude events;

(3) the proportion of total dissolved load carried by infrequent, high-magnitude events is probably less than that of suspended load.

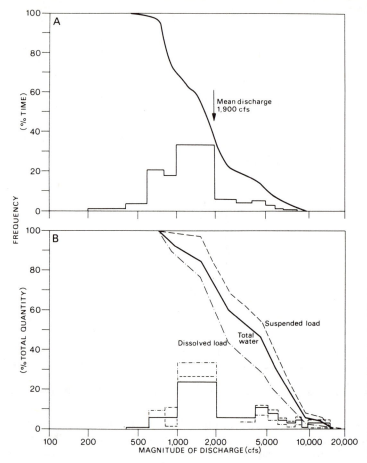

**Figure 3.15** Histograms and cumulative graphs of water and load, Bighorn River at Thermopolis, Wyoming: A. duration curve of water showing percentage of time various discharge rates are equalled or exceeded; B. cumulative graphs and histograms showing relative contribution of various discharge rates to total flow of water, suspended load and dissolved load.
*Sources:* L. B. Leopold *et al., Fluvial Processes in Geomorphology*, 1964, figure 3–19, p. 70, copyright © W. H. Freeman and Co., San Francisco, by permission.

**Table 3.10**  Suspended and dissolved loads related to runoff events

A. Suspended load: US rivers

| Basin area (km²) | % mean annual suspended sediment carried by runoff resulting from storms of large, moderate and small intensity | | |
|---|---|---|---|
| | *Large* | *Moderate* | *Small* |
| < 100 | 34 | 18 | 48 |
| 200–300 | 26·7 | 11·3 | 62 |
| 300–500 | 12·4 | 7·6 | 80 |

B. Dissolved load: carbonate rocks, south Canadian Rockies

| Discharge (m³/sec) | % daily mean discharges | % total basin runoff | % sulphate transported | % carbonate transported |
|---|---|---|---|---|
| Low (<113) | 60 | 16 | 32 | 23 |
| Medium (113–566) | 33 | 58 | 49 | 56 |
| High (>566) | 7 | 26 | 19 | 21 |

*Sources:* Gregory and Walling, 1973, table 4.7, p. 209, data from Priest; Drake and Ford, 1976, p. 167.

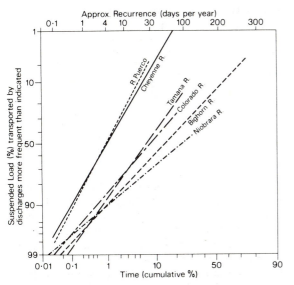

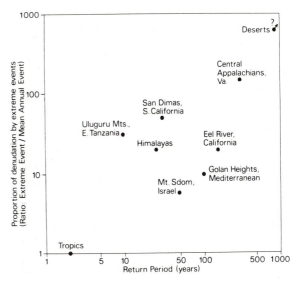

**Figure 3.16** Plot of cumulative percentage of time against cumulative percentage of total suspended load transported by six rivers, in Alaska (Tamana), Wyoming (Bighorn), New Mexico (Rio Puerco), South Dakota (Cheyenne), Arizona (Colorado) and Nebraska (Niobrara).
*Source:* Wolman and Miller, 1960, figure 2, p. 59.

**Figure 3.17** Relationship between denudation of hillslopes during large events of different recurrence intervals and mean annual denudation in various mountainous regions of the world.
*Source:* Wolman and Gerson, 1978, figure 7, p. 202. Data from a variety of sources, *Earth Surface Processes*, copyright © John Wiley and Sons, by permission.

Estimates such as those given above enable us to speculate regarding the total amount of work attributable to events of differing magnitude. It is clear from Figure 3.16 that, for all six rivers, even the two most fluctuating ones, 80–97 per cent of the total suspended load is carried during events which recur more frequently than once per year. The relative unimportance of infrequent, high-magnitude events is supported by the reasoning that, although rate of debris movement bears a strong exponential relationship to event magnitude, the skewness of the frequency of occurrences means that a product of these two curves gives a resulting maximum of total work over a long period in the middle range of event magnitudes (see Figure 20.19A).

The relative amount of work performed by events of given magnitude may not be an accurate measure of their contribution to the long-term development of landforms. The spatial concentration of the erosive effects of high-magnitude events which results in the cutting of new channels, slumping, valley-side steepening, channel widening and deepening may be locally of a permanent nature and, for example, once formed, a gully may continue to grow into a small valley under the action of events too small in magnitude to have cut it initially.

The observed persistence of geomorphic features produced by high-intensity events may give clues to the contribution of such events. On the Appalachian Piedmont, having low relief and thick soils, floods of greater than 100-year return periods have not been observed to cause slope failures, and their effects on many river channels (i.e. widening, bar deposition near bends, flushing out of all but coarsest bed material, destruction of vegetation) appear to be well on the way to recovery within one year. However, in temperate regions of higher relief, steeper slopes and thinner soils, high-magnitude events may have more lasting effects. In the Blue Ridge and Appalachian Plateaus slope failures accompany most high-magnitude events. The 150- to 250-year flood of August 1952, in Exmoor, England, produced valley-side scars and local channel scour apparent twenty-five years later, and these features may well survive until further exploited by subsequent events of similar magnitude.

Besides topography, climatic regime seems to influence the persistence of geomorphic features produced by infrequent events. Although river channel widening produced by 50- to 200-year events in temperate regions may recover in a few months or years, in arid regions this effect may be long-lived or even permanent. Slope scars resulting from greater than 100-year events in steep temperate regions may persist for more than twenty-five years, whereas in subtropical Japan the equivalent period may be much less. It is interesting that in Hawaii and Tanzania the effects of some one-to-ten-year events on slopes may persist longer than these return periods. Figure 3.17 attempts to express the proportion of denudation produced by extreme events as a function of return period for a variety of mountainous areas of the world.

### 3.3.3 Regional denudation

Despite problems both of measurement and of spatial and temporal generalization, it is instructive to attempt to identify current regional patterns of erosion and denudation which can be assembled into a world picture. Needless to say, estimates at these scales vary widely and there may be almost an order of magnitude of difference between the highest and lowest estimates for a given continental-scale region. Table 3.11 gives estimated averages of suspended and dissolved load for the continents (tonne km$^{-2}$ year$^{-1}$).

Denudation $D$ is calculated as follows:

$$D(m^3 \text{ km}^{-2} \text{ year}^{-1} = mm/1000 \text{ year})$$
$$= \frac{\text{total load (tonne)}}{\text{area (km}^2) \times \text{specific gravity of rock}}.$$

**Table 3.11** Estimated continental averages of suspended and dissolved load

|  | *Suspended load* | *Dissolved load* |
|---|---|---|
| Asia | 600 (95%) | 32 (5%) |
| North America | 96 (74%) | 33 (26%) |
| South America | 63 (69%) | 28 (31%) |
| Europe | 35 (45%) | 42 (55%) |
| Australia | 45 (95%) | 2·3 (5%) |
| Africa | 27 (53%) | 24 (47%) |

*Source:* Largely from Holeman, 1968.

The specific gravity of rock (2.5 or 2.64) is used on the assumption that, over the long term, soil thickness remains constant. Some authors purport to include measurements of suspended load, dissolved load and bedload; others the measures of only the first two, with an added 10 per cent to accommodate bedload; still others disregard bedload. Average measured denudation rates for a variety of major drainage basins are as given in Table 3.12.

To make sense of these gross figures, apart from the obvious relief effects, it has become the practice to assign to climate a major cause of regional differences in

**Table 3.12** Average denudation rates for major drainage basins

| Region | Mean Relief (m/20 × 20 km) | Present denudation rate (mm/1000 year) Mean | Range | Remarks |
|---|---|---|---|---|
| Himalayas |  | 1000 |  |  |
| Northern Alps |  | 610 |  |  |
| French and Swiss Alps | 2331 | 379 | 287–518 | Upper Rhine, Isère, Reuss, Kander, Rhône, Durance |
| Colorado River |  | 165* | 58– |  |
| Utah | 860 | 130 | 82–177* | Green, Escalante, Dirty Devil, Colorado |
| Hawaii |  | 130 |  | Oahu; mainly solution |
| California |  | 91* |  |  |
| Western Gulf (Texas) |  | 53* | 16– |  |
| Mississippi River |  | 51* |  | About twice Cenozoic average |
| SE USA |  | 41* | 28– |  |
| NE USA |  | 38 | 27–48* | Delaware, Juniata, Ohio, Potomac, Susquehanna |
| Columbia River |  | 38* |  |  |
| River Thames | 159 | 16 |  |  |

*Includes estimated 10 per cent bedload.
*Source:* Ahnert, 1970, table 1, p. 247.

**Table 3.13** Average rates of denudation (mm/1000 year)

| Temperature | Precipitation | | | | | |
| --- | --- | --- | --- | --- | --- | --- |
| | Arid (<200 mm) | | Normal (200–1500 mm) | | Humid (>1500 mm) | |
| | *Mountains* | *Plains* | *Mountains* | *Plains* | *Mountains* | *Plains* |
| Hot (15°N–15°S) | 1·0 | 0·5 | 25·0 | 10·0 | 30·0 | 15·0 |
| Tropical (15°–23½° N and S) | 1·0 | 0·5 | 30·0 | 15·0 | 40·0 | 20·0 |
| ExtraTropical (temperature >15°C) | 4·0 | 1·0 | 100·0 | 20·0 | 100·0 | 30·0 |
| Temperate (temperature 0°C–15°C) | 50·0 | 10·0 | 100·0 | 30·0 | 150·0 | 40·0 |
| Cold (temperature <0°C) | 50·0 | 15·0 | 100·0 | 30·0 | 180·0 | — |
| Polar | 50·0 | 15·0 | 100·0 | 30·0 | 150·0 | — |
| Glaciated polar | 50·0 | — | 1000 | | 2000 | |

*Source:* Corbel, 1964, p. 407.

sediment yield and denudation, and during the past quarter of a century a number of large-scale regional estimates have been made. Corbel prepared a table of rates of erosion ($m^3 km^{-2} year^{-1}$:mm/1000 year), showing the combined influence of precipitation, temperature and relief (Table 3.13). The major features of this scheme are that rates of erosion:

(1) vary inversely with temperature; in this respect Corbel's scheme differs from the other major estimates, because of his emphasis on supposedly high solution rates in high latitudes;

(2) vary directly with mean annual precipitation; Corbel recognized the oversimplification involved here, acknowledging locally high denudation rates in seasonally dry Mediterranean mountains (450 mm/ 1000 yrs):

(3) vary directly with relief; mountain denudation exceeding rates on the plains by a factor of 2 in the tropics, 4–5 in the temperate regions and 3–4 in cold regions.

Fournier studied the suspended sediment yield from seventy-eight major drainage basins (2460–1,060,000 km²) and showed that, for regions of differing relief, general relationships hold between sediment yields and a measure of seasonal precipitation ($p^2/P$: where $p$ = mean precipitation (mm) of wettest month: and $P$ = mean annual precipitation (mm)) (Figure 3.18). A plot of sediment yield against mean annual precipitation suggests maxima for sediment yield in the semi-arid regions and in the seasonally humid tropics (Figure 3.19). Finally, Fournier mapped the supposed world distribution of suspended sediment yields $S$ (tonne km⁻² year⁻¹) (Figure 3.20) and expressed it by the equation:

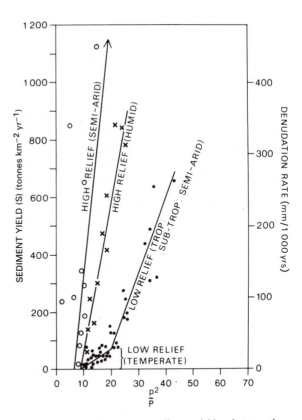

**Figure 3.18** Relation between sediment yield and seasonal precipitation ($p^2/P$) for four regions of differing climate and relief.

*Source:* Fournier, 1960, figure IX, pp. 124–7.

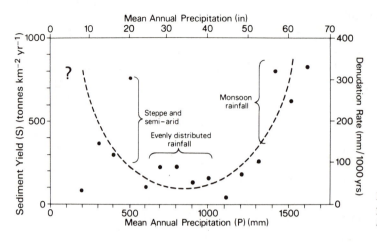

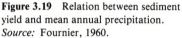

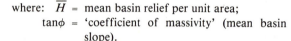

**Figure 3.19**   Relation between sediment yield and mean annual precipitation. *Source:* Fournier, 1960.

$$\log S = 2{\cdot}65 \log \frac{p^2}{P} + 0{\cdot}46 \log \overline{H}.\,\tan\phi - 1{\cdot}56$$

where:   $\overline{H}$ = mean basin relief per unit area;
$\tan\phi$ = 'coefficient of massivity' (mean basin slope).

A similar study by Strakhov (1967), however, yielded values about an order of magnitude less (Stoddart, 1969).

More detailed work relating sediment yields to climatic influences was carried out by Langbein and Schumm (1958) on gauging-station data relating to 94 catchments (mean area 3885 km²) and on reservoir sedimentation data from 163 catchments (mean area 78 km²). Correcting to a standard area of 3885 km² (1500 sq. mile), these were plotted against mean annual effective precipitation (i.e. the precipitation required to produce the known amount of runoff; avoiding problems of evapotranspiration, infiltration, etc.) of 203–2540 mm, producing a peak (198 mm/1000 year) at the semi-arid/grassland precipitation boundary (300 mm) and minima in very dry regions and in more humid ones (58 mm/1000 year) (Figure 3.21). At effective precipitations of less than 300 mm there appears to be insufficient runoff to produce maximum erosion, whereas above it the erosive effects of increased runoff are more than counteracted by the continuous vegetation cover. There are three additional points of note regarding this curve:

(1) Its lower end refines the problematic behaviour of Fournier's curve (Figure 3.19) at low precipitation values.

(2) As the mean annual temperature increases the maximum sediment yield occurs at higher amounts of mean annual effective precipitation (Figure 3.21). The 300-mm peak only relates to regions with a mean annual temperature of 10°C (50°F).

(3) With tropical climates having seasonal rainfalls (e.g. monsoonal), there may be another sediment yield maximum located above 1000 mm.

A follow-up study by Wilson (1973) employed 1500 worldwide basins with measured suspended sediment yields of more than three years, correcting sediment

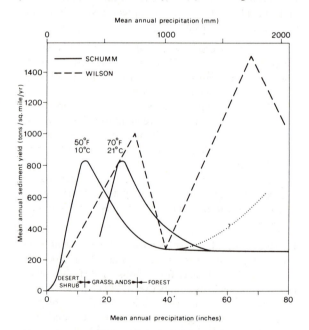

**Figure 3.21**   Relation between sediment yield and mean annual precipitation, estimated by Schumm and Wilson. *Sources:* Schumm, 1965, figure 2, p. 785, in H. E. Wright and D. G. Frey (eds), *The Quaternary of the United States*, copyright © 1965 by Princeton University Press, reprinted by permission of Princeton University Press; Wilson, 1973, figure 1, p. 336.

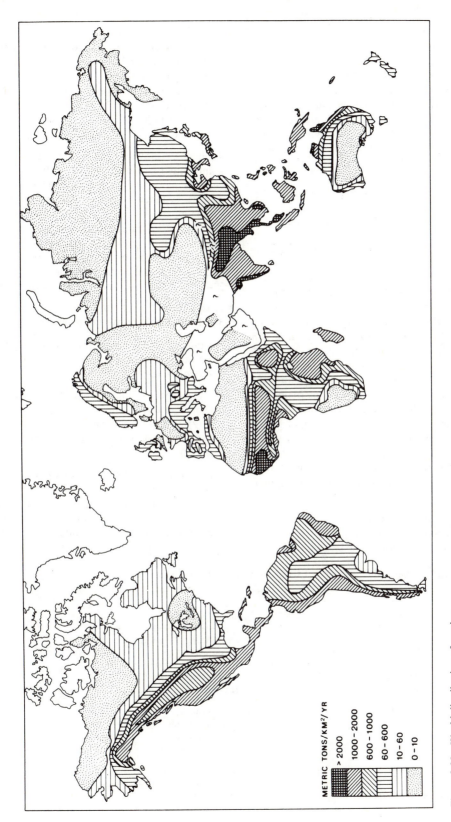

**Figure 3.20** World distribution of erosion.
*Sources:* Fournier, 1960, graphique XV, pp. 186–7, from Stoddart, 1969, figure 1.III.2, p. 45.

yields to a basin size of 259 $km^2$ (100 sq. mile). This produced a new curve with two pronounced peaks at 762 mm (30 in) and 1768 mm (70 in) (Figure 3.21). Wilson questioned the value of any one curve, however, pointing out that factors other than climate can cause a hundredfold variation in sediment yield, and reiterated the importance of seasonality effects in climatic regimen, with the following climates producing pronounced peaks of sediment yield:

| | | | |
|---|---|---|---|
| Mediterranean climates | 1270–1650 mm | (50–65 in) | together producing Wilson's second peak (Figure 3.21) |
| Tropical, seasonally wet | 1780–1905 mm | (70–75 in) | |
| Semi-arid | 250–500 mm | (10–20 in) | |

It is interesting that suspended sediment yields in Australia have shown peaks at mean annual precipitations of 40–50 mm (1·6–2 in), 305–381 mm (12–15 in) and at >2032 mm (>80 in), and although the last may be partly due to human activities, the first and last peaks correspond with peaks in drainage density (Abrahams, 1972). In terms of total denudation rates it should be pointed out that an emphasis on suspended load measurements is dangerous, even if estimates of dissolved load and bedload are included, because the humid tropics produce greater proportions of load in solution, the proportion of dissolved load decreases relatively with increasing relief and the proportion of bedload increases with relief. A recent study of seventy-nine worldwide basins (Jansen and Painter, 1974) used multiple regression techniques to obtain the control over mean annual suspended sediment yield (*S*) by combinations of:

*P* = mean annual precipitation (mm);
*T* = mean annual temperature (°C);
*G* = proneness to erosion (Paleozoic rocks = 3; Mesozoic = 5; Cenozoic = 6; Quaternary = 2);
*D* = mean annual discharge;
*H* = altitude (m) above sea level;
*R* = relief-length ratio (m/km);
*A* = basin area ($km^2$);
*V* = vegetation protection (forest = 4; grass = 3; steppe = 2; desert = 1).

Table 3.14 shows that, although *D,R,T* and *V* are dominant, the relative effects of these variables differ between major climatic regions. The most recent review of world sedimentation is by Walling and Webb (1983).

Finally, before turning to measurements and inferences regarding rates of diastrophic movements of the earth's crust, it is useful to recall the orders of magnitude which have been estimated by various workers for regional rates of denudation in the United States – mean 27–76 mm/1000 year, maximum 914 mm/1000 year.

## 3.4 Diastrophism: the geophysical cascade

For our purpose diastrophic movements, involving subterranean geophysical cascades, can be broadly divided into two classes:

(1) orogenic movements, those relating to the behaviour of plate margins,

(2) epeirogenic movements, those relating to the behaviour of continental platforms (plus ocean floors).

Geomorphic studies of rates of diastrophism have never been easy. On the one hand, precise measured rates of uplift and erosion have been conducted on a temporal scale of less than $10^2$ years, whereas mean rates of uplift and erosion inferred from geological evidence are on a scale of greater than $10^3$ years and give long-term

**Table 3.14** Variables controlling suspended sediment yields (S) in different climates

| *Climatic region* | *S = f(   )* | *Variance (%)* | *Remarks regarding S* |
|---|---|---|---|
| Tropical rainy | *D,A,R,T,* | 93·5 | Inverse with *T* because *V* increases with *T* |
| Dry | *D,H,P,T,V* | 86·0 | Inverse with *D*; shows effect of vegetation |
| Humid mesothermal | *D,R,T,V* | 62·8 | Inverse with *R*; unexpected!; perhaps due to human activity at low elevations |
| Humid microthermal | *H,P,V,G,* | 64·5 | |

*Source:* Jansen and Painter, 1974, p. 376, copyright © Elsevier Science Publishers.

averages only, masking shorter-term maximum rates which are the most geomorphologically significant. On the other hand, it seems clear that the interpretation of major landform features calls for a knowledge of maximum comparable rates of erosion and uplift on a temporal scale of $10^5$–$10^7$ years. It is of note in this connection that times estimated for peneplanation vary between 20 and 200 million years. A further difficulty is that tectonic uplift is rarely uniform either over large areas, giving a problem as to the spatial scale over which uplift can be generalized, or over long time periods, presenting great difficulties in extrapolating rates of uplift over time. In particular, intense rates of uplift (i.e. >10 mm/year) are very localized in space and time but probably contribute greatly to the surface features of the earth.

Lateral movements of the earth's surface are measured by triangulation, by tacheometer which uses stadia readings on a staff to give distance and, most recently, by laser beams employing satellites. Vertical movements can be inferred from the displacement of dated deposits or erosional features such as wave-cut platforms with respect to sea level, or measured exactly by precise levelling methods linked to long-term tidal records and, more recently, satellites.

### 3.4.1 Orogenic movements

Zones of spreading are associated with rift belts predominantly in mid-oceanic locations (see Chapter 5). These zones are characterized by uplift and the moving apart of rift boundaries, relative subsidence of the central graben and volcanic activity. Rates of spreading vary from a low of 10 mm per year in the Norwegian Sea to a maximum of 183 mm per year in parts of the East Pacific Rise. Projecting this latter rate backwards in time, it is possible that the whole floor of the Eastern Pacific could have been generated in little more than 130 million years (see Figure 5.13), although as Table 5.4 shows rates of spreading are very variable in time, for example, that of the Mid-Atlantic Ridge decreasing from an average of 20 mm per year in the Early Eocene to less than one-half this figure in the Late Eocene. In Iceland contemporary rates of rift-widening differ from place to

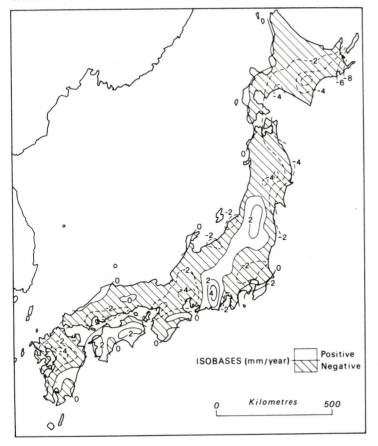

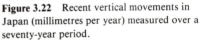

ISOBASES (mm/year) — Positive / Negative

0     Kilometres     500

**Figure 3.22** Recent vertical movements in Japan (millimetres per year) measured over a seventy-year period.
*Source:* Dambara, in International Union of Geodesy and Geophysics, 1975, figure 4, p. 160.

place with maxima ranging from 7 to 23 mm per year. The graben floor in north Iceland is subsiding at 4–7 mm per year, whereas the floor of the Ethiopian Rift Valley is subsiding at about 12 mm per year with the adjacent Somaliland coast rising at some 2 mm per year between 1935 and 1960.

Figure 3.22 gives a picture of vertical movements in Japan averaged over a seventy-year period, but these figures mask both the very detailed variations of movement over space and time and the importance of virtually instantaneous earthquake events. Figure 3.23B, for example, clearly shows the effect of the 1946 Nanki earthquake which followed a fifty-year period of vertical movements (1897–1946) free of major earthquakes in Shikoku. Longer-term evidence from dated raised beaches in the Greater Antilles suggests a maximum upwarping there of more than 400 m during the Quaternary (Figure 3.24).

Younger fold mountains cover some 5–10 per cent of the total land surface, and although like the island arcs they present a jumble of positive and negative movements in space and time, average maximum rates of uplift are some 6–7 mm per year. Best studied by precise levelling techniques, the California Coast Ranges have maximum uplift rates varying between 4 and 12 mm per year (Santa Monica Mountains, 4; Cajon Pass, 5; San Gabriel Mountains, 6; North Baldwin Hills, 9; North Peninsular Ranges, 10 (1897–1934); and southern San Joaquin Valley, 12). Uplift recorded over these short periods is clearly indicative of longer-term movements and, for example, the Ventura Avenue Anticline has been denuded at a rate which must average 1·6 mm per year for at least the past 1 million years. General paleontological evidence points to an average rate of uplift of the Coast Ranges of 5–8 mm per year during the past 36,000 years. As in island arcs, uplift occurs both steadily

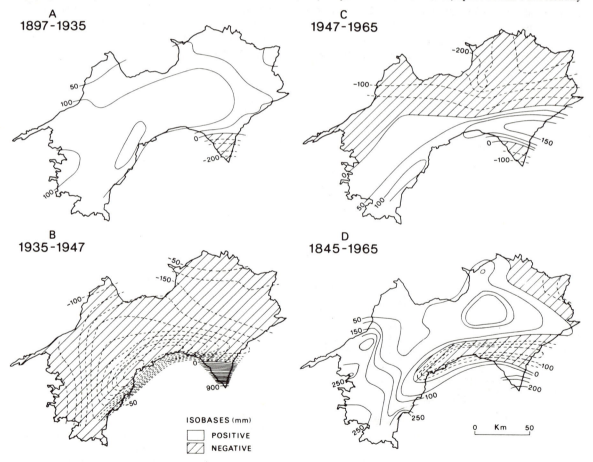

**Figure 3.23**   Vertical displacement of Shikoku, Japan, during three time periods (A–C), and for the total period 1845–1965, D. Values partly based on precise levelling are shown by 50-mm isobases; the effect of the 1946 earthquake is particularly apparent.
*Source:* Yoshikawa, 1970, figures 4 and 8, pp. 10, 11, 16.

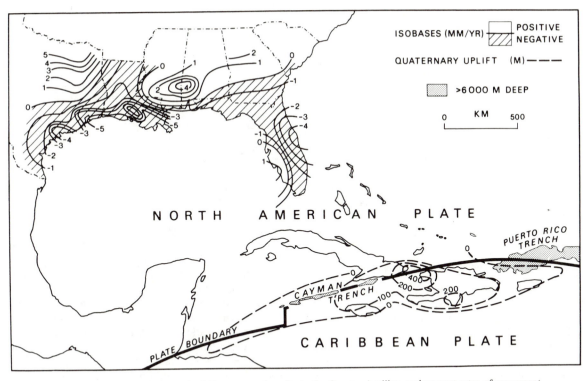

**Figure 3.24** Maximum uplift (in metres) of Quaternary deposits in the Greater Antilles, and current rates of movement (millimetres per year) in the southern United States.
*Sources:* Horsfield, 1975, figure 3, p. 934; Pavoni and Green, 1975, figure 4, p. 110.

over time and occasionally episodically. The Long Beach earthquake of 10 March 1933 caused a sudden uplift of a 4-mile (6·5 km) arch by as much as 178 mm and the San Fernando earthquake of 9 February 1971 (strength 6·4) a local uplift of as much as 2000 mm (Pavoni and Green, 1975). This pattern is repeated in the circum-Pacific belt with the Chilean earthquake (22 May 1960: magnitude 8·5) affecting an area of 1000 × 200 km and leading to a land subsidence of 2000 mm and a sea bed rise of 3000 mm; the Alaskan earthquake (27 March 1964: magnitude 8·4) raising an area of 1000 × 400 km by an average of 2000 mm and a maximum of 12,000 mm; and the Costa Rican earthquake (7 January 1953: magnitude 8·4) uplifting its Caribbean coastal plain by 200 mm.

Movements in the European Alps are of less magnitude and appear to have averaged only 0·199 mm per year from the Late Miocene onward. Present rates of measured uplift are 1 mm per year for the southern Swiss Alps (1918–70), decreasing towards the north (Figure 3.25). The Carpathians, however, are much more active, thrusting north-west towards the Bohemian Massif at some 3 mm per year along a foredeep showing oscillations of − 1·2 to + 2·0 mm per year. The western Car-

pathians have blocks rising at 1·5 mm per year and others subsiding nearby at − 5·3 mm/year, whereas the eastern Carpathians are rising generally at about twice this rate, perhaps reaching 10 mm per year in places.

Long-continued uplift has been demonstrated from geological and archaeological evidence in parts of the Mediterranean. Uplifted shorelines in southern Italy suggest average rates of 0.24 mm per year for as long as 230,000 years; shorelines in western Crete appear to have risen at 4–6 mm per year for the last 2000 years; and the Troodos Mountains of Cyprus have been uplifted 2000 m during the Pleistocene (average 1 mm per year for 2 million years) (Robertson, 1977). The Middle East is similarly unstable with the present uplift rates in the Persian Gulf being 3–10 mm per year, the southern Iranian coast of the Arabian Sea averaging 2 mm per year during the Holocene, the western Caucasus rising at 6–8 mm per year and the eastern Caucasus at about 12 mm per year. Further east overthrusting is taking place at 20–25 mm per year along the Peter the Great Fault Zone separating the Pamirs and the Tien-Shan Mountains. Figures for younger fold mountains suggest present average maximum rates of orogenic uplift of

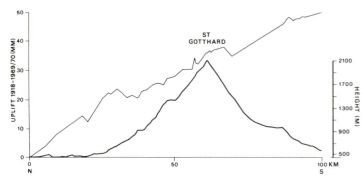

**Figure 3.25** Uplift (thin line) during the period 1918–70 along a 100-km surveyed line through St Gotthard (full line) in the Alps from Brunnen to Lavargo.
*Source:* Jeanrichard, in International Union of Geodesy and Geophysics, 1975, figure 2, p. 53.

some 7·6 mm per year, which could produce a mountain range 3048 mm (10,000 ft) high in less than 400,000 years, and extreme maximum values of about twice this figure.

### 3.4.2 Isostatic and epeirogenic movements

The crust of the earth is distinctly weak under vertical loading such that not only large lakes depress the crust, but also relatively small manmade reservoirs, such as Lake Mead in Arizona. During the past 16,000 years or so the Lake Bonneville region of Utah has rebounded up to a maximum of 64 m, giving an average rate of about 4 mm per year (see Figure 5.3). Large icecaps which developed during the Pleistocene depressed areas on a subcontinental scale to a maximum of several hundred metres. On the melting of the ice sheets an isostatic rebound resulted which is still continuing at the present time and reflects, in particular, the extent of icecaps in Fennoscandia and North America (see Chapter 19). The former area around the Baltic Sea began to updome following progressive deglaciation about 10,000 years BP. A characteristic of isostatic rebound is that its rate at any point decreases exponentially with time, giving very great rates at an early stage (possibly as high as 40 mm per year) which quickly decrease. Figure 3.26 shows the pattern of present isostatic uplift in Europe, which reaches a maximum of 10 mm per year or greater at the head of the Gulf of Bothnia, as determined by precise levelling since about 1886. Even so, the present rate is only some two-thirds the average over the past 7000 years, and about one-third of the total uplift to date probably took place during the first 1000 years of movement. Dated raised beaches indicate that a maximum uplift of over 275 m (approx. 900 ft) has taken place during the past 10,000 years, and it is estimated that significant rebound will continue for a further 5000 years at least. A feature of isostatic depression is a lateral displacement of material in the earth's mantle giving peripheral upbulging. Subsequent isostatic rebound results in peripheral subsidence, as in much of Denmark, the south Baltic coast and north-western Russia (parts of the east shore of the White Sea are subsiding at about 4 mm per year). This effect is clearly seen from comparisons of levelling, made in about 1916 and 1954, in England and Wales (Figure 3.27A). A contemporary rise over much of England increases northward to reach about 4 mm per year over northern Scotland, where the ice was thickest, whereas south-west England is subsiding at more than 1 mm/year. However, this detailed map shows that isostatic rebound may be complex in detail and may combine with other tectonic and epeirogenic mechanisms. More recent work, employing dated deposits and assumed uniform eustatic sea-level rise of 0.73 m/1000 year, has produced the pattern of deformation shown in Figure 3.27B.

Fold mountains show evidence of considerable bodily vertical uplift, separated in time from the major compressional phases, which has been effected by a number of episodic movements evidenced by erosion surfaces, straths, breaks of stream gradients, and so on. Although these movements may be the result of local mantle convection or of renewed compression, a lagged isostatic recovery of a thickened mountain root appears the most plausible explanation. Parts of the summit areas of the Sierra Nevada Mountains in California appear, on the basis of paleobotanical evidence, to have been elevated some 600 m (approx. 2000 ft) since the Early Pliocene (an average rate of 0.05 mm per year), and a similar behaviour has held for the Alps, Pyrenees and Atlas Mountains. These amounts, together with geomorphic evidence of periods of intervening stability, present problems of interpretation along Davisian lines when very young mountain ranges are examined in detail. For example, during the Pleistocene alone, the Gabilan Range in southern California has apparently suffered traces of two cycles of erosion and the Santa Lucia Range seems to have been peneplained, uplifted about 1370 m (4500 ft) and deeply dissected all during the same short period!

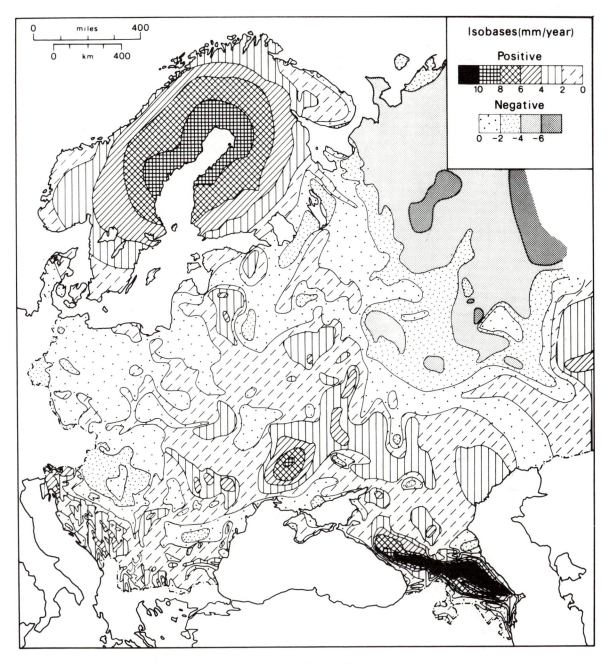

**Figure 3.26** Present rates of crustal movement in Northern and Eastern Europe.
*Source:* International Union of Geodesy and Geophysics, 1975, figure 1, p. 58.

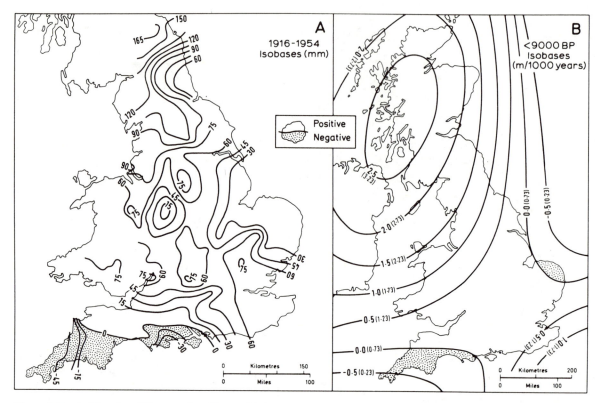

**Figure 3.27**   A. Elevation differences (in millimetres) inferred to have occurred between the second (1916) and third (completed 1954) Geodetic Levellings of England and Wales. B. Cubic trend surface fitted to some 100 locations containing dated deposits ranging in age from approximately 118 to 9961 years BP, and in elevation from − 27·2 m to + 9·3 m; the numbers indicate the rates of predicted vertical movement (m/1000 year) after assuming a uniform eustatic sea-level rise of 0·73 m/1000 year; the figures in brackets do not include this eustatic assumption.
*Sources:* Edge, 1959, figure 4 (Crown Copyright Reserved); Flemming, 1982, figure 8, p. 122.

True epeirogenic movements involving cratons are probably due to mantle plumes or to the crushing together of plates containing continental blocks. Figure 3.26 shows contemporary movements over the western Russian Platform, as well as glacial-isostatic ones. Local circular elevations occur at rates of as much as 8 mm per year (see south-central Ukraine and the southern Appalachians) which are probably associated with mantle convective plumes. West of Yellowstone Park an area of 8000 km² is rising at a rate of 3–5 mm per year, and during the period 1955–73 the central Adirondacks in New York State rose 40 mm, whereas their northern margin subsided 50 mm. The Rhine Massif has been elevated 300 mm since the Pliocene (probably episodically, as two major terraces at 200 m and 225 m near Bingen may imply). Long-term triangulation and levelling shows a contemporary uplift of the massif of 0·35 mm per year, maximum slipping along the faults of 0·23–0·5 mm per year and subsidence of the central

Rhine graben of 0·5 mm per year, and of the upper Rhine graben of 0·9 mm per year. A characteristic behaviour of the craton is the long-continued rise of localized plateaus. The Colorado Plateau appears to have risen some 550 m in the Late Cenozoic at an average rate of about 0·1 mm per year, and the uplift of the Deccan Plateau of India during the Tertiary and Quaternary continues at present at a rate of 0·36 mm per year.

### 3.5  Diastrophism and erosion

Considerations of the possible relationships between diastrophism and erosion have played a surprisingly small part in the development of geomorphology in the English-speaking world. The common assumption, adhered to most strictly by W. M. Davis, is that uplift has been of a predominantly discontinuous character and of short-lived duration. This type of uplift was thought to initiate different cycles of erosion, or parts of cycles, but not to have any more than a fleeting influence over the detailed development of individual landforms during the earliest part of youth. Davis acknowledged that slower uplifts were possible, although unusual, and that they would simply advance the progress of the cycle by producing landforms which were 'mature from birth'. This is, of course, in marked contrast to Penck's model. Nevertheless, Davis's view is broadly supported by a comparison between rates of orogeny (7·62 mm per year) and maximum regional rates of denudation, which show that the former are about eight times the latter. Such rapid elevation would not allow patterns of uplift to be reflected in patterns of slope form, and Penckian equilibrium landforms would occur only if there was a slow uplift of an area of low relief continued over a long period of time (i.e. *primärrumpf* conditions). Even this is very unlikely, however, because long-continued uplift is uncommon and because the effects of erosion do not affect all parts of the landscapes equally, with stream-channel erosion rates $D_c$ differing from mean slope denudation rates $D$ and from rates of summit lowering $D_s$ (Figure 3.28). In practice values of $D_c$ may approach uplift rates but seldom would values of $D_s$.

Variations in slope form usually reflect differences between channel incision and hillslope processes – rather than between rates of uplift and erosion. Figure 3.13A suggests that maximum denudation rates do not reach parity with rates of orogeny (i.e. 7·62 mm per year) until a relief of much more than 6000 m (20,000 ft) has been attained. As we have seen, Davis proposed a timespan of 20–200 million years for the peneplanation of fault blocks in Utah, and more recent estimates – even including isostatic recovery (see p. 103) – are tending towards the lower extreme – i.e. 15 or 33 million years, with a possible maximum of 110 million years. It is interesting that the general denudation rate, shown in Figure 3.13 B, of $D = 0.0001535H/1000$ years even when increased by only one-half (i.e. $D_s = 0.00021H/1000$ years) permits 90 per cent of initial relief $H$ to be reduced in 11 million years, and 99 per cent in 22 million years

(Figure 3.29) (Ahnert, 1970). These periods are short enough for peneplains to have developed during known periods of tectonic quiescence and it is of note that most known peneplains date from no earlier than the Tertiary.

Of course, large-scale denudation is accompanied by crustal responses which complicate calculations of rates of relief reduction. Such responses involve either the release of elastic compression or isostatic recovery. Erosion and unloading of compressed elastic rocks, such as mica-rich granite or mica schist or overconsolidated sedimentary rocks, leads to elastic rebound. In the former rocks this results in the development of exfoliation joints which affect surface form (as in the Sierra Nevada of California) and allow for more effective local erosion, such as may occur in the overdeepening of cirques, glacial troughs and fjords by ice scour. An example of the elastic rebound of valleys cut in overconsolidated clay-shales, siltstones and sandstones following the removal of 610 m (2000 ft) of overlying sediments by erosion during the Tertiary and Pleistocene has occurred in central Alberta, where upwarping of 6 m of the floors of 62–123-m-deep valleys is common (i.e. 3–5 per cent of valley depth) and as much as 10 per cent has been recorded.

The amount, rate and pattern of isostatic adjustment of the earth's surface to denudation depends upon the inherent strength of the crust. The inferred existence of summit peneplains and polycyclic surfaces points to a discontinuous post-orogenic vertical uplift of folded mountain belts. Figure 3.30 suggests a pattern of erosion and phases of uplift following an initial uplift of 4572 m (15,000 ft) with orogenic, isostatic and other uplift at the rate of 7·62 mm per year. Clearly, the weaker the crust the more sensitive isostatic recovery is to erosion, and it is noteworthy that young folded mountain ranges (e.g. Himalayas, Alps and Andes) are more perfectly isostatically adjusted than older ones (e.g. Appalachians), suggesting that isostatic adjustment may be more immediate in the weaker root zone of the younger ranges. Returning to Figure 3.29, the inclusion of realistic

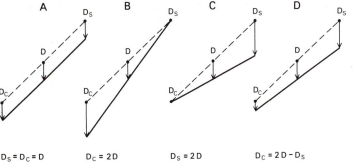

**Figure 3.28** Development of valley-side slopes assuming different rates of summit $D_s$, mean slope $D$ and stream channel $D_c$ erosion rates.
*Source:* Ahnert, 1970, figure 4, p. 252.

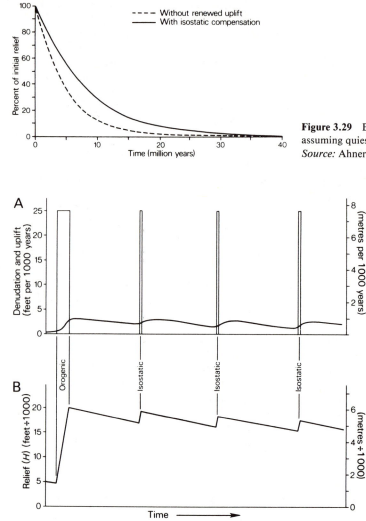

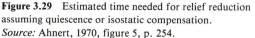

**Figure 3.29**   Estimated time needed for relief reduction assuming quiescence or isostatic compensation. *Source:* Ahnert, 1970, figure 5, p. 254.

**Figure 3.30**   Hypothetical relations of A. rates of orogenic and isostatic (plus other) uplift and denudation (thick line) to time, and B. drainage basin relief to time as a function of uplift and denudation shown in A. *Source:* Schumm, 1963, figure 3, p. 11.

isostatic compensation increases the time required for 90 and 99 per cent of the original relief to be removed by erosion to 18.5 and 37 million years, respectively.

It is of especial note that rapid rates of uplift and subsidence can readily affect the behaviour of alluvial rivers. A river flowing on alluvium can adjust to changes of valley-floor gradient by aggradation, degradation, or stream-pattern changes and this may be of considerable practical significance for the explanation of certain unstable alluvial river reaches (see Section 12.6).

# References

Abrahams, A. D. (1972) 'Drainage densities and sediment yields in eastern Australia', *Australian Geographical Studies*, vol. 10, 19–41.

Ahnert, F. (1970) 'Functional relationships between denudation, relief, and uplift in large mid-latitude drainage basins', *American Journal of Science*, vol. 268, 243–63.

Anderson, E. W. and Finlayson, B. (1975) 'Instruments for measuring soil creep', *British Geomorphological Research Group Technical Bulletin* 16.

Anderson, M. G. and Calver, A. (1977) 'On the persistence of landscape features formed by a large flood', *Transactions of the Institute of British Geographers*, n.s., vol. 2, 243–54.

Andrews, J. T. (1968) 'Pattern and cause of variability of postglacial uplift and rate of uplift in Arctic Canada', *Journal of Geology*, vol. 76, 404–25.

Bandy, O. L. and Marincovich, L. (1973) 'Rates of Late Cenozoic uplift, Baldwin Hills, Los Angeles, California', *Science*, vol. 181, 653–5.

Barry, R. G. and Chorley, R. J. (1982) *Atmosphere, Weather and Climate*, 4th edn, London, Methuen.

Bovis, M. J. and Thorn, C. E. (1981), 'Soil loss variation within a Colorado alpine area', *Earth Surface Processes*, vol. 6, 151–63.

Burmister, D. W. (1951) *Soil Mechanics*, New York, Columbia University Press, Vol. 1.

Carson, M. A. (1969) 'Soil moisture', in R. J. Chorley (ed.), *Water, Earth and Man*, London, Methuen, chapter 4. II, 185–95.

Carson, M. A. and Kirkby, M. J. (1972) *Hillslope Form and Process*, Cambridge University Press.

Chorley, R. J. (1963) 'Diastrophic background to twentieth-century geomorphological thought', *Bulletin of the Geological Society of America*, vol. 74, 953–70.

Chorley, R. J. (ed.) (1969) *Water, Earth and Man*, London, Methuen.

Chorley, R. J. (1978) 'The hillslope hydrological cycle', in M. J. Kirkby (ed.), *Hillslope Hydrology*, Chichester, Wiley, chapter 1, 1–42.

Chorley, R. J. and Kennedy, B. A. (1971) *Physical Geography: A Systems Approach*, London, Prentice-Hall.

Clark, S. P. and Jager, E. (1969) 'Denudation rate in the Alps from geologic and heat flow data', *American Journal of Science*, vol. 267, 1143–60.

Corbel, J. (1964) 'L'érosion terrestre, étude quantitative (Méthodes – techniques – résultats)', *Annales de Géographie*, vol. 73, 385–412.

Costa, J. E. (1974) 'Response and recovery of a piedmont watershed from tropical storm Agnes, June 1972', *Water Resources Research*, vol. 10, 106–12.

Derbyshire, E. (ed.) (1976) *Geomorphology and Climate*, London, Wiley.

De Sitter, L. U. (1952) 'Pliocene uplift of Tertiary mountain chains', *American Journal of Science*, vol. 250, 297–307.

Dickinson, W. T. and Wall, G. (1977) 'The relation between source-area erosion and sediment yield', *Hydrological Sciences Bulletin*, vol. 22, 527–30.

Dole, R. B. and Stabler, H. (1905) 'Denudation', *US Geological Survey Water Supply Paper* 294, 78–93.

Douglas, I. (1969) 'The efficacy of humid tropical denudation systems', *Transactions of the Institute of British Geographers*, no. 46, 1–16.

Douglas, I. (1973) 'Rates of denudation in selected small catchments in Eastern Australia', *University of Hull Occasional Papers in Geography* 21.

Douglas, I. (1976) 'Erosion rates and climate: geomorphological implications', in E. Derbyshire (ed.), *Geomorphology and Climate*, London, Wiley, 269–87.

Drake, J. J. and Ford, D. C. (1976) 'Solution erosion in the southern Canadian Rockies', *Canadian Geographer*, vol. 20, 158–70.

Edge, R. C. A. (1959) 'Some considerations arising from the results of the second and third geodetic levellings of England and Wales', *Bulletin Géodesique*, no. 52, 28–36.

Embleton, C. and King, C. A. M. (1975) *Periglacial Geomorphology*, London, Arnold.

Flemming, N. C. (1982) 'Multiple regression analysis of earth movements and eustatic sea-level change in the United Kingdom in the past 9000 years', *Proceedings of the Geologists Association*, vol. 93, 113–25.

Fournier, M. F. (1960) *Climat et érosion*, Paris, Presses universitaires de France.

Gage, M. (1970) 'The tempo of geomorphic change', *Journal of Geology*, vol. 78, 619–26.

Geikie, A. (1868) 'On denudation now in progress', *Geological Magazine*, vol. 5, 249–54.

Gilluly, J. (1949) 'Distribution of mountain building in geologic time', *Bulletin of the Geological Society of America*, vol. 60, 561–90.

Gilluly, J. (1964) 'Atlantic sediments, erosion rates, and the evolution of the continental shelf: some speculations', *Bulletin of the Geological Society of America*, vol. 75, 483–92.

Goudie, A., (ed.) (1981) *Geomorphological Techniques*, London, Allen & Unwin.

Gregory, K. J. and Walling, D. E. (1973) *Drainage Basin Form and Process*, London, Arnold.

Gupta, A. and Fox, H. (1974) 'Effects of high magnitude floods on channel form: a case study in Maryland piedmont', *Water Resources Research*, vol. 10, 499–509.

Haigh, M. J. (1977) 'Use of erosion pins in the study of slope evolution', *British Geomorphological Research Group Technical Bulletin* 18, 31–49.

Harvey, A. M. (1974) 'Gulley erosion and sediment yield in the Howgill Fells, Westmorland', *Transactions of the Institute of British Geographers, Special Publication* 6, 45–58.

Hathaway, J. C., Wylie Poag, C., Valentine, P. C., Miller, R. E., Schultz, D. M., Manheim, F. T., Kahout F. A., Bothner, M. H. and Sangrey, D. A. (1979) 'US Geological Survey core drilling on the Atlantic Shelf', *Science*, vol. 206, 515–27.

Holeman, J. N. (1968) 'The sediment yield of major rivers of the world', *Water Resources Research*, vol. 4, 737–47.

Horsfield, W. T. (1975) 'Quaternary vertical movements in the Greater Antilles', *Bulletin of the Geological Society of America*, vol. 86, 933–8.

Hudson, F. S. (1960) 'Post-Pliocene uplift of the Sierra Nevada', *Bulletin of the Geological Society of America*, vol. 71, 1547–74.

International Union of Geodesy and Geophysics (1975) *Problems of Recent Crustal Movements*, Tallinin, Valgus.

Janda, R. J. (1971) 'An evaluation of procedures used in computing chemical denudation rates', *Bulletin of the Geological Society of America*, vol. 82, 67–80.

Jansen, J. M. L. and Painter, R. B. (1974) 'Predicting sediment yield from climate and topography', *Journal of Hydrology*, vol. 21, 371–80.

Judson, S. and Ritter, D. F. (1964) 'Rates of regional denudation in the United States', *Journal of Geophysical Research*, vol. 69, 3395–401.

King, C. A. M. (1966) *Techniques in Geomorphology*, London, Arnold.

Langbein, W. B. and Schumm, S. A. (1958) 'Yield of sediment in relation to mean annual precipitation', *Transactions of the American Geophysical Union*, vol. 39, 1076–84.

Laronne, J. B. and Carson, M. A. (1976) 'Interrelationships between bed morphology and bed-material transport for a small, gravel-bed channel', *Sedimentology*, vol. 23, 67–85.

Ledger, D. C., Lovell, J. P. B. and McDonald, A. T. (1974) 'Sediment yield studies in upland catchment areas in southeast Scotland', *Journal of Applied Ecology*, vol. 11, 201–6.

Leopold, L. B. and Emmett, W. W. (1972) 'Some rates of geomorphological processes', *Geographia Polonica*, vol. 23, 27–35.

Leopold, L. B., Wolman, M. G. and Miller, J. P. (1964) *Fluvial Processes in Geomorphology*, San Francisco, Calif., Freeman.

Lustig, L. K. and Busch, R. D. (1967) 'Sediment transport in Cache Creek drainage basin in the Coast Ranges west of Sacramento, California', *US Geological Survey Professional Paper* 562-A, 3–36.

McGetchin, T. R. and Merrill, R. B. (1979) 'Plateau uplift: mode and mechanism', *Tectonophysics*, vol. 61, nos 1–3.

Maner, S. B. (1958) 'Factors affecting sediment delivery rates in the Red Hills physiographic area', *Transactions of the American Geophysical Union*, vol. 39, 669–75.

Marchand, D. E. (1971) 'Rates and modes of denudation, White Mountains, Eastern California', *American Journal of Science*, vol. 270, 109–35.

Matheson, D. S. and Thomas, S. (1973) 'Geological implications of valley rebound', *Canadian Journal of the Earth Sciences*, vol. 10, 961–78.

Meade, R. H. (1969) 'Errors in using modern stream-load data to estimate natural rates of denudation', *Bulletin of the Geological Society of America*, vol. 80, 1265–74.

Melton, M. A. (1960) 'Intravalley variation in slope angles related to microclimate and erosional environment', *Bulletin of the Geological Society of America*, vol. 71, 133–44.

Menard, H. W. (1961) 'Some rates of regional erosion', *Journal of Geology*, vol. 69, 154–61.

Miller, D. H. (1965) 'The heat and water budget of the earth's surface', *Advances in Geophysics*, vol. 11, 175–302.

Moberly, R. (1963) 'Rate of denudation in Hawaii', *Journal of Geology*, vol. 71, 371–5.

Oke, T. R. (1978) *Boundary Layer Climates*, London, Methuen.

Ollier, C. D. and Brown, M. J. F. (1971) 'Erosion of a young volcano in New Guinea', *Zeitschrift für Geomorphologie*, vol. 15, 12–28.

Pain, C. F. and Bowler, J. M. (1973) 'Denudation following the November 1970 earthquake at Madang, Papua New Guinea', *Zeitschrift für Geomorphologie*, Supplement 18, 92–104.

Pannekoek, A. J. (1961) 'Post-orogenic history of mountain ranges', *Geologische Rundschau*, vol. 50, 259–73.

Pavoni, N. and Green, R. (1975) *Recent Crustal Movements*, Amsterdam, Elsevier.

Peel, R. F. (1974) 'Insolation and weathering: some measures of diurnal temperature changes in exposed rocks in the Tibesti region, central Sahara', *Zeitschrift für Geomorphologie*, Supplement 21, 19–28.

Rapp, A. (1960) 'Recent development of mountain slopes in Kärkevagge and surroundings, northern Scandinavia', *Geografiska Annaler*, vol. 42, 65–200.

Reed, H. H. and Watson, J. (1975) *Introduction to Geology*, Vol. 2, *Earth History*, Pt II: *Later Stages of Earth History*, London, Macmillan.

Robertson, A. H. F. (1977) 'Tertiary uplift history of the Troodos massif, Cyprus', *Bulletin of the Geological Society of America*, vol. 88, 1763–72.

Ruxton, B. P. and McDougall, I. (1967) 'Denudation rates in northeast Papua from Potassium–Argon dating of lavas', *American Journal of Science*, vol. 265, 545–61.

Schumm, S. A. (1963) 'The disparity between present rates of denudation and orogeny', *US Geological Survey Professional Paper* 454-H.

Schumm, S. A. (1965) 'Quaternary paleohydrology', in H. E. Wright and D. G. Frey (eds), *The Quaternary of the United States*, Princeton, NJ, Princeton University Press, 783–94.

Schumm, S. A. (1973) 'Geomorphic thresholds and complex response of drainage systems', *Binghamton Publications in Geomorphology*, 299–310.

Schumm, S. A. (1975) 'Episodic erosion: a modification of the geomorphic cycle', *Binghamton Publications in Geomorphology*, 299–310.

Schumm, S. A. (1977) *The Fluvial System*, New York, Wiley.

Schumm, S. A. and Chorley, R. J. (1964) 'The fall of Threatening Rock', *American Journal of Science*, vol. 262, 1041–54.

Selby, M. J. (1974) 'Rates of denudation', *New Zealand Journal of Geography*, vol. 56, 1–13.

Slaymaker, H. O. (1972) 'Patterns of present sub-aerial erosion and landforms in Mid-Wales', *Transactions of the Institute of British Geographers*, no. 55, 47–68.

Smith, D. I. and Atkinson, T. C. (1976) 'Process, landforms and climate in limestone regions', in E. Derbyshire (ed.), *Geomorphology and Climate*, London, Wiley, 367–409.

Sparks, B. W. (1972) *Geomorphology*, 2nd edn, London, Longman.

Stocking, M. A. and Elwell, H. A. (1976) 'Rainfall erosivity over Rhodesia', *Transactions of the Institute of British Geographers*, n.s., vol. 1, 231–45.

Stoddart, D. R. (1969) 'World erosion and sedimentation', in R. J. Chorley, (ed.), *Water, Earth and Man*, London, Methuen, 43–64.

Strakhov, N. M. (1967) *Principles of Lithogenesis*, Edinburgh, Oliver & Boyd, Vol. 1.

Sukachev, V. and Dylis, N. (1968) *Fundamentals of Forest Bio-geocoenology*, Edinburgh, Oliver & Boyd.

Thornes, J. B. (1979) *River Channels*, London, Macmillan.

Thornes, J. B. and Brunsden, D. (1977) *Geomorphology and Time*, London, Methuen.

Tricart, J. and Macar, P. (eds) (1967) 'Field methods for the study of slope and fluvial processes', *Revue de Géomorphologie dynamique*, vol. 17, no. 4 (October), 147–88.

Trimble, S. W. (1975) 'Denudation studies; can we assume stream steady state?', *Science*, vol. 188, 1207–8.

Trimble, S. W. (1977) 'The fallacy of stream equilibrium in contemporary denudation studies', *American Journal of Science*, vol. 277, 876–87.

Wager, L. R. (1937) 'The Arun river drainage pattern and the rise of the Himalaya', *Geographical Journal*, vol. 89, 239–49.

Walling, D. E. and Webb, B. W. (1983) 'Patterns of sediment yield', in K. J. Gregory (ed.), *Background to Palaeo-hydrology*, Chichester, Wiley, 69–100.

Whitten, C. A., Green, R. and Meade, B. K. (eds) (1977) 'Recent crustal movements', *Tectonophysics*, vol. 52.

Wilson, L. (1973) 'Variations in mean annual sediment yield as a function of mean annual precipitation', *American Journal of Science*, vol. 273, 335–49.

Winkler, E. M. (1970) 'Errors in using modern stream-load data to estimate natural rates of denudation: discussion', *Bulletin of the Geological Society of America*, vol. 81, 983–4.

Wolman, M. G. and Gerson, R. (1978) 'Relative scales of time and effectiveness of climate in watershed geomorphology', *Earth Surface Processes*, vol. 3, 189–208.

Wolman, M. G. and Miller, J. P. (1960) 'Magnitude and frequency of forces in geomorphic processes', *Journal of Geology*, vol. 68, 54–74.

Yoshikawa, T. (1970) 'On the relations between Quaternary tectonic movement and seismic crustal deformation in Japan', *Bulletin of the Department of Geography, University of Tokyo*, no. 2.

Young, A. (1969) 'Present rate of land erosion', *Nature*, vol. 224, 851–2.

Young, A. (1974) 'The rate of slope retreat', *Institute of British Geographers Special Publication* 7, 65–78.

# Part Two  *Geological geomorphology*

Primary landforms are those formed by endogenic processes, igneous and tectonic. Exposed rocks (Chapters 4 and 8) and structures (Chapter 7) are immediately acted upon by exogenic weathering and erosional processes to form secondary landforms, which reflect geologic controls at both a global and a local scale. Rates of diastrophism and isostatic adjustment have been considered in Chapter 3, but in order to understand the first- and second-order landforms (see Table 1.3) of continents and ocean basins, consideration of global structures and endogenic processes is also necessary (Chapter 5).

Volcanoes are, of course, smaller primary features that are formed by endogenic constructional processes. Vast areas of the earth, moon and the planets are volcanic in origin, and both intrusive and extrusive igneous bodies and landforms are discussed in Chapter 6.

Secondary landforms are modified by erosional or depositional processes that are considered in Parts III and IV. A major factor determining the character of the secondary landform is the geologic character of a landscape, both structural and lithologic. A major practical application of geomorphology has been the interpretation of structure and lithologic mapping based on landform characteristics as viewed on aerial photographs. Therefore, variables 2 and 3 in Table 1.1 are worthy of separate treatment and this is, in fact, the classical geological geomorphology that is considered in Part II.

# **Four** *Minerals, rocks and sediments*

Basic to the understanding of geological controls over landforms (variable 3, Table 1.1) is a background in petrology. This deals with the physical and chemical properties of rocks, their mineral constituents and the resulting sedimentary weathered products. In this chapter the essentials of mineralogy and petrology are presented, primarily to provide the non-geologist with a brief introduction to the terminology involved.

## 4.1 Igneous minerals and rocks

### 4.1.1 Minerals

Minerals possess definable chemical compositions and characteristic crystal structures. There is an immense number of different minerals, but it is possible to classify those which are most important for igneous petrology into seven groups, the first five of which are dominated by silica structures.

(1) Quartz – the basic building-block of the silicate minerals, of which most igneous rocks are largely composed, is the silica tetrahedron ($SiO_4$) in which the silicon ion ($Si^{++++}$) fits snugly between four oxygen ions ($O^{--}$), covalently bonding them by sharing one electron with each (Figure 4.1A). This structure is both chemically efficient and geometrically compact, making the tetrahedra strong and very difficult to break up chemically, and, therefore, slow to weather.

(2) Ferromagnesian minerals – these are complex silicates of iron, magnesium and calcium forming a discrete series (i.e. exhibiting no continuous chemical gradations between the members) in which the silica tetrahedra are joined by $Fe^{++}$, $Mg^{++}$, $Ca^{++}$ and other ions. These structures are chemically weaker than quartz and, as we shall see in a subsequent chapter, this has far-reaching

implications in respect of the weathering of rocks containing ferromagnesian minerals.

Olivine ( $(Mg,Fe)_2SiO_4$) is a small greenish mineral composed of isolated tetrahedra linked on all sides by $Mg^{++}$ and $Fe^{++}$ ions, giving a silica–oxygen proportion of 1:4 (Figure 4.1A).

Pyroxenes represent a rather more structurally robust form of silica structures, being mainly represented by the minerals hypersthene ( $(Mg,Fe)SiO_3$) and augite ($Ca(Mg,Fe)Si_2O_6$) in which the silica tetrahedra form single chains by sharing one oxygen atom (i.e. having a silica–oxygen ratio of 1:3) but being weakly linked on four sides by other cations (Figure 4.1B).

Hornblende ( $(OH)_2Ca_2(Mg,Fe)_6Al_2Si_7O_{22}$) is a complex mineral which is the most common representative of the amphibole group. It is formed by double silica tetrahedron chains (Figure 4.1C) linked by other ions (silica-oxygen ratio approximately 4:11).

Biotite ($K(Fe,Mg)_3(AlSi_3O_{10})(OH)_2$), or black mica, is a complex alumina-silicate in which the silica tetrahedra form plates (Figure 4.1D), between which perfect cleavage occurs.

(3) Plagioclase feldspars – these are complex alumina-silicates ranging in composition from the high-temperature calcic plagioclase anorthite ($Ca(Al_2Si_2O_8)$) to the low-temperature sodic variety albite ($Na(AlSi_3O_8)$), linked by a continuous series of different varieties of plagioclase exhibiting all proportions of these two end-members. In the plagioclases (Greek: *plagio* – 'oblique', *klastos* – 'broken') varying proportions of the $Si^{++++}$ ions of the silica tetrahedra have been replaced by $Al^{+++}$ ions, together with additional cations to compensate for the resulting loss of charge. The more substitutions of $Al^{+++}$, the chemically weaker the structure becomes such that anorthite, in which $Al^{+++}$ replaces every other $Si^{+++}$, is less resistant to weathering than albite, where

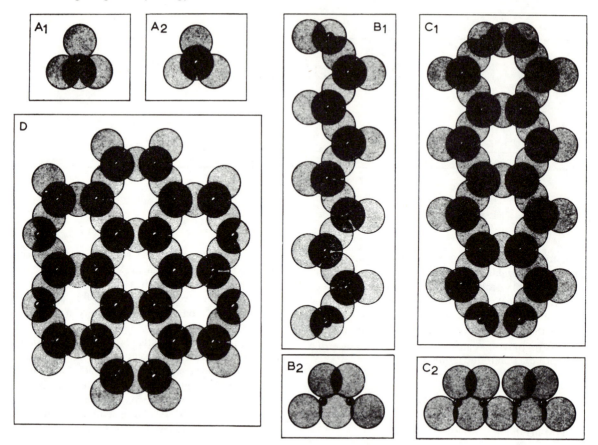

**Figure 4.1**   The structure of silicate minerals: A. silicon–oxygen tetrahedron viewed from the side (1) and above (2); B. single chain of tetrahedra viewed from above (1) and side (2); C. double chain of tetrahedra viewed from above (1) and side (2); D. tetrahedral sheet viewed from above.
*Source:* Leet and Judson, from Chorley, 1969, figure 3.II.3, p. 139.

$Al^{+++}$ replaces only every fourth $Si^{++++}$.

(4) Potassium (K) feldspars – potassium-rich alumina-silicates of which orthoclase ($K(AlSi_3O_8)$) (Greek: *orthos* – 'right' (angled), *klastos* – 'broken') is the most important member.

(5) Feldspathoids (foids) - composed of a number of unrelated sodic and potassic alumina-silicates (chiefly nepheline ($Na_3K(Al_4Si_4O_{10})$)), being feldspar-like in structure and occurring where $SiO_2$ is lacking (i.e. in undersaturated rocks).

(6) Muscovite or white mica ($KAl_2(AlSi_3O_{10})(OH)_2$) – a potassium-rich alumina-silicate similar in structure to biotite except that it has more $Al^{+++}$ ions, instead of the latter's $Fe^{++}$ and $Mg^{++}$ ions, giving it a more chemically resistant structure.

(7) Hydrous silicates of aluminium and magnesium – complex minerals formed by metamorphic alteration of other minerals in the presence of water. Two of the most important are serpentine ($Mg_3SiO_5(OH)_4$) formed from olivine, and chlorite (Greek: *chloros* – 'green') a mica-like, green mineral formed from foids.

### 4.1.2 Igneous rocks

Although there are characteristic associations of minerals which permit the identification, naming and classification of igneous rock types (Figure 4.2), there are sufficient variations among these associations of the large number of minerals involved to make any given classification incomplete, ambiguous and unsatisfactory to some degree. The most obvious variations stem from the original composition of the magma, and from its subsequent history of cooling and crystallization, although other circumstances such as magma reheating or its digestion of surrounding 'country' rocks may be important. Magmas are seldom, if ever, 'ideal' in composition

| TEXTURE | QUARTZ ABUNDANCE | DOMINANT FELDSPAR | | | MAINLY FERROMAGNESIAN MINERALS |
|---|---|---|---|---|---|
| | | K – FELDSPAR | ALKALI–CALCIC | CALCIC | |
| COARSE | MUCH (SATURATED) | GRANITE[1] | TONALITE | QUARTZ GABBRO | —— |
| COARSE | LITTLE OR NONE (UNDERSATURATED) | SYENITE (NEPHELINE–SYENITE)[2] | DIORITE | GABBRO | PERIDOTITE DUNITE PYROXENITE |
| FINE | MUCH (SATURATED) | RHYOLITE[3] | DACITE | THOLEIITE | —— |
| FINE | LITTLE OR NONE (UNDERSATURATED) | TRACHYTE (PHONOLITE)[2] | ANDESITE | BASALT[4] (OLIVINE–BASALT)[2] | —— |

**Figure 4.2** A classification of igneous rocks based on amount of quartz, dominant feldspar type, presence of foids, presence of ferromagnesian minerals and texture. 1, termed *pegmatite* when very coarse-grained – grades into *granodiorite* as more calcic feldspars are present; 2, rocks in parentheses contain significant proportions of foids; 3, especially rapid cooling produces *obsidian* (volcanic glass) and *pumice* (frothy glass); 4, when sodic feldspar is dominant, the rock is termed *spillite* – some basalts may contain olivine.
*Source:* Partly after B. W. Sparks, *Rocks and Relief*, 1971, figure 3.27, p. 123, copyright © Longman, London.

and, in view of its importance as the basic building-block of most igneous minerals, the amount of available silica (much = oversaturated or saturated; little = undersaturated) is especially vital. The amount of silica which is present in the original magma controls not only the amount of quartz in the resulting rock, but also the presence or absence of foids. The importance of the history of cooling can be illustrated in three respects:

(1) The proportions of calcic to sodic plagioclases are, to some degree, influenced by the rapidity of the cooling of the melt.

(2) The discontinuous reactions producing the ferromagnesian minerals are very much influenced by the rate of cooling such that, for example, a slowly cooled granite may contain hornblende and biotite, whereas a rapidly cooled rhyolite (Greek: *rheo* – 'I flow') of equivalent original composition may be rich in pyroxenes and even contain some olivine.

(3) The texture of igneous rocks relates to the following main attributes:

(a) crystal size – coarse, mean grain size > 5 mm (e.g. pegmatite); medium, 5–1 mm; fine, < 1 mm;

(b) degree of crystallinity – e.g. holocrystalline (entirely crystalline) v. glassy (non-crystalline; e.g. obsidian);

(c) crystal shape – e.g. euhedral (near-perfect crystal forms) v. anhedral (crystal faces absent, no regular crystal form);

(d) general characteristics – e.g. granular (equal-sized grains); porphyritic (more than 50 per cent of rock composed of large crystals enclosed by smaller groundmass); vesicular (containing gas holes, e.g. pumice and scoria), etc.

The texture of igneous rocks is determined partly by their history of cooling, but mainly by the viscosity (i.e. resistance to flow; measured in poises) of the magma as the crystallization sequence takes place. In general, the higher the ratio $Si:Fe,Mg,Ca$, the higher the viscosity, as is shown by the orders of magnitude for the viscosity of magmas appearing on the surface given in Table 4.1.

**Table 4.1** Viscosity of common magmas (poises)

| | |
|---|---|
| Olivine basalt | $10^2$ |
| Basalt | $10^3$–$10^5$ |
| Andesite | $10^5$–$10^7$ |
| Dacite | $10^{10}$ |
| Rhyolite | $10^{12}$ |

Basalt is very mobile at 1000°C, whereas at the same temperature granite appears as an incompletely melted, highly viscous glass. Viscosity is inversely related to temperature and depends on the amount of solid load which is included. An important factor is the amount and condition of contained gasses because the more gas that is dissolved in the magma, the higher its viscosity. The effect of temperature and viscosity upon igneous rock texture is that in magmas which are mobile, of low viscosity and being slowly cooled, the constituent ions have sufficient time to mobilize themselves into coarsely crystalline rocks, whereas with high viscosity and rapid cooling, crystallization is hindered and volcanic glasses may form. Where magma is sprayed into the air and cooled very rapidly, gas bubbles may be unable to escape from highly viscous material and pumice may form from granitic magma. Having said this, it should be recognized that the view of igneous rocks being divisible into coarse-grained plutonic (i.e. assumed to have crystallized slowly when intruded at depth) and fine-grained volcanic (i.e. assumed to have crystallized rapidly when extruded on the earth's surface) is too simplistic in that, other factors apart, some minerals (e.g. olivine) tend never to form large crystals.

As most medium- and small-scale landforms evolve by means of their superficial breakdown by the processes of chemical weathering (see Chapter 9), it is particularly useful for the geomorphologist to concern himself with a chemical and mineralogical classification of igneous rocks. Figure 4.2 illustrates five main criteria for classification:

(1) Amount of quartz – i.e. oversaturated and saturated v. undersaturated.
(2) Type of feldspar – i.e. alkali v. calcic.
(3) Presence of foids.
(4) Type and proportion of ferromagnesian minerals.
(5) Textures – i.e. coarse or fine (instrusive or extrusive).

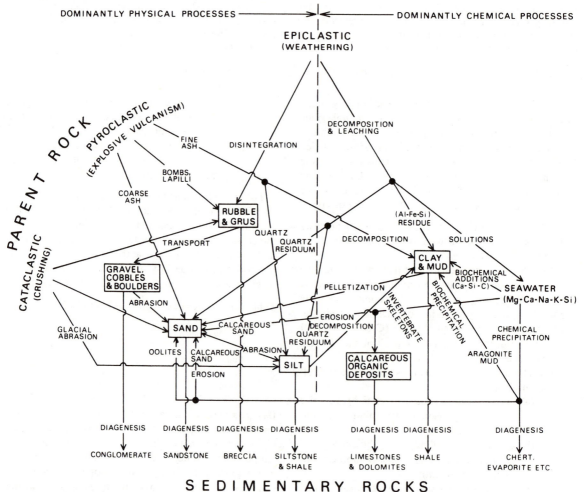

**Figure 4.3**   The provenance of sedimentary rocks.
*Sources:* F. J. Pettijohn, *Sedimentary Rocks*, 1975, 3rd edn, figures 8-7, 10-24, 10-50, pp. 226, 349, 369, copyright © 1949, 1957 by Harper & Row, Publishers, Inc., copyright © 1975 by Francis J. Pettijohn, reproduced by permission of Harper & Row, Publishers, Inc.; Pettijohn *et al.*, 1972, figure 8-1, p. 295.

## 4.2 Sediments

### 4.2.1 Provenance

The concept of provenance (French: *provenir* – 'to originate', 'to come forth') involves the origin of sediments and their delivery to the depositional environment. Figure 4.3 shows the provenance of six major types of sediments: rubble and grus; gravel, cobbles and boulders; sand; silt; clay and mud; and calcareous organic deposits. These originate from parent rocks by three major sets of processes – epiclastic (weathering), pyroclastic (explosive vulcanism) and cataclastic (crushing and grinding by differential earth movements along faults, slump planes, etc.). The processes of weathering (Chapter 9) are clearly the most important.

(1) Rubble and grus are accumulations of angular rock fragments larger than sand size (i.e. >2 mm diam.). Figure 4.4 shows the conventional size limits for the various classes of clastic, or fragmental, sediments (Greek: *klastos* – 'to break'). Rubble is produced by crushing and fracture associated with movement along faults, slumps and slides, glacial grinding and the like; by explosive vulcanism; and by the disintegration accompanying weathering. Grus is a term reserved for very immaturely disintegrated igneous rocks, particularly granite.

(2) Gravel (2–64 mm), cobbles (64–256 mm) and boulders (>256 mm) have the size equivalent of rubble and grus but are to some degree rounded, showing signs of transport by water action. Like the latter, the individual particles are commonly composed of rock fragments, rather than of individual minerals which make up the grains of the finer clastic rocks. Gravel, cobbles and boulders occur in contemporary environments, such as beaches, river channels and alluvial fans, and are almost exclusively derived from the transportation of rubble and grus.

(3) Sand ($\frac{1}{16}$–2 mm) consists of non-coherent grains of a variety of minerals, but most commonly quartz. Quartz sand usually derives from the weathering decomposition of parent sedimentary and igneous rocks, and from the

abrasion of gravel during its transportation. Angular sands of varying composition result from cataclastic and pyroclastic activity, and small particles of clay and mud may form into sand-size pellets by aggregation. Some sand is calcareous and originates by the weathering and erosion of calcareous skeletons, coral-reef deposits and chemical precipitates. Oolitic sand derives from the chemical association of calcareous oolites and other sands from the erosion of chemical precipitates (e.g. gypsum sand). Sand deposits are found in a very wide variety of environments, except in the deep ocean basins. They occur most commonly on the continental shelves, in shallow near-shore seas, beaches, deltas, floodplains, river channels, alluvial fans, glacial outwash deposits and aeolian sand dunes. Sandstones (i.e. rocks composed predominantly of sand-sized particles) form one-quarter to one-third of the sedimentary rocks of the earth, and of these rocks it has been estimated that 65 per cent is quartz, 15 per cent feldspar and 18 per cent other rock fragments (chiefly clay).

(4) Silt ($\frac{1}{256}$–$\frac{1}{16}$ mm) is overwhelmingly composed of fine grains of quartz and derives partly from fine volcanic ash and from the intense weathering of quartz-rich rocks. Its main sources, however, are the result of abrasion of water and wind-borne sand and the grinding of rocks by glaciers. Silt may be viewed as the end-product of abrasion and is found as widespread marine deposits, in floodplains and deltas (the Mississippi delta is composed of 60 per cent silt, 23 per cent very fine sand, 11 per cent clay and 6 per cent fine sand) and as wind-borne loess (0·06–0·01 mm), an aeolian silt partly derived from the floodplains of glacial outwash rivers. Most shale rocks normally contain one-third to one-half silt.

(5) Clay (<$\frac{1}{256}$ mm, 0·004 mm, or 4 $\mu$m) consists of minute particles of complex hydrous aluminosilicates formed predominantly by the weathering of feldspars, ferromagnesian minerals, micas and volcanic ash (Figure 4.5). Clay minerals are composed of covalent tetrahedral layers of oxygen and silicon ($Si_2O_5$) and octahedral layers of hydroxyls, aluminiums, magnesiums, etc. (covalent or ionic bonding) with, on occasion, additional layers of potassium, water, etc. (see Figure 9.1 and Table 9.6).

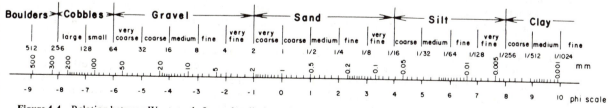

**Figure 4.4**  Relation between Wentworth–Lane class limits and the phi-scale.
*Source:* F. J. Pettijohn, *Sedimentary Rocks*, 1975, 3rd edn, figure 3-6, p. 35, copyright © 1949, 1957 by Harper and Row, Publishers, Inc., copyright © 1975 by Francis J. Pettijohn reproduced by permission of Harper & Row, Publishers, Inc.

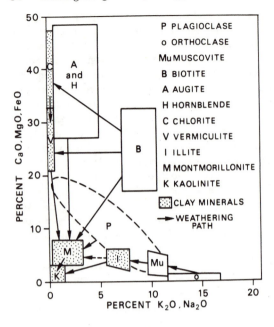

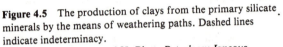

**Figure 4.5** The production of clays from the primary silicate minerals by the means of weathering paths. Dashed lines indicate indeterminacy.
*Source:* E. G. Ehlers and H. Blatt, *Petrology: Igneous, Sedimentary and Metamorphic*, 1982, figure 11-6, p. 280, copyright © W. H. Freeman & Co., San Francisco, by permission.

*Kaolinite* derives from the mature weathering of granitic or feldspar-rich rocks under moist, well-drained conditions. It has low plasticity, a relatively high permeability for a clay and absorbs little water. It is a major component of most shales and is abundant in lower-latitude marine sediments. *Chlorite* is a magnesium-rich clay deriving from the immature weathering of biotite mica, ferromagnesian minerals and sometimes from pre-existing montmorillonite clay. It is common in marine sediments, particularly in those of higher latitudes. *Illite* is the most abundant clay mineral, but when mixed with other clays its small crystal size and intermediate properties of permeability, water absorption and plasticity often result in its presence being masked. The origin of illite is from the weathering of feldspars, muscovite mica and montmorillonite clay, but there is considerable debate as to whether it may represent a stage in the development of kaolinite (particularly under mid-latitude marine conditions) or whether it may be formed by the alteration of kaolinite. *Montmorillonite*, the most impermeable, plastic and water-absorptive of the common clay minerals, has a chemically weak and highly expandable crystal lattice. It develops from ferromagnesian minerals, volcanic ash and gabbroic rocks under moist and poorly drained conditions, particularly in the tropics, and has been deposited over large areas of the tropical ocean floors. Besides being common in marine conditions, clays occur in river deltas and floodplains (forming a major part of the sediment load of present rivers) and as glacial clays resulting from the weathering

of finely comminuted subglacial debris. Some clays are formed by weathering after they have reached a major depositional basin (i.e. they are authigenic), and it is believed that much of the clay content of greywacke is of this origin.
(6) Calcareous organic deposits will be treated in Section 4.3.3.2.

### 4.2.2 Particle size and sorting

The sizes of clastic particles are given with reference to the length of three axes (or 'diameters') at right angles to one another, where $a$ = the long axis, $b$ = the intermediate axis and $c$ = the short axis. Obviously, the more irregular the particle shape, the more the values of $a$, $b$ and $c$ diverge, and the more difficult it is to characterize the particle size by the length of any one diameter. Thus although the average particle diameter may be the mean of $a$, $b$ and $c$, the sieve-mesh which just traps the particle is marginally smaller than $b$. Figure 4.4 gives the standard size nomenclature for clastic material. Terms such as gravel, sand and clay create certain problems because of their common association with given mineralogical components (i.e. rock particles and minerals, quartz and clay minerals), and some workers replace them by the purely textural terms rudite, arenite and lutite, giving them compositional prefixes (e.g. quartz arenite, calcilutite, etc.). Particle sizes are determined by individual measurement for those larger than 10 mm diameter by sieving ($10-\frac{1}{10}$ mm) and by settling velocity ($<\frac{1}{10}$ mm).

Despite the strong tendency for moving currents to sort clastic material into single size ranges, natural clastic aggregates seldom occur within narrow size ranges. There are many reasons for this, such as sudden and drastic decreases in current velocity, the trapping of finer particles between coarser ones, the reworking and mixing of earlier sediments, and transportation by such non-selective processes as slumps, earthflows and glaciers. This lack of sorting makes it difficult to classify sediments in terms of an overall diagnostic grain size. One method of classification and naming of mixed sediments is illustrated in Figure 4.6, using a triangle divided into percentages parallel to each side, so that any combination of three constituents (in this case sand, silt and clay) totalling 100 per cent may be represented by a point within it. Each clastic combination may be precisely named with reference to standard divisions of the triangle.

A standard method of depicting sedimentary grain sizes is by means of histograms (Figure 4.7) and cumulative frequency distributions (Figure 4.8) by plotting weight percentages within standard grain size ranges. The characteristics of histograms are conventionally described by such parameters as the arithmetic mean ($\bar{x}$) and the standard deviation ($\sigma$) by approximating the distribution by a bell-shaped, 'normal' distribution (A1 in Figure 4.9). Unfortunately, most natural sedimentary assemblages have non-normal, right-skewed size distributions (B1 in Figure 4.9) because of an excessive composition of fine material. Many of these right-skewed distributions can be normalized by plotting their size ranges on a logarithmic scale (B2 in Figure 4.9), and this is the conventional manner of plotting sedimentary histo-

grams (Figure 4.7), allowing logarithmic normal distributions to display themselves (e.g. disintegrated granite and shale). Even with logarithmic plots, most sediments have 'skewed' distributions (e.g. left-skewed, river sand; right-skewed, glacial outwash gravel, weathered gneiss, dune sand, loess), and the following additional parameters are commonly employed to express their size characteristics:

(1) The mode (*Mo*) – central diameter of the largest size class.
(2) The median (*Md*) – the size for which 50 per cent by weight is coarser.
(3) Quartiles (*Q*) – the sizes for which 75 per cent ($Q_1$), 50 per cent ($Q_2 = Md$) and 25 per cent ($Q_3$) by weight is coarser (Figure 4.8).
(4) Percentages finer (*D*) the sizes for which given percentages of the sediment are finer (e.g. $D_{50} = Md = Q_2$). These parameters are useful because many of the dynamic properties of sediments (e.g. porosity, permeability, shearing resistance, etc.) are largely controlled by the proportion of fine material. $D_{10}$, the so-called Hazen's coefficient, is especially diagnostic in this regard (Figure 4.8).
(5) Sorting (*So*) – an expression of the range of sizes present $= \sqrt{(Q_1/Q_3)}$.
(6) Skewness (*Sk*) – a measure of the departure of a distribution from normality $= \sqrt{(Q_1 Q_3/Md^2)}$.

It should be noted that, as is shown in Figure 4.7, some distributions are bimodal (e.g. pyroclastic fall) and others multimodal (e.g. glacial till), which is often evidence of the interaction of two or more different processes of sediment generation, transportation, or deposition. Normal distributions or log normal distributions can be plotted in cumulative form on arithmetic or

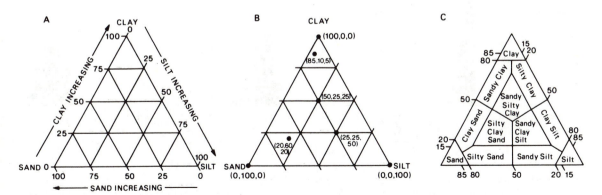

**Figure 4.6** Triangular classification for mixtures of sand, silt and clay showing A. its general basis, B. the percentages at some sample locations; and C. the domains for each class of mixed sediment.
*Source:* Trefethen, 1950, figure 4, p. 60.

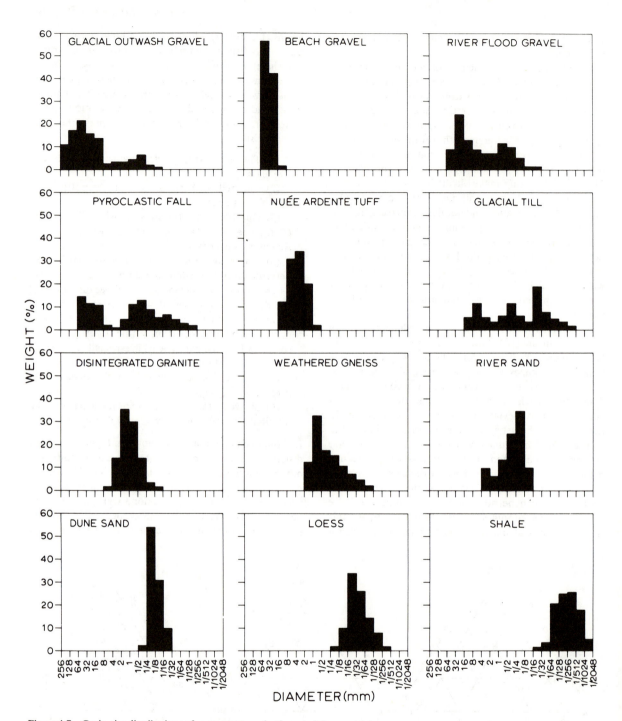

**Figure 4.7** Grain-size distributions of some common clastic materials.

*Sources:* F. J. Pettijohn, *Sedimentary Rocks*, 1975, 3rd edn, composite, copyright © 1949, 1957 by Harper & Row, Publishers, Inc., copyright © 1975 by Francis J. Pettijohn, reproduced by permission of Harper & Row, Publishers, Inc.; Krumbein and Sloss, 1963, figure 4-3, p. 98.

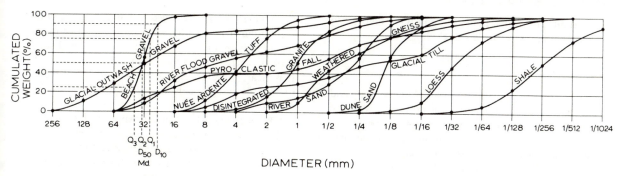

**Figure 4.8** Cumulative frequency grain-size curves from the common clastic materials shown in Figure 4.7.
*Sources:* F. J. Pettijohn, *Sedimentary Rocks,* 1975, 3rd edn, copyright © 1949, 1957 by Harper & Row, Publishers, Inc., copyright © 1975 by Francis J. Pettijohn, reproduced by permission of Harper & Row, Publishers, Inc.; Krumbein and Sloss, 1963.

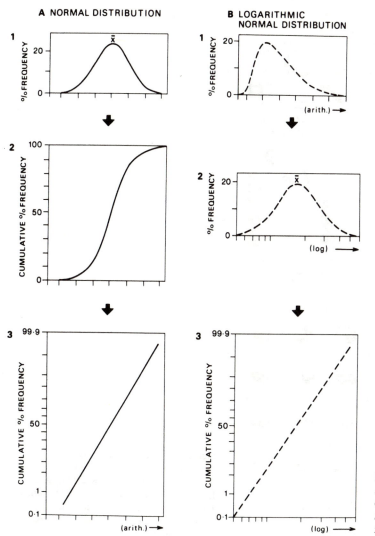

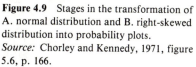

**Figure 4.9** Stages in the transformation of A. normal distribution and B. right-skewed distribution into probability plots.
*Source:* Chorley and Kennedy, 1971, figure 5.6, p. 166.

logarithmic probability paper (A3 and B3 in Figure 4.9) to yield readily identifiable straight lines.

### 4.2.3 Particle shape

The shape of clastic particles depends on the strength and structure of the parent material; the process of weathering to which it was subjected; the abrasion, corrasion and breakage suffered during transportation; and on such post-depositional processes as surface chemical precipitation. The shape of even an individual clastic particle is a composite quality, requiring at least three parameters to describe it:

(1) Form. This is achieved by comparing the grain in three dimensions with regular forms. One method of this type is by means of the Zingg diagram (Figure 4.10A), where the axes ratios $b/a$ and $c/b$ are plotted. It is clear, however, that apparently identical forms under this system may have considerably different shapes.

(2) Sphericity. This measure expresses the extent to which the longest particle diameter ($a$, as distinct from the median and shortest diameters $b$ and $c$) compares with the diameter of a sphere with the same volume as that of the particle ($d_n$). Sphericity = $d_n/a$, ranging from 0 to 1·0.

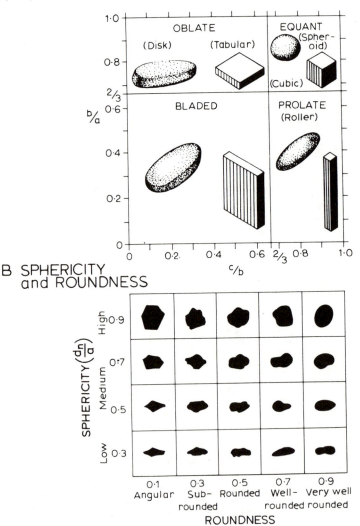

Figure 4.10  Particle shape classifications: A. form; B. sphericity and roundness. *Sources:* T. Zingg; W. C. Krumbein and L. L. Sloss, *Stratigraphy and Sedimentation*, 1963, figure 4-10, p. 111, copyright © W. H. Freeman and Co., San Francisco, by permission.

**Table 4.2** Sediment roundness and sphericity

| | *Average sphericity* | *Average roundness* | *Description* |
|---|---|---|---|
| Recent dune sand | 0·75 | 0·70 | Medium sphericity, well rounded |
| Recent beach sand | 0·83 | 0·64 | High sphericity, well rounded |
| Recent beach gravel | 0·64 | 0·61 | High medium sphericity, well rounded |
| Recent stream gravel | 0·71 | 0·34 | High medium sphericity, subrounded |
| Glacial till pebbles | 0·72 | 0·54 | High medium sphericity, rounded |

*Source:* W. C. Krumbein and L. L. Sloss, *Stratigraphy and Sedimentation*, 1963, table 4-4, p. 113, copyright © W. H. Freeman and Co., San Francisco, by permission.

(3) Roundness. This is a complex dimensionless measure referred to the two-dimensional particle shape, ranging from 0 to 1·0, expressing the ratio of the average radius of the corners and sides to the radius of the maximum inscribed circle. It is clear from Figure 4.10B that particles of a given sphericity may possess very different roundness, and vice versa.

The effects of transportation processes on particle shapes are particularly complex in that changes in shape due to abrasion also involve changes in size; particles of the same size but different shapes may be transported with different facility; different transportation processes produce sharp changes on given particles at different rates; etc. Table 4.2 gives average sphericity and roundness values for a range of sediments (compare with Figure 4.10B).

Experimental and observational evidence supports the view that, although the rounding of gravel by transport is relatively rapid, the rounding of sand is much slower and that selective transportation of different grain sizes is a more important factor than abrasion in explaining textural variations in sediments with distance of transportation. During sedimentary transport most of the actual decrease in particle size is effected by an increase in rounding, rather than by an increase in sphericity. Experimental stream transport studies have suggested that angular limestone pebbles become rounded (0·6) after 18–37 km, quartz gravels after 161 km and granite cobbles after 125 km. Abrasion has little effect on fine quartz gravel and is very slow for quartz sand. Certain processes are more capable of causing abrasion, however, wave action producing flattened beach cobbles and wind action being 100 to 1000 times more effective than water transport in rounding sand grains.

#### 4.2.4 Sedimentary fabric

The fabric of a sediment is a term which describes the spatial arrangement of its constituent particles. The two most important aspects of fabric are particle *orientation* and *packing*. Orientation relates to the position of the particle in space and, in particular, to the direction and dip of the long axis. An important orientation structure is the imbricate (Latin: *imbrex* – 'a gutter tile') or gravel fabric in which the particles are arranged somewhat like the tiles on a roof, overlapping and dipping in the opposite direction to that of the current flow at 15°–30° in fluvial deposits and 2°–15° in marine beach gravels. Another is the sub-glacial till fabric expressed by the long axes of large pebbles being orientated parallel to the direction of the flowing ice.

Packing describes the spacing of sedimentary particles, controlling the density and porosity of the deposit. Porosity

$$\left[ = 100 \left( \frac{\text{bulk volume} - \text{grain volume}}{\text{bulk volume}} \right) \right]$$

is the percentage of pore space in a given volume of sediment. Loosely packed spheres of equal size have a porosity of 47·64 per cent, tightly packed ones 25·95 per cent. Beach and dune sand porosities average about 45 per cent. Porosity is linked with density by engineers who describe the density of a loosely packed sediment as its 0 per cent *relative* density and that of its tightly packed equivalent as its 100 per cent relative density. Porosity is a static, dimensionless parameter which increases with goodness of sorting and decreases as the range of grain sizes increases; however, it has considerable influence over the dynamic sediment properties, such as permeability and shearing resistance, which will be examined in Chapter 10.

### 4.3 Sedimentary rocks

Sedimentary environments impose a variety of primary structures on sedimentary rocks. Of these, grain characteristics, mineral composition, bedding, biological structures (see especially Section 15.6.1), diagenesis and facies are the most important. Many of these characteristics are

## FUNDAMENTAL PROPERTIES

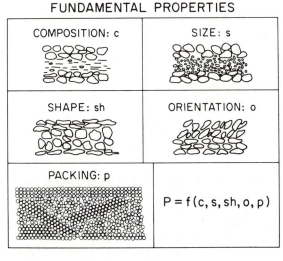

**Figure 4.11** Bedding properties (P) as a function of different combinations of composition, size, shape, orientation and packing of sediments.
*Source:* Pettijohn *et al.*, 1972, figure 4-1, p. 102, modified after Griffiths.

involved in the classification of sedimentary rocks, which is also introduced here (Pettijohn *et al.*, 1972; Pettijohn, 1975).

### 4.3.1 Bedding

Clastic sedimentary deposits are commonly arranged in beds, which are layers built up parallel to the surface of deposition and which are distinguishable by upper and lower discontinuities of grain composition, size, shape, orientation and packing (Figure 4.11), as well as colour, minor structures, etc. Most sediments exhibit some type of bedding, but an exception is found in some types of subglacial lodgement tills. Thickly bedded sediments are termed 'massive', whereas beds less than 1 cm thick are referred to as 'laminations'. Bagnold has recognized three types of bedding mechanisms:

(1) Simple sedimentation – settling from suspension.
(2) Accretion – deposition due to changing circumstances (e.g. of velocity, bed roughness, etc.).
(3) Encroachment – lateral outbuilding of the deltaic type.

For general purposes, the two major types are graded bedding and cross-bedding. Graded bedding (Figure 4.12A) is a very common form of stratification in which grain size increases towards the base of each bed. Such vertical grading may be produced in a number of ways, such as by differences in the settling velocities of sedimentary particles in a given sedimentary 'flush', by rapid pulsating sedimentation under stable conditions, settling from turbidity currents within water or air (e.g. nuées ardente), or a general decrease in grain size in the provenance during slower sedimentation over a longer

## A GRADED BEDDING    B CROSS-BEDDING

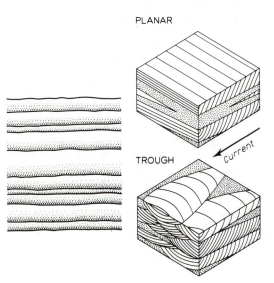

**Figure 4.12** Common types of bedding: A. graded bedding; B. planar and trough cross-bedding, with current direction indicated by arrow.
*Source:* F. J. Pettijohn, *Sedimentary Rocks*, 1975, 3rd edn, figures 4-4 and 4-5, pp. 105, 107, copyright © 1949, 1957 by Harper and Row, Publishers, Inc., copyright © 1975 by Francis J. Pettijohn, reproduced by permission of Harper & Row, Publishers, Inc.

period of time. Such a consideration of sedimentation mechanisms shows the wide range of sedimentation rates which are possible, varying from the virtually instantaneous deposition by floods, mudflows, or sand dune invasion to much slower rates under offshore marine conditions. Some thin beds and laminations can be directly related to individual seasonal (e.g. varves) or tidal variations in sedimentation.

Cross-bedding or current bedding results from lateral deposition building forward in a unidirectional current producing bedding lying at angles of some 30° or less to the major planes of stratification. Cross-bedding may be planar or trough, depending on the form of the surface upon which the bedding is deposited (Figure 4.12B). Thick sequences of cross-bedded deposits may assume very complex structures if directions of current flow vary through time. There are three major sources of cross-bedded sediments: sand dunes (Chapter 16), deltas (Section 14.3) and beach deposits (Section 15.2). The erosion of complex, massive cross-bedded rocks may produce a bizarre assemblage of niches and arches, as does the 200–600-m-thick Navajo Sandstone in Utah which is composed of Jurassic sand dunes.

### 4.3.2 Diagenesis

After deposition, sediments are commonly subjected to a group of processes which, to a greater or lesser extent, are responsible for transforming them into harder, less porous sedimentary rocks. This distinction between sediments and rocks is a fine one, but the production of sedimentary rocks is usually associated with the action of the processes of *diagenesis*, which has been defined by Pettijohn *et al.* (1972) as 'all processes, chemical and physical, which affect the sediment after deposition and up to the lowest grade of metamorphism, the greenschist facies'. In terms of the diagenic processes and of the conventional classification of sedimentary rocks it is convenient here to deal separately with the diagenesis of non-carbonate and of carbonate rocks.

Non-carbonate clastic sediments are subjected to a sequence of possible changes as they are progressively more deeply buried over longer periods of time. These changes involve four major steps:

(1) Compaction, crushing and the decrease of porosity.

(2) Syneresis, or dewatering under pressure.

(3) Cementation, a complex set of cavity and pore-space filling mainly by carbonate, silica, or iron-oxide cements, plus the growth of individual grains by precipitation. In some heavily cemented sandstones cementation may comprise one-quarter to one-third of the total rock volume. In some chemical and pressure environments, however, solution, rather than cementation, may occur.

(4) Incipient metamorphism. This occurs under higher temperatures and confining pressures and is responsible for the alteration of kaolin and montmorillonite to illite and, especially, chlorite. Under

extreme diagenesis clays may be partially recrystallized to produce micas and schistose textures.

Carbonate clastic and non-clastic sediments are also subjected to the diagenic processes of compaction, syneresis and, especially, cementation, together with some recrystallization associated with incipient metamorphism. Cementation often takes the form of deposition of, and replacement by, crystalline calcite associated with the action of groundwater or of the cementation of clastic material by the secondary growth of aragonite in the tidal zone (e.g. to produce beach rock).

### 4.3.3 Classification

The most obvious division of sedimentary rocks is into mechanical or clastic sediments (excluding chemical clastics) on the one hand, and chemical and biochemical sediments (including chemical clastics) on the other. Mechanical sediments may then be subdivided on the basis of the size of 'significant' components into coarse clastics (rudites), sandstones and mudstones. The coarse clastics are further divisible into transported, rounded conglomerates and mechanically fractured, angular breccias. Chemical sediments fall into the two classes of carbonate rocks and other less important chemical deposits (e.g. evaporites, phosphates and iron deposits). Carbonate rocks are most obviously divisible into boundstones, the organic structures of biochemical origin (e.g. coralline growths), and calcareous clastics which range from calcareous breccia and conglomerates into lime mudstones (Krumbein and Sloss, 1963; Pettijohn *et al.*, 1972; Pettijohn, 1975).

#### 4.3.3.1 Clastic rocks

Rudites are made up of rocks containing more than 10–30 per cent of material coarser than 2 mm, depending on the classifier. The primary subdivision into conglomerates and breccias identifies the latter as composed of coarse material which is dominantly angular (i.e. roundness $< 0.2$), resulting from a variety of mechanical fracturing associated with movement along fault zones, vulcanism, scree production, etc. Conglomerates are composed of material which shows evidence of having been transported by water or ice for considerable distances. The rocks are further divisible into:

(1) orthoconglomerates (grain-supported), which are composed of a framework of more or less rounded material (usually of resistant quartzitic composition) of fluvial or marine transportation; such orthoconglomerates, which may form rock beds hundreds of metres thick in places, are well sorted when laid down by the action of waves or of high-velocity streams, or

more poorly sorted when composed of flood gravels or alluvial fan deposits;

(2) paraconglomerates (mud supported), which are very poorly sorted and contain a proportion of as much as one-third to one-half of silt or clay; these conglomerates are not deposited by fluvial means, but made up of glacial tills (unstratified clays plus angular glacially derived material) as well as by debris derived from slope processes, such as solifluction and slumping (the latter landslide material occasionally assuming the characteristics of breccia).

Sandstones are classified on the basis of 'end-members'; predominant components possessing or achieving some degree of long-term chemical stability under conditions of weathering, transportation, deposition and diagenesis, between 0·06 and 2 mm in size. These components are the clays and silts of the matrix, and the particles of quartz (ubiquitous and commonly <0·6 mm in diam.), feldspar (less common than quartz in sedimentary rocks because of its lower chemical stability under processes of weathering, transportation and deposition) and of rock fragments made up of more than one mineral (principally forming the coarse sand fraction). The division of sandstones into arenites and wackes, determined by the percentage of the gross bulk made up by the matrix, is shown in Figure 4.13 and this is reflected in the provenance of sandstone (Figure 4.14). The chemical composition of the gross bulk of sandstones (Figure 4.15) is specified in terms of the ratios of $SiO_2/Al_2O_3$ and $Na_2O/K_2O$, to which mineralogical significance is clearly ascribable. Arenites contain a gross bulk of less than 15 per cent matrix and, depending on the relative framework proportions of quartz, feldspar and rock fragments, are classified as orthoquartzites, arkose, or lithic arenites.

Orthoquartzites contain a well-sorted and rounded framework normally well in excess of 50 per cent quartz particles, indicative of a very mature sandstone formed by prolonged, multicycle marine, fluvial, or aeolian activity. Most contemporary one-cycle beach sands are not pure quartz arenites. Orthoquartzites vary from the almost pure quartz arenites (90–95 per cent quartz) to arkosic (subarkose: 5–25 per cent feldspar, exceeding rock fragments) and lithic (sublithic arenite: 5–25 per cent rock fragments, exceeding feldspar) varieties (Figure 4.13) and are characteristic of cratonal and marginal miogeosynclinal sedimentary sequences.

Arkoses have a framework composed of more than 25 per cent feldspar particles (although not commonly exceeding 60 per cent) and having a smaller proportion of rock fragments. The grains are usually angular or sub-rounded, and the rock derives most characteristically from material eroded rapidly from granitic uplands of high relief which is not subjected to much chemical decomposition (i.e. is rapidly transported and buried under subhumid or arid conditions) or to extreme diagenesis. Arkoses are thus produced in local specialized cratonal environments (e.g. in zeugogeosynclines ('yoked' geosynclines) which are subsiding adjacent to rising granitic blocks and receiving material from them under arid conditions).

Lithic arenites (framework >25 per cent rock fragments), the most abundant of all sandstones, are mixtures of grey sands and rock particles derived from the erosion of pre-existing sedimentary rocks. These rocks are characteristic of post-orogenic molasse facies and

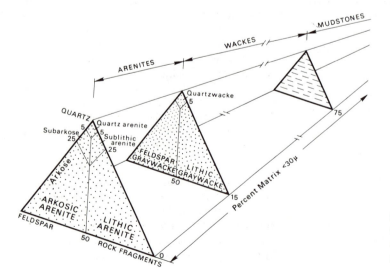

**Figure 4.13**   Classification of terrigenous sandstones.
*Source:* Pettijohn *et al.*, 1972, figure 5-3, p. 158, modified after Dott.

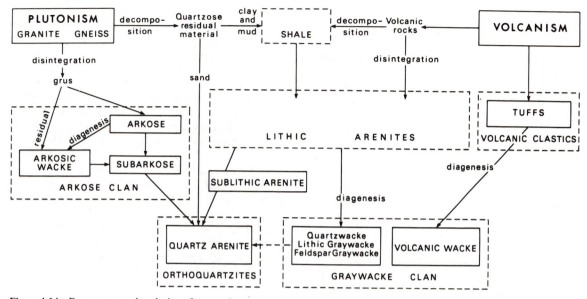

**Figure 4.14** Provenance and evolution of non-carbonate sands.
*Source:* F. J. Pettijohn, P. E. Potter and R. Siever, 1972, *Sand and Sandstone*, 1972, figure 6-1, p. 176, copyright © Springer-Verlag, New York.

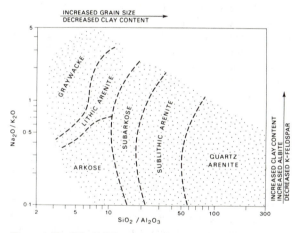

**Figure 4.15** $SiO_2/Al_2O_3$ v. $Na_2O/K_2O$ ratios in sandstones, emphasizing differences in chemical composition.
*Source:* Analyses largely by Pettijohn, modified from Pettijohn *et al.*, 1972, figure 2-11, p. 62.

originate particularly within miogeosynclines. The wackes contain 15–75 per cent matrix and are hard, indurated rocks composed of immature sands and muds which made up one-fifth to one-quarter of all sandstones. The most abundant of the wackes is lithic graywacke, a poorly sorted mixture of sand grains (rock fragments exceed feldspar) and clay–slate material exhibiting all types of bedding structures, and considerably hardened by physical and chemical diagenesis.

Lithic greywacke is a one-cycle deposit related to lithic arenites characteristic of eugeosynclinal flysch facies, commonly associated with deposition by turbidity currents. It is an abundant constituent of Paleozoic and older stratigraphic sequences. Less common are feldspar greywackes and quartzwackes, the latter containing less than 10 per cent feldspar and rock fragments and being a reworked, polycyclic greywacke. There are other types of specialized sandstones, of which an important one is greensand, which is rich in the mineral glauconite derived from biotite by submarine weathering.

Mudstones are dominated by a greater than 75 per cent silt–clay matrix. Dominantly siltstone is uncommon, except in the form of loess, but shale, by far the most abundant mudstone, makes up about one-half of the total sedimentary rocks. Shales are usually fissile, thinly bedded and composed of 40–100 per cent clay material with up to 40 per cent of silt scattered through it. Modern muds average 15 per cent sand, 45 per cent silt and 40 per cent clay. Shales are formed in shallow cratonal conditions, as well as in miogeosynclines where they form part of molasse deposits. Important varieties of shales are the carbonaceous black shales formed within vegetated swamps and red shales derived from oxidizing subaerial environments (red shale is now being deposited offshore by the Amazon River).

### 4.3.3.2 Chemically precipitated rocks

By far the most important chemical sedimentary rocks

are the carbonates, which contain in excess of 50 per cent carbonates of calcium and magnesium and make up more than 10 per cent of exposed sedimentary rocks the oldest of which date back 2.7 billion years. Secondary dolomitization (i.e. replacement of Ca by Mg to yield $CaMg (CO_3)_2$) has affected about one-half of all carbonate rocks, which largely originated as coralline and other near-shore carbonate sedimentation in warm, clear, shallow marine waters of cratonal and miogeosynclinal environments.

Boundstones are autochthonous (i.e. formed in place) coral-reef (i.e. bioherm) or platform biochemical deposits exhibiting growth bedding and other growth structures. The calcareous clastics are allochthonous (i.e. formed elsewhere and mechanically broken up and transported). Coarse clastics are either breccias, especially that formed by wave attack on the coral-reef front, or conglomerates, formed as the result of marine and fluvial transportation of coarse limestone material. Carbonate sandstones include wide admixtures of carbonate sand and muds, which were originally deposited on reef flats to the seaward of reef faces and in back-reef zones. A particularly pure and well-sorted carbonate sand is composed of oolites, which are spherical marine carbonate deposits around nuclei developed especially in offshore areas of strong bottom currents. Carbonate mudstones, characteristic of lagoons, tidal flats and deep-sea environments, are mainly developed from the abrasion of carbonate sand particles or pieces of coral or from the silt and clay-sized particles produced by calcareous planktons and algae (e.g. chalk).

### 4.3.4 Facies

A facies is a spatially distinguishable and limited part of a stratigraphic time unit, differing from other areal parts of the same unit in some physical and/or chemical (i.e. lithofacies) and/or organic (i.e. biofacies) sense. *Time units* are bodies of sedimentary rock, perhaps with areal

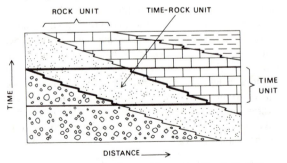

**Figure 4.16** Schematic diagram illustrating the distinction between rock units, time units and time–rock units.

variations in facies, laid down within a given time interval (Figure 4.16). *Rock units* are continuous bodies of uniform facies not necessarily all deposited within a given restricted time interval. A *time–rock unit* is a rock body of uniform facies laid down within a given time interval. Because of regular lateral changes in the loci of given types of sedimentation (e.g. migration of shorelines or river channels), sedimentary facies changes which are observed laterally are often found to form a vertical sequence through time (Figure 4.17). This principle is expressed as Walther's Law which implies that each distinctive sedimentary environment possesses its own characteristic and ordered suite of facies, recognizable both spatially and in vertical section. In Figure 4.17 a regression of the sea has produced a prograding (forward-building) shoreline and an offlapping sedimentary sequence of which certain vertical sections mirror the character and order of the facies changes observed in the spatially disposed surface outcrops.

## 4.4 Metamorphic rocks

Sedimentary, and even igneous, rocks can be *metamorphosed* (Greek: *meta* – 'change', *morphe* – 'form') by the action of heat and pressure. There are three classes of metamorphic processes: large-scale regional metamorphism which occurs in belts of mountain-building; contact metamorphism around the boundaries of igneous intrusions; and dislocation metamorphism which occurs during active frictional movement along fault or thrust planes. Geomorphically significant metamorphic effects are of three kinds:

(1) Production of new, often higher-temperature, minerals (e.g. garnet, pyroxenes, sillimanite, serpentine) or the recrystallization of existing minerals. In some instances these may be less resistant to the processes of weathering in either an absolute or relative sense. In other instances extremely resistant rocks may be produced, such as quartzites from orthoquartzites.

(2) The production of new structural features of jointing, banding and foliation which influence erosional processes.

(3) The most important effect is that of regional metamorphism whereby huge masses of existing rocks are chemically altered to produce such small-scale variety that a metamorphic outcrop may behave as a uniform lithological unit. This is the case in some extensive areas of the Coast Ranges of California.

Regional metamorphism occurs in belts of orogenic deformation, producing changes in zones of rock measured in kilometres or tens of kilometres wide broadly

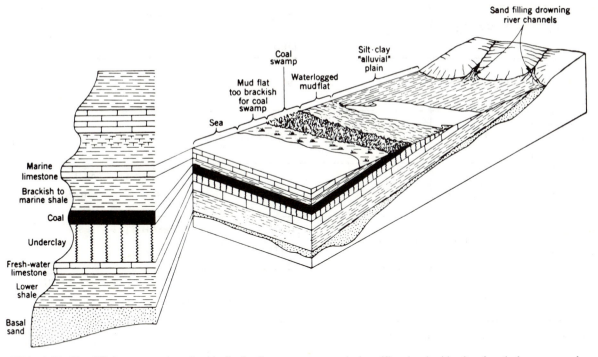

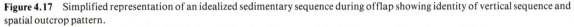

**Figure 4.17** Simplified representation of an idealized sedimentary sequence during offlap showing identity of vertical sequence and spatial outcrop pattern.
*Source:* J. M. Weller, *Stratigraphic Principles and Practice*, 1960, copyright © Harper and Row, by permission.

unrelated to igneous heat sources. These geomorphically significant changes involve varying degrees of recrystallization, the production of new structures (e.g. cleavage, schistosity and banding) by the preferred orientation of minerals or by mineral recrystallization, and the emergence of new secondary minerals diagnostic of the intensity, or *grade*, of metamorphism.

Metamorphic grade depends broadly on the temperature and pressure to which the rock has been subjected, which generally increase towards the geographical axis of an orogenic belt and with depth of burial of the rocks within a geosyncline, together with the presence of chemically active volatiles, notably steam. The processes of regional metamorphism are immensely complex but the production of different types of metamorphic rocks are mainly dependent upon the parent rock and the metamorphic grade (Figure 4.18). Some important metamorphic rock types are as follows:

| | |
|---|---|
| marble | – banded, recrystallized calcite or dolomite; |
| quartzite | – recrystallized quartz sandstone; |
| gneiss | – coarse-grained, irregularly banded, high-grade rock composed mainly of quartz and feldspar, deficient in |

mica; produced from granitic or arkosic rocks;

| | |
|---|---|
| granulite | – medium-grained, even-textured rock rich in quartz and feldspar, together with pyroxene and garnet, exhibiting some banding; very high-grade metamorphic rock related to gneiss; |
| slate | – fine-grained, low-grade (i.e. temperatures 200°C–400°C) metamorphic product of mudstone, shale, or siltstone, with regular slaty cleavage developed parallel to axial planes of folding; |
| phyllite | – slightly higher metamorphic grade of slate with recrystallization giving a coarser texture and parallel banding of mica and chlorite; these introduce some features of a schistose texture; |
| schist | – well-laminated, mica-rich, medium-coarse-grained, medium–high-grade metamorphic product of argillaceous rocks; there are many types of schist each named after its characteristic high-temperature minerals (e.g. |

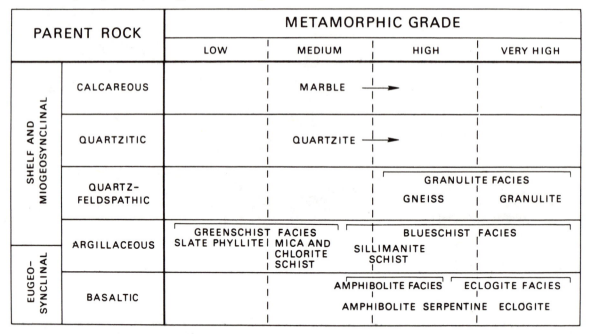

| PARENT ROCK | | METAMORPHIC GRADE | | | |
|---|---|---|---|---|---|
| | | LOW | MEDIUM | HIGH | VERY HIGH |
| SHELF AND MIOGEOSYNCLINAL | CALCAREOUS | | MARBLE → | | |
| | QUARTZITIC | | QUARTZITE → | | |
| | QUARTZ-FELDSPATHIC | | | GRANULITE FACIES<br>GNEISS | GRANULITE |
| EUGEO-SYNCLINAL | ARGILLACEOUS | GREENSCHIST FACIES<br>SLATE PHYLLITE | MICA AND CHLORITE SCHIST | BLUESCHIST FACIES<br>SILLIMANITE SCHIST | |
| | BASALTIC | | | AMPHIBOLITE FACIES<br>AMPHIBOLITE SERPENTINE | ECLOGITE FACIES<br>ECLOGITE |

**Figure 4.18**    A classification of metamorphic rocks

sillimanite schist is a high-grade variety);

amphibolite – medium-coarse-grained, hornblende- and plagioclase-rich rock produced by the medium–high-grade metamorphism of basic igneous rocks;

serpentine – greenish rock, characterized by the mineral serpentine, produced by the metasomatism of peridotite;

eclogite – medium-grained, very high-grade metamorphic product of basalt, rich in pyroxene and garnet.

Of course, it is difficult to separate high-grade metamorphic rocks from their igneous equivalents, for under conditions of plastic flowage and recrystallization the distinction between igneous and metamorphic rocks disappears.

The term 'metamorphic facies' is applied to regional groups of rocks having the same metamorphic grade:

granulite facies – associated with very high temperatures and pressures in water-deficient environments; this assemblage of gneiss, granulites and granites develops in the deepest parts of the geosyncline and examples are presently found outcropping in such areas as the Pre-Cambrian shields, the Adirondack Mountains of New York and the upfaulted San

Gabriel Mountains of southern California;

greenschist facies – associated with moderate pressures, low temperatures and the presence of water; this assemblage of slates, phyllites, chlorite and mica schists (together with quartzite and marble) is characteristic of miogeosynclinal deformation and is sometimes associated with batholithic intrusion; much of New England and the Appalachian Piedmont Crystalline Province (Figure 7.14) exemplifies these facies (e.g. Carolina Slates and Brevard Schists);

blueschist facies – associated with very high temperatures and pressures in the presence of abundant water; as the result of deformation of subduction zones, and characteristic of eugeosynclines and the most intensely deformed central parts of mountain belts, these fine- to medium-grained schistose rocks occur in the California Coast Ranges and, as sillimanite schists, in an axial belt of the Appalachian Crystallines some 70 km (45 miles) east of the Brevard Belt; the deep erosion of the Appalachians since their orogeny at the beginning of the Permian has exposed parts of their high-grade metamorphic core; in contrast, the metamorphic rocks exposed in the more recent Alps are restricted to narrow near-vertical bands of marble, schists and basic intrusions in the zone of roots;

amphibolite facies – associated with blueschist facies and containing amphibolites developed from the high-grade metamorphism of ophiolites and other eugeosynclinal volcanics, together with micaceous schists and quartzites; small outcrops of these rocks occur in the Appalachians from Maine to Georgia and, in particular, there is a belt of scattered serpentine (ultramafic) outcrops in the southern Piedmont (see Figure 7.14);

eclogite facies – the highest grade of metamorphosed basic igneous rocks, even more restricted in extent but similarly associated with blueschist facies. Examples are found in the California Coast Ranges.

This chapter has shown how characteristic rock types develop, later to be exposed at the surface of the earth as the result of diastrophic and igneous forces (Chapters 5 and 6) and, through their structure and lithology (Chapters 7 and 8), to exert profound influences over the landforms produced from them.

# References

Berner, R. A. (1971) *Principles of Chemical Sedimentology*, New York, McGraw-Hill.

Blatt, H., Middleton, G. and Murray, R. (1972) *Origin of Sedimentary Rocks*, Englewood Cliffs, NJ, Prentice-Hall.

Carmichael, I. S. E., Turner, F. J. and Verhoogen, J. (1974) *Igneous Petrology*, New York, McGraw-Hill.

Chorley, R. J. (1969) 'The role of water in rock disintegration', in R. J. Chorley (ed.), *Water, Earth and Man*, London, Methuen, 135–55.

Chorley R. J. and Kennedy, B. A. (1971) *Physical Geography: A Systems Approach*, London, Prentice-Hall.

Ehlers, E. G. and Blatt, H. (1982) *Petrology: Igneous, Sedimentary and Metamorphic*, San Francisco, Calif., Freeman.

Gilluly, J., Waters, A. C. and Woodford, A. O. (1968) *Principles of Geology*, 3rd edn, San Francisco, Calif., Freeman.

Holmes, A. (1978) *Principles of Physical Geology*, 3rd edn, London, Nelson.

Hyndman, D. W. (1972) *Petrology of Igneous and Metamorphic Rocks*, New York, McGraw-Hill.

Krumbein, W. C. and Pettijohn, F. J. (1938) *Manual of Sedimentary Petrography*, New York, Appleton-Century-Crofts.

Krumbein, W. C. and Sloss, L. L. (1963) *Stratigraphy and Sedimentation*, 2nd edn, San Francisco, Calif., Freeman.

Leet, L. D. and Judson, S. (1965) *Physical Geology*, 3rd edn, Englewood Cliffs, NJ, Prentice-Hall.

Pettijohn, F. J. (1975) *Sedimentary Rocks*, 3rd edn, New York, Harper & Row.

Pettijohn, F. J., Potter, P. E. and Siever, R. (1972) *Sand and Sandstone*, New York, Springer-Verlag.

Selley, R. C. (1976) *An Introduction to Sedimentology*, London, Academic Press.

Sparks, B. W. (1971) *Rocks and Relief*, London, Longman.

Streckeisen, A. (1976) 'To each plutonic rock its proper name', *Earth Science Reviews*, vol. 12, 1–33.

Trefethen, J. M. (1950) 'Classification of sediments', *American Journal of Science*, vol. 248, 55–62.

# Five *Diastrophism*

During the past twenty-five years geomorphology has ceased to be a rather isolated science and has become part of the main current of the earth sciences. Nowhere is this more apparent than in the debt which 'megageomorphology' owes to recent work by geophysicists and geologists relating to global diastrophism. It must always be remembered that landform development is essentially related to the *initial relief* (see Table 1.1) provided by diastrophic processes.

Diastrophic movements (Greek: *diastrophos* – 'turned', 'twisted', 'distorted') are those which result directly or indirectly in relative or absolute changes of position, level, or attitude of the rocks forming the earth's crust. Diastrophic movements are mainly the result of *endogenic processes* (Greek: *endo* – 'within', *geneia* – 'origin') (classes 1–4, below) which originate within the earth and are responsible for the production of *primary landforms*. Primary landforms can vary in size from continental (first order) to miniature earthquake fault scarps (seventh order) and are subsequently modified by the *exogenic processes* (Greek: *exo* – 'outside') described later in this book. There are five major classes of diastrophic movements, all interrelated:

(1) *Orogenic* movements (Greek: *oros* – 'mountain'), involving intense folding, thrusting, faulting and uplift of narrow belts subjected to severe lateral stress.

(2) *Epeirogenic* movements (Greek: *epeiros* – 'land', 'continent'), involving broad, gentle upwarping of relatively large crustal areas, sometimes associated with faulting. (Classes 1 and 2 are collectively termed *tectonic* movements (Greek: *tekton* – 'a builder').)

(3) *Isostatic* movements (Greek: *isos* – 'equal', *stasis* – 'standing'), involving vertical movements under the action of 'floatation' displacements between rock layers of differing density and mobility to achieve balanced crustal columns of uniform mass above a level of compensation in which topographic elevation is inversely related to the underlying rock density.

(4) *Igneous* movements (Latin: *ignis* – 'fire'), involving the movement and recrystallization of molten rock between various levels, sometimes appearing as surface volcanic rocks and sometimes as masses intruded or emplaced in other rocks which may cause surface deformation.

(5) *Eustatic* movements (Greek: *eu* – 'well', 'good'), involving worldwide movements of sea level resulting from changes in the total volume of liquid sea water or in the capacity of the ocean basins. Obviously, on the one hand orogenic, epeirogenic and igneous movements are important in affecting the capacity of the ocean basins, and on the other eustatic changes affect the weight of crustal columns and are usually accompanied by compensatory isostatic movements.

## 5.1 Earth structure

The direct evidence of the character and behaviour of the outer layers of the earth is limited to the comparatively shallow depths of our deepest well-borings (approx. 5 km), and recourse has to be made to indirect topographic, seismic, gravitational, heatflow and magnetic evidence.

### 5.1.1 Global topography

Topographically the earth is most inhomogeneous. The oceans comprise 70·8 per cent of its surface (Pacific 35·4 per cent, Atlantic 18·4 per cent, Indian 14·5 per

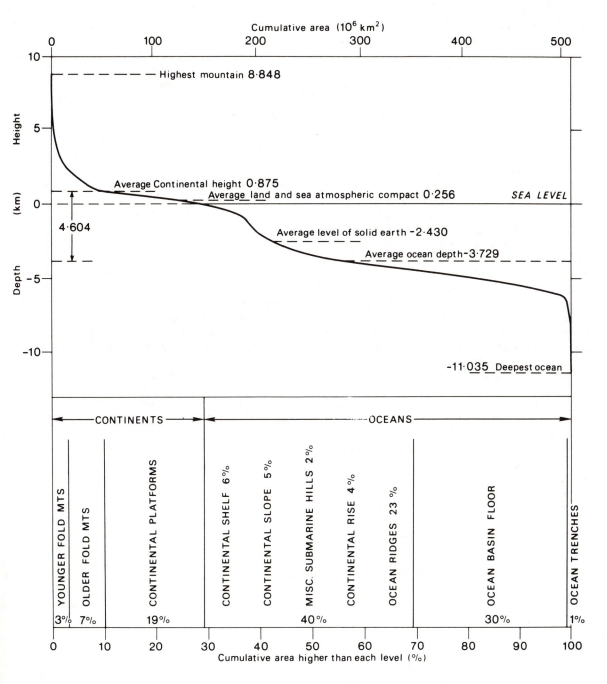

**Figure 5.1** Hypsometric curve of surface of the solid earth.
*Source:* P. J. Wyllie, *The Way the Earth Works*, 1976, figures 3-8 and 3-11, pp. 32, 38, copyright © John Wiley and Sons, by permission.

cent) and the lands 29·2 per cent. The distribution of elevations suggests that our present shoreline is of little importance, and that the outer edge of the continental shelf (approx. 0·135 km below sea level) significantly delimits the 34·8 per cent of the global surface composed of continental rocks. In this connection it should be remembered that only 18,000 years ago sea level fell more than 100 m at the maximum of the last glacial period. However, this would have exposed much less than three-quarters of the present-day continental shelves because the smaller volume of ocean water would have resulted in less isostatic depression of the ocean floor, and consequently a somewhat smaller ocean basin capacity, than is the case today. If all the existing world ice were melted, sea level would rapidly rise eustatically by 50–70 m, which would finally stabilize at 35 m after isostatic load adjustments. Land and sea are generally antipodally arranged, about two-thirds of the land being in the northern hemisphere, and only 1·5 per cent of the earth's surface having land antipodal to land. It is thus convenient to identify an ocean hemisphere, centred near New Zealand, of which 89 per cent is ocean and 11 per cent land, and a continental hemisphere, centred on Spain, of which 47 per cent is land and 53 per cent ocean. Figure 5.1 gives the vertical distribution of global elevations (i.e. a hypsometric curve), showing that only 1·6 per cent of the global surface lies above +3 km (max. Mount Everest: 8·848 km) and 1 per cent lies below −6 km (max. Nero deep in the Marianas trench: −11·035 km). This represents a maximum relief of the solid earth of

some 20 km which, as we shall see, is more than one-half the thickness of the average continental crust. Of more significance is the 4·604 km separating the mean elevation of the continental platforms from that of the ocean basin floors, and the existence of these two distinct levels is the most valuable structural evidence provided by the hypsometric curve. Care should be taken not to regard the hypsometric curve as anything resembling a topographic cross section, for as the east–west section along the zone 20°–25°S given in Figure 5.2 shows, the ocean deeps are often located close to high mountains and the middle depths of the ocean are made up both of ridges situated far out in the oceans, as well as the continental rises marginal to the lands (Wyllie 1971 and 1976).

There are eight major topographic elements of the earth.

(1) *Ocean ridges or rises.* These are broad, transversely-fractured linear swells, often lying near the centres of oceans. The ridges are tectonically unstable, being the sites of shallow earthquake sources, high heatflow and volcanic activity, and covered with a very thin sedimentary blanket. Ocean ridges are a major feature of terrestrial topography, totalling more than 80,000 km in length, having a width of 2000–4000 km and rising 1–3 km or more above the ocean floor. A distinction can be made between the irregular mid-Atlantic ridge (and that of the Indian Ocean) and the broad smoother arch of the East Pacific ridge. The crest of the former is a 500–1000 km-wide arch which is broken by an axial rift

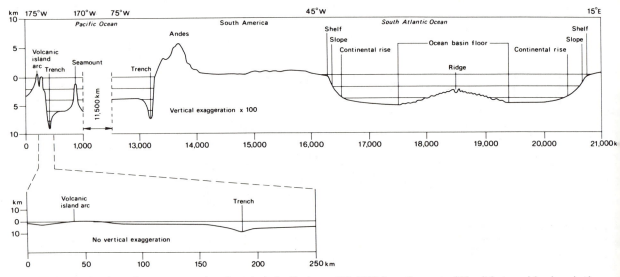

**Figure 5.2** A schematic east–west section along the latitudinal zone 20°–25°S from the coast of Namibia to an island arc in the vicinity of Fiji (see Figure 5.5).
*Source:* P. J. Wyllie, *The Way the Earth Works*, 1976, figure 3-10, pp. 36–7, copyright © John Wiley and Sons, by permission.

valley 25–50 km wide and 2000 m below the flanking submarine mountains. The outer flanks are each 500–1500 km wide, falling away from the flanking heights (lying at some 2000 m below sea level) in a series of highly dissected fractured steps (local relief up to 1000 m, adjacent peaks some 10–30 km apart) to the ocean floor some 3–3·5 km below sea level. A feature of this and all ocean ridges are huge tear (transform or transcurrent) faults situated at right angles to the ridge axis and exhibiting by terrain offsets recent differential lateral movements of up to 4000 km. The East Pacific rise is, in contrast, a less irregular feature exhibiting no prominent central rift valley and flanked by lower and less-continuous faulted steps, ridges and troughs. The crest of the mid-Atlantic ridge emerges above sea level to form the island of Iceland, and the Indian Ocean and East Pacific ridges extend on to the present continents as the rift valley system of East Africa and the Red Sea, and that of the Gulf of California and the faulted basin and range area of California and Nevada, respectively.

(2) *Ocean basins.* These flank the ridges and are characteristically made up of abyssal plains interrupted by tectonically inactive rises (e.g. Bermuda rise, 300–500 km wide and up to 5 km high), seamounts and abyssal hills. The abyssal plains are themselves tectonically inactive areas of low slope (gradients < 1:1000) and virtually no relief, being up to 1000 km wide and situated at depths of 3–6 km. The plains are covered at their centres by some 300 m of sediment which thickens landward towards the continental rise, slope and shelf. The numerous abyssal hills are up to 100 km wide and generally rise 50–1000 m above the sea floor; they appear to be associated with the ocean ridges which they, in some places, parallel (notably in the Atlantic), and give the impression of being subsided parts of the ridges either due to depression or to the lateral movement of the sea bed into deeper water. The many seamounts are isolated submarine volcanic features 2–100 km wide and more than 1000 m high, whose surfaces lie less than 2000 m below sea level, many of which form volcanic islands and of which Hawaii and the Azores are the most notable. Many seamounts are sharply pointed but others are flat-topped hills whose summits lie more than 200 m below sea level. These so-called 'guyots', of which more than 1400 so far have been identified in the Pacific alone, have steep sides (12°–35°) and give the impression of being wave-bevelled platforms submerged largely by sea floor subsidence, partly by the post-glacial rise of sea level. This view is supported by their being commonly capped by coral growth in the tropics which could not have been built at depths greater than 150 m or so.

(3) *Continental shelf, slope and rise.* Where the submerged continental edge is not bounded by steep slopes of 10° or more leading down to deep ocean trenches, the floors of the ocean basins rise to above sea level in a characteristic profile well exhibited by the east coast of North America. Off such coasts a thick wedge-shaped shelf of young sedimentary rocks in varying degrees of consolidation forms a continental shelf (slope < 1:1000; relief > 20 m), ranging in width up to more than 300 km (averaging 78 km, but in places approaching 1500 km), being generally widest opposite larger rivers.

From its outer edge, ranging between 20 m and 550 m below sea level (average 130 m), the continental slope descends at a gradient of 1·3°–6° (opposite major deltas 1·3°, stable coasts 2°–3·5°, young mountain range coasts 4·6°, fault coasts 5·6°), for widths of up to 150 km to depths of about 2 km. At the base of the continental slope the sea bed appears to consist of coalescing fans and aprons of continental sediment (produced by turbidity currents and slumps) which form the continental rise. This is up to more than 300 km wide and slopes gently down to the ocean floor at gradients of 1:700 to 1:2000 between depths of 2 km and 5 km, having a low relief (20–40 m) except where slumped steps occur (e.g. at 3 km and 4·2 km depths off the eastern United States).

In places, especially opposite large rivers (e.g. the Congo and Hudson), the continental shelf and slope are cut by submarine canyons exhibiting seaward slopes, bifurcations and sinuosities analogous to subaerial rivers. The steep-sided canyons are of many sizes (widths 1–15 km; lengths 20–2000 km; gradients < 1:40), and although they may be partly drowned features, they are probably the result of erosion by density or turbidity currents. The continental shelf and slope seem to have been produced by sedimentation in a subsiding belt, and in some tropical areas (e.g. the Great Barrier Reef off eastern Australia) the former is capped by thick coralline limestone.

(4) *Ocean trenches, deeps, or troughs.* Along some coasts (e.g. the western Pacific and the Caribbean) a narrow or complex continental shelf drops steeply into trenches forming the deepest parts of the oceans. The trenches are 30–100 km wide and 300–5000 km long and their sides slope, first, at angles of about 4°–8°, and then at 10°–16° to depths of more than 10,000 m. The walls of the deeper trenches may reach 25°–45° in angle and sometimes show a series of steps, perhaps due to slumping on a massive scale. Ocean trenches parallel island arcs or young volcanic orogenic zones on their seaward side, exhibit strong negative gravity anomalies and are V-shaped, with narrow flat floors up to several kilometres

wide, in many places being floored by only thin layers of sediment.

(5) *Island arcs.* To the landward most ocean trenches are paralleled by arcuate festoons of islands, individually ranging in size from less than 1 km² to the size of New Guinea, Luzon, or Hokkaido. These belts are seismically active, exhibit strong negative gravity anomalies (with strong positive ones on their continental sides), and are topographically and structurally continuous with some continental belts of young folded mountains (e.g. Malaya, Kamchatka, Alaska), which they strongly resemble.

Each island arc is generally made up of two parallel arcs of islands 50–150 km apart, the outer one comprising folded and thrust sediments lacking volcanic activity and the inner one characteristically composed of volcanic islands. In places the outer arc is submerged (e.g. south of Java), poorly developed (e.g. the Antilles, with Barbados the main example), dominant (e.g. Ryukyu Islands between Japan and Formosa), or topographically and structurally fused with the inner arc (e.g. Japan).

(6) *Marginal sea basins.* Between the island arcs and the continents lie the topographically varied marginal sea basins (e.g. Sea of Japan, Sea of Okhotsk), which in places separate parallel island arcs (e.g. the Philippine Sea). They are 500–1000 km wide and some include abyssal plains extending down more than 5 km below sea level. Generally, however, they exhibit a complex and rugged topography of a 'continental character', broken by faults and interrupted by undulations, small seamounts and other irregularities reflecting complex histories and varied sediment sources. Some are tectonically active, others are not.

(7) *Folded mountains.* These are curvilinear belts of pressure which have led to folding, thrusting, compression and uplift of sedimentary rocks, and are associated with volcanic activity, deep igneous intrusions and large-scale deep metamorphism of sedimentary and igneous rocks. Belts of folded mountains may be several hundred to several thousand kilometres in width and thousands of kilometres in length, and some of the younger belts characteristically connect with island arcs. Folded mountains may be roughly divided into younger and older groups, the age of the former being measured in terms of millions of years and that of the latter in hundreds of millions. The younger fold mountains include the highest terrestrial elevations in the Himalayas, Alps, Rockies, Andes, Alaska Mountains, Sierra Madre of Central America, Atlas, Carpathians, Pyrenees, Rhodope Mountains, Apennines, Dinaric Alps, Hindu Kush, Pamirs, Elburz-Zagros Mountains,

Verkhoyansk Mountains, and so on. The highest of the ranges have crests at elevations of 5–8 km and the largest of them include large areas of uplifted, undeformed crustal rocks (e.g. Tibetan Plateau, 4500 m, and the Andean Alti Plano, 3000–3800 m). In general, the younger folded mountains are tectonically unstable (experiencing shallow, medium and deep earthquakes) and exhibit strong magnetic and gravity anomalies, whereas the older folded mountains are much more tectonically stable, less gravitationally anomalous and have medium-scale elevations and exposed geological structures suggestive of considerable erosion.

Topographically the crests of the older folded mountains are smoother and more accordant than those of the younger folded mountains, and their present relief appears to be more the result of their bodily uplift *en bloc* than of crustal thickening by the initial orogenic compressive forces. Some older folded mountains are high (Tien Shan, about 7000 m, connecting eastward with the Mongolian Plateau, 1000–1600 m; Altai, about 5000 m), but most are either of medium relief (Appalachians, more than 800 m; Scandinavian Highlands, 700 m; Iberian peninsula, 600–700 m; Scottish Highlands; Black Forest; Bohemian Massif; Massif Central of France; large areas of Africa; the Urals; the eastern Australian Alps; and the Queen Maud Mountains of Antarctica) or persist as low relief (e.g. Brittany, the Cape Mountains of South Africa and the tectonically continuous Buenos Aires ranges of South America).

(8) *Continental platforms.* These are areas of low relief and commonly low elevation, of central continental location, composed of ancient metamorphosed and intruded rocks, in places overlain by wedges of undeformed sedimentary rocks up to 3 km in thickness which in general thicken seaward. The core areas, or shields, are not covered with sediments and are tectonically stable, and have positive gravity anomalies (e.g. northern Canada). Most continental platforms have been uplifted, including Antarctica (although much of its average height of 2·2 km is made up of glacial ice), eastern Siberia (1–1·5 km), the African plateaus (1–1·5 km), the Deccan Plateau of southern India (800 m), the western Russian Platform, eastern South America and Western Australia. Some parts of the continental platforms have suffered local subsidence of thousands of metres at periods in the past (e.g. the Michigan region of North America and the Aral Sea), sometimes accompanied by significant faulting.

### 5.1.2 Geophysical evidence

The evidence provided by topography regarding the

structure of the outer layers of the earth, although important, is limited and at best secondhand. There is, however, an extensive body of much more direct geophysical evidence, only the barest of outlines of which are given below.

Accurate contemporary measuring techniques, particularly topographic levelling, together with historical inferences regarding the vertical displacements of datum levels (e.g. sea or lake levels) are useful in determining both the amount and rate at which the earth's surface subsides under vertical loading and rises when unloaded. Recent observations are showing that the earth's surface, even the supposed stable continental platform, is universally much more sensitive to loading by thicknesses of rock, water, or glacial ice than was hitherto thought. In the period 1935–50 the weight of water in Lake Mead behind Boulder Dam on the Colorado River depressed the surface to a maximum of 170 mm and to a lesser

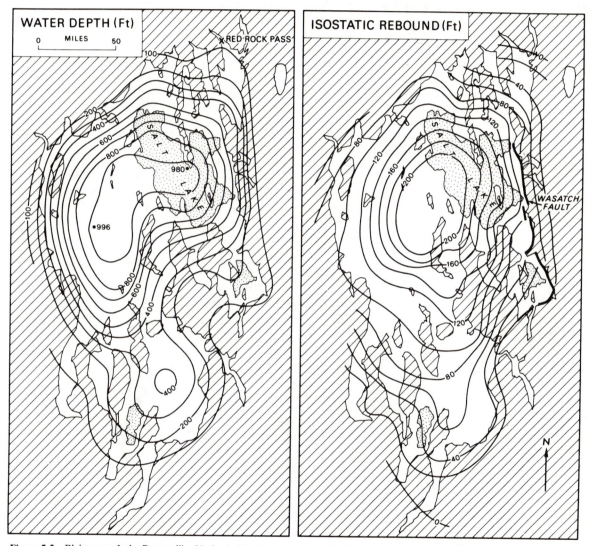

**Figure 5.3** Pleistocene Lake Bonneville, Utah. A. Maximum water depth (in feet) averaged over circles of twenty-five-mile radius. The white area shows the extent of the lake at the Bonneville shoreline and the stippled areas indicate the present lakes. Drainage of the Pleistocene lake occurred as the result of an outlet being cut at Red Rock Pass. B. Isostatic rebound (in feet) resulting from the removal of lake water, as measured by the deformation of the Bonneville shoreline. Recent displacement along the Wasatch Fault is shown.

*Source:* Crittenden, 1963, figures 3 and 4, pp. 9–10.

extent for tens of kilometres around. Figure 5.3 shows the maximum water depth attained by Lake Bonneville (the Pluvial ancestor of the present Great Salt Lake) during the period 25,000–11,000 BP (i.e. before the present time), and because old lake shorelines were cut as horizontal lines on the flanks of the surrounding mountains, the observed upward displacement of these originally horizontal shorelines gives a measure of the crustal (isostatic) rebound which has occurred following the draining away of most of the water. Figure 5.3 shows that water depths of less than 30 m (100 ft) caused the surface to depress under its weight and that the depression, as measured by subsequent rebound, was generally about 20 per cent of the thickness of the water layer, of which some 2 per cent was probably due to the elastic compression of the earth's solid crust.

By far the most important source of crustal information is that provided by seismic observations of earthquake waves. Earthquake shocks originate as sudden failures along planes of relative movement centred on foci which occur at depths in the earth down to about 700 km below their epicentres, or projections on the earth's surface. These shockwaves, measured in terms of their primary or compressional (P) and secondary or

transverse (S) components of movement, travel outwards around and through the earth. Their travel times to different surface recording-stations enable estimates to be made of the velocities of the P and S waves at different depths below the earth's surface and, from these, information is gained regarding the strength, elasticity, density, solidity and other properties of earth material. Table 5.1 summarizes some of the major findings of seismic studies, together with their implications regarding rock density and behaviour.

The spatial distribution of earthquake foci and epicentres is also instructive (Figure 5.4). More than 80 per cent of shallow earthquakes (less than 70 km in depth) are located in the young mountain and island arc circum-Pacific zone and the rest are mainly associated with the ocean ridges, the Mediterranean–Himalayan belt, and the East African and other rift valley systems. A small proportion of intermediate shocks (70–300 km deep) originate under young fold mountains, but 90 per cent of the intermediate shocks and almost all the deep shocks (300–700 km deep) originate in the circum-Pacific zone. In this latter zone earthquake foci at all depths appear to be in a zone (the Benioff Zone) 250 km wide dipping at about 30° continentward beneath the

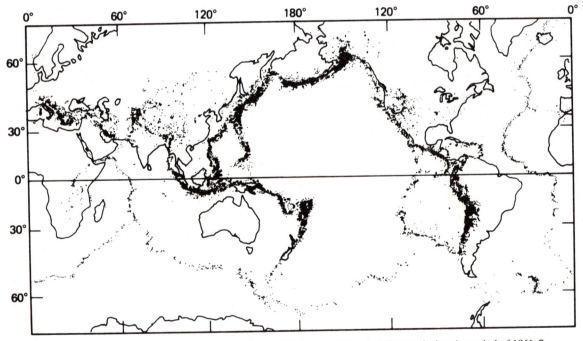

**Figure 5.4** Plot of all earthquake epicentres recorded by the *US Coast and Geodetic Survey* during the period of 1961–7. Shallow focus earthquakes (< 100 km depth) comprise those recorded beneath the main continental areas and ocean ridges, whereas intermediate (100–300 km) and deep focus earthquakes (300–700 km) are confined mainly to the island arcs and newer mountain ranges of the Pacific, East Indies and South and Central America.
*Source:* Compilation from Barazangi and Durman, 1969. (See Wyllie, 1971, figure 2-5a, p. 14.)

island arcs (Figure 5.5), and at 30° (0–300 km deep) and 60° (300–700 km deep) under young continental folded mountains flanked by ocean trenches (e.g. the Andes).

The gravitational attraction of the earth, measured at its surface by the period of swing of a pendulum, exhibits considerable spatial variation. Gravity measurements express the mass of the unit column of the earth material lying between the point of measurement and the earth's centre. The average gravity (corrected for the non-sphericity of the earth and for topographic variations) for the earth's surface is 980 gal (where 1 gal is an acceleration of 1 cm/s) and the minute variations, or anomalies, of gravity from this value are expressed in milligals (i.e. thousandths of a gal). Strong positive anomalies are characteristic of ocean rises (+ 200 to + 250 mgal) and of oceanic volcanic islands, which because of their low absolute elevation suggests that dense material has been brought near the surface in these locations. The ocean floors are also positively anomalous, although strong negative anomalies occur under the ocean trenches and island arcs. Such negative anomalies, which are also characteristic to a lesser degree of younger

folded mountains (– 300 to + 100 mgal), indicate that, assuming that there is a tendency for rock material to flow at depth to equalize the mass of adjacent crustal columns (i.e. the existence of an isostatic tendency), forces are (or have been) at work decreasing the column mass by lowering its elevation by forcing less-dense material to displace more-dense material at depth. Over young folded mountains and island arcs, however, the existence of both positive and negative gravity anomalies suggests a measure of structural complexity at depth, and the positive anomalies (0 to + 200 mgal) characteristic of marginal sea basins suggest a rise of more-dense material beneath them. Continental shields and platforms exhibit small negative anomalies (– 10 to – 50 mgal) which are only partly due in some instances to glacial isostasy, but the clear negative anomalies over the still thickly ice-covered Antarctic continent present a problem in explanation.

Dating by radioactive methods is particularly useful in assisting the development of ideas regarding the gross structural features of the outer layers of the continents and oceans. The rate of radioactive decay is measured by

**Table 5.1** The composition of the earth's interior

| DEPTH (km) | WAVE VELOCITIES (km s⁻¹) P AND S | SPATIAL VARIATIONS OF WAVE VELOCITIES | DENSITY (g.cm⁻³) | BEHAVIOUR[5] |
|---|---|---|---|---|
| SURFACE 0 / 10² Oceans | P AND S INCREASE WITH DEPTH | CONSIDERABLE VARIATION | 2·7 [4] | SOLID |
| MOHO DISCONTINUITY[1] 33 / Continents / 50[3] | 6·6 / 8·1(4·0) | LOW P VELOCITIES UNDER MOUNTAINS, ISLAND ARCS AND OCEAN RIDGES | 2·9 / 3·32 | |
| 250 | LOW VELOCITY LAYER 50–100 KM THICK IN THIS RANGE 7·8 | HIGH P VELOCITIES UNDER OCEAN FLOORS | | VERY WEAK |
| 400 | RAPID INCREASE OF P AND S WITH DEPTH 8·97 | | 3·64 | WEAK |
| 700 | 10·70 | SOME VARIATIONS OF S VELOCITIES | 4·29 | VARYING SOLIDITY |
| 1000 | 11·42(6·5) | | 4·64 | |
| | P INCREASES RAPIDLY WITH DEPTH 14·0(7·5) | | 5·66 | SOLID |
| GUTENBERG 2900 DISCONTINUITY | 8·10 / P RECOVERS SLOWLY WITH DEPTH / S DISAPPEARS | | 9·71 | LIQUID |
| LEHMANN 5200 DISCONTINUITY | 10·31 / 11·23 / P REMAINS CONSTANT | | 14·0 | SOLID |
| EARTH'S CENTRE 6370 | | | 16·0 | |

*Notes:*
1. Identified by Mohorovičić in 1910.
2. Reaches a minimum of 3 km under some parts of the ocean floor.
3. Reaches a maximum of 80 km under some younger fold mountains.
4. 2·7 beneath the continental surface; 2·9–3·0 beneath the ocean floor.
5. Inferred from earthquake wave velocities, wave 'attenuation' (i.e. rate of damping of wave amplitude with distance), existence of the S-wave, etc.

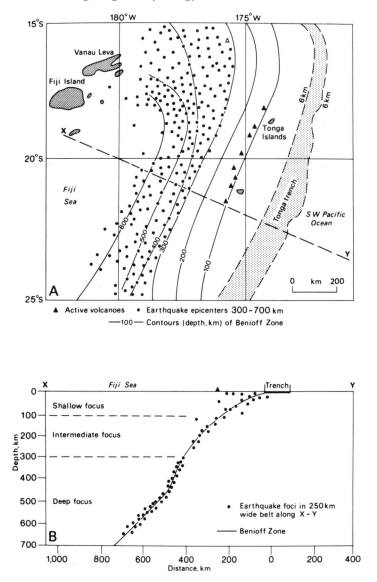

▲ Active volcanoes   • Earthquake epicenters 300–700 km
—100— Contours (depth,km) of Benioff Zone

**Figure 5.5** Distribution of earthquakes between the Tonga Trench and the Fiji Islands, north of New Zealand: A. map of deep-focus epicentres, active volcanoes and contours of the Benioff Zone; B. cross section of all earthquake foci along XY. *Sources:* A. P. J. Wyllie, *The Way the Earth Works*, 1976, figure 4-9, p. 58, copyright © John Wiley and Sons, by permission; B. data from L. R. Sykes.

the halflife, which is the unit time required for half the existing unstable radioactive isotopes to decay to stability (Table 5.2). Many of the older rocks are dated by the radioactive ratios exhibited by the granitic rocks formed or intruded within them during the major orogenic episodes. By this means it is possible to examine the development of the continental shields which give the impression that some of the present continental areas have been growing by the accretion of successive peripheral orogenic belts over long periods of time. More than 50 per cent of the present North American continent was in existence $2 \cdot 5 \times 10^9$ years ago. It appears

to have grown irregularly at an average rate of 7000 km²/10⁶ year with the outer, younger provinces tending to have been thrust successively over the inner, older ones by some 20–60 per cent of the latters' width. The maximum age of the earth's surface, as estimated by these and other methods, appears to be approximately $4 \cdot 6 \times 10^9$ year.

Most of the earth's heat originates from the decay of the radioactive isotopes, and 40 per cent of the world's heatflow loss at the surface comes from heat produced within the continental crustal rocks (i.e. above the Moho), whereas the remaining 60 per cent originates

**Table 5.2** Decay of major radioactive isotopes

| Isotope | Name | Half life (years) | Decay product | Name |
|---------|------|-------------------|---------------|------|
| $^{238}U$ | Uranium | $4 \cdot 5 \times 10^9$ | $^{206}Pb$ | Lead |
| $^{235}U$ | Uranium | $713 \times 10^6$ | $^{207}Pb$ | Lead |
| $^{232}Th$ | Thorium | $13 \cdot 9 \times 10^9$ | $^{208}Pb$ | Lead |
| $^{40}K$ | Potassium | $1 \cdot 3 \times 10^9$ | $^{40}Ar$ | Argon |
| $^{887}Rb$ | Rubidium* | $4 \cdot 7 \times 10^9$ | $^{87}St$ | Strontium |
| $^{14}C$ | Carbon* | $5710 \pm 30$ | $^{12}C$ | Carbon |

*Particularly important.

from the mantle which underlies both the shallow oceanic crust and the thicker continental crust. Global variations in the heatflow provide important information regarding the structure of the earth's outer layers. A relatively high heatflow suggests either a pronounced upward flow of heat (or of heated material) from the mantle underlying the crust or a high rate of heat generation in the crust itself; conversely, a low heatflow implies a cool mantle or low heat production in the crust. The average heatflow from the earth's solid surface of about 61 mW per square metre ($mW/m^2$) derives almost equally from the surface of the continents and the ocean bed. The highest values over the sea floor occur on the ocean ridges ($80\ mW/m^2$ – reaching over $100\ mW/m^2$ for extensive areas of the eastern Pacific rise and locally over $200\ mW/m^2$ along the 200-km-wide crest of the mid-Atlantic ridge). Areas of continental rifting such as the Red Sea, East Africa and the south-western United States also have high values. The continental marginal zones present a more complex pattern, with low values of heatflow in the ocean trenches (average $49\ mW/m^2$) but high values (approaching $200\ mW/m^2$ in places) in a belt to the landward side of the island arcs over the marginal seas (Figure 5.6). The low trench heatflow values suggest descending currents of viscous rock in the mantle and/or the existence of a thick wedge of relatively cool rock at depth. On the continents the younger fold mountains have above-average values of heatflow (averaging $74\ mW/m^2$), although like the island arcs, their pattern is locally complex. The older folded belts and the continental platforms average $60–62\ mW/m^2$ and the oldest shield areas have a low average of $41\ mW/m^2$.

The magnetic field of the earth approximates that of a dipole magnet with its axis offset from the rotational axis, and 90 per cent of the present magnetic field is thought to be accounted for by the liquid outer core which acts in the same way as a dynamo. Crystals of minerals containing much iron show a 'frozen' statistical preference to align with the earth's magnetic field, when cooled to a critical temperature (Curie point), as do fragments of these crystals to a much lesser extent during processes of sedimentation in the ocean. Such natural remnant paleomagnetism exists in many ancient rocks and the reconstruction of the past positions and orientations of parts of the earth's surface with reference to these antique magnetic fields has cast much light on the diastrophic evolution of the earth. A most important type of secular change is the sudden, complete reversal of polarity, possibly due to catastrophic overturnings in the liquid core. These have been organized into a timescale spanning more than the last 4 million years by use of associated potassium–argon dates (Table 5.3).

**Table 5.3** Recent magnetic polarity timescale

| | | |
|---|---|---|
| Brunhes | $0–0\cdot7(\times 10^6)$ years BP, | normal polarity |
| Matuyama | $0\cdot7–2\cdot4(\times 10^6)$ years BP, | reversed polarity |
| Gauss | $2\cdot4–3\cdot5(\times 10^6)$ years BP, | normal polarity |
| Gilbert | $>3\cdot5(\times 10^6)$ years BP, | reversed polarity |

In the late 1950s such belts of weak magnetic differences 5–50 km wide, hundreds of kilometres long and regularly offset by transverse faults were identified by airborne magnetometer in the rocks overlying the eastern Pacific rise and the mid-Atlantic rise south-west of Iceland, more or less symmetrically arranged parallel to the crests of the rises (Figure 5.7). In 1963 the important suggestion was made that these belts of magnetic anomalies could be interpreted as resulting from polar reversals of the earth's magnetic field and thus correlated with the reversal timescale (Figure 5.8). Over the next few years this interpretation was confirmed, giving rise to the interesting conclusion that, whereas the cores of superficial continents appear to have been growing by external accretion over a long geological timescale, the ocean ridges appear to have grown by the addition of belts of new sea floor formed centrally along their crests (perhaps since the Triassic).

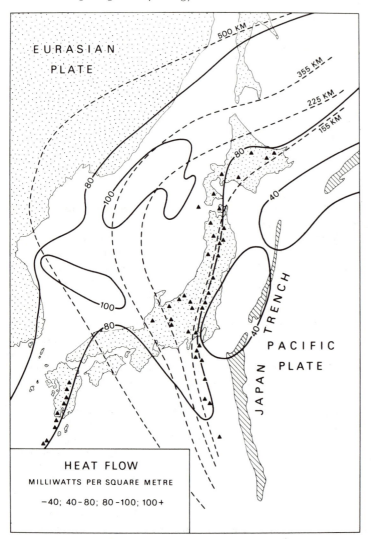

Figure 5.6 Heatflow (in milliwatts per square metre) (full lines) and contours of the Benioff Zone (depth in kilometres) in the vicinity of Japan.
*Source:* Pollack and Chapman, 1977, p. 67, copyright © *Scientific American.*

### 5.1.3 Interpretations of the evidence

Direct and indirect topographic and geographical evidence, although to some extent ambiguous and allowing of considerable differences of opinion, permits a relatively consistent picture of the structure of the earth's outer layers to appear (Wyllie, 1971 and 1976).

Since 1955 the earth's crust has been defined as that superficial layer occurring above the Moho (i.e. at which depth $P$-wave velocities increase to greater than 7·7 km/s). Below this lies the mantle with $P$-wave velocities of 8–8·5 km/s and density 3·4. However, the sharpness of the Moho is not everywhere well marked,

and even its existence may be questionable in places. This is due in part to the arbitrary definition of the Moho, but much more to the processes which are taking place at the base of the crust. The lower layer of the crust under the continents, and most of the total oceanic crust, is composed of material exhibiting the properties of basalt (or gabbro, a coarsely crystalline equivalent of it) and the abrupt increase of $P$-wave velocities at the Moho can be interpreted in two main ways:

(1) as a chemical discontinuity, with the basaltic–gabbroic crustal base overlying a more-dense rock called peridotite (Figure 5.9);

(2) as a phase discontinuity below which gabbroic

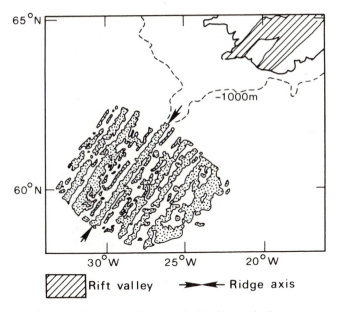

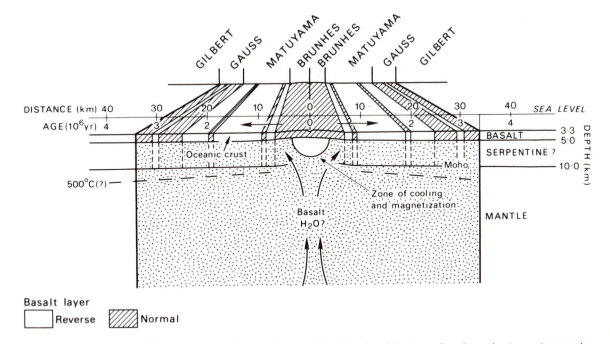

**Figure 5.7** Belts of magnetic anomalies (positive – stipples; negative – white) arranged symmetrically about the crest of mid-Atlantic Ridge, south-west of Iceland. Note the landward continuation of ridge crest as a rift valley complex in Iceland. *Source:* Heirtzler *et al.*, *Deep Sea Research*, 1966.

**Figure 5.8** Schematic model illustrating the mechanism for generating belts of positive (normal) and negative (reverse) magnetic anomalies on either side of an ocean ridge crest. *Source:* F. Vine, *Jour. Geol. Edn.*, 1969, vol. 17.

material becomes changed, or metamorphosed, by increased heat and pressure into eclogite, a rock of density 3·4-3·5 which when fully formed would transmit *P*-waves at 8·1-8·5 km/s (Figure 5.9).

Under parts of the Pacific floor the Moho is sharply defined (0·1 km thick), over the stable continental areas it is at about 0·5 km, increasing to several kilometres under many other regions and becoming most indeterminate under young fold mountains and especially under island arcs, oceanic ridges and continental rift zones. Under the last two belts anomalous transition layers (*P*-wave velocities 7·2-7·8 km/s) occur at the base of the crust. The most spectacular global variations in the crust exist between the continents and oceans (Figure 5.10). The continental crust is characteristically composed of a layer of variably deformed sedimentary rocks overlying a

thick layer of granite resting on the basaltic/gabbroic layer, with the Moho averaging 33 km below sea level under the shield areas, somewhat more under the continental platforms (Asia 37 km, and the central United States 40 km over a transition layer) and the older folded mountains (e.g. Appalachians 37 km), whereas *locally* the Moho may lie more than 60 km below sea level under the younger folded mountains forming a crustal 'root'. Under the ocean floors, in contrast, a variable layer (1) of sediments (mean thickness 0·3 km, *P*-wave velocities 1·5-3·4 km/s, density 1·7-2·5) overlies a thin layer (2) of granitic behaviour (less than 1 km thick, median *P*-wave velocity 5·1, density 2·6), floored by some 5 km of basaltic material (3) (*P*-wave velocity 6·5-6·8, density 2·9), with the Moho situated on average some 12 km below sea level and 5-7 km below the ocean floor. Under oceanic ridges conditions are complex, with the Moho

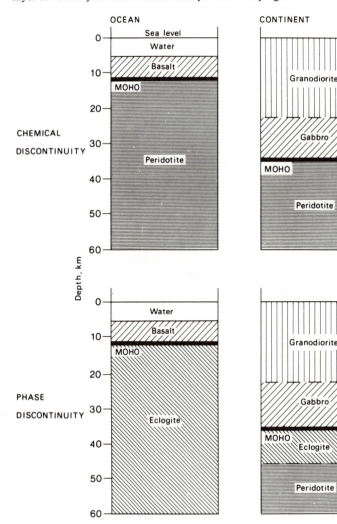

**Figure 5.9**  Possible crustal and upper mantle structures under oceans and continents, under assumptions of chemical discontinuity (above), and phase discontinuity (below).
*Source:* P. J. Wyllie, *The Dynamic Earth*, 1971, figures 5-5 and 5-6, pp. 69-70, copyright © John Wiley and Sons, by permission.

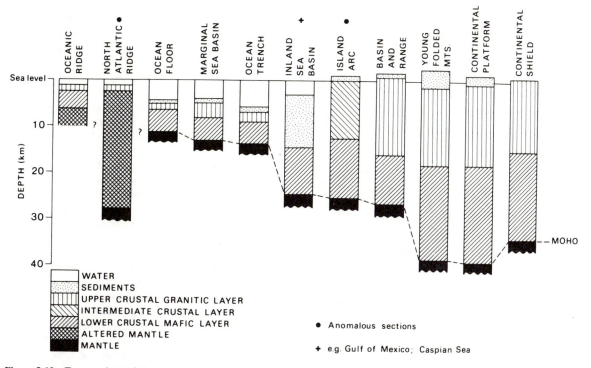

**Figure 5.10** Type sections of the crust and upper mantle for eleven locations.
*Source:* Reprinted by permission from Condie, 1976, figure 4-4, p. 77, copyright © Pergamon Press.

possibly absent and altered mantle lying close to the surface.

It is clear that the distinction between the crust and the mantle, separated by the Moho seismic discontinuity, is not as diastrophically significant as that between the more rigid outer layer of the earth and the weaker underlying layer of which the low-velocity layer forms the upper part (Table 5.1). The low-velocity layer is 50–100 km thick, the top of which is 60–70 km below sea level under the oceans and more than 100 km below sea level under the continents; the base being situated at depths of less than 200 km under the oceans and down to 300 km under some active continental margins. Thus the low-velocity layer lies well below the Moho and is not radically different in composition from the rocks immediately above and below, but is weaker and capable of slow plastic flow under the action of long-continued forces. It will fracture under sudden stress to produce earthquake foci and will transmit both *S* and *P* waves. The condition of the low-velocity layer, together with that of the underlying layer, gives a complex range of reactions to stress, probably because its material is at a temperature and pressure which makes it susceptible to incipient melting, particularly when the presence of small

amounts of water lowers its melting point. It is diastrophically useful, therefore, to distinguish the following three outer layers of the earth:

(1) the lithosphere, made up of the crust and the upper rigid part of the mantle above the low-velocity layer; this behaves as a relatively cool, rigid shell some 50–150 km thick;

(2) the asthenosphere, made up of the hot and very weak low-velocity layer, together with an underlying layer of variable but generally low strength; also capable of plastic behaviour and extending locally down to several hundred kilometres;

(3) the mesosphere, a zone of greater heat but also of greater strength.

The concept of isostasy can be applied to columns of the lithosphere being buoyed up in the plastic asthenosphere in direct proportion to their thickness and in inverse proportion to their general density (Figure 5.11). As Wyllie (1971, p. 38) has put it, 'all large land masses on the earth's surface tend to sink or rise so that, given time for adjustment to occur, their masses are hydrostatically supported from below, except where total stresses are acting to upset equilibrium'. Thus topographic

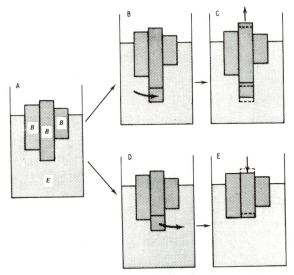

**Figure 5.11** Isostatic equilibrium illustrated by crustal basalt blocks (**B**) floating in mantle eclogite (**E**): A. isostatic equilibrium of assumed initial condition: B. and C. effect of insertion of additional crustal material beneath central block causing it to rise; D. and E. effect of removal of crustal material at root of the central block causing it to subside. *Source:* P. J. Wyllie, *The Dynamic Earth*, 1971, figure 10-1, p. 234, copyright © John Wiley and Sons, by permission.

heights are supported by roots in which less-dense material displaces more dense at depth, and the erosion of these mountains leads to decaying compensatory uplift which continues until the net result of erosion and isostatic recovery is the production of a surface of low relief near sea level. As we have seen with Lake Mead, even small applied stress loads cause the earth's surface to suffer relatively rapid isostatic adjustments. Figure 5.11 illustrates another important mechanism producing alterations in the elevation of the earth's surface, in which phase changes in the lower lithospheric rocks cause a density increase (e.g. gabbro changed to eclogite) leading to a general isostatic subsidence without any large-scale plastic deformation of the asthenosphere.

## 5.2 Global tectonics

### 5.2.1 Continental drift and plate tectonics

The development of the crustal floatation ideas of isostasy in the late 1880s subsequently became associated with those of intercontinental geological similarities which developed between 1910 and the 1950s, leading to a minority but persistent view among some earth scientists that earth history had been dominated by large-scale lateral continental movements. These movements were thought to have resulted in the splitting apart of a super-continent (Pangaea), first, into two major hemispherical continents (Laurasia in the north and Gondwanaland in the south), and then into smaller units whose movements across the surface of the globe left strings of detached islands in their rear and threw up mountain ranges along their leading edges or, as in the case of the Alpine–Himalayan system, between colliding blocks. There were four main classes of intercontinental geological evidence in support of this model of continental drift:

(1) Structural and stratigraphical. This involved the continuation of truncated structural zones and the matching of closely correlated stratigraphic successions across now wide oceans; for example, between north-west Scotland and Newfoundland and between south-west Africa and the east coast of South America.

(2) Paleontological. This stemmed from similarities of fossils at present widely spaced and difficult to explain in the absence of continental drift, except by clumsy hypotheses involving land bridges which appeared and disappeared at convenient times.

(3) Paleoclimatological. Of this, the anomalous distribution of Late Paleozoic glacial deposits over parts of South Africa, South America, India and Australia was the most important.

(4) Geometrical fit of continental edges. This evidence (e.g. that of Western Africa and eastern South America) was the least strong.

The mechanisms proposed for continental drift included lunar attraction, the earth's rotational forces, gravity sliding of continental blocks and the drag applied by thermally driven convection currents in the quasi-solid mantle (sima) beneath the continental crust (sial). Before the later 1950s, the overriding majority of earth scientists rejected the concept of continental drift partly because they found the geological evidence ambiguous, but mainly because the mechanisms proposed for it seemed completely inadequate (Hallam, 1973).

In the late 1950s and early 1960s the recent revolution in the earth sciences began when it was discovered that natural remnant magnetism in rocks suggested that systematic changes in the direction of the earth's magnetic polarity had occurred since Pre-Cambrian times (i.e. for much more than the last 600 million years) and, more important, that a globally coherent picture of these changes could only be reconstructed by assuming that different parts of the earth's surface (e.g. Europe and North America) had drastically changed positions relative to each other during this time. Subsequent

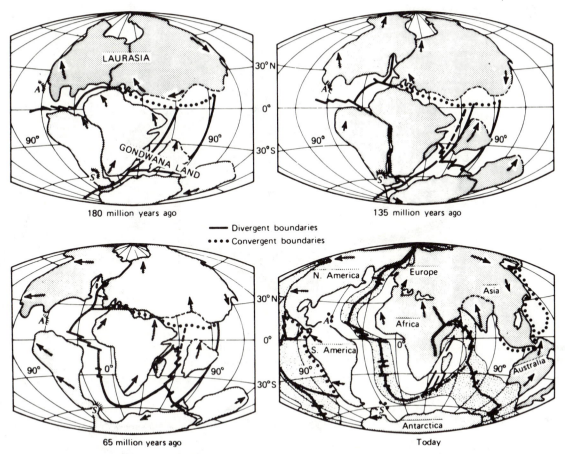

180 million years ago          135 million years ago

—— Divergent boundaries
•••• Convergent boundaries

65 million years ago          Today

**Figure 5.12** Break-up of the supercontinent Pangaea during the past 200 million years as inferred from the positions of fossil magnetic poles. **A** and **S** show the positions of the Antilles and Scotia arcs.
*Source:* Dietz and Holden, 1970, compilation figures 3-6.

developments have led to a very radical view of the relative movements of large parts of the earth's surface, involving continents, marginal seas and ocean floors (Figure 5.12). The most important of these developments was the previously described discovery of a chronology of magnetic reversals and its linkage in the early 1960s with the magnetic anomalies present within symmetrical belts of the ocean ridge surface on either side of the ridge crests. This provided proof of the reality of ocean floor spreading in the manner of a conveyor-belt away from ridge-crest zones of new crustal formation, but also a measure of the past rates of spreading and amounts of total movement (Figure 5.13). These axes of spreading were observed to extend on to the recently faulted belts of the continents (e.g. the Red Sea and East Africa, also the south-western United States), and these were assumed to represent early stages of continental break-up by spreading.

The crustal units involved in these movements were in some instances part oceanic and part continental, and in others entirely oceanic, so that the previous concept of continental movement was replaced with that of the movement of crustal plates. On these lithospheric plates the granitic continents (as well as the thin granitic 'scum' covering the oceanic lithosphere) are carried passively, not moving independently as previously postulated under continental drift theories. Before the end of the 1960s, the generalities of a global tectonic model had found broad acceptance wherein superficial plates of the lithosphere are moving in association with the creation of new crust along zones of spreading. The plates, varying in size from $10^6$–$10^8$ km² and from 70 km thick under the oceans to some 150 km under the continents, can be divided into seven major plates ($10^8$ km²) (e.g. African and Pacific) and eight intermediate plates ($10^6$–$10^7$ km²) (e.g. Arabian and Cocos) (Figure 5.14), as well as more

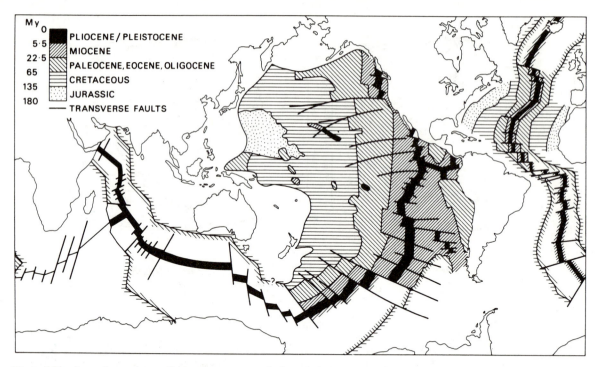

**Figure 5.13** Axes of crustal spreading, major transverse faults and the age of the sea floor.
*Sources:* Gass *et al.*, 1972, figure 16.10, p. 243; Rona, 1980; Stoddart, personal communication.

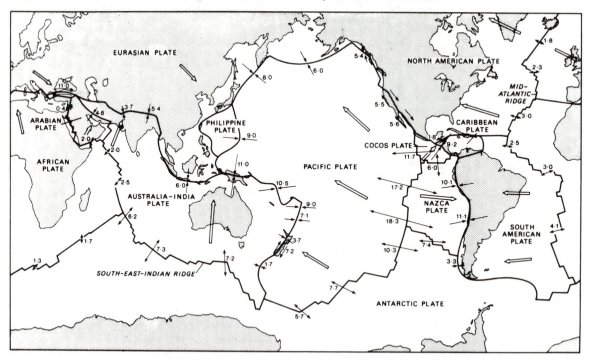

**Figure 5.14** Rates of plate divergence and convergence and directions of major plate movements; figures indicate relative velocity (in centimetres per year).
*Source:* McKenzie and Richter, 1976, p. 74, copyright © *Scientific American*.

than twenty smaller plates ($10^5$–$10^6$ km²) (e.g. Turkish–Aegean and Iran plates) which generally occur near continent–continent or arc–continent collision boundaries. Plate boundary zones can thus be divided into three broad groups (Figure 5.15):

(1) Zones of spreading, where a conveyor-belt process is creating new lithosphere along belts of tension and vulcanism.

(2) Transform (transverse or transcurrent) faults, along which plates slide differentially without great changes in level and without crust being significantly created or destroyed.

(3) Subduction zones of convergence and destruction along which the leading-edges of lithospheric plates are being thrust down into the asthenosphere to form the Benioff Zone and consumed by melting and attendant igneous activity in association with island arcs and ocean trenches or with young cordilleran mountain belts. Perhaps the most complex local situations exist where triple junctions occur between plates (e.g. in the north-east Pacific and in Asia Minor). Under this model the trailing-edges of the continental parts of the plates form more stable margins often characterized by subsidence and sedimentation (e.g. the Atlantic coastal margin of the United States) (see Figure 5.18B).

With the evidence for movement being so much more secure, the problem of mechanism seems less insuperable to plate tectonics than to continental drift, especially as at least five classes of mechanisms are possible:

(a) Viscous drag of the solid lithospheric plates by convection currents in the asthenosphere, rising along the zones of spreading and subsiding along the subduction zones. Earthquake evidence along the Benioff Zone suggests that the mantle below 700 km is too resistant to allow the penetration of subduction plates, which further suggests that convection is in all probability limited to depths above this, and that there is a hierarchy of convection cells of differing size.

(b) The plate being pulled by the weight of the relatively cool subducting slab, possibly assisted by an increase in the density of the slab as the gabbro metamorphoses into eclogite at depth.

(c) The gravitational sliding of the lithospheric slab away from the oceanic ridge zone raised by rising material in the asthenosphere. It has been estimated that a surface slope of only 1:3000 would produce a movement of the lithosphere of 4 cm/year over the low-velocity top layer of the asthenosphere.

(d) Lithospheric plates being pushed apart by magma rising along the axis of the zone of spreading to form wedges of new lithosphere along the trailing-edge of the plate (see Figure 6.2).

(e) Mantle plume mechanisms (see Chapter 6) in which magma rising from some twenty-one major hot spots in the mantle and spreading out under the lithosphere plates might provide a horizontal driving-force.

### 5.2.2 Zones of spreading

It is important to recognize that activity along the zones of spreading comprising the oceanic ridges and certain of the rifted intracontinental belts is neither constant in space nor time, and that this activity may be closely related to widespread eustatic, geosynclinal, epeirogenic

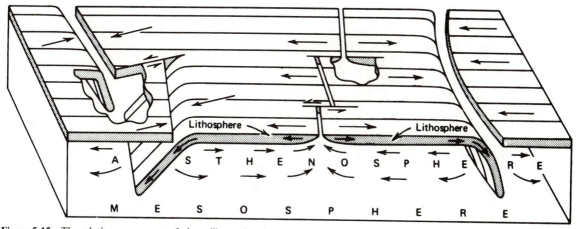

**Figure 5.15** The relative movements of plates illustrating the possible roles of the lithosphere, asthenosphere and mesosphere; the relations of zones of spreading, subduction and transform faulting.
*Source:* Isacks *et al.*, 1968, figure 1, p. 5857.

and orogenic events. It has been shown that the rate of spreading of the floor of the north Atlantic has experienced three maxima (each lasting 10–70 million years) and three minima (lasting 10–20 million years) since the beginning of the Cretaceous (i.e. 135 million years BP), and that the former were associated with increases in elevation of the mid-Atlantic ridge leading to eustatic rises of sea level, marine transgressions and corresponding increases of rates of sedimentation on the Atlantic shelves (controlled by water depth). The reverse situation occurred with low spreading rates (Table 5.4).

The greater protuberance of the zones of spreading during their more active phase has the effect of decreasing the capacity of the ocean basins and creating a tendency for a eustatic rise of sea level. Provided that this

is not accompanied by widespread orogeny – which it may be – this will lead to a marine transgression of the continental edges and a greater depth of coastal water in which a greater rate of sedimentation can occur. These sedimentary wedges along the more stable 'trailing' continental margins form one type of downbuckling, the *paraliageosyncline* (coastal geosyncline) (see Figure 5.18B). Subsidence along these zones was probably initiated by the foundering of the raw continental edge following its original fracture and this may have been aided subsequently by the thermal contraction of the lithosphere as it moved away from the hot zone of spreading, but it is probable that continued subsidence along paraliageosynclines has been assisted by the thicknesses of sediments laid down, particularly during the

**Table 5.4** Ocean floor spreading and shelf sedimentation

| Time (million years BP) | Geological period | Ocean floor spreading rate (cm/year) | Shelf sedimentation rate (cm/10³ year) | |
|---|---|---|---|---|
| | | | SE USA | Central-Eastern USA |
| 135–65 | Cretaceous | 1·7–2·0 max. | 1·9–4·0 | 2·3–2·9 |
| 65–53·5 | Paleocene | 1·3 min. | 0·4–2·6 | 0·4–1·4 |
| 53·5–45 | Early–Middle Eocene | 2·0 max. | 3·3 | 1·5–6·5 |
| 45–22·5 | Late Eocene–Oligocene | 0·9 min. | 0·3–0·6 | 0·6–1·7 |
| 22·5–5·5 | Miocene | 1·3 max. | 1·3–3·8 | 1·7–2·9 |
| 5·5–0* | Post-Miocene | ? min. | 0·1–1·0 | — |

*Plus glacial eustatism.
*Source:* Rona, 1973, pp. 2858–9.

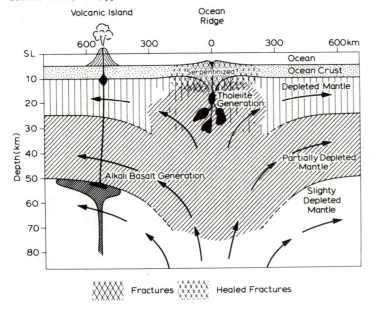

**Figure 5.16** Model of a mantle convection cell supporting an ocean ridge. See Chapter 6, for explanation of the rock types.
*Source:* Condie, 1976, figure 7-12, p. 174, after Gast.

transgressive phases. Continued subsidence may eventually so thin and weaken the lithospheric plate that failure, subduction and cordilleran type orogeny may result.

Orogenic belts are associated with subduction zones along island arcs or where the lithospheric 'conveyorbelt' has carried the leading-edge of a continent to an equilibrium position where it is being consumed along a convergence zone of subduction. The strongest compressive pulses producing orogenic buckling of the crust are probably associated with times of maximum rates of spreading. The maximum compression tends to arch up the crust such that major orogenies are often locally associated with epeirogenic movements leading to marine regressions. This is why for a time of increased rate of crustal spreading to be correlated with a eustatic rise of sea level, it must be accompanied by a *net* worldwide orogenic quiescence.

The upper part of the sea floor in ocean ridges may be composed of peridotite, which has been metamorphosed and hydrated by rising water to form serpentine (Figure 5.16) (see Section 4.4), and covered in places by a thin veneer of granitic scum and unconsolidated sediments. The low *P*-wave velocities characteristic of these ridges may be attributed to the high temperatures and excessive fracturing which are probably present. A zone of spreading may develop as the result of a mantle plume or convective upwelling causing the lithosphere to stretch, to thin and fracture under tension into blocks separated by high-angle faults. If this zone develops under a continent, it may result in a sequence of developmental stages which include (Figure 5.17):

(1) Intracontinental rifting, resulting from spreading and updoming. Possible present examples of this stage are the East African rift valley system, the Rhine trough and that of Lake Baikal. Work in East Africa is showing that the major faulting there is much more recent in origin than was previously supposed (i.e. starting in the Miocene or Pliocene – see Ollier, 1981, p. 80), and clearly post-dating the extensive erosion surfaces which are displaced by the faults.

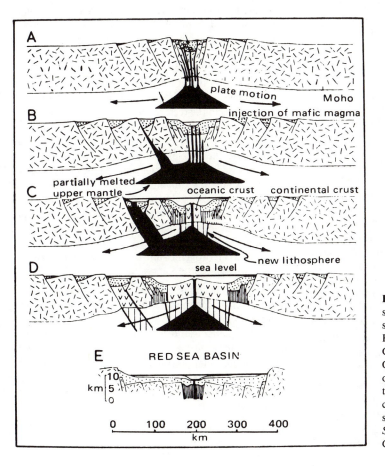

**Figure 5.17** Stages in the rupture and separation of a continent along a zone of spreading: A. analogous to condition of the Basin and Range Province of eastern California and Nevada at the end of the Oligocene; B. analogous to present condition of the East African Rift Zone; C. analogous to present Red Sea Basin; see E. for comparison; D. is a somewhat later stage of separation.
*Sources:* Dewey and Bird, 1970; and Condie, 1976, figure 9-1, p. 213.

(2) Interplate thinning during which the crust thins and the surface sinks. The Red Sea trough may be an example of this stage.

(3) Oceanic rises, which are large-scale features at a lower elevation. The Carlsberg Ridge in the Indian Ocean, apparently continuous with the Red Sea structures, is one such. A major Icelandic graben has appeared to have spread laterally some 70 km in 9000 years. At present laser beams are being used to make accurate measurements of spreading rates in Iceland and elsewhere (Condie, 1976).

### 5.2.3 Subduction zones and orogeny

Where lithospheric plates are converging, it is usual for the oceanic plate to be subducted beneath the edge of the continental one due to a combination of descending convection currents, metamorphism, gravity and underthrusting, so that crustal material is forced down into the mantle and consumed by complex processes of remelting and metamorphism. The edge of the continental plate is fractured and buckled, and the subsiding slab produces the earthquakes associated with the Benioff Zone, the negative gravity anomalies, the andesitic vulcanism and the excessive heatflow characteristics of the island arc–trench belts (Wyllie, 1976).

The downbuckling of the continental margin allows thick belts of shallow-water sedimentation to occur (i.e. geosynclines), as well as ocean trenches. The latter, however, may only exist at certain tectonic stages in the development of a subduction zone dependent upon the rates of downbuckling, faulting, sedimentation and erosion from nearby island arcs. The dips of the subducting plates appear to be variable both in space and time (range $30°-90°$, with a mean of $45°$), and inversely related to the associated rates of sea floor spreading. For the first 200 km of its descent, the upper part of the oceanic lithosphere plate is in tension, but at greater depths it meets increasing resistance from the more solid mantle and the whole plate is strongly compressed. Some existing subducting plates have penetrated to less than 300 km (e.g. below the Aleutian Arc), but most have been forced to a maximum limiting depth of 700 km into the resistant mantle.

The buckling of the continental edge and the attendant vulcanism causes a succession of rising and subsiding belts to be initiated. These belts are dynamic features and develop through time as the convergence zone deforms and collapses. Rising belts, or geanticlines, may occur due to upbuckling of the continental plate, forcing up by subduction underthrusting, the upwelling of magma from depth to form batholiths, or other processes. They may be associated with surface vulcanism, as with inner volcanic island arcs – or not, as with outer tectonic arcs or welts in the marginal seas. These geanticlines are flanked by geosynclines which are belts of long-continued subsidence and sedimentation, the latter mostly in shallow water, indicating a rapid supply of material keeping pace with subsidence (Figure 5.18). Geosynclinal subsidence may be caused by lateral buckling, subduction, removal of lava from depth, mantle convection, weight of overlying sediments, phase changes at depth, or any combination of these. Patterns of geosynclinal sedimentation vary in space and time in two important respects:

(1) On the flanks of tectonically active geanticlinal arcs a rapid supply of ill-sorted sands and muds results in thick sequences of greywackes (see Section 4.3.3), together with interspersed andesitic igneous rocks which form an outer *eugeosynclinal* (truly geosynclinal) belt. On the landward side is the non-volcanic *miogeosynclinal* (lesser geosynclinal) belt with thick sequences of limestone, well-sorted sandstones and shales. This, in turn, may be backed by a stable foreland on which very much slower sedimentation is occurring, which can be considered as an extension of the continental *craton*. The craton is the antithesis of the geosynclinal belt in that it is that central portion of a continent which appears to have been stable over long periods of time.

(2) The production of flysch and molasse sequences of sedimentary rocks in geosynclines are the result of differences in the tectonic behaviour of the associated geanticlines. When the flanking geanticline is tectonically active and rising more rapidly than erosion can remove it, the coarse material deposited adjacent to it extends further and further out into the geosyncline (a *flysch* 'suite' of sediments). When the geanticlinal belt is inactive, erosion progressively reduces its elevation and the coarse sedimentary material extends less and less far out into the associated geosyncline with time, producing a *molasse* suite. A flysch is, therefore, viewed as an orogenic suite of rocks and a molasse as a post-orogenic suite.

The orogenic collapse of the subduction zone and the production of a complex mountain belt can probably occur either due to activation or collision. Activation takes place under island arc/cordilleran conditions when the zone adjacent to a thermally driven subducting oceanic plate collapses (Figure 5.19). Although it is difficult to generalize regarding the typical history for such a belt, a number of stages have been suggested (Condie, 1976; Wyllie, 1976):

(a) Geosynclinal:
  (i) generative;

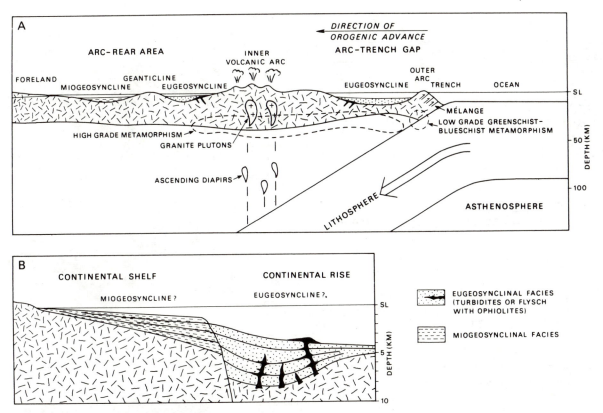

**Figure 5.18** The locations of eugeosynclines and miogeosynclines with respect to: A. an active subduction zone; a modern example might be the section NNW–SSE from Vietnam to the Indian Ocean: Foreland – South Vietnam; Miogeosyncline – South China Sea; geanticline – Borneo: inner eugeosyncline – Java Sea; inner volcanic arc – Java-Flores; outer non-volcanic arc – Sumba; trench – Java Trench; B. a more stable shelf.
*Sources:* Seyfert and Sirkin, *Earth History and Plate Tectonics*, 1973, figure 6.7, p. 73, copyright © 1973 by Carl K. Seyfert and Leslie A. Sirkin, reproduced by permission of Harper & Row, Publishers, Inc.; Condie, 1976, figure 8-12, p. 196; Dietz, 1963, figure 3B, p. 320.

(ii) development – e.g. the present condition of the western Pacific margins;

(iii) orogenic – the collapse of the mobile belt appears to begin normally with the inner slope of the trench in the vicinity of which a *mélange* (French: *mélange* – 'medley', 'a jumble') of sheared and thrust slices form at shallow depths in the subduction zone; the orogeny thus begins in the eugeosyncline but then migrates into the miogeosyncline; flysch deposits form some 35 million years after the start of the orogeny and in the latter organic phases great *nappes* (French: *nappe* – 'a cover', 'table cloth') of eugeosynclinal rocks and associated intrusive igneous rocks are thrust continentward for tens of kilometres over the deformed meogeosynclinal rocks (and even over the foreland), in association with intense and widespread metamorphism at depth.

(b) Late geosynclinal: alongside the rising cordillera

(e.g. the present Andes) trenches and deeps are formed. After some 50 million years from the beginning of orogenesis, the complete width of the marginal belt is at a late geosynclinal stage, which is associated with the deposition of molasse suites in marginal downbuckles.

(c) Post-geosynclinal: this is characterized by vertical movements, arching, fracturing and possibly the emission of basaltic lavas on various parts of the belt and foreland.

Mountain-building by collision occurs when arcs and continents are present on both sides of the subduction zone, as was the case in the orogenies involving the Alps, Himalayas and the central Urals (Figure 5.20). In this circumstance particularly complicated structures arise, subduction may be reversed, and *ophiolites* (Greek: *ophis* – 'a serpent') and gabbroic and other rocks forming the oceanic lithosphere may be squeezed up and

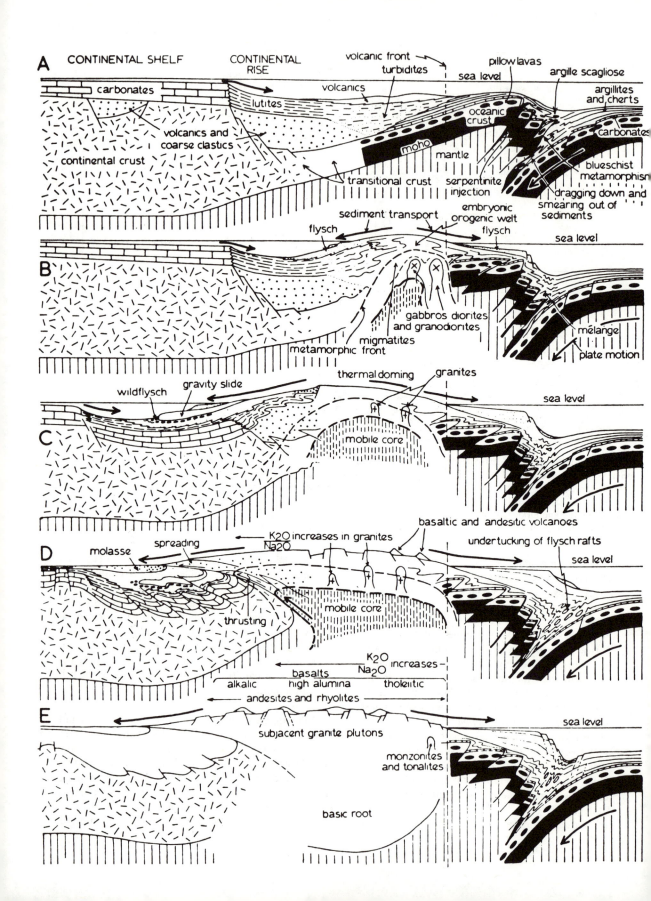

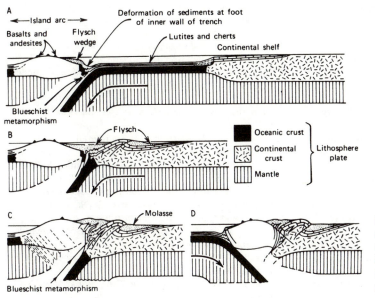

Basalts and
andesites
Flysch
wedge
Island arc
Deformation of sediments at foot
of inner wall of trench
Lutites and cherts
Continental shelf
A
Blueschist
metamorphism
B
Flysch

Oceanic crust
Continental crust
Mantle
Lithosphere plate

C
Molasse
D
Blueschist metamorphism

**Figure 5.20** Schematic sequence of sections showing the collision of a continental margin with an island arc, followed by a change in direction of plate descent.
*Source:* Dewey and Bird, 1970, figure 13, p. 2642.

intruded on a large scale along a central *suture zone* (Latin: *sutura* – 'a stitch') where the two land masses become welded together. Two possible means of continental growth through such orogenies are by the sweeping together by subduction of adjacent island arcs or by the seaward migration of the continental edge by the welding on to it of successive orogenic belts composed of deformed, metamorphosed and intruded geosynclinal material.

It is thus apparent that the plate theory of crustal development is capable of explaining the major features of first-order continents and oceans, and many second-order mountain ranges and third-order folded and faulted features, as well as third-/fourth-order features, such as the alignment of some of the world's major rivers along lines of structural weakness (e.g. the Ganges, Mississippi, Niger and Rhine). Indeed, the continued adjustment of the earth to major endogenic forces may explain much of the morphology and stability of these large-scale fluvial features. However, it is also clear that the mechanisms of global diastrophism which have been described are also responsible for characteristic patterns of small-scale activity, which also make important contributions to geomorphology. For example, recent studies show that rivers crossing areas of active uplift show pattern, gradient and depth changes that are clearly associated with the decrease and increase of valley slope as the river approaches and crosses the axis of uplift (Burnett and Schumm, 1983).

### References

Barazangi, M. and Durman, J. (1969) 'World seismicity maps 1961–67', *Bulletin of the Seismological Society of America*, vol. 59, 369–80.

Boot, M. H. P. (1971) *The Interior of the Earth*, London, Arnold.

Brander, J. (1976) 'The measure of plate tectonics', *New Scientist*, vol. 69, 110–13.

Bullard, E. (1969) 'The origin of the oceans', *Scientific American*, vol. 22, 66–75.

Bullard, F. M. (1976) *Volcanoes of the Earth*, Austin, Texas, Texas University Press.

Burk, C. A. and Drake, C. L. (eds) (1974) *The Geology of Continental Margins*, New York, Springer-Verlag.

Burnett, A. W. and Schumm, S. A. (1983) 'Alluvial-river

**Figure 5.19** Schematic sequence of sections showing the evolution of a cordilleran-type mountain belt developed as result of the active underthrusting of a continent by an oceanic plate. This should be contrasted with the situation when two continents collide along a subduction zone producing more pronounced thrust, nappe and mélange structures, as in the Alps and Himalayas.
*Source:* Dewey and Bird, 1970, figure 10, p. 2638.

response to neotectonic deformation in Louisiana and Mississippi', *Science*, vol. 222, 49–50.

Condie, K. C. (1976) *Plate Tectonics and Crustal Evolution*, Oxford, Pergamon.

Crittenden, M. D. (1963) 'New data on the isostatic deformation of Lake Bonneville', *US Geological Survey Professional Paper* 454-E.

Dewey, J. F. (1973) 'Plate tectonics and the evolution of the Alpine system', *Bulletin of the Geological Society of America*, vol. 84, 3137–80.

Dewey, J. F. and Bird, J. M. (1970) 'Mountain belts and the new global tectonics', *Journal of Geophysical Research*, vol. 75, 2625–47.

Dietz, R. S. (1963) 'Collapsing continental rises', *Journal of Geology*, vol. 71, pp. 314–33.

Dietz, R. S. and Holden, J. C. (1970) 'Reconstruction of Pangaea', *Journal of Geophysical Research*, vol. 75, pp. 4939–56.

Gass, I. G., Smith, P. J. and Wilson, R. C. L. (1972) *Understanding the Earth*, Chichester, Artemis Press.

Hallam, A., (1973) *A Revolution in the Earth Sciences: From Continental Drift to Plate Tectonics*, Oxford University Press.

Isacks, B., Oliver, J. and Sykes, L. (1968) 'Seismicity and the new global tectonics', *Journal of Geophysical Research*, vol. 73, pp. 5855–99.

McKenzie, D. P. and Richter, F. (1976) 'Convection currents in the earth's mantle', *Scientific American*, vol. 235, 72–89.

Ollier, C. D. (1981) *Tectonics and Landforms*, London, Longman.

Pollack, H. N. and Chapman, D. S. (1977) 'The flow of heat from the earth's interior', *Scientific American*, vol. 237, 60–76.

Rona, P. A. (1973) 'Relations between rates of sediment accumulation on continental shelves, sea-floor spreading, and eustacy inferred from the central North Atlantic', *Bulletin of the Geological Society of America*, vol. 84, 2851–72.

Rona, P. A. (1980) *The Central North Atlantic Ocean Basin and Continental Margins*, NOAA Atlas 3 (National Oceanographic and Atmospheric Administration).

Seyfert, C. K. and Sirkin, L. A. (1973) *Earth History and Plate Tectonics*, New York, Harper & Row.

Wyllie, P. J. (1971) *The Dynamic Earth: Textbook in Geosciences*, New York, Wiley.

Wyllie, P. J. (1976) *The Way the Earth Works*, New York, Wiley.

# Six  *Igneous activity and landforms*

Primary landforms are produced by igneous activity, which may be either intrusive or extrusive. Igneous activity always attracts popular attention, whether as a result of the recent activity of Mount St Helens in Washington State or of the aesthetic satisfaction afforded by such undissected primary landforms as the volcanoes Fujiyama and Cotopaxi. However, the geomorphologist must view igneous phenomena more dispassionately and recognize, for example, that the Mount St Helens eruption had an insignificant topographic effect compared with that of Krakatoa in 1883, and that basalt plateaus and batholith surfaces are by far the most areally extensive igneous features of the globe.

## 6.1  Igneous activity in space and time

Magma intruded into the earth's crust or extruded upon its surface appears to have been mobilized at relatively shallow depths. For example, basalt which erupts at the surface at a temperature of 1200°C probably became molten at a depth of less than 100 km – at a depth of 500 km the confining pressure of 150,000 bar (1 bar ≃ surface atmospheric pressure) would impose a pressure melting point in excess of 2000°C on this rock. Under atmospheric pressure solidification of lava occurs between 600°C and 900°C, depending on the chemical composition and gas content. The melting of magma within the earth is the result of a complex interaction of increases of temperature, decreases of pressure and addition of water. Volcanic heat contributes only about one-hundredth of the total terrestrial heat loss but it is very localized along belts of rising convective currents or at the more than 120 hot spots associated with the rising 'plumes' which will be described later. As was suggested earlier, the convecting layer probably occurs at depths of

less than 700 km and contains a hierarchy of different convection cell sizes. About half of the heat in the convecting layer is generated within the layer itself and half comes from the mantle below, the latter source explaining why localized volcanic activity can continue at a given location despite the movement of a lithospheric plate over it. Magma rises within the earth as coherent through-piercing blobs or *diapirs* (Greek: *diapeiro* – 'pierce through'), or as columns of partly molten rock by a complex set of processes including the 'elbowing aside' of superincumbent rocks by mechanisms of slow creep and their partial 'digestion' (anatexis). At shallow depths a combination of diapir pressure, thermal expansion and gas pressure (partly from groundwater turned to steam) may lead to surface eruptions.

Generalizations regarding the occurrence of eruptions in space and time are made more difficult because a combination of the following processes often causes a variety of magmas to be erupted at different locations within a given volcanic belt, or at different times at the same geographical location:

(1) A given magma may be subjected to a variety of geochemical conditions which may generate a wide spectrum of lavas.

(2) Fractionation and differentiation take place within the magma chamber, which might successively yield peridotite, gabbro and granite, for example.

(3) Different parts of the magma chamber may be tapped at different times.

(4) Anatexis involving a variety of country rocks.

Thus the volcanic chains forming the Puys of Auvergne in central France exhibit a close association of basalt, andesite and trachyte, while the chain of large composite andesitic volcanoes in western Guatemala is paralleled by a belt of smaller basaltic ones somewhat to the east. A given volcano can produce different material at different times – Krakatoa in the East Indies yielded alternately rhyodacites and andesites – and these differences can be manifested even within a single eruptive phase. In the longer term many volcanoes seem to become more acid with time, although this is by no means a universal rule as demonstrated by Krakatoa, which has become more basic, and by a change from andesitic to more basic flows from Etna in Sicily at about 5400 BP. During the entire span of earth history there appears to have been a relative global increase of basic igneous activity.

Despite the foregoing variations, each zone of distinctive behaviour within the global plate tectonic model possesses a characteristic assemblage of igneous intrusive and extrusive rocks (Figure 6.1), although the mechanisms of magma generation include some of the greatest unresolved contemporary geochemical problems. These igneous assemblages are, in general, made up of the following major classes of rocks:

(1) tholeiite – a subalkaline, olivine-deficient basaltic rock; it has low viscosity, forming intrusives, flows and floods, as well as pillow lavas (i.e. rocks composed of rounded masses of 10–100 cm diam.) when extruded beneath the sea; the most extensive lava flows (e.g. forming the Deccan Plateau in India and the Snake River Plateau in the north-western United States) are composed of tholeiite; the majority of the energy output of tholeiite eruptions is in the form of heat;

(2) calc-alkaline rocks – these include highly viscous andesitic extrusive rocks with associated granodiorite plutons; also smaller amounts of rhyolite, dacite and even tholeiite commonly occur, causing this group to embrace a wide range of acid to basic magmas; the andesitic and rhyolitic extrusive rocks are associated with explosive releases of gas, ash, ejecta and pumice which form symmetrical conical volcanoes (e.g. Parícutin), and on occasions great amounts of frothy gas-charged rhyolite (ignimbrite) are extruded locally to build substantial lava plateaus (e.g. in Sumatra); the high viscosity of the magma helps to produce threshold effects which lead to periodic tendencies in the eruptions;

(3) alkali basalt – a low-saturation olivine basalt. This is the least abundant of the three classes and commonly associated with smaller amounts of trachyte, latite and phonolite; some of this class may result from the assimilation of limestones and other calcareous sedimentary crustal rocks by basalt.

Along the tensional rift systems forming the divergent zones of crustal spreading tholeiite is intruded as dyke swarms (Figure 6.2) or is extruded as flows and pillow lavas, the process of spreading resulting in its ultimately forming much of the oceanic crust. Ocean ridges appear to be associated with lines of long-lived hot spots, and it is thought that peridotite diapirs may originate in the low-velocity zone at depths of about 100 km. These rise by plastic deformation, melting to produce tholeiite as the confining pressure decreases such that they may be 25 per cent molten at 40–60 km (mostly at the base) and much more at about 20 km. Tholeiites are thus formed by the partial melting of partially depleted mantle peridotite, and they may be associated with the production of some serpentine by the metamorphism of peridotite. The situation along continental tensional rift systems is more complex because of the melting, fractionation and anatexis of crustal rocks (including sedimentary ones at shallow depths). In these belts tholeiite floods occur only locally, in areas such as East Africa, Ethiopia and the Rhine Valley, and they are associated with flows and ring complexes of alkali basalt and trachyte.

Mantle plumes occur not only along zones of spreading, but also within continental and oceanic plates. They are varied both in size and magmatic production, have little obvious spatial organization at a global scale, but are clearly associated with local updoming and excessive crustal heatflow. Tholeiite plumes occur more exclusively in the ocean basins, one of the most notable examples of which are the Hawaiian Islands which have existed for the past 70 million years, apparently tapping magma chambers which extend from depths of more than 60 km to within 8 km of the surface. A persistent mantle hot spot may exert a continued effect over successive portions of a lithospheric plate as it moves above it, giving rise to a line of volcanic peaks decreasing in age in the direction of the centre of spreading. In the Pacific Ocean these plumes have formed many seamounts and guyots (See Figures 6.3 and 6.15) which appear to be older in the west and younger in the east, supporting their place in the general scheme of sea floor spreading. Those which are truncated, submerged and coral-capped give further support to Darwin's theory of coral-atoll formation by successive volcanic action, marine erosion, subsidence and coral growth (see Section 15.6.1). Plumes of alkali basalt are widespread in both continental and oceanic locations, occurring with latite, trachyte and phonolite. In the case of Easter Island even associated rhyolites are found. In other instances early-stage eruptions of tholeiite are followed by alkali basalt as a result of magma differentiation and even by andesites, presumably developed by the anatexis of shallow crustal rocks. The pronounced tendency for tholeiitic plumes in marginal sea basins has suggested to some

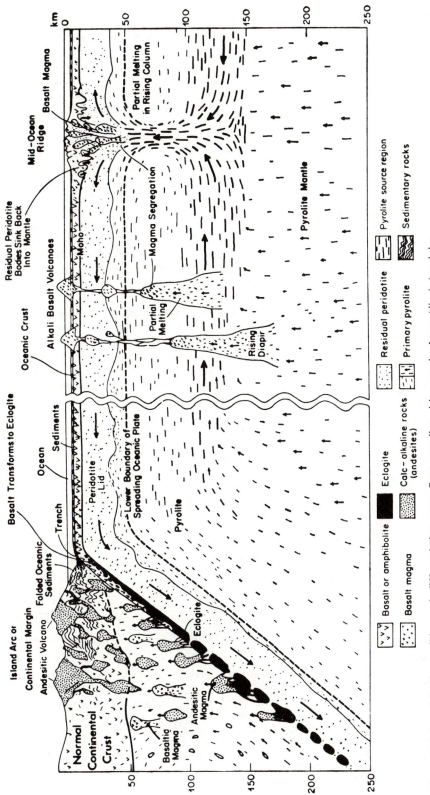

**Figure 6.1** A model of irreversible mantle differentiation, ocean floor spreading, rising mantle plumes (diapirs) and the evolution of the continental edge.

*Source:* Ringwood, 1974, figure 11, p. 201.

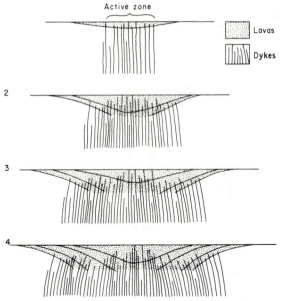

**Figure 6.2** Diagram illustrating spreading of the earth's crust beneath Iceland, with the intrusion of vast numbers of dykes and gradual accumulation of volcanic rocks in the sinking graben at the surface.
*Source:* G. Bodvarsson and G. P. L. Walker, *Royal Astronomical Society Geophysical Journal*, 1964, vol. 8, figure 5, p. 291.

authorities that these belts may be subjected to tension and spreading.

Igneous activity associated with the subduction zone is very complex in character. The subducting plate is subject to frictional heating, mineralogical phase changes (leading to additional heating), increasing pressure and metamorphism; these processes giving rise to a dominantly calc-alkaline suite of igneous rocks and being responsible for more than 80 per cent of the world's present active volcanoes. These magmas are generated by a complex of processes involving partial melting of wet ultramafic rocks, melting and metamorphism of the descending oceanic crustal slab, fractionation of tholeiite, partial crustal melting and anatexis. Within the subduction zone diapirs of tholeiite are produced at shallow depths (80–100 km) and of andesite at greater depths (100–150 km). Pulses of subduction activity may result in variations of diastrophic and igneous activity in the subduction belt. For example, the lack of active volcanoes above a subduction zone (e.g. as in Ecuador and southern Peru at present) may imply a stationary subduction plate segment and a lack of magma generation. When subduction ceases, a basin may open up behind the volcanic belt and be filled with sediments and lava which become deformed on the resumption of subduction (Figure 6.4).

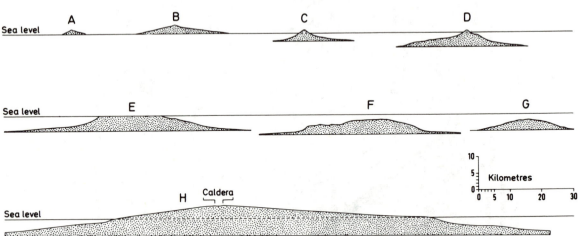

**Figure 6.3** Comparison of sizes of volcanic features: A. Vesuvius, Italy, composite strato land volcano, relief 1281 m; B. Mount Etna, Sicily, composite strato land volcano, relief 3290 m; C. Stromboli, Aeolian Islands, Mediterranean, composite strato volcanic island, relief 2885 m above sea floor; D. Gunungapi, Lesser Sunda Islands, East Indies, composite strato volcanic island; E. Bikini Atoll, Pacific, basaltic guyot with reefcap; F. Guyot 29°N 153°E, Pacific, large guyot far removed from the eastern Pacific axis of spreading; G. Scripps Seamount, small basaltic guyot in the north-east Pacific near to axis of spreading; H. Mauna Loa, Hawaii, NE–SW section across this huge basaltic volcanic island (part of a larger group) which rises 4170 m above sea level and has a total relief of more than twice this above the ocean floor; even so, the largest Martian shield volcano, Olympus Mons, is some two and a half times this size.
*Source:* Menard, 1964, figure 4.8, p. 68.

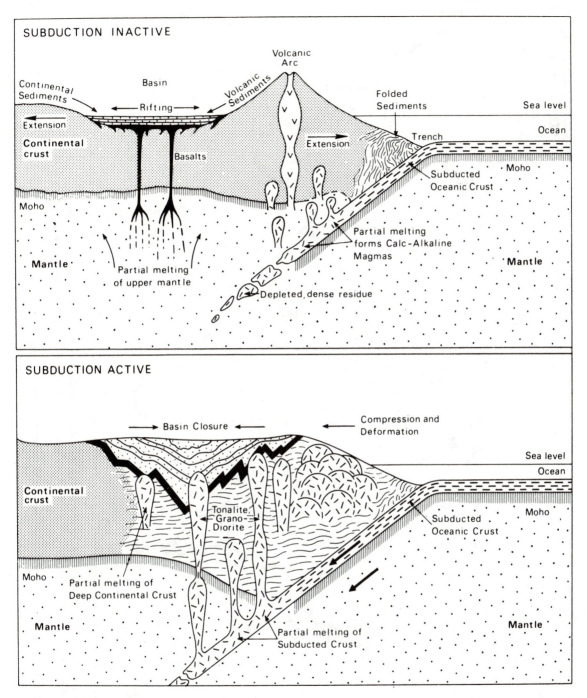

**Figure 6.4** Possible stages in the evolution of the subduction zone along the continental margin. Subduction weak or inactive; opening up of basin behind the island arc under tension which fills with sediments and basaltic lavas derived from the melting of the upper mantle; subduction actively resumes causing intense deformation of the basin, welding it to the old crust, together with massive intrusions of tonalite and granodiorite from the melted subducted crust.
*Source:* Moorbath, 1977, p. 103, copyright © *Scientific American*.

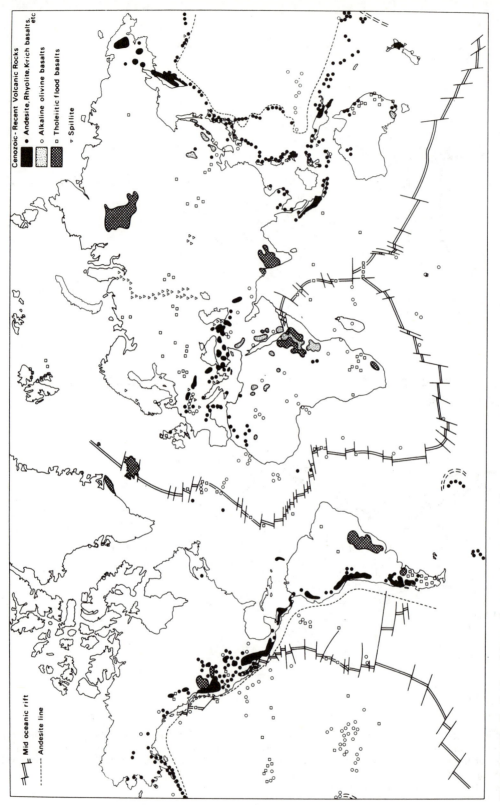

**Figure 6.5** The distribution of Cenozoic (Tertiary) to Recent volcanic extrusive rocks of the world, divided into the andesite group, alkaline olivine basalts, tholeiitic flood basalts and spillites. Centres of oceanic spreading and the andesite line are also shown.

*Source:* D. W. Hyndman, *Petrology of Igneous and Metamorphic Rocks*, 1972, frontis., copyright © McGraw-Hill, New York, by permission.

Figure 6.5 shows the present distribution of igneous extrusive rock bodies of Cenozoic – Recent age. This pattern bears the stamp of the distribution of Cenozoic (i.e. Tertiary plus Quaternary) orogenic activity and plate tectonics and foreshadows much of the present distribution of active vulcanism, although it has been estimated that in the last 100 million years there have existed hundreds of thousands of volcanoes on the continents alone (also all the Pacific Ocean volcanoes and seamounts are younger than 100 million years).

## 6.2 Intrusive constructional forms

### 6.2.1 Plutons

The largest and most common plutons are the diapiric granite and granodiorite batholiths, having volumes in excess of 100,000 km³, which have risen to levels of

hydrostatic equilibrium within mountain ranges during their period of folding. In some of these ranges erosion has exposed the broadly domed heads of these plutons as extensive outcrops in positive structural locations elongated along the trend of the folded anticlinoria. The exposed batholith of the Coast Ranges of Alaska and British Columbia is 1650 km × 160 km in extent, and that of the Sierra Nevada 640 km × 88 km. In some, but by no means all, folded ranges the highest elevations are supported by batholiths, the Alaska Range rising to well over 6000 m and the Cordillera Blanca of Peru to 5346 m.

The internal structure of granitic plutons dominates the landforms which are developed on their exposures. Figure 6.6 shows the three dominant sets of initial joints and faults – a less steeply dipping cross-set P resulting from tension and stretching of the pluton head, and near-vertical cross Q and longitudinal S joints cutting each

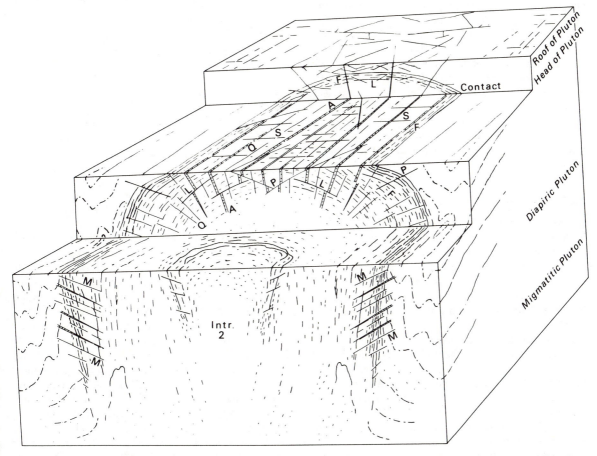

**Figure 6.6**  The major structural features near the top of a batholith: **A** = Aplite dikes: **F** = foliation and pseudo-bedding; **P** = tension joints; **Q** = cross-joints: **S** = longitudinal joints; **L** = exfoliation joints; **M** = marginal thrusts.
*Sources:* H. Cloos, combined from Rittmann, 1962, figure 101, p. 215 and Hills, 1972, figures XII-19 and XII-21, pp. 375, 377.

other almost at right angles. Within many of these vertical joints dykes of aplite (a fine-grained granite deficient in ferromagnesian minerals) have been intruded. The lateral·flow of the granite at the head of the pluton has resulted in a crystal orientation exhibiting primary flow structures giving a 'pseudo-bedding' forming dome structures. Granite is emplaced under enormous pressure, and because it is about twice as elastic as basic igneous ·rocks when erosion begins to remove the roof of the pluton, the uppermost 100 m or so of the granite expands to produce a secondary set of very low-angle exfoliation joints L varying in spacing from a few centimetres to tens of metres and taking advantage of the pseudo-bedding. The exfoliation joints, pseudo-bedding and the sets of vertical joints are dominant in controlling the production of domes and tors which, as will be seen, are so characteristic of granitic landforms. The cross- and longitudinal joints often impart a right-angular or trellis tendency to drainage patterns developed on exposed batholiths.

Basic plutons are represented by the lopolith (Greek: *lopos* – 'a basin'), which is a large saucer-shaped gabbroic intrusion ranging in size up to batholithic proportions. The gabbro tends to be layered and to give rise to a series of outward-facing small erosional scarps. The Duluth lopolith in Minnesota is 200 km long, some 40,000 km$^2$ in area, up to 16 km thick and has a volume of 200,000 km$^3$, but the largest of such features is the Bushfeld Complex in the Transvaal which covers 55,000 km$^2$.

### 6.2.2 Smaller intrusions

On the margins of larger intrusions, near the surface and in the roots of volcanoes smaller-scale intrusions occur, the dimensions and geometry of which depend partly on that of primary sedimentary and on secondary stress fracture structures, and partly on the viscosity of the intruded magma. It is convenient to divide these intrusives into those which are concordant and those which are discordant (Figure 6.7).

Concordant intrusions occur when igneous rocks have been intruded as layers within the existing bedding. *Sills* are thin sheetlike intrusions of igneous rock injected between bedding planes, being most commonly associated with less competent rocks such as shales (i.e. rocks which will deform relatively easily under stress). In places sills may cut across more competent beds (i.e. more resistant rocks) as *dykes* (US: dikes) and may be subsequently folded or faulted. Sills vary greatly in size and composition. Smaller sills are usually composed of acidic rocks (e.g. quartz porphyry) but some composed of basic

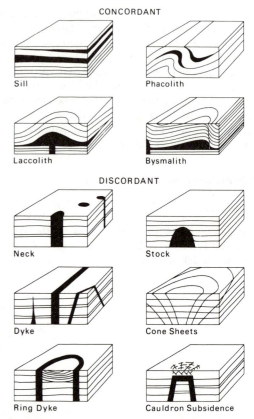

**Figure 6.7** The main types of igneous intrusive rock bodies divided into concordant and discordant types (the block diagrams are not necessarily drawn to scale).
*Sources:* Ollier, 1969, figure 23, p. 8; B. W. Sparks, *Rocks and Relief*, 1971, figure 3.9, p. 92, copyright © Longman, London.

rocks of low viscosity (e.g diabase) may be very extensive and may be looked upon as the intrusive equivalents of large-scale basic lavaflows. The quartz-diabase Whin Sill (Figure 6.8) intruded into Carboniferous rocks underlies an area of some 40,000 km$^2$ in northern England, varying in thickness from 5 m to 60 m and averaging 30 m. Because most intrusive igneous rocks are more resistant to erosion than the country rocks in which they are intruded, when eroded they produce terrain features reflecting their structures. The exposed vertically jointed northern edge of the Whin Sill produces an extensive north-facing scarp along the crest of which Hadrian's Wall was constructed by the Romans to protect northern England from Scottish invasion. An even more extensive system of basic sills forms part of the Karroo Series of South Africa underlying 500,000 km$^2$, with individual sills exceeding 500 m thick and together with swarms of smaller sills giving an overall intrusive rock volume of

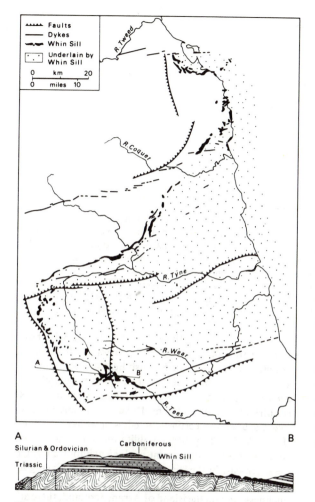

**Figure 6.8** Extent and outcrop of the Great Whin Sill of northern England, together with its four *en échelon* series of contemporaneous dykes of Late Carboniferous age; the line of cross section AB is shown on the map.
*Source:* Holmes, 1978, figures 11.5 and 11.6, p. 164, copyright © Nelson, London.

200,000–400,000 km³. Sills, in common with most smaller intrusives, are often bounded by zones of contact thermal metamorphism, where the heat has recrystallized a zone of country rock varying in width from 1 cm to 100 + m. These zones also find topographic expression due to the excessive resistance of rocks, such as quartzite, marble and hornfels – the thermal metamorphic equivalents of sandstone, limestone and shale. A special type of sill is the *phacolith*, a lens-shaped concordant intrusive occupying a crest or trough in folded sedimentary rocks and elongated in the direction of the fold axis, as at Corndon Hill in the Welsh borderlands (Figure 6.9).

Another variation of the sill is the *laccolith* (Greek: *laccos* – 'a cistern'), which is thicker and commonly less extensive than the former, generally of an inverted-bowl or mushroom shape, above which the overlying rocks have been updomed. Laccoliths, which have a huge range of sizes, commonly result from the injection of more viscous acid magmas which tend to solidify rapidly and thicken up close to the magma source. They may be considered to be the intrusive equivalent of the acid lava dome. The type example of laccoliths are the Henry Mountains, Utah (Figure 6.10), formed by the intrusion of diorite porphyry into the shales of a flat-lying shale sandstone sequence of the Colorado Plateau of which the largest (Mt Ellen) rises some 1500 m above the plateau surface and has a basal diameter of 24 km. In fact, each peak is an intrusive complex consisting of up to a dozen smaller laccoliths radiating from a central *stock*, a small diapirlike discordant igneous intrusion punched up through the country rocks (Figure 6.11). Mt Hillers, the central stock in the group, has a partly exposed stock some 6 km in diameter. On the north-west side of Mt Ellen, Table Mountain is an example of a marginally faulted variant of the laccolith, the *bysmalith* (Plate 4).

The terminology of intrusions and other geologic features has at times, bordered on the grotesque. A wonderful example of the tongue-in-cheek use of jargon is provided by C. B. Hunt, who, during his investigations in the Henry Mountains of Utah, discovered an intrusive that resembled a saguaro cactus, and he describes it as follows:

> The feeder to the Trachyte Mesa laccolith has a distinctive form *and some may wish it named*. Because the form has certain resemblances to the woody structure of the cane cactus the name cactolith might be used and defined as a quasi-horizontal chonolith composed of anastomosing ductoliths whose distal ends curl like a harpolith, thin like a spenolith, or bulge discordantly like an akmolith or ethmolith.

Hunt managed to pass this beauty through the entire *US Geological Survey* editorial system!

A characteristic discordant intrusive rock body is the *neck*, or *plug*, which commonly formed the feeder-pipe for a volcano. These are circular or oval in section, range in diameter from a few metres to 1–2 km, and although some have near-vertical walls, the majority tend to get narrower with depth. Volcanic necks are composed of a variety of rocks but commonly exhibit columnar jointing which forms a more regular inverted-fan shape near the original top of the intrusion, tending to radiate outwards from the vertical axis normal to the cooling walls at lower depths (the Hopi type). Near the margins there may be

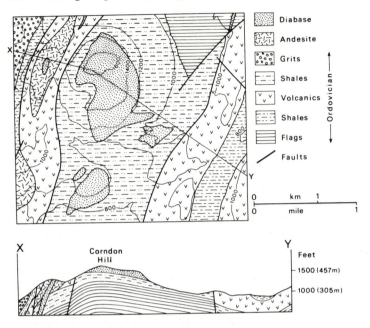

**Figure 6.9** The Corndon Hill phacolith near Montgomery in the Welsh borderland; the diabase phacolith is shown by the dotted symbol, and the line of cross section is shown on the map.
*Source:* Cross section adapted from B. W. Sparks, *Rocks and Relief*, 1971, figure 3.27, p. 94, copyright © Longman, London.

evidence of the slumping and stoping of country rocks (see later). Erosion commonly results in volcanic rocks standing up as striking residual landforms, in some cases as much as 600 m above the surrounding countryside. Hopi Buttes in Arizona comprise some 200 alkali-rich basic necks of near-surface crater fill and breccia, whereas the so-called Navajo type of neck is a more deeply eroded feature. Shiprock, New Mexico, is a classically jagged, dyke-ridden, tuff-breccia (Navajo type) neck with radial dykes, as is the nearby Agathla neck which is some 300 m high and of 1000 m diameter at its base. In the Old World, Castle Rock, Edinburgh, and the spectacular Le Puy at Velay, France, are notable necks.

*Dykes* are sheet-like intrusions of igneous rock filling mechanical fracture patterns (tensional or shear) and usually cutting the bedding of the country rocks at a high angle. Although dykes are seldom more than 60 m wide (an exception being the Great Dyke of Rhodesia, which is 3–13km wide at the outcrop and 540 km long), being commonly only 1–6 m wide, they may be very extensive as a result of highly fluid magma and/or great injection pressures. The thinness of dykes in comparison with sills, combined with their high angles, means that they have much less control over terrain than the latter. Most dykes are 'dilitational', due to the forcing apart of fracture walls under tensional stretching of the country rock, the sides of which are seldom greatly displaced vertically, except for the ring dykes which will be described shortly. Above updoming associated with igneous intrusion and volcanic activity radial tension fractures are formed

which may be filled with *radial dykes*. Figure 6.12 shows some of the 500 dykes radiating from small stocks of granite and diorite in the Spanish Peaks area, Colorado. These steeply dipping dykes are 1–30 m wide and have been eroded to form walls 20–30 m high extending across country for up to 16–24 km. Dykes commonly occur in *swarms*, as do the Tertiary tholeiites of western Scotland (Figure 6.13). The process of dilitation may, under these circumstances, be of extreme geological importance. On the south-east of the island of Mull 375 dikes (averaging only 1·75 m thick) in a distance of 20 km total 0·75 km in thickness, and on the island of Arran, Scotland, there are 525 diabase dykes in 24 km totalling 1600 m. However, in the east Greenland region thousands of dykes extend in a belt hundreds of kilometres long and the dyke swarms of the mid-oceanic divergence zones have added hundreds of kilometres to the width of the ocean floors in recent geological times.

Finally, two related systems of dykes, *cone sheets* and *ring dykes*, often occur in association as *ring complexes*. These are believed to form the roots of volcanoes (Figure 6.14) and to result from:

(1) upward percussion pressure of magma injection, nested conical fracturing of the overlying country rocks, uplift and emplacement of magma in the resulting cone sheets;

(2) subsequent subsidence of the central plug forming steeply outward-dipping cylindrical faults, the cauldron subsidence of the pistonlike masses and the emplacement of ring dykes in the fracture spaces.

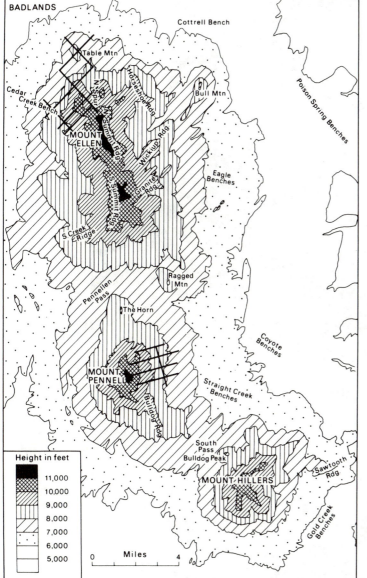

**Figure 6.10** The three most northerly of the Henry Mountains, Utah (Mt Ellen, Mt Pennell and Mt Hillers). The lines north-west of Mt Ellen and east of Mt Pennell indicate the isometric fence lines in Figure 6.11.

*Source:* Hunt, 1953, plate 17.

Cone sheets are nests of inward-dipping inverted cones averaging 45°, but varying from 70° near the centre to 30° at the margins. Individual sheets of intruded dykes are very thin (commonly less than 2 m and seldom more than 5 m thick) and are never complete around a given centre. Ring dykes are much thicker than cone sheets, measuring from a few hundred metres to some 2 km in width and mostly being composed of granitic or alkalic-silicic rocks. They do not commonly form a complete ring (see Figure 6.14).

## 6.3 Extrusive constructional forms

Although the intrusion of magma may deform the surface and the exposure of intruded material by later erosion lead to the development of distinctive igneous-rock landforms, the most spectacular primary igneous landforms are those associated with the breakout of igneous rocks to produce extrusive volcanic features (Plate 5). There are more than 800 volcanoes which are either active or dormant (as distinct from inactive or

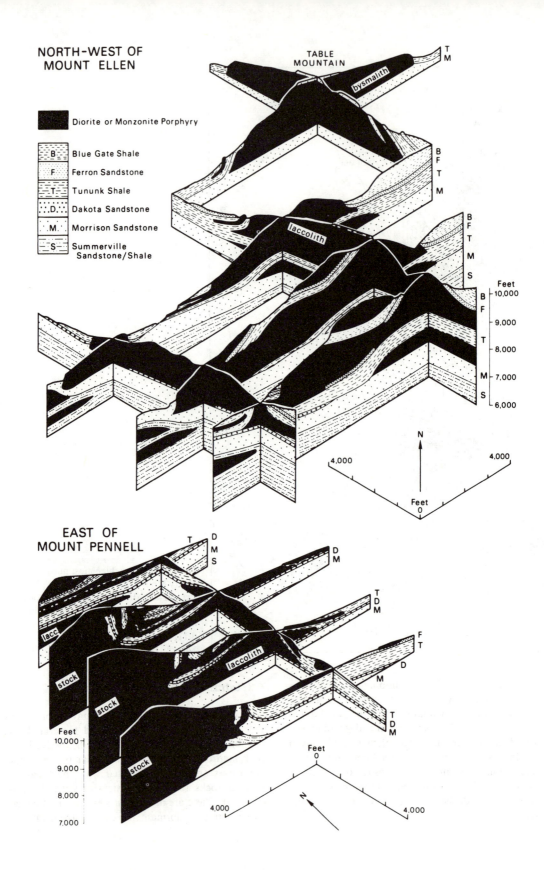

NORTH-WEST OF
MOUNT ELLEN

Diorite or Monzonite Porphyry

B   Blue Gate Shale
F   Ferron Sandstone
T   Tununk Shale
D   Dakota Sandstone
M   Morrison Sandstone
S   Summerville
    Sandstone/Shale

TABLE
MOUNTAIN

bysmalith

laccolith

Feet
10,000
9,000
8,000
7,000
6,000

N

4,000          4,000

Feet
0

EAST OF
MOUNT PENNELL

lacc

stock

stock

stock

laccolith

Feet
10,000
9,000
8,000
7,000

Feet
0

N

4,000          4,000

**Figure 6.11** Isometric fence diagrams showing the laccolithic structures of diorite or monzonite porphyry intruding the shales and sandstones north-west of Mt Ellen and east of Mt Pennell.
*Source:* Hunt, 1953, figures 39 and 47, pp. 108, 122.

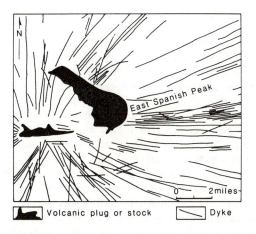

**Figure 6.12** Spanish Peaks, Colorado. These consist of two stocks of granite and granodiorite porphyry, rising to some 4180 m (13, 623 ft) from which radiate between 300–500 dykes around 360°. These dykes give the impression of walls which interrupt normal drainage. The Spanish Peaks lie on the north side of the Park Plateau on the Colorado–New Mexico boundary.
*Source:* Hills in Macdonald, 1972.

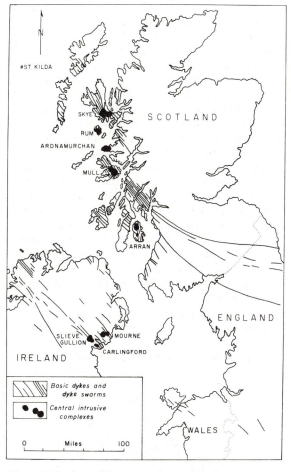

**Figure 6.13** Some dyke swarms form huge subparallel regional complexes around intrusive stocks, such as the Tertiary dyke swarms of western Scotland and northern Ireland.
*Source:* J. E. Richey, *Scotland: The Tertiary Volcanic Districts*, Geological Survey, Crown Copyright Reserved.

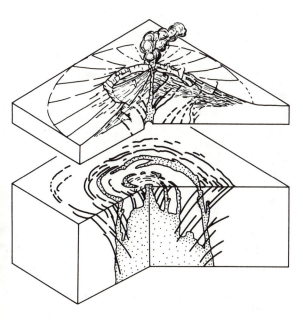

**Figure 6.14** Structure of a volcano, caldera and ring dyke complex. Surface structure – central cone, crater, caldera, vent, radial dykes, cone sheets and sills. Structure at depth – cone sheets and ring dykes.
*Source:* Hans Cloos.

extinct) (Figure 6.15), of which 80 per cent are located in island arcs or in recent folded mountain chains. Of these, 62 per cent are circum-Pacific (West Pacific Islands, 45 per cent; west coasts of North, Central and South America, 17 per cent), 14 per cent in Indonesia and 4 per cent in the Mediterranean and South-west Asia; 17 per cent of active volcanoes occur in the ocean basins (3 per cent in the central Pacific; 1 per cent, Indian Ocean; 13 per cent, Atlantic Ocean). Apart from the active volcanoes, the Pacific has some 10,000 seamounts (Figures 6.3 and 6.15) greater than 1 km in height and 100,000 smaller abyssal hills, which tend to lie in lines reflecting plate movements and transverse fractures. The active volcanoes in the Atlantic occur either along the mid-Atlantic Ridge zone of spreading (e.g. Iceland, Azores, Ascension, Tristan da Cunha) or in isolated groups (e.g. Canary Islands, Cape Verde Islands, St Helena). A mere 3 per cent of the active volcanoes lie in the continental interiors, particularly associated with the zone of spreading of the East African Rift system (Macdonald, 1972; Bullard, 1976).

It is possible to group the active, or recently active, volcanoes into four spatial classes.

(1) Volcanic arcs. These are probably early-stage sub-duction zones (some twenty-two in the west Pacific and East Indies: and four in the Lesser Antilles, Scotia, Tyrrhenian and Aegean Seas). The shape of each arc depends on the curvature of the tension fractures in the subducting plate and the line of volcanoes along each arc or chain may be separated into segments by gaps, offsets, or changes in the volcanic products. This latter feature suggests that the subducting plates may have been fractured into segments, some still actively descending and capped by active volcanoes, the others stationary and capped by inactive ones. Within the more or less straight volcanic segments varying from 100 km to 1100 km in length active volcanoes tend to be rather evenly spaced at an average of 50–60 km.

(2) Volcanic chains. These are straight lines of volcanoes within post-tectonic-stage folded mountains (e.g. the Andes). The volcanoes seem least active where subduction is most moribund (e.g. western North America) but more active in other regions (e.g. Central and South America). The volcanic belts are strongly segmented and, for example, seven clearly defined segments have been recognized in the region between Guatemala and Costa Rica. Active volcanoes tend to occur singly within belts of inactive ones, with an average spacing of some 80 km, although the spacing tends to be variable (in southern

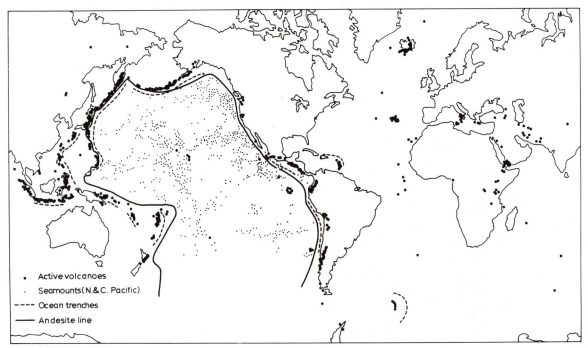

**Figure 6.15**  Distribution of active volcanoes (excluding one in Antarctica), major seamounts (north and central Pacific only), ocean trenches and the Andesite Line.
*Sources:* Holmes, 1978, figure 29.36, p. 662; G. A. Macdonald, *Volcanoes*, © 1972, figure 14-1, facing p. 346, and figure 14-4, p. 349, reprinted by permission of Prentice-Hall, Inc., Englewood Cliffs, NJ.

Peru it varies between 20 km and 80 km in as little a distance as 300 km).

(3) Volcanic clusters. These appear to be associated with the action of thermal plumes rising under relatively stationary plates, for example, groups of Atlantic islands. Each group of dominantly extinct volcanoes displays a tendency for some regularity of volcanic arrangement (e.g. Madeira, radial; Galapagos, rectangular) but their spacing may reflect the pattern of lithospheric rupture above the plume, which is dependent upon the lithospheric thickness – and, indeed, may be a direct reflection of it (e.g. Galapagos Islands 42 km spacing, Azores 52 km, Canary Islands 80 km, Ethiopia 9–130 km, East African Rift Valleys 30–53 km).

(4) Volcanic lines. These mainly include lines of dominantly extinct volcanoes and seamounts the linearity of which possibly reflects the influence of one persistent mantle plume on a lithospheric plate as the latter moves across it. The spacings of individual volcanoes in these lines are dependent both upon lithospheric thickness and the intensity of diapiric upwelling. Plate movements may thus explain the spatial development of volcanic centres through time in so far as some lines of volcanoes in the Pacific tend to be inactive at one end and active at the other.

Volcanoes exhibit significant differences in composition. Those of the oceanic ridges and sea floors are characterized by mainly non-explosive outflows of molten basaltic material, basalt being a dark-coloured, fine-grained crystalline rock of density 2·9–3·0. Vulcanism on the continents is much more varied but includes significantly more explosive eruptions of granitic material, granite being a light-coloured crystalline rock of density 2·7. On the continental margin of belts associated with island arcs and younger folded mountains with ocean trenches (e.g. the Andes) the vulcanism is explosive and continental in character but dominantly composed of andesitic material, the composition of andesite lying between that of granite and basalt. Seaward of the outer limit of this belt, termed the Andesite Line (Figure 6.15), there is a sharp transition to the characteristically basaltic rocks of the ocean bottoms, reinforcing the idea that the surface rocks of the continents and of the ocean basins are broadly different in composition.

### 6.3.1 Types of eruption

The solid products of volcanic activity can be broadly classified into fragmental material (tephra or pyroclastics) and lava. The percentage relation between the amounts of these two products expresses the degree of explosiveness of the eruption (Table 6.1). The explosiveness of the eruption depends largely on the viscosity of the lava which, in turn, is related to the rate of release of gas ($CO_2$, $SO_2$, etc.), water vapour and other volatiles from the magma near the surface. As the pressure of the magma drops less volatile material can be held in solution, and at a critical depth this material is released in the form of explosive gas bubbles. The rate of expansion of the bubbles is, reciprocally, controlled by the viscosity of the magma such that, for example, highly viscous rhyolitic material allows the pressure to build up to critical points at which especially intense explosions

**Table 6.1** Percentage of acid lavas

| | | (%) | |
|---|---|---|---|
| Effusive basalts and trachytes dominant | Central Pacific and Hawaii | 3 | Oceanic islands, primarily basic lavas |
| | Atlantic and Indian Ocean Islands | 16 | |
| | Canary Islands | 20 | |
| | Iceland | 39 | |
| Intermediate | Southern Italy, East-Central Africa, Ryukyu Islands | 40–45 | |
| | Kamchatka, Azores | 60 | |
| | Mexico | 70· | |
| | Celebes, Greek Islands | 80 | |
| Explosive Pacific type, andesites, dacites and rhyolites dominant | New Guinea, New Britain, Japan, Marianas, Andes | 90 | Island arcs and young fold mountains, primarily acid lavas |
| | Solomons, Aleutians, Lesser Antilles, New Hebrides | 95 | |
| | Indonesia, Central America | 99 | |

occur, whereas the eruptions of lower-viscosity basic magmas are characterized by low explosive activity. A geomorphic result of these complex pressure relationships is that extrusive forms produced by the eruption of acid lavas appear to be limited in height to 3–4 km, whereas landforms of basic lava may exceed 6·5 km in height.

Table 6.2 classifies eruptions in terms of the magma type, explosive activity, products and associated constructional landforms. The eruptions of low-viscosity basalts from fissure systems (Icelandic) or central vents (Hawaiian) are primarily responsible for producing lava plains or plateaus and large shield volcanoes (or lava domes), respectively. A variety of more explosive eruptions involving acidic material (Strombolian, Vulcanian, Vesuvian, Plinian and Peléean) may produce large stratified conical construction forms, which may from time to time suffer explosive destruction and collapse associated with Plinian- or Krakatoan-type eruptions. When the acid lava consists of solid particles suspended

**Table 6.2**  A classification of types of volcanic eruption

| | Eruption type | Physical nature of the magma | Character of explosive activity | Nature of effusive activity | Nature of dominant ejecta | Structures built around vent |
|---|---|---|---|---|---|---|
| Basic | Basaltic flood (Icelandic) | Fluid basic | Very weak ejection of very fluid blebs; lava fountains | Voluminous widespreading flows of very fluid lava from systems of fissures | Cow-dung bombs and spatter; very little ash | Spatter cones and ramparts; very broad flat lava cones; broad lava plains with cone building along fissures in terminal phases. |
| | Hawaiian | Fluid basic | Weak ejection of very fluid blebs; lava fountains | Thin, often extensive flows of fluid lava from dominant central vents | Cow-dung bombs and spatter; very little ash | Spatter cones and ramparts; very broad flat lava domes and shields. |
| Composite or strato | Strombolian | Moderately fluid, part acid, part basic | Weak to violent ejection of pasty fluid blebs | Thicker, less extensive flows of moderately fluid lava; flows may be absent | Spherical to fusiform bombs; cinder; small to large amounts of glassy ash | Cinder cones and lavaflows |
| | Vulcanian | Viscous acid | Moderate to violent ejection of solid or very viscous hot fragments of new lava | Flows commonly absent; when present, they are thick and stubby; ashflows rare | Essential, glassy to lithic, blocks and ash; pumice | Ash cones, block cones, block-and-ash cones, explosion craters |
| | Vesuvian (strong Vulcanian) | Viscous acid | Moderate to violent ejection of solid or very viscous hot fragments of new lava | Flows commonly absent; when present, they are thick and stubby; ashflows rare | Essential, glassy to lithic, blocks and ash; pumice | Ash cones and flows, explosion calderas |
| | Plinian (exceptionally strong Vulcanian) | Viscous acid | Paroxysmal ejection of large volumes of ash, with accompanying caldera collapse | Ashflows, small to very voluminous; may be absent | Glassy ash and pumice | Widespread pumice lapilli and ashbeds; generally no cone-building |
| | Peléean | Viscous acid | Like Vulcanian, commonly with glowing avalanches | Domes and/or short very thick flows; may be absent | Like Vulcanian | Ash and pumice cones; domes |
| | Krakatoan | Viscous acid | Highly explosive | Widespread ashfalls | Glassy ash and pumice | Huge explosion calderas |
| Acid | Rhyolitic flood | Viscous acid (fluid when not degassed) | Relatively small amounts of ash projected upward into the atmosphere | Voluminous widespreading ashflows; single flows may have volume of tens of cubic km | Glassy ash and pumice | Flat plain, or broad flat shield, often with caldera |

*Source:* G. A. Macdonald, *Volcanoes,* © 1972, table 10-1, p. 211, reprinted by permission of Prentice-Hall, Inc., Englewood Cliffs, NJ.

in gas, the resulting material is of very low viscosity and repeated eruptions of this type produce lava plateaus of rhyolite and related rocks.

### 6.3.2 Basaltic magmas

Basaltic lavas are of low viscosity, having no free silica, and may flow in a very fluid manner at extrusion temperatures of 1000°C–1100°C and velocities in excess of 30 m/s. As a consequence, individual flows may be very extensive (up to 100 km or more in length) but comparatively thin (normally a few metres, but occasionally very much thicker). The steplike eroded edges of thick sections of superimposed flows led to basaltic rocks being originally termed 'trap' (German: *trappen* – 'steps'). The surface of such flows presents a twisted or rolled crust (pahoehoe), resulting from the continual movement of the lava beneath the cooled surface skin. This internal movement may cause partial drainage of the interior of flows to produce lava caves and tubes, as well as the deformation of the surface into pressure ridges of up to 20 m high. Submarine eruptions of basaltic lava (mostly spillite – i.e. basalt with excessive sodium) produces a 'pillow' structure of sacklike ellipsoids due to the rapid surface cooling and continued movement of small individual parts of the flow. The slower terrestrial cooling of basaltic lava commonly results in the production of vertical columnar jointing, with the joints set at angles of approximately 60° and separating four-to eight-sided columns 1–20 m thick, as in the Giant's Causeway, Northern Ireland, and the Devil's Postpile in the Sierra Nevada.

The most extensive features associated with basaltic extrusion are the lava plains and plateaus resulting from the superimposition of vast series of individual flows of tholeiitic and olivine basaltic lavas from systems of surface fissures, some of regional extent. Such combinations of superimposed thin lava streams, individual flows being normally much less than 10 m thick but very rarely reaching 100 m, can form extensive plains by drowning areas of low relief and disrupting drainage. Vast and persistent regional flows can produce lava plateaus thousands of metres thick composed of superimposed flows and intruded sills and dykes (see later). Some especially important basaltic lava plateaus are:

(1) The Columbia River Plateau (Figure 6.16). This is composed of hundreds of individual flows of Miocene basalts, covering an area of more than 130,000 km² up to 2000 m thick in places. The plateau has been built on a relatively flat surface by extensive flows some of which exceed 160 km in length. One single flow near Roza in eastern Washington State covered 45,000 km².

(2) The Snake River lava plain (Figure 6.16) resulted from 50,000 km² of Quaternary basalts first filling a 200-m-deep canyon, and then overflowing a plain. A characteristic later feature of such eruptive phases was the building of lines of spatter cones and scoria mounds along large fissures, as in the Craters of the Moon National Monument.

(3) The Deccan Plateau in peninsular India consists of more than 500,000 km² of Eocene basalts, accumulating to a depth of 2000 m by flows of 1–60 m thick. Although the second largest basaltic plateau in the world, it is thought to have been originally almost twice as extensive as at present.

(4) The Paraná Plateau of southern Brazil and Uruguay is made up of Early Jurassic basalts, covering 750,000 km² to depths of up to 3000 m.

(5) Siberia, where Early Jurassic basalts cover 250,000 km².

(6) Iceland. Here more than 55,000 km² of young basalts are up to 5000 m thick in places. A most recent eruption by the 30-km-long Laki fissure in 1783 produced almost 10 km³ of lava, ending with the production of a line of lava cones.

Shield volcanoes or basaltic domes are huge structures made up of many flows originating from a location on the oceanic floor or mid-ocean ridge, which have formed a broad lava dome with lower slopes of 2°–4° and only 5°–6° near the summit plateaus on which are located unrimmed subsidence craters (small calderas) consisting of a sink or lava lake which periodically overflows with a minimum of explosive activity. Shield volcanoes are composed of wide effusions of alkaline olivine basalts, together with tholeiites, which produce large convex domes. Mauna Loa, Hawaii, rises 10,000 m from its 200-km-diameter base on the ocean floor to a broad plateau with an elliptical pit crater (dimensions 4 × 5·6 km) (see Figure 6.3H), and is only one of a complex of five superimposed domes (e.g. the flank volcanoes of Kilauea and Mauna Kea). Even these structures are quite small compared with the Martian shield volcano of Olympus Mons, which is 20 km high and has a basal diameter of 400 km. Smaller domes, seldom more than 1000 m and sometimes less than 100 m high (basal diameters of about twenty times their height and craters 100–2000 m diameter) and lacking summit plateaus, are found in other island locations such as Iceland. Smaller conical basalt lava cones with long concave lower flanks and upper slopes of more than 20° on scoria sometimes occur near the vents of basalt domes, as well as along fissure eruptions on basaltic lava plateaus (Macdonald, 1972).

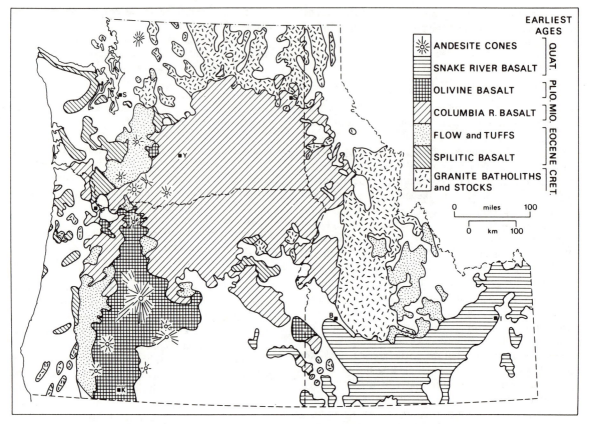

**Figure 6.16** The basalts of the Washington, Oregon and Idaho area (simplified from the *Geologic Map of North America*, 1965).
*Sources:* Hyndman, 1972, figure 4-18, p. 167, and Geological Society of America.

### 6.3.3 Acidic magmas

Granite magmas, extruded at 700°C–900°C, are much more viscous than basaltic ones and form relatively short and thick lava flows (coulées) with blocky clinker or slaggy surfaces (aa lava), having a local relief of 5 m or more between hollows and ridges. Individual flows may be more than a hundred metres thick and advance at rates of up to 10 m/h on flat surfaces, but occasionally reaching 300 m/h on slopes. The toe of an aa flow may be 20 m or more high, grinding along by slumping like a pile of clinker pushed by a bulldozer. Pyroclastic material (tephra) is commonly derived from lava shattered by the explosive production of gas bubbles and is either deposited as foreset breccias dipping at about 25° away from the central vent or more widely distributed as ash falls (ash <4 mm diam.; lapelli 4–32 mm; bombs >32 mm). Andesitic material in particular is associated with the production of much fine ash, together with lapelli and bombs. Dacitic and rhyolitic material pro-

duces much pumice, together with very short and thick lavaflows – one rhyolite flow near Mono Craters, California, is some 230 m thick and only 1.5 km long. Trachyte is less viscous than dacite and rhyolite because it is deficient in free silica and contains more dissolved gas; it may even exhibit pahoehoe surface structures.

Cumulo lava domes are produced by the extrusion of highly viscous rhyolite and dacite lavas from a central vent to generate steep-sided forms which are commonly broader than they are high. The crest of Lassen Peak, California, is a cumulodome 800 m high and 2·5 km in basal diameter. Similar features are Montagnone Maschiatta on the island of Ischia in the Bay of Naples (300 m high, 600 × 1200 m in elliptical plan, with a subsidence crater 400 × 550 m diam. and 80 m deep) and the Grand Sarcouri, Auvergne (260 m high and 800 m diam.). Some cumulodomes take the form of thick, convex, mesalike lavaflows (Figure 6.17), as with obsidian flows at Mono Craters, California. An example of the development of a cumulodome is that of Showa Sin-zan

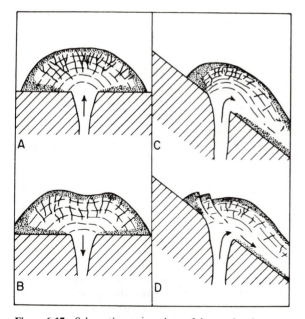

**Figure 6.17** Schematic cross sections of domes showing flowlines and brecciated margins (stippled); the arrows indicate the direction of lava movement: A. dome built on a nearly horizontal surface; B. the top of dome has sagged due to drainage of magma back into the vent; C. dome built on a sloping surface; D. summit of dome broken by faults due to downslope movement.
*Sources:* Modified after van Bemmelen, 1949, 1970 (Netherlands Government Printing Office), from Macdonald, 1972, figure 6-1, p.109.

dome at Usa volcano in Hokkaido, Japan, where earthquakes began in December 1943. By the end of June 1944 the local land surface had risen by 50 m when the dome broke the surface, and by September 1945 a feature more than 300 m high had developed. Where domes grow in a pre-existing crater, the term 'tholoid' (Greek: *tholos*, – 'domed building') is employed, and successive flows of trachyte generate steep-domed 'mamelons'. Where the magma is particularly viscous, it may be extruded from a central vent as an upstanding pistonlike 'plug dome' up to more than 300 m high and having an onion-like structure (e.g. Mt Edgcombe, New Zealand) – these are related to intrusive laccolithic structures. Smaller plugs form 'spines', such as the Puy de Dôme, Auvergne (more than 500 m high), and the famous spine of Mt Pelée, Martinique, which grew at a rate of up to 25 m per day in 1902 reaching a height of more than 300 m in nine months, after which it was rapidly eroded away.

Pyroclastic features occur in a wide variety of sizes with cinder, ash, tephra, pumice, or (more rarely) scoriae cones ranging from groups 15 m high to single conical

structures approaching 500 m high. At the latter height slumping appears to limit the continued increase of height of purely pyroclastic cones, but many much larger dominantly pyroclastic strato or composite cones (i.e. partly supported by interlayered lava flows) occur, like those in Java which have a lava/pyroclastic ratio of 1: 370. Mt Agua in Guatemala (rising to more than 3 km) is an example of a large stratovolcano with a very high ash content. Ash and cinder cones have concave repose slopes of 30°–35°, steepening to about 40° near the central vent where the coarsest material is found. These cones can grow initially at a rapid rate and Monte Nuovo, in the Bay of Naples, grew to 140 m during a few days in 1538, since when it has been quiescent. A recent example

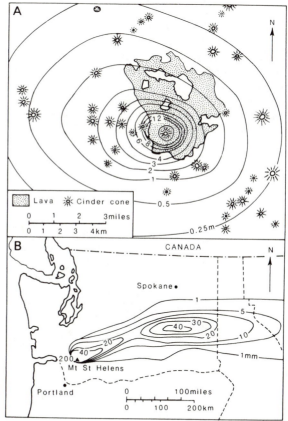

**Figure 6.18** Distribution of volcanic ash. A. Ash growth (in metres) around the Parícutin Volcano from its inception in February 1943 until October 1946, by which time the central ash cone was some 350 m high. It ceased to grow in 1952 when it was 410 m high. The extent of the lava flows is shown in August 1947. B. Ash fall (in millimetres) resulting from the eruption of Mt St Helens, Washington, on 18 May 1980.
*Sources:* H. Williams, *U.S.G.S. Bull.* 965-B, 1950, plate 9; Foxworthy and Hill, 1982, figure 35, p. 68.

of such growth is that of Parícutin, Mexico, which grew to about 140 m in one week and to about 300 m in two months (Figure 6.18A). The finer products of pyroclastic eruptions may be carried considerable distances by air circulations depending on the force of the eruptions and the wind velocity (Figure 6.18B).

The large strato or composite volcanoes are conical features with concave slopes of 20°–30° built of rhyolitic or andesitic lava and pyroclastics round a central vent (Figure 6.19). The most perfect forms are represented by Mt Mayon, Luzon (2421 m high) and El Misti, Peru (3450 m). After the eruption of some 450–500 m of volcanic rocks around a single vent, there is a tendency for the main cone to continue to grow by the addition of superimposed parasitic cones. Most of the huge stratovolcanoes, measuring up to 30 km in diameter and 3650 m in height, consist of such superimposed structures: Mt Etna, Sicily (3250 m high), has more than 200 parasitic cones, and Mt Fuji, Japan (3750 m), consists of at least three superimposed volcanoes (Figure 6.20). A notable exception is Cotopaxi, Ecuador (having a relief of 3000 m and an absolute elevation of 5896 m), which has erupted through a single vent and is the world's highest active volcano. A characteristic feature of the stratovolcano is the central crater formed by explosion or by the collapse of the magma chamber which is of the order of 400,m (e.g. Ngauruhoe, New Zealand) or 500 m in diameter (e.g. Mt Mayon, Luzon) (see Section 6.4). More violent explosions cause the upper parts of the magma chamber to be blown clear, and general collapse to occur, producing a summit caldera, such as is evidenced by Monte Somma, encircling the central peak of Vesuvius (Figure 6.21).

The last major constructional feature of acid extrusive activity is that produced by the ashflow. These consist of solid particles of volcanic glass, crystals and pumice suspended in gas which flows like a fluid of low viscosity due to excessive gas expansion. The most striking example of this is the nuée ardente, a glowing avalanche, one of which flowed for some 5·6 km at a speed of up to 160 km per hour following the Mt Pelée eruption, on Martinique, on 8 May 1902. The resulting rock of rhyolite, dacite, andesite, or trachyte composition is termed *ignimbrite*. Ignimbrites can be relatively soft, occurring as bedded or chaotic tuffs, but are often found in the form of welded tuffs. Welding occurs when the glassy particles are so hot when they come to rest that they become welded together, expelling the gas, and compact to produce a very dense, homogeneous rock of low porosity, susceptible to marked columnar jointing,

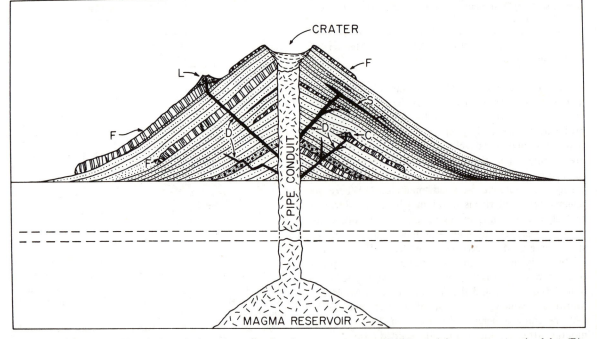

**Figure 6.19** Diagram of a typical composite volcano showing the cone, crater, central pipe conduit, magma reservoir, dykes (**D**), a dyke conduit feeding a lateral cone (**L**) and lava flow (**F**), a buried cinder cone (**C**) and a sill (**S**). Tephra layers are shown by broken lines, breccia layers are marked with small triangles and lava flows (**F**) are irregularly cross-hatched.
*Source:* G. A. Macdonald, *Volcanoes,* © 1972. Reprinted by permission of Prentice-Hill, Inc., Englewood Cliffs, N.J.

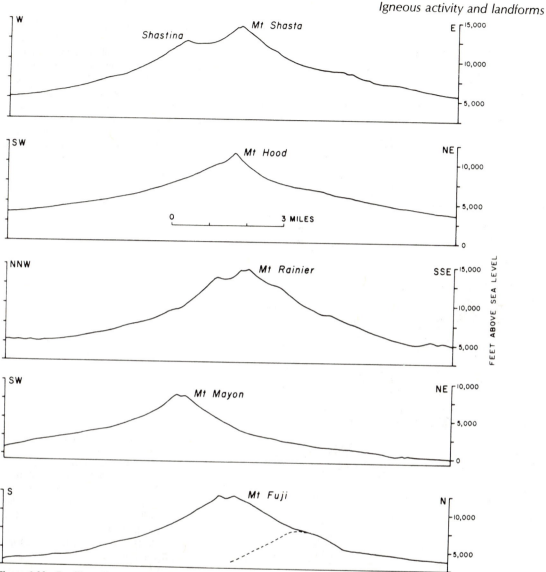

**Figure 6.20** Profiles of large symmetrical composite volcanoes. Shastina is a younger cone on the flank of Mt Shasta. The hump on the north-west side of Mt Rainier is part of the rim of an old crater largely buried by the younger summit cone. The bulge on the north side of Mt Fuji is caused by an older volcano now largely buried by Fuji. Note that the relief of Mt Rainier compares with that of Mt Etna (see Figure 6.3 B).
*Source:* G. A. Macdonald, *Volcanoes,* © 1972, figure 11-11, p. 284. Reprinted by permission of Prentice-Hall, Inc., Englewood Cliffs, N.J.

especially in the lower layers. Ashflows are erupted in great quantity from vents or systems of fissures – one emission of Katmai, Alaska, produced 28 km³ of rock in sixty hours (Figure 6.22). Individual flows may be many tens or even hundreds of metres thick. Some individual Tertiary ignimbrite flows in Nevada covered up to 16,000 km² and were as much as 60 m thick. Repeated ashflows characteristically produce flat-topped plateaus

with abrupt eroded edges. Two of the largest existing ignimbrite plateaus are in the North Island of New Zealand (20,000 km²; 8000 km³) and in Sumatra (20,000 km²; 2000 km³) (see Figure 6.5) (Macdonald, 1972).

### 6.4 Igneous tectonism

Igneous tectonism occurs when igneous activity at depth

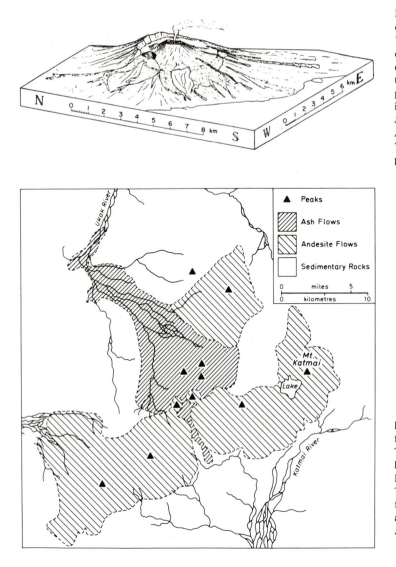

**Figure 6.21** Monte Somma–Vesuvius, a composite stratovolcano. The actual cone of Vesuvius has been built up in the summit caldera of Monte Somma since the Plinian eruption of the year AD 79. To the north, the crater edge of Monte Somma is still well preserved; to the south, on the other hand, it is buried under the lava flows of Vesuvius. *Source:* A. Rittmann, *Volcanoes and Their Activity*, 1962, copyright © Ferdinand Enke Verlag and John Wiley and Sons, by permission.

**Figure 6.22** Relatively minor rhyolitic flood eruption in the Valley of Ten Thousand Smokes, north-west of Mt Katmai, Alaska, in 1912; 25 cm of ash fell at Kodiak, a hundred miles from the vents. This was followed by ashflow deposits which filled the valley to a length of 22 km and an average depth of 30 m. *Source:* G. H. Curtis, *G.S.A. Mem.* 116, 1968, figure 1, p. 160.

results in the fracturing and displacement of surface country rocks or of previously deposited volcanic material. The most characteristic of such features are *calderas* (from Spanish, 'cauldrons').

Basic shield calderas are large circular or elliptical sunken craters on the crests of volcanic domes bounded by high-angle faults without raised rims. Examples of basic volcanic calderas are those of Mauna Loa (diam. 6 × 3 km; depth 200 m) and Kilauea (diam. 4 × 3 km; depth 120 m), while on the Martian shield of Olympus Mons the largest of the intersecting summit calderas is 65 km across.

The calderas of composite stratovolcanoes are larger than those of shield volcanoes (commonly 10–20 km diam.: e.g. Aso, Japan, 20 km, but La Garita, San Juan Mountains, Colorado, is 45 km diam.) and are formed by collapse of a magma chamber drained by the eruption of magma and pyroclastics, particularly in the form of nuées ardentes, ashflows and ignimbrites, although a lesser amount of material may be removed by explosions. Calderas are usually formed late in the history of a composite volcano due to collapse of a magma chamber near the surface, producing a series of inward-dipping normal faults (Figure 6.23) sometimes involving almost the whole of the original volcano (e.g. Krakatoa). They exhibit strong negative gravity anomalies due to the deep fills of collapsed lighter surface material and some of them (e.g. Messum Complex, South-West Africa) are clearly associated with cauldron subsidence. Some 'resurgent' calderas subsequently become updomed, refaulted and the seat of renewed volcanic activity as, for example, Valles, 50 km north-west of Santa Fé, New Mexico (diam. 19 × 24 km; floor 600 m below the rim) (Figure 6.24). Macdonald (1972) classifies composite volcano calderas into:

(1) Krakatoa type – the collapse of a mature volcano due to eruptions of pumice falls and flows from the main vent.

(2) Katmai type – due to magma drainage through adjacent conduits, rather than through the main vent itself.

(3) Valles type – collapse following the discharge of large volumes of ash and pumice flows from fissures unrelated to the pre-existing volcano.

Five examples of such calderas will serve to show their range of characteristics. Monte Somma, Vesuvius, with a diameter of 3 km and formed by the eruption of 79 AD has already been depicted in Figure 6.21. Crater Lake, Oregon, was formed by the eruption of a 3600-m cone in about 6000 BP, leaving a caldera of 10 km diameter now partially filled by a lake some 600 m deep in places above the level of which is a 150–600-m high rim and a more recent circular volcanic cone of Wizard Island rising 230 m above lake level (Figure 6.23E). Three well-documented calderas were formed in the nineteenth century: the Krakatoa caldera of 6 km diameter was formed from the 1883 eruption of an 1800-m-high Pleistocene volcanic cone; the Coseguina caldera, Nicaragua, of 4 km diameter and 450–600 m deep was formed by the collapse of about one-half of the mountain in 1835; and the Bandai San caldera in Japan was formed by a 400-m subsidence of 2·5 km diameter in 1888 accompanied by violent explosions, but no solid or liquid products.

Beside the circular collapse calderas associated with crests of volcanoes, there are other tectonic faulted features resulting from vulcanism. At a smaller scale

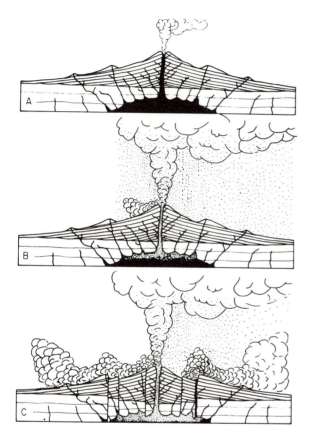

**Figure 6.23** Diagrams illustrating the formation of the Crater Lake caldera, Oregon: A. before the eruption; B. during the early stage of the caldera-forming eruption, showing a volcanian ash cloud and a small ashflow from the central vent: C. during the climax of the eruption, with big ashflows issuing from the central vent and from ring fractures part-way down the slope while the summit starts to sink as a series of big blocks: D. after the eruption; E. in its present state, with new eruptions on the floor and the caldera partly filled with water. The present Crater Lake is 8–10 km across and 1130 m deep from the caldera rim to the lake bottom. Wizard Island is a small cinder cone formed in Post-Pleistocene times.
*Source:* Macdonald, 1972, figure 12-5, p. 301, modified after H. Williams, Carnegie Inst. Washington, Pub. 540, 1942.

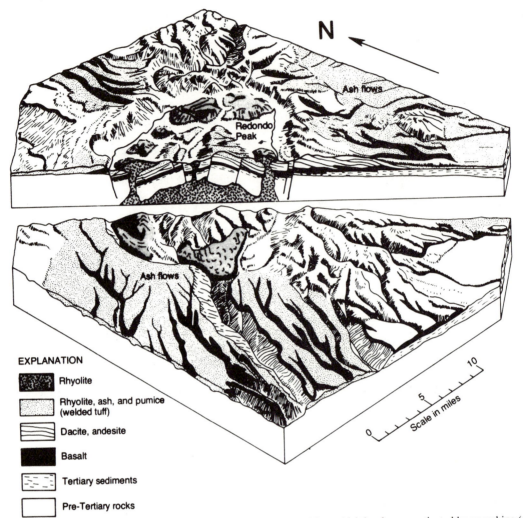

**Figure 6.24** Valles caldera, north-central New Mexico, an example of a caldera which has been reactivated by uparching (e.g. resurgent caldera) with the secondary formation of some ten large rhyolitic domes (up to 3 km in diameter and 60 m in relief) surrounding Redondo Peak.
*Source:* F. M. Bullard, *Volcanoes of the Earth*, 1976, figure 14, p. 107, copyright © University of Texas Press, modified after Smith and Bailey.

there are faulted troughs produced by the stretching of expanded shield volcanoes leading to local subsidence which is sometimes difficult to estimate because of subsequent lava filling. On Kilauea these troughs are up to 2 km wide and several kilometres long, on Fayal in the Azores a similar trough is 16 km long, 6–10 km wide and 150 m deep, whereas the whole volcano of Renggit in east Java is completely cut by a fault trough. Larger-scale intrusion at depth may updome an area to produce a faulted jumble of surface rocks, as did the Early Quaternary intrusion of trachyte beneath the island of Ischia in the Bay of Naples. Linear volcanic updoming is believed

to have produced the long faulted troughs of Rotorua–Taupo, North Island, New Zealand (almost 100 km long and 20–30 km wide) and that of Lake Toba, central Sumatra (100 km long, 13 km wide, and 425 m deep).

At this point it is convenient to draw a parallel between volcano–tectonic features and those associated with the removal or emplacement of other mobile earth materials. Subsidence due to mining, the tapping of oil and the pumping of groundwater has been on a sufficient scale to produce effects of geomorphic importance. For example, in western Fresno County, California, some

2850 km² has subsided at least 0·3 m (1–1·5 m common, locally up to almost 7 m) due to the withdrawal of groundwater leading to subsidence and collapse of the upper 60 m of sediments with cracks 6 m deep. Between 1933 and 1946 parts of Long Beach Harbour, California, subsided as much as 1·4 m due to oil extraction (see Section 10.3). Salt domes, in particular, provide small-scale analogies with certain volcanic phenomena. These are diapirs of mobile salt punched up through several kilometres of overlying sediments, which are especially common in the western Gulf Coast of the United States (over 300 located) and in the Persian Gulf region. The domes are roughly circular with diameters of 1–10 km (mean 3 km) which commonly produce updoming of up to 30 m of the overlying rocks, together with intersecting concentric and radial terrain fractures giving 70°–30° normal faulting. In places secondary solution of the salt near the surface has led to closed surface collapse hollows, but such collapse is more spectacularly associated with folded anticlinal structures involving thick salt beds. Such a region lies on either side of the La Sal Mountains laccolith group on the Utah–Colorado

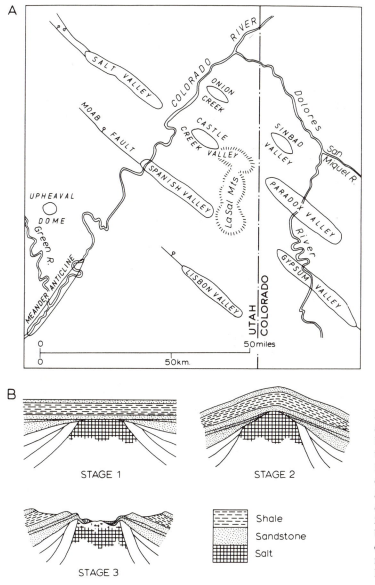

**Figure 6.25** The development of salt anticlines in western Colorado and eastern Utah: A. map of the principal structures due to salt flowage and solution (after Stokes); B. three stages in the proposed development of salt anticlines (after Stokes).

*Source:* A. J. Eardley, *Structural Geology of North America*, 1962, 2nd edn, figures 26.12 and 26.13, p. 417, copyright © 1951 by Harper and Row, Publishers, Inc., copyright © 1962 by A. J. Eardley, reproduced by permission of Harper & Row, Publishers, Inc., after Stokes.

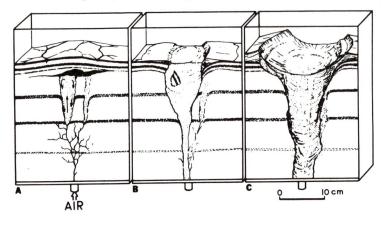

**Figure 6.26** Diagrammatic series showing principal features developed in model diatremes. A. Compressed air entering base of container propagates fractures in cohesive clay or marble-dust mixture: doming of the surface with ring and radial fractures results from uplift. B. Breakthrough to surface occurred above main fluidization cell; surface layers display inward dip on right and detachment and descent of a surface slab on left; a ring of bedded ejecta has begun forming. C. Continued bubbling circulation has built a larger bedded ejecta ring with rotational slumping into crater; bedded ejecta is carried downward during continued bubbling fluidization accentuating saucerlike form by drag along the pipe walls. *Source:* Woolsey *et al.*, 1975, figure 7, p. 40.

border (Figure 6.25), comprising eight elongate anticlines modified by collapse and faulting due to salt solution at depth. The largest of these features is the Paradox Basin, which forms a topographic valley 240 km long and up to 110 km wide.

Cryptovolcanic features are circular low domes or depressions, some 3–12 km in diameter, with little or no ejecta, believed to result from explosions of volcanic gases near the surface. Eight have been identified in the United States. Some consist of a central uplift with a marginal depression cut by a radial and concentric pattern of normal tension faults, but others like the Rieskessel basin in Germany have a calderalike expression. Smaller structures are the roughly circular explosion craters or *maars* (German, 'lake') which generally have diameters of 500 m to about 1 km, are less than 150 m deep and have low raised rims of 4° outward slope. Maars are quite widespread in the Eifel district of Germany, New Zealand, Luzon and elsewhere, but an especially large one is Hole-in-the-Ground, Oregon (1500 m diam. and 130 m deep). It has been suggested that some maars may have been formed by the explosive mixing of rising magma and groundwater, or by degassing to produce diatremes and flat-floored fluidization craters (Figure 6.26).

Some depressions are the result of meteorite impacts.

Fifty-three impact craters have been identified on earth (Canada 15; United States 14; Africa 10; Europe 7; Australia 5; rest 2), ranging in diameter from 2 m to 64 km. The majority of these are located on Pre-Cambrian shields or on stable Paleozoic structures because of their long exposure to potential impact events and to their tectonic inactivity. However, Meteor Crater, Arizona, is a 1·2-km-diameter, 120–140 m-deep hole with a raised rim that was punched into Kaibab Limestone and Coconino Sandstone during the Pleistocene. It is floored with more than 200 m of breccia and meteorite material. The Guelb er Richât structure in the Dhar Adrar, Mauritania, is more than 50 km in diameter, forming a nest of concentric rings in Early Paleozoic rocks. This is similar to the larger multiringed concentric lunar impact craters which are generally in excess of 200 km diameter, an example of which is Orientale (620 km diam.) and from which material was ejected as far as 300 km by the impact which caused the melting of 75,000–255,000 km³ of lunar rocks. Smaller lunar impact craters, such as Copernicus, Aristarchus and Tycho, exhibit concentric rings due to crater wall slumping. We are, however, only at the earliest stages in attempts to draw helpful analogies between the geomorphic features of the earth and those of extraterrestrial landscapes.

# References

Bullard, F. M. (1976) *Volcanoes of the Earth*, Austin, Texas, Texas University Press.

Clapperton, C. M. (1977) 'Volcanoes in space and time' *Progress in Physical Geography*, vol. 1, 375–411.

Cotton, C. A. (1944) *Volcanoes as Landscape Forms*, Christchurch, NZ, Whitcombe & Tombs.

Dennis, J. G. (1972) *Structural Geology*, New York, Ronald Press.

Eardley, A. J. (1962) *Structural Geology of North America*, 2nd edn, New York, Harper & Row.

Foxworthy, B. L. and Hill, M. (1982) 'Volcanic eruptions of 1980 at Mount St Helens: the first 100 days', *US Geological Survey Professional Paper* 1249.

Francis, P. (1976) *Volcanoes*, Harmondsworth, Penguin.

Gass, I. G., Smith, P. J. and Wilson, R. C. L. (1972) *Understanding the Earth*, 2nd edn, Chichester, Artemis Press.

Hills, E. S. (1972) *Elements of Structural Geology*, 2nd edn, London, Chapman & Hall.

Holmes, A. (1978) *Principles of Physical Geology*, 3rd edn, London, Nelson.

Hunt, C. H. (1953) 'Geology and geography of the Henry Mountains region, Utah', *US Geological Survey Professional Paper* 228.

Hyndman, D. W. (1972) *Petrology of Igneous and Metamorphic Rocks*, New York, McGraw-Hill and Stuttgart, Ferdinand Enke Verlag.

Macdonald, G. A. (1972) *Volcanoes*, Englewood Cliffs, NJ, Prentice-Hall.

Menard, H. W. (1964) *Marine Geology of the Pacific*, New York, McGraw-Hill.

Moorbath, S. (1977) 'The oldest rocks and the growth of continents', *Scientific American*, vol. 236, no. 3, 92–104.

Ollier, C. D. (1969) *Volcanoes*, Canberra, Australian National University Press.

Ringwood, A. E. (1974) 'The petrological evolution of island arc systems', *Journal of the Geological Society, London*, vol. 130, 183–204.

Rittmann, A. (1962) *Volcanoes and their Activity*, New York, Wiley.

Sparks, B. W. (1971) *Rocks and Relief*, Longman, London.

Vogt, P. R. (1976) 'Plumes, subaxial pipe flow, and topography along the mid-oceanic ridge', *Earth and Planetary Science Letters*. vol. 29, 309–25.

Woolsey, T. S., McCallum, M. E. and Schumm, S. A. (1975) 'Modeling of diatreme emplacement by fluidization', *Physics and Chemistry of the Earth*, vol. 9, 29–42.

Wyllie, P. J. (1971) *The Dynamic Earth: Textbook in Geosciences*, New York, Wiley.

Wyllie, P. J. (1976) *The Way the Earth Works*, New York, Wiley.

# Seven *Structure and landforms*

The distinctive characteristics of landscape are commonly a complex response to variations in rock type (lithology), to primary structures within the rock units (see Chapters 4 and 6), to secondary structures involving groups of rock units mainly due to diastrophic processes, to the effects of different exogenic processes and to the geomorphic history. This chapter is devoted to the production of large-scale secondary rock structures and to the influence exercised by these structures over the resulting terrain. For example, the twenty-three major physiographic provinces of the United States, together with many of the seventy-seven secondary sections (Figure 7.1), are mainly distinguished on the basis of their secondary structural characteristics.

The deformation of rocks which produces secondary structures ranges from gentle uplift and doming to intense lateral deformation. Each set of such secondary structures finds its place in the tectonic framework set out in Chapter 5 and is explainable not only in terms of the applied secondary stresses, but in the light of the previous sedimentary history of the area. Uplift mechanisms can be categorized conveniently as eustatic, isostatic, tectonic and orogenic (see Chapter 5), while recognizing that a given movement cannot always be unambiguously ascribed to any one of them. Of these mechanisms, only the tectonic produces well-marked structural deformation; isostatic and orogenic are clearly related in the post-tectonic history of orogenic belts; and only the eustatic can be said to occur without some structural deformation.

## 7.1 Horizontal and domed structures

When a valley or canyon is cut into uplifted, flat-lying, sedimentary rocks, stepped valley sides appear as the steep, often vertical, outcrops of the more resistant sand-

stones and limestones separated by the less steep shale slopes. The classic example of this is the Grand Canyon of the Colorado River (see Figure 3.13A). The retreat of these exposed sedimentary edges takes place by a variety of processes which we will return to on several occasions in this book, but these processes are commonly focused on or near the bedding planes separating the overlying sandstones and limestones from the underlying shales. In this zone where groundwater is moving horizontally in the overlying permeable rock, chemical weathering and spring sapping is concentrated and, in addition, surface erosion, flowage and slumping of the shales may cause the more competent overlying beds to be undermined and to collapse.

The retreat of the edges of horizontal sedimentary sequences produce, first, terraces and structural benches, such as the Tonto Platform (due to stripping back to Tonto Shales) and the Esplanade (supported by the Redwall Limestone) in the Grand Canyon, and then widening stripped and structural plains (Figure 7.2) which may be dotted with isolated residuals of larger (*mesa* – Spanish, 'table') or smaller (*butte*, from butt of a tree) size. Such residuals or outliers occur, albeit in a less stark form, in humid regions, such as Brent Knoll, Somerset, and Bredon Hill, Worcestershire. Some of these structural surfaces are very extensive, for example, covering thousands of square kilometres in Southern Africa, and it is a matter of current dispute as to whether this scale of structurally controlled stripping is possible of accomplishment at elevations well above sea level (see subject of pediplains, Section 2.4.2). Three examples of structural surfaces are the Edwards Plateau of Texas supported by the Edwards Limestone (about 300–1000 m above sea level and 130–400 m above the surrounding countryside); Salisbury Plain, England, supported by the flinty and permeable Echinoid Chalk from which the

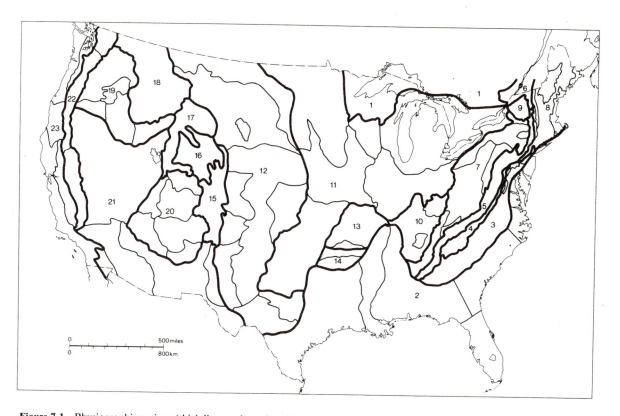

**Figure 7.1**  Physiographic regions (thick lines and numbered) and subregions (thin lines) of the coterminous United States:

1, Superior Upland – hilly area of erosional topography on ancient crystalline rocks;

2, Coastal Plain – low, hilly to nearly flat terraced plains on soft sediments;

3, Piedmont Province – gentle to rough, hilly terrain on belted crystalline rocks becoming more hilly towards mountains;

4, Blue Ridge Province – mountains of crystalline rock 3000–6000 ft high, mostly rounded summits;

5, Ridge and Valley Province – long mountain ridges and valleys eroded on strong and weak folded rock strata;

6, St Lawrence Valley – rolling lowland with local rock hills;

7, Appalachian Plateaus – generally steep-sided plateaus on sandstone bedrock, 3000–5000 ft high on the east side, declining gradually to the west;

8, New England Province – rolling hilly erosional topography on crystalline rocks in south-eastern part to high mountainous country in central and northern parts;

9, Adirondack Province – subdued mountains on ancient crystalline rocks rising to over 5000 ft;

10, Interior low plateaus – low plateaus on stratified rocks;

11, Central lowland – mostly low rolling landscape and nearly level plains; most of area covered by a veneer of glacial deposits, including ancient lake beds and hilly lake-dotted moraines;

12, Great Plains – broad river plains and low plateaus on weak stratified sedimentary rocks; rises towards Rocky Mountains at some places to altitudes over 6000 ft;

13, Ozark plateaus – high, hilly landscape on stratified rocks;

14, Ouachita Province – ridges and valleys eroded on upturned folded strata;

15, Southern Rocky Mountains – complex mountains rising to over 14000 ft;

16, Wyoming Basin – elevated plains and plateaus on sedimentary strata;

17, Middle Rocky Mountains – complex mountains with many intermontane basins and plains.

18, Northern Rocky Mountains – rugged mountains with narrow intermontane basins;

19, Columbia Plateau – high rolling plateaus underlain by extensive lava flows; trenched by canyons;

20, Colorado Plateau – high plateaus on stratified rocks cut by deep canyons;

21, Basin and Range Province – mostly isolated ranges separated by wide desert plains; many lakes, ancient lake beds, and alluvial fans;

22, Cascade–Sierra Nevada Mountains – Sierras in southern part are high mountains eroded from crystalline rocks; Cascades in northern part are high volcanic mountains;

23, Pacific Border Province – mostly very young steep mountains, includes extensive river plains in California portion.

*Source: US Geological Survey.*

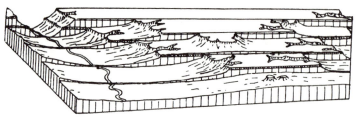

**Figure 7.2** Dissection of an area of flat-lying rocks under arid conditions shown progressively by five blocks, the oldest at the front and the youngest at the back.
*Source:* Cotton, 1948, figure 115, p. 140, after W. M. Davis.

Belemnite Chalk has been stripped; and the summit levels of the Appalachian Plateaus (650–750 m; 2000–2250 ft) supported by the Pottsville and Pocono Sandstones which have been dissected into a hierarchy of regular drainage basins of 300–700 m (1000–2000 ft) relief (see Chapter 13.2).

However, even in its apparently most simple instances, regional uplift always produces structural deformation. The 7000-ft uplift of the Colorado Plateau, which commenced at the end of the Cretaceous in association with lateral compressive forces of the Rocky Mountain (Laramide) orogeny, produced a number of broad domes or upwarps separated by basins (Figure 7.3) which remain today with characteristically eroded central parts (e.g. Kaibab uplift: Figure 7.4). Almost all these upwarps are characterized by gentle western and steeply dipping eastern limbs (Figure 7.5), except for the Zuni uplift which has a steep western flank. The asymmetry was not only associated with lateral compression forces, but also with differential uplift of a block-faulting nature, as evidenced by the fact that some of the steeply dipping flanks (monoclines) extend laterally into normal faults (Figure 7.6). Much of the present relief of the Colorado Plateau is due to bodily uplift in Later Tertiary times, but this was also accompanied by deformation, in particular, by block-faulting in its western parts which are transitional with the Basin and Range Province, to be discussed later in this chapter.

The deformation of surface rocks depends not only upon the nature of the applied forces, but also upon the resistance to deformation presented by the older basement rocks. Indeed, as we have seen, deformation patterns and basement structures are clearly related to each other. The deformation of cratonal areas takes the form of broad upwarps and downwarps, one example of the latter, the autogeosyncline, having been introduced in Chapter 5. The cratonal domes of North America (e.g. Nashville and Cincinnati) are characteristic upwarps, having diameters in excess of 150 km, dips of 2 m/km and eroded central parts which form topographic depressions.

An example of uplift and moderate deformation on the cratonal edge of a geosyncline is provided by the upheaval, doming and minor folding of south-east England beginning in the Cretaceous and culminating in the movements associated with the Alpine (Oligocene–Early Miocene) orogeny. Possibly the Chalk was gently deformed into the Wealden dome, then truncated by fluvial and marine action into the Sub-Eocene plain, followed by intense Alpine deformation in the Mid-Tertiary (Figure 7.7), together with further peneplanation, which was succeeded by uplift of 200 ft (60 m), the cutting of a marginal Calabrian marine platform and then intermittent uplift of a further 600 ft (180 m). The sections in Figure 7.8 show the homoclinal structures developed north of the River Thames (Figure 7.8A) where the resistant Paleozoic basement is covered by less than 300 m of later rocks, contrasting with the greater deformation of the Weald anticlinorium (Figure 7.8B) and the south coast of England (Figure 7.8C) where the basement lies more than 2000 m deep (see Section 2.3).

## 7.2 Homoclinal structures

The most common homoclinal structures are sequences of sedimentary rocks having a general dip in one direction. These structures are produced in two main ways, either by the uplift of a sequence of off-lapping coastal plain sediments or as part of one limb of a large dome or fold. Figure 7.9 suggests a sequence of development for landforms produced on sedimentary rocks offlapping an old land (see Figure 4.17), as in the Coastal Plain of the eastern United States or in south-eastern England. The retreat of the sea is accompanied by the extension of consequent rivers from the old land (i.e. consequent on the original surface slope of the sedimentary sequence) and lateral tributaries (i.e. subsequent rivers) are later developed exploiting the outcrops of less-resistant rocks (e.g. clays and shales), etching out lines of asymmetrical hills (cuestas) consisting of steeper, landward-facing scarp slopes and gentler, seaward-facing, dip slopes, controlled in part by the original dip of the beds. The dip slopes merge into the valley floors formed by the overlying less-resistant rocks to produce a characteristic terrain termed 'scarp and vale'. The innermost vale (i.e. inner lowland) is commonly separated from the higher

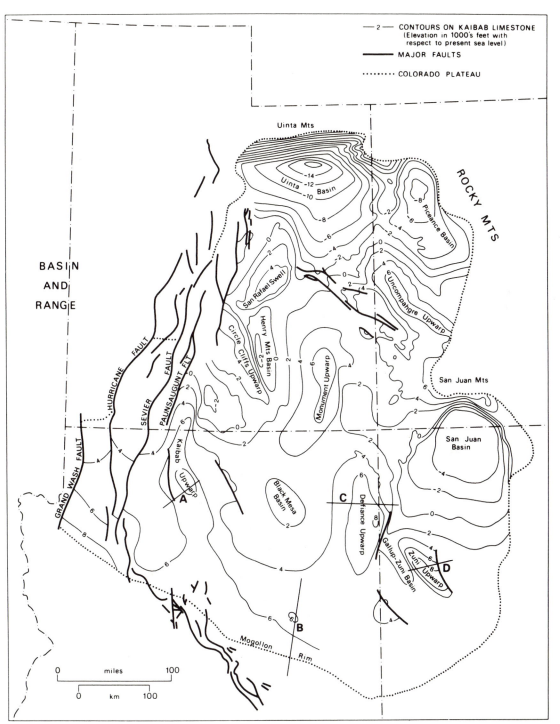

**Figure 7.3** Structure contour map of the Colorado Plateau; contours are those on top of the Permian Kaibab limestone formation at present, with elevations in thousand feet above sea level; major faults are shown; the four sections refer to Figure 7.5.
*Source:* Hunt, 1956, figure 37, p. 52.

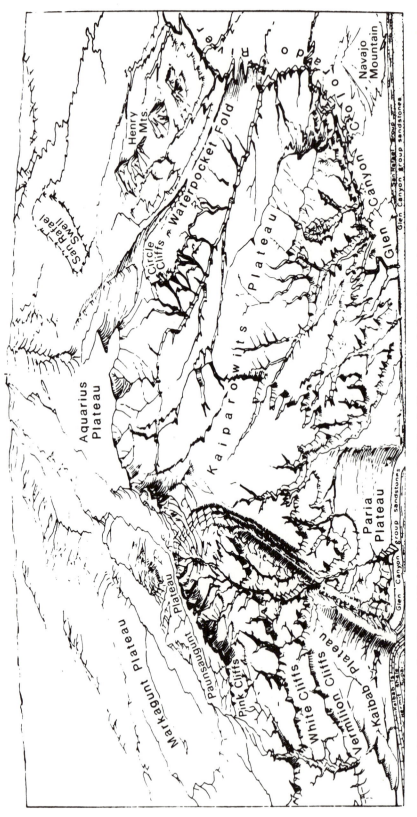

**Figure 7.4** Generalized view of south-central Utah, looking north across the Colorado Plateau from the Utah-Arizona boundary line.

*Source:* Hunt, 1956, figure 50, p. 69.

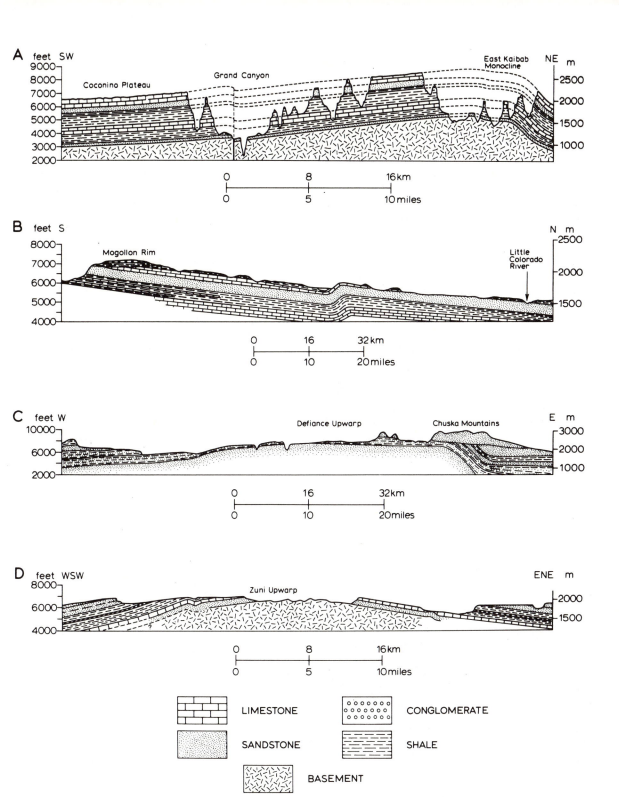

**Figure 7.5** Four structural sections in the southern Colorado Plateau shown in Figure 7.3: A. Kaibab Upwarp; B. Mogollon Rim; C. Defiance Upwarp; D. Zuni Upwarp.
*Sources:* Geological Society of America and U.S. Geological Survey.

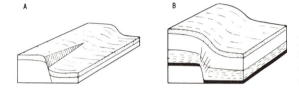

**Figure 7.6**  Relation of monoclines to normal faults: A. monocline passing laterally into a fault; B. monocline passing downwards into faults.
*Source:* Hills, 1972, figure VIII-9, p. 223.

terrain of the old land by a break of slope or 'fall line', especially apparent in the longitudinal profiles of the extended consequent streams. The extension of the subsequent streams along the vales allows resequent (i.e. reconsequent) and obsequent (i.e. opposite consequent) rivers to develop, and eventually causes extended and other consequent rivers to be beheaded by river piracy at an elbow of capture, perhaps leaving a former watergap bereft of drainage, forming a windgap. Captures may occur wherever extending rivers have a headward-cutting advantage, due to lack of rock resistance (e.g. subsequent rivers) or to topographic slope (e.g. obsequent rivers). Along with this drainage development goes slope development, which is dominated by scarp recession producing 'homoclinal shifting', the erosional displacement of the scarp and cuesta crest in the direction of stratigraphic dip. The mechanisms of scarp slope retreat are akin to cliff-recession processes (see Chapter 11). In

south-east England the deepening of subsequent valleys has led to much shallow slumping in the clays forming the lower parts of the scarp slopes and to the squeezing out of these clays from beneath the overlying more 'competent' beds (i.e. non-plastic). These processes, which were especially active prior to the last interglacial period, have given rise to the bulging of incompetent clays in the valley bottoms by lateral creep (i.e. valley bulges), the draping of the overlying competent rocks around the upper cuesta slopes (i.e. cambering) to such an extent that in places the original dip may be reversed near the scarp edge, and the consequent fracturing (i.e. gulling) of the deformed competent beds (cambering, gulling and valley bulging are shown in Figure 10.5E).

The major competent beds which in south-east England are associated with cuestas and watergaps are the limestones of the Inferior Oolite (Jurassic), having low dips of a few degrees and reaching thicknesses of

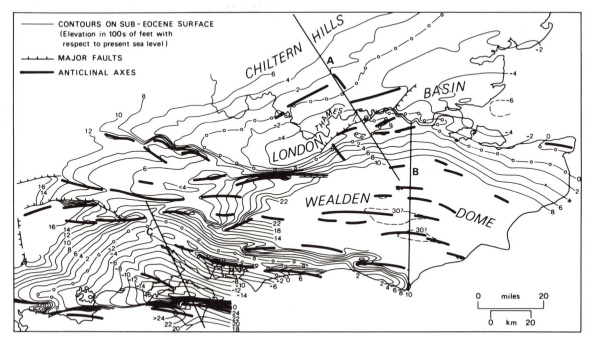

**Figure 7.7**  Major Post-Alpine structures of south-east England, as illustrated by structure contours on the Sub-Eocene surface, major faults and anticlinal axes. The three lines of section are shown in Figure 7.8: **A.** across the London Basin; **B.** across the Wealden Dome; **C.** across the Hampshire Basin.
*Source:* Wooldridge and Linton, 1955, figures 5 and 6, pp. 12–13, 17.

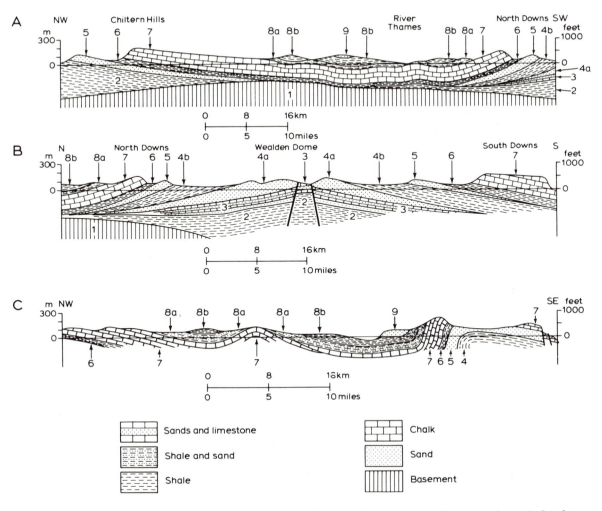

**Figure 7.8** The three geological and topographical sections shown in Figure 7.7; generalized rock types are shown. A. London Basin; B. Wealden Dome; C. Hampshire Basin:

| | | |
|---|---|---|
| 9 | Eocene and Oligocene sands | |
| 8b | Eocene (London) clay | Tertiary |
| 8a | Lower Eocene sands and clays | |

| | | |
|---|---|---|
| 7 | Chalk | |
| 6 | Gault clay | |
| 5 | Lower Greensand | |
| 4b | Weald clay | Mesozoic |
| 4a | Weald sands | |
| 3 | Portland and Purbeck beds | |
| 2 | Oxford Clay and other Jurassic rocks | |

| | | |
|---|---|---|
| 1 | Paleozoic platform | Paleozoic |

*Source:* Stamp, 1946, figures 64, 67 and 66, pp. 179, 185, 182.

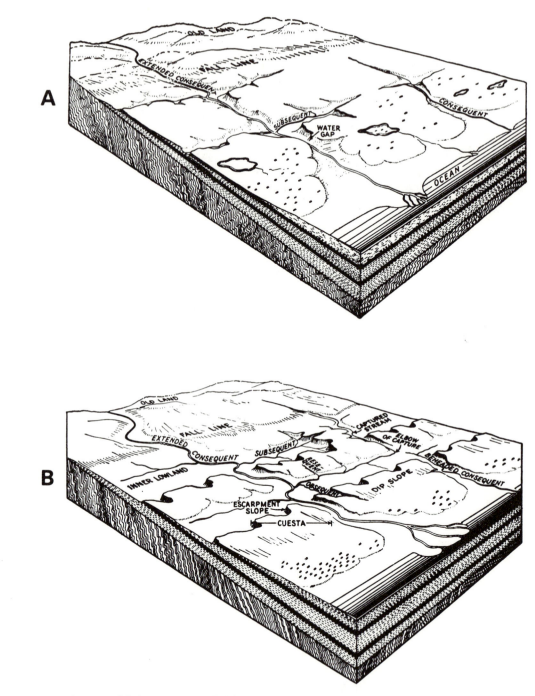

**Figure 7.9** Development of drainage on a coastal plain of homoclinal strata. A. Early stage: development of consequent and extended consequent drainage with water gaps, together with subsequent drainage. B. Late stage: etching out of cuestas, retreat of scarp slopes, river capture and beheading, and the development of resequent and obsequent drainage.

*Source:* Von Engeln, 1948, figures 47 and 48, pp. 122, 123. (Drawings by Elizabeth Burckmyer.) Copyright © The Macmillan Co., New York, by permission.

more than 100 m in places, forming the Cotswold Hills; the Cretaceous sandstones of the Upper and Lower Greensand and the Hythe Beds; and most important, the Upper Chalk which forms the topographic framework of much of the region (i.e. the Downs and the Chilterns) through which many watergaps have been cut. Ignoring for the moment these effects of erosional history, the relief of cuestas can be ascribed to a number of features of the scarp-forming rock (Plate 6):

(1) thickness;

(2) resistance, including permeability and shearing resistance of the rock and its weathered debris;

(3) angle of dip. For example, the Chalk of the Isle of Wight produces hills of some 215 m (700 ft) high where dipping at 4° and hills of only 150 m (475 ft) where dipping at 70°–80°. Where dips approach 45°, cuestas have more or less symmetrical 'hog-backs'. The dip of a cuesta formation has also been shown to influence the morphometry of its dip slope and on Clinch Mountain, a cuesta of quartzite, sandstones and shales in the folded Appalachians, stream lengths, basin areas and hypsometric integrals (see Section 13.2) bear significantly negative relationships to the dip which varies from less than 20° to more than 60°;

(4) The amount of recession. This is influenced by the rate of operation of mass movements on the scarp slope, together with spring sapping, stream incision and basal weathering.

A number of studies have been carried out to evaluate the relative importance of some of the above factors which determine the relief of scarp and vale topography. An investigation of the soil shearing resistance and permeability of four formations south of Oxford, England, showed that the highest terrain is supported by the Shotover Sand which possesses a combination of high shearing resistance and high permeability, whereas the lowest

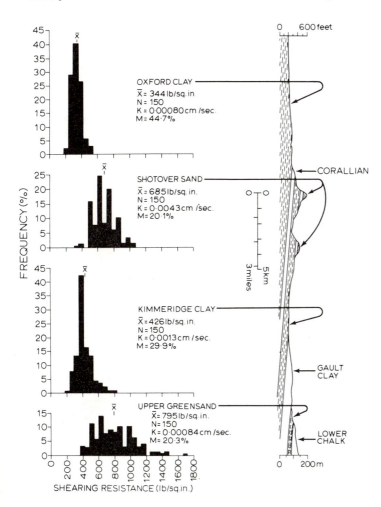

**Figure 7.10** North–south geological cross section near Oxford, England, showing the following information relating to the surface soil of four formations: histograms of shearing resistance (lb/in²), mean shearing resistance $\overline{X}$; number of samples $N$; coefficient of permeability (cm/s) $K$ (see Chapter 5); and moisture content $M$. *Source:* Chorley, 1959.

elevations lie on the Oxford Clay which possesses low shearing resistance and low permeability (Figure 7.10). Shearing resistance is the sum of the frictional and cohesive forces within a mass of soil particles which opposes failure by shear. A further study was conducted on an 80-km (50-mile) Lower Greensand ridge forming a belt of hills 5–6 km (3–4 miles) wide of friable sandstone, rising from 13 m (42 ft) to 169 m (556 ft) across south-central England. Its elevation was correlated with a number of variables including thickness of the formation (varying from 3 m to 72 m), percentage silt and clay, and the projected altitude of the upper contact of the Lower Greensand. Of these, Greensand thickness was found to explain fully 98·92 per cent of variation in the ridge elevation and the other two factors a significant percentage but well below 0·50 per cent (Chorley, 1969).

## 7.3 Folded structures

### 7.3.1 Simple folding

A common mode of rock failure is by folding into a series

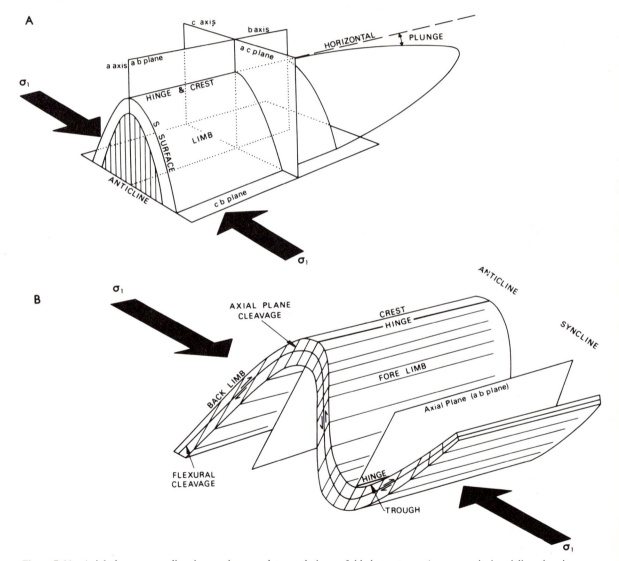

**Figure 7.11** Axial planes, stress directions and nomenclature relating to folded structures: A. symmetrical anticline plunging out; B. asymmetrical anticline and syncline.
*Sources:* De Sitter, 1964, figure 202, p. 272; Whitten, 1966, figure 104, p. 127.

of *anticlines* and *synclines*. Folding is most prevalent where stresses are long-continued and where high confining pressures permit rocks to bend and flow, rather than fracture. Intense compression, particularly close to the surface or subparallel to incompetent beds (e.g. shales and silts), is commonly accompanied by failure along low-angle thrusts. Figure 7.11 shows the geometrical features, and associated terminology in respect of folds, in which the axial (*ab*) plane is vertical in A and inclined in B. Anticlines are commonly terminated by 'plunging out', and Figure 7.12A and B shows a characteristic fold assemblage as is exemplified by parts of the folded Appalachians and the Jura Mountains on the Franco-Swiss border.

Folds are formed at some depth and are usually exposed at the surface as the result of erosion which has commonly breached the anticlines along their crests (see Figure 7.16C) or even truncated a series of folds. Figure 7.12C shows the surface outcrop of such a truncated series of plunging anticlines and synclines, and the zigzag ridges produced by the outcrops of the more resistant beds are characteristic of much of the folded Appalachian region of the eastern United States. Sometimes two different alignments of folding, of different date, are superimposed on each other, and this sometimes results in the later folds being arranged *en échelon*, like the wings of a theatrical stage, or in much more complex structures.

Simpler folds, together with their associated fractures, fall into three groups of increasing complexity.

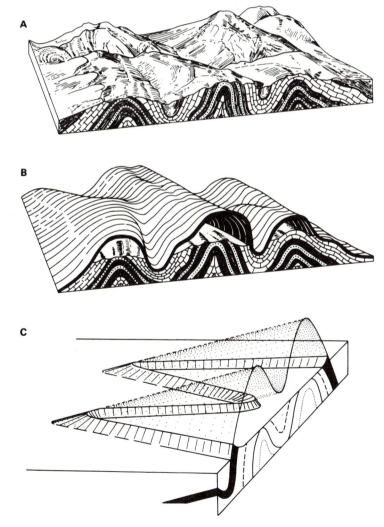

**Figure 7.12** Examples of simple folded structures: A. relief of part of the Jura Mountains, Switzerland; B. sketch of restored top of the resistant limestone bed shown in A.; C. zigzag pattern of eroded plunging folds.
*Sources:* Gilluly *et al.*, 1968, figure 9-11, p. 143, after Albert Heim; Lobeck, 1939, figure 1, p. 596. Copyright © McGraw-Hill, New York.

(1) Parallel folding. This involves a more or less simple concentric bending or flexing of the layers of the sedimentary rock. The lack of internal space causes upward stress within the anticlines leading to a pattern of crestal joints and fractures (see Section 7.4) contrasting with those of the synclines. The crests of parallel folds are commonly cut by a series of longitudinal normal faults (throws < 30 m) and joints intersected by shear fractures.

(2) Similar folding. In this type successive beds have similar curves which preserve the folded form with depth. Similar folding usually involves more than simple flexure in that flowage may occur in the less competent beds which tend to thicken symmetrically at the hinges of folding and to thin on the limbs. Slippage and flowage of rocks, together with the preferred reorientation of minerals produces *flexural cleavage* parallel to the bedding and *axial plane cleavage* parallel to the axial (*ab*) plane (Figure 7.11B).

(3) Disharmonic folding. This involves abrupt changes in geometry in passing from one lithic unit to another in a folded sequence such that the folds in one bed may be unrelated to those in successive layers. Disharmonic folding is mainly due to flowage or failure within incompetent beds. Examples of the latter are widespread in the folded Appalachians (Figure 7.13) where thrusting has occurred along the shale beds (Figure 7.14B). The best case of this is the Pine Mountain overthrust (Figure 7.14C) which developed in the Rome and Chattanooga Shales and, in passing from one to the other, generated the Powell Valley anticline. This thrust is 30 km wide and was subjected to 3 km of movement at the leading (western) edge and 13 km at the rear. More extreme

examples of disharmonic folding occur in the Jura Mountains (Figure 7.15) where, in addition to thrusts in the Liassic shales, the whole Mesozoic sequence has been thrust and folded over a rigid basement along Triassic anhydrite and salt beds to produce *décollement*. Differential thrusting produces a series of near-vertical *strike-slip* or *tear faults* (see Figure 7.21C) which are very common in the Jura, and which clearly delimit the north and south ends of the Pine Mountain overthrust.

The initial development of landforms in a region of simpler folding is less easy to postulate than that of homoclinal, coastal plain regions. This is because rising folds are rapidly attacked by erosive forces, especially along their tensional crests, and because many folded structures only develop at depth where rock bending and flowage is possible under superincumbent weight, while their surface expression may be more characterized by faulting, an analogous situation to the deformation of glaciers. However, direct and indirect measurements suggest that relatively rapid surficial folding does occur. For example, a monoclinal flexure in the Chilean Andes has risen 19 m in the last 1700 years, distorting underground water conduits, and the Long Beach earthquake in California on 10 March 1933 raised an arch 18 cm high and 6 km across. It is also true to say that most folded mountains, except the high ranges of the Tertiary thrust mountains, have undergone prolonged and complex erosional histories. This is true of the Flinders Range of Australia, and especially of the folded Appalachians, which were treated in detail in Chapter 2. Regions of more recent and, possibly, more simple erosional histories may include the Jura Mountains of the Franco-

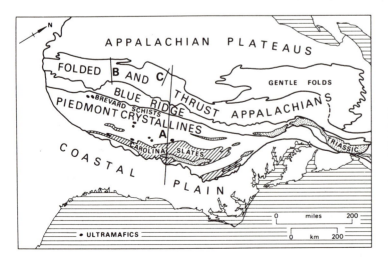

**Figure 7.13** The major structural systems of the eastern United States; major provinces are in bold lettering, subprovinces in small lettering; the three cross sections relate to Figure 7.14.
*Source:* A. J. Eardley, *Structural Geology of North America*, 1962, 2nd edn, figure 7.1, p. 92, copyright © 1951 by Harper and Row, Publishers, Inc., copyright © 1962 by A. J. Eardley, reproduced by permission of Harper & Row, Publishers, Inc.

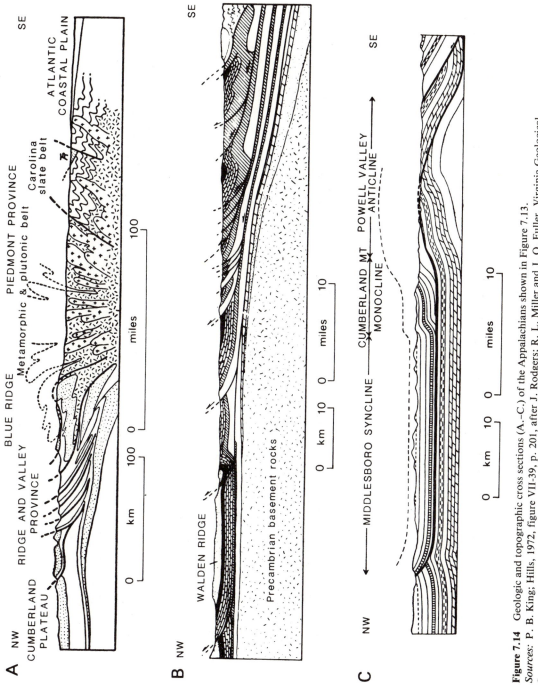

**Figure 7.14** Geologic and topographic cross sections (A.–C.) of the Appalachians shown in Figure 7.13.
*Sources*: P. B. King; Hills, 1972, figure VII-39, p. 201, after J. Rodgers; R. L. Miller and J. O. Fuller, Virginia Geological Survey.

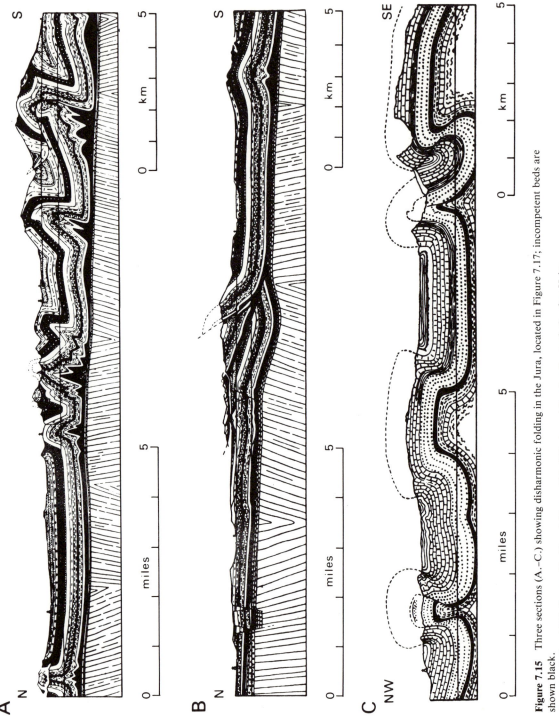

**Figure 7.15**  Three sections (A.–C.) showing disharmonic folding in the Jura, located in Figure 7.17; incompetent beds are shown black.

*Sources*:  Collet, 1927, figures 43, 44 and 47, pp. 133, 136; A. and B. after Buxtorf, C. after Heim.

Swiss border and, more certainly, the very recent folds of the Zagros Mountains of south-western Iran.

The hypothetical development of drainage and erosional forms on a series of simple folds is depicted in Figure 7.16. The initial terrain (Figure 7.16A) is characterized by anticlinal mountains, and synclinal valleys (French: *val*), as in east Hampshire, England, where ESE–WNW anticlinal ridges of Upper Chalk occur. The longitudinal consequent drainage of the synclines crosses the ridges at their low points to form watergaps (French: *cluse*), and these primary consequents are fed by secondary lateral consequents (French: *ruz*) draining the anticlinal flanks. The development of lateral consequent erosion (Figure 7.16B) leads to the widespread breaching of anticlinal crests (Figure 7.16C), with erosion exploiting the weak crests and anticlinal cores. Extended anticlinal breaching is effected by the extension of longitudinal subsequents on the exposed softer outcrops, producing breached anticlines with inward-facing escarpments (French: *crêtes*), such as the Wealden anticline and the smaller vales of Wardour, Pewsey and Blackmore in southern England (Figure 7.16D). River piracy begins as the subsequents extend along the strike and master transverse rivers cut down to produce the deepest watergaps. The continued development of erosion by subsequent rivers produces a complex pattern of captures and reversal of ruz drainage leading to an inversion of relief with synclinal mountains and anticlinal valleys (Figure 7.16E) (e.g. Burkes' Garden, Virginia). Ultimately (Figure 7.16F) there is a complete adjustment of the integrated drainage to structure with a zigzag ridge pattern of eroded pitching folds (see Figure 7.12C), as in the Flinders Range of Australia from which some 6000 m of rock have been eroded, leaving a skeleton of contorted limestone ridges. In detail such relief is similar to that produced by homoclinal structures with ridges of resistant sandstone and limestone, separated by subsequent weak clay–shale vales.

### 7.3.2 Complex folding

Complex folding involves large-scale disharmonic features and the production of large asymmetrical or *recumbent folds*, often broken along their axial planes by thrusts of kilometres or tens of kilometres displacement, of which the European Alps are the prime example (Figure 7.17). The *root zone* of such folds (i.e. where the core of the recumbent anticline is seen in relation to the older basement) has usually been subjected to regional metamorphism (see Section 4.4). *Autochthonous* folds are connected with their appropriate root zone, whereas *allochthonous* ones have been separated from their roots by large-scale thrusting. These thrust sheets, commonly composed of the back limbs of recumbent folds, are termed *nappes* (French, 'cover'; German equivalent: *Decke*), parts of which isolated by erosion are called *klippe* (German, 'reef') and erosional depressions through which are *windows* (German: *Fenster*) (Figure 7.18). Most mountain ranges which developed from the violent collapse of geosynclines exhibit complex folding with thrusts oriented outwards from the geosynclinal axis. However, most ranges show a dominant thrusting towards the marginal craton (i.e. foreland), and this is particularly so in the case of the European Alps which were formed by the collision of two crustal plates (Figure 7.19A). North of the root zone (Figures 7.17 and 7.19B) are, successively, the pile of recumbent folds of the Pennine nappes (i.e. Simplon, Great St Bernard, Monte Rosa and Dent Blanche); the Helvetic nappes, a complex thrusted mass of calcareous Mesozoic and Eocene rocks pierced by the massifs which are exposed upthrusted splinters of the crystalline basement; and the Pre-Alps, which are the leading-edges of the huge thrusted recumbent folds, where a series of smaller closely spaced subparallel imbricate *thrusts* characteristically occur. The huge size and distance of travel of the allochthonous Alpine nappes has led to their explanation, at any rate in part, by the concept of *gravity tectonics* (see also Chapter 10). This concerns the sliding of huge rock masses down the flank of a squeezed-up crustal belt under the action of gravity, rather than primarily by lateral thrusting from the rear.

### 7.4 Faulted structures

#### 7.4.1 Faulting

Crustal stress fields can be resolved into three resultant compressive stresses ($\sigma_1$ maximum; $\sigma_2$ intermediate; and $\sigma_3$ minimum). The minimum compression is usually negative, being a tensional extension force. Besides producing rock bending, deformation and flowage, crustal stresses commonly result in strain fractures in the form of joints and faults. *Joints* are small-scale fractures with little relative movement on either side, whereas *faults* are larger-scale fractures with significant movement (i.e. *slip*) resulting in the offsetting of the two sides. *Sets* of fractures comprise a group of parallel failures, *systems* of fractures are two or more sets of fractures intersecting at a more or less uniform angle. Different orientations of the stress field produce sets and systems

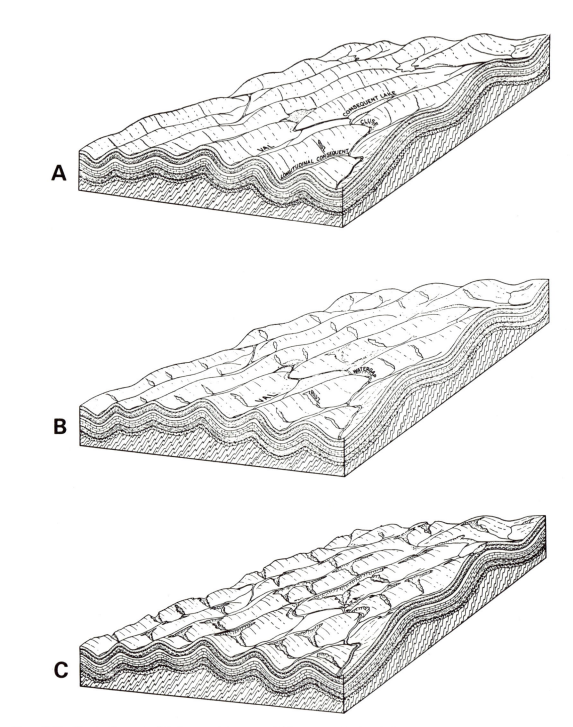

**Figure 7.16** The development of drainage on intact, open folds; see text, for description: A. dominantly consequent drainage; B. ruz development; C. anticlinal breaching;

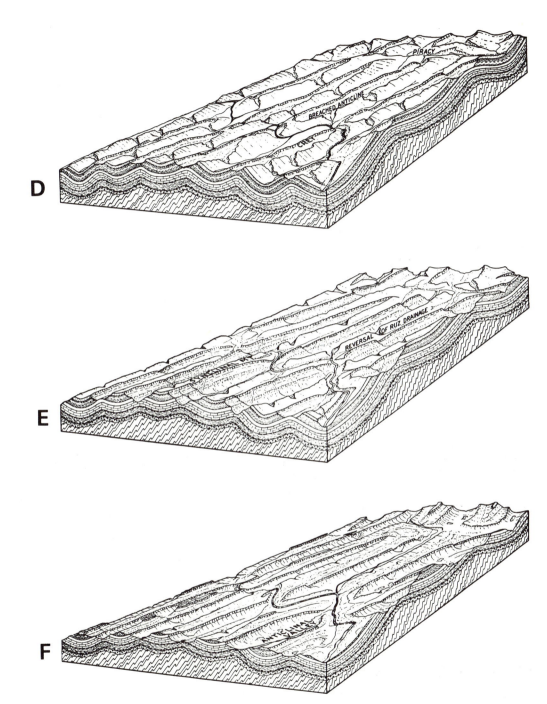

Figure 7.16 (cont.) D. subsequent drainage development, piracy and evolution of master transverse river with deepest cluses (water gaps); E. complex pattern of capture and drainage reversal in some ruz and subsequents; inversion of relief now well marked due to anticlinal breaching (bounded by crets); F. complete adjustment of terrain to structure, drainage integrated; topography dominated by synclinal summits and anticline valleys.

*Source:* Von Engeln, 1948, figures 178-81, 183 and 185, pp. 317, 318, 320, 321, 323, 325. (Drawings by Donald Rockwell and Thomas Chiswell.) Copyright © The Macmillan Co., New York, by permission.

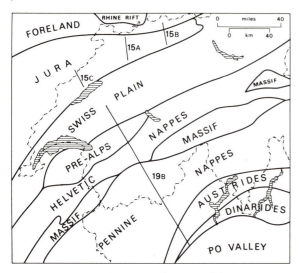

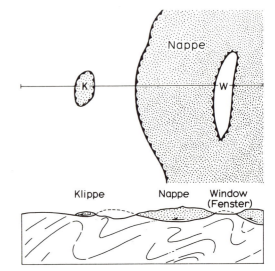

**Figure 7.17** The major structural systems of the western Alps and adjacent areas; the four cross sections relate to Figures 7.15 and 7.19.

**Figure 7.18** Schematic map (above) and cross section (below) illlustrating the structural relations of a nappe, a klippe and a window or fenster.
*Source:* J. G. Dennis, *Structural Geology*, 1972, figure 12-18, p. 277, copyright © John Wiley and Sons, by permission.

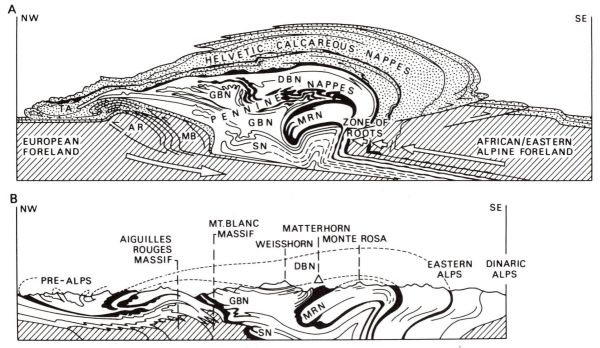

**Figure 7.19** Sections across the European Alps. A. The collision of two plates (forelands) which are splintering in places (**AR**, Aiguilles Rouges Massif; **MB** Mont Blanc Massif) squeezing the Helvetic Calcareous nappes over the Pennine nappes (**GBN** Great St Bernard; **SN**, Simplon; **DBN**, Dent Blanche; **MRN**, Monte Rosa) and constricting the zone of roots. B. Tectonic cross section across the western Alps, according to E. Argand and R. Staub. This is section 19B on Figure 7.17 and the abbreviations are the same as in Figure 7.19 A.
*Sources:* Collet, 1927, figure 62, p. 208; Holmes, 1965, figure 826, p. 1151.

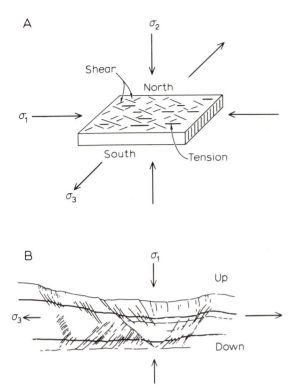

**Figure 7.20** Patterns of rock fracture. A. Shear and tension joints (faults) in an unfolded sheet of rock; directions of compression ($\sigma_1$ and $\sigma_2$) and tension ($\sigma_3$) are shown; see text, for description. B. Tension fault pattern developed in clay stretched on a rubber plate, by H. Cloos; tension ($\sigma_3$ and resultant compression ($\sigma_1$).
*Sources:* L. U. De Sitter, *Structural Geology*, 1964, figure 59, p. 109, copyright © McGraw-Hill, New York; Hans Cloos.

of fractures differently oriented with respect to the earth's surface. The two most common are sets of tension fractures oriented parallel to $\sigma_1$ and normal to $\sigma_3$, and systems of two conjugate sets of shear fractures oriented parallel to $\sigma_2$ and making acute angles of 7°–45° with $\sigma_1$ (Figure 7.20). The angles between pairs of shear joints may vary greatly (i.e. 15° to approx. 90°) with differences in rock strength and in the confining pressure under which the stress field is operating. The term *competence* is applied to the brittleness of a rock such that rocks (like shale) which flow rather than fracture are termed less competent, whereas those which strain by fractures (like many limestones) rather than by bend and flow are termed more competent. The attitude of a plane of failure is described in terms of its *strike*, the direction of the line formed by the intersection of the plane with the horizontal, and its *dip*, the maximum angle of inclination (measured from the horizontal) of the fracture plane

measured at right angles to the strike direction. A somewhat related geological term is *plunge*, the angle of inclination of a linear structural element (i.e. fold axis). Various methods are employed to plot the frequency of fractures of differing dip and strike (De Sitter, 1964).

Near the surface the intermediate stress ($\sigma_2$) direction is usually either horizontal or vertical, and this orientation produces the ideal cases of fracture (Figure 7.21):

(1) $\sigma_2$ and $\sigma_3$ horizontal, and $\sigma_1$ vertical. This stress field orientation is associated with zones of major tension and the production of dip–slip or *normal faults* (i.e. dipping towards the block which has been relatively lowered).

(2) $\sigma_2$ and $\sigma_1$ horizontal, and $\sigma_3$ vertical. This stress field orientation produces *reverse*, or *thrust*, *faulting* (i.e. the fault plane dipping towards the block which has been relatively raised). Apart from some high-angle reverse faults or relatively small slip associated with local uplift of the basement, or with the compensating jostling of a minority of a group of normally faulted blocks, thrust faults are restricted to zones of compression and folding.

(3) $\sigma_2$ vertical, and $\sigma_1$ and $\sigma_3$ horizontal. This produces *strike–slip*, or *wrench*, *faults* which are vertical

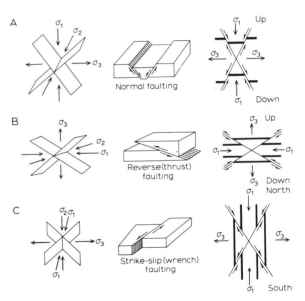

**Figure 7.21** Initial stress distribution causing A. normal, B. reverse and C. strike–slip faulting. $\sigma_1$ is the direction of maximum compressive stress (i.e. compression), $\sigma_3$ is the direction of minimum compressive stress (i.e. tension) and $\sigma_2$ is the direction of mean compressive stress; see text, for description.
*Source:* Hills, 1972, figure VII-19, p. 183.

and commonly oriented in one of the two shear directions. These failures are also characteristic of compressive folding. Across strike–slip fault zones there commonly occur sets of parallel *en échelon* fractures which are vertical tension fractures making angles of 12°–17° with the strike–slip direction (Hills, 1972).

In an ideal situation, where a uniform stress field is operating on a rock body of uniform properties, sets and systems of joints and faults would be accordant and distinguishable only by their magnitude and displacement. In practice, few locations exhibit such accord between joints and faults because of local variations in the regional stress field and of rock competence. A graphic example of this is given in Figure 7.22, where differently oriented fracture sets and systems have developed on an adjacent anticline and syncline because the convex and concave bending of competent limestone beds was associated with different orientations of local stress.

The features of dip–slip or normal faulting were simulated by Hans Cloos, who fractured damp clay by stretching it on a rubber base (Figure 7.20B). Normal fault systems are usually formed of one dominant set of shear fractures with the other set subordinate. Normal faults may dip between 30° and 90°, although they are most frequently found to lie at dips of 45°–70°, depending partly on rock competence, being more steep in more competent beds, and partly on confining pressure, being less steep at increasing depths. Systems of normal faults are always associated with dominant tension parallel with the earth's surface producing one dominant set of shear planes parallel to $\sigma_2$, and relative displacements ranging from a few centimetres to several kilometres. Such crustal extension occurs mainly under three circumstances:

(a) Large-scale crustal divergence zones. The major zones of normal faulting are associated with such tectonic environments, the Tanzanian graben (Figure 7.23) being 700 km long, 12–65 km wide, with a subsidence of 1400 m, and the Upper Rhine graben being 300 km long, 35 km wide, with a subsidence of 4000 m bounded by faults dipping at 60°–65°.

(b) Zones of updoming and stretching during orogeny and emplacement of large intrusive igneous rock bodies – e.g. the Rocky Mountain trench of Canada, the Central Valley of Chile and the Barisan rift valley of Sumatra.

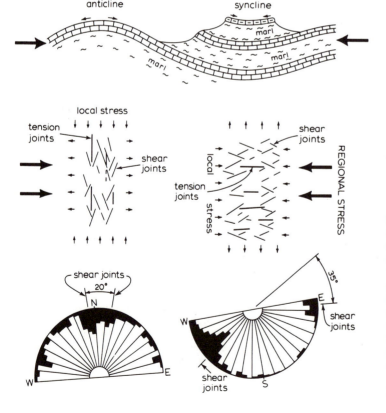

**Figure 7.22** An adjacent anticline and syncline in south-east Algeria, composed of marls and limestones, showing the direction of regional compression by the large arrows. The local tension direction on the anticlinal crest and the corresponding compression in the synclinal trough are shown by small arrows and produce contrasting patterns of shear fractures which appear in the strike frequency diagrams. Note that the axial planes of the two folds are not quite parallel. *Source:* L. U. De Sitter, *Structural Geology*, 1964, figure 56, p. 104, copyright © McGraw-Hill, New York.

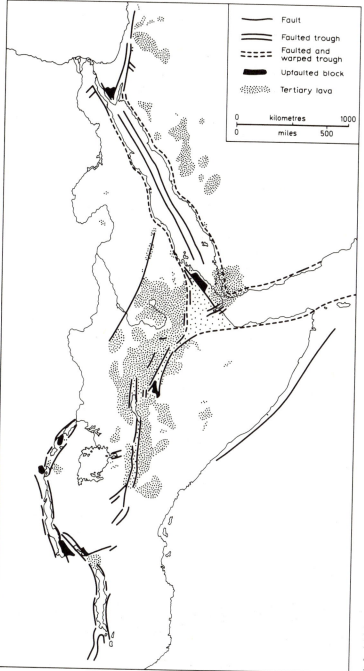

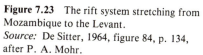

**Figure 7.23** The rift system stretching from Mozambique to the Levant.
*Source:* De Sitter, 1964, figure 84, p. 134, after P. A. Mohr.

(c) Uplift and broad doming of a previously folded orogenic belt, perhaps as the result of post-orogenic isostatic compensation. Examples of normal fault systems are the Connecticut Valley (Figure 7.24) and the Basin and Range province of Nevada, California, Utah and Arizona. The Connecticut Valley consists of a series of blocks of Triassic sandstone intruded by sills with beds dipping 10°–20° towards the major faults which have a maximum displacement of some 5000–12,000 m.

Compression and doming in two dominant directions, or alternating compression and tension, may produce a group of differentially moving basement fault blocks, draped with overlying sediment. Examples of this type of complex structure are the folded and faulted blocks of Saxony and the California Coast Ranges.

Large-scale transcurrent faults are associated with gross global tectonic processes, such as have produced the large displacements between tectonic plates or which are aligned at right angles to the crests of the oceanic ridges (see Chapter 5). Transcurrent faults (*dextral* if the opposite side of the fault has moved to the right, and *sinistral* if to the left) are shear failures developed in one dominant direction as the result of regional compression. Transcurrent movement between plates is exemplified, first, by the sinistral Dead Sea fault which moved 62 km in the Miocene and a further 45 km beginning in the Pliocene; and, second, the sinistral Great Glen fault of Scotland which moved more than 130 km in the Devonian and Carboniferous, to be followed in the Tertiary by a reverse movement of some 30 km. To link this scale of movement with that of modern earthquakes, it has been estimated that it would require the combined displacement of 20,000 average earthquakes to produce these results. The best-studied transcurrent fault system is that of the San Andreas in southern California (Figure 7.25), where north–south compression has produced a dominant set of NW–SE faults. The main San Andreas

fault, which is about 1000 km long, appears to have originated in the Mid-Oligocene due to movement between plates, and this has continued until the present day. Older rocks have been displaced some 550 km; the Lower Miocene of southern California more than 300 km; the Middle Miocene some 250 km and the Upper Miocene 120 km. The most recent disturbances have produced topographic scarps and the offset of stream channels, and the earthquake of 18 April 1906 caused some 4 m of lateral movement along 435 km of strike, with a maximum movement of 6 m in places.

### 7.4.2 Faulted landforms

Landforms associated with thrust and transcurrent faulting are referred to elsewhere but it is necessary here to discuss the landforms of high-angle faulting. As has been suggested, such faulting results from crustal stretching such that most of the faults dip normally (i.e. are dip–slip) into the centre of the stretching zone at angles of 45°–80° (see Figure 7.21), with the occasional reverse fault occurring in the jumble of blocks. This stretching, due to crustal divergence or to post-orogenic uplift, is reinforced by the extrusion of volcanic material, such as the rhyolitic ignimbrites in the Basin and Range Province of the United States and the volcanics of East Africa. Such flows are of great value in providing absolute or relative dating markers for subsequent fault movements.

Hans Cloos has provided a classification to aid in the explanation of the geomorphic expression of faults (Figure 7.26), depending on the relative rates of displacement and erosion (A, B and C), on the possibility of renewed (posthumous) faulting (E) and the differential erosion on either side of a stable fault plane (F and G). Thus (Dennis, 1972):

A is a faulted sequence of considerable faulting rate and negligible erosion producing, first, a monoclinal flexure (1) which breaks into a fault (2) (e.g. the East

W             CONNECTICUT VALLEY             E

Upland                                                        Upland

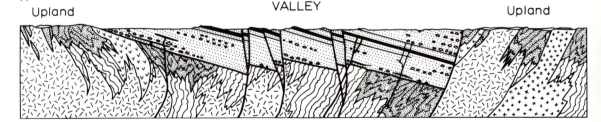

**Figure 7.24** Topographic and geologic cross section of the Connecticut Valley where faulting has affected the Triassic clastic rocks (with intruded lavas shown in black), as well as the closely folded Paleozoic metamorphic rocks and the underlying Pre-Paleozoic complex.
*Source:* Fenneman, 1938, figure 106, p. 374, interpretation by Joseph Barrell.

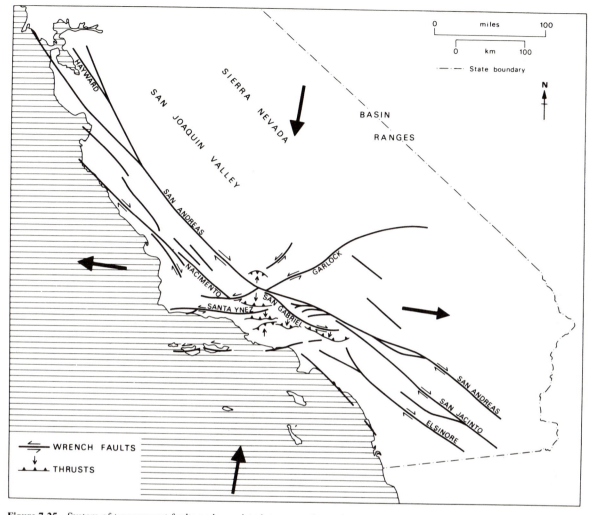

**Figure 7.25** System of transcurrent faults and associated structures in southern California. Major compression and tension directions are shown, but the major movement is occurring along one predominant shear direction between the adjacent plates. *Source:* M. L. Hill, *Calif. Div. Mines Bull.* 170, 1954.

Kaibab Monocline, Colorado Plateau: see Figure 7.6), and then becomes a significant fault scarp (3) whose height represents the vertical throw of the displacement and slope length the dip-slip of the fault.

B is an intermediate condition of erosion and uplift occurring together. The rising block is eroded by headward valley-cutting, at the margins of which alluvial fans are built and other deposits are laid down on the subsiding block (1–2). Continued erosion produces characteristic triangular truncated spur facets on the fault scarp (3), as does the advent of active erosion on an existing fault scarp (A3) (see Figure 7.27A).

C has an erosion rate which greatly exceeds the uplift rate of the rising unit such that there is a general flattened surface across the fault zone (1), except where differential erosion can form small scarps (2 and 3).

D represents the condition after cessation of fault movement and the completion of erosional truncation and depositional filling.

E condition D interrupted by renewed (posthumous) faulting and the production of a new fault scarp (2) (see Figure 7.27B).

F stability along the fault plane may continue but renewed erosion produces a further excavation of the fault to produce a *resequent fault-line* scarp (i.e. fault

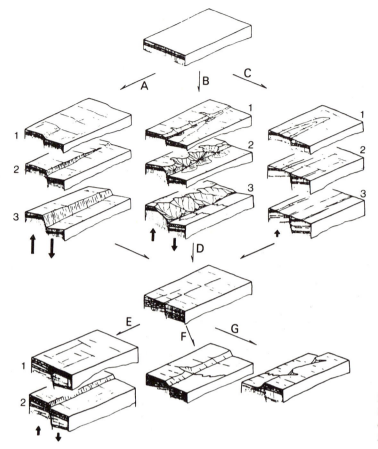

**Figure 7.26** Physiographic expression of faults, by Hans Cloos; see text, for detailed description.
*Source:* Dennis, 1972, figure 12.2, p. 258.

line because it has been produced entirely by erosion, and resequent because it faces the original downthrow side).

G On the other hand, if erosion during stability excavates a scarp facing towards the original upthrow side, due to chance lithological associations on either side of the original fault plane, an obsequent fault-line scarp is produced.

As erosion proceeds, of course, the distinction between a long-eroded fault scarp and a fault-line scarp becomes blurred – indeed, the two types may exist side by side. Fault scarps are most readily identifiable in areas of recent or contemporary crustal activity or where the original 'throw' (i.e. vertical displacement: see Figure 7.21A and B) of the fault was very large. The American Basin and Range Province (Figure 7.1) is characterized by a series of discontinuous, north–south-oriented blocks which, for the most part, have been normally faulted and tilted (Plate 7). This contrasts with a horst and graben terrain where the mountains are composed of

upthrown blocks and the intervening valleys by separate downthrown blocks.

Some fault scarps appear relatively fresh (e.g. on the west side of Upper Klamath Lake, Oregon, and the west face of the Panamint Range, California) and many exhibit triangular facets (e.g. on the east side of Deep Springs Valley, California; see 2 in Figure 7.26B2 and C in Figure 7.27A); but most fault scarps are considerably dissected, like the west face of the Wasatch Range, Utah; and some (particularly in the southern Arizona section) have been much eroded back into low pediment margins considerably removed from the original fault line. Naturally the relief of most fault scarps has been decreased by erosion and valley filling, but some scarps have been increased in height by renewed faulting (e.g. the base of the Wasatch Front shows recent fault scarps up to 12 m high) or by subsequent erosion of the downthrown block. For example, the north-east face of the Ruby–East Humboldt Range in Nevada is 1200 m high, of which the top 900 m is fault scarp and the bottom

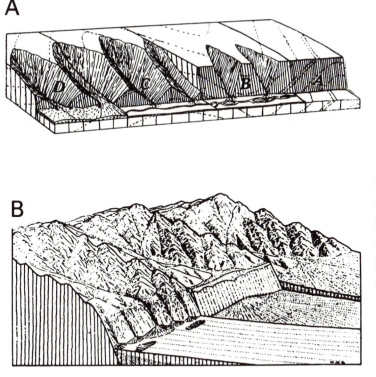

**Figure 7.27**   Some forms associated with faulting: A. the sequential dissection of a fault scarp (**AD**); **A** is the initial form; in **B** and **C** basal debris has been removed by a river; in **D** debris has accumulated as alluvial fans;  B. three stages in the evolution of a greatly dissected fault scarp with fans (back block); which has been subject to renewed normal faulting (middle block); and then to renewed erosion and deposition.
*Sources:* Cotton, 1948, figure 295, p. 396; W. M. Davis, 1930, 'Physiographic Contrasts, East and West', *Scientific Monthly*, vol. 30, pp. 394–415, figure 4. Copyright © 1930 by the AAAS.

300 m is resequent fault-line scarp. Striking fault scarps are not limited to the Basin and Range Province in the western United States, for in the Rockies the steep east face of the Bitterroot Range in Montana exhibits triangular fault facets. On a larger scale the East African divergence zone horst and graben system extends some 5500 km from Mozambique north to the Levant (see Figure 7.23), having a bifurcating system of grabens averaging 30–70 km wide (Dead Sea, 5–20 km; Red Sea, 200–400 km). Fault-line scarps are also ubiquitous in tectonically long-stable areas, such as Britain and New England, but are capable of producing very striking topographic features, such as the east side of the Vale of Clwyd in North Wales where a relief of more than 300 m has been produced by the relative erosion of Triassic sandstones lying adjacent to resistant, scarp-forming Upper Silurian Grits.

The foregoing discussion has shown clearly the effects of geological structures on landforms at all scales, from the continental and Alpine on the large scale to the control of stream patterns on the small scale (see also Figures 8.2 and 13.4). It is also apparent that much structural control manifests itself through the influence of the absolute and relative characteristics of the rock units involved. It is to this matter that we now turn our attention in Chapter 8.

## References

Chorley, R. J. (1959) 'The geomorphic significance of some Oxford soils', *American Journal of Science*, vol. 257, 503–15.

Chorley, R. J. (1969) 'The elevation of the Lower Greensand ridge, south-east England', *Geological Magazine*, vol. 106, 231–48.

Collet, L. W. (1927) *The Structure of the Alps*, London, Arnold.

Cotton, C. A. (1948) *Landscape*, 2nd edn, Christchurch, NZ, Whitcombe & Tombs.

Davis, W. M. (1930) 'Physiographic contrasts, east and west', *Scientific Monthly*, vol. 30, 394–415, 501–19.

Dennis, J. G. (1972) *Structural Geology*, New York, Ronald Press.

De Sitter, L. U. (1964) *Structural Geology*, 2nd edn, New York, McGraw-Hill.

Eardley, A. J. (1962) *Structural Geology of North America*, New York, Harper & Row.

Fenneman, N. M. (1938) *Physiography of Eastern United States*, New York, McGraw-Hill.

Gilluly, J., Waters, A. C. and Woodford, A. O. (1968) *Principles of Geology*, 3rd edn, San Francisco, Calif., Freeman.

Hills, E. S. (1972) *Elements of Structural Geology*, London, Chapman & Hall.

Holmes, A. (1978) *Principles of Physical Geology*, 3rd edn (2nd edn, 1965), London, Nelson.

Hunt, C. B. (1956) 'Cenozoic geology of the Colorado Plateau', *US Geological Survey Professional Paper* 279.

Hunt, C. B. (1967) *Physiography of the United States*, San Francisco, Calif., Freeman.

Lobeck, A. K. (1939) *Geomorphology*, New York, McGraw-Hill.

Stamp, L. D. (1946) *Britain's Structure and Scenery*, Edinburgh, Collins.

Thornbury, W. D. (1965) *Regional Geomorphology of the United States*, New York, Wiley.

Twidale, C. R. (1971) *Structural Landforms*, Cambridge, Mass., MIT Press.

Von Engeln, O. D. (1948) *Geomorphology*, New York, Macmillan.

Whitten, E. H. T. (1966) *Structural Geology of Folded Rocks*, Chicago, Rand McNally.

Wooldridge, S. W. and Linton, D. L. (1955) *Structure, Surface and Drainage in South-East England*, London, George Philip.

# Eight *Lithology and landforms*

In Chapter 7 the geomorphic importance of both geologic structure and stratigraphic variability were considered. In the present chapter the specific effect of lithology on landforms will be discussed. Lithological controls over landforms produce a large number of variations and, more important, these variations may be associated with a wide range of discrete regions varying in size from a distinctive outcrop of a few square metres to areas of uniform rock type extending over hundreds of square kilometres. It is convenient in this chapter to examine some of the major geomorphic features associated with arenaceous, argillaceous, calcareous, igneous and metamorphic rocks, while recognizing the many variations within each rock type.

## 8.1 Arenaceous landforms

Arenaceous rocks, including siltstones, exhibit a wide range of geomorphic behaviour when exposed to the processes of denudation. Although the climatic environment within which erosion occurs exercises a strong influence over the resulting landforms (see Chapter 18), the intrinsic rock properties are also important. Among these properties, the most significant are cementation, jointing and bedding, and permeability.

Silica-cemented sandstones are the most resistant arenaceous rocks, and they usually provide landforms of low drainage density and striking relief on account of their resistance to chemical weathering, particularly where jointing is sparse. The differences in drainage density (i.e. total channel length per unit area: see Chapter 13) between lithologically similar resistant sandstones (Exmoor (Hangman Grits) 3·0 mile/sq. mile (1·9 km/km²); Central Appalachian Plateau (Pocono and Pottsville) 5·0 (3·1 km/km²); Northern Alabama (Pottsville) 8·5 (5·3 km/km²)), such as are shown in Figure 8.1, are probably mainly attributable to significant differences in rainfall intensity (at present the mean monthly maximum precipitation intensities are, respectively, 1·72, 2·95 and 4·81 in. (44, 75, 122 mm) per 24 hours). The following examples, important in other contexts, will serve to illustrate the topographic resistance of silica-cemented clastic rocks:

(1) Conglomerate: Shinarump Member (Triassic), a 10–60-m-thick conglomerate capping shales and sandstones, forming the Chocolate Cliffs north of the Grand Canyon (up to 330 m high).

(2) Greywacke: Coniston Grits (Silurian), a formation 750–1700 m thick, giving relief in excess of 400 m in the southern Lake District, in north-west England.

(3) Arkose: Torridonian Sandstone (Pre-Cambrian), the Applecross Group, 2000–2600 m (quartz 60 per cent, feldspar 30 + and 5 + per cent iron oxide, etc.) of almost horizontal, cross-bedded arkosic grits and conglomerates, produces striking, steep-sided hills of relief up to 700 m in the north-west Highlands of Scotland.

(4) Sandstone: Pocono Sandstone (Mississippian), 230 m of massive sandstone (quartz 63 per cent, rock fragments 28 per cent and matrix 6 per cent) forms pronounced ridges in the Appalachian Ridge and Valley Province (see Figure 7.1) and, with the overlying similar Pottsville Sandstone (Pennsylvanian: 130 m), supports the elevations of 650–750 m in the Appalachian Plateau. Hangman Grits (Devonian), up to 1650 m of fine-grained fluviatile deltaic sandstones, dipping south at angles of up to 30°, supports high relief and elevations of up to some 550 m in Exmoor, south-west England.

Iron oxide provides a less resistant cement, allowing locally high rates of infiltration and rapid weathering rates. The New Red Sandstone (Triassic) of England is associated with low rolling terrain, but a more resistant iron cementation characterizes the lower part of the Navajo Sandstone (Triassic/Jurassic) of Zion Canyon, Utah, where near-vertical cliffs rise some 650 m. In some instances the iron cementation can be extremely weak, both mechanically and chemically, as in the case of the incoherent sandy facies of the Lower Greensand (Cretaceous) rocks of south-east England which, because of their high permeability, support pronounced escarpments. A third important cement is calcium carbonate, characteristic of the massive, medium–fine-grained, cross-bedded, strongly jointed aeolian sandstones of the Colorado Plateau. The chief examples of these are the upper parts of the Navajo Sandstone, the Wingate Sand-

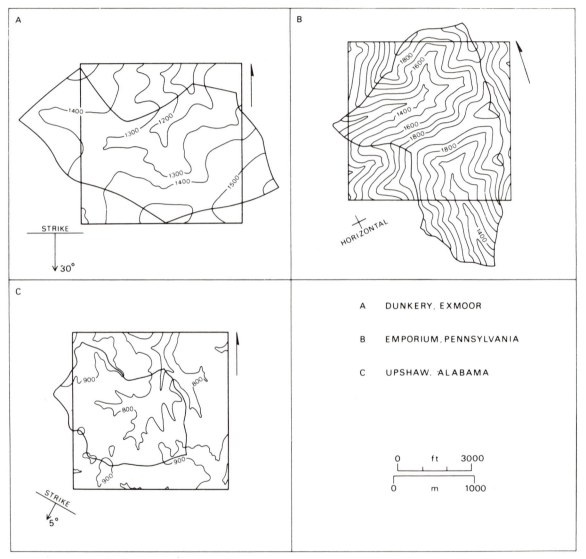

**Figure 8.1** Drainage basins developed on three suites of arenaceous formations: A. a basin on the Ilfracombe Beds and Hangman Grits of Exmoor, England (4600 ft of southward-dipping sandstones with some shales of Devonian age), having a drainage density of about 3 mile/sq. mile; B. two basins on the Pottsville and Pocono Sandstones of the Appalachian plateaus of Pennsylvania (1100 ft of horizontal coarse, massive sandstones of Carboniferous age), having a drainage density of about 5 mile/sq. mile; C. a basin on the Pottsville Formation of Northern Alabama (350–500 ft of siliceous sandstone and arkose dipping south-west of Upper Carboniferous age), having a drainage density of 8–9 mile/sq. mile. (Drainage densities 1·9, 3·1 and 5–5·6 km/km².)
*Source:* Chorley, 1957, figure 3, p. 633.

stone (100–150 m, forms the Vermilion Cliffs, north of Grand Canyon) and the cliff-forming Entrada Sandstone (up to 200 m thick, 95 per cent quartz, 90 per cent of grains average about 0·08 mm diam.) (Table 8.1).

The more gross influences of bedding and jointing on topography have already been referred to. Suffice it here to note, for example, the influence of the east–west strike on the basin shape and stream alignment in Figure 8.1A, and the striking control of the two vertical joint sets (N20°–30°W and N70°–80°E) in the Navajo Sandstone, where the very continuous joints may be spaced up to 160 m apart, over the alignment of canyons (Figure 8.2).

The permeability of many older, silica-cemented arenaceous rocks is very low and their power to transmit water is almost entirely governed by their jointing, as is also the case with granite. In addition, there is usually a great difference between permeability parallel to the bedding and at right angles to it. In a Pennsylvanian sandstone, in Illinois, containing only 12–14 per cent thin shaly partings, the vertical permeability can vary between $10^1$ and $10^{-1}$ darcy in 2 m (see Section 3.2). Types of bedding facies are also important controls over per-

meability, as evidenced by the Bethel Sandstone (Mississippian) of Illinois:

Ripple-bedded facies – poorly sorted; clay 30 per cent, sand 55 per cent, porosity 7–22 per cent; permeability $10^{-2}$ darcy.

Cross-bedded facies – well sorted; cement 12–14 per cent (silica and iron), porosity 22–27 per cent; permeability $10^0$–$10^{-1}$ darcy.

The combined effects of porosity, permeability, cement and other properties on observed weathering rates for some of the important cliff-forming sandstones in the Colorado Plateau are given in Table 8.1, suggesting that susceptibility to weathering is very much a function of high permeability plus weak cementation (see Figure 9.15).

Before leaving this consideration of arenaceous terrain, it is most important to stress that where landforms are developing in association with a stable soil and vegetation cover, the surficial geomorphic forms and processes tend to be dominated by the characteristics of the weathered debris, rather than those of the bedrock.

**Table 8.1** Physical properties of five sandstone formations in the Colorado Plateau

| Formation (location) | Unconfined compressive strength (p.s.i.) | Schleroscopic hardness | Porosity (%) | Cement | Horizontal permeability (Darcy) | Experimental depth of wetting ($\frac{1}{2}$ in. precipitation in 15 min.) (in) | Observed weathering weight loss* (%) |
|---|---|---|---|---|---|---|---|
| Entrada (Gallup, New Mexico) | 5600 | 15 | 30 | Calcium Carbonate | $10^4$ | $1\frac{3}{4}$ | 23·0 |
| Cliff House (Chaco Canyon, New Mexico) | 2500 | 10 | 28 | — | $10^3$–$10^4$ | $1\frac{1}{4}$ | 3·1–3·7 |
| De Chelly (Monument Valley, Arizona–Utah) | 7200 | 21 | 28 | Poor | $10^3$ | $1$–$1\frac{1}{4}$ | 3·2 |
| Navajo (Teec Nos Pass, Arizona) | 2200 | 16 | 30 | Calcium carbonate and iron oxide | $10^3$–$10^4$ (?) | $1\frac{1}{4}$ (Zion Canyon) | 1·5–1·7 |
| Point Lookout (Mesa Verde, Colorado) | 4500 | 11 | 34 | — | $10^2$–$10^3$ | $\frac{7}{8}$ (Fruita, Colorado) | 0·9–1·8 |
| Comments | Moderate mechanical strength | Very low hardness | High porosity | Poorly cemented | Variable (poor vertically) | | |

\* December 1962 – March 1964: See Figure 9.15.
*Sources:* Data from Jobin, 1962; Schumm and Chorley, 1966, tables 2 and 3, pp. 24–5.

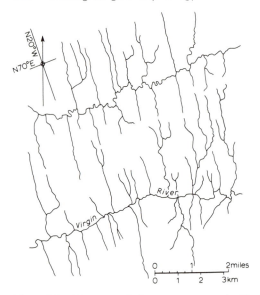

**Figure 8.2** Alignment of tributary canyons to the Virgin River and Clear Creek following pronounced joint patterns (especially N 20° W) in the Navajo Sandstone just east of Zion Canyon, Utah.
*Source:* Gregory, 1950, figure 96, p. 159.

This being so, landforms may frequently appear as the offspring of debris characteristics and only the grand-children of bedrock attributes such that, under otherwise similar soil, climatic, hydrological and vegetational conditions, different types of bedrock may exhibit similar landforms. A striking example of this is the almost indistinguishable difference between the slope forms on the Hangman Grits and the overlying Ilfracombe Slates of Exmoor, south-west England. Shear tests on soil (55 per cent silt, 40 per cent sand, 5 per cent gravel) from valley-side slopes yield a value of $\phi_{sf}$ (i.e. saturated, semi-frictional angle of internal friction) of 25·8° under saturated conditions, which is close to the modal value of stable valley-side slopes surveyed in the field (see Section 11.6). This compares with the 28° stable Hangman Grit slopes. Slopes at 33° on the Ilfracombe Slates appear to be unstable and compare directly with the unstable Hangman Grit slopes.

## 8.2 Argillaceous landforms

Landforms underlain by clay and shale universally present low relief and, for this reason, tend to receive only passing mention by geomorphologists. However, argillaceous landforms and their associated processes are of considerable geomorphic interest for two major reasons; first, because in conjunction with other sedi-

mentary rock layers their properties exert an important control over denudational processes, and, second, because argillaceous rocks support differing slope forms depending upon their moisture environment. The properties of clay minerals (e.g. shearing resistance, volume, density, etc.) are strongly influenced by the presence of water, and it is with the influence of moisture regime on the morphometry of clay and shale rocks that this section is mainly concerned. Although an oversimplified view, it is convenient to divide argillaceous landforms into two groups:

(1) Those which evolve under humid conditions of abundant moisture throughout the year with water tables at shallow depths, significant moisture content (except for limited periods) within a few centimetres of the surface and a more or less continuous vegetation cover.

(2) Those developing under semi-arid or arid conditions where the moisture intake and changes of moisture are limited to a surface layer a few tens of centimetres thick and the vegetational cover is sparse or non-existent.

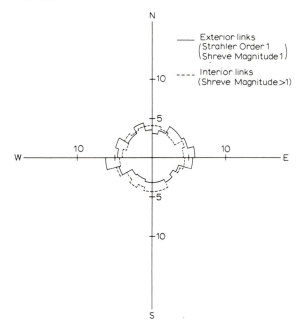

**Figure 8.3** Percentage frequency of stream links in 15° absolute azimuth classes for streams on the Culm Measures (shales), Devon, England; comparing exterior (i.e. fingertip) links (Shreve magnitude = 1: full line) with interior links (Shreve magnitude > 1: broken line). This shows the greater west–east orientation of the exterior (fingertip) links which are most controlled by rock jointing.
*Source:* Jarvis, 1976, figure 3, p. 571.

Between these two extremes lie the argillaceous land-forms of the summer-seasonally dry areas of Mediterranean climate.

Argillaceous landforms in humid regions commonly present slopes of very low angle ($< 8°$) with moderate drainage densities which are prevented from being very high by the resistance to surface erosion and channel formation provided by the vegetation cover. Drainage patterns are dendritic with the orientation of drainage lines dominantly controlled by the regional gradient of the terrain. In some shales, however, the direction of jointing and bedding does exercise some control over the orientation of the smallest fingertip tributaries, as with the west–east striking Carboniferous Culm Measures of Devon, England (Figure 8.3) (see Chapter 13).

The contrasting conditions controlling the development of clay–shale landforms in subhumid regions have been investigated in the western United States. The average rainfall of 8–15 in occurs mostly as short-lived storms of small amount in the period autumn–spring (e.g. 0·3 in in storms spaced about seven to ten days apart) with the occasional heavier summer storm (e.g. 3–4 in in 24 hours). These rainfall characteristics, together with the lack of vegetation and high evaporation, cause the moisture changes to be limited to a surface layer of the argillaceous rocks the differing behaviour of which is controlled by differences in mineral composition. These conditions produce a characteristic 'bad-lands' terrain with very high drainage density and a low relief of a few tens of metres made up of short, steep (i.e. 30°–45°) slopes, separated from basal slopes of very low angle by sharp breaks and rising to narrow divides (see Figure 11.20).

## 8.3 Calcareous landforms

Because of their susceptibility to solution, calcareous rocks commonly present the most distinctive of the land-form assemblages associated with a given rock type, so

much so that the term *karst* (German, 'bare stony ground') has been applied to it. 'True' karst (holokarst) terrain is characterized by the development of underground solution networks, disorganized surface drainage and a surface terrain characterized by a hierarchy of holes and closed depressions resulting from solution and collapse. The classical karst terrain, such as covers some 25 per cent of Yugoslavia, develops under an ideal combination of circumstances:

(1) A considerable thickness of pure limestone ($> 4000$ m in parts of Yugoslavia; 700 m in Jamaica and 200 m in Yorkshire) with a high crushing strength, low permeability and intersected by a massive organized joint system (in parts of northern England joints are spaced 0·5–1 m apart, but tens of metres are common in some areas). These joints focus surface and subsurface solution, as do bedding planes which may, however, be absent throughout considerable thicknesses of limestone (especially forereef facies).

(2) Considerable uplift above sea level, giving high relief and great initial height of the surface above any local or regional water tables which may develop at depth. Parts of the Yugoslavian karst region are 2000 m above sea level.

(3) A high annual precipitation, associated with a measure of vegetational cover and soil development (some parts of the Yugoslavian karst receive $> 5000$ mm/year). Characteristic karst landforms tend to be absent where the mean annual precipitation has been less than 300 mm (12 in) for a long period.

The solution of calcareous minerals is governed by a complex series of reversible reactions between the gas phase (carbon dioxide ($CO_2$) in the atmosphere, soil air, cave air, etc.), the liquid phase (e.g. carbonic acid ($H_2CO_3$) in natural water) and the solid phase (notably calcite ($CaCO_3$), as a naturally occurring mineral or reprecipitated). The solid-to-liquid reactions are (Pricknett *et al.*, 1976, pp. 216–17):

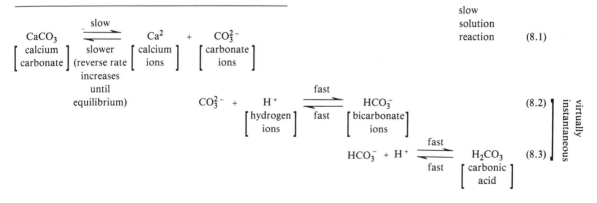

$$\underset{\substack{\text{[calcium} \\ \text{carbonate]}}}{CaCO_3} \underset{\substack{\text{slower} \\ \text{(reverse rate} \\ \text{increases} \\ \text{until} \\ \text{equilibrium)}}}{\overset{\text{slow}}{\rightleftharpoons}} \underset{\substack{\text{[calcium} \\ \text{ions]}}}{Ca^2} + \underset{\substack{\text{[carbonate} \\ \text{ions]}}}{CO_3^{2-}} \qquad \begin{array}{l}\text{slow} \\ \text{solution} \\ \text{reaction} \end{array} \quad (8.1)$$

$$\underset{\substack{\text{[hydrogen} \\ \text{ions]}}}{CO_3^{2-} + H^+} \underset{\text{fast}}{\overset{\text{fast}}{\rightleftharpoons}} \underset{\substack{\text{[bicarbonate} \\ \text{ions]}}}{HCO_3^-} \qquad (8.2)$$

$$HCO_3^- + H^+ \underset{\text{fast}}{\overset{\text{fast}}{\rightleftharpoons}} \underset{\substack{\text{[carbonic} \\ \text{acid]}}}{H_2CO_3} \qquad (8.3) \quad \left.\right\} \text{virtually instantaneous}$$

The gas-to-liquid reactions are (Pricknett *et al.*, 1976, p. 217):

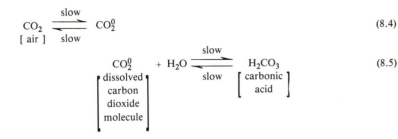

$$CO_2 \underset{\text{slow}}{\overset{\text{slow}}{\rightleftharpoons}} CO_2^0 \qquad (8.4)$$
[ air ]

$$CO_2^0 + H_2O \underset{\text{slow}}{\overset{\text{slow}}{\rightleftharpoons}} H_2CO_3 \qquad (8.5)$$

[ dissolved carbon dioxide molecule ] · [ carbonic acid ]

The dissolving of $CO_2$ depends differentially on a number of factors (Jennings, 1971; Sweeting, 1972):

(1) The amount of $CO_2$ in the free atmosphere. This varies slightly with location and time but its average value is expressed as a partial pressure of 0·00033 atm (i.e. 33/100,000 of atmospheric pressure is due to the contribution of $CO_2$).

(2) The amount of $CO_2$ in the soil air with which percolating water may be in contact. The $CO_2$ content of soil air may be greatly in excess of that in the free atmosphere (reaching at times a partial pressure of 0·1 atm – i.e. some 300 times that of the free atmosphere), especially in tropical soils. The production of soil $CO_2$ is due to a complex of biological processes, including vegetational decay, root respiration and bacterial action (Figure 8.4). The following figures of $CO_2$ production (milligrams per 24 hours produced by the equivalent of 1 g of dry vegetation or bacterial matter) show, in particular, the importance of the latter action: sugar beet 3, barley 70, wheat 90, oats 120 and four types of bacteria 480–13,000.

(3) The amount of $CO_2$ in the confined air of subterranean fissures and caves.

(4) The temperature of the fluid. This exerts a powerful inverse control over $CO_2$ solubility such that, for example, at 20°C the solubility of $CO_2$ is half that at 0°C.

(5) The temperature of the atmosphere. This is much less important than factor 4.

(6) The hydrostatic pressure of the fluid. A complex control that is operative at depth in confined openings.

(7) The time of fluid–gas contact. This is not greatly important as $CO_2$ equilibrium is reached in a matter of minutes or hours, the time commonly involved in rain falling, running across the surface and percolating through the soil.

(8) The addition of a chemical base to the fluid,

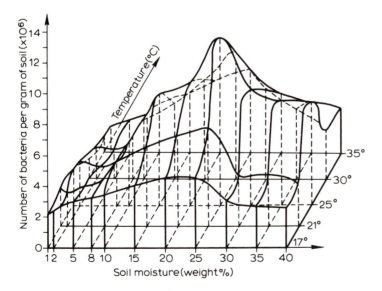

**Figure 8.4** Three-dimensional block diagram representing the biological activity of soil (in terms of number of bacteria) as a function of soil temperature (°C) and moisture (% by weight).
*Source:* Jakucs, 1977, figure 54, p. 148; based on Fehér.

**Figure 8.5** Saturation equilibrium curves for solution of calcium carbonate as a function of pH and equilibrium carbon dioxide in solution; temperatures are in deg C: A. climatic regions are only roughly indicated after the work of Corbel. For instance, both points **H** and **I** are in north Yorkshire, the first from sparry, crystalline limestone and the second from water derived from permeable limestones. **A** = saturated at 30°C, **A**′ = **A** cooled by 20°C would dissolve a further 110 p.p.m. of $CaCO_3$; **B** = supersaturated (deposition?); **C** = undersaturated (solution?); B. saturated water at **D** when cooled (i.e. displaced in direction c) can dissolve more limestone, if it is warmed (direction w) it will precipitate, if it comes into contact with air with less $CO_2$ (direction l) it will lose $CO_2$ and precipitate $CaCO_3$, if the adjacent air has more $CO_2$ (direction h) it will take in more $CO_2$ and dissolve $CaCO_3$. A 1:1 mixture of two equilibrium solutions (**E** and **F**) produces an undersaturated aggressive solution (**G**).

*Sources:* B. W. Sparks, *Rocks and Relief*, 1971, figure 5.11, p. 208, copyright © Longman, London; Jennings, 1971, figure 4, p. 26; based on Trombe.

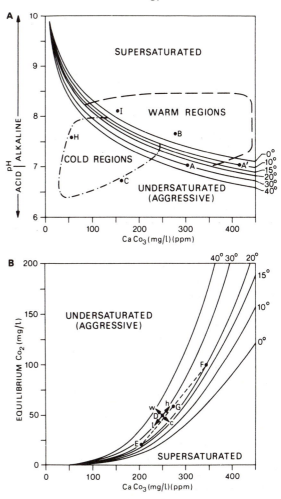

destroying $H_2CO_3$ and allowing renewed solution of $CO_2$.

The amount and rate of limestone solution thus depends directly upon:

(1) the amount of $CO_2$ present in the fluid;

(2) the amount of water in contact with the rock – i.e. the discharge;

(3) the degree of turbulence, which is related to velocity and temperature;

(4) the time of contact between fluid and limestone, which operates counter to controls 2 and 3;

(5) the temperature of the fluid; higher temperatures give higher rates of solution but lower equilibrium amounts of dissolved $CaCO_3$;

(6) the presence of organic, humic and bacterial acids, due to vegetational decomposition, etc. (Table 8.2 gives average figures for dissolved calcite and shows the influence of soil on solution);

**Table 8.2** Dissolved $CaCO_3$ (mg/l)

|                     | Bare rock | Soil–rock interface |
|---------------------|-----------|---------------------|
| Cold climates       | 45        | 100                 |
| Temperate climates  | 50 +      | 114                 |
| West Indies         | 90        | 85                  |

(7) the presence of lead, iron sulphides, sodium, or potassium in ground or soil water. For example, water from pyritic shales contains weak sulphuric acid.

The amount of dissolved $CaCO_3$ can be measured against the optimum amount for that temperature, expressed as pH (see Section 9.2.2) or equilibrium $CaCO_3$ (mg/l) (Figure 8.5). It is interesting that many natural limestone waters are supersaturated, especially in warm regions, showing the effect of other solution controls, notably the presence of organic compounds. The effect of solution of limestone surfaces, notional average rates of which will be discussed later, produce systems of surface flutings (*Karren*, German; *lapiès*, French; 'grikes', northern English), which are usually small but which may occasionally reach 20 m long and 1 m deep. These are best developed on non-porous limestones with steep surface slopes (especially 60°–80°) and are commonly associated with joint or bedding planes. A karren complex forms a limestone pavement which many authorities believe to have developed most favourably under snow, peat, or glacial till covers which yielded chemically aggressive water at their base. Where the calcareous rock is impure, a residual cover may form of acid clay, clay-with-flints (possibly derived from the dissolved chalk of southern England and northern France, and possibly also from the overlying Tertiary beds), terra rossa in Mediterranean regions and bauxite in the tropics (see Chapter 9). This material fills depressions; blocks fissures; encourages surface drainage; and commonly provides much of the scanty arable land in karst regions.

The development of calcareous terrain is profoundly influenced by surface and subsurface hydrology, which is itself very much a function of the susceptibility of the rock to discrete chemical attack and the time during which this attack has been taking place. At one end of a possible hydrological spectrum is the weak, finely fissured, porous and permeable chalk which presents the classically simple, coherent model of little surface drainage except during occasional floods, high and universal surface infiltration rates, a vadose zone within which flow is occurring throughout the rock under gravity, a distinct and continuous water table which fluctuates widely in level with lags of days or weeks after changing rainfall inputs, and a saturated groundwater zone of slower flow under hydrostatic pressure (see Section 3.2). At the other extreme is a less-organized system of discrete subterranean flows developed over short periods in relatively impermeable, massively jointed, crystalline limestones. Because of the difficulties of underground hydrological investigation and because of the many, often conflicting, variables controlling subsurface solu-

tion, two somewhat conflicting models of karst hydrology have polarized. First, there is the model which finds particular support in the United States which is based on the view that, after a relatively short time interval, subterranean solution along joints, faults and bedding planes will produce a generally integrated, interconnected fissure system within which it is possible to distinguish discrete vadose and groundwater zones. These linked fissure systems connect to form an organized regional water table, albeit structurally interrupted in places, of varying inclination. Just below this regional water table the maximum discharges of subterranean water occur. The second model, more favoured in Europe, is that of a more disarticulated subsurface flow involving multiple aquifers, independent conduits, a mixture of gravity and hydrostatic flow, the existence of siphons, with the majority of underground flow confined to a few major cave systems even in the late stages of karst development (Figure 8.6). There is little doubt that subterranean flow in karst areas is discrete, at least in the vadose zone, and perhaps even at levels far below any possible general water table. Water entering the surface through a number of openings tends to unite at depth and appear as a smaller number of springs under gravity (mainly at the base of more grossly permeable limestone outcrops, at junctions with underlying impermeable beds, along faults, etc.) or as risings under hydrostatic pressure (vauclusian springs) such that it has been suggested that a three-dimensional stream ordering pro-

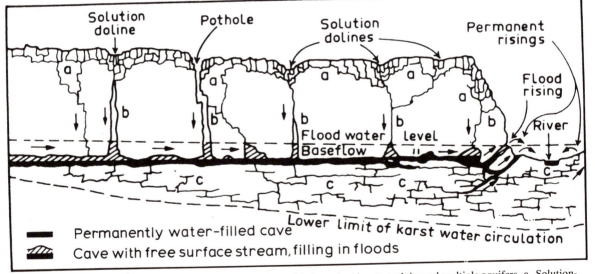

**Figure 8.6** Karst hydrologic circulation based on the concept of independent karst conduits and multiple aquifers. a. Solution-opened joints and bedding planes with seepage water; b. Potholes and joints with seepage and stream flow; c. Solution-opened joints and bedding planes permanently water filled.
*Source:* Jennings, 1971, figure 24, p. 92, after Cavaillé.

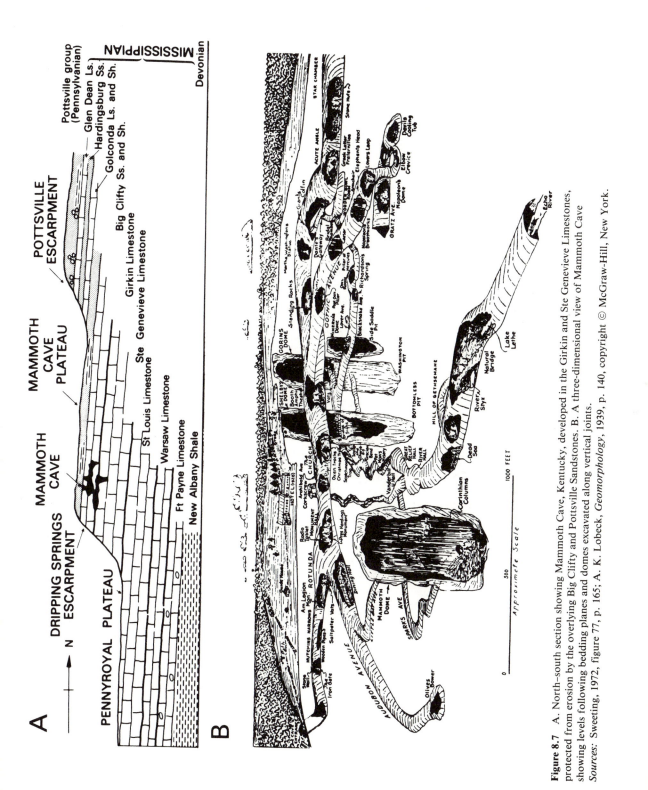

**Figure 8.7** A. North–south section showing Mammoth Cave, Kentucky, developed in the Girkin and Ste Genevieve Limestones, protected from erosion by the overlying Big Clifty and Pottsville Sandstones. B. A three-dimensional view of Mammoth Cave, showing levels following bedding planes and domes excavated along vertical joints.

*Sources:* Sweeting, 1972, figure 77, p. 165; A. K. Lobeck, *Geomorphology*, 1939, p. 140, copyright © McGraw-Hill, New York.

cedure might be applied to the subterranean flow cavities. Long subterranean rivers flowing under gravity are occasionally encountered, such as the River Timavo which disappears into a Yugoslavian cave to reappear almost 50 km away on the Italian coast. The geometry of such rivers is of interest in that they frequently present concave-up longitudinal profiles unrelated to lithology or bedding, and even meandering patterns and evidence of stream capture; in other words, features common in surface rivers. The possible significance of this fluvial morphometry will be returned to later.

The problems of interpreting the disposition and flow of subterranean water, together with uncertainties regarding its chemical aggressiveness have led to considerable theorizing as to the development of limestone caves. The location and form of cave systems is clearly influenced by lithology and structure. They develop best in pure, resistant, impermeable, well-jointed limestone having a high crushing strength; and are located with particular reference to joint systems (e.g. on the crests of anticlinal ridges in the Jura Mountains), along bedding planes (e.g. Mammoth Cave, Kentucky: Figure 8.7), at contacts of pure and impure limestone and at shale contacts. Subterranean caves vary greatly in size from embryonic openings to huge chambers. When the slow solutional enlargement of planes of parting has caused a large-scale interconnected system of openings to develop with widths exceeding some 5 mm, turbulent flow may occur, and this drastically increases the rate of solution such that large cave systems may develop at favoured locations with water flowing at rates of 4–1000 m/h or more. It is now generally agreed that most subterranean enlargement is by corrasion and that corrasion is only important where resistant exotic bed material has been introduced by allochthonous rivers rising outside the karst region. It has been suggested that some cave systems may have developed quite rapidly, and the lack of glacial till material in some caves of western Ireland indicates that caves 1000 m long may have been formed in post-glacial times. One of the most taxing problems is to try to relate levels of preferential cave formation to the groundwater disposition, and four zones have been suggested:

(1) the top 10–20 m of limestone just below the surface, as being the level of particularly aggressive water;

(2) the vadose zone of free gravity flow with a gas phase possible with cave air. Most cave systems lie in this zone, but it is by no means sure that, at the time of formation, their location was in this zone. Some authorities believe that uplift may have drained cave systems dissolved below the water table;

(3) at or just below the water table because of the huge quantites of relatively fast-moving groundwater here. Many cave systems lie at this level (including Mammoth Cave, Kentucky, which is partly a vadose system, and partly at the water table), but again there are complications in that there is a complex feedback effect because the existence of a cave level itself influences the elevation of the water table. It has also been suggested that large cylindrical cave shafts (e.g. Mammoth Dome: Figure 8.7) develop by solution and collapse along master vertical joints where water table fluctuations cause an alternation between vadose and groundwater conditions;

(4) below the water table in the true groundwater zone. The amounts of water involved here, together with the possibility that the mixing of saturated waters may result in an aggressive, undersaturated solvent (Figure 8.5, see points E, F and G) and that of local high-velocity flow under considerable hydrostatic pressure has led to the view that many large cave systems, subsequently elevated and drained, may be of true groundwater origin. For example, Carlsbad Caverns in the Permian reef complex of west Texas extend well below the level of the adjacent plain and may have been partly of groundwater zone development.

The largest explored cave system is in the Swiss Alps (Hölloch); Mammoth Cave, Kentucky, is the second largest, with 72 km (45 miles) of explored passages; the largest single chamber is Big Room at Carlsbad Caverns, being 500 m (1500 ft) long, 90 m (285 ft) high and covering 5·6 ha (14 acre); and the deepest known cave is the Gouffre Berger, near Grenoble, France, being 1130 m (3750 ft) below the surface. It is clear that the subterranean development of integrated fissure and cave systems influence both surface drainage and topography, causing surface rivers to disappear by means of a swallow hole (American, 'sink hole'; Slavic, *ponor*) or vertical shaft (English, 'pot hole'; Slavic, *Jama*) and surface collapse to occur (Sweeting, 1972).

As the underground cave system develops and the karst surface is lowered by solution and weathering more and more of the surface drainage becomes diverted underground, leaving a very specialized subaerial valley system, and the terrain becomes increasingly dominated by a hierarchy of closed depressions. A feature of some holokarst regions (as well as of chalk terrains, which will be referred to later) is the existence of variably integrated systems of dry-surface valleys incised into a terrain the bare slopes of which are dominated by the stepped forms resulting from the outcropping of the limestone bedding (e.g. the Derbyshire area of the English Pennines and the

Carlsbad region of Texas and New Mexico). The origin of these valley systems, parts of which may be temporarily inundated during very wet weather, has been variously ascribed to the existence of an earlier water table at or near the surface, an earlier more-pluvial climate, or to the initial superimposition of drainage developed on overlying non-calcareous rocks which have been eroded away. The efficacy of the last mechanism would seem to depend very much upon the character of the overlying rock, for if it were relatively impermeable, an integrated surface drainage network might be delivered to the limestone surface to be later progressively disorganized by subsurface abstraction of drainage. On the other hand, if superimposition took place from a permeable stratum, it might be on to a limestone already riddled with caves which would abstract the surface drainage before substantial fluvial valley-cutting could be effected. Other kinds of more characteristic holokarstic valleys are:

(1) through-valleys, commonly steep-sided and gorgelike (especially where the limestones are horizontal: e.g. the Tarn gorge of south-western France and the Dove and Manifold valleys in Derbyshire, England), produced by rivers rising outside the karst region carrying exotic, resistant bed material to aid corrasion;

(2) gorges produced by the collapse of the roofs of cave systems near the surface (e.g. Cheddar Gorge in the Mendip Hills of south-western England); partial collapse may produce natural bridges;

(3) blind valleys, closed at the downstream end by the abstraction of surface drainage by a swallow hole;

(4) steepheads, which are valley heads of steep inclination formed by the sapping back of a cliff due to accelerated erosion at its base by a spring or rising.

The most characteristic feature of karst terrains is the doline, a closed depression occurring in a wide range of depths (2–100 m) and diameters (10–1000 m) and at densities of up to 2460/km². The two main mechanisms of doline formation appear to be accelerated solution and settling in a location of especially close jointing (Figure 8.8A) or roof collapse of near-surface caves (Figure 8.8B). The former mechanism produces slopes of 30°–40°, although they may vary widely (e.g. dolines in the Causses region of south-western France commonly have slopes of <5°), whereas the latter mechanism may give more narrow and deep dolines with almost vertical walls. Dolines occur most characteristically in well-jointed, pure, resistant limestones, especially on flatter surfaces or along dry valleys. These depressions occur in a hierarchy of sizes and appear to be susceptible of enlargement and deepening, especially as a result of the aggressive chemical action of water deriving from doline soils, residual material and snow accumulations. The uniting of several adjacent dolines produces a more complex uvala, but the largest class of karst closed depressions, the polje, cannot be attributed to such growth. Poljes are large, more or less closed depressions (up to 50 km long and 6 km wide in Yugoslavia) with steep sides and flat alluvial floors, frequently dotted with conical residual hills (hums), with slopes of 20°–30° (Figure 8.8C). They appear to be grossly associated with major structural depressions, such as fault and synclinal troughs, but there is ample evidence for their deepening

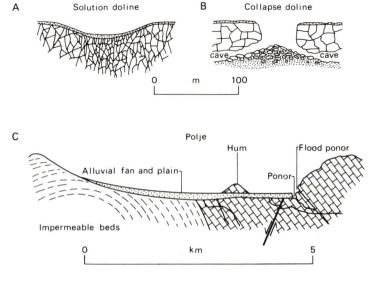

**Figure 8.8** Schematic cross sections of karst depressions: A. solution doline; B. collapse doline; C. theoretical cross section to show polje formation by lateral corrasion on an alluvial floor.
*Sources:* Williams, 1969, figure 6.II.3, p. 273; Jennings, 1971, figure 46D, p. 141, based on Roglić and Louis.

and widening by chemical processes. These depressions are frequently subject to seasonal inundations, being imperfectly drained by ponors, and this action probably produces high local rates of solution.

The lithological and structural importance of widely spaced open joints, thick bedding, high crushing strength and the purity of crystalline limestone has been stressed as important to the development of holokarst land-scapes. In this respect the massive deeply jointed and relatively unbedded forereef deposits are of especial importance (as distinct from the bedded backreef lime-stone with more closely spaced and shallow jointing). Where magnesium carbonate ($MgCO_3$) is present with $CaCO_3$ in only 2 or 3 per cent, solution is strikingly inhibited. The calcareous rock whose lithological and structural characteristics militate most strikingly against the development of karst landforms is chalk, most classically present in the Cretaceous sequence of southern England and northern France. In southern England the Chalk is up to 450 m (1500 ft) thick of which the bottom third (the Lower and Middle Chalk) is more argillaceous, impure and impermeable, forming generally lower ground, whereas the thick Upper Chalk is a soft, permeable, firmly jointed rock. Table 8.3 com-pares the properties of such a rock with those of others. The chalk is too weak and permeable to form true karst, with solution being spread uniformly throughout the rock mass, which is too weak to sustain many large cavities. Caves and sink holes are rare but solution pipes filled with residual clay and flints are found, especially towards the top of the formation where aggressive water has percolated down from the overlying Tertiary beds, now extensively stripped off. The chalk is thus associated with striking escarpments rising to nearly 275 m (900 ft) in places, composed of rolling terrain with broad convex divides. This surface, which may be complex in age (see Chapter 2), is intersected with a low-density network of largely dry valleys (southern England 2·26 km/km$^2$ or 3·66 mile/sq. mile; Picardie, northern France, 0·74

km/km$^2$ or 1·30 mile/sq. mile), parts of which appear never to be inundated under present conditions. Various origins have been proposed for these valleys, including climatic changes, the decrease of permeability by soil freezing during periglacial times and regional lowering of the well-developed chalk water table by recession of the scarp front.

Various estimates have been made for the theoretical rates of karst surface lowering by solution, ignoring solu-tion at depth and the presence of organic acids (i.e. based on the theoretical Henry–Dalton law of gas absorption). One of the most popular is (Sweeting, 1972):

$$X = fQTn/(10^{12}AD)$$

where:  $X$ = solution in millimetres during a given time;
$Q$ = drainage basin discharge (m$^3$);
$T$ = hardness of the discharged water (p.p.m.);
$A$ = area of drainage basin (km$^2$);
$D$ = density of rock;
$\frac{1}{n}$ = fraction of basin with limestone out-cropping;
$f$ = a constant, depending on the units used (1000 for metric).

Such formulae accord with the very rough estimates of surface lowering in Table 8.4. These very approximate theoretical estimates are based on the assumption that amount of water and lowering of temperature produce the highest solution rates. However, Jakucs (1977) takes strong issue with these assumptions and gives the follow-ing estimate of relative (%) karst denuding agencies in different regions (Table 8.5).

The two sets of figures in the tables demonstrate a wide divergence of views, particularly regarding the relative aggressiveness of karst waters in cold and wet v. hot and wet climates. There is, however, general agreement on the virtual absence of karst processes in desert areas and on the relatively high rates in seasonally wet Medi-terranean highlands, like the Yugoslav karst region. Wet tropical karst is coming to be regarded as a zone of

**Table 8.3**   Comparative properties of some limestones

|  | Specific gravity | Porosity (%) | Compressive strength (lb/in$^2$) |
|---|---|---|---|
| Chalky limestone (USA) | 1·8 | 26·0 | 3,000 |
| Older crystalline limestones (USA) | 2·6 | < 6·0 | 17,000–37,000 |
| Modern reef breccia (Eniwetok) | 2·35 | 14·0–16·0 | 4,960 |
| Sandstones (USA) | 2·06–2·33 | 3·0–16·0 | 6,000–32,000 |

*Source:* Sweeting, 1972, table 1b, p. 11.

**Table 8.4** Rates of limestone denudation

| Region | mm/1000 year |
|---|---|
| Cold and wet | 450 |
| Southern Alps | 300 |
| Periglacial | 200 |
| Cold and damp | 160 |
| Temperate mountains | 100 |
| Yugoslavia | 80 |
| Derbyshire | 80 |
| Hot and wet | 80 |
| Jamaica | 72 |
| West Ireland | 55 |
| Warm and wet | 45 |
| East England Chalk | 25 |
| Cold arid | 14 |
| Warm arid | 6 |

higher surface limestone solution than temperate karst because of the great amounts of surface runoff aided by vegetal and biological activity, particularly by organic acids in swampy lowlands. It certainly appears that the corrosive action of tropical waters is largely limited to the surface and near-surface where they become rapidly saturated. This view is supported by the relative lack of large limestone cave systems in the tropics, where caves tend to be more usually networks of small tunnels, and of collapse dolines (an exception being in the Mulu Hills, Sarawak).

Two related forms of wet tropical karst landforms have been identified:

(1) Cone karst (German, *Kegelkarst*) (Figure 8.9A), exemplified by the 'cockpit country' of Cuba and Jamaica. This is an assemblage of steep-sided (60°–90°), rounded hills 100–160 m (300–500 ft) high

and 0·75–1 km in diameter, separated by star-shaped closed depressions with inner slopes of 30°–40°. These hollows appear to follow the intersections of major joint systems and to be formed by a combination of rapid-surface and near-surface solution, assisted by soil and vegetation chemical effects, together with minor near-surface collapse. The star shape of the depressions is most probably produced by the erosional valley effects associated with the rapid runoff and surface erosion resulting from the intense tropical rainstorms.

(2) Tower karst (Figure 8.9B), as found in an area of some 600,000 km² in the monsoonal area of south China (especially Kwangsi, where it has stimulated much classical Chinese landscape art) and Vietnam. These landforms are dominated by groups of more or less isolated very steep-sided hills up to some 300 m high surrounded by flat alluvial plains. The hills appear to be controlled by strong vertical jointing and to have been subjected to pronounced basal stream sapping and lateral corrosion associated with special conditions of seasonal lowland inundations. Some authorities believe that the limestone residuals have evolved over very long periods (e.g. perhaps during the 65 million years since the Cretaceous), but others view them as resulting from exceptionally rapid solution (Figure 8.9B).

An earlier chapter was partly devoted to the notion that landforms may pass through a sequence, or cycle, of recognizable and well-marked stages, and such a cycle has been proposed for the theoretically ideal development of holokarst in wet temperate latitudes (Figure 8.10). At the beginning of youth (1 in Figure 8.10) drainage is assumed to have initiated on a thick, pure, massively jointed limestone sequence either by its uplift

**Table 8.5** Relative (%) karst denuding agencies

| Agency | High mountains and periglacial | Temperate | Mediterranean | Desert | Wet tropics | Worldwide |
|---|---|---|---|---|---|---|
| Atmospheric $CO_2$ | 2·70 | 0·63 | 0·48 | 0·30 | 0·36 | 4·47 |
| Inorganic $CO_2$ | 0·30 | 0·81 | 0·96 | 0·15 | 1·80 | 4·02 |
| Biogenetic $CO_2$ | 1·80 | 4·86 | 6·60 | 0·00 | 36·00 | 49·26 |
| Inorganic acids | 0·30 | 0·45 | 0·96 | 0·55 | 2·88 | 5·14 |
| Organic acids | 0·90 | 2·25 | 3·00 | 0·00 | 30·96 | 37·11 |
| Total | 6·00 | 9·00 | 12·00 | 1·00 | 72·00 | 100·00 |

*Source:* Jakucs, 1977, table 14, p. 111.

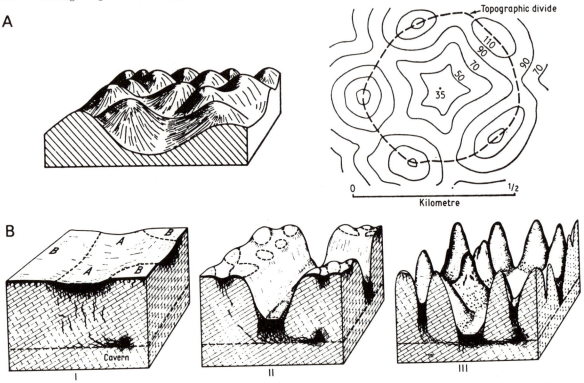

**Figure 8.9** Tropical karst features: A, cone karst with associated map of 'cockpit', B. three stages in the development of tropical tower karst: I, preferential solution and debris accumulation in area **A**, as compared with **B**; II, intense chemical corrosion under the soil cover increases the disproportionate elevations of areas **A** and **B**; III, the **B** areas are reduced to small, steep towers whose rate of erosion may be only one-tenth that of the intervening, chemically active lowlands.
*Sources:* Williams, 1969, figure 6.II. 9, p. 281; Jakucs, 1977, figure 45, p. 133.

or by superimposition from overlying rocks. During youth solution proceeds in the upper limestone levels, caves form, subsurface drainage achieves some measure of integration and dolines begin to form. Maturity begins when most of the original limestone surface has been destroyed by coalescing doline development and most drainage is rapidly conducted underground into a well-developed cave system (2 and 3 in Figure 8.10). In old age erosion has thinned the limestone, the water table lies close to the surface, general collapse has reintroduced surface drainage and residual hills are evident (4 in Figure 8.10). Today there is less support than hitherto for this simplified picture of karst evolution because of doubts regarding the development of coherent bodies of groundwater, the possible effects of recent climatic changes (for example, Carlsbad Caverns were probably formed during a wetter phase than at present) and the individualistic behaviour of different types of limestone. However, the possible evolutionary significance of certain limestone cave levels (e.g. in Czechoslovakia: Figure 8.11,

and in the Craven area of northern England) has been proposed as perhaps giving evidence of previous water table levels or baselevels of erosion (Sweeting, 1972).

## 8.4 Igneous destructional landforms

The forms assumed by igneous rock bodies at the surface of the earth commonly mirror their original geometry both because of the greater resistance of many igneous rocks to erosion, compared with that of the associated sedimentary rocks, and because of the comparative youth of many extrusive igneous forms. The geometry of volcanic landforms depends on the type of magma and its variations during the history of eruption, the morphology of the extrusive vent or of the intrusive bodies, and the denudation history of the feature (Chapter 6).

Basalt and ignimbrite flows produce flat-topped plateaus with cliffed and stepped edges controlled by the vertical jointing and flow bedding, respectively. There is commonly an absence of river erosion on the plateau tops

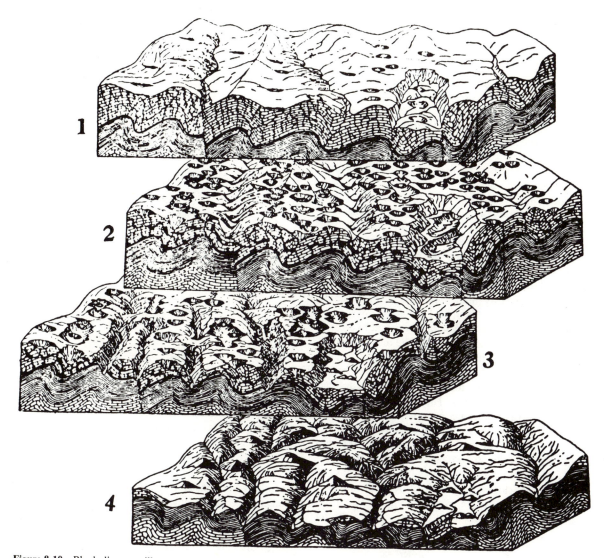

**Figure 8.10** Block diagrams illustrating four stages in the karst cycle; see text, for description.
*Source:* Cotton, 1948, figure 365, p. 468, after Cvijić.

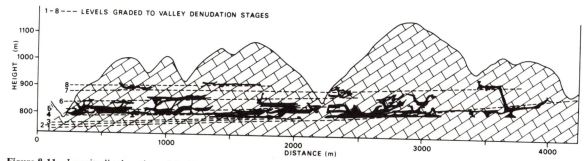

**Figure 8.11** Longitudinal section of the Demänová Caves near Liptovsky Mikulaš, Czechoslovakia.
*Source:* Sweeting, 1972, figure 72, pp. 152–3, after Droppa.

because the drainage is conducted underground by the joint systems, permeable ash and flow cavities, but deep weathering of the basalts (especially where closely jointed in the humid tropics) and areas of poorly welded tuffs may lead to considerable piecemeal reduction of volcanic plateaus by erosion. Normally the most rapid and spectacular erosion takes place around the plateau margins, especially by means of landslides produced by weathering and spring sapping at the base of the flows. Basalt plateaus, for example that of the Snake River, are bordered by a number of deep, steep-sided, theatre-headed 'alcoves' bordered by talus at the sides, but not at the heads, which are aggressively backcutting at a zone of springs and rapid weathering (Figure 8.12). The advanced erosion of a lava plateau produces an assemblage of mesas and buttes, and may also lead to the superimposition on the underlying rocks of some drainage lines developed on the lava surface. Smaller-scale flows can influence landform development by diverting surface drainage or by filling pre-existing valleys leading to 'inversion of relief' (Figure 8.13).

Scoria and ash cones are generally resistant to erosion during their early history because of the relatively small catchment area which they present to runoff and because of their high permeability. When surface weathering produces a clay-rich soil, however, these cones become ribbed with radiating gullies. The readily datable character of extrusive volcanic material allows accurate weathering rates to be established, such as on andesitic ash in St Vincent, West Indies, where clay soil is forming at the rate of 0·45–0·60 m/1000 year. Larger stratovolcanoes are very susceptible to fluvial erosion because of their long steep slopes, interbedded flows and considerable relief. This erosion is commonly initiated by mudflows (lahars) which may produce gullies during

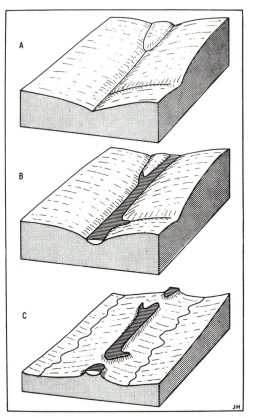

**Figure 8.13** Stages in the production of inversion of relief due to the preservation of a valley form by a resistant lava flow.
*Source:* Ollier, 1969, figure 45, p. 124.

eruptions or subsequently when the unconsolidated volcanic debris becomes saturated by rain. The continued erosion of stratovolcanoes results in a series of radiating valleys (barrancas) with theatre-shaped heads, which are enlarged by lateral spring sapping and mass movements undercutting along ash layers dipping in the direction of the valley gradient (Figure 8.14). These theatre heads may breach the central crater and coalesce to form a cliff-encircled erosion caldera, such as the Caldera of Las Palmas, Canary Islands, where 1000 m of resistant lava is underlain by less-resistant flows. The radial valleys isolate triangular facets, or planèzes, which are especially well developed on the volcano of Cantal, Massif Central of France, and on Mt Rainier, Washington State. Figure 8.15 shows the stages characteristic of the erosion of a stratovolcano, culminating in the exposing of the skeleton of its intrusive structure, in the form of necks (Figure 8.16) and dykes. The deep erosion of multiple volcanic cones may produce a much more complex assemblage of landforms. The erosion of

**Figure 8.12** Headward development of a characteristic steep-sided tributary valley (alcove) in the basalt of the Snake River Plateau.
*Source:* Cotton, 1944, figure 181, p. 345.

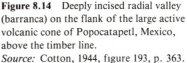

**Figure 8.14** Deeply incised radial valley (barranca) on the flank of the large active volcanic cone of Popocatapetl, Mexico, above the timber line.
*Source:* Cotton, 1944, figure 193, p. 363.

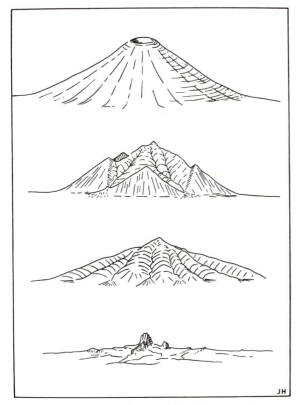

**Figure 8.15** Four stages in the erosion of a stratovolcano: intact volcano, planèze stage, residual volcano, volcanic skeleton with necks and dykes.
*Source:* Ollier, 1969, figure 41, p. 118.

**Figure 8.16** The volcanic neck of Rocher Saint Michel, Le Puy, Massif Central, France.
*Source:* Cotton, 1944, figure 202, p. 371.

the huge basalt domes also results in the formation of barrancas, planèzes and erosion calderas (e.g. the Enclos, Réunion, and Huakeine, Society Islands). However, the theatre valley heads are much more marked than on stratovolcanoes due to accentuated chemical weathering at the foot of cliffs of basic rock and to water-

fall plunge pool erosion, both of which processes are especially well marked in such humid tropical localities as Hawaii.

The eroded tops of granitic plutons cover more than 10 per cent of the present land surface and their associated landforms are dominated by the igneous structures and fabric which have been described previously. Domes, produced by erosion being guided by foliation, horizontal joints and unloading structures, are widespread features of granite terrains in all climates (e.g. Sugar Loaf, Rio de Janeiro; Yosemite Park, Sierra Nevada; Stone Mountain, Georgia). They are particularly well developed in subhumid climates where they are accentuated by the processes of pedimentation. The near-vertical and horizontal jointing of granitic outcrops affects the pattern of erosion, particularly as closely spaced jointing results in more rapid and deep weathering than does widely spaced jointing. In some tropical localities closely jointed granite has been weathered to depths of more than 50 m and the relief at the base of the weathered mantle is considerable and related to the degree of jointing. This is because granite has a low permeability and the intrusion of water into the rock mass is controlled by the frequency of joints. Weathering along joint planes produces concentrically weathered corestones, and the removal of weathered material may leave piles of partially weathered corestones in the form of *tors*. It has been suggested that granite tors (e.g. those of Dartmoor, England) may have been formed in two stages, first, by deep chemical weathering (perhaps associated with more tropical Pliocene climates) irregularly penetrating the jointed rock down to the water table in places, and, second, by the mechanical removal of the weathered material (perhaps by Pleistocene solifluction) (see Section 9.4). An alternative theory proposes a one-cycle formation involving the removal of differentially metamorphosed, less-resistant granite. Inselbergs (German, 'island mountains') are larger-scale widely jointed granite residuals which are particularly well marked in drier climates where basal weathering and pedimentation processes maintain their sharp outlines. The considerable height of some of these residuals argues against a simple history of weathering and excavation and they may be the result of successive phases of weathering on polycyclic landscapes.

## 8.5 Metamorphic landforms

The significance of metamorphism for geomorphology is that this process commonly converts rocks of lower resistance (e.g. shale and sandstone) to those of higher resistance (e.g. slate and quartzite). Like igneous rocks, ease of weathering depends on mineralogical composition and minor structures (e.g. jointing).

Although metamorphic rocks generally present more resistance to erosion than do their sedimentary counterparts, it is not easy to identify a separate class of distinctly metamorphic landforms. In some instances, however, the geometry of metamorphism is translated into distinctive landforms, as is the resistant contact aureole of metamorphosed Ordovician greywackes and argillites (the Rhinns of Kells and Merrick) around the Loch Doon composite laccolith (18 × 10·5 km) of granite and relatively less-resistant tonalite in south-west Scotland (Figure 8.17). For the most part, the complex layering, banding and cleavage of gneisses, schists, quartzites, slates and marbles is on such a small scale that landforms resulting from their erosion present more or

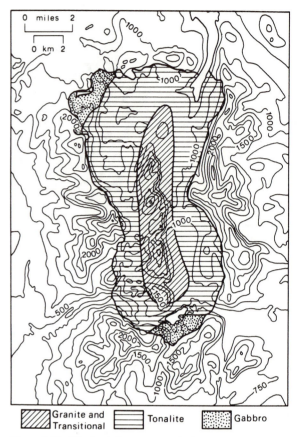

**Granite and Transitional** ▨   **Tonalite** ▤   **Gabbro** ▨

**Figure 8.17** The Loch Doon composite laccolith (granite, tonalite and gneiss), Dumfries and Galloway, south-west Scotland. The contact metamorphism of the Ordovician argillites and greywackes has produced the resistant rock forming the encircling rim of high land, commonly reaching in excess of 2000 ft. The rim to the east of the intrusion is termed the Rhinns of Kells. Contours in feet.
*Source:* B. W. Sparks, *Rocks and Relief*, 1971, figure 3.4, p. 77, copyright © Longman, London, after Gardiner and Reynolds.

less homogeneous terrains of dendritic drainage, differing only in the degree to which rock fracture and composition affects the surface debris characteristics and, thereby, drainage density, slope angles and rates of denudation. In addition, generalizations are difficult in that the majority of metamorphic rocks are of great age and were formed at considerable depth such that their present eroded exposures (e.g. in the Canadian shield, Fenno-Scandia, north-west Scotland and the Appalachian piedmont) commonly outcrop at low elevations. Exceptions occur when basement rocks have been involved in later orogenies, as in the Coast Ranges of California (Verdugo Hills and San Gabriel Mountains), the Rockies and the Blue Ridge Mountains of the older Appalachians.

The banded acidic Lewisian Gneiss of north-west Scotland provides a good example of a terrain of low elevation and low relief developed by long-continued Pre-Cambrian erosion before the overlying Torridonian Sandstones were laid down, and by Pleistocene glacial scour after the partial removal of the latter. Where zones of regional metamorphism have been peneplained and subsequently uplifted, it is possible to identify differences in terrain associated with different metamorphic rock types. The Appalachian Piedmont Province (see Figure 7.1) reaches elevations of 70–100 m (200–300 ft) in New Jersey, rising to 600 m in parts of Georgia. The dominant gneissic and schistose bedrock support a local relief of 15–100 m (50–300 ft), consisting of a rolling terrain with low knobs and ridges sloping eastward at about 20 ft/mile (3 m/km). Above this general level rise erosional remnants (monadnocks ?) of quartzite (e.g. Pine Mountain, a 60-mile ridge in Georgia) and granite

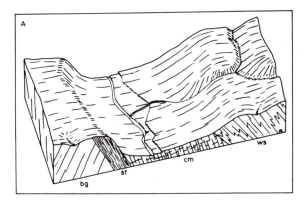

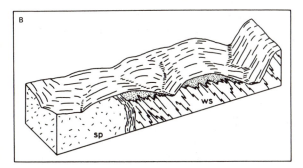

**Figure 8.18** Block diagrams showing the differential erosion of complex metamorphic terrain of the Piedmont Province in Baltimore County, Maryland., **ws** – schist; **cm** – marble; **sf** – quartzite, etc.; **bg** – gneiss; **sp** – serpentine; the band of wavy lines in B is chlorite schist.
*Source:* Cleaves *et al.*, 1974, figures 6 and 5, p. 443.

(e.g. Stone Mountain, Georgia, rising 650 ft (200 m) above the surface), whereas belts of marble, more susceptible to solution, form valleys in Pennsylvania and Maryland (Figure 8.18A). Although the calculated solution rate of the serpentine outcrops in Maryland (lowering rate 2·2 m/10[6] year) is almost twice that of the neighbouring biotite–feldspar–quartz schist, there is no elevation difference between them (except that the schist has deeper valleys and fewer ephemeral streams: Figure 8.18B), which suggests that their long-term total denudation rates are about equal (2·4 m/10[6] year). The surface runoff from the deep weathered mantle of quartz, gibbsite and kaolin covering the schist provides enough solid load to redress the erosional balance between the two rock types.

In the Fall Zone of southern Connecticut a striking relationship has been shown between metamorphic rock type and average terrain elevation (Table 8.6). About 20 per cent of the province is composed of the Carolina Slate Belt which forms slightly lower ground than the schists and gneisses. Under humid conditions it is, therefore, possible to make some generalizations regarding the resistance to erosion of certain metamorphic rocks:

**Table 8.6** Metamorphic rock type and average terrain elevation in Southern Connecticut (data from Flint, 1963)

| Low | Marble | 710 (ft) | (elevations in New Preston Quad., Connecticut) |
|---|---|---|---|
| | Mica schist | 820 | |
| | Gneiss relatively poor in quartz | 900 | |
| | Amphibolites Gneiss rich in quartz Granites | 1050 | |
| High | Quartzites | | |

(1) Quartzite – this is most mechanically and, especially, chemically resistant and is responsible for the highest relief, except after prolonged erosion. For example, sampling of solutes in streams draining granite, silica-cemented arkosic sandstone (conglomerates 15 per cent; sandstones 60 per cent; shale 25 per cent) and Pre-Cambrian quartzite (97 per cent quartz) in part of the Sangre de Cristo Range in New Mexico suggests that the last-named is being denuded at a rate many times less than either of the other two (Figure 8.19), explaining its higher elevation.

(2) Slates – these low-grade metamorphic rocks are relatively susceptible to erosion, compared with other rocks in the group, because of their impermeability

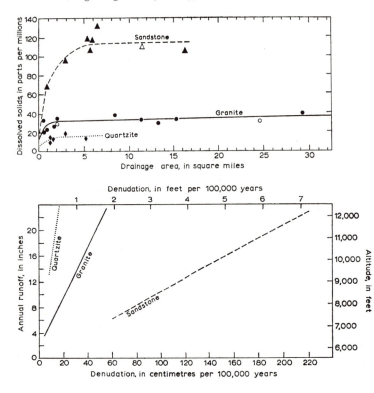

**Figure 8.19** Solution in the Sangre de Cristo Range, New Mexico: A. relation of dissolved solids in streams to drainage area; open symbols represent mean values below tributary junctions; B. estimated denudation as a function of altitude, calculated from the annual runoff of solutes, showing the relative resistance of quartzite, granite and (arkosic) sandstone.
*Source:* Miller, 1961, figures 3 and 4, pp. 18, 22.

and high degree of cleavage, especially where the cleavage planes intersect the surface at a steep angle, as is most usual. In the mountains of North Wales, the (Cambrian) slates support some of the lower elevations and in the northern part of the English Lake District the Ordovician Skiddaw Slates present a subdued glaciated upland relief, very much in contrast with the sharp relief of the andesitic lavas, tuffs and agglomerates of the Borrowdale volcanics in the central Lake District.

(3) Schists – with gneisses comprise the majority of the rocks of regional metamorphic belts and the cores of certain mountain ranges. Schists vary greatly in their resistance to weathering and erosion, depending on their composition, and resistant schists form the majority of the Grampian Highlands of central Scotland. A study by Melton (1957) of the characteristics of several dozen small drainage basins on many different lithologies in the south-western United States suggested that schistose rocks produce more clayey debris than that of gneisses, acid volcanics and arenaceous sedimentary rocks, leading to a significantly lower surface infiltration capacity (1·0 in/h (25 mm/h) v. 2·7–3·7 in/h (69–94 mm/h) for gneiss debris), and that they are associated with higher drain-

age densities (up to 80 mile/sq. mile (50 km/km²) v. 5–30 mile/sq. mile (3·1–18·6 km/km²) for gneiss) and less steep valley-side slopes (some 12° v. 21°–27°).

(4) Gneissic rocks – behave topographically very much like granites, especially when layered and massively jointed, exhibiting tors and dome structures. Gneisses weather to produce a relatively permeable rubble and sand, and the morphometric constraints of the landforms are commonly determined by the characteristics of the surface debris, which may be indistinguishable from that of many other resistant granitic and sedimentary rocks.

## 8.6 Rock strength

Geomorphic literature is filled with qualitative expressions regarding the 'strength', 'hardness' or 'softness', or 'resistance to erosion' of different rock types. One of the few attempts to quantify this central concept has been made by Selby (1980), who set up a fivefold rating scale in respect of seven variables assumed to be important in determining geomorphic rock strength (Table 8.7). From this table, a total rock mass strength rating ($\Sigma r = 25$–$100$) was obtained which was shown to exercise a high prediction over angles of slope exhibited by sedimentary rocks and dolerites in the Britannia Range of Antarctica and in

**Table 8.7** Geomorphic rock mass strength classification and ratings

| Variable | 1 Very strong | 2 Strong | 3 Moderate | 4 Weak | 5 Very weak |
|---|---|---|---|---|---|
| Intact rock strength* | 100–60 | 60–50 | 50–40 | 40–35 | 35–10 (Schmidt R) |
|  | r: 20 | r: 18 | r: 14 | r: 10 | r: 5 |
| Weathering | Unweathered | Slightly weathered | Moderately weathered | Highly weathered | Completely weathered |
|  | r: 10 | r: 9 | r: 7 | r: 5 | r: 3 |
| Spacing of joints | > 3 m | 3–1 m | 1–0·3 m | 300–50 mm | < 50 mm |
|  | r: 30 | r: 28 | r: 21 | r: 15 | r: 8 |
| Joint orientations | Very favourable; steep dips into slope, cross-joints interlock | Favourable; moderate dips into slope | Fair; horizontal dips or nearly vertical (hard rocks only) | Unfavourable; moderate dips out of slope | Very unfavourable; steep dips out of slope |
|  | r: 20 | r: 18 | r: 14 | r: 9 | r: 5 |
| Width of joints | < 0·1 mm | 0·1–1 mm | 1–5 mm | 5–20 mm | > 20 mm |
|  | r: 7 | r: 6 | r: 5 | r: 4 | r: 2 |
| Continuity of joints | None continuous | Few continuous | Continuous, no infill | Continuous, thin infill | Continuous, thick infill |
|  | r: 7 | r: 6 | r: 5 | r: 4 | r: 1 |
| Outflow of groundwater | None | Trace | Slight < 25 l min$^{-1}$ 10 m$^2$ | Moderate 25–125 l min$^{-1}$ 10 m$^2$ | Great > 125 l min$^{-1}$ 10 m$^2$ |
|  | r: 6 | r: 5 | r: 4 | r: 3 | r: 1 |
| Total rating ($\Sigma r$) | 100–91 | 90–71 | 70–51 | 50–26 | < 26 |

* Measured according to a corrected value of R obtained from the Schmidt–Hammer test, involving the rebound of a 'controlled, spring-loaded mass, impacted on a rock surface' (Selby, 1980, p. 34).

*Source:* Selby, 1980, table 6, pp. 44–5.

the warm humid environment of northern New Zealand (77 per cent of slope angle variance explained) (Figure 8.20). The strength of slope materials will be examined further in Chapters 10 and 11.

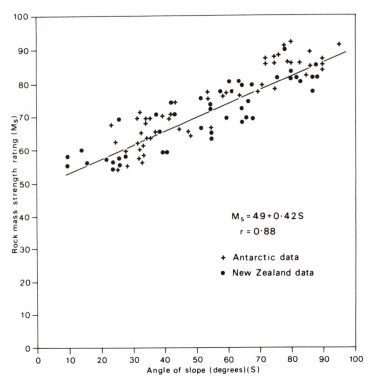

**Figure 8.20** The relationship between mass strength and profile angle for rock units studied in Antarctica and New Zealand. *Source:* Selby, 1980, figure 4, p. 48.

# References

Chorley, R. J. (1957) 'Climate and morphometry', *Journal of Geology*, vol. 65, 628–38.

Chorley, R. J. (ed.) (1969) *Water, Earth and Man*, (London, Methuen), especially chapter 2.II, 'The role of water in rock disintegration', 135–55.

Cleaves, E. T., Fisher, D. W. and Bricker, O. P. (1974) 'Chemical weathering of serpentinite in the Eastern Piedmont of Maryland', *Bulletin of the Geological Society of America*, vol. 85, 437–44.

Cotton, C. A. (1944) *Volcanoes as Landscape Forms*, Christchurch, NZ, Whitcombe & Tombs.

Cotton, C. A. (1948) *Landscape*, 2nd edn, Christchurch, NZ, Whitcombe & Tombs.

Flint, R. F. (1963) 'Altitude, lithology and the Fall Zone in Connecticut', *Journal of Geology*, vol. 71, 683–97.

Gregory, H. E. (1950) 'Geology and geography of the Zion Park region, Utah and Arizona', *US Geological Survey Professional Paper* 220.

Gregory, H. E. (1951) 'The geology and geography of the Paunsaugunt region, Utah' *US Geological Survey Professional Paper* 226.

Hack, J. T. (1960) 'Interpretation of erosional topography in humid temperate regions', *American Journal of Science*, vol. 258A, 80–97.

Jakucs, L. (1977) *Morphogenetics of Karst Regions*, Bristol, Adam Hilger.

Jarvis, R. S. (1976) 'Stream orientation structures in drainage networks', *Journal of Geology*, vol. 84, 563–82

Jennings, J. N. (1971) *Karst*, Cambridge, Mass., MIT Press.

Jobin, D. A. (1962) 'Relation of the transmissive character of the sedimentary, rocks of the Colorado Plateau to the distribution of uranium deposits', *US Geological Survey Bulletin* 1124.

King, P. B. (1959) *The Evolution of North America*, Princeton, NJ, Princeton University Press.

Legrand, H. E. and Stringfield, V. T. (1973) 'Karst hydrology – a review', *Journal of Hydrology*, vol. 20, 97–102.

Lobeck, A. K. (1939) *Geomorphology*, New York, McGraw-Hill.

Martonne, E. de (1947) *Géographie Universelle. Tome VI(1), France Physique*, 2nd edn, Paris, Armand Colin.

Melton, M. A. (1957) 'An analysis of the relations among elements of climate, surface properties, and geomorphology', *Columbia University, Department of Geology, Office of Naval Research Technical Report* 11.

Miller, J. P. (1961) 'Solutes in small streams draining single rock types, Sangre de Cristo Range, New Mexico', *US Geological Survey Water Supply Paper* 1535-F.

Miller, V. C. (1953) 'A quantitative geomorphic study of drainage basin characteristics in the Clinch Mountain area, Virginia and Tennessee', *ONR Geography Branch, Project*

089-042, *Contract N6 ONR 271, Task Order 30, Columbia University*.

Morgan, R. P. C. (1971) 'A morphometric study of some valley systems on the English Chalklands', *Transactions of the Institute of British Geographers*, no. 54, 33–44.

Ollier, C. D. (1969) *Volcanoes*, Cambridge, Mass., MIT Press.

Pricknett, R. G., Bray, L. G. and Stenner, R. D. (1976) 'The chemistry of cave waters', in T. D. Ford, and C. H. D. Cullingford (eds), *The Science of Speleology*, London, Academic Press, 213–66.

Schumm, S. A. and Chorley, R. J. (1966) 'Talus weathering and scarp recession in the Colorado Plateau', *Zeitschrift für Geomorphologie*, vol. 10, 11–36.

Selby, M. J. (1980) 'A rock mass strength classification for geomorphic purposes: with tests from Antarctica and New Zealand', *Zeitschrift für Geomorphologie*, vol. 24, 31–51.

Sparks, B. W. (1971) *Rocks and Relief*, London, Longman.

Sweeting, M. M. (1972) *Karst Landforms*, London, Macmillan.

Thornbury, W. D. (1965) *Regional Geomorphology of the United States*, New York, Wiley.

Williams, P. W. (1969) 'The geomorphic effects of ground water', in R. J. Chorley (ed.), *Water, Earth and Man*, London, Methuen, 269–84.

# Part Three    *Geomorphic processes and landforms*

Acting upon the earth's surface are a variety of erosional and depositional processes. These processes may modify existing landforms or produce landforms (e.g. sand dunes, landslides). In Part III the significant processes will be described and resulting landforms and their evolution will be considered, and an attempt made to illustrate both the historical and functional approaches to landforms. Topics considered are weathering (Chapter 9), mass movement (Chapter 10), hillslopes and hillslope erosion (Chapter 11), fluvial processes and landforms (rivers, valleys, drainage basins and depositional landforms: Chapters 12–14), and coastal, aeolian and glacial processes and landforms (Chapters 15–17).

The effects of climate and climate changes on both the morphologic and cascading systems will be reserved for Part IV.

# Nine  *Weathering*

Weathering represents the response of minerals which were in equilibrium at a variety of depths within the lithosphere to conditions at or near the earth–atmosphere interface. Here they are in contact with the atmosphere, hydrosphere and biosphere giving rise to their largely irreversible change to a more clastic or plastic state, involving increases in bulk, decreases in density and particle size, and in the production of new minerals which are more stable under the interface conditions. Obviously, landform evolution depends to a large extent upon these weathering processes. We have already

examined the solar energy budget and the hillslope hydrological cycle (see Section 3.2), and it is now necessary to see how these, in turn, result in physical and chemical exchanges between the surface and bedrock resulting in weathering.

## 9.1 The earth–atmosphere interface

Chemical activity involved in near-surface nutrient cycling has two primary sources; a small atmospheric source and a major source from the breakdown of soil and rock minerals (see Figure 3.1C). Rainfall is chemically active

**Table 9.1**  The biological cycle involving the soil and litter in different biomes

| Biome | Total biomass $(kg\ ha^{-1})$ | Mineral elements in biomass $(kg\ ha^{-1})$ | Net primary production $(kg\ ha^{-1}year^{-1})$ | Net mineral uptake $(kg\ ha^{-1}year^{-1})$ | Total litter fall $(kg\ ha^{-1}year^{-1})$ | Minerals returned in litter $(kg\ ha^{-1}year^{-1})$ |
|---|---|---|---|---|---|---|
| Tropical rainforest | 517,000 | 11,081 | 34,200 | 2,029 | 27,500 | 1,540 |
| Subtropical forest | 410,000 | 5,283 | 24,500 | 993 | 21,300 | 795 |
| Beech forest, Central Europe | 370,000 | 4,196 | 13,000 | 492 | 9,000 | 352 |
| Dry savanna, India | 26,800 | 978 | 7,300 | 319 | 7,200 | 312 |
| Northern taiga spruce | 260,000 | 970 | 7,000 | 118 | 5,000 | 100 |
| Semi-shrub desert, USSR | 4,300 | 185 | 1,220 | 59 | 1,200 | 59* |
| Arctic tundra | 5,000 | 159 | 1,000 | 38 | 1,000 | 37 |
| Harsh desert, USSR | 1,400 | 143 | 610 | 84 | 700 | 84* |

\* Mostly root residues; a 5 per cent surface vegetation cover in deserts may provide a *complete* near-surface root
   mat, which is important in resisting surface erosion.
*Source:* Rodin and Basilevic, 1967.

due to some of its water molecules being split into $H^+$ and $OH^-$ ions by the dissolving of $CO_2$ in the atmosphere; due to its dissolving of sea salts carried into the air; and due to its dissolving of other atmospheric chemicals, many resulting from the effects of human pollution. $H^+$ ions are particularly chemically active, due to their small size and relatively high electrical charge. The negative logarithm of the $H^+$ ion concentration is expressed by the value of pH, where neutral solutions have a pH of 7 and acid solutions pH values of less than 7.

The breakdown of soil and rock minerals is largely effected by the chemical activity of solutions associated with the decomposition of biomass. The total biomass in the surface canopy, wood and roots of vegetation varies markedly between different biomes (i.e. vegetational types), as do their annual mineral uptakes and litter falls (Table 9.1). The discrepancy between mineral uptake and that returned in litter fall tends to vary directly with total biomass and is made up by the weathering of soil and rock minerals. However, some massive tropical ecosystems achieve almost closed nutrient cycles with rapid and almost complete decay of organic material on a forest floor, where a stable microclimate is dominated by the vegetation and with little leaching losses such that primary weathering of soil and rock minerals is almost at a standstill. In such instances the organic soil (A horizon) layer may only be about 1 m thick and contain about 20 per cent of the total mineral nutrient pool, compared with as much as 80 per cent for temperature biomes (the rest being in the growing vegetation). Table 9.2 shows the turnover times (i.e. average length of storage, in years) of calcium in the soil, tree wood, canopy and litter (i.e. surficial organic layer) stores in different forest biomes. One of the striking features of these figures is the rapidity of nutrient turnover in tropical forests and, in particular,

the rate at which fallen litter is decomposed and its nutrients passed, via infiltration, to the soil storage solutions. The breakdown of organic material in the litter and upper soil layers is effected by macrofauna (e.g. earthworms, which may have a mass of up to 250 $g/m^2$, and termites in the tropics), but mostly by microfauna (e.g. bacteria, with a mass generally less than 1 $g/m^2$) and especially by microflora (e.g. fungi, with a mass of up to 10 $g/m^2$). The respiration of the microflora and fauna may raise the percentage of $CO_2$ in soil air to 0·5–20 per cent (atmospheric average = 0·003 per cent), especially in humid tropical soils, and this produces $H^+$ ions, as follows:

$$CO_2 + H_2O \rightarrow H_2CO_3 \rightarrow H^+ + HCO_3^-.$$

The same process can occur in non-organic weathering material.

Both complex acids (e.g. fulvic and humic) and more simple ones (e.g. oxalic, citric and tartaric) are directly produced by organic activity; and chelates, which are complex solutions carrying mineral cations, are also generated by organic decay. The reaction of active soil solutions with soil and rock minerals provides the major original source of cycling nutrients by solution of elements at the mineral–liquid interface, by differential removal of ions (e.g. $K^+$ and $Si^{++++}$ ions may be successively removed from potassium feldspar to leave kaolin; see pp. 211–13), and by the transport of ions by soil water flow. The efficiency of these reactions depends on the activity of the soil solutions, their rate of movement (i.e. drainage) and the reaction times of the minerals. It is known from field evidence that the presence of organic acids in the soil can increase the rate of silica solution by up to at least five times, and that silica solubility (2·8–51 p.p.m.) is greater in tropical regions. Recent

**Table 9.2**  Turnover times (years) for calcium cycling

| Climate | Biome | Soil | Tree wood | Tree canopy | Forest-floor litter |
|---|---|---|---|---|---|
| Tropical | Montane tropical rainforest (Puerto Rico) | 3·0 | 6·4 | 0·9 | 0·2 |
| | Moist tropical forest (Ghana) | 8·2 | 6·8 | 1·5 | 0·2 |
| Temperate | Mixed oak (Belgium) | 108·2 | 12·6 | 0·4 | 0·9 |
| | Scotch pine (UK) | 11·2 | 6·1 | 0·8 | 3·4 |
| | Douglas fir (Washington State) | 57·4 | 20·2 | 5·5 | 10·2 |
| Taiga | Spruce (USSR) | 7·6 | 18·3 | 3·2 | 13·6 |

*Source:* Jordan and Cline, 1972, p. 47.

laboratory tests have indicated that in a seven-day period the microflora *penicillum* mobilized 31 per cent Si, 11 per cent Al, 64 per cent Fe and 59 per cent Mg of the constituents of a sample of basalt. Residence time of soil water depends on its rate of flow through soil openings (fissures 0·5 mm – pipes of >3 cm) and, naturally, maximum solution, weathering and leaching rates are usually associated with high intensity and high frequency of precipitation, except where the surface is protected by layers of tall vegetation. Under the latter conditions leaching losses to rivers may be generally small (e.g. in undisturbed tropical rainforests), but otherwise they are at a maximum in hot and wet conditions. Temperate biomes have generally small leaching losses, except for calcium (up to 4–12 kg ha$^{-1}$year$^{-1}$). Root reactions involve the movement of H$^+$ ions from the roots through the clay colloids to the minerals, replacing cations of Ca, Mg, K, etc. which move to the roots, together with dissolved nutrients from soil solutions. The nutrient uptake involves a wide range of elements including silica which comprises a significant percentage of dry weight of certain biomasses (e.g. bamboo 10 per cent, steppe plants 5 per cent, tropical hardwoods 3·5 per cent). Assuming a 2·5 per cent silica content and no silica return to the nutrient uptake (which is rather unrealistic), it has been calculated that some tropical forests consume 800 kg ha$^{-1}$year$^{-1}$ of silica, which would be sufficient alone to lower a basalt (49 per cent silica) surface at a rate of 60 mm/1000 year.

## 9.2 Processes of weathering

To the student of geomorphology the study of weathering presents a picture of complexity which appears in many ways divorced from the rest of the discipline. Complex because of the apparent multitude of causative factors, and isolated because of the apparent dominance of chemistry over physics and of the subordination of gravitational forces to intermolecular ones. Before beginning a treatment of the processes of weathering, it may be helpful to make three points:

(1) Although it is convenient for the textbook writer to separate systematically apparently different weathering processes (in our case physical, chemical and biochemical), most rock disintegration is effected by a complex interplay of processes. No chemical weathering takes place without the production of physical stresses; disintegration of rock by thermal expansion probably does not occur in the absence of the chemical processes associated with the presence of water; in the company of even the sparsest vegetation

chemical weathering is replaced in part by biochemical,

(2) The dominance of chemical and biochemical processes has made the study of weathering more the province of the geochemist than of the geomorphologist. This has meant the sole object of weathering studies has too often been the investigation of processes for their own sake, whereas the geomorphologist is concerned with weathering merely as a stepping-stone to the understanding of the resulting landforms. Except where the chemistry and physics of the weathered material provide a direct indication of the environmental conditions under which previous weathering occurred, and thus a clue to denudation chronology, the attention of the geomorphologist is most commonly directed away from the detailed processes of weathering. The characteristics of the *weathered products* condition to a great extent the denudational processes to which they are susceptible; areal *rates of weathering* place limits on the speed of change of landscape geometry and, in particular, on the manner of slope development; *the structure of the weathered mass* influences its hydrology and stability, as well as the form of the resulting land surface should the weathered material be quickly evacuated.

(3) Weathering is commonly described as 'the breakup of rock *in situ*'. This is manifestly impossible, as even the most subtle chemical changes in the crystal lattice involve absolute and relative movement. All movement which occurs in a gravity field partakes of a net downslope component and it is thus strictly impossible to distinguish between the mechanisms of weathering and those of mass movements (see Chapter 10). The increase in volume of a feldspar on being changed to a clay mineral is both a result of weathering and a mechanism of debris creep; the fall of a section of cliff face is a manifestation of macroweathering. When one considers even the most subtle of chemical changes to which near-surface bedrock is susceptible one embarks on a chain of thought which runs unbroken through slope and stream processes to the destruction of continents.

### 9.2.1 Physical weathering

Notwithstanding the foregoing remarks, it is convenient to identify a set of weathering processes which result from a clearly identifiable set of physical stresses which find expression close to the earth's surface.

Geological stresses are set up when crystalline rocks (e.g. granites and marbles) have been crystallized or recrystallized and sedimentary rocks (e.g. massive, poorly jointed sandstones, arkoses and limestones) have been lithified under high confining pressures of tectonic or superincumbent origin. Surface erosion and unloading may lead to pressure release and the formation of joints and sheeting structures subparallel to the surface, from a few centimetres to more than 30 m thick, within a hundred metres or so of the surface. The great elasticity of granite makes this rock particularly susceptible to pressure release jointing which erodes into surface domes, as in the Sierra Nevada, California. Similar features occur at depths of up to 30 m in many important Colorado Plateau sandstones (e.g. Navajo, Wingate, De Chelly and Entrada), for example, in Monument Valley and in the vicinity of Glen Canyon Dam.

Solar stresses are induced by the thermal gradients generated just below rock surfaces by solar heating (see Section 3.1). As rocks are generally poor thermal conductors, quite high thermal gradients are achieved, especially in deserts or at high altitudes. Average gradients through the top 30 cm are $0.5°$ C/cm for quartz monzonite (Mohave Desert) and of $0.85°$ C/cm for sandstone (Sahara), but near the surface these gradients are considerably greater. Experiments involving purely thermal changes, in the best-known case the equivalent of diurnal change of $110°$ C over 244 years, have produced no effect, however, although the intervention of water during the cooling phase soon causes surface spalling. Purely thermal stresses seem insufficient alone to break up surface rock even in desert conditions, but the long-term effects of fatigue have not been investigated. The existence of pronounced chemical rotting, particularly in shady sites where more moisture is available, shows that chemical weathering is dominant even in arid areas, as does the expansion of exfoliation shells on boulders, which is obviously due to expansion accompanying chemical alteration. Occasional rainstorms and, particularly, nocturnal dew form a significant supply of desert moisture, the presence of which can be detected several feet below the surface of boulders. Observations on a quartz monzonite boulder in the Mohave Desert have shown that a diurnal temperature range of $24°$ C would cause a significant linear expansion of $0.0084$ per cent, but that the temperature gradients within the rock are sufficiently uniform to allow the whole mass to expand and contract with little differential stress within it. The products of desert weathering, however, differ from those of more humid areas, being on the average rather coarser and not possessing such a high proportion of clay or organic material. These characteristics partly explain differences between arid and humid geomorphic processes and forms, in that creep of the organic- and clay-rich humid soils often

contrasts with sheet erosion in desert areas, and that repose slopes in humid areas tend to be of lower angle (see Chapter 11).

Although hydration and the associated stresses due to swelling of the crystal lattice accompany a wide range of chemical weathering processes, the most striking mineral swelling due to absorption of water is associated with certain of the clay minerals. This swelling is responsible for much surface disintegration of shales, mudstones and greywackes, particularly in respect of montmorillonite (see Chapter 4) (Table 9.3).

Crystal growth stresses involved in weathering generally stem from two sources – ice crystals and salt crystals. Ice crystallization is usually the more effective because in this case all the solution enters into the solid phase and crystallization begins from the outside and may form a closed, high-stress system. Ice expands some 9 per cent on freezing and in a closed system of sealed voids can reach a maximum stress of 2115 kg/cm² at a temperature of $-22°$ C. This maximum pressure requires rapid freezing and about 80 per cent moisture saturation, and is easily able to exceed the tensile strength of rocks (marble 100, granite 70, limestone 35 and sandstone 7–14 kg/cm²), especially when the rocks are layered or foliated. Even where the optimum conditions are not present, considerable frost damage can result from repeated cycles of freezing and thawing and because ice crystals often align normal to the freezing rock surfaces in cracks and grow rapidly and in unison. Under certain conditions the crystallization of salts (sodium chloride, gypsum, calcite, etc.) is important in rock disintegration, granulation and cavernous weathering. Salt crystallization exhibits preferred orientations and growing stresses similar to ice, and with 1 per cent supersaturation calcite may crystallize against a pressure of 10 atm – of the same order as the tensile strength of rocks. Permeable rocks such as chalk and sandstone are particularly susceptible to disintegration by salt crystal growth, whereas igneous rocks are much less so. Crystallization occurs near the surface of an outcrop, where the rock is porous, where both salts and water are abundant, and where evaporation permits crystal growth. This type

**Table 9.3** Clay swelling in presence of unlimited moisture

|  | % |
|---|---|
| Sodium montmorillonite | 1400–1600 |
| Calcium montmorillonite | 45–145 |
| Illite | 5–120 |
| Kaolinite | 5–60 |

*Source:* California Div. Mines, Bull. 169.

**Table 9.4**  Mineral susceptibility to chemical weathering

| | | |
|---|---|---|
| Most susceptible to chemical weathering | Olivine | Anorthite (31,935) |
| | Augite (30,728) | |
| | Hornblende (31,883) | |
| | | Albite (34,335) |
| | Biotite (30,475) | |
| Least susceptible | Orthoclase (34,266) | |
| | Muscovite (32,494) | |
| | Quartz (37,320) | |

of weathering occurs in many environments, but is particularly effective in polar and desert areas. In high latitudes the nuclei of snowflakes provide salt, which because melting and runoff are small, tends to accumulate near rock surfaces and disintegrate them. In arid regions the excessive evaporation causes salts to be drawn up from depth in capillary films and crystallization to occur at the surface, producing weathering which is particularly effective in shady locations. In general, the present desert areas have tended to be arid in the past, and therefore many of the underlying rocks are sources of salts which migrate to the surface.

Biological stresses which induce physical weathering are of two main classes, faunal and floral. Burrowing earthworms ingest particles up to 1 mm diameter, burrow to depths of 1.5 m and over large parts of the earth bring an average of 43 tonne/ha/year to the surface (i.e. a 5-mm layer). In parts of the tropics this is exceeded and measurements in Nigerian forests have given figures in excess of 50 tonne/ha/year, added to which the activity of termites may be two and a half times as great. Plant root growth also generates important physical stresses. Lichen roots can penetrate diorite, and tree tap-roots are commonly 3 m deep and fine tree roots extend to more than twice this depth. However, as will be

seen in Section 9.2.3, the major organic weathering processes are biochemical.

### 9.2.2 Chemical weathering

A measure of mineral susceptibility to chemical weathering is given by the bond strengths (Kcal/mol) between oxygen and the following common cations:

$$K^+ 299; \; Na^+ 322; \; H^+ 515, \; Ca^{2+} 839; \; Mg^{2+} 912; \; Fe^{2+} 919; \; Al^{3+} 1793; \; Al^{4+} 1878; \; Si^{4+} 3110\text{--}3142.$$

From these, approximate total bond strengths (Kcal/mol) can be theoretically deduced for the common igneous rock-forming minerals (Table 9.4). These theoretical bond strengths do not completely reflect mineral susceptibility to chemical weathering because the mineral structure may be sustained by a robust architecture of silicon–oxygen bonds despite the removal of the weaker $K^+$ and $Na^+$ ions. In general, minerals rich in silicon–oxygen bonds are the most resistant to chemical weathering. Reiche (1950) has assigned a weathering potential index to many common minerals (Table 9.5).

The susceptibility of the silicates to weathering depends on the number and weakness of the cation links $(K^+, \; Na^+, \; Mg^{++}, \; Ca^{++}, \; Fe^{++}, \; Fe^{+++}, \; Al^{+++})$ between the structure of silica tetrahedra. In quartz these

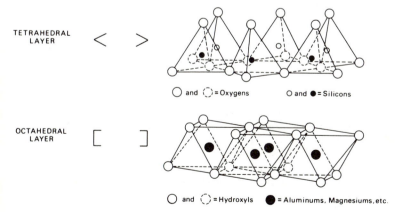

TETRAHEDRAL LAYER

○ and ◌ = Oxygens    ○ and ● = Silicons

OCTAHEDRAL LAYER

○ and ◌ = Hydroxyls    ● = Aluminums, Magnesiums, etc.

**Figure 9.1**  The atomic structure of tetrahedral and octahedral layers in clays, after Grim.

**Table 9.5**   Weathering potential indices

| Group | Mineral | Formula | Weathering potential index |
|---|---|---|---|
| Silicates | Olivine | $(Mg, Fe)_2 Si O_4$ | 54 |
| | Serpentine | $Mg_3 Si O_5 (OH)_4$ | – |
| | Augite (pyroxene) | $Ca (Mg, Fe) Si_2 O_6$ | 39 |
| | Hornblende (amphibole) | $Ca_2 (Mg, Fe)_6 Al_2 Si_7 O_{22} (OH)_2$ | 36 |
| | Nepheline (foid) | $Na_3 K (Al_4 Si_4 O_{16})$ | 25 |
| | Anorthite (calcic plagioclase) | $Ca (Al_2 Si_2 O_8)$ | 25 |
| | Biotite (mica) | $K (Mg, Fe)_3 (Al Si_3 O_{10}) (OH)_2$ | 22 |
| | Albite (sodic plagioclase) | $Na (Al Si_3 O_8)$ | 13 |
| | Orthoclase (potassium feldspar) | $K (Al Si_3 O_8)$ | 12 |
| | Quartz | $Si O_2$ | 0 |
| | Muscovite (mica) | $K Al_2 (Al Si_3 O_{10}) (OH)_2$ | $-10 \cdot 7$ |
| Clays | Montmorillonite (Na or Ca) | $Na_{0.5} Al_{1.5} Mg_{0.5} Si_4 O_{10} (OH)_2$ | |
| | Illite | $K Al_2 (Al Si_3) O_{10} (OH)_2$ | |
| | Vermiculite | | |
| | Chlorite | $(Mg, Fe)_3 (Al, Fe)_3 (Si Al_3) O_{10} (OH)_8$ | |
| | Kaolinite (+ halloysite) | $Al_2 Si_2 O_5 (OH)_4$ | $-67$ |
| Metallic residua | Boehmite ⎫ Gibbsite ⎭ Bauxite | $Al O (OH)$<br>$Al_2 (OH)_6$ or $Al (OH)_3$ | $-100$<br>$-300$ |
| | Goethite (Laterite: ferric hydroxide) | | |
| | Iron oxides | $Fe_2 O_3$ or $Fe O H_2 O$ | |

*Source:* Loughnan, 1969, table 16, p. 60, after Reiche, 1950.

tetrahedra are completely interlocked; in orthoclase the framework of silica tetrahedra is tight enough seriously to inhibit the escape of $K^+$ ions; in albite the silica tetrahedra structure is weakened by $Al^{+++}$ replacing every fourth $Si^{++++}$; in biotite sheets the silica tetrahedra are sandwiched between $Al^{+++}$, $Mg^{++}$, $Fe^{++}$ and $K^+$ ions; in anorthite $Al^{+++}$ replaces every other $Si^{++++}$; in the foids (e.g. nepheline) the Si:Al ratio is still lower; in hornblende the silica tetrahedra form only double chains; in augite there are single silica tetrahedra chains loosely bound on four sides by $Mg^{++}$, $Fe^{++}$, $Fe^{+++}$ and $Al^{+++}$ ions; and in olivine isolated silica tetrahedra are loosely bonded on all sides by $Mg^{++}$ and $Fe^{++}$ ions. The susceptibility of clays to further weathering is also a function of their structure of tetrahedral layers of oxygen and silicon; octahedral layers of hydroxyls, aluminiums,

**Table 9.6** Properties of clay minerals

| Clay minerals | Structure* | Ideal composition* | Maximum particle size (μm) | Cation exchange capacity (meg/100 g) | Relative water adsorption | Relative permeability | Relative plasticity | Derived from | Change in components | Possible environments of formation |
|---|---|---|---|---|---|---|---|---|---|---|
| Kaolinite | T O T O | $Al_2Si_2O_5(OH)_4$ | 4.0–0.3 | 1–10 (slight) | Slight | Large | Slight | Feldspars Muscovite | $-(K^+, SiO_2)$ | Mature weathering Moist conditions Good drainage Granitic source |
| Chlorite | T O T O T O T | $[Mg,Al]_3(OH)_6$ $[Mg,Al]_3$ $(Si,Al)_4O_{10}(OH)_2$ | 0.3–0.1(?) | 5–30 (moderate) | Moderate | Moderate | Moderate | Ferromagnesians Biotite Montmorillonite(?) | $+(Fe^{++}, Mg^{++})$ $-(SiO_2, H_2O, Na^+, Ca^{++})$ | Immature weathering Mafic source |
| Illite | T O T K O T | $K_{1-x}[Al_2]$ $(Al_{1-x}Si_{3+x})$ $O_{10}(OH)_2$ | 0.3–0.1 | 10–40 (moderate) | Moderate | Moderate | Moderate | Feldspars Muscovite Kaolinite Montmorillonite | $\begin{bmatrix} +(K^+, SiO_2) \\ -(Al_2O_3, H_2O) \\ +(K^+) \\ -(SiO_2, H_2O, Na^+, Ca^{++}, Mg^{++}, Fe^{++}) \end{bmatrix}$ | Immature weathering Marine conditions |
| Montmorillonite | T O T H₂O Ex H₂O T O T | $Ex_x[Al_{2-x}Mg_x]$ $(Si_4)$ $O_{10}(OH)_2$ | 0.2–0.02 | 80–140 (large) | Very large | Small | Large | Ferromagnesians Volcanic ash (Bentonite) Volcanic rocks | $+(H_2O)$ $-(Na^+, K^+ Ca^{++})$ | Waterlogged conditions Poor drainage (alternatively, very dry) Gabbroic source |

* ⟨T = tetrahedral layer); [O = octahedral layer] (see Figure 9.1).
*Sources:* Krumbein and Sloss, 1963, table 6–2, p. 195; Berner, 1971, Figure 9–2, p. 166; and Grim.

magnesiums, etc.; and of linking $K^+$, $Na^+$, $Mg^{++}$, $Ca^{++}$ ions (Table 9.6 and Figure 9.1). Increased weathering selectively removes these ions, together with Si and O, adding $(OH)^-$ until metallic residua of aluminium or iron may result. Montmorillonite and illite have the loosest structures, based on two tetrahedral layers linked to one octahedral layer, the former very subject to absorption of water and expansion (especially Na – montmorillonite) and the latter (expanding little) very characteristic of alkaline marine conditions and other situations where leaching is not excessive. Chlorite has a structure involving two tetrahedral and two octahedral layers (vermiculite lying between chlorite and montmorillonite in structure), and kaolinite, a tight, little-expanding structure of one tetrahedral and one octahedral sheet from which all linking ions have been leached. Continued leaching may remove all but $Al^{+++}$, $Fe^{++}$ and $(OH)^-$ ions in extreme conditions to yield either bauxite minerals (bohemite and gibbsite) or laterites (goethite) and iron oxides (haematite and limonite). Whether gibbsite or goethite is produced depends on the nature of the original weathered rock (goethite derives most usually from basic rocks and gibbsite from silicic undersaturated rocks, like nepheline syenite, or from alumina-rich rocks), on the environment of leaching (wet conditions – gibbsite; wet–dry conditions – goethite) and on the rate of destruction of soil organic matter. Iron residuum, like silcretes and calcretes, can act topographically as a resistant caprock. The nature of the clay or metallic residuum produced by continued weathering thus depends on:

(1) The nature of the original weathered minerals, of which feldspars are the most important source. Illite and kaolinite are mostly produced from the breakdown of feldspars and micas under alkaline conditions; vermiculite from the partial weathering of micas; and montmorillonite often from the weathering of ferromagnesian minerals.

(2) Time. Continued leaching under humid, well-drained conditions may cause montmorillonite to weather to a kaolinite.

(3) Composition of soil water (see p. 211).

(4) Amount of soil water making contact with the weathered surface (see later discussion of leaching).

The mechanisms of chemical weathering represent a complex association of the following types of chemical reactions. The *solubility* of given minerals is related to the presence of dissociated $H^+$ and $OH^-$ ions in water, which is expressed by the pH value (i.e. pH = 7 implies the existence of $10^{-7}$ mole/l each of $H^+$ and $OH^-$ ions). The $H^+$ ions are small, carrying relatively large electrical charges and readily penetrate mineral lattices to displace

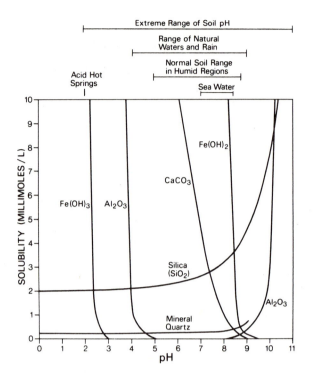

**Figure 9.2** Solubility (in millimoles per litre) in relation to pH for some components released by chemical weathering. *Source:* Loughnan, 1969, figure 15, p. 32.

cations which combine with $OH^-$ ions to form solutions which are leached away. This reaction between chemical elements and dissociated water is termed *hydrolysis*. $K^+$, $Na^+$, $Ca^{++}$ and $Mg^{++}$ ions are readily replaced and, for example, the solubility of $CaCO_3$ is strongly controlled by the natural pH (Figure 9.2), where $Si^{++++}$ is slowly leached and $Al^{+++}$ is virtually immobile within the natural range of pH values (4·5–9·5). The solubility of silica from the breakdown of silicate minerals is relatively unaffected by natural variations in pH and that of pure mineral quartz is only about one-tenth as great. The presence of $H^+$ ions depends on such factors as the composition of atmospheric and soil water, the action of organisms, the presence of organic acids and *cation exchange* (the substitution of ions in roots for those in minerals – especially in clay minerals). Environmental pH is thus a complex function of reactions with minerals, rate of leaching of cations, and the nature and cation exchange capacity of the residual mineral products. It must be remembered that chemical equations indicate only the probable course of weathering, but not the rate, which is both a function of the solubility of the mineral and the amount of water passing. Water may also enter the mineral lattice by *hydration* and this may prepare the way for subsequent weathering by oxidation and carbonation. *Oxidation* occurs when elements lose electrons to an oxygen ion in solution. The ease of oxidation depends on the redox potential (Eh: inversely related to pH), which is controlled by the amount of organic matter and the accessibility of free oxygen. *Carbonation* involves the action of carbonic acid on calcite (see Section 8.3) and on feldspars as a stage in their breakdown. A final means of extracting ions (usually metallic) from minerals is by the complex chemical process of *chelation*, with decomposing plant matter being rich in chelating agents.

The importance of the amount, rate of circulation and periodicity of water availability to the rate and products of weathering has been particularly stressed by Trudgill (1976). Weathering rates are adjusted to the amount of water input to the weathering mass, to the water output, to the water chemistry, to organic factors (e.g. organic products with solution or chelatory capacity: see Section 9.2.3) and to the susceptibility of given minerals to weathering. Climate is clearly important in a complex manner – for example, although for each rise of temperature of 10° C the rate of chemical reactions almost doubles, if these higher temperatures decrease the amount of infiltration this can operate to decrease weathering rates. In arid regions the small amount of available water, the return of water to the surface by capillary rise accompanying evaporation and the paucity of organic matter slows down weathering rates. Here, and in waterlogged situations, montmorillonite, illite and chlorite are the resulting clay products. In humid regions, with good drainage, intensive leaching and abundant vegetation, deep weathering results leading to the production of kaolinite or, in tropical regions, gibbsite or goethite. The amount of free percolation in the weathering mass is crucial in determining the rate and completeness of removal of chemically weathered constituents and, in this respect, evenly distributed persistent showers are more effective than occasional high-intensity storms or alternating wet/dry seasons. However, the rate of weathering depends on a trade-off between the rate of chemical solubility at a given mineral face and the rate of waterflow against that face (Trudgill, 1976):

> High solubility and slow flow – chemical equilibrium is achieved and the weathering rate is controlled by the *solubility level* (i.e. the saturation value of the solution).
>
> Lower solubility and fast flow – the weathering rate is controlled by the rate of flow and the *solution velocity* (i.e. the rate of achievement of saturation value).

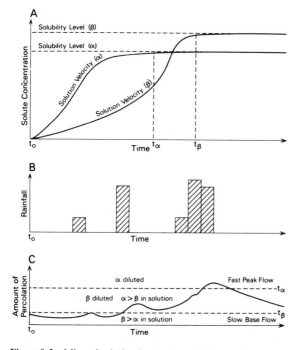

**Figure 9.3** Mineral solution in relation to the dynamic properties of water in the soil and weathered mantle.
A. solution velocities of minerals $\alpha$ and $\beta$, $\beta$ having the greater solubility but the lower solution velocity.
B. and C. model of the influence of rainfall on soil water percolation, and its relation with solution velocity.
*Source:* Trudgill in E. Derbyshire (ed.), *Geomorphology and Climate*, 1976, figures 3.10 and 3.11, pp. 77–8; copyright © John Wiley and Sons, by permission.

Figure 9.3A compares two minerals having different solubility levels and solution velocities, and Figure 9.3B and C suggest that, under free drainage, infrequent and intense rainstorms will cause both minerals to be more or less equally weathered over time, whereas frequent and low-intensity rainstorms will result in mineral $\beta$ being more weathered than $\alpha$ over time. The weathering of albite can be used to show the effects of velocities of percolation on the products of weathering (Trudgill, 1976, pp. 66–7).

(1) Bad drainage – stagnant conditions, or low percolation velocities; equilibrium quickly reached; not all $Na^+$ and $Mg^{++}$ cations flushed out and those remaining react with $Al^{+++}$ and $Si^{++++}$ to give montmorillonite:

$$\underset{\boxed{\text{albite}}}{Mg^{++} + 3NaAlSi_3O_8} + 4H_2O \rightarrow \underset{\boxed{\text{montmorillonite}}}{2Na_{0.5}Al_{1.5}Mg_{0.5}Si_4O_{10}(OH)_2} + \underset{\boxed{\text{in solution}}}{2Na^+ + H_4SiO_4.}$$

(In the same way, the initial weathering of anorthite produces montmorillonite and 3 Ca(OH)$_2$ in solution.)

(2) Better drainage – higher percolation rates; Na$^+$ and Mg$^{++}$ dissolved and removed; more slowly dissolving Al$^{+++}$ and Si$^{++++}$ remain to form kaolinite:

$$\underset{\boxed{\text{albite}}}{2H^+ + 2NaAlSi_3O_8} + 9H_2O \rightarrow \underset{\boxed{\text{kaolinite}}}{Al_3Si_2O_5(OH)_4} + \underset{\boxed{\text{in solution}}}{2Na^+ + 4H_4SiO_4.}$$

(3) Rapid drainage – velocities of percolation fast compared with solution velocities; Si$^{++++}$ lost as well as Na$^+$, Mg$^{++}$ and Ca$^{++}$ cations; Al$^{+++}$ remains to give bauxite:

$$\underset{\boxed{\text{albite}}}{H^+ + NaAlSi_3O_8} + 7H_2O \rightarrow \underset{\boxed{\text{gibbsite}}}{Al(OH)_3} + \underset{\boxed{\text{in solution}}}{Na^+ + 3H_4SiO_4.}$$

Thus, given suitable weathering material, tropical rugged relief, intense rainfall and rapid drainage tends to produce gibbsite.

In a similar way orthoclase can be weathered to illite:

$$\underset{\boxed{\text{orthoclase}}}{5KAlSi_3O_8} + 4H^+ + 4HCO_3^- + 16H_2O \rightarrow \underset{\boxed{\text{illite}}}{KAl_5Si_7O_{20}(OH)_4} + 8H_4SiO_4 + 4K^+ + 4HCO_3^-.$$

and continued leaching may yield kaolinite:

$$\underset{\boxed{\text{illite}}}{2KAl_5Si_7O_{20}(OH)_4} + 2H^+ + 2HCO_3^- + 13H_2O \rightarrow \underset{\boxed{\text{kaolinite}}}{5Al_2Si_2O_5(OH)_4} + 4H_4SiO_4 + \underset{\substack{\boxed{\text{soluble potassium}}\\\boxed{\text{carbonate}}}}{2K^+2HCO_3^-.}$$

$$2KAl_2(AlSi_3)O_{10}(OH)_2 + 5H_2O \rightarrow 3Al_2Si_2O_5(OH)_4 + 2KOH.$$

The weathering of acid igneous and metamorphic rocks well illustrates both the relative susceptibilities of common minerals to weathering and the influence of climate on the products of weathering. The classic work of Goldich (1938) on thick sections of weathered gneiss in Minnesota indicated that plagioclase had weathered faster than orthoclase, but that both, with biotite, had been changed into kaolinite by prolonged leaching (Figure 9.4A). These relative changes are expressed in Figure 9.4B, where it should be noted that the initial apparent increases in quartz and potash feldspar are due to *constant* amounts of these minerals being expressed as a percentage of a weathering mass which is at the time being rapidly depleted of leached plagioclase constituents. The influence of climate on the mechanisms and products of igneous rock weathering is shown by recent work on a weathered quartz diorite in Antarctica, which although apparently considerably decomposed, was found to be little changed chemically. No clay had been produced, the only chemical change being the oxidation of some iron, and the disintegration seemed mainly due

to physical weathering by the growth of ice and sea-salt crystals. In an area of Colorado, probably persistently dryer than Minnesota, a 25-m-thick weathered profile (now fossil) developed. This profile indicates the rapid weathering of hornblende and alteration of biotite to vermiculite by the loss of ferrous iron; the destruction of albite by the removal of sodium and the later destruction of 50 per cent of the orthoclase, both resulting in replacement by clay minerals and iron oxides; the removal of 15 per cent of the original silica content by the weathering of hornblende and augite; and the development of least-stable montmorillonite in the lower, less-weathered layers, more-stable illite above and most-stable kaolinite throughout the weathered section. Figure 9.5C illustrates clearly how in California (winter rainfall; mean annual temperature 10°C–16°C) montmorillonite and illite are produced by the weathering of acid rocks in drier areas (where not all K$^+$, Mg$^{++}$ and Ca$^{++}$ ions have been leached away), and kaolinite and vermiculite in wetter areas. The transformations associated with extreme tropical weathering of a granite are illustrated in Figure 9.6

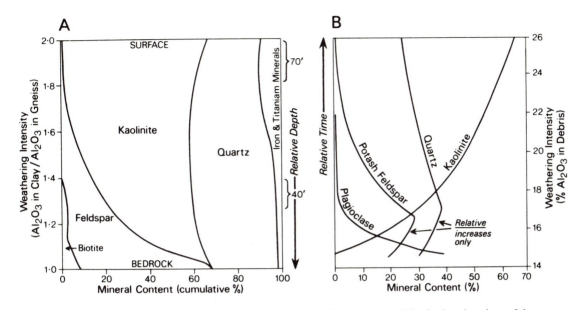

**Figure 9.4** Weathering intensity of a granite gneiss showing decrease with depth: A. variation in the mineralogy of the weathered Morton Gneiss, Minnesota; B. variation of individual mineral composition in a granite gneiss weathering under humid temperate conditions.
*Sources:* Loughnan, 1969, figure 48, p. 101; Goldich, 1938.

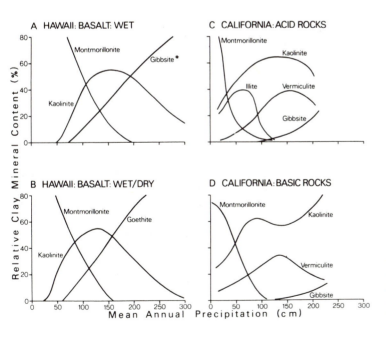

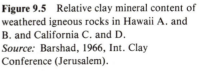

**Figure 9.5** Relative clay mineral content of weathered igneous rocks in Hawaii A. and B. and California C. and D.
*Source:* Barshad, 1966, Int. Clay Conference (Jerusalem).

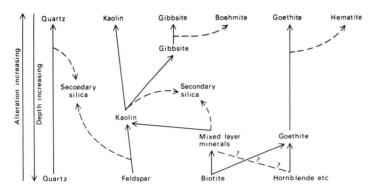

**Figure 9.6**  Lateritic weathering of a granite.
*Source:* Gilkes *et al.*, 1973, figure 4, p. 529.

where, under prolonged leaching, a variety of clay and metallic residues are possible.

The chemical weathering products of basic igneous rocks also reflect a complex interaction of rock type and degree of leaching. Even in dry areas of California, no illite is produced by the breakdown of basic rocks (Figure 9.5D). Moderate leaching of olivine basalt in Hawaii leads to the rapid production of montmorillonite from olivine and pyroxene, and some kaolinite from the breakdown of plagioclase (Figure 9.5A and B). Optimum leaching in a continuously wet tropical climate yields predominantly gibbsite because of iron losses through leaching (Figure 9.5A), whereas that in a seasonal wet–dry climate results in the retention of iron and the production of goethite (Figure 9.5B) (basalt contains some 12 per cent iron, as against 4 per cent or less in an average granite). Weathered profiles associated with the breakdown of basic rocks in differing tropical rainfall regimes are shown in Figure 9.7. Figure 9.7D shows the results of the weathering of an olivine basalt (plagioclases 30 per cent; olivine 25 per cent; augite 20 per cent; glass 20 per cent; magnetite iron 5 per cent) at Bathurst, New South Wales, in a monsoonal wet–dry climate (rainfall 22 in (56 cm); mean annual temperature 57°F (13·9°C)). The leaching of the olivine, augite and plagioclases produced, first, montmorillonite, and then kaolinite, but the continued presence of unleached $Fe^{++}$ has led to the production of goethite and iron oxides, particularly towards the top of the weathered zone. The source of the quartz is obscure. The breakdown of a serpentine in the continuously wet climate of Kalimantan, Borneo, is given in Figure 9.7C. Here montmorillonite and chlorite are formed initially, but continued leaching destroys, first, the former, and then the latter. The loss of $Mg^{++}$ and $Si^{++++}$ from the chlorite leads to the production of gibbsite; the abundance of original $Fe^{++}$ to goethite; and secondary addition of dissolved silica near the surface to the generation of kaolinite there.

Being composed of previously weathered material, most sedimentary rocks are not notably susceptible to further chemical weathering, except where changed circumstances permit less-stable clays to be leached into more-stable ones, partially weathered feldspars (commonly orthoclase) to be weathered into clay, or silica to be dissolved. In Figure 8.19 we have already seen the relatively high chemical solution associated with the breakdown of feldspar in an arkosic sandstone, compared with that of a recrystallized quartz sandstone. Figure 9.7A gives a weathered section of clay–shale-slates at Goulburn, New South Wales (rainfall 21–46 in (53–117 cm); temperature 57°F–65°F (13·9°C–18·3°C)), where the 20 per cent of original montmorillonite was further leached into kaolinite and later, within the top 3–5 m, about half the original 20 per cent illite composition was changed to kaolinite. A kaolinitic sandstone at Weipa, Queensland, in a tropical monsoonal wet–dry climate, has been leached to give mainly gibbsite, together with some goethite and kaolinite, and only 5 per cent quartz (Figure 9.7B) (Loughnan, 1969).

Chemical weathering produces three classes of products:

(1) Solutes of sodium, potassium, magnesium, calcium, strontium, etc., together with some silica particularly in humid tropical regions, which are flushed into lakes and seas to be reprecipitated as limestones, dolomites and other chemical sedimentary rocks.

(2) Clays, complex hydrous aluminosilicates in the form of minute platy crystals largely derived from the weathering of feldspars and ferromagnesian minerals. These clays produce a large proportion of shales and other common argillaceous rocks.

(3) Mineral residua, largely silica and as-yet-unweathered feldspars and micas, together with small proportions of heavy minerals. These residua largely comprise the sandstones and other clastic sedimentary rocks.

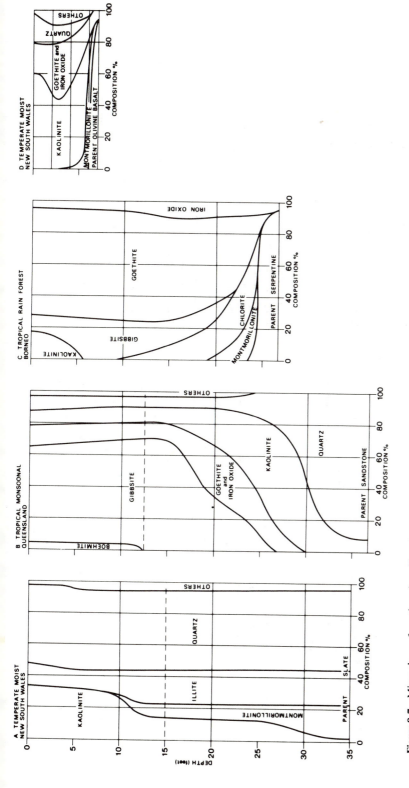

**Figure 9.7** Mineralogy of weathered profiles: A. slate, New South Wales, Australia (after Bayliss and Loughnan); B. sandstone, north Queensland, Australia (after Bayliss and Loughnan); C. serpentine, Borneo (after Schellmann); D. olivine basalt, New South Wales, Australia (after Craig and Loughnan).

*Source:* Loughnan, 1969, figures 54, 55, 32 and 42, pp. 110, 111, 77, 88, copyright © Elsevier, Amsterdam.

**Table 9.7** Resistance to weathering related to rock properties

| Rock Properties | Physical weathering | | Chemical weathering | |
|---|---|---|---|---|
| | Resistant | Non-resistant | Resistant | Non-resistant |
| Mineral composition | High feldspar content<br>Calcium plagioclase<br>Low quartz content<br>$CaCO_3$<br>Homogeneous composition | High quartz content<br>Sodium plagioclase<br>Heterogeneous composition | Uniform mineral composition<br>High silica content (quartz, stable feldspars)<br>Low metal ion content (Fe–Mg), low biotite<br>High orthoclase, Na feldspars<br>High aluminum ion content | Mixed/variable mineral composition<br>High $CaCO_3$ content<br>Low quartz content<br>High calcic plagioclase<br>High olivine<br>Unstable primary igneous minerals |
| Texture | Fine-grained (general)<br>Uniform texture<br>Crystalline, tightly packed clastics<br>Gneissic<br>Fine-grained silicates | Coarse-grained (general)<br>Variable textural features<br>Schistose<br>Coarse-grained silicates | Fine-grained dense rock<br>Uniform texture<br>Crystalline<br>Clastics<br>Gneissic | Coarse-grained igneous<br>Variable textural features (porphyritic)<br>Schistose |
| Porosity | Low porosity, free-draining<br>Low internal surface area<br>Large pore diameter permitting free draining after saturation | High porosity, poorly draining<br>High internal surface area<br>Small pore diameter hindering free-draining after saturation | Large pore size, low permeability<br>Free-draining<br>Low internal surface area | Small pore size, high permeability<br>Poorly draining<br>High internal surface area |
| Bulk properties | Low absorption<br>High strength with good elastic properties<br>Fresh rock<br>Hard | High absorption<br>Low strength<br>Partially weathered rock (grus, honeycombed)<br>Soft | Low absorption<br>High compressive and tensile strength<br>Fresh rock<br>Hard | High absorption<br>Low strength<br>Partially weathered rock (oxide rings, pitting)<br>Soft |
| Structure | Minimal foliation<br>Clastics<br>Massive formations<br>Thick-bedded sediments | Foliated<br>Fractured, cracked<br>Mixed soluble and insoluble mineral components<br>Thin-bedded sediments | Strongly cemented, dense grain packing<br>Siliceous cement<br>Massive | Poorly cemented<br>Calcareous cement<br>Thin-bedded<br>Fractured cracked<br>Mixed soluble and insoluble mineral components |

**Table 9.7** — *continued.*

| Rock Properties | Physical weathering | | Chemical weathering | |
| --- | --- | --- | --- | --- |
| | *Resistant* | *Non-resistant* | *Resistant* | *Non-resistant* |
| Representative rocks | Fine-grained granites<br>Some limestones<br>Diabases, gabbros, some coarse-grained granites, rhyolites<br>Quartzite (metamorphic)<br>Strongly cemented sandstone<br>Slates<br>Granitic gneiss | Coarse-grained granites<br>Poorly cemented sandstone<br>Many basalts<br>Dolomites, marbles<br>Soft sedimentary (poorly cemented)<br>Schists | Igneous varieties (acidic)<br>Metamorphic (other than marbles, etc.) varieties<br>Crystalline rocks<br>Rhyolite, granite, quartzite (metamorphic), gneisses<br>Granitic gneiss | Calcareous sedimentary<br>Poorly cemented sandstone<br>Limestones, basic igneous, clay-carbonates<br>Slates<br>Marble, dolomite<br>Carbonates (other)<br>Schists |

*Source:* Lindsey *et al.*, 1982, table 4.

Weathering seldom proceeds sufficiently far *in situ* for the weathered material to achieve a new chemical stability before the partly weathered material is removed by erosion and transported on its interrupted journey towards the basin of deposition.

Although chemical breakdown and physical attrition of sediments continue during their transportation, sediments transported long distances by large rivers show relatively few changes during transport and up to 90 per cent of the apparent changes are due to the selective transportation of different sediment types. The mechanical wear which does occur, resulting in the reduction in size and in the increased rounding of sediments during transport, operates differentially, as the following mineral resistances to mechanical wear show (the figures relate to an arbitrary resistance of haematite iron ore equal to 100): orthoclase 150; apatite 275; olivine 290; augite 420; tourmaline 817. The result of these observations leads to the conclusion that many sediments exhibiting the effects of extreme physical attrition and of decomposition to advanced stages of new chemical stability are those which have been subjected to two or

more cycles of sedimentation. These properties are those expressive of the *maturity* of a sediment – the degree to which the sediment 'approaches the ultimate end product to which it is driven by the formative processes that operate upon it' (Pettijohn, 1975). Because of the common occurrence of both quartz and feldspar, and of the relatively universal chemical stability of the former, a simple measure of the chemical maturity of a sediment is expressed by the ratio of quartz to feldspar. Although the sediment now transported by the Ohio River at Cairo (quartz 86 per cent, feldspar 5 per cent, rock fragments 8 per cent – mostly clay), and by the Mississippi River at its mouth (quartz 64 per cent, feldspar 11 per cent, rock fragments 19 per cent) shows a low feldspar percentage, the average content for all rivers in the United States is 22 per cent (quartz 58 per cent). This figure compares with a feldspar content of 11·5 per cent for all ancient reworked clastic sedimentary rocks and with 10 per cent for present beach sands which have probably been reworked under conditions particularly favourable to the breakdown of all but the most stable sediments. Conversely, sedimentary environments characterized by low-

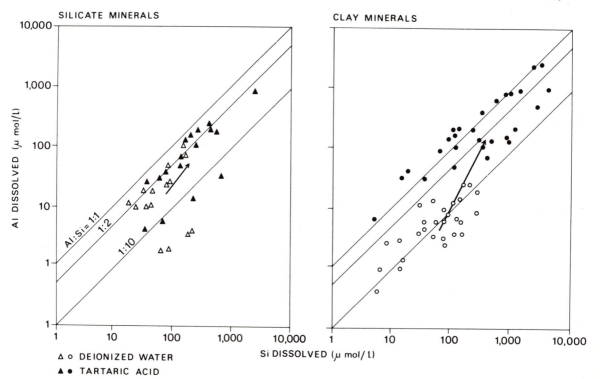

**Figure 9.8**  Amounts of Al and Si dissolved from primary rock-forming silicate and clay minerals by deionized water and tartaric acid.
*Source:* Huang and Keller, 1973, figure 2, p. 150. Reprinted by permission from *Nature*, vol. 239, copyright © 1973 Macmillan Journals Ltd.

intensity chemical weathering, to which sediments have been delivered from their sources with great rapidity (e.g. arid alluvial fans), have much higher proportions of feldspars and of other minerals of high susceptibility to chemical and physical breakdown.

Table 9.7 summarizes the resistance to physical and chemical weathering presented by the properties of composition, texture, porosity, bulk and structure, together with a generalized classification in terms of representative rock types.

### 9.2.3 Biochemical weathering

Organic material, both growing and decaying under the action of microfauna and flora, operates within almost all weathering zones to produce a very complex set of biochemical processes including cation root exchange, chelation, solution by root exudates and the production of organic acids. By these means, weathering rates may be increased up to ten times those expectable from distilled water by weak complexing acids (e.g. acetic and aspartic) and up to more than 100 times by strong complexing acids (e.g. citric and tartaric). In some circumstances organic acids dissolve $Al^{+++}$ preferentially to $Si^{++++}$, but this ratio is a complex matter depending on the type of acid, rate of decay and the rate of circulation of the dissolving solutions. Figure 9.8 shows that tartaric acid dissolves both silicate minerals and clay many times

faster than deionized water; clay minerals faster than silicate minerals; and $Al^{+++}$ more rapidly than $Si^{++++}$ in clay.

Humic acids (e.g. $C_{40}H_{24}O_{12}$) are active in chelation, decompose silicates and, in particular, amphiboles. Fulvic acids, (i.e. humic acids derived from peats) are especially important agents of weathering. Humic acid is produced most effectively by moderate soil moisture of neutral pH under high soil temperatures (Figure 9.9) such that the following soil zones exhibit different humus characteristics:

tundra soils – low humic acid content; low microbiological activity; low cation exchange capacity;

chernozem soils – high humic acid content; high microbiological activity; high cation exchange capacity;

podzol soils – intermediate humic acid content; very high microbiological activity.

Table 9.8 shows the experimental results of distilled water (pH = 5·5) percolating slowly through 1-m columns of acid and basic rock fragments for one month (i.e. approx. 200 mm total percolation) with and without a humus cover.

**Table 9.8**  Losses (in milligrams) to percolating water

|  | *Basalt* | | *Trachyte* | |
|---|---|---|---|---|
|  | *Humus* | *No humus* | *Humus* | *No humus* |
| CaO | 1300 | 180 | 270 | 240 |
| MgO | 310 | 29 | 45 | 45 |
| $K_2O$ | 230 | 65 | 260 | 150 |
| $Na_2O$ | 480 | 270 | 320 | 180 |
| $SiO_2$ | 3070 | 1920 | 1870 | 1550 |
| $Al_2O_3$ | 270 | 360 | 165 | 174 |
| Organic material | 1230 | 0 | 570 | 0 |

*Source:* Kerpen and Scherpeusul 1967, *Proc. Int. Atomic Energy Symp.*

Here the basalt is more affected than the trachyte, and the weak humic acid is more effective in dissolving all the constituents except $Al_2O_3$, which is more susceptible to solution by distilled water. However, there is some evidence that, as weathering proceeds, increasing amounts of Al may be dissolved by humic acids. Another set of experiments suggests that humic acid initially dissolves anorthite (calcic plagioclase) faster than does acetic acid (Figure 9.10), although in the long run in a closed system the effects of the two do not differ significantly from that of distilled water. However, in natural weathering fresh weak acid is constantly coming into contact with the minerals, and humic acids may be

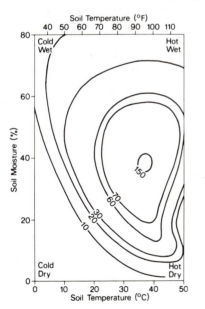

**Figure 9.9**  $CO_2$ production (milligrams per litre of soil air), a measure of organic activity, as a function of soil temperature and moisture content.
*Source:* Kononova, 1966, figures 54–6, p. 248.

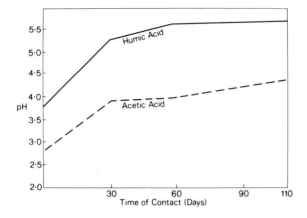

**Figure 9.10**   Change of pH of acid–anorthite mixtures with time.
*Source:* Loughnan, 1969, figure 18, p. 38, after Graham.

particularly effective under these conditions.

Bacterial acids (e.g. lactic, acetic, oxalic $(COOH)_2$, and gluconic) are also effective in weathering. They attack a wide range of minerals including magnesium carbonates, Ca and Mg silicates, feldspars and kaolinite. Aluminium-rich silicates are not particularly susceptible to their action, but in waterlogged, reducing situations bacterial acids produce sulphides, oxidize iron and assist in the solution of silica. Basic minerals are particularly susceptible to the development of, and destruction by, bacteria, as is shown by the following numbers of organisms per gram found in the interior of weathered rocks in north-east Scotland:

| | |
|---|---|
| hornblende–chlorite schist | 1,780,000 |
| serpentine | 860,000 |
| basic gneiss | 263,000 |
| granite | 20,960 |

Practical experiments produce similar results. The following percentages refer to the weight of ground-up minerals ($<0.3$ mm diam.) dissolved in a number of days by 2-ketogluconic acid:

| | |
|---|---|
| pyroxene | 43% |
| serpentine | 21% |
| nepheline | |
| (foid) | 5% |

Microfloral acids (e.g. oxalic and citric), produced by fungi and lichens, attack silicate minerals and clays, and produce $CO_2$. They also form insoluble precipitates of $Ca^{++}$, $Fe^{++}$ and $Al^{+++}$ to which humus substances are linked, and thus help to convert humus substances into a soluble state.

## 9.3  Rates of weathering

Rates of weathering are primarily of interest to the student of landforms in relation to the rates at which weathered material is removed from its location of break-up. Thus two contrasting weathering environments can be identified.

(1) Weathering limited: where transport processes (e.g. gravity fall, surface wash, wind action, etc.) are more rapid than weathering processes. Here little or no soil is free to develop and surface form is dominated by structural and lithological controls. Direct measurement of weathering rates (e.g. by erosion pins, microerosion meters, etc.) can only be really effective under weathering limited conditions, and extrapolation of rates over time can be reasonably secure only in such instances if constant transport rates are assumed.

(2) Transport limited: where weathering rates are more rapid than transport processes. Here a soil or debris cover develops and surface form is dominated by the action of mass movements (see Chapter 10). Rates of transport limited weathering are highly variable and non-linear, particularly under early weathering conditions where a combination of mechanical and chemical break-up is occurring. Figure 9.11 shows a series of theoretical relations between rate of basal rock weathering and waste cover thickness for pure mechanical (where weathering rate decreases exponentially with increasing waste thickness) and pure chemical weathering (where an optimum rate is associated with a cover thick enough to retain significant amounts of water percolating for long periods at rates compatible with mineral solubility). Of

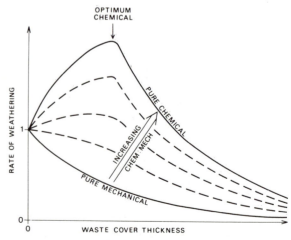

**Figure 9.11**   Model illustrating relative rates of combinations of chemical and mechanical weathering.
*Source:* Ahnert, 1976, figure 2, p. 34.

course, both chemical and mechanical processes normally operate together and the family of broken-line curves in Figure 9.11 is broadly appropriate. During the early stages of weathering, rock breakup by mechanical means exposes an exponentially increasing internal rock surface to the internal debris processes of chemical weathering (Figure 9.12). As the waste cover thickness increases, combined weathering rates decrease, until an equilibrium (steady state or mature) debris or soil cover may evolve where the surface is being eroded at the same rate as subdebris rock weathering, so that a constant weathering profile is maintained. The geomorphic significance of the weathering processes under transport limited conditions is much more complex than under weathering limited conditions. Rates of chemical change, leaching and soil formation have both a direct erosional significance – in that they are associated with the physical disaggregation of the debris cover – and an indirect significance in the light which their progress throws on the relative and absolute dating of exposed surfaces and the inferences which may be drawn regarding paleoclimates.

Field observations of weathering limited (Table 9.9) and transport limited (Table 9.10) weathering rates present a bewildering variety (Ollier, 1969; Brunsden, 1979).

Clearly, moisture environment, vegetational cover, aspect and similar local conditions exercise roles which are in some respects comparable to the gross influences of climatic regime and rock type. It has been suggested that an inch of soil can be formed in any time between 10 min and 10 million years and, for example, whereas several centimetres of crude soil was formed on Krakatoa pumice in forty-five years, and beneath granite building foundations in Rio de Janeiro a 45-cm rotted layer formed in only twenty years; no soil horizons have developed on 1200-year-old mudflows on Mt Shasta, California, and many glacially scoured rock surfaces show little evidence of 10,000 years of post-glacial weathering. Of course, transport limited weathering is particularly difficult to assess because of the problems of defining an immature soil. Nevertheless, considerable field evidence of weathering rates has been assembled from dated archaeological structures, industrial tips, engineering works, volcanic depositional surfaces, glacially exposed surfaces, moraines and limestone surfaces. Problems of generalization arise from the many regional and local controls over weathering rates, the difficulty of quantitatively defining weathering under transport limited conditions with initially incoherent material (i.e. sands, gravel, volcanic ash, etc.), the non-linearity of weathering rates and their great variability

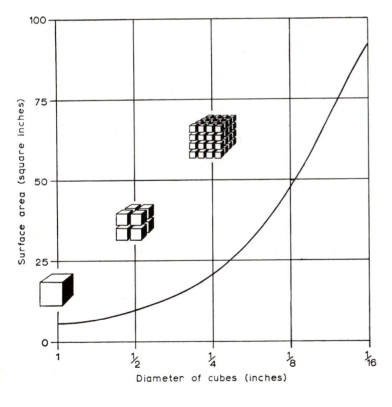

**Figure 9.12** Relation of exposed surface area of a cube to the dimensions of smaller cubes into which it may be decomposed. *Source:* Chorley, 1969, figure 3.II.9, p. 152, adapted from Leet and Judson.

**Table 9.9** Weathering rates – weathering limited

| Rock type | Location | Surface | Period (years) | Result | Average (mm/year) |
|---|---|---|---|---|---|
| Sandstone | Edinburgh | Tombstones | 200 | Little | |
| Slate | | | 90 | Engraving clear | |
| Marble | | | 90 | Crumbling | |
| Limestone | Yorkshire | Tombstones | 500 ⎫ | Based on 2·5 cm | 0·051 |
| Limestone | | | 300 ⎪ | of weathering | 0·085 |
| Limestone | | | 250 ⎬ | in period | 0·102 |
| Limestone | | | 240 ⎭ | | 0·106 |
| Granite | Aswan | Ancient structures | 4,000 | Fresh | |
| Granite | Giza | Ancient structures | 5,400 | Flaking 0·5–0·8 cm | 0·0009– 0·0015 |
| Hard grey limestone | | Great Pyramid | 1,000 | Little | |
| Soft grey limestone | | | 1,000 | 1–2 cm pits | 0·01–0·02 in part |
| Grey shale | | | 1,000 | 20 cm niches | 0·2 |
| Yellow limey shale | | | 1,000 | Rubble | 0·2 (average over whole surface) |
| Carboniferous limestone | N. England | Erratic block | 12,000 | 30–50 cm | 0·025–0·042 |
| Carboniferous limestone | | Glacial striae | 13 | 3–5 cm | 2·2–3·8 |
| Carboniferous limestone | | Runnels in glacial surface | 13 | 15 cm | 11·5 |
| Limestone | Austrian Alps | Bare surface | 1,000 | 9–12·5 mm | 0·009–0·0125 |
| Limestone | Norfolk Island | Jetty intertidal | 16 | 28 mm | 0·5–1·0 |
| Limestone | Point-Perm, Aust | Coastal notch | — | — | 1·0 |
| Limestone | La Jolla, California | Inscriptions | — | — | 0·5 |
| Limestone | Puerto Rico | Intertidal notch | 155 | — | 1·0 |
| Granite | Narvik | Glacial surface | 10,000 | Variable Max. at 120 m | 0·00105 |
| Limestone chert | Spitsbergen | Glacial surface | 70 | — | 0·02–0·2 |
| Sandstone | | Glacial surface | 10,000 | — | 0·34–0·5 |

*Source:* Brunsden, 1979, table 4.12, p. 102 and others.

over space and time which inhibits the use of the ergodic method of comparative reasoning. A crude climatic generalization has been made that, with the exception of solution rates of calcium carbonate, rates of weathering are slower in periglacial, arid and mountainous areas, and faster in temperate and tropical ones.

Experimental investigations of weathering rates commonly involve simulated attempts to speed up the effects of natural processes by increasing the rates of heating–cooling and wetting–drying cycles, the amounts of heating and wetting, the concentration of chemical solutions, etc. This time compression introduces scale problems which distort the effects of natural lag and long-term fatigue effects and the increase in the intensity of the processes further removes them from their supposed real-world counterparts. Thus experimental work tends to throw more light on relative, rather than absolute, rates of weathering. A coarse granite block was exposed to an accelerated equivalent of 244 years of diurnal temperature ranges of 110°C with little effect, but after a period equivalent to only two and a half years where cooling was effected by a water spray, the surface

**Table 9.10** Weathering rates – transport limited

| Rock type | Location | Evidence | Period (years) | Result and rate |
|---|---|---|---|---|
| Andesite ash | St Vincent, West Indies | Soil depth | | 1·83 m soil in 4000 years<br>Clayey soil 0·45–0·6 m/1000 years<br>Glass decomposed 15 g cm$^{-2}$/100 years |
| Andesite stratovolcano | Papua | Depth of weathering profiles | 650,000 | 58 mm/1000 years |
| Granite | Ivory Coast | Leaching rate of Si, Ca, Mg, K, Na | | Ferrallitization of 1 m granite 22,000–77,000 years |
| Volcanic | New Guinea | Nature of mantles | | Skeletal soil 5000 years<br>Immature 5000–20,000 years<br>Mature 20,000 years |
| Limestone | Ukraine | Soil on Kammetz Fortress | 230 | 30·48 cm soil<br>1·32 mm/year<br>possible blown sand addition |
| Limestone | Austrian Alps | Soil under dwarf pine | 1000 | 28 mm soil<br>0·028 mm/year |
| Andesite stratovolcano (pumice) | Krakatoa | Soil depth | 45 | 35 mm soil, Si, K, Na leached<br>Al and Fe O accumulated at surface |
| Tin spoil heaps | Malaya | Soil | 25 | Little development – no clay to assist |
| Iron spoil heaps | Northamptonshire, England | Soil | few decades | Soil on south-facing slopes; not on north-facing |
| Moraines | Tyrol | Soil | 80 | Significant soil |
| Coal spoil heaps | Somerset, England | Soil | up to 178 | Stratification after 21 years;<br>in 178 years 54% coal parent material reduced to 2% of regolith |

*Source:* Brunsden, 1979, table 4.10, pp. 98–9 and others.

exhibited significant cracking and loss of polish. Although ignoring the longer-term fatigue factor and lack of rock confinement, this experiment showed the efficacy of wetting and drying in association with thermal changes. The effect of crystal growth, hydration and thermal expansion in association with salt crystallization was investigated, where a number of rocks were immersed in a salt solution at 17°C – 20°C for 1 h, dried at 60°C for 6 h and then dried at 30°C for 17 h. The resulting changes of weight by surface fragmentation are shown in Figure 9.13 against number of daily cycles of wetting and drying, indicating that chalk, sandstone and limestone are more susceptible to salt weathering than are igneous rocks and black shale. The most effective salt used was $Na_2SO_4$ and rates of disintegration appeared to be related to water absorption capacity. This weathering mechanism is particularly applicable in deserts and arctic coastal areas. Other controlled experiments have involved the exposure of different rock types to natural weathering processes, although choices of initial particle size have been arbitrary. Figure 9.14 gives the percentage decreases of particles of the initial 10–20 mm diameter

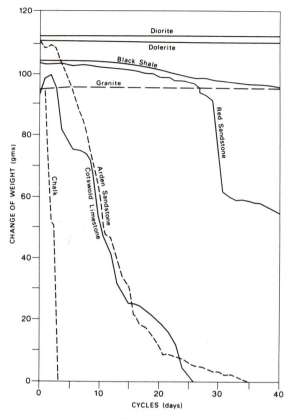

**Figure 9.13** Examples of salt weathering of eight rock types when treated with sodium sulphate in laboratory experiments. *Source:* Goudie *et al.*, 1970, figure 3, p. 45.

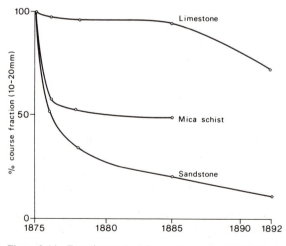

**Figure 9.14** Experiments by Hilger, who exposed 10–20-mm particles of three types of rock to natural weathering for up to seventeen years. *Source:* Jenny, 1941, figure 15, p. 32.

range when exposed to seventeen years of weathering in Central Europe, with sandstone and mica schist being more susceptible to weathering than limestone. In a series of observations between 9 December 1962 and 19 March 1964 at Denver, Colorado, the natural weathering of a piece of Entrada Sandstone was measured against that for two sets of other Colorado Plateau cliff-forming sandstones (Figure 9.15). During the total eight periods the rocks were subject to surface granulation and cracking under the influence of 21·53-in precipitation (including 3·87 in of snow in periods 1, 2, 6, 7 and 8) and 124 freeze–thaw cycles (period 1–33; period 2–15; period 6–13; period 7–22; period 8–41). These observations indicated the high present susceptibility of all these sandstones to weathering, the especial importance of freeze–thaw effects particularly after periods of wetting, and the marked vulnerability of the Entrada Sandstone which exhibited a rate of weathering more than four times the average of the rest.

## 9.4 The weathered mantle

The thickness of the weathered mantle on transport limited slopes depends on the balance between the depth of the equilibrium weathered profile (depending on rock resistance, jointing, permeability, climate, etc.) and the degree of transport limitation (depending on angle of slope, amount of surface runoff, rate of creep, frequency of slides, frequency of wetting/drying and frost heaving, etc.) (see particularly Ollier, 1969). Thus the thickest weathered mantles occur in areas of moderate to low relief, gentle slopes, good drainage, unimpeded through-flow and rapidly weathering bedrock. It is also possible to generalize regarding climatic controls over soil thickness, although differences resulting from rock type may be as important locally. Figure 9.16 gives an impression of the range of climatically controlled weathering depths, which vary from the shallow Arctic soils to 1–3 m in the temperate regions and to sometimes over 100 m in the humid tropics. Of course, depth of weathering may be due to relict conditions, and some deep soils in temperate regions may be relict from tropical climates which occurred there during Tertiary times. The variety of controls, both contemporary and relict, make generalizations regarding the thickness of the weathered mantle very difficult, especially as definitions of significant weathering are not clear-cut, as we have seen. However, the following examples of deep weathering are of interest (Ollier, 1969) (Plate 8):

granite – 300 m New South Wales; 100 m Czechoslovakia; 80 m Victoria, Australia; 50 m Queensland; 50 m Jos Plateau, Nigeria; 40 m Western Australia;

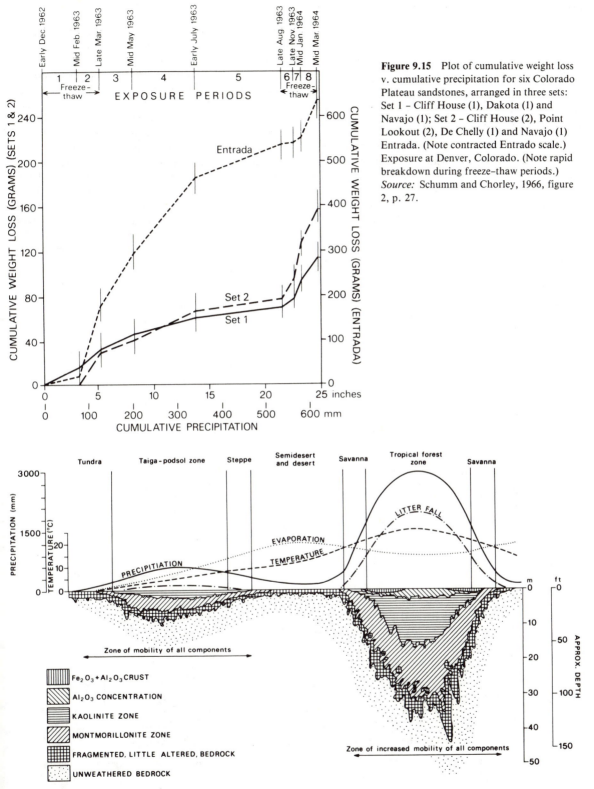

**Figure 9.15** Plot of cumulative weight loss v. cumulative precipitation for six Colorado Plateau sandstones, arranged in three sets: Set 1 – Cliff House (1), Dakota (1) and Navajo (1); Set 2 – Cliff House (2), Point Lookout (2), De Chelly (1) and Navajo (1) Entrada. (Note contracted Entrado scale.) Exposure at Denver, Colorado. (Note rapid breakdown during freeze–thaw periods.) *Source:* Schumm and Chorley, 1966, figure 2, p. 27.

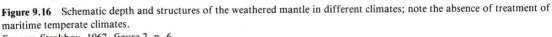

**Figure 9.16** Schematic depth and structures of the weathered mantle in different climates; note the absence of treatment of maritime temperate climates.
*Source:* Strakhov, 1967, figure 2, p. 6.

> 30 m Georgia, United States.
  acid metamorphics – 130 m Victoria, Australia;
> 100 m Brazil.
  shales – > 130 m Brazil.

The debris mantle–bedrock contact exhibits considerable variety, however it is convenient to classify it into
four groups (Figure 9.17):

(1) Transitional – where chemical rotting is deep
and the contact with unweathered bedrock indeterminate. Commonly found in rocks of uniform structure and texture which exhibit differences of weathering susceptibility only at the mineralogical level.
Weathered gneisses in Uganda exhibit such a transition, with the weathered mantle depth exceeding 30 m.

(2) Sharp – where there is a sharp transition, sometimes only a few millimetres thick between the
weathered debris mantle and the bedrock. Such a
contact is sometimes termed a 'weathering front' by
analogy with the advance of some metamorphic processes – indeed, weathering itself is sometimes referred
to as 'katamorphism'. Sharp contacts are variously
associated with dense, mineralogically and structurally
uniform rocks; rocks susceptible to solution with an
insoluble surficial residuum remaining; basic rocks of
low resistance to weathering in the tropics; rocks
composed of unresistant minerals, so that the rock
weathers uniformly layer-by-layer from the top down;
and rocks which are of low permeability or those
supporting a constant water table close to the surface.
Sharp contacts can form under thin mantles, as with
the Chalk of southern England, or at depth as in the
quartz diorite of the Columbian Andes beneath some
90 m of weathered debris.

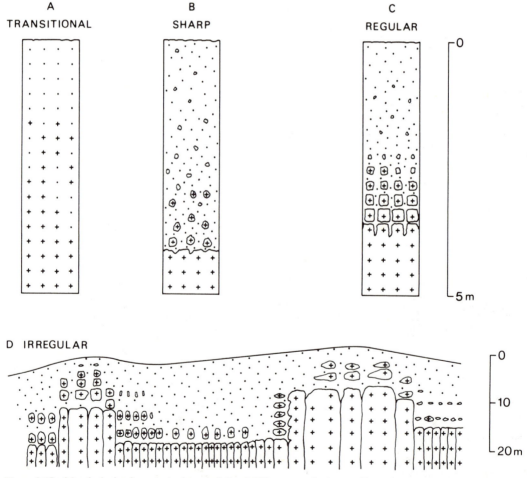

**Figure 9.17** Mantle–bedrock contacts characteristic of different weathering configurations (A.–D.); it is clear in D. that the
basal weathering front may not necessarily be pseudo-horizontal.

(3) Regular – where the contact is not unduly transitional, being characterized by jointed corestones of increasing size and fairly regular in depth. Regularly jointed rocks of uniform resistance to chemical weathering commonly exhibit this form of contact.

(4) Irregular – where the contact may vary in depth by up to tens of metres. This occurs where the bedrock characteristics of susceptibility to weathering (e.g. amount of quartz) and degree of jointing vary spatially in a gross way. It may also occur by irregular weathering below a water table due to hydrolysis and reduction. Although such a contact may occur in variably jointed conglomerates, sandstones and limestones, it is most characteristic of granites where little jointed areas coincide with resistant quartz-rich facies, and alternate with large areas of highly jointed feldspar-rich granite. The base of weathered granite in Hong Kong is highly irregular with a maximum relief of 100 m; in the Snowy Mountains, Australia, where the surface relief is >1000 m, weathered rock is found locally at depths of 300 m below the surface; and crystalline rocks in Nigeria have been shown to have a weathered contact of 50-m relief where the surface relief is only 25 m. The geomorphic significance of such mantle/bedrock surfaces may emerge when a period of deeply irregular weathering may be succeeded by stripping of the mantle due to a climatic change inaugurating accelerated sheetwash, solifluction, etc. This two-cycle origin has been suggested both for granite tors (Figure 9.18) and for tropical inselbergs, although both these forms may have an incremental origin of progressive weathering and removal.

Although vertical cracking is a common feature of weathered debris mantles, their most characteristic structural feature is a systematic layering, subparallel to the surface, involving variations in grain size, shape and sorting, organic and chemical content, permeability,

shearing and frost heaving susceptibility. These variations are found to some degree even on steep slopes, except where exceptionally rapid solifluction, earthflows, or other mass movements have inhibited them. This layering, of which soil horizons are an example, is produced by the vertical mobility of cations, differential weathering with depth, surface additions of vegetable litter, washing down of fine particles, near-surface frost heaving, vertical fluctuations of the debris water table, root layering zones, and the lateral downslope movement

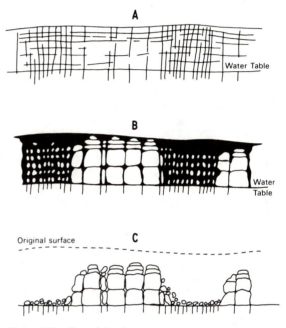

**Figure 9.18**   The origin of tors by A. jointing and the establishment of a water table; B. weathering down to water table and the production of corestones; C. evacuation of the finer material.
*Source:* Linton, 1955, figure 2, p. 475.

**Table 9.11**   Influence of debris layering

| Soil depth (cm) | Character | Grain Sizes (%) | | | Density | Permeability (K) (cm/h) | % water transmitted laterally |
| | | Sand 0·1–0·05 mm | Silt 0·05–0.002 mm | Clay <0·002 mm | | | |
|---|---|---|---|---|---|---|---|
| 0–5 | Surface litter | — | — | — | — | High | 1 |
| 5–56 | Sandy loam | 64 | 28 | 8 | 1·33 | } 28·6 | 16 |
| 56–90 | Sandy loam | 67 | 18 | 15 | 1·41 | | 64 |
| Hiatus | | | | | | | |
| 90–120 | Loam | 42 | 38 | 20 | 1·78 | 1·7 | 16 |
| 120–150 | Clay loam | 31 | 40 | 29 | 1·80 | 0·2 | 3 |

*Source:* Whipkey, 1965.

of debris more or less maintaining its relative vertical position. The geomorphic effects of this layering lie mainly in its control over soil moisture movements, over the differential shearing resistance of various layers and in the frost heaving susceptibility of fine-grained layers. An example of the control of weathered mantle layering over lateral movement of soil water has been experimentally measured on a 17-m-long, 16° slope on sandstone in the Allegheny Plateau having a sixty-year-old mixed oak cover. The figures in Table 9.11 show the influence of debris layering on the lateral transmission of rainwater downslope.

When a surface which has been under the influence of deep, transport limited weathering is subjected to a sudden increase in erosion, the irregular basal weathering surface may be exposed to form the land surface, littered with the larger residual corestones. In this manner the tors of temperate regions have been ascribed to periglacial debris evacuation by solifluction (see Figure 9.18) and the inselbergs or bornhardts of the wet–dry tropics to accelerated sheetwash erosion. Extensive landscapes (etchplains) may have been formed by the etching action of deep weathering followed by the wholesale removal of the finer weathered debris.

## References

Ahnert, F., (1976) 'Description of a comprehensive three-dimensional model of landform development', *Zeitschrift für Geomorphologie,* Supplement 25, 29–49.

Berner, R. A. (1971) *Principles of Chemical Sedimentology,* New York, McGraw-Hill.

Blatt, H., Middleton, G. and Murray, R. (1972) *Origin of Sedimentary Rocks,* Englewood Cliffs, NJ, Prentice-Hall.

Brunsden, D. (1979) 'Weathering', in C. Embleton and J. Thornes (eds), *Process in Geomorphology,* London, Arnold, 73–129.

Carson, M. A. and Kirkby, M. J. (1972) *Hillslope Form and Process,* Cambridge University Press.

Chesworth, W. (1973) 'The residua system of chemical weathering: a model for the breakdown of silicate rocks at the surface of the earth', *Journal of Soil Science,* vol. 24, 69–81.

Chorley, R. J. (1969) 'The role of water in rock disintegration', in R. J. Chorley (ed.), *Water, Earth and Man,* London, Methuen, 135–55.

Chorley, R. J. and Kennedy, B. A. (1971) *Physical Geography: A Systems Approach,* London, Prentice-Hall.

Douglas, I. (1968) 'The effects of precipitation chemistry and catchment area lithology on the quality of river water in selected catchments in eastern Australia', *Earth Science Journal,* vol. 2, 126–44.

Gilkes, R. J., Scholz, G. and Dimmock, G. M. (1973) 'Lateritic deep weathering of granite', *Journal of Soil Science,* vol. 24, 523–36.

Goldich, S. S. (1938) 'A study in rock weathering', *Journal of Geology,* vol. 46, 17–58.

Goudie, A. S., Cooke, R. U. and Evans, I. S. (1970) 'Experimental investigation of rock weathering by salts', *Area,* no. 4, 42–8.

Griggs, D. (1936) 'The factor of fatigue in rock exfoliation', *Journal of Geology,* vol. 44, 781–96.

Huang, W. H. and Keller, W. D. (1973) 'Organic acids as agents of chemical weathering of silicate minerals', *Nature, Physical Science,* vol. 239, 149–51.

Jenny, H., (1941) *Factors of Soil Formation,* New York, McGraw-Hill.

Jordan, C. R. and Kline, J. R. (1972) 'Mineral cycling: some basic concepts and their application in a tropical rain forest', *Annual Review of Ecology and Systematics,* vol. 3, 33–50.

Kelly, W. C. and Zumberge, J. H. (1961) 'Weathering of a quartz diorite at Marble Point, McMurdo Sound, Antarctica', *Journal of Geology,* vol. 69, 433–46.

Kirkby M. J. (ed.) (1978) *Hillslope Hydrology,* Chichester, Wiley.

Kononova, M. M. (1966) *Soil Organic Matter,* 2nd edn, New York, Pergamon.

Krumbein, W. C. and Sloss, L. L. (1963) *Stratigraphy and Sedimentation,* 2nd edn, San Francisco, Calif., Freeman.

Lindsey, C. G., Doesburg, J. M. and Vallario, R. W. (1982) 'A review of long-term rock durability', *Proceedings of the 5th Symposium, Uranium Tailing Management, Colorado State University,* 101–15.

Linton, D. L. (1955) 'The problem of tors', *Geographical Journal,* vol. 121, 470–87.

Loughnan, F. C. (1969) *Chemical Weathering of the Silicate Minerals,* New York, Elsevier.

Ollier, C. D. (1969) *Weathering,* Edinburgh, Oliver & Boyd.

Pettijohn, F. J. (1975) *Sedimentary Rocks,* 3rd edn, New York, Harper & Row.

Reiche, P. (1950) *A Survey of Weathering Processes and Products,* Albuquerque, NM, University of New Mexico Press.

Ritter, D. F. (1978) *Process Geomorphology,* Dubuque, Iowa, W. C. Brown.

Rodin, L. E. and Basilevic, N. I. (1967) *Production and Mineral Cycling in Terrestrial Vegetation,* Edinburgh, Oliver & Boyd.

Schumm, S. A. and Chorley, R. J. (1966) 'Talus weathering and scarp recession in the Colorado Plateaus', *Zeitschrift für Geomorphologie,* vol. 10, 11–36.

Sparks, B. W., (1971) *Rocks and Relief,* London, Longman.

Strakhov, N. M. (1967) *Principles of Lithogenesis,* Edinburgh, Oliver & Boyd.

Thomas, M. F. (1974) *Tropical Geomorphology,* London, Macmillan.

Trudgill, S. T. (1976) 'Rock weathering and climate: quantitative and experimental aspects', in E. Derbyshire (ed.), *Geomorphology and Climate*, London, Wiley, 59–99.

Trudgill, S. T. (1977) *Soil and Vegetation Systems*, Oxford University Press.

Watts, D. (1971) *Principles of Biogeography*, London, McGraw-Hill.

Whipkey, R.Z. (1965) 'Subsurface stormflow from forested slopes', *Bulletin of the International Association of Scientific Hydrology*, vol. 10, 74–85.

Witkamp, M. (1971) 'Soils as components of ecosystems', *Annual Review of Ecology and Systematics*, vol. 2, 85–110.

# Ten  *Mass movement*

Mass movement is the detachment and downslope transport of soil and rock material under the influence of gravity. The sliding or flowing of these materials is due to their position and to gravitational forces, but mass movement is accelerated by the presence of water, ice and air. This definition of mass movement permits consideration of the movement of earth materials at all scales and at all rates. Therefore, the slow downslope creep of soil and rock fragments, as well as the rapid movement of large landslides over long distances, are both mass movement.

Mass movement occurs everywhere in varying degrees and, indeed, landslides of major dimensions have been identified on the moon, on Mars and beneath the Atlantic Ocean on continental margins. For example, one Martian landslide as described by Lucchitta (1978) is about 60 km long and 50 km wide; it contains about 100 km³ of rock debris, and it moved about 2 km from an escarpment that is 6 km high (Plate 9).

Large translational slides and slumps on the continental margins of the Atlantic Ocean have travelled hundreds of kilometres on slopes of less than a half-degree. Many slides occur in water depths greater than 2000 m. According to Embley and Jacobi (1978) the largest is off the Spanish Sahara where the area of disturbance is 18,000 km². The origin of these slides may be sediment overloading near the edge of the continental shelf, when sea level was lower during Pleistocene time.

## 10.1 Significance

Frequently it is considered that landslides occur without warning and yet no slide can take place unless the factor of safety or the ratio between the average shearing resistance of the ground and the average shearing stresses on the potential surface of sliding decreases from an initial value greater than one to unity at the instant of the slide.

Most landslides that are preceded by an almost instantaneous decrease of this ratio are those due to earthquakes; usually a gradual decrease of the ratio precedes failure. Hence, if a landslide comes as a surprise to the eyewitnesses, it would be more accurate to say that the observers failed to detect the phenomena which preceded the slide. A slide at Goldau, Switzerland, took the villagers by surprise, but the horses and cattle became restless several hours before the slide, and the bees deserted their hives.

As the population of this planet increases, areas of potential landslide danger will be utilized for recreational purposes or simply as places to live, and in many instances changes of land use have contributed to the frequency and magnitude of landslides.

The more rapid forms of mass wasting have been and continue to be of concern to land managers, highway engineers and, in fact, to anyone living in landslide-prone areas. The destruction of highways, dwellings and loss of life associated with mass movement through historic times has been very significant (Table 10.1). It is estimated that landslide damage in California will total $10 billion between 1979 and 2000 (Leighton, 1976). Of greater significance is the loss of life associated with landslides. For example, in 1963 a large landslide displaced the water behind the Vaiont Dam in northern Italy causing a flood which killed almost 2000 people in the valley downstream. Perhaps the most disastrous landslide in North America was that which occurred at Frank, Alberta, Canada, in 1903, in which a large segment of Turtle Mountain was displaced into the valley, burying part of the town, industrial area, railroads and taking seventy lives (Figure 10.17B). Many examples can be cited, but the 1970 disaster in Peru will suffice. On 31 May 1970 at 3·23 p.m. an earthquake of Richter magnitude 7·7 occurred twenty-five miles off the coast of

Peru at latitude 9·2°S. One effect of this event was the detachment of an 800-m-wide slab of ice and rock from the west face of Mt Huascarán. This mass, with volume greater than 25 million cubic metres, moved down valley at velocities up to 400 km/h, totally destroying two towns and killing at least 20,000 people. Although a catastrophe of this magnitude is unusual, deposits related to a similar type of event cover 130 km² of the Puget Sound lowland 32 km east of Tacoma, Washington State (Figure 10.1). This was a volcanic mudflow, probably associated with an eruption of Mt Rainier. Crandell (1971, p. 71) recommends that in the event of another eruption, valley floors within a radius of at least 40 km from the volcano should be evacuated immediately.

The individual landslide event may be only part of a larger problem, and downstream river aggradation and flooding as a result of the deposit of mass movement debris into the river systems is a problem of national importance in Japan, New Zealand and Taiwan. For example, Sheng (1965) calculates that 83 per cent of the sediment produced by the Tsengwen Watershed in Taiwan is derived from landslides. In this drainage basin relief is high, slopes are steep and mean annual precipitation is 2·5 m. During one 24-hour period, 1·3 m of rain fell during a typhoon. These conditions understandably lead to mass movement on a large scale and downstream aggradation.

The importance of rapid slope failures is undoubtedly great in certain environments. On 15°–45° mica schist and amphibolite slopes at Kärkevagge, Sweden, it has been estimated that current erosion is apportioned as follows:

rockfalls 7% ⎫
debris avalanches 8% ⎬ rapid failures 49%
earthslides 34% ⎭

scree movement 1% ⎫
solifluction 2% ⎬ slow failures 51%
solution 48% ⎭

In the Upper Rhine Valley of Switzerland rapid failures are responsible for about 84·6 per cent of the mean annual displacement of material, as against 15·4 per cent by creep and 0·001 per cent by river action. The earthquake of 1970 (magnitude 7) in north Papua New Guinea caused rapid slope failures over 240 km², including debris avalanches, which affected 60 km² and which removed an average of 6·7 cm of debris from the whole area. Of this amount, 50 per cent reached the sea in less

**Table 10.1** Landslide disasters

| | Date | People killed | Remarks |
|---|---|---|---|
| Brenno Valley, Switzerland | 1512 | 600 | Rockslide-dammed valley; dam broke in 2 years causing destruction |
| Tour d'Ai, Switzerland | 1584 | 300 | Landslide-devastated village of Yvorne in Rhône Valley |
| Mount Conto, Switzerland | 1618 | 2,430 | Rockslide |
| Goldau, Switzerland | 1806 | 457 | Landslide-destroyed village |
| Mt Ida, Troy, New York | 1843 | 15 | Sediment slump and flow |
| Elm, Switzerland | 1881 | 115 | Rock avalanche also demolished 83 houses |
| Trondheim, Norway | 1893 | 111 | Liquefaction flow in marine clays |
| Frank, Canada | 1903 | 70 | Rock avalanche destroyed most of town |
| Kansu Province, China | 1920 | 100,000–200,000 | Earthquake caused loess flows |
| Nordfjord, Norway | 1936 | 73 | Rockfall created 74-m wave |
| Kobe, Japan | 1938 | 461 | Rocky mudflows |
| Kure, Japan | 1945 | 1,154 | Rocky mudflows |
| Yokohama, Japan | 1958 | 61 | Rocky mudflows |
| Madison, Montana | 1959 | 28 | Rock avalanche buried campers |
| Vaiont, Italy | 1963 | 2,000 | Rockslide into reservoir created wave that flooded below dam |
| Anchorage, Alaska | 1964 | 114 | Combined toll from landslides and earthquake |
| Aberfan, Wales | 1966 | 144 | Manmade mining spoiled hill; landsliding buried mostly children |
| Brazil | 1966–7 | 2,700 | Combined toll from landslides and floods |
| Nelson County, Virginia | 1969 | 150 | Combined toll from debris avalanches and floods |
| Huascarán area, Peru | 1970 | 21,000 | Combined rock avalanche and debris-flow buried two cities |
| St Jean Vianney, Canada | 1971 | 31 | Slab flows buried people and houses |

*Source:* Coates, 1977, table 2, p. 19: various sources.

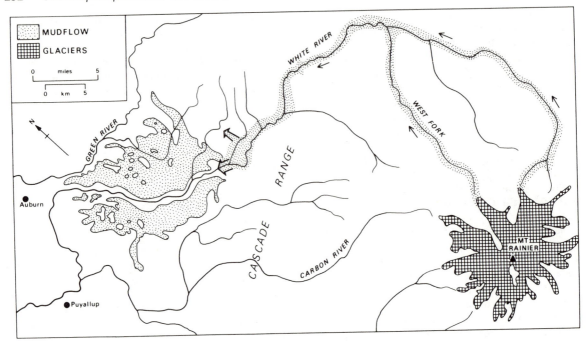

**Figure 10.1** Map showing path and distribution of the Osceola mudflow, Washington State.
*Source:* Crandell and Waldron, 1956, figure 1, p. 350.

than six months. It has been estimated that in the very tectonically active Adelbert Range of Papua New Guinea the average vertical erosion rate is 0·8–10 m/1000 year, of which 60–70 per cent is triggered by earthquakes.

## 10.2 Gravity tectonics

The relation between mass wasting and tectonics is a relatively clear one. Where rocks are shattered and relief is high, this is where mass movement is common and, in fact, the denudation of high mountains may, as indicated above, be the result of mass wasting rather than fluvial or glacial processes.

In some cases the mountain mass itself is significantly shaped by mass movement. For example, it was recognized in the Apennine Mountains of northern Italy that rocks had moved, apparently by thrust faulting, at least 100 km and perhaps as far as 200 km. The rocks are relatively incompetent, and they cannot maintain their continuity under the stress required for this type of movement. Italian geologists concluded that the movement was due to uplift and massive gravitational sliding. Not only did large blocks of the earth's surface move at velocities of 0·5–1 cm/year, but as they moved they exposed an underlying surface by tectonic denudation. Figure

10.2 shows an example of gravity tectonics in an area about 50 km south of Florence, Italy, where there is evidence of massive overthrusting and folding due to mega-mass wasting.

In the United States the Heart Mountain fault in western Montana has been ascribed to gravity tectonics. Heart Mountain is a piece of a much larger fault plate that became detached and moved independently for a distance of at least 20 km on a 2° slope. In the same area the McCulloch Peaks, fragments of the fault, travelled 105 km. This movement over such a flat slope, partially on a fault plane and partially over the preexisting land surface, apparently coincided with a major period of earthquake activity and volcanic intrusion in that region of Wyoming (Pierce, 1973).

The geomorphic features involved in gravity tectonics are far larger than those usually associated with landslides and other forms of mass wasting, although significant areas can be covered with modern landslide debris (Table 10.2). For example, about 10,000 years ago a paleoslide of exceptional dimensions occurred in Iran. The Saidmarreh landslide (Figure 10.14), which consists of a portion of a limestone hogback 15 km long, 5 km wide and 300 m thick, slid into the adjacent valley where it dammed the Karkheh River. The moving debris rose 600 m from the valley floor, and it actually crossed the

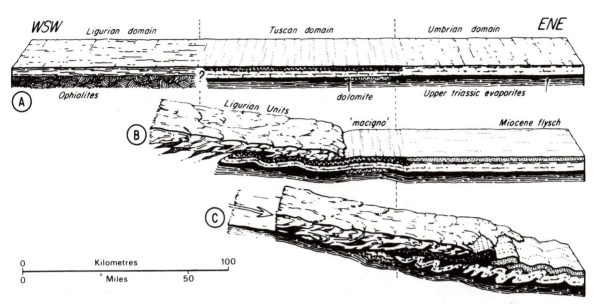

**Figure 10.2** Mid-Miocene thrusting and gravity tectonics in the northern Apennines, 50 km south of Florence: A. original basins of deposition; B. thrusting from the WSW; C. associated gravitational sliding along Triassic evaporites (black).
*Source:* Elter and Trevison, 1973, figure 15, p. 187, in K. A. De Jong and R. Scholten, *Gravity and Tectonics*, 1973, copyright © Wiley International, by permission.

**Table 10.2** Landslide terrain properties

| Locality | Date | Volume $(10^4 m^2)$ | Area covered $(km^2)$ | Vertical displacement or fall $(m)$ | Horizontal movement $(km)$ | Runup distance $(m)$ | Velocity (est) $(km/h)$ | Landslide type |
|---|---|---|---|---|---|---|---|---|
| Huascarán, Peru | 31 May 1971 | 10 | | 4,000 | 14·5 | | 400 | Rock avalanche–debris flow |
| Little Tahoma Peak, Washington | 14 December 1963 | 10·7 | | 1,890 | 6·9 | 90 | 152 | Rock avalanche |
| Elm, Switzerland | 11 September 1881 | 12·7 | 0·6 | 610 | 1·4 | 103 | 160 | Rock avalanche |
| St Albans, Canada | 27 April 1894 | 19·1 | | | | | | Liquefaction flow |
| Madison, Montana | 17 August 1959 | 28·3 | 0·5 | 400 | 1·6 | 130 | 180 | Rock avalanche |
| Sherman, Alaska | 27 March 1964 | 28·3 | | 600 | 5·0 | 137 | 185 | Rock avalanche |
| Frank, Canada | 29 April 1903 | 36·5 | 2·5 | 870 | 4·0 | 120 | 175 | Rock avalanche |
| Gros Ventre, Wyoming | 23 June 1925 | 38·2 | 0·8 | 640 | 1·9 | 106 | 164 | Rock avalanche |
| Goldau, Switzerland | 2 September 1806 | 40 | | 550 | 1·7 | | | Rock avalanche |
| Apollo 17, Moon | — | 200 | 21 | 2,000 | 5·0 | | | Rock avalanche |
| Silver Reef, California | Prehistoric | 226 | | 790 | 5·3 | 46 | 104 | Rock avalanche |
| Vaiont, Italy | 9 October 1963 | 260 | | | 1·8 | 240 | | Rockslide |
| Blackhawk, California | Prehistoric | 283 | | 1,220 | 8·0 | 64 | 120 | Rock avalanche |
| Gohna, India | September 1893 | 290 | | 1,470 | 1·6 | | | Rock avalanche |
| Martinez, California | Holocene | 382 | | 2,000 | 7·6 | | | Rock avalanche |
| Ticino River, Switzerland | During glacial retreat | 500 | | | | | | Rock avalanche |
| Upper Garhwal, India | 22 September 1893 | 566 | | 1,520 | 3·2 | | | Rock avalanche |
| Sawtooth Ridge, Montana | Prehistoric | 650 | | 360 | | | | Rock avalanche |
| Klonsee, Alps | Interglacial | 770 | | | | | | Rock avalanche |
| Tin Mountain, California | Prehistoric | 1,795 | | | | | | Rock avalanche |

**Table 10.2** — *continued.*

| Locality | Date | Volume (10⁴m²) | Area covered (km²) | Vertical displacement or fall (m) | Horizontal movement (km) | Runup distance (m) | Velocity (est) (km/h) | Landslide type |
|---|---|---|---|---|---|---|---|---|
| D'Onsoi, Pamir Mountains | 18 February 1911 | 2,080 | 10·3 | 1,160 | 3·6 | | | Rock avalanche |
| Flims, Switzerland | Interglacial | 12,000 | 41·4 | 1,980 | 16·1 | | | Rock avalanche |
| Lake Tahoe, California–Nevada | — | 10,000 | | | | | | Subaqueous slide |
| Saidmarreh, Iran | Holocene | 20,000 | 165·7 | 1,650 | 14·5 | 457 | 338 | Rock avalanche |
| Sàmar Island, Philippines | 2000 years BP, | 135,000 | | | | | | Block glide |
| Gulf of Alaska, Alaska | 27 March 1964(?) | 590,000 | | | | | | Subaqueous slide |

*Source:* Coates, 1977, table 1, p. 18; various sources.

divide to the north, finally coming to rest in the next valley as much as 20 km from the point of origin (Watson and Wright, 1969).

## 10.3 Classification

To the layman the term 'landslide' means a rapid downward movement of a mass of rock, soil, or earth on a slope. It is a spectacular and hazardous geomorphic event. However, mass movement includes many other types of mass wasting phenomenon than the landslide. The term landslide is actually a misnomer because in most cases the material flows rather than slides. Classifications of landslides and mass movement phenomena (Figures 10.3 and 10.4) are usually based on the type of movement (slide, flow, heave), on the velocity of movement and on the water content (dry rock materials or

earth materials with varying proportions of ice and water). In Table 10.3 the important types of mass movement are classified according to direction of movement, type of movement and the presence or absence of mobilizing agents (water, air, ice). In all cases gravity is the prime-mover.

*Vertical movement.* Under vertical displacements there are two types of movement: the fall of earth material from a cliff face (figure 10.5A), and subsidence due usually to subsurface removal of material. The material can be dry and no transporting agent need be present, although frost action is an important factor causing rock falls. Rapp (1960), while working in northern Sweden, found that when the temperature remains below 0°C there is little or no mass wasting, but when temperatures rise above 0°C rockfalls are numerous (Figure 10.6).

**Table 10.3** Mass movement – mass wasting phenomena

| Direction of movement | Vertical | | Lateral | | Diagonal | | |
|---|---|---|---|---|---|---|---|
| Type of movement | Fall | Subsidence | Slide | Spread | Creep | Slide | Flow |
| Presence of transporting agent | No | No | Minor in basal layer or on sliding surface | Moderate in basal layer plastic | Minor | Minor to moderate | Moderate     Major |
| Types of mass movement | Rockfall, earthfall, debris fall, topple | Collapse settlement | Block slide | Spread cambering sackung | Soil creep, rock creep, talus creep | Rock slide, debris slide, soil slip, slump | Earth flow, debris flow     Solifluction mudflow, sturzstrom, (rock avalanche), rock glacier |

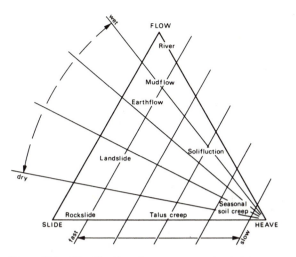

**Figure 10.3** Classification of mass movement processes.
*Source:* Carson and Kirkby, 1972, figure 5.2, p. 100.

There are three types of falls based on the type of material involved. *Rockfalls* involve bedrock fragments, *earthfalls* involve alluvium, colluvium and soil, and *debris-falls* involve all of these materials plus vegetation, houses, cars and people. Although the fall is a simple process, it can lead to other types of rapid mass movement, which will be considered under the category of diagonal flow features. A *topple* (Figure 10.5B) is the rotational fall of a slab of material, such as the section of riverbank.

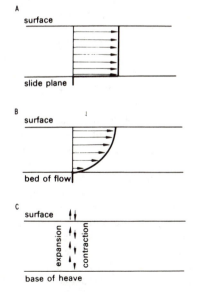

**Figure 10.4** Velocity profiles for ideal mass movement types in Figure 10.3: A. pure slide; B. pure flow; C. pure heave.
*Source:* Carson and Kirkby, 1972, figure 5.1, p. 100.

If the falling rocks are not removed from the base of a slope, they accumulate to form a *talus* or scree slope. The material accumulates at an 'angle of repose', which reflects the shape and roughness of the particles. Figure 10.7 shows the effect of particle size and shape on the angle of repose. Obviously, the rounder and smoother the material, the gentler is the talus slope. Sorting is also important, and the presence of fine sediments among the coarse increases the angle of repose.

Carrigy (1970) and others have made a distinction between the critical angle of repose, which is the maximum angle at which sediments can stand, and the angle of rest, which is defined as the inclination of the slope after some adjustment has occurred. For some dune sand, the angle of repose is about 35°, whereas the angle of rest is about 31°, which is the angle at which such material is normally found in the field.

The other category of vertical mass movement is *subsidence* or sinking of the ground, and within this category there are two types of displacement. The first involves collapse of the roof of a subterranean cavity to form a circular sink hole, or a linear depression forms if the roof of a limestone cavern or a lava tube collapses. Limestone regions are characterized by such collapse features, and the term karst is given to this topography (see Section 8.3). Collapse can be man-induced and many mining regions display closed depressions or pseudo-karst features resulting from collapse of underground workings.

*Settlement*, on the other hand, can be a result of the withdrawal of fluids from permeable material, thereby permitting rearrangement of the grains of this material and its compaction. This is common throughout the world due to groundwater and petroleum withdrawal, and it has been described in southern California as a result of oil extraction and in Japan as a result of groundwater pumping, where subsidence of up to 29 cm per year has been measured in the Tokyo area. The city of Venice is being subjected to inundation primarily as a result of the pumping of both groundwater and oil from beneath the city. In addition, poorly compacted materials can subside and be compacted by the addition of water (hydrocompaction) or by vibrations.

*Lateral movement.* Under the heading of lateral movement there are two types of movement, sliding and spreading. Sliding is translational in nature with movement occurring along a plane that is generally horizontal or slightly inclined. Movement can be along a horizontal fracture or most likely along an interface between strata of two different compositions, for example, sandstone overlying shale. In this category is the *block glide* which is the movement of a large block of massive rock or sediment on a surface that has been lubricated, or over a

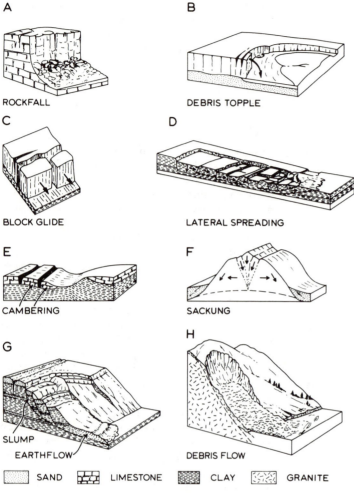

A ROCKFALL

B DEBRIS TOPPLE

C BLOCK GLIDE

D LATERAL SPREADING

E CAMBERING

F SACKUNG

G SLUMP EARTHFLOW

H DEBRIS FLOW

| SAND | LIMESTONE | CLAY | GRANITE |

**Figure 10.5** Types of vertical and lateral mass movement (A.–H,): A., B., D., G. and H. are very rapid; C. and F., slow; E., very slow.
*Sources:* Varnes, 1978; also Selby, 1982, figure 7.18, p. 165.

surface below which the underlying material is moving plastically and transporting the block (Figure 10.5C).

A fine example of a blockglide and *topple* is provided by a 90 m-long, 30 m-high and 9 m-thick block of sandstone which separated from the cliff face and finally fell on the Indian ruin of Pueblo Bonito in Chaco Canyon National Monument, New Mexico in 1941 (Plate 10). In this arid region (208 mm/year) mass movement might be expected to be non-existent; nevertheless, the sandstone monolith had two names. It was known as Threatening Rock by the National Park Service personnel who viewed it as a threat to the Pueblo Bonito ruins, and Braced-Up Cliff by the Navajo Indians who observed pine logs that had been wedged beneath the rock by the residents of Pueblo Bonito to prevent its movement. The logs, as dated by tree-ring count (dendrochronology), were cut in AD 1057 and 1004. Thus concern about the

stability of Threatening Rock existed about 1000 years ago.

During the late 1930s the custodian of the Chaco Canyon National Monument became concerned that the rock would topple on to the ruins and destroy them. In order to convince his superiors that some mechanical means of stabilizing the rock was required, the custodian climbed to the top of the cliff and established a measuring-point on the top of Threatening Rock and on top of the adjacent cliff. Each month he measured the distance between the two points and included this information in his monthly report. The data he collected are presented on Figure 10.8, which shows an exponential increase in movement between 1935 and January 1941 when the rock finally toppled (Figure 10.5B), with in fact only minimal damage to Pueblo Bonito. An interesting aspect of the movement data is that they show a

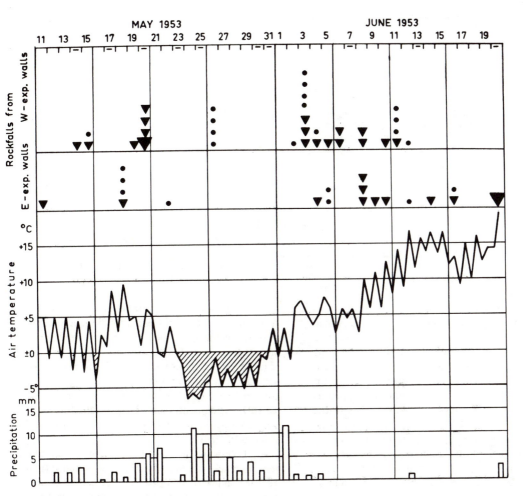

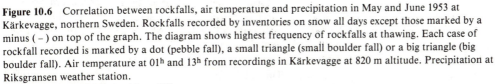

**Figure 10.6** Correlation between rockfalls, air temperature and precipitation in May and June 1953 at Kärkevagge, northern Sweden. Rockfalls recorded by inventories on snow all days except those marked by a minus ( – ) on top of the graph. The diagram shows highest frequency of rockfalls at thawing. Each case of rockfall recorded is marked by a dot (pebble fall), a small triangle (small boulder fall) or a big triangle (big boulder fall). Air temperature at 01ʰ and 13ʰ from recordings in Kärkevagge at 820 m altitude. Precipitation at Riksgransen weather station.
*Source:* Rapp, 1960, figure 18, p. 105.

markedly seasonal pattern, with most of the movement occurring during winter months. It is unlikely that freeze–thaw affected this huge rock, but the low temperatures of winter reduced evaporation and perhaps the collection of snow behind the rock, and its subsequent melting wetted the sandstone–shale interface and permitted movement, due to plastic deformation of the shale. During the hot summer months the interface was probably dry and little or no movement was recorded.

Extrapolation of the movement data suggests that the rock began to slide away from the cliff about 2300 years

ago, and 1000 years ago when it was 'braced up', it was separated from the cliff by 0·3 m. This combination block glide and topple is significant because it indicates that even large mass movements can be monitored, and that the rate of movement may increase exponentially through time.

Spreading is the lateral movement of a series of blocks away from a centre (Figure 10.5D). The distinction between a spread and a glide is that the spread has multiple blocks, and there is no well-defined basal slip surface. The overlying blocks float on an underlying

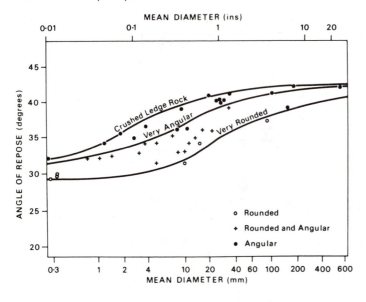

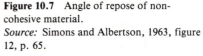

**Figure 10.7** Angle of repose of non-cohesive material.
*Source:* Simons and Albertson, 1963, figure 12, p. 65.

plastic or liquefied zone.

Under this category are *cambering* and *sackung*, which are features of geologic importance. During cambering (Figure 10.5E) there is a draping of sedimentary units over the hillsides, as the underlying sediments creep towards the valleys. In an area of cambering, as a result of this deformation, the hillsides could be mapped as anticlinal axes and the valleys as synclinal axes. In the case of sackung (Figure 10.5F) lateral spreading away from a centre results in grabenlike features in the centre of the spreading mass. Both cambering and sackung can confuse the structural geologist and lead him to map pseudo-structural features that are, in fact, due to mass wasting.

Rapid spreading may be the result of weakening of materials by seismic waves. This was obviously the case during the 1964 Alaskan earthquake. The epicentre was 129 km east of Anchorage, and the Richter magnitude was 8·5. In Anchorage 21-m-high bluffs failed and spreading occurred over an area of 52 hectometres with the destruction of seventy-five houses (Figure 10.9). Within the slide area the original ground surface was broken up into a complex system of ridges and depressions (Plate 11). The Turnagain Heights landslide occurred where a flat plain of sand and gravel, that was underlain by clay, formed the coast. The slide resulted from a loss of strength in the clay. The slide was simulated by a model study (Figure 10.10) in which a 10-cm layer of very weak clay was overlain by layers of stronger clay. Support along one side of the mass was removed, and the resulting sliding was described by Seed and Wilson (1967) as follows:

(1) A series of several rotational slides caused outward and downward movement of the ground surface, with extrusion of the weak layer from the toe of the slide (Figure 10.10A–D).

(2) After several such slides, the upper layers of stronger clay sank and disrupted the weak layer, so that no further extrusion from the toe occurred (Figure 10.10D). Then continued sliding resulted in the outward movement of a block without change in elevation (Figure 10.10E).

(3) Following the lateral movement of the block, tension cracks developed in the clay behind it (Figure 10.10F), and settlement of the area behind the block formed a depressed zone in which tension cracks separated the clay into a series of blocks that moved laterally and subdivided and sealed off the weak layer (Figure 10.10G).

The model provided an explanation for the complex spread topography observed in the Turnagain Heights area, and it gives an example of how simple physical models can provide information on complex natural phenomena.

*Diagonal movement.* The diagonal movement of materials is downslope and away from their point of origin, and as compared to lateral movement, the gravitational forces involved are much larger. There are three types of movement under this category: creeping, sliding and flowing, with an increase of velocity from the former to the latter. *Creep* is the slow, imperceptible downslope movement of earth materials under the influence of gravity. Relatively minor amounts of water

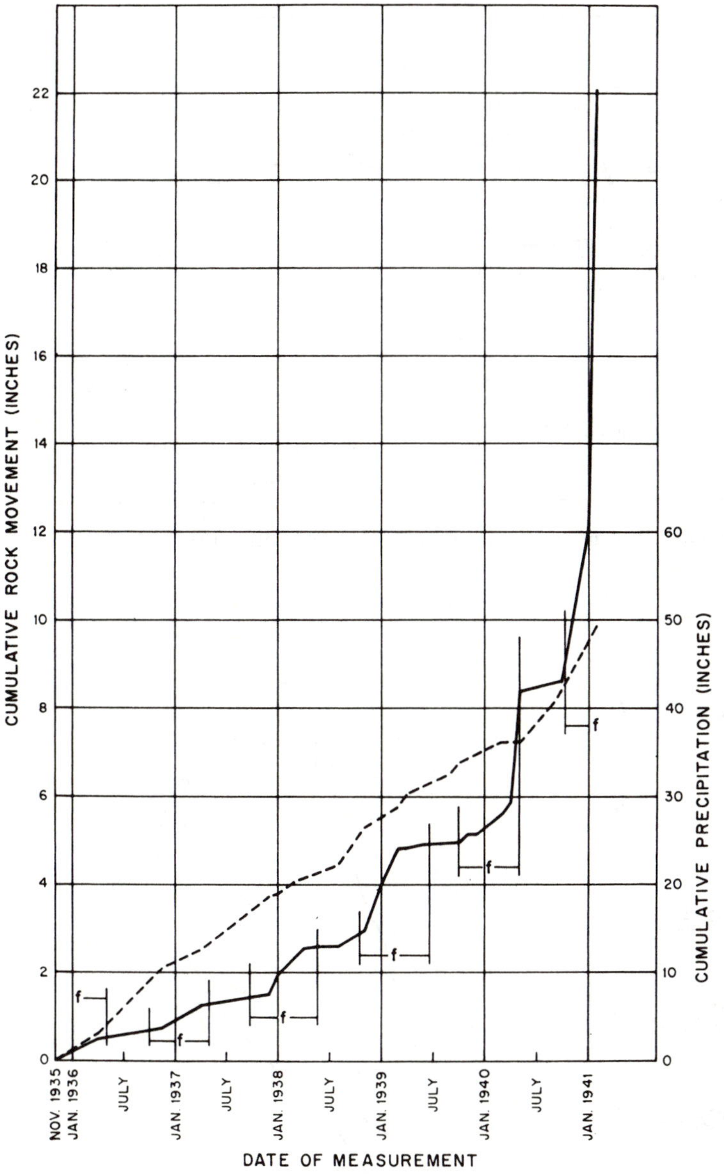

**Figure 10.8**  Cumulative movement of Threatening Rock (full line) and cumulative precipitation (broken line) plotted against date of measurement or time. Vertical lines, drawn at dates of first killing frost of the autumn and the last killing frost of the spring for the years of record, indicate cumulative movement which occurred during periods of freeze and thaw. Freeze–thaw periods are designated f.
*Source:* Schumm and Chorley, 1964, figure 2, p. 1047.

and ice cause creep not because they act as a transporting medium, but due to the disruptive effects of freeze–thaw and swelling. Later, during a discussion of hillslopes (p. 258), we shall discuss the importance of creep to the appearance of the landscape as a whole. Soil creep is one type of mass movement that can operate over the entire interfluve area of a landscape. The other forms of mass movement are restricted areally and are landforms in themselves (Figure 10.5G and H). Soil creep, however, modifies the landscape, without intruding in any signifi-

cant way upon the regularity of the landscape. It involves the movement of soils as a viscous fluid, or as a plastic, and the vertical velocity profile shows that soil movement is usually greatest near the soil surface (Figure 10.11).

Rock creep is the movement of rock upon rock. If a rock fragment is lying on an inclined bedding or joint plane, there will be movement of that rock due to thermal creep, which is the expansion and contraction of the rock due to heating and cooling. Due to the influence of gravity, any disturbance of this type produces downslope

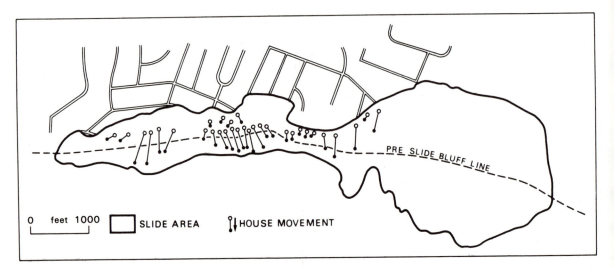

**Figure 10.9** Movements of houses in the slide area of Turnagain Heights landslide, Anchorage, Alaska, 1964.
*Source:* Seed and Wilson, 1967, figure 6, p. 328.

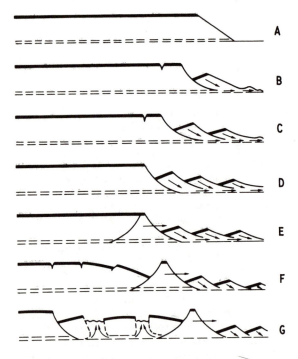

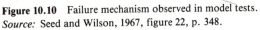

**Figure 10.10** Failure mechanism observed in model tests.
*Source:* Seed and Wilson, 1967, figure 22, p. 348.

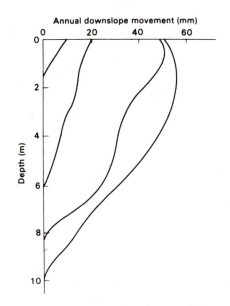

**Figure 10.11** Four representative soil creep velocity profiles from the Berkeley Hills area, California. Note the different horizontal and vertical scales.
*Source:* Kojan, U.S. Forest Exp. Station Berkeley, 1967.

movement. This movement, although very slow , can be particularly important with regard to the safety of travellers in canyon areas. During construction of a canyon road unstable rock materials are removed, but during a period of years rocks that were considered to be stable creep towards a position of instability and eventually fall on to the roadway.

During the last century the effect of diurnal temperature changes on the expansion and contraction of materials and their downslope movement was demonstrated clearly by Canon Moseley, who recorded the 45-cm creep of lead plates in less than two years on the roof of Bristol Cathedral. The coefficient of expansion of rocks is less than that of metal, but measurable downslope movement of rock on rock by thermal creep has been recorded. Tamburi (1974) concluded that large blocks of rock will move as much as 1 cm per year on slopes of about 30°. He found by experimentation that the velocity of movement increased with slope inclination and the maximum grain size of the minerals forming the rock surface. Rock creep decreases with an increase of roughness of the rock surfaces, but with very large mineral grains the number of contact points between the surfaces is reduced and movement is relatively rapid.

*Talus creep* is the slow rearrangement and downslope movement of the surface fragments of talus. Movement is due to the frost action, thermal creep and perhaps to the tracking of animals on the slope.

Typical *landslides*, rapid movements of large quantities of earth material that can be very destructive and hazardous to human life, comprise the last two categories of diagonal movement (sliding and flowing)(Table 10.3). It is difficult to identify a single type of movement that characterizes these mass movements. The initial movement may be a rock or debris slide or a soil slip, which then is converted to a debris-flow or an earthflow. What begins as a fall or slide may be converted by disruption of material and the incorporation of transporting agents, such as air and water, into a high-velocity debris or mudflow.

*Rock slides, debris slides* and *soil slips* are the detachment of earth materials along a bedding plane or weathering front or any zone of weakness with subsequent sliding of the mass downslope. As noted above, the continuity of the sliding mass may be almost immediately disrupted upon the beginning of movement.

A very distinctive mass movement landform is the *slump* (Figure 10.5G) which involves sliding along a curved fracture surface. The initial slump frequently is converted to a flowing type of movement, and it becomes an *earthflow* or *debris-flow* (Figure 10.5G and H).

In all previous noted types of mass movement, water, air and ice have played a minor or moderate role by aiding in the development of conditions that are favourable for slope failure, but in the final diagonal-flow category the transporting agent may be very important indeed. For example, a rock glacier will not move without an icecore, and solifluction will not occur without saturation of the soil by water. Indeed, glaciers themselves are an example of diagonal-flow mass movement (see Chapter 17).

One type of solifluction or soil-flow is that associated with periglacial conditions and it is termed gelifluction. Gelifluction is the result of soil saturation that occurs because permafrost prevents downward percolation of the water. The melting of ice and snow saturates the surface zone and soil flowage results (Section 20.2.2).

A *mudflow* is the result of a high water content and flowage in a preexisting valley or depression. A mudflow or debris flow can move at a high velocity and be destructive. For example, during the years 1962–71 twenty-three people in the greater Los Angeles area were killed as a result of debris flows that probably originated as soil slips. Soil slips are shallow failures of alluvial soil and ravine fill (Campbell, 1975). The debris flows are generated when the initial slipping movement of slabs of soil and wedges of ravine fill cause reconstitution of the sliding masses into viscous, debris-laden mud, which then flows down available drainage courses, accelerating to avalanche speed until reaching gradients gentle enough for deposition. Soil slips require a combination of three conditions: (1) a mantle of colluvial soil or a wedge of colluvial ravine fill; (2) a steep slope; and (3) soil moisture sufficient to permit flowage. In southern California the most common range of slopes for soil slips that give rise to destructive debris flows is from about 45° to about 27°. The steeper slopes generally do not have a continuous mantle of soil and the slips cannot take place. On the gentler slopes of less than 27° the few debris flows that form do not accelerate downslope.

Very heavy rains fell during 18–26 January 1969 in the Los Angeles area, and on 25 January 1969 eight debris flows of this soil slip origin caused twelve deaths among the residents of the Santa Monica Mountains. The soil slips occurred during two periods of highest-intensity precipitation (Figure 10.12). Apparently the strong correlation between debris-flow activity and rainfall of moderate to high intensity can be attributed to the buildup of water in the regolith, as infiltration at the surface takes place at a greater rate than deep percolation. This situation is illustrated in Figure 10.13, which shows shallow-rooted vegetation with a thin mulch of dead leaves and grass growing in a regolith of colluvial soil, the upper part of which contains abundant living

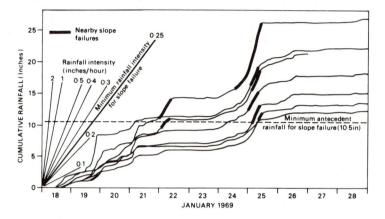

**Figure 10.12** Cumulative rainfall at selected continuously recording gauges in and near the Santa Monica and San Gabriel Mountains showing times of nearby slope failures, 18–28 January 1969.
*Source:* Campbell, 1975, figure 9, p. 13.

and dead roots as well as animal burrows. When the rate of infiltration into and through the upper layers is equal to or less than the capacity of the bedrock to remove it by deep percolation, the water moves down towards the permanent water table in the bedrock, and the stability of the surficial material on the slope is not affected. However, when infiltration through the regolith exceeds the transmissive capacity of the rocks below, a temporary perched water table is formed. When infiltration continues at a high rate, the whole surficial zone will become saturated, at which time all the rainfall in excess of the transmissive capacity of the bedrock becomes surface runoff and downslope seepage within the saturated surficial zone. This reduces the cohesion of the slope materials, and seepage and porewater pressures lead to a reduction in shearing resistance. The soil slips occur as a relatively rigid slab of soil, and vegetation becomes detached at an underlying slip surface. Following initial movement, the slab changes to a viscous fluid and the movement changes from sliding to viscous flow. This change helps to explain how on a given slope the mass that has just become unstable can accelerate to avalanche speed (12 m/s), rather than move at a steady slow velocity.

The eventual disposition of the slipped hillslope material is of considerable significance because it is an important aspect of the geomorphology of the area, and the debris-flow deposits indicate where mass movement and debris flows occur most frequently within a watershed. Where steep low-order drainages on the hillsides descend abruptly to slopes of relatively gentle gradient, extensive fans have been built at the mouths of some remarkably small drainages. Some first- and second-order drainages are so small that they show no bare stream channels even after very severe rainstorms. This condition suggests that surface runoff in many of the small drainage basins is not sufficiently powerful even

during severe rainstorms to have eroded and transported all the debris now deposited in the fans. Fans of this kind, therefore, indicate a history of recurring debris flows. Moreover, their abundance indicates that at least within the recent geologic past, erosion by slab failure and transport of the debris by mud flow are major elements in the geomorphic processes that form the present landscape.

The relatively small soil slips in California are mobilized by the addition of water. Another example of the introduction of a transporting medium into landslide debris is described by Shreve (1968). The Blackhawk landslide is a prehistoric landslide lying at the western base of the San Bernadino Mountains in southern California where a landslide lobe of marble breccia 2 miles (3·2 km) wide and 30–100 ft (9–30 m) thick can be recognized. The Blackhawk landslide is notable not only because it involves $283 \times 10^6 \, m^3$ of crushed rock, but because it also extends 8 km horizontally from its obvious source. Because of its form and structure, it can be compared with the smaller but well-known historic landslides that occurred in 1903 at Frank, Alberta, Canada; in 1964 on the Sherman Glacier in Alaska; and the great prehistoric landslide at the Saidmarreh in south-western Iran (see Figure 10.14 and Table 10.2). Geologic evidence, and in the case of the Elm and Frank slides eyewitness reports, suggest that the Blackhawk slide and similar slides started as high rock falls, which were launched into the air and then traversed gently inclined and relatively smooth gentle slopes at high speeds on a lubricating layer of air. Landslides of this type acquired such high speeds in their descent that at a sudden steepening of the slope they leave the ground, override and compress trapped air upon which they traverse the gentler slopes with little friction. This air-layer lubrication readily accounts for the low friction, high speed and non-flowing motion of these large landslides (sturzstrom) and explains many other puzzling details of their former structure. This

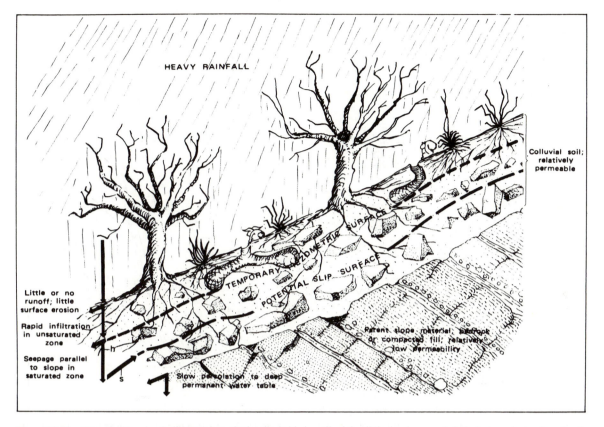

HEAVY RAINFALL

Colluvial soil; relatively permeable

TEMPORARY PIEZOMETRIC SURFACE

POTENTIAL SLIP SURFACE

Little or no runoff; little surface erosion

Rapid infiltration in unsaturated zone

Seepage parallel to slope in saturated zone

Slow percolation to deep permanent water table

Parent slope material; bedrock or compacted fill; relatively low permeability

**Figure 10.13** Diagram showing build-up of perched water table in colluvial soil during heavy rainfall. A sharp bedrock–subsoil junction is shown, but soilslips also occur where the transition from parent material to colluvial soil is more gradational. The significance of the piezometric head (h) in reducing shearing resistance S is suggested.
*Source:* Campbell, 1975, figure 16, p. 18.

mechanism is also involved in certain high-speed snow avalanches, particularly slab avalanches and the fallback from nuclear explosions and impact cratering.

Assuming a launching speed of 275 km/h, a straight slope of 3.2° and negligible drag of leading-edge of the main lobe, the Blackhawk landslide travelled for 8 km from the launching-point to its present position in approximately 80 s, accelerating to a maximum speed of about 435 km/h. In some cases observers of slides of this type felt a strong outrush of air along the margins of the slide, and indirect evidence strongly suggests windblast at many points around the sliding mass of the Frank landslide. As such a flow leaves the confinement of a canyon it advances and spreads. Therefore, the layer of trapped air must support an ever-increasing area, and at some point it escapes completely, and the slide comes in contact with the ground. The area of first contact generally is at the front of the slide, and then the contact expands rearward until the slide halts.

These high-velocity mass movement forms deserve a term of their own, and the name sturzstroms (fall flows) is suggested by Hsu (1975) for phenomena that show clear flowage behaviour, but Coates (1977) uses the more familiar term 'rock avalanches' (Table 10.3). Sturzstroms are a very dangerous mass movement phenomenon, and the Mt Huascarán event may have been of this type.

Of course, the rapidly moving forms of mass movement are not amenable to detailed study during movement, but the slower forms, such as creep, earth flows and debris flows, can be investigated over a period of years. D. B. Prior and N. Stephens and their associates (1968, 1971) have monitored the movement of several of the slower forms of debris and earth flow in north-east Ireland. These flows consist of three parts, a crescentic depression at the head, the main track and the depositional toe. Precipitation in the area is 1100 mm per year and potential evapotranspiration is only 590 mm. Figure

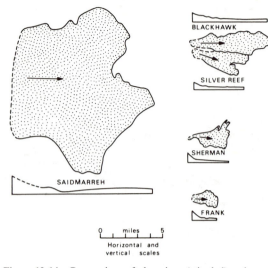

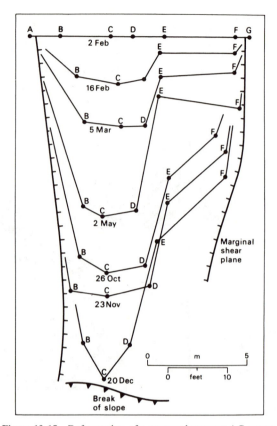

**Figure 10.14** Comparison of plan views (stippled) and profiles of five large landslides: Saidmarreh, south-west Iran; Blackhawk and Silver Reef, San Bernadino Mountains, California; Frank, Alberta, Canada; Sherman Glacier, Alaska.
*Source:* Shreve, 1968, figure 1, p. 3.

**Figure 10.15** Deformation of a surveyed transect AG across the Minnis North debris flow, Northern Ireland, during the period 2 February–20 December 1966.
*Source:* Prior *et al.*, 1968, figure 10, p. 77.

10.15 shows the displacment of a line of pegs across the surface of one debris flow during a year. The movement is not unlike that of glacier ice with low velocity at the margins of the flow. Note also the differential movement between the right and left sides of the flow. There is more of the swelling clays, montmorillonite and illite, on the left side of the slide which indicates the effect of material properties on velocity of slide movement.

Significant relations between slide movement and rainfall amount were obtained for a one-week period of record which illustrates the influence of precipitation on slide movement and the manner in which the slide moves. A plot of cumulative movement against time provides a graph which shows changing rates of movement through time (Figure 10.16).

For some periods, the rate of movement was fairly constant. For example, between noon on Wednesday 28 October and 6 a.m. on Thursday 29 October the rate was approximately 1·7 cm/h. However more generally, the trace shows both concave and convex curves. The former indicates gradual acceleration of movement (between 6 a.m. on Sunday 1 November and 6 a.m. on Monday 2 November), whereas the convexities indicate gradual deceleration. Clearly, the mudflow moves in a complex manner, involving constant rates, acceleration and deceleration.

Short periods of very rapid movement are represented by steeply inclined or vertical sections of the graph. Each of these is followed by a short period of relative stability where the trace is flatter. Thus it seems that a short-lived surging movement is also an important component of the mudflow activity. The time of occurrence of surge periods may be compared with a record of rainfall (Figure 10.16). Preliminary results show that the mudflow surges frequently accompany periods of rainfall or closely follow them. For example, the 19·8-cm surge (3 November) followed a total of 20 mm of rainfall in the previous 11 h. Within this rainstorm there were two periods of high rainfall intensity (4 mm between 2 and 3 a.m., and 9 mm between 7 and 10 a.m.). The largest surge occurred 1 h after the intense rainfall. Type of material and precipitation events strongly influence the type and intensity of mass movement.

## 10.4 Location of mass movement

The description and classification of mass movement

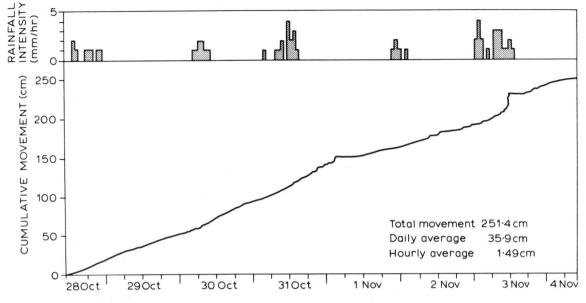

**Figure 10.16** Cumulative movement of a debris flow in Northern Ireland during the week 28 October–4 November 1970, correlated with rainfall data.
*Source:* Prior and Stephens, 1971, figure 4, p. 243, copyright © Elsevier, Amsterdam.

phenomena is a necessary part of the study of these features, but from a practical point of view the prediction of mass movement is an ultimate objective. It must be emphasized that, although mass wasting can occur anywhere, in its most significant forms it occurs in preferred locations. These locations consist of areas of high relief, shattered rocks, high precipitation and tectonic activity, where earthquakes can disrupt and trigger landslides. In such areas the mass wasting processes have a great influence on the landscape. For example, in New Zealand, the Andean countries of South America, California, Taiwan and Japan not only is mass wasting a significant component of landscape denudation, but it also strongly influences the morphology of the rivers and the nature of the sedimentary deposits that form in piedmont areas and along the coast. Delivery of large quantities of sediment to the stream channels by mass wasting produces aggradation, with the result that thick accumulations of mass-wasted sediments can be stored within valleys.

The regional nature of landslide problems is clear for the United States. They occur where relief is high in the Appalachian, Ozark, Columbia and Colorado Plateaus Physiographic Provinces and in the northern and southern Rocky Mountains, the Sierra–Cascade Mountains and the coastal ranges of the Pacific Border Physiographic Province.

Within a given region the distribution of mass movement phenomena can be related to lithologic and structural variations. For example, shale outcrop areas are more susceptible to failure than are sandstones. Layered rocks dipping towards a cliff face are more susceptible to failure than rocks dipping into the cliff (Figure 10.17A). Of course, the orientation of joints will act in the same manner as bedding planes (Figure 10.17B).

Weathering and rock character are important factors in determining mass movement type and frequency. For example, Durgin (1977) relates mass movement to the weathering of granitic rocks during the erosional evolution of a granitic terrain. He recognises four stages of granitic weathering: (1) fresh rock; (2) corestones: (3) decomposed granitoid; and (4) saprolite. Fresh rock contains less than 15 per cent weathered material, the corestone stage is characterized by 15–85 per cent weathered granite with unweathered boulders *in situ*. Decomposed granitoid consists of 85–100 per cent disintegrated rock composed of granules (grus). Saprolite is the fine-grained weathered product of the chemical decomposition of granitic rock. According to Durgin fresh rock is subject to rockfalls, rockslides and block glides most of which are controlled by fractures. The corestone stage is characterized by rockfall avalanches. Decomposed granitoid is characterized by debris-flows, debris-slides and saprolite is vulnerable to slumps and earthflows. Durgin states that failures in granitic rocks are most common in stage 4, which is achieved in areas of intense chemical

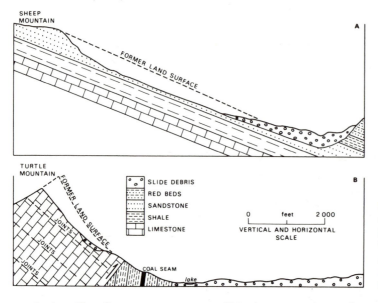

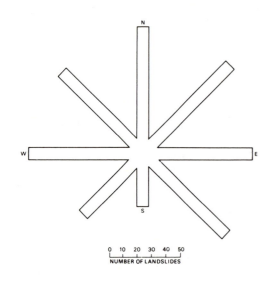

**Figure 10.17**   Reconstructed geological cross sections through two important landslides: A. Gros Venture, Wyoming; B. Turtle Mountain, Frank, Alberta.
*Source:* Leggett, 1939, figures 10.7 and 10.15, pp. 227, 243 after Alden and Canadian Geol. Surv.

weathering. Therefore, mass movement will be important in the humid tropical areas of the world.

The effects of microclimatic variations are also important. In most areas mass movement is associated with moist soils. For example, in semi-arid regions the higher moisture content of north-facing slopes renders them more susceptible to mass wasting than drier south-facing slopes (Figure 10.18) Elsewhere in the Arctic greater

**Figure 10.18**   Rose diagram illustrating number of slope exposures of landslides in Utah. This shows the paucity of landslides on slopes with a south or south-west aspect.
*Source:* Shroder, 1971, figure 5, p. 8.

freeze–thaw activity can cause south-facing slopes to be the most active. However, studies by Rice and others (1969) in southern California reveal that the greatest frequency of slides occurs on north-facing slopes because in this area of Mediterranean climate the north-facing slopes are brush-covered, whereas the south-facing slopes are grass-covered. The deeper root systems of the brush render the north-facing slopes less resistant to mass movement. Therefore, the effect of aspect changes in the different climatic zones.

Seepage hollows and hillslope depressions are very likely mass movement sites, and the California soil slips are frequently located in existing slope concavities. At these locations soil moisture is greatest and added precipitation will trigger soil slippage and debris-flows. During the 1969 storm in Virginia 85 per cent of the debris-slides occurred in such positions, and the most active were on north, north-east and east-facing slopes (Williams and Guy, 1971).

Another finding of interest is that most slides do not occur on the steepest slopes; rather, slopes of intermediate inclination are most susceptible. For example, in the coastal range of British Columbia O'Loughlin (1972) determined that most landslides originate on slopes between 31° and 39° in the mountains. Slopes steeper than 40° are rocky and lack soils that are involved in mass movement. In the Los Angeles mountains the soil slips occur most frequently between 26° and 45°. On gentler slopes the gravitational forces are not adequate to produce slippage, and on steep slopes rainwash and other erosion processes have kept the soils thin and rocky, thereby preventing mass movement.

In summary, where relief, slope inclination, rock type and climatic conditions are favourable, mass movements can dominate the landscape. However, within that landscape mass movement will occur at positions dependent on aspect, slope configuration, slope angle and, of course, zones of weak rock and closely jointed rocks.

These general statements provide a basis for geomorphic investigation that could lead to the identification of the most landslide-prone areas within a region. The identification of such areas should lead to restrictions on land use and especially construction on or near these sites.

## 10.5 Causes of mass movement

Mass movement occurs when the stresses acting on a hill or valley-side exceed the strength of the material involved. This can be expressed as a factor of safety $F_S$:

$$F_S = \frac{\text{strength or shear resistance of material}}{\text{magnitude of stress.}}$$

It is clear, then, when $F_S$ is less than $1 \cdot 0$, the slope will fail. This can come about by the application of increased stress due to changes in conditions external to the slope, which increases the denominator, or by internal processes that weaken the slope material, thereby decreasing the numerator (Table 10.4). In most cases failure will result from a combination of both processes, an increase in shearing forces and a decrease in the resistance to shear. Varnes (1978) has summarized the causes of mass movement under two headings; (1) factors that contribute to increased shear stress, and (2) factors that contribute to reduced shear strength. The major categories are listed in Table 10.4.

Obviously, any steepening, undercutting, increase of height, or removal of underlying support of a slope increases the shear stress as does loading of the slope (surcharge), and vibrations due to earthquakes, blasting and heavy traffic can weaken the internal structure of the earth materials forming the slope, as at Anchorage.

An increase of lateral pressure towards the slope face by hydrostatic pressure, freezing, and swelling and pressure release will lead to rockfalls and slab failure (see Figure 10.6). The creeping of earth materials can be due to any factor that disturbs the soil. This can include a number of events and processes as follows: wedging and prying by the growth of plant roots, swaying of trees in the wind, expansion of water by freezing in joints and cracks, the expansion of soil and rock materials due to heating, and the expansion of soil due to wetting and contraction due to drying. All these involve expansion and contraction, and downslope movement occurs during

**Table 10.4** Causes of mass movement

I Factors that contribute to increased shear stress:
 (1) Removal of lateral support (undercutting–steepening of slope)
   (a) erosion by rivers
   (b) erosion by glaciers
   (c) wave action
   (d) weathering
   (e) previous rockfall or slide, subsidence, or faulting (steepens slope)
   (f) construction of quarries, pits, canals, roads
   (g) alteration of water levels in lakes and reservoirs
 (2) Surcharge (loading of slope)
   (a) weight of rain, snow, water from pipelines, sewers, canals
   (b) accumulation of talus
   (c) vegetation, trees
   (d) seepage pressures of percolating water
   (e) construction of fill, waste piles, buildings
 (3) Transitory earth stress
   (a) earthquakes
   (b) vibrations, blasting, traffic
   (c) swaying of trees in wind
 (4) Removal of underlying support
   (a) undercutting by rivers and waves
   (b) solution at depth, mining
   (c) loss of strength of underlying sediments
   (d) squeezing out of underlying plastic sediments
 (5) Lateral pressure
   (a) water in cracks
   (b) freezing of water in cracks
   (c) swelling (hydration of clay or anhydrite)
   (d) mobilization of residual stress (pressure release)

II Factors that contribute to reduced shear strength:
 (1) Weathering and other physiochemical reactions
   (a) softening of fissured clays
   (b) physical disintegration of granular rocks (frost action, thermal expansion, etc.)
   (c) hydration of clay minerals causing decreased cohesion, swelling
   (d) base exchange (changes physical properties)
   (e) drying of clays and shales – racking, loss of cohesion – entrance of water
   (f) removal of cement by solution
 (2) Changes in intergranular forces due to water content (porewater pressure)
   (a) saturation – buoyance, decreased intergranular pressure and friction, and capillary cohesion destroyed
   (b) softening of material
 (3) Changes of structure
   (a) fissuring of shale and consolidated clays
   (b) remoulding of loess, sand and sensitive clay
 (4) Organic
   (a) burrowing animals
   (b) decay of roots

*Source:* Varnes, 1978.

both phases of the cycle. Tracking by man and animals can be very important, in fact, studies in western Colorado show that the very significant difference in sediment yield from small drainage basins relates to movement of animals through unfenced areas in the spring of the year, when the surface soil is relatively loose and uncompacted. Burrowing and excavating by animals and the decay of plant roots leave voids, which are filled primarily from the upslope side.

Reduction of strength occurs by weathering, swelling, fracture development, removal of cement and softening of material by increased water content. Most unconsolidated earth materials become susceptible to failure when water is added. For example, a dry mixture of silt, clay and sand may be stable, but with the addition of water it will eventually become plastic when the water content exceeds the 'plastic limit', and then a liquid when the 'liquid limit' is exceeded. These limits of water content are the Atterberg limits, so-called after the Swedish soil scientist, who developed a method of describing quantitatively the effect of varying water content on fine-grained soils (Table 10.5). Figure 10.19 shows how, as the liquid limit is exceeded, the type of mass movement changes from slides and slumps to flows.

Over long spans of time the weathering of the slope materials will produce a progressively less-stable con-

**Table 10.5** Atterberg limits

| Stage | Description | Limit |
|---|---|---|
| Liquid | Slurry, viscous liquid, mudflow | |
| | | Liquid limit |
| Plastic | Deforms, will not crack | |
| | | Plastic limit |
| Semi-solid | Deforms permanently, will crack | |
| | | Shrinkage limit |
| solid | Fails upon deformation | |

*Source:* Sower and Sower, 1970, p. 27.

dition as more clay is produced and formerly coherent materials become increasingly disaggregated and even plastic. Chemical weathering requires water, and it is water that is a major factor in internal causes of mass wasting. However, water rarely acts as a lubricant. Terzaghi (1950) makes clear that the effect of the introduction of water to earth materials, and particularly in those cases where a block of rock is moving over a surface of rock, is that the surfaces are not lubricated by the water. In fact, the coefficient of static friction between smooth, dry quartz surfaces is increased when the surfaces are wetted.

Attewell and Farmer (1976) point out that most surfaces are irregular and the contact between two rocks, for example, is limited to 'asperities' or the roughness elements on the surfaces. The water would, therefore, not be at the points of contact. However, when the voids are saturated, a porewater pressure is generated that can reduce the pressure of contact of the asperities or grain contact. The porewater pressure $v$ at depth $h$ below the water table is given by $v = \gamma h$ where $\gamma$ is the density of water. The porewater pressure will act in all directions in the porespace, and it will exert an uplift or buoyant effect that can produce failure.

When the groundwater is stationary, the above equation is sufficient, but when there is a piezometric gradient the water flows, and as it does, a frictional drag will be exerted at the water–solid interface. This 'seepage force' acts on the soil or rock in the direction of flow. For example, when the flow is upward in sand, quicksand can be produced. In fact, in some cases water under pressure acts as a hydraulic jack in which the overlying material is floated by water pressure. This is the mechanism whereby some major overthrust faults occur, but it also can be a significant factor in the development of soil slips, as described for the Los Angeles area, and in glacier sliding (see Chapter 17).

In the words of Hubbert and Rubey (1959) an 'elegant demonstration' of this process can be made with a piece of glass and an empty beer can (Figure 10.20). Any open-ended metal container will serve the purpose but an

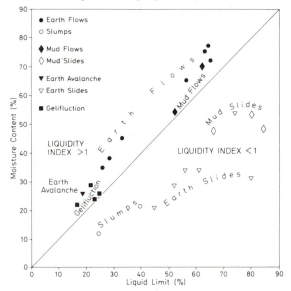

**Figure 10.19** Types of mass movement, especially flows (solid) v. slips (open) for twenty-five locations, ranging from Greenland to Brazil plotted in terms of a liquidity index (**LI** = moisture content percentage/liquid limit percentage). *Source:* Carson in E. Derbyshire (ed.) *Geomorphology and Climate*, 1976, figure 4.4, p. 109, copyright © John Wiley and Sons, by permission.

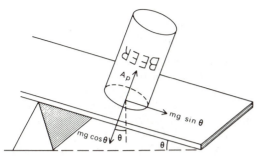

**Figure 10.20** The beer can experiment: an analogue for the effect of pore water pressure on the angle of repose of fragmented material. A description is given in the text: $\mathbf{A_p}$ = air pressure.
*Source:* Hubbert and Rubey, 1959, figure 32, p. 161.

empty beer can always appears to be available to the earth scientist:

A piece of plate glass a meter or so in length is cleaned with a liquid or nonabrasive detergent so that it will retain a continuous film of water. On this glass, which is first wet with water, is placed in upright position an empty beer can. The glass is then tilted until the critical angle is reached at which the can will slide down the surface. This angle of approximately 17 degrees gives for the coefficient of friction of metal on wet glass a value of about 0·3.

Next, the beer can is chilled, either by being placed in the freezing compartment of a refrigerator or in a container of solid carbon dioxide, and the experiment is repeated. The can is first placed on the glass with its open end upward and the angle of sliding redetermined. It is found to be the same as that previously, indicating that the coefficient of friction of the metal on wet glass is not temperature sensitive within this range of temperature.

Finally, with the angle of slope of the glass fixed at about 1 degree, the can (rechilled if necessary) is placed on the glass with its open end downward. In this case it will slide down the slope the full length of the plate, but will stop abruptly at the edge.

The physical reason for this behaviour is exactly the same as that which we have deduced for the case of overthrust faulting. As the cold can becomes warm the air inside expands and causes the pressure to increase. This, in turn, partially supports the weight of the can and so reduces the normal component of force between the metal and the glass without affecting the tangential component. The can stops at the edge of the glass because the pressure is released. (Hubbert and Rubey, 1959, p. 161)

The variety of materials and types of motion involved in mass movement render any comprehensive discussion of causes of mass movement very difficult. In the simplest sense gravity is the prime moving-force, and the strength of the materials or the friction between particles resists the downslope force of gravity.

In Figure 10.21 a slab of rock rests on slope $\theta$, its weight $W$ equals its mass $m$ times the acceleration of gravity ($g$ = 980 mm/s$^2$):

$$W = mg.$$

The vertical force is of less interest in this case than its components which act normal ($W_n$) and parallel ($W_p$) to the slope:

$$W_n = W \cos \theta = mg \cos \theta$$
$$W_p = W \sin \theta = mg \sin \theta.$$

This means that the sine of the angle of inclination of a slope is proportional to the component of gravity acting along the slope surface. Hence any attempt to relate rates of movement to the slope should be expressed as the sine of the slope angle.

If the surface of the slope in Figure 10.21 is frictionless, the rock slab will slide downslope, accelerating as it does so, but actually it is held on the slope by friction. For a given block weight $W$, a critical force $F_s$ is required to set it in motion on that surface. The downslope force $F$ equals the weight times a constant $\mu$ which is the coefficient of sliding friction.

$$F = W\mu.$$

As $\mu$ is constant, an increase of $W$ means an increase in $F$. On a sloping plane a block may be stable (i.e $F < F_s$), but as the slope $\theta$ is increased the block will slide at a critical angle $\theta$, when $F = F_s$ such that:

$$\mu = \frac{\sin \theta}{\cos \theta}$$
$$\mu = \tan \theta$$

The critical angle $\theta$ is also equal to the angle of internal sliding friction $\phi$.

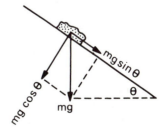

**Figure 10.21** The resolution of gravity forces acting on a slab of rock of mass $m$ lying on a slope of angle $\theta$.

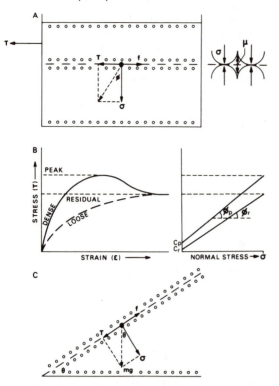

$S = C + \sigma \tan \phi$

where: $S$ = shear strength (kgf/cm$^2$)
$C$ = cohesion (kgf/cm$^2$).

For non-cohesive materials: $S = \sigma \tan \phi$. Figure 10.22B shows that when deformation (i.e. strain $\epsilon$) is occurring, $\tau$ (and therefore $\phi$ and $S$) assumes a range of values for a given granular material depending mainly on its relative density (i.e. dense or loose). The analogous situation to the shear box test is shown for a slope of granular material in Figure 10.22C, where at the moment of failure:

$f = \tau = \sigma \tan \theta$

where: $\theta$ = the angle of repose of the material (for dry medium sand: $\theta = 33°$).

Put in the simplest terms, $\phi$ is approximated by $\theta$, which varies depending on size, sorting, shape, relative density, etc., of the granular material as well as with cohesion and moisture content, particularly for clay soils.

The slump is a failure along an arcuate shear surface. Figure 10.23 shows the forces involved in a large-scale slump (Swedish break). A number of circumstances may lead to such slope failure (see Table 10.4). For example, a rise in the water table will relatively increase $W_1$ (i.e. more than $W_2$). Although such a rise will increase pore-water pressure $v$ and therefore decrease the effective normal stress $\sigma'$ on the potential surface of failure, it will also decrease the interparticle cohesion and the frictional resistance $f$. Changes of slope geometry involving increasing the height or steepness of the slope by erosion or the loading of the slope crest by deposition will, of course, result in the relative increase of $W_1$. When any of the above changes causes the driving-force to exceed the resistive force (i.e. the safety factor to fall below unity) a slope failure will occur.

The work of Skempton and Hutchinson and Chandler has done much to clarify humid slope development on the argillaceous rocks of England, particularly in respect of overconsolidated clays (i.e. those that have been geologically compressed by superincumbent strata, then unloaded by erosion, expanded and increased in water content), like the London and Lias Clays, which have greater shear strength than normally consolidated clays. Where slopes are being increased in height and steepened by lateral erosion greatly in excess of vertical degradation by weathering, such as where coastal wave action is causing the base of 30–50-m-high cliffs of London Clay to recede at average rates of 1–2 m/year, failure by a single Swedish break occurs almost immediately and repeated lateral erosion leads to other deep failures at intervals of some thirty to forty years. Under subaerial conditions of free slope degradation with no lateral erosion, any clay

**Figure 10.22**   The behaviour of cohesionless material under the application of a shear stress $\tau$:
A. conditions in a shear box along the shear plane at the moment of shearing ($f$ = frictional resistance to shear; $\sigma$ = stress normal to the plane; $\phi$ = angle of internal friction; $v$ = porewater pressure); B. comparison of deformation of loose and dense material (i.e. strain $\epsilon$) with the application of stress $\tau$, showing strain softening of dense material from a peak strength to a residual strength; the related patterns of normal stress $\sigma$ during increasing deformation are given on the right related to the peak $\phi_p$ and residual $\phi_r$ angles of internal friction; C. forces operating on a particle (weight $W$) lying on a slope of cohesionless material at the angle of repose $\theta = \phi$, showing the analogy with the conditions in the shear box.
*Source:* Carson in E. Derbyshire (ed.), *Geomorphology and Climate*, 1976, figure 4.8c, p. 115, copyright © John Wiley and Sons, by permission.

The shearing resistance or strength of a solid is measured by applying a shear stress $\tau$ to material in a shear box (Figure 10.22A) such that at the moment of failure:

$f = \tau = \sigma \tan \phi$;

where: $f$ = frictional resistance to shear;
$\sigma$ = normal stress (kgf/cm$^2$);
$\phi$ = angle of internal friction.

Coulomb's equation expresses the shear strength $S$ of clastic material:

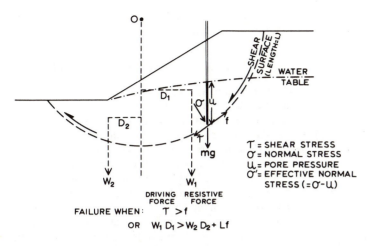

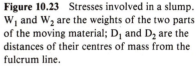

T = SHEAR STRESS
$\sigma$ = NORMAL STRESS
$u$ = PORE PRESSURE
$\sigma'$ = EFFECTIVE NORMAL STRESS $(=\sigma-u)$

FAILURE WHEN: $T > f$

OR $W_1 D_1 > W_2 D_2 + Lf$

**Figure 10.23** Stresses involved in a slump. $W_1$ and $W_2$ are the weights of the two parts of the moving material; $D_1$ and $D_2$ are the distances of their centres of mass from the fulcrum line.

slope of initial angle ($\beta_0$) about 15°, is immediately susceptible to failure by a series of shallow rotational (i.e. Swedish-type) slips, producing an upper degradation zone ($\beta = 13°$ for London Clay and 12·5° Lias Clay) and a more stable lower accumulation zone ($\beta_u = 8°$ and 8·5°). Thereafter the lower zone encroaches towards the head of the slope, and the whole slope is progressively reduced to the stable $\beta_u$ value, first, by a series of successive rotational slips working headward from the lower part of the degradation zone (giving values of $\beta = 8·5°$ and 11·5°), and then by undulations and soil waves (Figure 10.24A). When the whole slope has been reduced to the $\beta_u$ angle, it is thereafter slowly decreased in angle by creep and, particularly, by overland flow erosion. Surficial creep of clays is quite rapid, even on low slopes, where seasonal wetting alternates with seasonal drying. In California a 1·5-m (5-ft) surface layer of silty clay on a 7°–8° slope has been observed to creep during winter wetting at a maximum rate of 8–13 mm/year (0·3–0·5 in/year). Table 10.6 gives characteristic values for properties of some British argillaceous rocks, showing an increase of the residual angle of internal friction $\phi_r$ with decreasing clay content. Of particular interest are the time scales for the degradation of argillaceous slopes. It has been observed that a 6·5-m (20-ft) high vertical slope in London Clay may stand for several weeks without failure, a 26° slope of the same height for ten to twenty years and a 17° slope for fifty years. Railway cuttings of 18°–25° in the Lias Clay have not failed for 90–130 years, and it has been estimated that a 50-m-high Lias Clay slope formed some 100,000 years ago (Ipswichian interglacial) was first lowered to 17° by periglacial processes and now lies almost entirely at about 8° except for a small upper degradation zone at 11°. Figure 10.24A suggests a rate of reduction in mean angle $\beta$ for London Clay slopes of differing initial angles $\beta_0$ towards the angle of ultimate stability $\beta_u$ under free degradation, and Figure 10.24B gives a ratio of hilltop loss for 30-m-high slopes in London Clay after being abandoned by basal wave-cutting (i.e. free degradation) and where lateral wave erosion exceeds free degradation weathering (Hutchinson, 1973).

**Table 10.6** Properties of clays

| | % clay | % sand | Peak angle of internal friction ($\phi_p$) | Residual angle of internal friction ($\phi_r$) | Peak cohesion ($C_p$)(lb/ft²) | Liquid limit (%) | Plastic limit (%) |
|---|---|---|---|---|---|---|---|
| Carboniferous Mudstone | 70 | 30 | 21° | 13° | 320 | — | — |
| London Clay | 55 | 45 | 20° | 16° | 320 | — | — |
| Lias Clay (E. Midlands) | 55 + | 45 − | 24° | 17°–18° | — | 56–68 | 28–36 |
| Glacial Drift (Yorkshire) | 17 | 83 | 32° | 30° | 180 | — | — |

*Sources:* Chandler, 1969; Hutchinson, 1973; Skempton and Hutchinson, 1976.

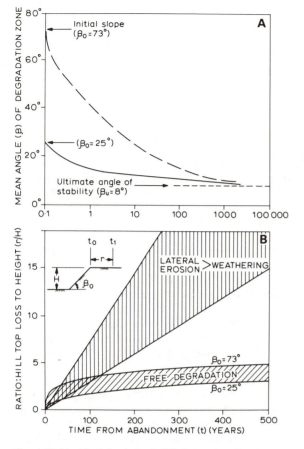

**Figure 10.24** The behaviour of cliffs in London Clay: A. the rate of decline of initial marine-cut cliffs by free degradation (i.e. by weathering and mass movements after the abandonment of the cliff base by wave action); B. crest recession (calculated assuming cliff height = 30 m) under free degradation for an initial slope angle range of 73°–25°; the range of values when lateral wave erosion continues at a greater rate than free degradation processes is also shown. *Source:* Hutchinson, 1973, figures 12 and 14, pp. 234, 236.

Some argillaceous slopes do not become immune to rapid mass failure after they have reached their stable lower angle of about 8°. So-called 'quick clays' are subject to spontaneous liquefaction, causing them to flow on slopes as low as 4° producing bowl-shaped depressions hundreds of metres in diameter and slopes after failure of about 3°. Quick clays are commonly platy, open-structured marine clays (not overconsolidated) whose strength has been decreased by the leaching of salt and by other weathering processes (i.e. liquid limit probably reduced from 34 to 26 per cent) such that an increase in water content and porewater pressure

will cause failure in the form of a thick sludge, possibly triggered off by construction work, earthquakes, etc.

## 10.6 Mass movement and landform evolution

Several examples of prehistoric landslides have been mentioned above and in many areas of the world the climatic changes of the Pleistocene epoch have been responsible for periods of significant mass movement. Particularly in arid and semi-arid regions an increase in precipitation and soil moisture will trigger mass movement. Large slump blocks of Pleistocene age at the base of the Vermilion and Echo Cliffs in Arizona (Strahler, 1940) and elsewhere in the dry Colorado Plateaus provide clear evidence of climate change and what can be expected if there is an increase in the water content of cliff faces. Those slump blocks range from 400 m to 4 km in length and some lie at distances up to 5 km from the cliffs (Figure 10.25). They are evidence of episodic cliff retreat by slumping during periods of higher precipitation.

The fact that mass wasting in many areas of the world appears to be a relatively insignificant component of the denudational processes does not mean that mass movement during the early stages of the erosion cycle does not play a predominant role. With relatively rapid uplift and dissection of terrain it can be expected that, unless the rocks are very strong and massive, a significant part of the denudation of the high relief will be the result of mass wasting. It has been noted that in Taiwan 80 per cent of the sediment derived from small drainage basins is produced by mass movement, and it has been suggested that the relief of mountains could reach a critical maximum value, and continued uplift would then be offset by slope failure and mass movement.

Obviously, the entire spectrum of mass wasting processes requires further study, especially from the geomorphic perspective. This will require an analysis of landslide distribution in space and time. For example, a very important point has been made by Terzaghi (1950) that most slopes fail during periods of exceptionally heavy rainfall, or in the spring when snow melts. However, exposure to rain or melting snow belongs to the normal experience of a slope. Hence, if a slope has not formed recently, heavy rainstorms or rapidly melting snow cannot be the sole cause of slope failure because it is unlikely that these events are without precedent in the history of the slope. With this in mind, a search for geomorphic controls on landslide occurrence seems warranted. That is studies leading to the identification of those components of the landscape that are most susceptible to failure seem to be the most worthwhile geomorphic approach to this problem.

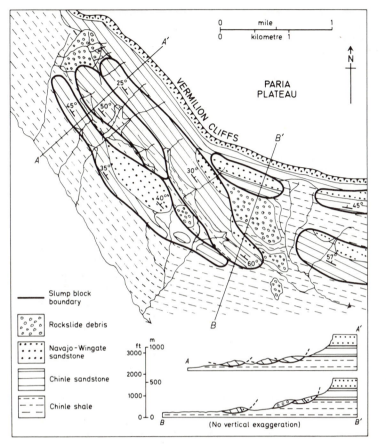

**Figure 10.25** Geologic sketch map of landslides along the Vermilion Cliffs at the southern end of House Rock Valley, Arizona (see Figure 7.4).
*Source:* Strahler, 1940, figure 3, p. 290.

# References

Attewell, P. B. and Farmer, I. W. (1976) *Principles of Engineering Geology*, London, Chapman & Hall,

Campbell, R. H. (1975) 'Soil slips, debris flows and rainstorms in the Santa Monica Mountains and vicinity, southern California', *US Geological Survey, Professional Paper* 851.

Carrigy, M. A. (1970) 'Experiments on the angle of repose of granular materials', *Sedimentology*, vol. 14, 147–58.

Carson, M. A. (1971) *The Mechanics of Erosion*, London, Pion.

Carson, M. A. (1976) 'Mass-wasting, slope development and climate', in E. Derbyshire (ed.), *Geomorphology and Climate*, Chichester, Wiley, chapter 4, 101–36.

Carson, M. A. and Kirkby, M. J. (1972) *Hillslope Form and Process*, Cambridge University Press.

Chandler, R. J. (1969) 'The degradation of Lias clay slopes in an area of the east Midlands', *Quarterly Journal of Engineering Geology*, vol. 2, 161–81.

Coates, D. R. (1977) 'Landslide perspectives', *Geological Society of America, Reviews in Engineering Geology,* vol. 3, 3–28.

Crandell, D. R. (1971) 'Postglacial lahars from Mount Rainier volcano, Washington', *US Geological Survey Professional Paper* 677.

Crandell, D. R. and Waldron, H. H. (1956) 'A recent volcanic mudflow of exceptional dimensions from Mt Rainier, Washington', *American Journal of Science*, vol. 254, 349–62.

Durgin, P. B. (1977) 'Landslides and the weathering of granitic rocks', *Geological Society of America, Reviews in Engineering Geology*, vol. 3, 127–32.

Elter, P. and Trevison, L. (1973) 'Olistostromes in the tectonic evolution of the northern Apennines', in K.A. De Jong and R. Scholten (eds), *Gravity and Tectonics*, New York, Wiley Interscience, 175–88.

Embley, R. W. and Jacobi, R. D. (1978) 'Distribution and morphology of large, submarine sediment slides and slumps on Atlantic continental margins', *Marine Geotechnology*, vol. 2, 205–28.

Ericksen, G. E. and Plafker, G. (1970) 'Preliminary report on the geologic events associated with the May 31, 1970 Peru earthquakes', *US Geological Survey Circular* 639.

Hsu, K. J. (1975) 'Catastrophic debris stream (Sturzstoms) generated by rockfalls', *Bulletin of the Geological Society of America*, vol. 86, 129–40.

Hubbert, M. K. and Rubey, W. W. (1959) 'Role of fluid pressure in mechanics of overthrust faulting', *Bulletin of the Geological Society of America,* vol. 70, 115–60.

Hutchinson, J. N. (1973) 'The response of London Clay cliffs to differing rates of toe erosion', *Estratto da Geologia Applicata e Idrogeologia*, vol. VIII, pt 1, 221–39.

Jones, F. O. (1973) 'Landslides of Rio de Janeiro and the Serra das Araras Escarpment, Brazil', *US Geological Survey Professional Paper* 697.

Leggett, R. F. (1939) *Geology and Engineering*, New York, McGraw-Hill.

Leighton, F. B. (1976) 'Urban landslides: targets for land-use planning', in D. R. Coates (ed.), *Urban Geomorphology*, *Geological Society of America Special Paper* 174, 57–60.

Lemoine, M. (1973) 'About gravity gliding tectonics in the western Alps', K. A. De Jong and R. Scholten (eds), *Gravity and Tectonics*, New York, Wiley-Interscience, 201–18.

Lucchitta, B. K. (1978) 'A large landslide on Mars', *Bulletin of the Geological Society of America*. vol. 89, 1601–9.

O'Loughlin, C. L. (1972) 'A preliminary study of landslides in the Coast Mountains of southwestern British Columbia', in O. Slaymaker and H. J. MacPherson (eds), *Mountain Geomorphology*, Vancouver, BC, British Columbia Geographical Series 14, 101–12.

Pierce, W. G. (1973) 'Principal features of the Heart Mountain fault and the mechanism problem', in K. A. De Jong and R. Scholten (eds), *Gravity and Tectonics*, New York, Wiley-Interscience, 457–71.

Prior, D. B. and Stephens, H. (1971) 'A method of monitoring mudflow movements', *Engineering Geology*, vol. 5, 239–46.

Prior, D. B., Stephens, N. and Archer, D. R. (1968) 'Composite mudflows on the Antrim coast of north-east Ireland', *Geografiska Annaler*, vol. 50(A), 65–78.

Prior, D. B., Stephens, N. and Douglas, G. R. (1971) 'Some examples of mudflow and rockfall activity in north-east Ireland', *Institute of British Geographers Special Publication* 3, 129–40.

Rapp, A. (1960) 'Recent development of mountain slopes in Kärkevagge and surroundings, northern Scandinavia', *Geografiska Annaler*, vol. 42, 71–200.

Rice, R. M., Corbett, E. S. and Bailey, R. G. (1969) 'Soil slips related to vegetation, topography and soil in southern California', *Water Resources Research*, vol. 5, 647–59.

Schumm, S. A. and Chorley, R. J. (1964) 'The fall of Threatening Rock', *American Journal of Science*, vol. 262, 1041–54.

Seed. H. B. and Wilson, S. D. (1967) 'The Turnagain Heights landslide, Anchorage, Alaska', *Journal of the Soil Mechanics and Foundations Division, American Society of Civil Engineers, Proceedings*, vol. 93, SM4, 325–53.

Selby, M. J. (1982) *Hillslope Materials and Processes*, Oxford University Press.

Sharpe, C. F. S. (1938) *Landslides and Related Phenomena*, New York, Columbia University Press.

Sheng, T. C. (1965) *Development of a landslide classfication for mountain watersheds of Taiwan, China*, Unpublished MS thesis, Colorado State University.

Shreve, R. L. (1968) 'The Black Hawk landslide', *Geological Society of America Special Paper* 108.

Shroder, J. F. Jr (1971) 'Landslides of Utah', *Utah Geological and Mineralogical Survey Bulletin* 90.

Simons, D. B. and Albertson, M. E. (1963) 'Uniform water conveyance channels in alluvial material', *American Society of Civil Engineers, Transactions*, vol. 128, 65–108.

Skempton, A. W. and Hutchinson, J. N. (eds) (1976) 'A discussion on valley slopes and cliffs in southern England: morphology, mechanics and Quaternary history', *Philosophical Transactions of the Royal Society of London*, series A, vol, 283, 421–631.

Sower, G. B. and Sower, G. F. (1970) *Introductory Soil Mechanics and Foundations*, New York, Macmillan.

Strahler, A. N. (1940) 'Landslides of the Vermilion and Echo Cliffs, northern Arizona', *Journal of Geomorphology*, vol. 3, 285–301.

Tamburi, A. J. (1974) 'Creep of single rocks on bedrock', *Bulletin of the Geological Society of America*, vol. 85, 351–6.

Terzaghi, K. (1950) 'Mechanism of landslides', S. Paige (ed.), *Application of Geology for Engineering Practice*, New York, Geological Society of America, 83–123.

Varnes, D. J. (1978) 'Slope movement types and processes', in R. L. Schuster and R. J. Krizek (eds), *Landslide Analysis and Control*, Washington, DC, National Academy of Sciences, Special Report 176, 11–33.

Watson R. A. and Wright, H. El, Jr (1969) 'The Saidmarreh landslide, Iran', *Geological Society of America Special Paper* 123, 115–39.

Williams, G. P. and Guy, H. P. (1971) 'Debris avalanches – a geomorphic hazard', in D. R. Coates (ed.), *Environmental Geomorphology*, Binghamton, NY, State University of New York, Publications in Geomorphology, 25–48.

## 11.1 Introduction

The study of the form and erosional evolution of slopes has been a focus of geomorphic studies for a very long time. Not only do slopes comprise the greater part of the landscape, but as an integral part of the drainage system they provide water and sediment to streams. Therefore, hillslopes are an important component of the complex landscape that forms a drainage basin.

The term *slope* has two commonly used meanings, one referring to the angle of inclination of the surface, expressed in degrees or a percentage, and the other is the inclined surface itself. In order to avoid confusion the inclined surface is frequently referred to as a hillslope or a valley-side slope, and the inclination of that slope is referred to as the slope angle or slope. However, taking the word slope in its most general sense, one may consider the entire earth to be composed of both sub-aerial and subaqueous slopes of greatly varying inclination, orientation, length and shape. Of the diverse types of slopes, only those that comprise the surface between a drainage divide and a valley floor, in other words, the hillslope, will be discussed here.

With great global variations of climate, vegetation, lithology and structure, the processes of hillslope erosion are very different. Therefore, hillslopes have a variety of forms, and it is difficult to formulate a general model for their development and evolution.

The study of slopes is not only of academic, but also of practical, significance. Agricultural engineers study runoff and erosion on slopes of low inclination. They have collected the largest body of data on erosion in these situations, but unfortunately their sloping fields are far too gentle to be of major concern to the geomorphologist.

The manner in which erosion progresses on a slope is of importance not only to the agriculturalist, but to those who wish to use slopes for recreational or other practical purposes. In fact, the design of slopes for long-term stability is an important facet of land management. The stability of mine tailing piles, especially those that contain radioactive materials, and the reclamation of strip-mined areas raises questions concerning stable slopes. The length, shape and angle of inclination chosen may be a compromise between the erodibility of the material and the needs and the cost of designing the new landscape. A complete understanding of the long-term erosional evolution of hillslopes is the minimal background information required for this purpose.

## 11.2 Characteristic slopes

In a given area many investigators have indicated that there is a characteristic slope that gives the terrain its general appearance (Young, 1972). This characteristic hillslope is that which is most frequently identified under particular conditions of rock type and climate (Figure 11.1). A. N. Strahler stated this concept as his law of the constancy of slopes as follows:

> The tendency of slopes to group closely about a mean value may be taken as an expression of a morphologic law relating slope to other form factors. This law may be stated as follows: In an area of essentially uniform lithology, soils, vegetation, climate and stage of development, maximum slope angles tend to be normally distributed with low dispersion about a mean value determined by the combined factors of drainage density, relief and slope profile curvature. (Strahler, 1950, pp. 684–5).

Later we shall discuss each of these variables and their influence on slope morphology, but note that it is only the straight, steep slope angles that were measured by Strahler, as a sample of the valley-side slope angle most typical of the area. Frequently investigators have not studied or measured the same thing, and they have worked under different climatic and lithologic conditions. It is no wonder that there has been disagreement among geomorphologists concerning hillslope development and evolution.

## 11.3  Classification of hillslopes

In the simplest way we can think in terms of three types of slopes or slope segments. These are straight, concave and convex. Most slopes are composed of more than one of these segments, which form a compound slope (Figure 11.2A). It has been suggested that there is a standard or normal hill slope that is composed of four segments (Figure 11.2B). It is more likely that a slope of three segments, upper convex, straight and lower concave is the dominant hillslope form, with the vertical straight segments or cliffs reflecting undercutting by streams or the outcropping of resistant strata. Figure 11.3 shows a nine-unit classification of hillslopes based not only on profile characteristics, but on planar form. A slope, of course, is not simply a two-dimensional profile, but it is a three-dimensional feature. In this classification not only the profile that is normal to the contours of the slope is considered, but the slope form on both sides of that profile is also examined, and the surfaces can be classi-

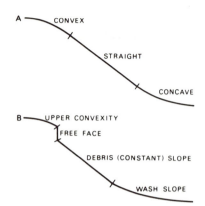

**Figure 11.2**  Terminology of slope segments associated with A. a convex-concave slope, and B. a slope formed by the consumption of a retreating free face.

fied as in Figure 11.4. Obviously, this is much more realistic as regards the erosional forces operating on the slope.

In addition to the profile form, another approach to the study of slopes is to evaluate what Jahn (1963) considers to be the denudation balance of the slope in which segments of the slope that are undergoing progressive erosion and those that are undergoing progressive deposition are identified. In this way the slope can be separated into its dynamic components, and a much more realistic view of hillslope behaviour is obtained. This would be the most valuable type of classi-

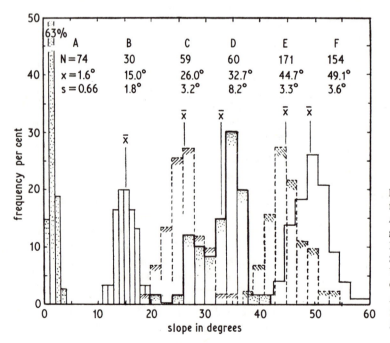

**Figure 11.1**  Histograms of six slope frequency distributions:
A. Steenvoorde, France; B. Rose Well gravels, Arizona; C. Bernalillo, New Mexico, Santa Fé formation; D. Hunter–Shandaken area, Catskill Mountains, New York; E. Kline Canyon area, Verdugo Hills, California; F. dissected clay fill, Perth Amboy, New Jersey; all data except in A. is from field readings.
*Source:*  Strahler, 1950, figure 2, p. 680.

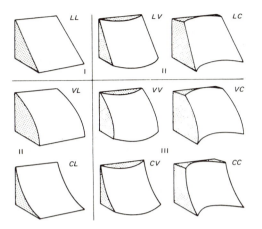

**Figure 11.3**  Geometric forms of hillslopes. Slope length is down the form; slope width is across the form: **L** = linear, **V** = convex, **C** = concave. The simplest form (I) is colinear (**LL**); group III forms, the most complex, are doubly curved; group II forms are linear in one dimension and curved in the other.
*Source:* R. V. Ruhe, 1975, *Geomorphology*, copyright © Houghton Mifflin, used with permission.

fication if, in fact, data on slope erosion and deposition through time were generally available, but the geomorphologist rarely has this luxury, and he is forced to draw conclusions from form alone.

## 11.4 Origin of hillslopes

The classification of slopes may be useful but of primary interest to a geomorphologist is the manner in which

slopes change through time and the erosional processes that act on the slopes. There are three ways in which slope evolution can be studied. It can be studied theoretically, which involves the development of mathematical models based on the laws of physics. It can be studied experimentally when an inclined surface is subjected to precipitation or other erosional forces, but perhaps best of all are the field studies of slope erosion.

The first theoretical analysis of hillslope evolution was by W. M. Davis, and a foremost exponent of this approach is Scheidegger (1970), who presents a number of mathematical and graphical models of slope retreat (see also Ahnert, 1976; Kirkby, 1971). Experimental studies consist primarily of work that is oriented towards agricultural problems and, in the United States, the Agricultural Research Service has been involved in the study of rainsplash, runoff and rill erosion on slopes of relatively gentle inclination.

In the field, of course, the problem is that erosional processes occur too slowly for one to obtain the necessary information on slope retreat over a relatively short period of time. Another alternative is to base studies on the ergodic hypothesis (Sections 2.5.1. and 13.4.1) that space can be substituted for time. For example, along a spur or ridge of declining altitude, slope profiles are surveyed and ordered in a sequence suggesting age. This is a simple technique for determining the manner in which the form of hillslopes might change through time.

However, before discussing hillslope evolution, let us consider the initial development of the slope. First of all, what is needed is relief, a surface elevated sufficiently above baselevel, so that stream incision can take place, and slope forms can develop adjacent to the incising

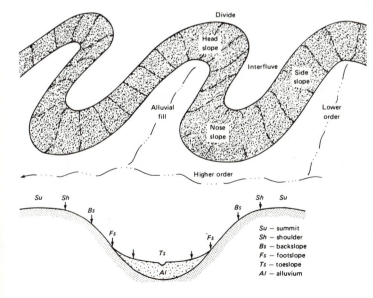

**Figure 11.4**  Components of hillslopes. Above: in plan; orthogonal (i.e. downslope) lines converge on the headslope, are parallel on the sideslope and diverse on the noseslopes. Below: in profile.
*Source:* Ruhe and Walker, 1968.

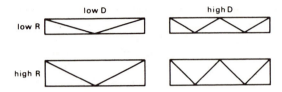

**Figure 11.5** Effects of drainage density **D** and relief **R** on hillslope inclination and length.

channel. In an area of uniform lithology, climate and stage of development characteristic slope angles will form. Strahler's California study showed clearly that the hillslopes cannot be considered separately from the drainage basin as a whole, for what is happening on the interfluve areas between the streams will have a dominant influence on the character of the streams themselves and on the hydrology of the drainage basin. Therefore, when Strahler found that 'maximum slope angles tend to be normally distributed with low dispersion about a mean value determined by drainage density, relief and slope profile curvature', he was concluding that the character of the slopes themselves is determined largely by the relief available. The greater the relief, the steeper the slopes. In addition, drainage density, which reflects the spacing of streams, is also going to affect the angle of inclination and length of slopes. The more closely the streams are spaced, the steeper will be the slopes (Figure 11.5). Strahler quantified this relationship as follows: $S_g = 4S_c^{0.8}$, where $S_g$ is a mean maximum slope angle and $S_c$ is stream channel gradient. This relation (Figure 11.6)

shows that the steeper the hillslope, the steeper are the streams that drain the area. This is the expected relation because the steep slopes provide a large amount of sediment to the channels which, in turn, must be steep to transport it.

## 11.5 Hillslope erosion

There are several erosional processes that act to modify hillslopes (e.g. creep, overland flow or rainwash, rainsplash and rilling). Creep has previously been discussed in Chapter 10 as a mass movement phenomenon, but it is such an important hillslope process that it is considered again here.

### 11.5.1 Creep

In well-vegetated humid temperate regions soil creep is five to ten times more important than sheet erosion, being of the order of 1–2 mm/year downslope in England. On a 20°–30° grass slope in the Pennines an average movement of 0·25 mm/year has been measured in the upper 10 cm of the soil. By contrast, in savanna and semi-arid regions sheet erosion plus soil splash moves 5–10 times more sediment than does soil creep. In periglacial regions mean rates of gelifluction plus frost creep are at least ten times faster than creep in humid temperate regions, and maximum rates more than 100 times faster. Maximum gelifluction rates of 10–300 mm/year have been widely observed on slopes of 15°–25°

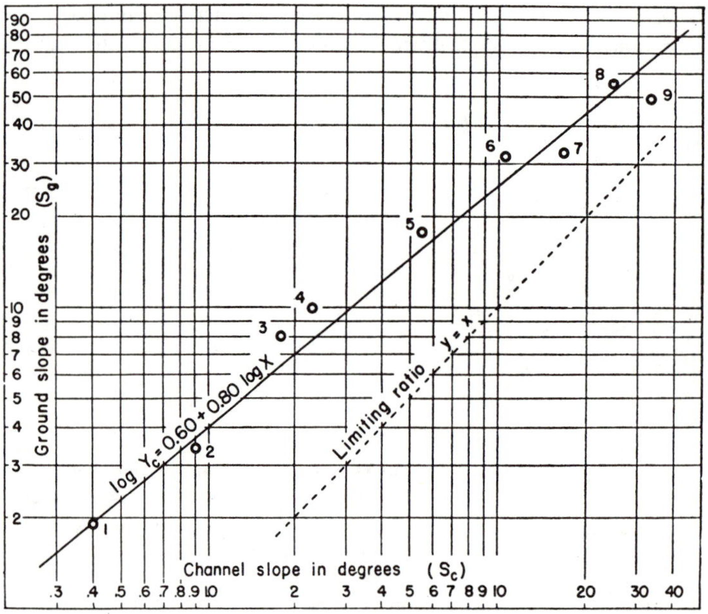

**Figure 11.6** Slope ratio $S_c/S_g$ for nine maturely dissected regions: 1,2, Grant, Louisiana; 3, Rappahannock Academy, Virginia; 4, Belmont, Virginia; 5, Allen's Creek, Indiana; 6, Hunter–Shandaken, New York; 7, Mt Gleason, California; 8, Petrified Forest, Arizona; 9, Perth Amboy (clay fill), New Jersey. All data from US Geological Survey, Army Map Service, or special field maps.
*Source:* Strahler, 1950, figure 8, p. 689.

and these rates vary greatly in space and time, a movement of 250 mm in twenty-one days being measured on a 19° slope at Dovrefjell, Norway. Gelifluction rates decrease rapidly with depth, as is shown by the following rates at Kärkevagge, Sweden:

| depth (cm) | rate (mm/year) |
|------------|----------------|
| 0 | 50 |
| 10 | 38 |
| 20 | 28 |
| 30 | 17 |
| 40 | 8 |
| 50 | 3 |

On some steep slopes in the humid tropics soil creep rates may exceed those in periglacial regions.

Creep rates can be highly variable within one area and a study on surface rock creep provides an illustration of the difficulty of obtaining hillslope erosion information. Figure 11.7 is the record of the movement of small painted sandstone fragments, the position of which was measured semi-annually, during seven years, on shale slopes in western Colorado. The results show the influence of gravity (size of slope angle) on creep rates. In addition, this particular figure illustrates extremely well the difficulties of geomorphic research. It is obvious that the data scatter widely about the regression line, and this is expected because the movement of a given particle is significantly influenced by the movement of animals over the slope, the position of the particle with reference to a clump of vegetation, the physical and chemical properties of the surficial materials and microclimate. Therefore, there are many reasons why the variability in the rate of rock creep should be high. Hence, if one is to become involved in such studies a large sample is required, and a considerable range of the independent

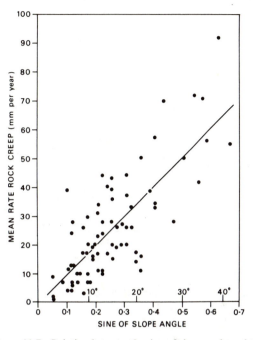

**Figure 11.7** Relation between the sine of slope angle and the rate of movement of rock fragments on Mancos Shale hillslopes in western Colorado over a seven-year period. The standard error is 11.8 mm and the correlation coefficient is 0·77.
*Source:* Schumm, 1967, figure 1, p. 561.

variables, in this case slope, is necessary. For example, a significant relationship could not have been developed from the data if only hillslopes with a range of inclination of 10° were studied, e.g. between 10° and 20°. In fact, if only a limited range of slope is sampled and only very few rock fragments measured, it is possible that an inverse

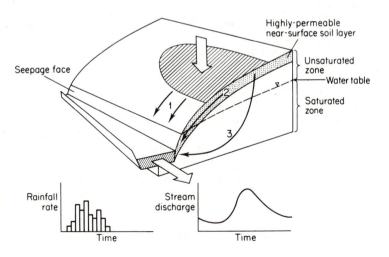

**Figure 11.8** Mechanisms of delivery of rainfall to a stream channel from a hillslope in a small tributary watershed: 1, overland flow; 2, throughflow; 3, groundwater flow.
*Source:* Freeze in M. J. Kirkby (ed.), *Hillslope Hydrology, 1976*, figure 6.1, p. 180, copyright © John Wiley and Sons, by permission.

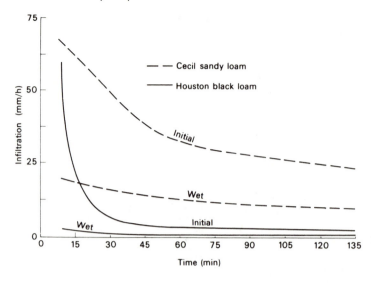

**Figure 11.9** The effect of initial dry moisture content on infiltration rates. *Source:* Carson and Kirkby, 1972, figure 3.12, p. 48, after Free, Browning and Musgrave, U.S. Dept. Agr.

relationship between creep and slope would result from such a small sample.

### 11.5.2 Overland flow

Another agent of slope erosion is flowing water (Figure 11.8). Precipitation falling on the surface of a hillslope runs off, is held on the surface by surface detention until it evaporates, or infiltrates into the soil, where it moves downward to the water table or parallel to the slope as interflow. Surface runoff will occur only when the rate at which water supplied to the slope exceeds the infiltration capacity of the soil, which is the maximum rate at which water can enter the soil. Figure 11.9 shows infiltration rates for a sandy and a loamy soil with wet and dry initial conditions. During the early period of a storm there should be no runoff unless the soil is originally very wet or the storm intensity is very high. Initially infiltration rates are high, but after about 15 min infiltration rates decrease in the loam to the point that all of the water cannot be accommodated by the soil, and runoff begins. This surficial runoff modifies hillslopes; and it is important to know how the water moves over the slope, how it erodes and how the soil resists erosion. Assume that a sheet of water is moving over a plane surface; the velocity flow of the water will depend on the slope of the surface, the depth of water and the roughness of the surface itself. The rougher the surface, the slower will be the velocity of water movement. Therefore, the smoother the surface, the greater the depth of water; and the greater the slope, the higher will be the velocity of flow. This can be expressed as a modified Manning equation (imperial units) as follows:

$$V = \frac{1 \cdot 49}{n}\ d^{2/3}S^{1/2}$$

where: $V$ is flow velocity; $n$ is a roughness index, $d$ is depth of flow; and $S$ is slope inclination. Velocity of flow is proportional to a depth–slope product; therefore, the velocity of flow of water over a surface is proportional to the shear force that is exerted on the surface because the equation for shear or tractive force also contains a product of depth and slope as follows: $\gamma = w\,d\,S$, where $\gamma$ is tractive force and $w$ is specific weight of water.

Although overland flow or surficial runoff may be referred to as sheet flow, the concept of a uniformly thick sheet of water moving down a slope is unrealistic. Precipitation may deliver water uniformly to the slope, but it will run off the slope as subdivided flow (Figure 11.10). Multiple flow paths form and the bulk of the runoff follows these paths because of the roughness elements, vegetation, rocks, etc., which occur on the slope surface.

### 11.5.3 Rainsplash

Another factor that is generally not considered important to the erosive effects of surface runoff is raindrop impact itself. Numerous studies have shown that rainsplash can be an important erosive agent, where the surface of the slope is not completely covered by the vegetation. Experimental studies by Mosley show (Figure 11.11) that on a 5° slope about 60 per cent of the material moved by raindrop impact is moved downslope, and 40 per cent is moved upslope. This ratio changes until at a 25° slope approximately 95 per cent of the sediment moved by rainsplash is moved in a downslope direction. Erosional reduction and modification of a slope wholly

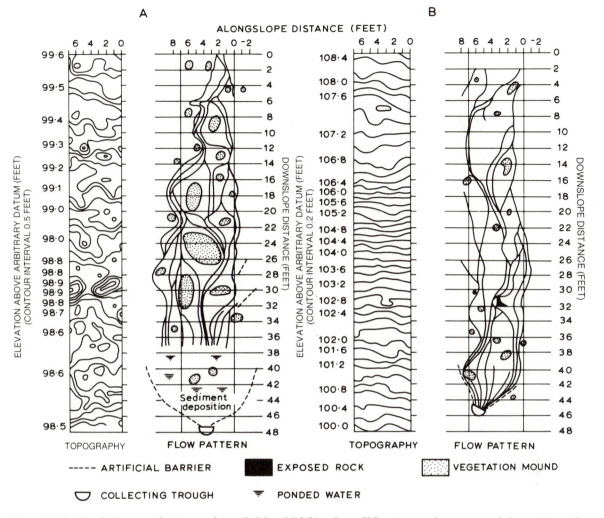

**Figure 11.10** Detailed topography (contour intervals 0·5 and 0·2 ft) and runoff flow patterns for two natural slope segments (I, New Fork River site 2; II, Boulder Lake site 1) on morainic sand, gravel and boulders in west-central Wyoming. *Source:* Emmett, 1970, figures 29 and 30, p. 33.

by splash is possible. For example, maximum rates of downslope rainsplash transport are 2·6 cm³ cm⁻¹year⁻¹ in south-western United States (mean rainfall 375 mm/ year). This is thirty times greater than the rates measured on densely vegetated areas in Britain.

Another experiment (Figure 11.12) further emphasizes the importance of rainsplash. When artificial precipitation is applied to a slope surface, the raindrop impact alone is important in disrupting the surface and causing erosion, as previously indicated. Similarly, water moving over a surface without raindrop impact will initially cause significant erosion, but when the loose surface material has been swept away, erosion due to moving

water alone is relatively insignificant. The combination of water on the slope plus raindrop impact produces an intermediate value of erosion. In other words, the sheet of water dissipates the raindrop energy to some extent but raindrop impact, nevertheless, provides sediment to the flowing water and so erosion continues.

### 11.5.4 Slope erosion

Experimental and field studies by the Department of Agriculture as reviewed by Smith and Wischmeier (1962) show that erosion on slopes is, of course, complicated. It depends on the eroding forces and resistance of the

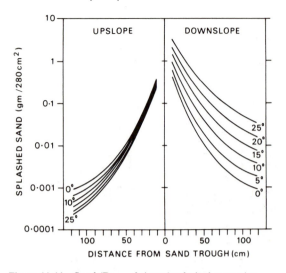

**Figure 11.11** Sand ($D_{50}$ = 0·4 mm) splashed at varying distances (units = $\frac{1}{10}$ m) upslope and downslope from a sand trough under the action of artificial precipitation applied for 30-min periods at an intensity of 35 mm/h with six different angles of slope.
*Source:* Mosley, 1973, figure 3, p. 20.

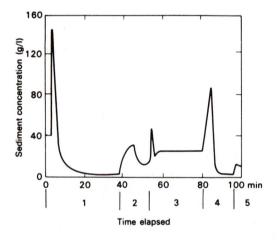

**Figure 11.12** Variations in soil loss during the course of an experiment, applying rainfall and overland flow to a silt loam in differing combinations: 1, 0–38 min; overland flow only at 50 cm/h; 2, 38–55 min: overland flow plus rain at 82 mm/h; 3, 55–80 min: rain at 82 mm/h only; 4, 80–96 min: overland flow at 50 cm/h only; 5, 96–100 min: overland flow plus rain.
*Sources:* Carson and Kirkby, 1972, figure 8.15, p. 214, after Ellison.

surface to erosion. A soil loss equation developed for agricultural fields can be used to consider the variables influencing hillslopes.

This universal soil loss equation, as it is called, can be summarized as follows: $A = RKLSCP$, where $A$ is the annual soil loss per unit area; the rainfall factor $R$ is a measure of the erosive potential of average annual rainfall; $K$ is a soil erodibility factor; $L$ and $S$ are the geomorphic factors of slope inclination ($S$) and length ($L$); $C$ is the cropping factor and relates to the type of crop, but for non-agricultural areas, it can be vegetational cover; $P$ is the soil conservation factor, and it can be considered to be man's influence on the erosion process.

This relation indicates that the more precipitation that falls and the higher its intensity, the greater is erosion. Offsetting this is the protecting influence of vegetation. Figure 11.13 shows the influence of vegetation, and Figure 11.14 presents the interaction between the two which indicates that maximum erosion will occur between 200 mm and 500 mm of annual precipitation. Above this precipitation level the vegetative cover decreases the effect of raindrop impact and runoff. Note that this precipitation range is also that at which sediment yield rates from drainage basins are at a maximum (see Section 20.2.1).

Another way of viewing the same relationship is to consider that as slope inclination increases, the length of slope exposed to vertically falling precipitation decreases, that is as the slope angle increases, the vertically falling water is dispersed over a larger surface area

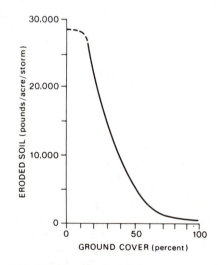

**Figure 11.13** Relation of soil eroded to percentage of groundcover by vegetation for an experimental watershed in Utah. The broken part of the curve is hypothetical.
*Source:* Noble, 1963, figure 4, p. 118.

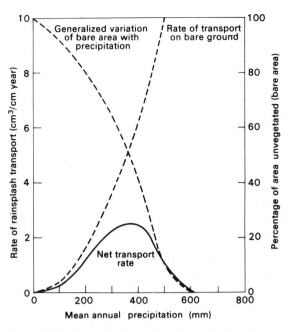

**Figure 11.14** Generalized variation of rainsplash erosion on unvegetated ground: percentage bare (unvegetated) area and net transport rate calculated as product of the other two; all expressed in terms of mean annual precipitation in the southern United States.

*Source:* Carson and Kirkby, 1972, figure 8.9, p. 206, after Kirkby.

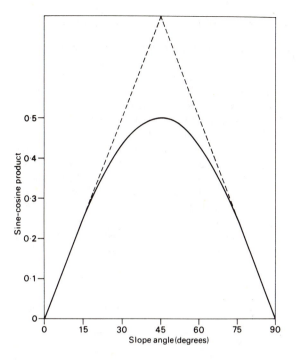

**Figure 11.15** The sine–cosine product for slopes between horizontal and vertical.

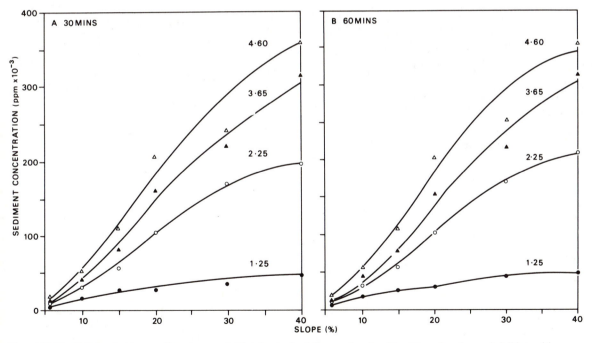

**Figure 11.16** Relation between sediment concentration in overland flow and angle of landslope for given rainfall intensities (inches per hour) after A. 30 min and B. 1 h of experimental erosion (1 in = 25·4 mm).
*Source:* Kilinc and Richardson, 1973, figures 12 and 14, p. 25.

and this effect can be expressed by the cosine of the slope angle. When sine is used to express the gravitational force a sine–cosine product can represent the effect of slope on hillslope erosion. This is plotted as Figure 11.15, which shows that slope erosion increases to a maximum at 45°. The most interesting aspect of this relationship is the rapid increase in the product to about 25° or 30°, when the rate of increase decreases. This suggests that a small increase of slope angle will produce a significantly greater change in erosion when slopes are flat than when slopes are steep. This seems improbable but experimental studies by Kilinc and Richardson (1973) show this type of relation for slopes of less than 40 per cent (Figure 11.16). Their results confirm the relationship shown by the sine–cosine product, when rain is falling vertically.

### 11.5.5 Rills

Another type of surface erosion that occurs on slopes,

particularly steep slopes with poor vegetative cover, is rilling. Rills are small linear, rectangular channels that have cut into a slope surface. They tend to be parallel, and they are most commonly displayed on new road cuts. An experimental study was performed to look into the development of rills on a planar surface to determine the effect of slope inclination under constant precipitation conditions, and Figure 11.17 is a series of maps showing the drainage patterns that developed as slopes were increased in inclination from about 2 to 7 per cent. The rill patterns change from a dendritic type of pattern at the low slopes to a parallel pattern at the higher slopes. In addition, the heads of the rills approach the upper divide more closely at higher slopes as one might expect. The drainage density of the channels on the experimental surface also changes with slope (Figure 11.18).

Mosley also found that sediment contributed from the rilled slope is closely related to slope inclination with a linear increase as the plane surface was steepened.

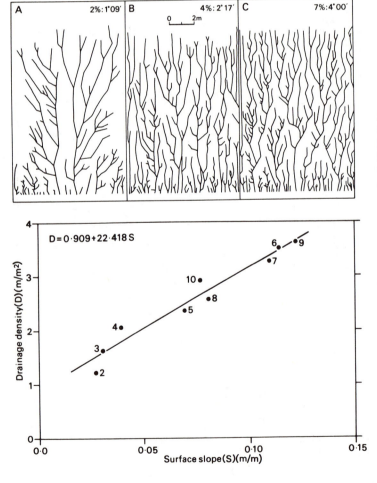

**Figure 11.17** Experimental development of equilibrium rill systems under artificial precipitation on planar surfaces of differing initial slope: A. 2 per cent (1° 9′); B. 4 per cent (2° 17′); C. 7 per cent (4°).
*Sources:* Mosley, 1972: Schumm *et al.*, 1984, figure 3-2.

**Figure 11.18** Mean drainage density as a function of surface slope for numbered surfaces 2–10
*Sources:* Mosley, 1972: Schumm *et al.*, 1984, figure 3-5.

However, different results emerge for converging and diverging surfaces of the same average inclination. The converging flow (head slope: see Figure 11.4) produces a significantly higher sediment contribution than the diverging flow (nose slope: see Figure 11.4) and this further emphasizes the need to consider slopes as three-dimensional landforms. The studies by Mosley and common sense indicate that if a slope is convex in plan, water cannot be concentrated and erosion will be significantly less than when flowlines converge on a concave slope.

With the previous information in hand, it is appropriate to ask what slope profile will be least susceptible to erosion by overland flow. The steeper the slope, the greater will be erosive forces acting on that material; but the shallower the depth of flow, the less will be erosion. Flow depths will increase from divide to slope base and, therefore, in any situation where water is to be led from a divide or slope crest to a slope base with minimal erosion, the steepest slope should be near the top of the slope where water flow velocities and quantities are less. Where the water has reached a high flow velocity and where flow

depth is greatest, slope inclination should be least. So a concave slope is most desirable when water is to be led from the divide area to a slope base with minimum erosional influence. If the slope is convex in plan, the dispersal of water will also keep slope erosion at a minimum.

### 11.5.6 Throughflow

The preceding discussion has emphasized the effect of surficial runoff on slope erosion and, in fact, the geomorphic literature places great emphasis on this process because the agricultural engineer has concentrated on rainsplash and surface runoff, and some of the most influential hillslope studies have been performed in semi-arid regions where surface runoff can be observed. However, in more densely vegetated regions major storm events may produce only minor surface runoff. The bulk of the rainfall infiltrates and moves to the water table or moves beneath the hillslope surface as interflow or throughflow (see Figures 3.4 and 11.8). Remember that it was this subsurface water flowing beneath the hillslope

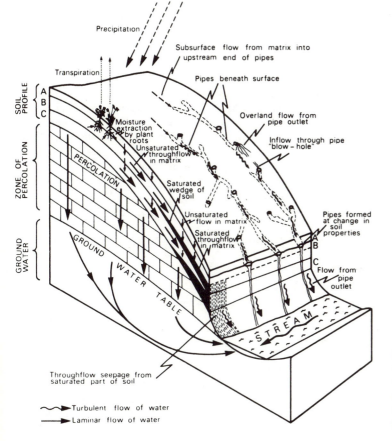

**Figure 11.19** Subsurface flow on soil-covered hillslopes.
*Source:* Atkinson in M. J. Kirkby (ed.), *Hillside Hydrology*, 1976, figure 3.4, p. 74, copyright © John Wiley and Sons, by permission.

surface and through the soil that triggered the California soil slips (see Figure 10.12).

It appears that, as in other geomorphic situations, there is a range of hillslope hydrologic conditions from predominant surface runoff to predominant interflow under conditions of high permeability and vegetation density (Figure 11.19). The differences are important not only to hillslope evolution, but to the hydrology of a drainage basin itself.

Tischendorf (see Chorley, 1978, p. 13) studied fifty-five storm events in a sixty-acre vegetated watershed in the south-eastern United States. He did not observe any overland flow, although nineteen storms produced channel flow. Of even greater interest, Kirkham (1947) concluded that water could infiltrate downwards near a hilltop, horizontally outwards over the mid-slope and vertically upwards near the slope base.

Soil moisture conditions will vary considerably within a drainage basin and along a slope profile and, therefore, certain areas will produce runoff, whereas others will not. The most favourable areas are:

(1) the slope base, which is wetter:
(2) concavities or topographic hollows (headslopes: see Figure 11.4), where surface flowlines converge;
(3) areas of thin soil cover.

This leads to the concept of partial area overland flow at those parts of the slope profile that have high moisture contents (see Figure 3.4). This concept is particularly important because it integrates certain types of mass movement and drainage pattern development with hillslope hydrology (Chapter 13).

Hillslope erosion can take many forms, the simplest two being erosion by creep and overland flow, but there can be combinations of the two processes and even seasonal changes in the intensity of each process. Throughflow carries away material in solution and also in suspension. The most striking morphologic influence of subsurface flow occurs along zones of varying permeability and it is manifested by piping. The water moving along an interface between an overlying relatively permeable and underlying relatively impermeable material will begin to develop subsurface channelways beneath the slope surface, and over a period of years these may enlarge to the extent that the roofs of the tunnels collapse during or following a large precipitation event. When this occurs, a previously smooth field can be converted to a deeply gullied field overnight. In the Wither Hills area of the South Island in New Zealand loess-covered hills developed very intricate gully patterns not through surface erosion, but through the collapse of pipes. In addition, Ruxton (1958) has described the flushing of fine sediments from decomposed granite debris at the foot of inselbergs in the Sudan. This process apparently maintains the sharp angle between the granite slope and pediment, as the presence of fines blurs this break.

## 11.6  The evolution of hillslopes

One of major concerns of geomorphologists is the evolution of hillslope profiles through time. In particular, attention was directed to two models of slope evolution, one involving the downwearing and decrease of slope angles through time, the other emphasizing the maintenance of steep angles and parallel retreat of slope segments (Figure 11.20). Geomorphologists working in humid regions where weathering is important favoured declining retreat, whereas those in dry regions or where caprocks were present supported parallel retreat.

Mathematical models of hillslope erosion have been developed in order to simulate hillslope erosion. The most common type of these deterministic geomorphic models involves the transformation of slope profiles under various assumptions regarding the original slope geometry and the process acting. One such family of models has been based by Kirkby (1971) on the transport law:

$$C = f(a) \left( -\frac{\partial y}{\partial x} \right)^n$$

where $C$ = transport capacity; $a$ = area 'drained' per contour length ($f(a)$ thus describes the influence of increasing distance from the divide); $x$ = horizontal distance from the divide; $y$ = elevation above baselevel; $n$ = a constant exponent (zero or positive) describing the influence of increasing gradient. From this equation can be derived the simple empirical transport capacity relationship:

$$C \propto a^m (\text{slope})^n.$$

The following values for $m$ and $n$ have been obtained from field measurement of forms associated with different transportational processes:

|           | $m$       | $n$       |
|-----------|-----------|-----------|
| soil creep | 0         | 1·0       |
| rainsplash | 0         | 1·0–2·0   |
| soil wash  | 1·3–1·7   | 1·3–2·0   |
| rivers     | 2·0–3·0   | 3·0       |

From this equation a range of equilibrium dimensionless curves has been generated to approximate characteristic-form slope profiles for a range of transportational processes (Figure 11.21) – assuming no chemical solution,

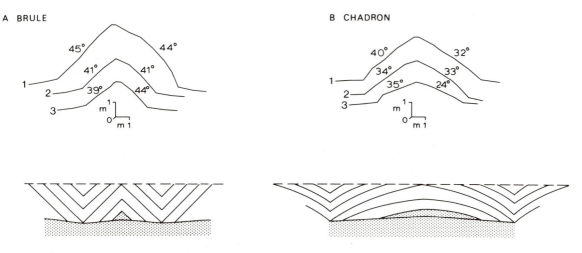

**Figure 11.20** Slope profiles. Above: two series of slope profiles of decreasing slope length measured on residuals of the Brule Formation (A) and the Chadron Formation (B) of the South Dakota shale badlands. The former is developing under the action of rainwash by parallel retreat, whereas the latter is covered by a creeping debris layer and is undergoing reclining retreat; below: the two contrasting, idealized models of slope development are shown.
*Source:* Schumm, 1956b, figures 6, 7 and 8, pp. 701–703.

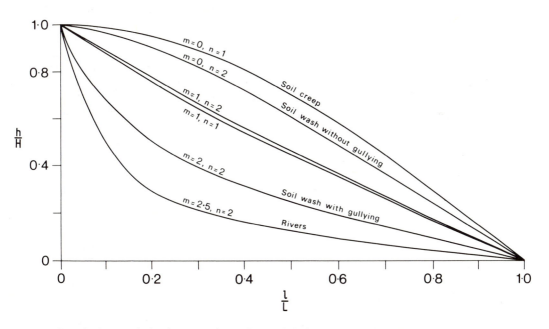

**Figure 11.21** Dimensionless graph showing approximate characteristic-form slope profiles for a range of transportational processes: see text, for description.
*Source:* Kirkby, 1971, figure 5, p. 26.

no lateral convergence or divergence downslope and that erosion is constant over time. These equilibrium characteristic-form curves are obtained by simulated changes through time, and Figure 11.22 shows the development of a slope under soil creep from an initially straight profile to a convex characteristic-form after approximately one-half of the elevation above baselevel has been removed. A similar, but rather more complex computer-based model has generated a set of characteristic-form slope profiles (Figure 11.23) from an initial slope profile (broken line) with a fixed baselevel for the following processes (Ahnert, 1976):

(1) rainsplash – i.e. $C\alpha$ sin (slope$^{0.75}$);
(2) suspended load wash;
(3) slope wash by rolling – i.e. no suspension;
(4) plastic flow – i.e. mass movements with a lower minimum threshold;
(5) viscous flow – i.e. mass movements at all slope angles (no lower threshold of action).

To return to the question of the manner in which slopes evolve through time, some studies of small-scale badland slopes reveal that within a small area with similar precipitation and vegetation cover, both parallel and declining slope retreat takes place depending on the erosional agent that is dominant within the area. In the Badland National Monument of South Dakota slopes on the Chadron formation are broadly convex as compared to the Brule formation slopes which are straight and steep and dominated by rainwash. The erosional evolution of these slopes is, in fact, shown in Figure 11.20. The Chadron slopes decline under the influence of creep, whereas the Brule slopes retreat parallel under the influence of rainbeat and overland flow (Plate 12).

The manner in which soil or rock fragments move downslope is the key to understanding the erosional evolution of these slopes.

Runoff or rainwash where observed on badland slopes occurs as surge or subdivided flow. Minute obstacles check and deflect the runoff in its downslope course; the movement of the eroded material is, therefore, not continuous. The runoff neither constantly accelerates nor becomes overloaded towards the slope base on straight slopes. Erosion measured on straight badland slopes is essentially uniform over the length of that slope. When rainwash acts on a slope, any particles A, B, C and D (Figure 11.24A) spaced equally along the slope may be removed from the slope during any one storm or during several storms, but in any case before the underlying particles W, X, Y and Z are removed. Rainwash is thus an eroding mechanism tending to remove a uniform thickness of material from a straight slope during a storm. It does this because as an eroding agent it attacks the steepest part of the slope with the greatest energy.

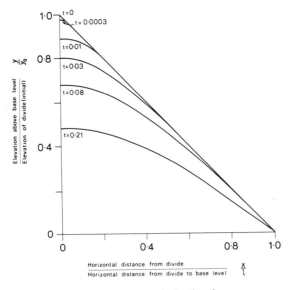

Figure 11.22   Dimensionless graph showing slope development by creep of an initially straight slope with a fixed divide (at $x = 0$; $dy/dx = 0$) and a fixed baselevel (at $X = \ell$, $y = 0$). This mathematical development resulted from a complex equation $y = f(t,X)$ with constants $K$ and $\ell$.
$K$ = a constant of the process rate ($d^2y/dx^2 = 1/K \, dy/dt$); $t$ = time elapsed; $\tau$ relative time elapsed ($\tau = Kt/\ell^2$). The rate of creep is assumed to be proportional to the tangent of the slope gradient which, in turn, is assumed to be proportional to the actual gradient for slopes in the field < 20°.
*Source:* Carson and Kirkby, 1972, figure 10.12, p. 298.

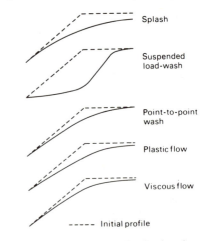

Figure 11.23   Characteristic profiles developed mathematically for five different denudation processes with a fixed baselevel.
*Source:* Ahnert, 1976, figure 14, p. 48.

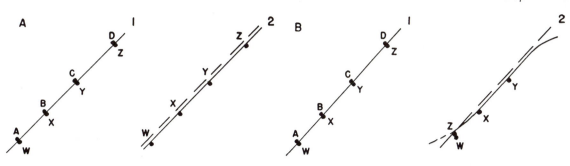

**Figure 11.24** A. Erosion of particles **A, B, C** and **D** from a slope surface under the action of rainwash: the grains are not necessarily removed in sequence, but each layer of grains is removed before erosion of the next can take place. B. Removal of particles **A–D** from a slope surface by creep; the grains were removed in sequence, followed by particle **Z**, causing maximum erosion on crest of the slope.
*Source:* Schumm, 1956b, figure 4, p. 698.

Thus, when particle A (Figure 11.24A) is removed from the slope, in effect, the baselevel for the particles above is lowered and they are quickly removed.

Creep is the dominant process active on the Chadron slopes. The individual aggregates move downslope by swelling from wetting, shrinking from drying and filling of the desiccation cracks from the uphill side, and possibly by rainbeat. Unlike the action of rainwash, in which a sheet of material moves on the entire slope surface, a particle D (Figure 11.24B) engaged in creep will not be removed from the slope at the same time or before a particle A, but must traverse the entire slope length behind the particle downslope from it. Thus the particles will be removed from the slope in the order A, B, C, D, Z. Since the movement of the creeping material is greatest on the steepest slopes, it seems that erosion would logically be greatest on these slopes. However, these slopes are principally slopes of transportation. The creeping material from upslope moves over these slopes, protecting the bedrock from rapid erosion. This means that the upper part of the slope provides the bulk of the moving material, as soil or rocky particles move downslope. The downslope segments are largely slopes of transportation and deposition occurs at the concave slope base. The net result is the development of an upper convexity and a lower slope concavity under the influence of creep.

Field studies show that with time slopes decrease after reaching a maximum value. Therefore, depending on the relative age of slopes, they can display different angles of inclination unless parallel retreat is the mode of slope erosion.

Figure 11.25A shows how the order of stream channels at the slope base might be used as an ergodic device. It was felt justifiable to assume that, as there seemed to be a general tendency for all streams within the basin to increase in order as the net extended, one could use the order of any individual channel segment to indicate the relative age of the adjacent valley-side slope (stream order is defined in Section 13.2). Figure 11.25B shows a plot of maximum valley-side slope angles for each of six stream orders, suggesting that initial channel incision is associated with the progressive steepening of this slope until it reaches a limiting angle close to its angle of repose. The size of the sample of highest-order slopes was too small, and the order of the whole network was too low, to permit significant inferences regarding the later stages of slope development when, perhaps, decline in angle might have become important.

A recent study of fault scarps in Nevada suggests that the processes that are dominant during one stage of slope evolution may be replaced by other processes later in their evolution. For example, Wallace (1977) studied fault scarps in alluvium which, of course, is a relatively erodible material. The original fault scarp was very steep, ranging between 50° and 90°. At this early stage (Figure 11.26) mass wasting is dominant and the top of the scarp retreats by debris fall. This material accumulates at the scarp base to form a slope controlled by mass wasting and debris. Later the scarp is controlled by the angle of repose of the debris, about 35°, but at this gentler slope water erosion becomes dominant and further slope decline takes place.

Wallace's observation that a steep slope will decline by slope failure to an angle of repose slope returns us to the characteristic slope concept introduced earlier and introduces the concept of limiting or threshold slope angles. Within a given area there is not only a characteristic slope angle, but a range of angles from an upper to a lower limit. For a meaningful analysis, measurements should be of the steep, straight segment of a profile.

A slope formed by river incision or steepened by

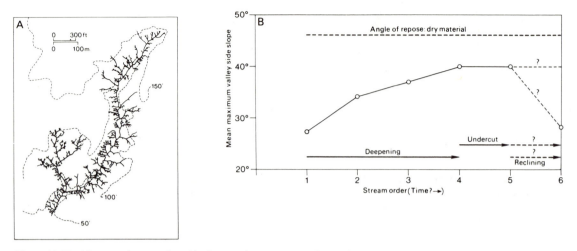

**Figure 11.25** Mean maximum valley-side slope angles v. stream order (B) for a small drainage system, actively cutting back into a terrace of the Farmington River, Connecticut (A).
*Source:* Carter and Chorley, 1961, figures 1 and 10, pp. 119, 128.

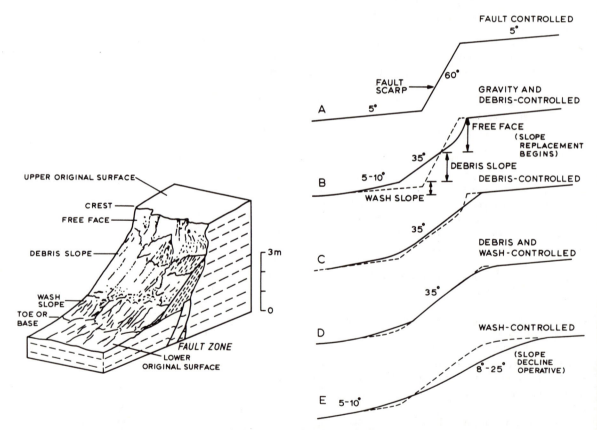

**Figure 11.26** Block diagram of a Nevada young fault scarp showing the terminology and a proposed sequence of fault scarp degradation; incremental change is shown by the broken line which represents the full line on the previous profile.
*Source:* Wallace, 1977, figures 2 and 3, p. 1269.

undercutting will decline by mass movement processes to the frictional threshold angle which is equal to the peak angle of internal friction $\phi_p$ (see Figure 10.22) or to the residual angle of internal friction $\phi_r$ after prolonged stress has tended to weaken the internal structure of the material (e.g. by the preferred alignment of platy minerals). When high porewater pressures regularly occur in the saturated slope debris, the threshold angle is reduced to the lower semi-frictional angle ($\phi_{sf}$). For shallow slides:

$$\tan\alpha = (1 - u/\gamma z \cos^2\alpha)\tan\phi$$

where: $\alpha$ = threshold angle of failure;
  $\gamma$ = bulk unit weight of the sliding material;
  $z$ = depth to the potential shear plane;
  $u$ = porewater pressure on the potential sliding surface.

(Cohesion is ignored in the long term due to weathering and unloading processes.) For dry material: $u = 0$ and $\alpha = \phi$. For complete saturation of the slope material: $u = \gamma_w z \cos^2\alpha$ (where $\gamma_w$ = unit weight of soil water) and $\tan\alpha = \frac{1}{2}\tan\phi$ ($= \tan\phi_{sf}$) (Carson, 1975). It has been noted that maximum valley-side slopes on debris of low cohesion in humid areas tend to be about half those in dry regions.

A good example of the relation between maximum valley-side slope angles and frictional threshold angles is provided by the gneissic terrain of the Verdugo Hills in the California Coast Ranges, which is characterized by steep, straight valley-side slopes covered by regolith up to 1 m thick. These slopes form two groups, one which is subjected to basal stream undercutting (mean 44·8°) and another with basal protection (mean 38·2°) (Figure 11.27). The undercut slopes are subject to more active erosion and are underlain by coarser debris ($d = 0\cdot1 - 30$ mm; $d_{50}$ estimated at 12 mm), whereas the more weathered protected slopes have a much finer sandy soil ($d < 0\cdot1$ mm). Experimental shear tests on debris samples yielded the following results:

(1) Undercut slopes: $\phi_p = 38° - 55°$, suggesting that the observed range of valley-side slope angles is controlled to a large extent by the shearing characteristics of debris at a range of densities.
(2) Protected slopes: $\phi_p = 35° - 44\cdot8°$ and $\phi_r = 32\cdot8° - 35\cdot9°$, which together bracket the range of observed slope angles.

On the Hangman Grits of north Devon, England, maximum valley-side slope angles have a mode of about 28°, and slopes of 24–30° have a continuous vegetation cover with the cover of regolith (i.e. soil plus weathered rock mantle) increasing in thickness from the top to the

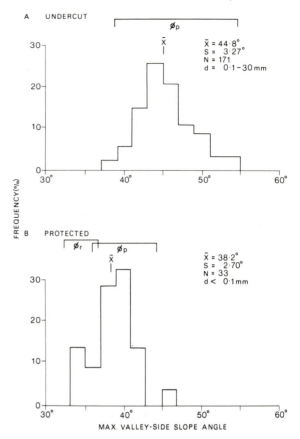

**Figure 11.27** Frequency distributions of maximum valley-side slope angles for (A) undercut and (B) protected slopes of gneissic debris in the Verdugo Hills, California, with the experimental ranges of angles of internal friction ($\phi_p$ *and* $\phi_r$) for the finer slope material shown.
*Source:* Strahler, 1950, figure 14, p. 812; Carson, 1975, figure 2, p. 24.

base of the slopes (Figure 11.28A), whereas those slopes increased in angle in excess of 30° by basal stream undercutting exhibit features of slumping (i.e. soil dislocation, terracettes, etc.) indicative of instability and transience. Young (1963) has suggested the following sequence of development of the steep slopes on the Hangman Grits (Figure 11.28B):

(1) River incision causing unstable slopes of 33°–40° to develop.
(2) Cessation of downcutting with a more stable 26°–29° slope developing at the base and working headward.
(3) The more stable 26°–29° segment becomes dominant.

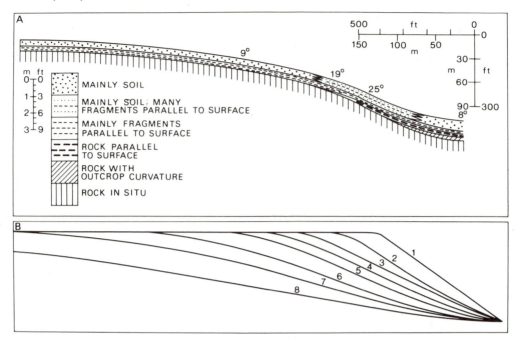

**Figure 11.28** A characteristic slope and soil profile on the Hangman Grits (Devonian) of the Heddon Basin, Exmoor, North Devon A., and a suggested sequence of slope development for non-calcareous Paleozoic rocks under humid temperate conditions B.
*Source:* Young, 1963, figures 8 and 11, pp. 17, 25.

(4) This segment retreats more or less parallel due to erosion producing a 25° maximum slope angle with a basal concavity and upper slope convexity.

(5)–(8) The decrease of the maximum slope angle and the extension of both concavity and convexity.

The effect of water content is further borne out by comparison of highly erodible Mancos Shale hillslopes in arid western Colorado with clay–shale slopes in Derbyshire, England, where during winter high porewater pressures have been measured close to the surface of the slope. Shearing tests on the Mancos Shale regolith show that the range of maximum angles observed for the 5–10-m-long slopes is encompassed by the experimental values of $\phi_r$ and $\phi_p$ whose range is dependent upon different conditions of void ratio, density and orientation of the platy fragments (Figure 11.29A). A link with the moist shale rubble (clay 33–54 per cent; silt 24–64 per cent; plasticity index 18–42 per cent) slopes of England is suggested by Figure 11.29B where the short (10 m) valley-head slopes ($>17°$) which are dry for much of the year contrast with the majority of the longer valley-side slopes which are saturated for much of the time, the latter material exhibiting experimental angles of internal friction when saturated ($\phi_{sf}$) about one-half the values of $\phi_r$.

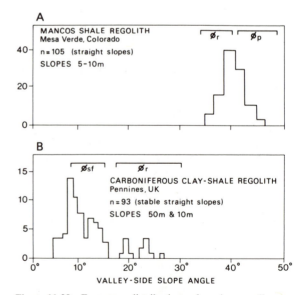

**Figure 11.29** Frequency distributions of maximum valley-side slope angles for Mancos Shale slopes at Mesa Verde, Colorado A., and for clay–shale slopes in the Pennines of Derbyshire, England B. Experimental ranges of angles of internal friction $\phi_p$ and $\phi_r$, and of semi-friction $\phi_{sf}$ of the regolith material, are shown.
*Source:* Carson, 1975, figure 6, p. 32.

It has been suggested that slope profiles could be used to interpret the earth's recent tectonic history (Chapter 2). For example, the ratio between rates of uplift and river incision should change as uplift becomes more rapid and consequently slope profiles will reflect this difference. If uplift of an area was relatively rapid or much greater than stream incision, broadly convex slopes would form. If uplift was relatively slow and stream incision was dominant, concave slopes would form. If there was a balance between uplift and stream incision, straight slopes would form. However, as appealing as this idea may be, it does not have general applicability because of *convergence* (equifinality). This concept states that similar forms can result from different causes. In other words, the profiles of slopes can be similar in appearance, but they can be formed by different erosional and depositional processes. For example, if we consider the three slope segments: – the convex, concave and straight – studies have shown that for each of these there are several processes that can develop the slope profile (Table 11.1).

Another type of landform that deserves attention is the escarpment. This is a line of cliffs generally formed by faulting or erosion, the most spectacular of which are controlled by lithologic variations. If the scarp is due to recent faulting or river incision or it is composed of homogeneous materials, it may quickly assume a typical hillslope form (see Figure 11.24), but if the scarp is protected by a resistant caprock or if its steepness is maintained by fractures, it can maintain its form as it retreats. In humid regions fluvial or marine erosion at the slope base commonly initiates multiple rotational slumps which are reactivated from time to time by mudflows and slides giving complex patterns of erosion and accumulation at the toe of the slope. In arid regions, such as the Colorado Plateau, there are characteristic mechanisms of slope recession where shales underlie more competent, permeable beds. There is also no doubt that such cliffs have retreated for long distances, in some cases many tens of kilometres, leaving behind surfaces of low relief which are either the stripped upper surfaces of resistant beds (e.g. the Great Sage Plain of south-east Utah and south-west Colorado, supported by the Dakota Sandstone) or surfaces cut at low angles across thick shale formations (e.g. the shale pediments of Utah, north of the Henry Mountains: see Section 18.4.1). The walls of the Grand Canyon are composed of scarps superimposed on scarps, and when more than two lithologic units are involved, a complex canyon wall results. We will discuss only two cases, the effect of a caprock and fractures. Figure 11.30 shows a case of scarp retreat downdip with progressively decreasing height of the scarp. Initially the

**Table 11.1** Mechanisms of slope segment production

I *Convex segment*:
(1) Erosion
  (a) creep
  (b) rainwash (on relatively permeable soils and upon moving from an upper flat segment to a lower steeper segment)
  (c) mechanical and chemical weathering of massive rocks (weathering is most effective on sharp edges)
  (d) pressure-release jointing (formation of domes)
  (e) basal erosion (lateral stream erosion, spring sapping, weathering – any process that steepens basal slope segments)
  (f) incision of widely spaced (low D) streams in areas of low or moderate relief
(2) Deposition
  (a) mantling of existing slopes with loess or volcanic ash

II *Concave segment*:
(1) Erosion
  (a) rilling (erosion greatest on middle segment)
  (b) creep (least efficient at slope base)
  (c) piping
(2) Deposition
  (a) accumulation of talus at slope base
  (b) accumulation of colluvium at slope base
  (c) accumulation of volcanic ejecta to form a cinder cone

III *Straight segment*:
(1) Erosion
  (a) rainwash (on relatively impermeable soils)
  (b) incision of closely spaced streams (high D) in areas of moderate or high relief
  (c) mass movement (separation along vertical joints including pressure-release jointing)
  (d) basal erosion (lateral stream erosion, wave action, erosion of weaker underlying rock)
(2) Deposition
  (a) talus, sand, volcanic ejecta at angle of repose

caprock is a small part of the cliff face and it is the erosion of the underlying weaker lithologic unit that controls the situation.

Arid cliffs are commonly composed of a near-vertical upper section varying from 10–75 per cent of the total cliff height, which may or may not be coincident with the edge of the caprock, and a lower shale slope (25–90 per cent of the height) of 32°–38° usually dissected into badlands (Figure 11.31). The erosion of the shale causes the caprock to be undermined and to collapse, and the height of the vertical cliff is generally controlled by the following resistive properties of the caprock:

(1) Mechanical strength. The greater the caprock strength, the higher the vertical cliff because more

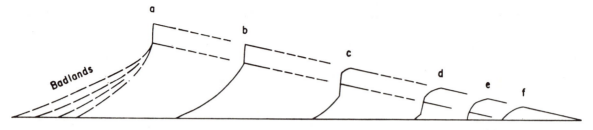

**Figure 11.30**  Schematic illustration of the change in scarp form as the downdip retreat causes the caprock to become an increasingly important component of the scarp face.
*Source:* Schumm and Chorley, 1966, figure 3, p. 32.

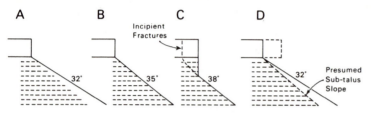

**Figure 11.31**  Idealized profiles showing stages in the retreat of an escarpment formed by shale capped by sandstone.
*Source:* Koons, 1955, figure 5, p. 49.

undermining is required to cause its failure. This is expressed by the following relation which was developed by Terzaghi (1962):

$$Hc = \frac{Sc}{W}$$

where $Hc$ is the critical height of the cliff at which failure occurs, $Sc$ is the compressive strength of the rock and $W$ is the unit weight of the rock. For example, if $Sc$ is 5000 lb/in² and the unit weight is 170 lb/ft² or 1·18 lb/in² per foot of height, the critical height $Hc$ is

then about 4200ft. As Terzaghi points out, no vertical slopes of such height exist and, therefore, the height of cliffs is more dependent on joints than on rock strength. It is interesting that gravel capping can give rise to small vertical cliffs (e.g. in the flight of gravel-capped shale pediments north of the Henry Mountains, Utah) in that their high permeability causes them to behave as a resistant rock.

(2) Jointing and bedding. The thinner the bedding and the closer the jointing, the more readily the cliff fails and the smaller the vertical slope. In the massive sandstones

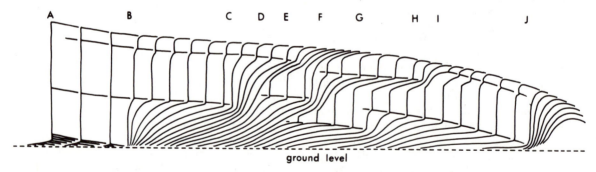

**Figure 11.32**  Development of scarp forms in massive sandstone through time and space. For simplicity, a constant ground level (broken line) is assumed during scarp retreat, along with equal thicknesses of removal from major slab walls in each unit of time. At **A** a cliff is present due to sapping above a thin-bedded substrate. At **B** the thin-bedded substrate passes below ground level, scarp retreat slows, and effective intraformational partings assume control of scarp form. Effective partings close at **C, E, F, G, H** and **J**, leading to local slab wall stagnation and rounding into slick rock. Partings that open at **D, E, F, G,** and **I** initiate growth of new slab walls. Note that the effect of former partings in rocks that have been removed continues to be expressed in the form of slick rock ramps and concave slope breaks. Lowering of ground level during backwearing would cause cliff extension upward from major contact.
*Source:* Oberlander, 1977, figure 8, p. 105.

of the Colorado Plateaus retreat is by granular disintegration of the sandstones forming the cliff face, and by slab failure as controlled by jointing (Figure 11.32). An important factor in the development of fractures is what is called pressure-release jointing (Figure 11.33). This is due to the expansion of the massive rocks, when erosion exposes rocks that were once buried. For example, when rivers incise deeply into massive rocks, joints develop parallel to the canyon walls and the retreat of the wall or scarp depends on these fractures. The rocks of the Colorado Plateau, although having considerable compressive strength, weather rapidly and the base of the scarp is relatively free of talus, for – as we have seen – the sandstones there are generally porous, of low strength, poorly cemented and susceptible to physical and chemical weathering. When this is not so, the system becomes clogged with debris, a talus slope results and the buried scarp face may assume a convex form as suggested in Figure 11.34.

(3) Direction of dip. Dip away from the scarp accentuates the slumping and sliding of the caprock.

(4) The thickness of the caprock. Thin caprocks give low cliffs which retreat rapidly and may leave behind extensive areas of shale badlands.

Figure 11.35 illustrates the effect of (A) a resistant caprock, such as the Shinarump Conglomerate overlying the Moenkopi Shale in the Chocolate Cliffs or the immensely thick Navajo and Wingate Sandstones overlying the Chinle Shales in the Zion Park Region; (B) a less resistant caprock, such as the Point Lookout Sandstone overlying the Mancos Shale at Mesa Verde, south-west Colorado; and (C) a thin, weak caprock overlying thick shales dissected into extensive badlands, such as the thin Wingate Sandstone over the Chinle Shales in the Painted Desert area of Arizona.

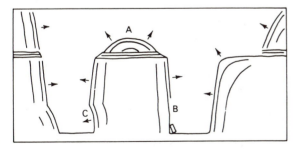

**Figure 11.33** Diagrammatic sketch of exfoliation joints in massive Colorado Plateau sandstones. Arrows show inferred directions of expansion: **A**, an exfoliation dome; **B**, an exfoliation cave; **C**, an overhanging exfoliation plate in a meander scar.
*Source:* Bradley, 1963, figure 2, p. 522.

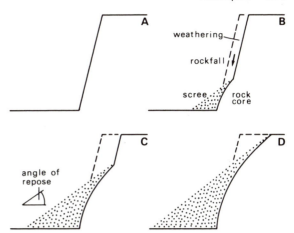

**Figure 11.34** Replacement of a cliff by a scree with the formation of a buried rock core.
*Source:* Young, 1972, figure 9, p. 31.

## 11.7 Summary

As the preceding material has demonstrated, one of the reasons that the understanding of the development and evolution of hillslopes has been slow is due to the complexity of the phenomena investigated. The effects of climate, lithology and vegetative cover and the differences between erosional and depositional slopes produce forms that cannot be generally related one to the other and, therefore, generalizations about hillslopes on a global basis are impossible to make. And yet, from a practical point of view, it is essential that we understand how slopes erode and what form they will take over long periods of time. In the past practical applications of hillslope erosion were related primarily to agricultural activities. However, more recently the disposal of tailings from oil shale and uranium mines has led to a need for slope stability of the order of 1000 years.

If slope evolution by the processes of creep and surface runoff is by declining and parallel retreat, respectively, then in a region where creep is dominant a peneplain will eventually develop, or the reduction in height of the divides by creep will produce a gently undulating surface. On the other hand, as slopes retreat parallel they leave pediments (Section 18.4.1) at their base, and as the steep slopes eventually are reduced the pediments will coalesce along the drainage divides to form a pediplain (see Figure 2.9D). Thus depending on the processes of hillslope retreat, which is dependent on the material being eroded and climate (Chapter 18), either the Davis or the King model (Chapter 2) will apply.

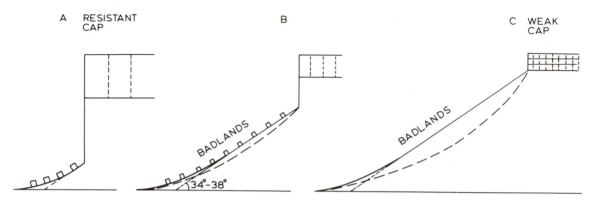

**Figure 11.35**   Retreat of an arid cliff under the influence of three types of caprock: see text, for description.

# References

Ahnert, F. (1976) 'A brief description of a comprehensive three-dimensional process–response model of landform development', *Zeitschrift für Geomorphologie*, Supplement 25, 29–49.

Atkinson, T. C. (1978) 'Techniques for measuring subsurface flow on hillslopes', in M. J. Kirkby (ed.), *Hillslope Hydrology*, Chichester, Wiley, 73–120.

Bradley, W. C. (1963) 'Large-scale exfoliation in massive sandstones of the Colorado Plateau', *Bulletin of the Geological Society of America*, vol. 74, 519–28.

Carson, M. A. (1975) 'Threshold and characteristic angles of straight slopes', *Proceedings of the 4th Guelph Symposium on Geomorphology*, Norwich, Geo Books, 19–34.

Carson, M. A. and Kirkby, M. J. (1972) *Hillslope Form and Process*, Cambridge University Press.

Carter, C. S. and Chorley, R. J. (1961) 'Early slope development in an expanding stream system', *Geological Magazine*, vol. 98, 117–30.

Chorley, R. J. (1978) 'The hillslope hydrological cycle', in M. J. Kirkby (ed.), *Hillslope Hydrology*, Chichester, Wiley, 1–42.

Chorley, R. J. and Kennedy, B. A. (1971) *Physical Geography: A Systems Approach*, London, Prentice-Hall.

Culling, W. E. H. (1963) 'Soil creep and the development of hillside slopes', *Journal of Geology*, vol. 71, 127–61.

Emmett, W. W. (1970) 'The hydraulics of overland flow on hillslopes', *US Geological Survey Professional Paper* 662A.

Finlayson, B. and Statham, I. (1980) *Hillslope Analysis*, London, Butterworth.

Freeze, R. A. (1978) 'Mathematical models of hillslope hydrology', in M. J. Kirkby (ed.), *Hillslope Hydrology*, Chichester, Wiley, 177–225.

Horton, R. E. (1945) 'Erosional development of streams and their drainage basins: hydrophysical approach to quantitative morphology', *Bulletin of the Geological Society of America*, vol. 56, 275–370.

Jahn, A. (1963) 'Importance of soil erosion for the evolution of slopes in Poland', *Nachrichten der Akademie der Wissenschaften in Göttingen Math.-Physik, Klasse*, vol. 15, 229–37.

Kilinc, M. and Richardson, E. V. (1973) 'Mechanics of soil erosion from overland flow generated by simulated rainfall', *Colorado State University Hydrology Paper* 63.

Kirkby, M. J. (1971) 'Hillslope process–response models based on the continuity equation', *Institute of British Geographers Special Publication* 3, 15–30.

Kirkham, D. (1947) 'Studies of hillslope seepage in the Iowan drift area', *Proceedings of the Soil Science Society of America*, vol. 12, 73–7.

Koons, D. (1955) 'Cliff retreat in the southwestern United States', *American Journal of Science*, vol. 253, 44–52.

Melton, M. A. (1960) 'Intravalley variation in slope angles related to microclimate and erosional environment', *Bulletin of the Geological Society of America*, vol. 71, 133–44.

Mosley, M. P. (1972) 'An experimental study of rill erosion', unpublished MS thesis, Colorado State Universtiy.

Mosley, M. P. (1973) 'Rainsplash and the convexity of badland divides', *Zeitschrift für Geomorphologie*, Supplement 18, 10–25.

Mosley, M. P. (1974) 'Experimental study of rill erosion', *Transactions of the American Society of Agricultural Engineers*, vol. 68, 909–16.

Noble, E. L. (1963) 'Sediment reduction through watershed rehabilitation', *US Department of Agriculture Miscellaneous Publication* 970, 114–23.

Oberlander, T. M. (1977) 'Origin of segmented cliffs in massive sandstones of southeastern Utah', in D. O. Doehring (ed.), *Geomorphology in Arid Regions*, Proceedings 8th Annual Geomorphology Symposium, Binghamton, State University of New York, 79–114.

Ruhe, R. V. (1975) *Geomorphology*, Boston, Mass., Houghton Mifflin.

Ruhe, R. V. and Walker, P. H. (1968) 'Hillslope models and soil formation. I, open systems', *Transactions of 9th International Congress of Soil Science*, paper 57, 551–60.

Ruxton, B. P. (1958) 'Weathering and sub-surface erosion in granite at the piedmont angle, Balos, Sudan', *Geological Magazine*, vol. 95, 353–77.

Scheidegger, A. E. (1970) *Theoretical Geomorphology*, 2nd edn., New York, Springer-Verlag.

Schumm, S. A. (1956a) 'The evolution of drainage systems and slopes at Perth Amboy, New Jersey', *Bulletin of the Geological Society of America*, vol. 67, 597–646.

Schumm, S. A. (1956b) 'The role of creep and rainwash on the retreat of badland slopes', *American Journal of Science*, vol. 254, 693–706.

Schumm, S. A. (1967) 'Rates of surficial rock creep on hillslopes in western Colorado', *Science*, vol. 155, 560–1.

Schumm, S. A. and Chorley, R. J. (1966) 'Talus weathering and scarp recession in the Colorado Plateaus', *Zeitschrift für Geomorphologie*, vol. 10, 11–36.

Schumm, S. A., Harvey M. D. and Watson, C. C. (1984) *Incised Channels*, Littleton, Colo, Water Resources Publications.

Schumm, S. A. and Mosley, M. P. (eds) (1973) *Slope Morphology*, Stroudsburg, Pa, Dowden Hutchinson & Ross.

Selby, M. J. (1982) *Hillslope Materials and Processes*, Oxford University Press.

Small, R. J. and Clark, M. J. (1982) *Slopes and Weathering*, Cambridge University Press.

Smith, D. D. and Wischmeier, W. H. (1962) 'Rainfall erosion', *Advances in Agronomy*, New York, Academic Press, 109–48.

Strahler, A. N. (1950) 'Equilibrium theory of erosional slopes approached by frequency distribution analysis', *American Journal of Science*, vol. 248, 673–96, 800–14.

Terzaghi, K. (1962) 'Stability of steep slopes on hard unweathered rocks', *Geotechnique*, vol. 12, 251–70.

Toy, T. J. (1977) 'Hillslope form and climate', *Bulletin of the Geological Society of America*, vol. 88, 16–22.

Wallace, R. E. (1977) 'Profiles and ages of young fault scarps, north-central Nevada', *Bulletin of the Geological Society of America*, vol. 88, 1267–81.

Young, A. (1961) 'Characteristic and limiting slope angles', *Zeitschrift für Geomorphologie*, vol. 5, 126–31.

Young, A. (1963) 'Some field observations on slope form and regolith, and their relation to slope development', *Transactions of the Institute of British Geographers*, no. 32, 1–29.

Young, A. (1972) *Slopes*, Edinburgh, Oliver & Boyd.

# Twelve *Rivers*

Mass movement and hillslope erosion were difficult to discuss because several mechanisms and processes were, and are, operating to produce landforms. In addition, the processes are difficult to observe. This is not true of rivers. A liquid is transported through an open channel composed of a solid. The flow is confined and, although the depth and velocity of flow in the channel can render the study hazardous, the liquid can be penetrated by various measuring devices and the system can be studied with relative ease in comparison, for example, to the complicated mix of overland flow, rainsplash and creep on hillslopes.

## 12.1 Significance

Although on a global basis the quantity of water in rivers is only a tiny fraction of the total (0.0001 per cent), when in flood rivers can be awesome phenomena. For example, the very destructive 1973 flood in the Mississippi River had a discharge of 855,000 cfs (24,210 m³/s) at St Louis. The greatest-known flood on the Mississippi was in excess of 2 million cfs (56,640 m³/s), but this is below the *average* discharge of the Amazon River, the mightiest river of all (Figure 12.1).

The energy of the hillslopes and drainage basin is concentrated in the water and sediment discharge and, therefore, the river channel will be the most dynamic component of the landscape. Dramatic changes in channel morphology and river behaviour occur as a result of upstream controls (Figure 1.3) and downstream influences such as baselevel change. As the rivers adjust, they affect not only their tributary streams, but also the adjacent hillslopes by undercutting or by aggradation. The effects of increased or decreased water discharge and sediment load are transmitted downstream to affect not only the downstream channels, but also depositional landforms such as alluvial fans and deltas. The influence can also be noted in the sedimentology and stratigraphy of the fluvial and shallow marine sedimentary deposits (Figure 1.3).

A river is a flowing body of water in an open channel. The characteristics of the flowing water are the domain of the hydraulic engineer, whereas the dimensions and pattern of the open channel (i.e the hydraulic geometry) are geomorphic problems. The river, perhaps best of all geomorphic features, integrates the cascading and morphologic systems because the forces that determine the dimensions and morphology of the channel are almost exclusively those exerted by the flowing water. The flowing water not only moves through and fashions the open channel, but it transports within it a variety of materials. The total sediment load of the channel is comprised of a chemical or dissolved load, a suspended load, fine particles that are kept in suspension by the turbulence of the flowing water, and a bedload composed of coarser particles that are moved on or near the bed of the stream. The sediments are moved by the transport capacity of water that is in the channel.

A major problem for the geomorphologist is to determine what fashions the channel. In most cases an observer will see the channel only partly filled with water. This is the case during dry periods of the year and in semi-arid regions, where the channel may be completely dry for much of the year. The dryland's channel carries an ephemeral flow only during and following storms, in contrast with perennial streams that are characteristic of humid regions. Intermittent streams flow seasonally for at least a month during each year.

It seems logical to assume that the quantity of water that just fills or nearly fills the channel (i.e. the bankfull discharge $Q_b$) is the dominant discharge that is responsible for establishing the characteristics of the channel as

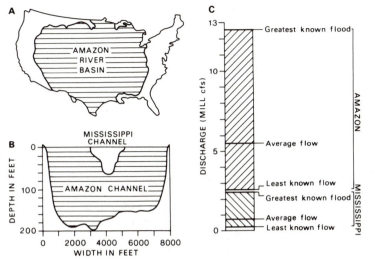

**Figure 12.1** The Amazon and the Mississippi: A. the Amazon river basin covers 6 million square kilometres, equivalent to nearly three-quarters the size of the coterminous United States; B. comparison of channel cross sections of the Amazon River at Obidos (more than 700 km upstream of the mouth!) and that of the Mississippi at Vicksburg (500 km upstream from the delta); C. comparison of discharges of the Amazon River at Obidos and the Mississippi River at Vicksburg.
*Source:* U.S. Geological Survey.

one sees it (but see Section 12.4). It is obvious that the flow velocity and the manner in which the water flows through the channel are important considerations in determining the morphology of that channel. The shear forces exerted against the channel boundaries will determine the final equilibrium form of that channel. Therefore, it is essential before considering the morphology of a river channel to consider hydraulics of flow in the channel and some aspects of the water and sediment discharge through it.

## 12.2 Open-channel hydraulics

Water in open channels is subject to two principal forces: the force of gravity which propels the water downslope, and frictional forces between the water molecules themselves and between the water and the channel boundaries which together resist the downslope movement. Hence a steep-gradient, smooth channel in which the gravitational forces are high and roughness is low will have a high velocity. A good example is a surface meltwater stream on a glacier (see Chapter 17). Velocity of flow in a channel can be estimated using Manning's equation:

$$\overline{V} = \frac{1 \cdot 49}{n} R^{2/3} S^{1/2} \text{ or } \frac{1}{n} R^{2/3} S^{1/2} \quad \text{(for metric units).}$$

$$(12.1)$$

This empirical equation shows that mean velocity of flow in an open channel ($\overline{V}$) is directly related to average depth of flow $R$ (the hydraulic radius, which is obtained by dividing the cross-sectional area of the channel by the wetted perimeter of the channel) (Figure 12.2) and to channel gradient $S$, and inversely related to a roughness

value $n$. Unfortunately, although $R$ can be obtained by direct measurement and $S$ measured or estimated relatively easily, the estimation of $n$ is often subjective and is based on experience.

Nevertheless, the coefficient of roughness represents a central concept which links hydraulics with hydraulic geometry. Roughness is directly controlled by the grain size of bed and bank material, the amount of channel vegetation and the irregularity of adjacent channel cross sections. It is indirectly controlled by the alignment of the channel plan and the discharge in that *increases* in the latter 'drown out' bed irregularities, make adjacent cross sections more comparable, and straighten the alignment of flowlines. A further complex control over roughness is exercised by the state of flow (see Figure 12.6). For simple, artificial, sand-bedded channels the roughness factor can be calculated from the median grain diameter $D_{50}$ as follows:

$$n = 0 \cdot 0342 D_{50}^{1/6} \quad (D \text{ measured in feet})$$
$$\text{or } n = 0 \cdot 016 D_{50}^{1/6} \quad (D \text{ measured in millimetres}).$$

However, for natural rivers, $n$ is much more difficult to estimate, but the following values are typical:

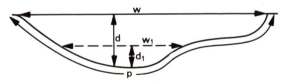

**Figure 12.2** Dimensions of the channel cross section: $w$ = bankfull width; $d$ = bankfull depth; $p$ = bankfull wetted perimeter; $w_1$ = width at mean annual discharge; $d_1$ = depth at mean annual discharge.

$$\frac{n \, (\mathrm{ft}^{1/6})}{}$$

sand-bedded  $\dfrac{n \, (\mathrm{ft}^{1/6})}{0.011 - 0.035}$   (depending on
channels                                          state of flow)

winding natural         0·035
streams

mountain         0.040 − 0·050.
streams with
rocky beds

Values of $n$ in the Manning equation assume that the ratio of channel depth $d$ to mean debris diameter $D$ exceeds 30, but where this value falls to 10 in shallow, rocky channels, the power of $R$ must be changed from 2/3 to 3/4.

Thus we have three linked aspects of roughness which enable us to progress from a simple, small-scale to a more complex, large-scale view of hydraulic geometry:

(1) Manning's definition of microscale roughness, based on grain diameter (i.e. $= D$).

(2) A definition for sand-bedded rivers based on the amplitude of mesoscale bed forms such as ripples and dunes (i.e $= 10^2 D$–$10^3 D$).

(3) Variations of overall macroscale channel roughness with variations of at-a-station discharge on a spatial scale of point-bar or riffle spacing (see later, especially Figure 12.24) (i.e. $= 10^4 D$–$10^5 D$).

Ignoring $n$, the Manning equation is essentially a depth–slope product, and the other common equation for velocity, the Chézy equation, relates mean velocity to the square root of the depth–slope product:

$$\overline{V} = C\sqrt{(RS)}, \text{ where } C \text{ is the Chézy constant.} \quad (12.2)$$

The Chézy equation can be related to the Manning equation when:

$$C = \frac{1\cdot49}{n} R^{1/6}. \quad (12.3)$$

The depth–slope product also appears in the equation for the tractive or shear force $\tau$ exerted on the bed of the stream, when the product of the specific weight of water $\gamma$, flow depth $d$ and slope $S$ equals the tractive force acting on the flow in the open channel:

$$\tau = \gamma \, d \, S. \quad (12.4)$$

One of the most important expressions of the hydraulics of flow in a channel is stream power $\Omega$, which is the work expended or energy loss of the stream:

$$\Omega = \gamma \, Q \, S \quad (12.5)$$
$$\text{or} \quad \Omega = \gamma \, w \, d \, \overline{V} \, S. \quad (12.5a)$$

Stream power per unit width of channel is:

$$\Omega = \frac{\gamma \, w \, d \, \overline{V} S}{w} = \gamma \, d \, S \, \overline{V} \quad (12.5b)$$

or, substituting equation (12.4):

$$\Omega = \tau \overline{V} \quad (12.6)$$

from which, because according to the Chézy equation (12.2), $\overline{V}^2 \alpha RS$;

$$\Omega \alpha \, \overline{V}^3 \quad (12.7)$$

Thus, if unit stream power varies as the cube of velocity, slight changes in velocity can significantly affect the work done in the channel. However, it should be noted that only 3–4 per cent of the work or energy of the stream is expended in the sediment transport, some 97 per cent being transformed into frictional heat by impacts between the water molecules themselves.

In any reach of a channel it is important to realize that, unless there are tributary contributions or diversions of water from the channel, the volume of water moving through one cross section should be essentially the same as that moved through another cross section in the same unit time. Considering the range in channel width and variations of the shape in the channel, this is a relatively important fact that must always be kept in mind. Thus the product of the mean velocity of flow $\overline{V}$ and cross sectional area of that flow $A$ at cross section 1 must equal product of flow at cross section 2 (i.e. $A_1 \overline{V}_1 = A_2 \overline{V}_2$). Therefore, when the cross-sectional area increases, velocity decreases. This simple equating of water discharge at adjacent cross sections is referred to as the *continuity equation*.

The Bernoulli energy equation indicates that the total energy of water in any streamline passing through a channel section is expressed as 'total head' of water $H$, which is equal to the sum of the elevation above a datum $h_2$, the pressure head $h_1$ and the velocity head ($V^2/(2g)$) (Figure 12.3). This can be rewritten as:

$$H = d + Z + V^2/(2g) \quad (12.8)$$

where: $d$ = depth of water;
         $Z$ = elevation above the datum plane.

The line representing the magnitude of the total energy at successive points along a channel is shown as $S_e$ and the slope of the line is known as the energy gradient (Figure 12.3). The slope of the water surface is denoted by $S_w$ and that of the channel bed by $S_b$.

Note that none of these slopes must be equal to the other. The slope of the channel bottom is sin $\theta$ (actually tan $\theta$ for low slopes). It is important to note that $S_e$ is the value of $S$ required by the Manning equation, $S_w$ is sometimes used as a surrogate for this but is difficult to

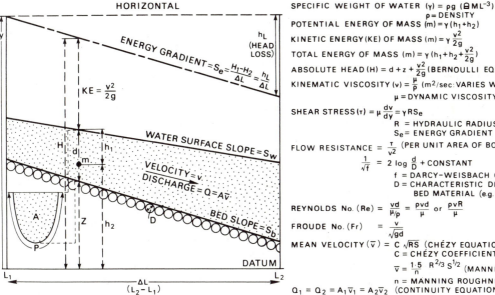

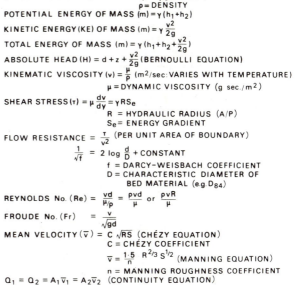

**Figure 12.3** The characteristics of open channel flow, together with some of the more important hydraulic equations.
*Source:* Simons, 1969, figure 7.1.6, p. 307.

measure accurately in the field, and $S_b$ is easy to measure but is of little value in the Manning equation except under simplified laboratory conditions. According to the principle of conservation of energy, the total energy at section $L_1$ upstream, should be equal to the total energy at the downstream section $L_2$ plus the head loss of energy $h_1$ between the two sections. However, a simplified view of stream energy is that it is related to the product of $D$, $S$ and $\overline{V}^2$. Because $R$ is a measure of $d$ and $\overline{V}^2 \alpha RS$

(equation (12.2)), it is convenient to consider stream energy as being proportional to $\overline{V}^4$.

Several types of flow are possible in open channels (Figure 12.4): uniform, non-uniform, steady, laminar, turbulent, tranquil and rapid flow. A basic understanding of these types of flow is essential to the hydraulician. With uniform flow, there is no change with distance either in magnitude or in velocity along a streamline. This is a condition that rarely exists. Non-uniform flow with

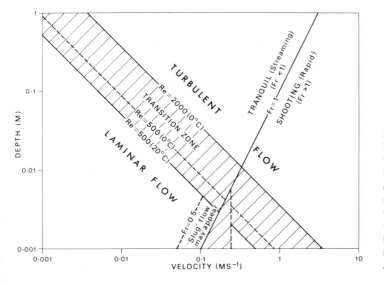

**Figure 12.4** The regimes of flow in a broad, open channel as a function of velocity and depth. The transition zone between laminar and turbulent flow is defined by a range of values of the Reynolds number (Re) of 500–2000, dependent largely upon temperature variations. The value of unity for the Froude number (Fr) separates tranquil (streaming) turbulent flow from rapid (shooting) turbulent flow.
*Source:* Sundborg, 1956, figure 1, p. 138.

its variations in velocity with time and space is more common. For steady flow, the velocity at a point does not change with time; for unsteady flow, the velocity at a point varies with time. Unsteady flow results when discharge changes with time and with travelling surges and floodwaves. Flow in alluvial rivers is generally non-uniform and unsteady.

The existence of laminar flow or turbulent flow depends upon the Reynolds number (Re) where:

$$Re = \rho \overline{V} d / \mu. \qquad (12.9)$$

This is the dimensionless ratio of the inertial to the viscous forces, where $\overline{V}$ = mean velocity, $d$ = depth of water, $\rho$ = water density and $\mu$ = dynamic viscosity (see Figure 12.3). When laminar flow occurs (i.e. Re < 2000 or so), the viscous forces predominate; when turbulent flow occurs, the inertial forces are large in comparison with the viscous forces. With laminar flow, the mixing in the stream is by molecular activity alone; whereas with turbulent flow, mixing is accomplished by random eddy motion of finite masses of molecules. Generally speaking, laminar flow rarely occurs in open channels except in very special instances in regions where the laminar sub-layer may exist, or perhaps in slack water.

The criterion for tranquil flow and rapid turbulent flow is the Froude number (Fr) which is the ratio of inertial to gravity forces (see Figure 12.3), where:

$$Fr = \overline{V} / \sqrt{(gd)} \qquad (12.10)$$

where $g$ = acceleration due to gravity, and $d$ = depth of flow. When the Froude number is less than 1, flow is tranquil; when it is equal to 1, flow is critical; and when it is greater than 1, flow is rapid. In the majority of rivers flow is usually non-uniform, unsteady, turbulent and tranquil, although rapid flow may occur in rare locations. Rapid flow is less common because the high velocity, the corresponding high rates of sediment transport and bank erosion usually alter the channel geometry, so that roughness increases, channel slope and velocity are reduced and flow becomes tranquil. For example, the Froude number on the Lower Mississippi

River is generally below 0·20 and it rarely exceeds 0·30, owing to great depths of flow.

The numerator of the Froude number is simply the mean velocity of flowing water in the channel; the denominator is the expression for the velocity of a gravity wave, which is a wave generated by a disturbance of the water surface. A rock dropped into a pond will create gravity waves and these will move across the surface at a velocity equal to $\sqrt{(gd)}$. This means that when the Froude number is less than 1 and a disturbance is created in the flow, the gravity wave has a higher velocity than the velocity of flow and the gravity wave will be propagated upstream. When the Froude number is equal to 1 (i.e. critical flow), the gravity wave will equal the velocity of the flow and it will not be propagated upstream. When the Froude number is greater than 1, the gravity wave will be swept downstream because the velocity of flow is greater than the velocity of the gravity wave. Above a Froude number of 1 pressure differences and downstream effects cannot be transmitted in an upstream direction. Therefore, at supercritical flow the flowing water cannot detect a downstream change of slope or other channel change, whereas at subcritical (tranquil or streaming) flows pressure differences are transmitted upstream, and the flow can adjust. For example, when approaching a slope change the velocity of supercritical flow will not change until it flows over the waterfall or rapids, whereas at subcritical flows the velocity of flow will increase as it approaches the slope change. In the latter case erosion above the slope break should eventually remove it. Figure 12.5 shows the idealized cross-sectional velocity distribution for tranquil flow in a straight channel.

The bed of a stream transporting sand may be characterized by the variety of ripple-scale patterns, and these patterns and the roughness of the channel can change dramatically as Froude number or stream power changes (Figure 12.6). As the velocity of flow increases in an open channel the essentially flat bed at the beginning of motion will be converted into a bed characterized by ripples, then by ripples and dunes, and at critical flow the

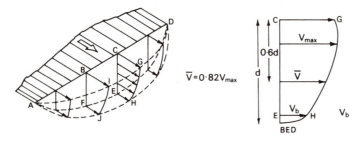

**Figure 12.5**  Idealized cross-sectional velocity distribution for tranquil (streaming) turbulent flow in an open channel. On the right the elevations of maximum velocity ($V_{max}$), mean velocity ($\overline{V}$) and velocity near the bed ($V_b$ i.e. that velocity associated with the tractive force responsible for bedload transport) are shown. The mean velocity occurs at about 0·6$d$ and is about 0·82 times the value of the maximum velocity.

bed again becomes plane, but with a high rate of sediment movement. In the range of supercritical flow antidunes develop. For the geomorphologist, these forms tend to be ephemeral because as flow depth and velocity decrease these bedforms disappear. Hydraulic changes from the edge of a channel to its centre can produce zones of different bedforms in a channel cross section (see Figure 12.30). Figure 12.6 shows these characteristic bedforms which are interesting from the hydraulic point of view because they markedly change the mesoroughness of the stream bed. For example, ripples and dunes create a rough channel, but at critical flow it becomes very smooth. The roughness increases as antidunes form. It has been demonstrated that these changes in bedform can significantly change water level and other characteristics of the stream.

Particles fall through water at velocities dependent on their size, the largest particles falling more rapidly than the smaller particles, and upon the viscosity of the water. However, temperature changes are sufficient to affect the viscosity of flow. Changes in the viscosity of water will, in turn, change the velocity at which a particle of a given size will fall, and hence its *effective* (hydraulic) diameter. Figure 12.7A shows sand bedforms as a function of stream power and sediment diameter. For example, an increase in viscosity with falling water temperature could have the effect of decreasing the effective diameter of a particle from *a* to *b* in figure 12.7A thereby decreasing channel roughness at a given stream power (i.e. the product of mean bed shear stress and mean flow velocity), as

well as increasing velocity and decreasing water depths. This has been documented for the Missouri River, where a significant fall in the water level, as a result of decreased roughness, caused grounding of barges and other river craft on sandbars as temperature decreased and flow velocity increased. Reservoir regulators were accused of not permitting sufficient water to flow through the channel in order to maintain river transport, and yet it was shown that discharge from the upstream reservoirs remained constant during the temperature change. It should be noted that antidunes do not appear in Figure 12.7A because their formation depends on a gravity-controlled, free-surface effect. In reality each grain size exhibits its own bedform regimes and Figure 12.7B shows these for a diameter of 0.19 mm in respect of variations in flow depth and mean velocity.

## 12.3 Sediment transport

From the point of view of the geomorphologist and the sedimentologist, the transport of sediment through an open channel is its most important attribute. Unfortunately, although it is relatively easy to obtain information on the water discharge through an open channel, it is far more difficult to sample the sediment load of a stream. This is due to the nature of the sediment itself, which has been described earlier as suspended-sediment load and bedload (see Section 3.3.1).

Although suspended sediment is easy to describe and to sample, its character changes as velocity increases. At

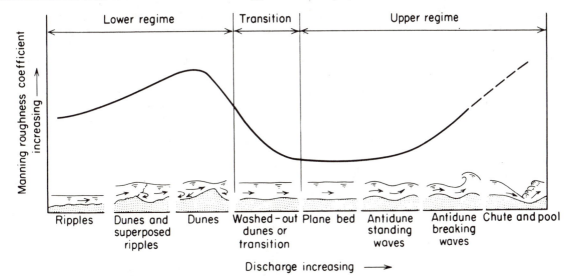

**Figure 12.6** Variations in bedforms and bed roughness as transport of a uniform sandy material increases with discharge. The upper part of the transition regime is associated with a Froude number of unity.
*Source: Coastal Eng. Div., Am. Soc. Civil Eng., vol. 99, pp. 231–43, after Simons et al.*

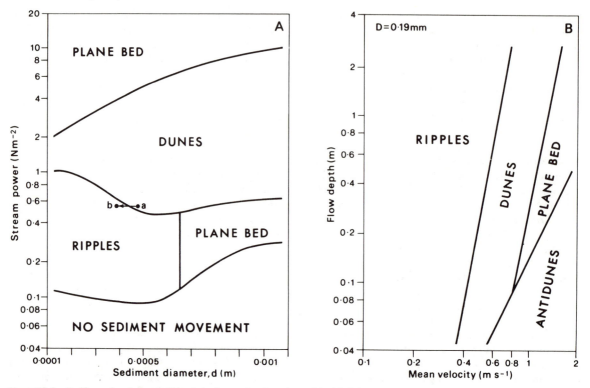

**Figure 12.7** Bedforms in relation to flow states for predominantly sand-bedded rivers: A. related to stream power and sediment diameter; the significance of the transition *a* to *b* is described in the text; B. related to depth and velocity for a uniform grain size of 0·19 mm.
*Source:* Allen, 1976, figure 1, p. 363, after Southard, copyright © John Wiley and Sons, by permission.

low velocity only silt and clay particles will be in suspension, but as the velocity and turbulence increase large particles are at least temporarily suspended. To avoid the problem of identifying large sedimentary particles as suspended load engineers refer to the fine fraction of the sediment load as wash load. Wash load moves through the channel at the velocity of the water, whereas bedload

moves more slowly by saltation (hopping) or, predominantly, by rolling on the bed. Bedload movement is related to velocity near the streambed ($V_b$: see Figure 12.5). The saltating sediments will be sampled with the suspended load, so engineers refer to the coarser sediment sands and gravels as bed-material load because it is the predominant sediment on the bed of the stream.

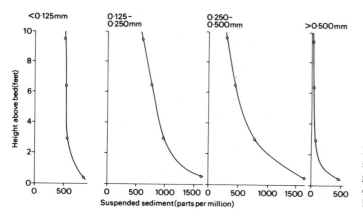

**Figure 12.8** Variations of suspended sediment concentration with depth for four grain-size ranges at a single vertical profile of the Niobrara River, near Cody, Nebraska.

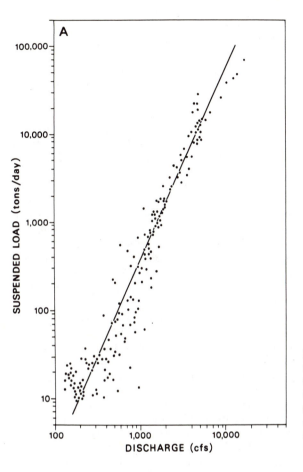

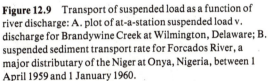

**Figure 12.9** Transport of suspended load as a function of river discharge: A. plot of at-a-station suspended load v. discharge for Brandywine Creek at Wilmington, Delaware; B. suspended sediment transport rate for Forcados River, a major distributary of the Niger at Onya, Nigeria, between 1 April 1959 and 1 January 1960.
*Sources:* Wolman, 1955, figure 21, p. 20; Allen, 1974, figure 14, p. 289, copyright © Elsevier, Amsterdam.

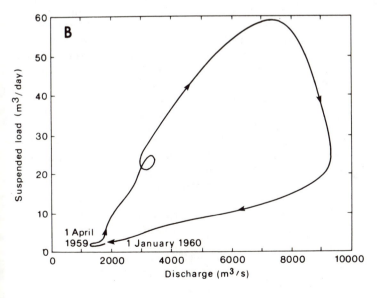

Figure 12.8 shows the vertical distribution of the various types of sediment in a river cross section. Silts and the clays are distributed relatively uniformly through the column of water, whereas the larger and heavier sands tend to be more concentrated near the floor of the channel, as expected. Although it is relatively simple to sample the sediment transported in suspension by taking water samples, it is very difficult to obtain measurements of bedload because any sampler placed on the bed disturbs the flow.

Logarithmic plots of suspended load transported as a function of discharge (Figure 12.9A) are rather misleading in that there are marked timelags between discharge and changes in rates of suspended-load transport and also because, even at a given station with a more or less constant debris size, a given discharge may at different times be associated with rates of suspended sediment transport which may differ by up to one order of

magnitude. These differences depend on the antecedent hydrological conditions, on whether discharge is increasing or decreasing, on state of the banks and on season of the year. Figure 12.9B shows this in detail during a nine-month period for a distributary of the River Niger at Onya, Nigeria. Numerous equations have been developed to permit calculation of sediment discharge in the stream channels. In Figure 12.10 a number of these have been compared with measured sediment transport in the Colorado River, and it is obvious that the results are not satisfactory. Usually the equations tend to be useful only for the channels for which they were developed. When there is a significant difference in the size of sediment transported, or perhaps in other characteristics of the fluid flow, the relationships are not readily applicable to other streams. However, Colby (1964) has shown that the simplest way of estimating transport of sand load through a channel is to express it in terms of the cube of velocity which, as will be recalled, is proportional to stream power.

The movement of pure bedload is much more difficult to estimate than is that of suspended load because attempts to measure it severely interfere with the movement itself and because bed movement is influenced by a complex association of bed variables, including grain size, sorting, shape, roundness and larger-scale bed environment configurations. Besides direct attempts to trap moving bedload in natural rivers, three approaches have been made towards the estimation of bedload movement: laboratory studies, the calculation of theoretical traction relationships and observations of the movement of individual bed particles in natural rivers. Controlled experiments on the critical tractive force required to initiate movement of particles of a given grain size forming part of a bed of similar sizes have produced the curve given in Figure 12.11. Hans Einstein (1950), the son of Albert, calculated a theoretical bedload function expressing bedload movement with respect to various debris sizes applied to Big Sand Creek, Mississippi (Figure 12.12). From a strictly pragmatic standpoint, Laronne and Carson (1976) painted and replaced 4823 pebbles (size range 4–256 mm) at four locations in Seale's Brook, Quebec, in April 1972. After the spring floods, a search for the particles was made in May and only 242 (i.e. some 5 per cent) were recovered. For each of these recovered particles, the size, the distance moved and detailed sedimentary environment of trapping was recorded and plotted (Figure 12.13). Allowing for the poor recovery rate, it seems likely that:

(1) Particle mobility (i.e. distance of transport) is greater for small particles than for larger ones.

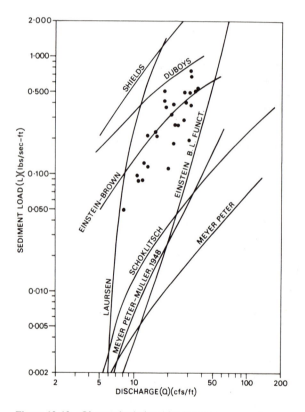

**Figure 12.10** Observed relationships between measured sediment load and discharge (dots) for the Colorado River at Taylor's Ferry (slope = 0·000217 ft/ft; mean sediment diameter = 0·320 mm; temperature = 60°F, 15·5°C). These values are compared with several standard rating curves.
*Source:* Vanoni, 1975, figure 2.113, p. 221.

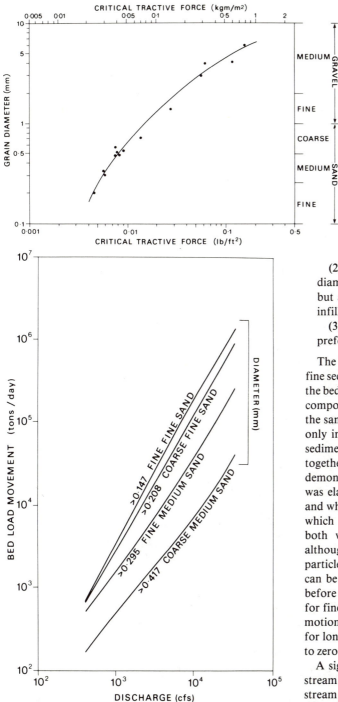

**Figure 12.11** The critical tractive force required to move grains of a given size on a bed of similarly sized material.
*Source:* Nevin, 1946, figure 2, p. 667.

(2) Particles less than 10 gm in weight (average diameter 20 mm) have a more or less equal mobility but are preferentially trapped by open and vertically infilled structures.

(3) Particles larger than 100 gm (i.e. 60 mm) are preferentially trapped by imbricate structures.

The banks of a river are usually composed of relatively fine sediments, such as fine sand, silts and clays, whereas the bed is composed of sands and gravels with some small component of silt and clay. These two types of sediment – the sands and gravels and the silts and clays – differ not only in size, but also in physical properties. The finer sediments tend to be cohesive in that the particles stick together and are difficult to erode. This was first demonstrated by Hjulström in a classic diagram, which was elaborated by Sundborg (1956) (see Figure 12.14), and which relates critical velocity to the sediment size at which erosion of sedimentary particles will begin for both wind and water (i.e. *competence*). Note that although there is a critical velocity required to set particles of a given size in motion, once in motion there can be a reduction in velocity below that critical value before deposition occurs. This is important, particularly for fine sediments, for once the fine sediments are set in motion they continue to be transported in turbulent flow for long distances, and velocity may decrease essentially to zero before the finest sediments begin to be deposited.

A significant characteristic of the sediment found in stream beds is that it normally decreases in size in a downstream direction. This is usually attributed to the wear of the particle as it moves along the stream bed, but of course, finer particles can be more readily transported in the flow and sorting also influences the size decrease. The downstream decrease in size can also be attributed to

**Figure 12.12** Estimated movement of different sand fractions as bedload under varying discharge at Big Sand Creek, Mississippi.
*Source:* Einstein, 1950, figure 23, p. 64.

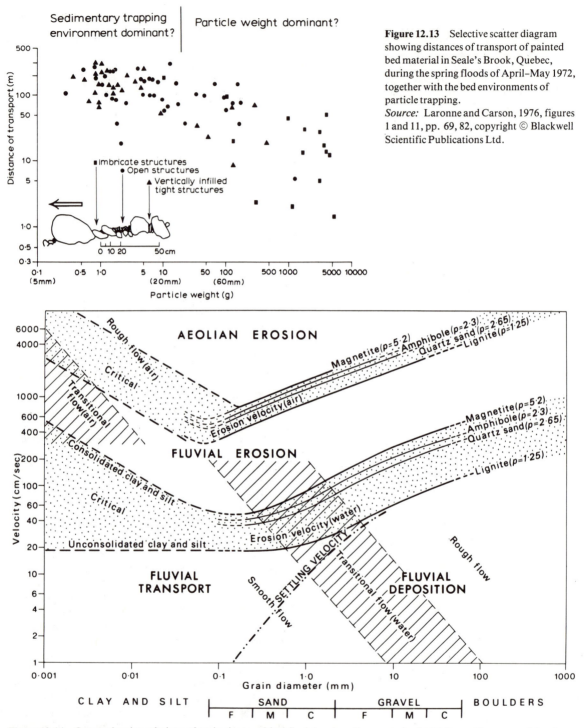

**Figure 12.13** Selective scatter diagram showing distances of transport of painted bed material in Seale's Brook, Quebec, during the spring floods of April–May 1972, together with the bed environments of particle trapping.
*Source:* Laronne and Carson, 1976, figures 1 and 11, pp. 69, 82, copyright © Blackwell Scientific Publications Ltd.

**Figure 12.14** Curves showing relations of grain size to critical fluvial and aeolian erosion velocity for uniform materials of differing densities. The critical fluvial erosion velocity refers to a height of 1 m above the river bed. The two critical zones around these curves and the settling velocity curve for particles in water delimit the four regimes of fluvial deposition, fluvial transport, fluvial erosion and aeolian erosion.
*Source:* Sundborg, 1956, figure 16, p. 197.

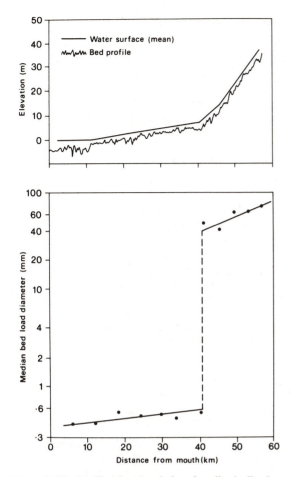

**Figure 12.15** Profile (above) and plot of median bedload diameter against distance from mouth (below) of the Kinu River, Kanto Plain, north of Tokyo.
*Source:* Yatsu, 1959, figures 3 and 7, pp. 225, 230.

weathering of the rock fragments as the sediment is stored in floodplains. Transport of a sedimentary particle can be episodic, and sediment can be stored on the floodplain and in bars in the channel for varying periods of time, where they weather. If the period of storage is sufficiently long, when the rock is reintroduced to the stream channel it may disintegrate to its constituent grains. This sort of dramatic reduction in size may lead to abrupt changes in sediment size in a stream channel, as described by Yatsu (1959) for some Japanese rivers where granule-size particles are missing in the bed (Figure 12.15).

## 12.4 Hydrology

The runoff from a drainage basin reaches a channel and,

together with amount and type of sediment load, this is one of the major variables determining the size and morphologic character of stream channels.

At gauging-stations the discharge is recorded and then analysed to provide information on annual discharge and flood characteristics. Discharges can vary from base-flow, which is the minimum discharge of a perennial stream (i.e. entirely due to the groundwater contribution to the channel flow), to the maximum probable flood. The latter is the largest flood for which there is a reasonable expectancy, and it is geomorphically significant because when it occurs, it should significantly alter channel characteristics. Nevertheless, channel morphology is closely related to average discharge $(\overline{Q})$, mean annual flood $(Q_{2.33})$ and bankfull discharge $(Q_b)$.

Considerable difficulty has been encountered in defining the dominant channel-forming discharge, 'the steady discharge which would form or maintain the average geometric properties of a mobile-bed channel' (Bray, 1975, p. 143), not least because channel-forming or maintaining discharges may not, of their nature, be steady! Definitions include the following:

(1) The discharge which just fills the channel to the level of the floodplain (i.e. the bankfull discharge $Q_b$). This is the most popular definition and is considered to be that discharge which occurs with an average recurrence interval of about 1·5 years. However, this recurrence interval is found to vary from less than 1 year to more than 3 years between different rivers (see Figure 14.13), with a tendency for tropical rivers to have shorter recurrence intervals than those in higher latitudes.

(2) That discharge which appears to correlate best with meander geometry. Suggestions here range from an upper value slightly in excess of $Q_b$ through a range of discharges (possibly occurring during falling stages) between the mean discharge of the month of maximum discharge and the mean annual discharge $(\overline{Q})$ – not to be confused with the mean annual maximum discharge (i.e., $Q_{2.33}$ the mean annual flood which has an average recurrence interval of about 2·33 years).

(3) The discharge at which the maximum sediment load is moved. Benson and Thomas (1966) believe this to be much less than $Q_b$. However, it is clear that the maximum sediment load is transported over a wide range of discharges.

(4) The discharge which occurs some eight to ten times a year in the north-eastern United States and attacks previously wetted and weakened banks.

Despite the problems of definition, it is clear that the hydraulic geometry of alluvial channels is, in general, adjusted to short-term runoff events having a recurrence interval of a few years only in humid regions and 30–100 years in more arid regions. Nevertheless, as Richards (1982, p .123) points out, 'the morphological impact of [discharge] events is partly a queueing problem, with inter-arrival times between peaks being as significant as peak magnitudes'. This is true of all landforms.

Obviously both water and sediment transport are very important for the understanding of channel morphology, and a simple diagram illustrates this fact. Lane's balance (Figure 12.16) shows how channel gradient $S$ is related to sediment load $L$ (i.e. *capacity*) and sediment size $D_{50}$, as well as mean water discharge $\overline{Q}$ as follows:

$$L D_{50} \alpha S \overline{Q} \qquad (12.11)$$

such that slope or gradient is directly related to sediment load and inversely related to discharge, as follows:

$$S \alpha \frac{L D_{50}}{\overline{Q}} . \qquad (12.12)$$

Either aggradation or degradation follows a change in these variables as slope is adjusted to altered conditions.

## 12.5  River morphology

Depending on perspective, a river can be described by many different terms. The layman usually describes it as a good fishing or boating stream, the hydrologist may think of the river in terms of flow characteristics, but for the geomorphologist and hydrologic engineer it is its form aspects, i.e. the channel (hydraulic) geometry and the spatial pattern of the river, that are of interest.

### 12.5.1  Hydraulic geometry

The variables involved in the geometry and dynamics of alluvial channels form a complex system of interacting feedback relationships in which cause and effect (i.e. independent and dependent variables) are difficult to distinguish (Figure 12.17). It is thus, like beaches and sand dunes, a complex process–response system with a short relaxation time (see Chapter 1). Changes in inputs lead to rapid equilibrium adjustments of channel form and dynamics, and evidence for long-term changes of channel form is commonly obliterated by the effects of short-term processes. However, the dynamic response of such a system is often difficult to analyse in that:

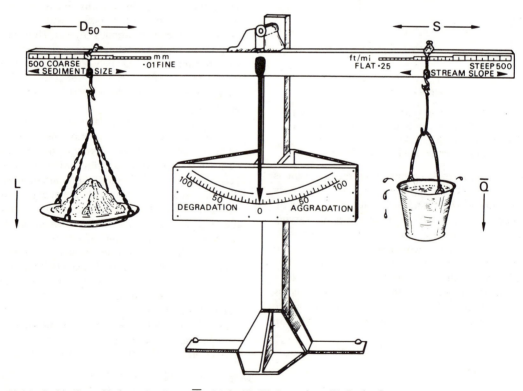

**Figure 12.16**  Stable channel balance $L \times D_{50} \propto \overline{Q} \times S$ (after E. W. Lane, from W. Borland).

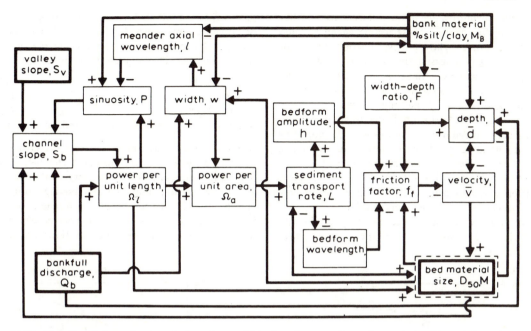

**Figure 12.17** The alluvial channel system. Independent variables have heavy outlines: direct relationships are shown by +, inverse by −; the arrow shows the direction of influence; double-headed arrows indicate reversible relationships.
*Source:* Richards, 1982, figure 1.8, p. 26.

(1) Although major independent variables can be identified (i.e. discharge, bed material size and valley slope, see Figure 12.17), their independence is not total. For example, bankfull discharge is defined by the geometry of the channel itself; bed material size is partly conditioned by the sorting effects of sediment transporting processes; and valley slope is quickly subordinated to the local channel gradients imposed by purely hydraulic and sediment factors.

(2) The detailed response of the system is indeterminate in that its complexity and richness of linkage make it difficult to specify the precise effects of a given change of input. For example, whether an increase in discharge is more easily accommodated by an increase in depth or of width depends on such little-known factors as the detailed stratigraphy of the buried bed material and the hydraulic state of the banks.

(3) Some given relationships are both direct and indirect, depending on their magnitude. For example, sediment transport rate exercises a complex control over bedforms (see Figure 12.6).

(4) Some relationships are completely reversible. For example, an increase in channel friction increases flow depth, but increasing depth 'drowns' channel bed resistance and reduces friction (i.e. there is negative feedback).

In order to establish relations between the dominantly dependent variables of channel morphology and the independent variables that influence the channel, it is important to distinguish between rivers that are stable and reflect these equilibrium controls and rivers that are unstable. Unstable channels are either depositing or eroding in response to some external influence. The stable or graded stream is neither progressively aggrading nor degrading, or changing its cross-sectional area through time. In addition, channels can be categorized by the types of sediments or materials forming their bed and banks. A very youthful stream incising into bedrock is bedrock-controlled. The dimensions of the channel and its morphologic character will be controlled by the nature of the bedrock and by the character of the rocks that are shed into the stream. The bedrock control in some streams can be local, with long reaches not being influenced by bedrock. Many rivers are largely alluvial and have channels whose geometry is entirely conditioned by the movement of water and debris within them. Alluvium is the sediment that is transported by the stream and which forms its bed and banks.

Channel stability is of great concern and, in fact, some of the earliest work associated with channel morphology was performed by engineers designing major irrigation systems in India, Pakistan and Egypt. It was necessary for the movement of water through the canals that they be stable or 'in regime'. These regime channels could aggrade or degrade slightly, but over a period of a year or through an irrigation season, the ideal situation would be for the channel at a given time to be similar in shape and dimensions and position to what it was during the previous irrigation season. Some simple quantitative relationships that related velocity of flow to channel depth and discharge to channel dimensions were developed by these engineers to assist in the design of the stable (regime) canals. Because this work was carried out in a number of locations with different flow and sediment characteristics, the equations tend to be applicable only in the locality for which they were developed.

More recently in the United States Leopold and Maddock (1953) took data from the *US Geological Survey* files and initiated studies of hydrologic geometry of stream channels which relate the dimensions and velocity of streams to at-a-station discharge variations (Figure 12.18A). Their equations are as follows:

$$w = aQ^b \qquad (12.13)$$
$$d = cQ^f \qquad (12.14)$$
$$\bar{v} = kQ^m \qquad (12.15)$$

(since $w\,d\,\bar{v} = Q$; $b + f + m = 1$ and $a\,c\,k = 1$).

It should come as no surprise that with an increase in the quantity of water moving through a channel, width $w$ and depth $d$ increase. This is necessary for the channel to accommodate the quantity of water moving through it. It is also apparent that an increase in discharge will lead both to an increase in depth and to a decrease in channel macroroughness (i.e. resistance to flow: see Section

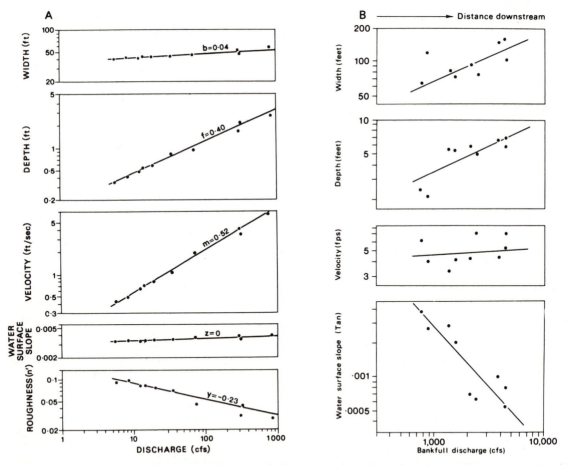

**Figure 12.18**   A. Relation of width, mean depth, mean velocity, water surface slope and roughness parameter ($n'$) to changing discharge at-a-station from Brandywine Creek at Cornog, Pennsylvania. B. Relation of width, mean depth, mean velocity and water surface slope to bankfull discharge along a 40-km (25-mile) stretch of Brandywine Creek, Pennsylvania.
*Source:* Wolman, 1955, figures 10 and 29, pp. 11, 31.

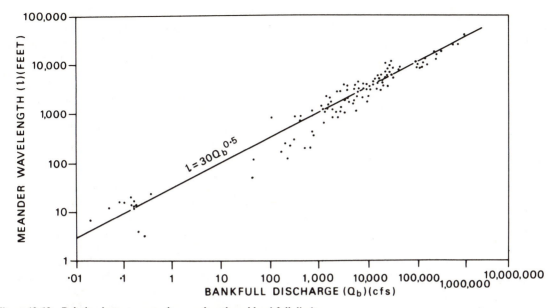

**Figure 12.19** Relation between meander wavelength and bankfull discharge.
*Source:* Dury, 1965, figure 5, p. 5.

12.2), both of which will increase at-a-station velocity, as the Manning equation demonstrates. What is more surprising perhaps is the fact that the average velocity $\bar{v}$ increases in a downstream direction with bankfull discharge (Figure 12.18B). This is because the greater the depth of water, then the smaller will be the frictional forces acting on the total flow and, in spite of a decrease in gradient in a downstream direction (mostly due to decreased bed calibre), average velocity will increase. One must be careful, however, in making the ergodic leap from the hydraulic variations associated with at-a-station changes in discharge (Figure 12.18A) *through time* to those changes *in space* resulting from differences in bankfull discharge at different stations along a river (Figure 12.18B). The essential difference between these lies in the downstream changes in the nature of bed material, and this explains the greater scatter of points about the regressions on the downstream graphs compared with those at-a-station.

It should be noted that Leopold and Maddock (1953) used the mean annual discharge in their downstream relationships because this could be readily obtained from the available records but, as we have seen, it is more likely that a discharge close to bankfull is the channel-forming discharge. Nevertheless, there is usually a good statistical relationship between mean annual discharge $\overline{Q}$, bankfull discharge $Q_b$ and mean annual flood $Q_{2.33}$, so that almost any of these hydrologic variables can be used in an equation which relates channel dimensions to discharge.

Not only are channel dimensions related to the quantity of water moving through the channel, but the dimensions of the stream pattern also have a similar relationship to discharge. The meander pattern itself, the amplitude of the meanders, meander wavelength and the width of the meander belt tends to be related to a quantity of water moving through the valley (Figure 12.19). These relationships, although highly significant, have a considerable standard error. That is for given discharge, the dependent variable ranges over a log cycle and there is an order of magnitude range of channel dimensions for a given discharge.

Discharge, of course, is only one of the two main independent variables influencing the stream channel. The other is the sediment load. It will be recalled that there are two types of sediment load, suspended-sediment load or wash load and bed-material load. Suspended sediment transport is dependent on the turbulence and velocity of the flowing water. For a given slope, the turbulence and velocity will be greater in a deeper channel. On the other hand, bed-material load is moved by shear on the bottom of the stream. Thus the most efficient channel for the transport of a suspended load is one that is relatively narrow and deep, whereas the channel required to move bedload with the same quantity of water will be wide and shallow. It seems likely, therefore, that in nature streams transporting large quantities of bed-material load will have a relatively large width–depth ratio, whereas those transporting a small

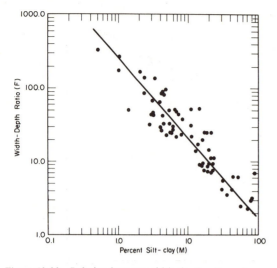

**Figure 12.20**   Relation between width–depth ratio and percentage of silt and clay in channel perimeter for stable alluvial streams.
*Source:* Schumm, 1960, figure 8, p. 21.

quantity of bed-material load will have a relatively small width–depth ratio. Figure 12.20 shows the relationship between width–depth ratio and percentage of silt and clay exposed in the perimeter of stream-channels. The bed and the banks of the channel are composed of the material transported by the stream and the percentage of silt–clay in the channel thus reflects the nature of the sediment load moved through that channel. That is if the percentage of silt–clay is large, the percentage of sand and gravel is small, and the channel will be narrow and deep because it will not be transporting large quantities of sand and gravel. On the other hand, when the percentage

of sand in the perimeter of the channel is large, this reflects a large sand and gravel bedload and the channel tends to be wide and shallow. Figure 12.21 shows the variation in channel width and depth in a downstream direction along the Smoky Hill–Kansas River system where tributaries introduce greatly different types of sediment into the channel, and instead of a progressive increase in width and depth in a downstream direction as is predicted by hydraulic geometry relationships (see Figure 12.18B), there is a great variation in the manner in which width and depth change with an increase in discharge. This clearly is an indication of the importance of sediment load on channel morphology (Plate 13).

Not only does sediment load affect channel dimensions and shape, but it influences stream gradient or bedslope. The larger the sediment load and the larger the size of sediment particles that must be transported through a stream, the steeper is the gradient of the channel required to transport this material. Unfortunately for the geomorphologist, the need to relate sediment size to the quantity of water moving through the channel requires a large number of stream-gauging stations which is rarely available. Hack (1957) circumvented this problem by using a general relationship in humid regions between drainage area and stream discharge. He developed an equation between median bed sediment size $D_{50}$, drainage area $A$ and stream-bed slope $S_b$ as follows:

$$S_b = \frac{18\,(D_{50})^{0.6}}{A}. \tag{12.16}$$

In Canada Bray (1973) found that there was a clear relationship between stream-bed slope, discharge $Q$ and sediment size for rivers in Alberta as follows:

$$S_b = 0.965 Q^{-0.334}(D_{50})^{0.586}. \tag{12.17}$$

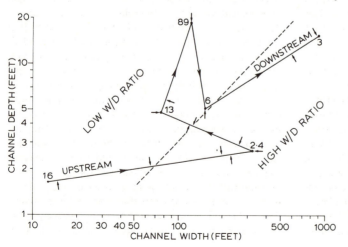

**Figure 12.21**   Relation of channel width **w** to depth **d** for the Smoky Hill–Kansas River. Numbers indicate the silt-clay percentage of the channel (bed and banks), arrows show the tributary entrance points and the broken line shows a constant width–depth ratio.
*Source:* Schumm, 1960, figure 14, p. 27.

It is apparent that gradient generally decreases with discharge and increases with sediment size and quantity of sediment. Therefore, in a downstream direction as $Q$ increases and $D_{50}$ decreases, $S_b$ will decrease and the normal longitudinal profile of a river is concave-up.

However, something of the complexity of the above relationships is illustrated by observations by Brush (1961) of stream channels in central Pennsylvania developed on sandstone, shale and limestone, respectively. On all three rock types the channel bedslopes decrease with distance downstream (Figure 12.22B) but with considerable scatter and rather poor correlation between the two. Even poorer correlations were obtained between mean bed particle size and distance downstream, and peculiarities of debris supply caused the size of limestone particles to show a downstream *increase* in size (Figure 12.22A). Nevertheless, the idealized longitudinal profiles were broadly concave-up, with sandstone channels being steepest and limestone least steep (Figure 12.22C).

## 12.5.2 Channel patterns

Although rivers are usually described as being straight, meandering, or braided, there is in fact a great range of channel patterns from straight through meandering to braided and anabranching or anastomosing (Figure 12.23). Straight and meandering channels are described by sinuosity $P$ which is the ratio of channel length $L_c$ to valley length $L_v$ or the ratio of valley slope $S_v$ to channel gradient $S_c$, as measured over the same length of valley:

$$P = \frac{L_c}{L_v} = \frac{S_v}{S_c}. \qquad (12.18)$$

In addition to sinuosity variation, channel width and meander-belt width variability, the meander pattern may be bimodal in that two meander patterns of different dimensions may be superimposed (see Figure 12.23).

The braided stream channel contains bars and islands, and the degree of braiding can be expressed by the percentage of reach length that is divided by one or more islands or bars. The anabranching channel, according to Brice (1975), is divided by islands that are three times the width of water at bankfull discharge. The degree of anabranching is the percentage of reach length that is occupied by large islands. Anabranched channels merge into the island-braided patterns.

Anastomosing channels are distinct from the anabranched channels, as they have major distributaries that branch and rejoin the main channel (Schumm, 1977). The individual branches of anabranching and anas-

tomosing channels can be meandering, straight, or braided and, therefore, they will not be considered separately from the three basic patterns, meandering, braided and straight.

Of major interest to both geomorphologists and engineers is the explanation of these patterns, and the question of why rivers meander has been one that has been reviewed from many different perspectives. The most coherent hydraulically based theory of meander development has been developed by Yalin and elaborated by Richards (1982). Above certain discharge thresholds systematic downstream variations in velocity occur which are associated with the formation of pools and riffles each of which are successively spaced at a distance of $2\pi\bar{w}$ (i.e. 5–7$\bar{w}$) (Figure 12.24B). At low, non-channel-forming flows the bed topography influences flow geometry, whereas at high, channel-forming discharges the reverse is true. At low discharges riffles form relatively steep, wide and shallow reaches with relatively steep water surface slopes, high velocities and coarse debris (although riffles can form in well-sorted sandy streams). Under these conditions pools are relatively flat, narrow and deep with relatively low water surface slopes, low velocities and fine debris. At channel-forming discharges of approaching bankfull and greater the flow in straight channels takes the form of roller eddies associated with reaches of accelerated and decelerated flow spaced at about $2\pi$ times the mean stream width (Figure 12.24A). Under these conditions the faster flow erodes the pools and the slower causes deposition on the riffles, which latter is also assisted by the trapping effect of the rougher particles there. Flowlines tend to converge in plan view over pools and to diverge over riffles. Relative scour acts to enlarge the pool cross sections and thus decreases average velocity there, whereas the relative riffle deposition decreases those cross sections and increases average velocity. This action accentuates the production of an undulating channel bed along which the systematic velocity variations at high discharges are minimized by variations in cross section. However, lateral oscillations of flow also develop which increase stream sinuosity as a result of diversions occurring from a uniform flow direction caused by sedimentological and topographical constraints which a river is unable to accommodate in any other way (Richards, 1982, p. 183). As Langbein and Leopold have suggested, the sedimentary character of pools causes them to possess less microroughness than the riffles, and an effect of increasing sinuosity is for bends to develop preferentially over pools and for straight reaches (i.e. crossings) to locate over riffles (Figure 12.24C). This also has the effect of evening out the continuity of channel roughness by relatively

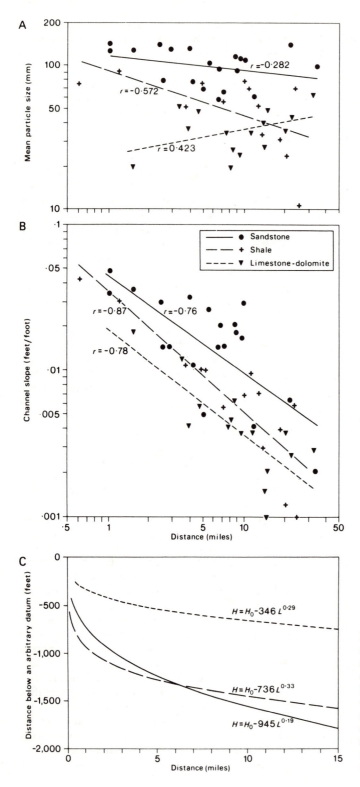

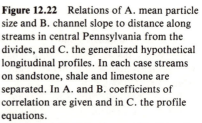

**Figure 12.22** Relations of A. mean particle size and B. channel slope to distance along streams in central Pennsylvania from the divides, and C. the generalized hypothetical longitudinal profiles. In each case streams on sandstone, shale and limestone are separated. In A. and B. coefficients of correlation are given and in C. the profile equations.

*Source:* Brush, 1961, figures 106, 104 and 105, pp. 167, 164, 166.

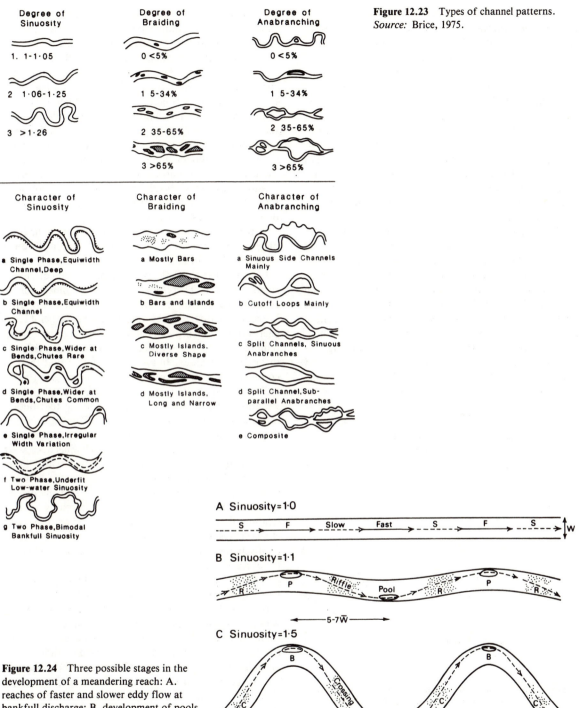

**Degree of Sinuosity**

1. 1-1.05
2. 1.06-1.25
3. >1.26

**Degree of Braiding**

0 <5%
1 5-34%
2 35-65%
3 >65%

**Degree of Anabranching**

0 <5%
1 5-34%
2 35-65%
3 >65%

**Figure 12.23** Types of channel patterns.
*Source:* Brice, 1975.

**Character of Sinuosity**

a Single Phase, Equiwidth Channel, Deep
b Single Phase, Equiwidth Channel
c Single Phase, Wider at Bends, Chutes Rare
d Single Phase, Wider at Bends, Chutes Common
e Single Phase, Irregular Width Variation
f Two Phase, Underfit Low-water Sinuosity
g Two Phase, Bimodal Bankfull Sinuosity

**Character of Braiding**

a Mostly Bars
b Bars and Islands
c Mostly Islands, Diverse Shape
d Mostly Islands, Long and Narrow

**Character of Anabranching**

a Sinuous Side Channels Mainly
b Cutoff Loops Mainly
c Split Channels, Sinuous Anabranches
d Split Channel, Sub-parallel Anabranches
e Composite

A Sinuosity = 1.0

B Sinuosity = 1.1

5-7$\overline{w}$

C Sinuosity = 1.5

10-14$\overline{W}$

**Figure 12.24** Three possible stages in the development of a meandering reach: A. reaches of faster and slower eddy flow at bankfull discharge; B. development of pools and riffles with a spacing of 5–7 $\overline{w}$; C. development of meanders with a wavelength of 10–14$\overline{w}$.
*Source:* Richards, 1982, figure 7.8d, p. 201.

increasing macroroughness in the pool reaches. It is significant that whereas water surface slope ($S_w$: a rough surrogate for the stream energy gradient $S_e$ in the Manning equation) is normally steeper over the riffles and flatter over the pools, at above $\frac{3}{4}Q_b$ Leopold noted that these differences are 'drowned out' in meandering rivers. This does not occur for straight rivers with pools and riffles which are, therefore, regarded as being inherently unstable in that there are uncompensated variations of energy dissipation along the flow direction. It should be stressed that the above explanation of the formation of pools and riffles and of their links with meander generation is largely theoretical and speculative, but it does accord with many field observations and forms a coherent whole.

The oscillating transport of bedload during meander development has been demonstrated by Ackers and Charlton (1970) in an experimental channel (Figure 12.25). With the constant conditions of flume discharge, debris calibre and rate of sediment inlet feed, the initial flume gradient (0·0020) was too small to transport the load and the channel was steepened by bed deposition beginning near the inlet. This caused an increase in bedload transport and after some 200 h meandering was initiated. Thereafter until 430 h the material moved along the channel in surges as the water surface slope increased and the meanders developed. After 430 h the

meanders were fully developed and the water surface slope stabilized in equilibrium at 0·00266 (Plate 14).

Field observations by Lewin (1976) of bars developing in an artificially straightened reach of the River Ystwyth in Mid-Wales over the period 1969–70 have provided further clues to the links between pools, riffles, point bars and meanders (Figure 12.26). The high discharges quickly produced mid-channel bars on one side of which the flow became dominant leading to a meander bend and bank erosion, and on the other deposition was dominant forming a chute. These primary bars became the cores of point-bar complexes which grew laterally towards the meander bends and downstream. Pools were located at the bends and riffles at the crossings at the downstream ends of the point-bar complexes.

Experimental studies have demonstrated that the complete range of river patterns from straight through meandering to braided is dependent on stream power ($\tau V$) which, in turn, reflects sediment load and discharge (Figure 12.27). Therefore, it is not so much simple fluctuations in flow or disturbances to the flow that influence the final equilibrium pattern, but rather it is the energy of the flowing water and the work that the water is required to do to transport the water and sediment through the channel. In addition, the fact that there is considerable pattern variability along a given river indicates that there is another explanation for pattern variations. Pattern differences can occur due to tributary influences, and the introduction of different types of sediment load into the channel may cause a relatively low-sinuosity, high width–depth ratio channel to convert to a relatively narrower deeper channel with higher sinuosity. Figure 12.28 shows the relationship between width–depth ratio and the sinuosity of streams on the Great Plains. Although there are exceptions to this relation, it suggests how channel slope and pattern could change in response to changes of sediment load, as well as to discharge changes resulting from climate change, tectonics, or man.

A factor influencing the variability of stream patterns is the Quaternary and Holocene history that produces variations of valley-floor slope. The slope of a valley can vary as a result of neo-tectonic deformation or as the result of large quantities of sediment introduced from tributaries. In order to maintain a constant gradient at a given location in the face of valley steepening, sinuosity will increase. Pattern changes can compensate for slope changes, as equation (12.18) indicates. In spite of the further variability caused by cutoffs and bedrock control, this effect can be demonstrated for the Mississippi River (Figure 12.29) where experimental work and the field study of patterns have identified two thresholds that

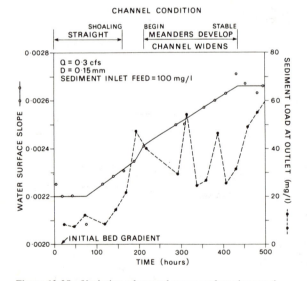

**Figure 12.25**  Variation of general water surface slope and sediment load at the outlet for a flume experiment with constant values of discharge $Q$, sediment size $D$, and sediment inlet feed rate.
*Source:* Yang, 1971, figure 6, p. 244, after Ackers and Charlton.

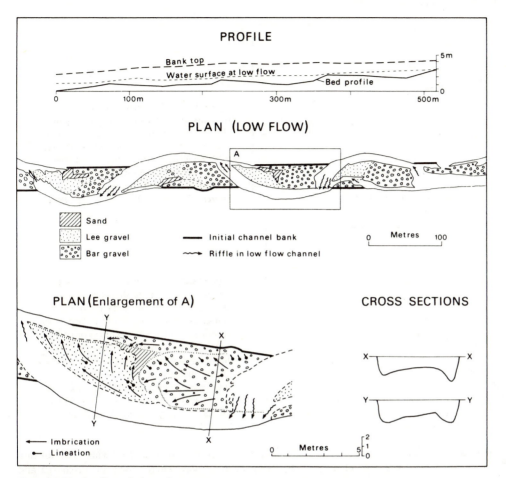

**Figure 12.26** Profile and plans of a reach of the River Ystwyth, Mid-Wales, showing a series of three bars which developed during the period 1969–70 after the channel had been artificially straightened.
*Source:* Lewin, 1976, figure 3, p. 283.

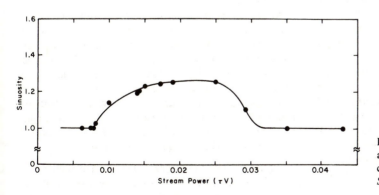

**Figure 12.27** Relation between sinuosity and stream power derived for experimental channels developed in a large flume.
*Source:* Khan, 1971.

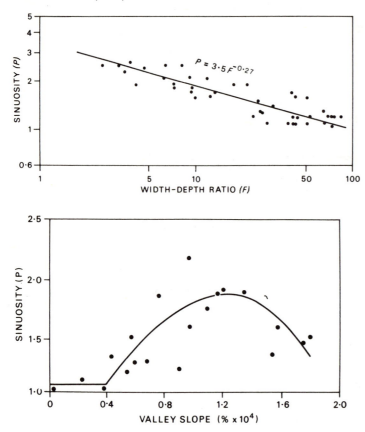

**Figure 12.28** The relation between sinuosity and width–depth ratio for some alluvial rivers on the Great Plains. This relationship does not appear to be affected by differences in mean annual discharge. The standard error is 0·064 log units and the coefficient of correlation is − 0·89.
*Source:* Schumm, 1963, figure 2, p. 1092.

**Figure 12.29** Relation between sinuosity and valley slope for the Mississippi River between Cairo, Illinois, and Head of Passes, Louisiana. Data from 1911–15 surveys which were carried out before modification of the channel patterns by artificial cutoffs. Two thresholds at valley slopes of 0·4 and 1·9 (‰ × $10^4$) are apparent.
*Source:* Schumm *et al.*, 1972, figure 2, p. 76. Reprinted by permission from *Nature*, vol. 237, copyright © 1972 Macmillan Journals Limited.

separate straight from meandering and braided patterns. This suggests that, in addition to changes of channel shape and dimensions, abrupt pattern changes can occur in response to small changes of sediment load or flood magnitude.

There is no one single explanation for the mechanisms of meandering. Certainly, disturbances and flow instabilities are required to set the meander patterns in motion, but these disturbances can only be effective at critical values of stream power (i.e. at certain stream gradients and velocities) and in the presence of critical amounts and sizes of available bedload.

Despite the problems of understanding the mechanisms of meander formation, it is clear that equilibrium (stable) meanders present predictable geometrical and dynamic characteristics. Figure 12.19 has suggested a relationship between meander wavelength $\ell$ and bankfull discharge. An observed relationship between meander wavelength and mean stream width has given support to the riffle theory of meander generation (see Figure 12.24):

$$\ell = 12.34\overline{w} \text{ (i.e. } \ell \approx 2 \text{ times } 2\pi\overline{w}). \tag{12.19}$$

Schumm has proposed the following relationship between meander wavelength, bankfull discharge and the silt–clay index $M$:

$$\ell = 618\,Q_b^{0.43}M^{-0.74}. \tag{12.20}$$

Later in this chapter the interrelationships between the variables involved in the meandering process are discussed in detail.

At this point it is instructive to examine Jackson's (1975) detailed survey of a Wabash River meander, which was conducted under low flow, near bankfull and overbank stages (mean depths 0·5 m, 4·3 m and 7 m, respectively) (Figure 12.30). Jackson surveyed channel cross sections, near-surface and near-bottom velocity vectors, active bedforms associated with increasing bedload transport (i.e. plane bed to active dunes > 50 cm high), and inactive dunes indicative of a very recent decrease of tractive force. These unique observations suggest:

(1) that flow conditions and bedload movement are most symmetrically organized at near-bankfull discharges, and that *overall* debris movement along the channel is probably at a maximum then;

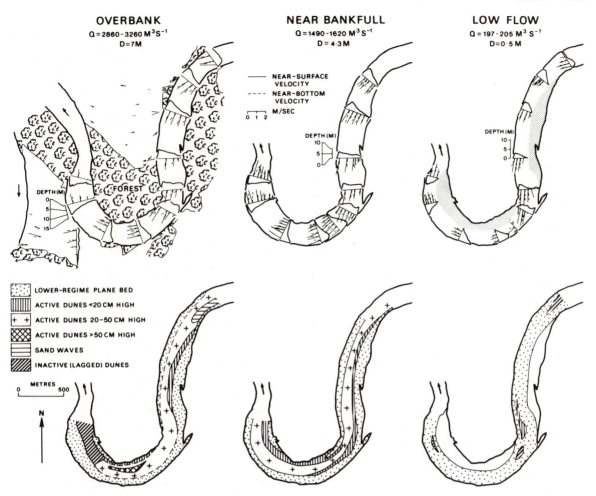

**Figure 12.30** Cross sections, velocity vectors and bed conditions for the Wabash River, near Grayville, Illinois, for low flow, bankfull, and overbank stages: **Q** = discharge (cubic metres per second); **D** = mean depth (in metres).
*Source:* Jackson, 1975, figures 5, 7, 10, 11, 13 and 14, pp. 1514, 1515, 1517, 1518.

(2) that despite the plan-view regularity, there are pronounced contrasts between the upstream (entry) and downstream (exit) limbs of the meander, particularly at bankfull discharge:

(a) the entry limb appears transitional, within which the thread of maximum velocity shifts from one bank to the other, cross sections are rather irregular, current velocities are strongest, and dunes and sand waves are most prominently developed;

(b) the exit limb shows the thread of maximum velocity against the left bank, the normal cross-sectional asymmetry, and pronounced helicoidal flow (i.e. a corkscrew effect with surface water moving towards the left bank and bottom water to the right).

The braided pattern can occur in two ways. When there is active aggradation, bedload deposits as bars, the flow is divided and the characteristic braided pattern develops. The other type of braided channel is associated with a steep-gradient channel where high stream power leads to the transport of large quantities of bedload (see Figure 12.27). In order to move this material it is necessary to have a wide, shallow steep-gradient channel with maximum shear forces acting on the bed. The braided channel develops on steep slopes.

Of considerable interest is the change of channel pattern through time (see Figures 2.14 and 2.15). Not only do bends grow and shift, but they can cutoff (Figure 12.31). This is of major concern to people who work with rivers and particularly, for example, highway engineers who construct bridges on meandering streams only to find them abandoned when cutoffs occur, or destroyed as the meander bends shift and migrate in a downstream direction. Perhaps the best example of the behaviour of a large meandering stream is the channel pattern changes along the Mississippi River which have been so beautifully documented by Fisk (1944, 1947) (see Figure 2.14A).

In addition to meander shift, there can be shifts of sand and gravel bars in channel; avulsive changes which include neck cutoffs, chute cutoffs and even total abandonment of the existing channel; and true avulsion (Figure 12.31). Of interest is the fact that if there is a political or administrative boundary line associated with the stream, the former is often held to move with the channel if the channel shifts slowly due to a process of bank erosion and lateral accretion. However, the boundary will often be held to remain fixed if the change is abrupt and avulsive in nature, for example, when a neck or chute cutoff occurs! This, of course, leads to disputes concerning the actual location of the boundary. A recent example of this involved the international boundary between the United States and Mexico. Meander cutoffs by the Rio Grande left meander cores both to the north and south of the river, for example, in Texas an old meander cutoff, which was Mexican territory but which was now separated from Mexico by the river, was occupied by Texans. Elsewhere along the river the reverse circumstance occurred, and it was only recently that a new treaty was signed which resolved this problem by awarding the cutoffs to the south to Mexico and those to the north to the United States.

## 12.6 Channel stability

The alluvial channel, because it is formed in erodible

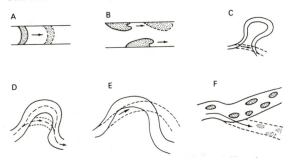

**Figure 12.31** Types of channel changes: A. transverse bar shift; B. alternate bar shift; C. neck cutoff; D. chute cutoff; E. meander shift; F. avulsion. Full lines indicate present status and broken lines the future potential changes. *Source:* Shen and Schumm, 1981, figure 5.

sediments and because the stress exerted by the flowing water often exceeds the strength of the sediment forming the bed and banks of the channel, will change naturally with time. In Figure 12.31 six types of natural channel changes are illustrated. Examples A and B are within-channel shifts of bars and islands. In case B alternate bars shift slowly downstream, and as they do the thalweg shifts position. Hence at one time a bank location is protected by the alternate bar, whereas at some later time the bar has shifted downstream, and the bank can be subjected to erosion. Examples C, D, E and F show a change of channel position gradually by meander shift or relatively quickly by neck or chute cutoffs or by avulsion (i.e. rapid channel diversion). In addition to meander shift (Figure 12.31E), meanders may also increase in amplitude. This is accomplished by erosion of the cutbank, or the outside of the bend, while sediment is deposited on the inside bank or point bar.

The continual reworking of sediment demonstrates that the floodplain of a meandering stream is not a permanent feature, but it is composed of material in temporary storage on its journey downstream. Rates at which the meanders migrate and rework the floodplain depend on many factors. Hickin and Nanson (1975) list discharge, water-surface slope, character of the boundary material, height of the concave bank, bank vegetation and the ratio of radius of channel curvature to channel width as important in controlling rates of meander migration. Sediment supply to the bend is also important.

Meander migration is inherent to sinuous rivers, and over a period of years a river will sweep back and forth across its floodplain, reworking floodplain deposits, destroying all surface features and replacing them with point-bar deposits. Hickin and Nanson (1975) report an average rate of migration of the Beatton River in British Columbia at 0·5 m per year, during approximately 250 years. The maximum rate during this time was 0·7 m per year. Nadler reported that the channel of the Arkansas River at Bent's Old Fort migrated at a rate of 8·0 m per year during 43 years. In addition to, and in part as a result of, meander migration, cutoffs also modify meandering channels. Meander cutoffs commonly occur in two different ways, by neck cutoffs and by chute cutoffs (Figure 12.31C and D). Both processes shorten the channel, thereby increasing local slope and causing increased scour upstream and deposition downstream.

Many rivers in the Great Plains region do not exhibit a meandering pattern, although their channels may be slightly sinuous. Straight channels are normally relatively stable, but braided channels (Figure 12.31F) continually change their appearance as bars and islands shift, disappear, or form during high flows. As bars grow flow is deflected towards the banks, eroding them and introducing coarse material into the channel. Therefore, the channel changes its appearance from one flood to the next as bars and islands shift, and the floodplain is eroded by the channel shift. However, a reduction of peak flow, which may be accompanied by increased base flow, allows vegetation to establish itself on areas previously inundated by floods. This vegetation can stabilize the banks, thereby cutting off much of the coarse sediment which feeds the growth of bars and islands.

### 12.6.1 Stable channels

Although bedrock and resistant alluvium can locally dominate channel morphology, nevertheless, a classification of alluvial channel patterns can be developed that, in addition, relates channel pattern to the relative stab-

ility of the channel. For simplicity and convenience of discussion, the range of channel patterns can be illustrated by only five patterns (Figure 12.32). These five patterns illustrate the overall range of channel pattern to be expected in nature, but of course they do not show the detailed differences within a pattern (Mollard, 1973). Nevertheless, Figure 12.32 is more meaningful than a purely descriptive classification of channels because it is based on cause and effect relations, and it illustrates the differences to be expected when the type of sediment load, flow velocity and stream power differ among rivers. It also explains why there are pattern differences along the same river (Schumm, 1977).

A classification of alluvial channels should be based not only on channel pattern, but also on the variables that influence channel morphology. This is particularly true if the classification is to provide information on channel stability. Numerous empirical relations demonstrate that channel dimensions are due largely to water

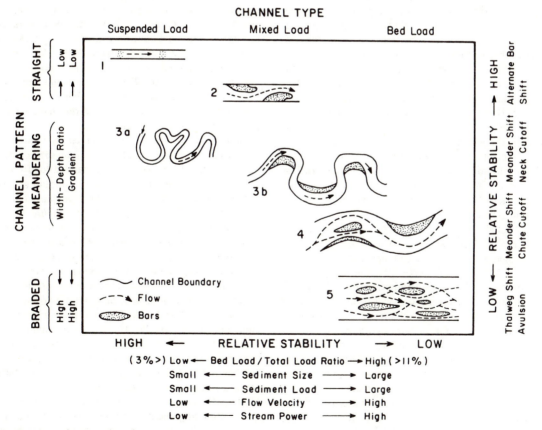

**Figure 12.32**  Channel classification based on pattern and type of sediment load with associated variables and relative stability indicated.
*Source:* Schumm, 1981, figure 4, p. 24.

discharge, whereas channel shape and pattern are related to the type and amount of sediment load moved through the channel. In nature there are large and small channels of each type illustrated in Figures 12.20 and 12.32, and in each case the large channel forms in response to a large water discharge, but the pattern itself and the shape of the channel depends on the proportion of the total sediment load (silt, clay, sand, gravel) that is bedload or bed-material load (sand and gravel). Geomorphic history is also important because it can determine the slope of the valley floor or alluvial plain upon which the stream flows. Some very straight rivers (the Illinois and the Mississippi below New Orleans) are flowing on alluvial surfaces that are relatively flat. The most sinuous reach of the Mississippi (Greenville Bends), before channel straightening, was localized on the steepest part of the valley floor below the confluence with the Arkansas River (Schumm, 1977).

When the proportion of bedload in a channel is small, the channel is a narrow and deep suspended-load channel (width–depth ratio less than 10). Depending on valley slope, the channel can be straight (Figure 12.32: pattern 1), or have a high sinuosity (pattern 3a). When the percentage of bedload is intermediate, the mixed-load channel has a lower width–depth ratio (between 10 and 40) and sinuosity is between about 2·0 and 1·3 (pattern 3b). This channel may also be relatively straight, but the thalweg or the deepest part of the straight channel may be sinuous (pattern 2). As the proportion of bedload increases, the width–depth ratio of a bedload channel increases (greater than 40), and sinuosity is low. There is a tendency for multiple thalwegs to form (pattern 4). The greatest development of channels and bars occurs in the braided channel (pattern 5) when the ratio of bedload to total load is high.

As indicated in Figure 12.32, not only does the channel pattern change from pattern 1 to pattern 5, but other morphologic aspects of the channel also change; for example, for a given discharge, gradient increases and width–depth ratio increases. In addition, peak discharge, sediment size and sediment load will probably increase from pattern 1 to pattern 5. Naturally with such geomorphic and hydrologic changes, hydraulic differences can be expected, and flow velocity, tractive force and stream power increase from pattern 1 to pattern 5. Therefore, channel stability decreases from pattern 1 to pattern 5, with patterns 4 and 5 being the least stable.

In nature there is a continuum of patterns between patterns 3 and 4 of decreasing sinuosity, increasing gradient and width–depth ratio, and decreasing bank and channel stability. A summary of the five basic patterns of Figure 12.32 is presented as a means of reviewing channel character as well as the nature of channel change and stability.

*Pattern 1*   The suspended-load channel is straight with relatively uniform width. It carries a very small load of sand and gravel. Gradients are low and the channel is relatively narrow and deep (low width–depth ratio). Banks will be relatively stable because of their high silt–clay content. Therefore, the channel will not be characterized by serious bank erosion or channel shift. Bars may migrate through the channel (Figure 12.31A), but this should not create undue instability. Pattern 1 channels are rare, but stable unless the straight channel has been artificially straightened and, therefore, steepened. A naturally straight channel will pose few problems, but an artificially straightened channel will be subject to degradation and scour, bank erosion and an increase of sinuosity.

*Pattern 2*   The mixed-load straight channel has a sinuous thalweg. It is relatively stable, but carries a small load of coarse sediment, which may move through the channel as alternate bars (Figure 12.31B). As these bars shift through the channel banks are alternately attacked and protected by the alternate bars. Hence at any one location the thalweg will shift with time. This means that apparent deposition or fill at one side of the channel will be replaced by scour as an alternate bar migrates downstream. Also at any time one side of the channel may be filling, while the other is scouring.

*Pattern 3*   This pattern is represented by two channel patterns that are only two of a continuum of meandering patterns. Pattern 3a shows a suspended-load channel that is very sinuous. It carries a small amount of coarse sediment. Channel width is roughly equal and the banks are stable, but meanders will cut off (Figure 12.31C). Pattern 3b shows a less stable type of meandering stream. Mixed-load channels with high bed loads and banks that contain low cohesive sediment will be less stable than the suspended-load channels. The sediment load is large and coarse sediment is a significant part of the total load. The channel is wider at bends, and point bars are large. Meander growth and shift (Figure 12.31E) and neck and chute cutoffs (Figure 12.31C and D) are characteristic. The channel, therefore, is relatively unstable but the location of the cutoffs and the pattern of meander shift can be predicted. The shifting of the banks and thalweg follows a more or less regular pattern.

The shift of a meander creates major channel problems as the flow alignment is drastically altered, and bank erosion may become very serious. The rate of a meander shift will vary greatly depending on where in the continuum of meandering patterns the river fits.

*Pattern 4* This pattern represents a meander-braided transition. Sediment loads are large, and sand, gravel and cobbles are a significant fraction of the sediment load. Channel width is variable. The channel is relatively wide and shallow (high width–depth ratio) and the gradient will be steep. Chute cutoffs and thalweg and meander shift and bank erosion are characteristic (Figure 12.31D and E). In addition to these problems, which are also characteristic of pattern 4, the development of bars and islands may modify flow alignments and change the location of bank erosion.

*Pattern 5* This bedload channel is a typical bar-braided stream. The bars and thalweg shift within the channel, and it is unstable. The sediment load and size are large. Braided streams are frequently located on alluvial plains and alluvial fans. Their steep gradients reflect a large and/or coarse sediment load. Bank sediments are easily eroded, gravel bars and islands form and migrate through the channel, and avulsion (Figure 12.3F) may be common.

The other type of braided stream is the island-braided stream. This is a much more stable channel, and it would appear to the left of pattern 5 on Figure 12.32. The Mississippi River above the junction of the Missouri River is of this type. Island formation, erosion and shift occur in these channels, but at a much slower rate than in a bar-braided channel.

The island-braided channel is not shown in Figure 12.32 because it may be a channel in transition from a bar-braided pattern to either a straight or meandering pattern (patterns 2 or 3b). The typically bar-braided South Platte, Platte and Arkansas rivers have, during the past century, changed to island-braided and to either straight, anabranching, or meandering patterns as a result of decreased discharge and flood peaks.

### 12.6.2 Unstable channels

The preceding discussion relates entirely to stable alluvial channels; that is the changes indicated for each channel type are typical and can be expected under all conditions. However, when the sediment load or discharge transmitted by the channels is altered, they respond by either eroding or depositing sediment. The channels become unstable, but because the channels are composed of sediments of different degrees of stability and because the manner of erosion, deposition and transport are different, the response of the channels can be different.

In contrast to the channels that are stable or 'graded', there are unstable channels that are usually responding to a change in an external variable, such as sediment load or discharge. During the major climatic changes of the

Quaternary rivers responded in dramatic fashion to changes in sediment load and discharge, as well as to changes in baselevel. A simple way of comprehending the effects of changing discharge and sediment load and sediment size on a stream channel is conveyed by the cartoon presented as Figure 12.16. However, it does not include all aspects of channel morphology and, based on empirical relations developed for stable channels, two simple equations can be presented which show the effect of water discharge $Q$ on channel width $w$, depth $d$, meander wavelength $\ell$ and slope $S$; and the effect of sediment load $L$ on these variables, as well as width–depth ratio $F$ and sinuosity $P$ as follows:

$$Q = \frac{w\,d\,\ell}{S}. \tag{12.21}$$

$$L = \frac{w\,\ell\,S\,F}{d\,P}. \tag{12.22}$$

$L$ relates to the bedload transport and, for the purpose of the following discussion it will be considered to be the quantity of sand and gravel moved through a channel.

To discuss in more detail the effects of changing discharge and sediment load on channel morphology, a plus or minus exponent will be used to indicate how, with an increase or decrease of discharge or bed-material load, the various aspects of channel morphology will change (see Figure 12.17). For the relatively straightforward cases of an increase or decrease in discharge or bed-material load alone, equations (12.23)–(12.26) are obtained:

$$Q^+ = w^+, d^+, \ell^+, S^-. \tag{12.23}$$

$$Q^- = w^-, d^-, \ell^-, S^+. \tag{12.24}$$

$$L^+ = w^+, d^-, \ell^+, S^+, P^-, F^+. \tag{12.25}$$

$$L^- = w^-, d^+, \ell^-, S^-, P^+, F^-. \tag{12.26}$$

An increase or decrease in discharge alone could be caused by diversion of water into or out of a river system or climate change. An increase of $L$ will result from increased erosion in the catchment area, which can be induced by deforestation, climate change, or by an increase in the area under cultivation. A decrease of $L$ will result from improved land use or a programme of soil conservation.

An increase or decrease in discharge changes the dimensions of the channel and its gradient, but an increase or decrease in bed-material load at constant mean annual discharge changes not only channel dimensions, but also the shape of the channel (width–depth ratio) and its sinuosity.

In nature, however, rarely will a change in discharge or sediment load occur alone. Generally any change in discharge will be accompanied by a change in the type of sediment load and vice versa.

Using the plus or minus exponents to indicate an increase or decrease in a variable, four combinations of changing discharge and sediment load can be considered. For example, if both discharge and bed-material load increase, perhaps as a result of increased agricultural activity, deforestation, or a climate change; equation (12.27) suggests the nature of the resulting channel changes:

$$Q^+, L^+, = w^+, d^\pm, \ell^+, S^\pm, P^-, F^+. \qquad (12.27)$$

Equation (12.27) indicates that, with an increase of both discharge and bed-material load, then width, meander wavelength and width–depth ratio should increase and sinuosity decrease. The influences of increasing discharge and of bed-material load on channel depth and gradient are in opposite directions, and it is not clear in what manner gradient and depth should change. However, by including width–depth ratio, an estimate of the direction of change of depth can be obtained. Width–depth ratio is predominantly influenced by type of load, equation (12.22), and, therefore, it increases in equation (12.27). This suggests that depth will remain constant or decrease because both width and width–depth ratio increase. Channel gradient will probably increase because sinuosity decreases, thereby straightening the channel and increasing its slope.

When both $Q$ and $L$ decrease, perhaps as a result of dam construction, the reverse of equation (12.27) pertains, as follows:

$$Q^-, L^- = w^-, d^\pm, \ell^-, S^\pm, P^+, F^-. \qquad (12.28)$$

When, as common in nature, the changes in $Q$ and $L$ are in opposite directions, the following relations are obtained:

$$Q^+, L^- = w^\pm, d^+, \ell^\pm, S^-, P^+, F^-. \qquad (12.29)$$
$$Q^-, L^+ = w^\pm, d^-, \ell^\pm, S^+, P^-, F^+. \qquad (12.30)$$

Equations (12.29) and (12.30) reflect changes expected with a change of climate from semi-arid to subhumid or subhumid to humid and vice-versa (see Figure 3.21). The result of the hydrologic changes described by the above equations is a complete change of channel morphology, which can be referred to as channel metamorphosis.

## 12.7 Examples of river metamorphosis

It should be possible to substantiate the conclusions concerning the effect of changed discharge and sediment load on river channels, and some information, although incomplete in many cases, is available for major river changes both during historic time and during the recent geologic past.

### 12.7.1 Historical river metamorphosis

First, a few examples of the conversion of what appear to be suspended-load or mixed-load channels to bedload channels will be cited. These changes occurred on rivers that were not subject to significant regulation of flow, and the formerly meandering rivers were converted to straight channels by a combination of high peak discharges and an influx of coarser sediment. For example, the highly sinuous, relatively narrow and deep Cimarron River channel of south-west Kansas was destroyed by a major flood in 1914. Between 1914 and 1939 the river widened from an average of 50 ft to 1200 ft (15–366 m), and the entire floodplain was destroyed. Large floods moved considerable sand and caused this transformation despite the fact that annual discharge was probably less during the drought of the 1930s. The hydrologic record is short, but an abrupt increase in annual discharge after 1940 was recorded at the Wyanoka, Oklahoma, gauging-station.

Precipitation data indicate that the years 1916–41 were generally a period of below-average precipitation. Thus during years of low runoff and high floodpeaks, the Cimarron River was converted from a narrow sinuous channel characterized by low sediment transport to a very wide, straight bedload river. These changes were apparently the result of climatic fluctuations, although agricultural activities within the basin may have increased the flood peaks and the sediment loads by destruction of the natural vegetation (Schumm and Lichty, 1963).

It appears that large floods override the effect of decreased mean annual discharge and that the change in Cimarron River morphology can be considered analogous to that caused by increased $Q$ and $L$, equation (12.27). An increase in bed-material load must have occurred as the channel was widened and as the gradient increased by straightening of the channel (decreased sinuosity), but there are no data to support this suggestion.

Equally great changes along some major rivers east of the Rocky Mountains can be documented. Especially impressive is the conversion of the broad North and South Platte rivers and the Arkansas River to relatively insignificant streams owing to flood-control works and diversions for irrigation. The width of these rivers, as shown on topographic maps published during the latter part of the nineteenth century, can be compared with the width shown on new maps of the same areas. For example, the North Platte River has narrowed from about one-half to three-quarters of a mile wide to a few hundred feet wide (Plate 15).

The South Platte River has always been cited as a classic example of a braided stream. About fifty-five miles above its junction with the North Platte River, the South Platte River was about a half-mile wide in 1897, but it narrowed to about 200 ft (60 m) wide by 1959 (Figure 12.33).

The tendency of both rivers is thus to form one narrow well-defined channel in place of the previously wide braided channels. In addition, the new channel is generally somewhat more sinuous than the old.

The narrowing of the North Platte can be attributed to a decrease in the mean annual flood from 13,000 to 3000 cfs and to a decrease in the mean annual discharge from 2300 to 560 cfs as a result of river regulation and the diversion of flow for irrigation. The major decrease in annual runoff occurred about 1930. A similar change occurred on the South Platte during the drought of the 1930s. However, the annual discharge of the South Platte increased after 1940, partly as a result of transmountain diversions, whereas because of upstream regulation, the discharge of the North Platte did not. A decrease in the magnitude of the annual momentary maximum discharge occurred for both rivers, and this decrease was undoubtedly the major factor determining the present channel size. Data on changes of sediment load are lacking, but it is probable that much less sand is being transported through these channels at present.

These major changes of river character have occurred without significant change in altitude of the channel. Adjustments to altered runoff and sediment load were accomplished primarily by changes in channel shape and pattern. Of course, where a major increase in the quantity and size of sediment load occurs, aggradation will occur and it may be very significant.

Although data on changes of sediment load are not available for the historic examples of channel metamorphosis, it may be assumed that as rivers are widened and steepened by destruction of their original channels, larger quantities of bed-material load are moved through the channels by the recorded high flood peaks. If this condition persisted, the channels would have remained wide and straight. However, with decreased flood peaks, deposition of sediment in the wide channel and the encroachment of vegetation into the channel caused narrowing of the channel, and a tendency towards development of meanders is evident.

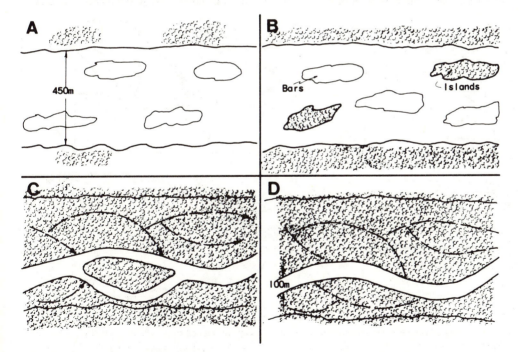

**Figure 12.33** Model of South Platte River metamorphosis: A. early 1800s, discharge is intermittent, bars are transient; B. late 1800s, discharge is perennial, vegetation is thicker on floodplain and islands; C. early 1900s, drought allows vegetation to establish itself below mean annual high-water level, bars become islands, single thalweg is dominant; D. modern channel, islands attached to floodplain, braided patterns on floodplain are vestiges of historic channels.
*Source:* Nadler and Schumm, 1981, figure 10, p. 107.

### 12.7.2 Geological river metamorphosis

The historic examples of river metamorphosis, although interesting, do not provide a complete picture of river response to changes of hydrologic regimen because they were in some cases local and temporary. However, the changes of climate and sea level during relatively recent geologic time produced examples of complete channel metamorphosis, although information concerning water and sediment discharge is indirect (e.g. the Murrumbidgee River: see Section 20.2.1.).

One of the most intensively studied rivers of the world is the Mississippi River. The pioneering work of Fisk (1944, 1947) and his colleagues provides insight into the effect of climate and sea-level changes on river morphology. The two factors which determined the behaviour of the Mississippi River between the maximum extent of the last glaciation (18,000 BP) and the present, were the presence of the ice sheet itself, which supplied tremendous quantities of meltwater and sediment to the upper Mississippi Valley, and the lowering of sea level to a maximum depth of about 130 m about 15,000 years ago. The fall of sea level lowered the base level of the Mississippi River system, permitting removal of earlier alluvial deposits and scour of the floor of the valley. The details of the history are summarized from Fisk's (1947) report as follows.

During the time of maximum lowering of sea level the Mississippi River was entrenched between 400 and 450 ft (120–137 m) (Figure 12.34A). In addition, the tributaries incised their channels to produce an irregular valley floor with an average slope of 0·83 ft/mile. When the continental ice sheet began to waste away, the introduction of water and sediment into the Mississippi Valley was greatly increased at the same time that sea level began to rise rapidly between 14,000 and about 4000 years BP. Deposition in the valley accompanied the rise of sea level, and deposits grading upward from coarse sands and gravels through clean sands were formed. The upward decrease in particle size resulted from the progressive decrease in slope and the northward retreat of the margins of the ice, which supplied the greater part of this material. In effect, a wave of alluviation moved slowly upstream and into the tributary valleys, so that the entire drainage system was affected. However, the tributary streams still transported coarse sediments, and alluvial

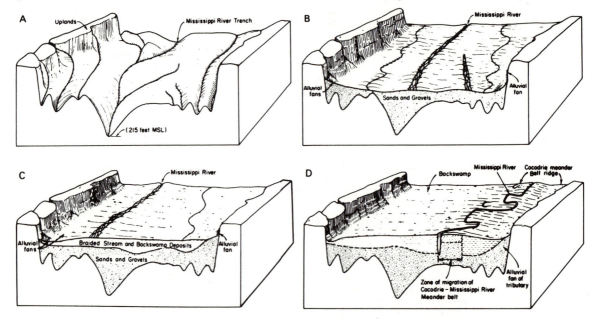

**Figure 12.34** Quaternary history of the Mississippi River valley: A. late Glacial entrenched stage: 18,000–14,000 years BP; sea level −400 ft; valley slope 0·83 ft/mile; river overloaded, carrying gravel; B. valley aggradation stage I: sea level −100 ft; valley slope 0·75 ft/mile; river overloaded, braided, carrying sand and fine gravel; C. valley aggradation stage II: sea level −20 ft; valley slope 0·68 ft/mile; river overloaded, braided, carrying silts and sands, build-up of thick backswamp silts and clays; D. valley aggradation stage III: sea level as at present; valley slope 0·60 ft/mile; development of Cocodrie meander belt; meandering deep-channel river with great discharge variation gradually replacing braided shallow channel (0·60, 0·68, 0·75 and 0·83 ft/mile = 0·11, 0·13, 0·14 and 0·16 m/km).
*Source:* Fisk, 1944, plate 5.

fans were built into the valley.

The lack of fine sediment (silts and clays) in this part of the alluvium indicates that these fine sediments were transported to the sea. According to the evidence of these deposits, the Mississippi River at this time was a braided stream shifting across the alluvial valley on a slope of about 0·75 ft per mile (Figure 12.34B).

With continued but slower rise of sea level and a decrease in the size of the sediments moving from the north and from the tributary valleys, the sediment-load was eventually reduced to fine sands, silts and clays. However, the quantity of material carried into the valley was sufficient to cause deposition of fine sediment on the reduced valley slope. According to Fisk, 'The braided streams, during this stage, wandered widely on the plain and built low alluvial ridges of sands and silts' (Figure 12.34C). As valley alluviation continued the gradient of the valley was reduced to about 0·60 ft per mile and 'the size and amount of sediment contributed to the river decreased'. At this stage the river was flowing on a slope only 0·15 ft per mile (0·03 m/km) less than that when it was braided. With essentially constant sea level and reduced sediment load, the river began to meander. According to Fisk, 'The change from a braided stream to a meandering one brought about the confinement of the Mississippi flows through a single deep channel'. No longer did the river wander 'freely in shallow channels across the alluvial surface as did the braided stream' (Figure 12.34D). The history of the Mississippi River strongly supports the idea that meandering is largely the result of a river's attempt to reduce its gradient in response to changed hydrologic regimen.

According to the general relations presented in the preceding section, it is likely that during the erosional evolution of a landscape river morphology will change, as relief is reduced and as the character of the sediment load is changed from bedload to suspended-sediment load. Therefore, in the early stages of steep relief and high energy the channels should be braided. As the sediment load is reduced in grain size and quantity, and as it becomes more of a suspended load, the character of the channels should change to meandering or even straight, depending on valley-floor slope.

Also a change of channel morphology, for example by incision, leads to a period of instability during which the channel evolves to a new and predictable condition. For example, very sinuous streams have been channelized in northern Mississippi in order to reduce flooding by reducing the sinuosity of the channels from about two to one and doubling the gradient. The channels responded by incising and then widening. For one channel, Oaklimiter Creek, when the width–depth ratio increased

to six the channel cross section had evolved to a new condition of relative stability (Figure 12.35). This information is of considerable value in planning channel stabilization works in Oaklimiter Creek (Schumm *et al.*, 1984) and arroyos and gullies elsewhere, because a similar evolution of these incised channels has been recognized.

## 12.8 Rivers and valley morphology

Most valleys, except some that have been subjected to recent uplift, contain rivers that are flowing on alluvial valley-fill deposits. This means, of course, that the morphology of the valley itself is significantly obscured by these deposits. It is very difficult to obtain information on the configuration of the bedrock floor of a valley, and yet the configuration of the valley floor and the valley sides, as well as the character of the alluvium at depth, is geologically important. Unlike stream channels that are visible and available for study, the valley fill and the valley floor can be studied only where penetrated by wells, where excavated at construction sites, at quarries and mines, or where studied by seismic methods.

When sketches of a valley profile or cross-section are made, a relatively smooth-floored profile is drawn. The smoothness of the valley floor, as sketched, is really a reflection of our ignorance of its true condition. This simplistic view of the bedrock floor of valleys ignores, for example, the great variability that may be due to the variations in the resistance of the materials exposed within the valley walls and floor. If little is known about valley cross-sections and dimensions, there is even less known about their longitudinal profile.

Valleys are fluvial features that are eroded into resistant older material. This means that the floor of the valley and the valley sides are more resistant to erosion than the materials that the valley contains. For this reason, variability of valley morphology will reflect variations in the resistance to erosion of the material into which the river has eroded. Although several types of erosive mechanisms may be investigated as they occur, incision of a river into bedrock is not one of these.

As with stream channels, the dimensions of a valley should reflect the quantity of water that moved through the valley. Despite exceptions dependent on the resistance of the bedrock to erosion, major river systems occupy rather large valleys. However, it is also true there are cases where small misfit rivers occupy large valleys. There are two explanations for this: (1) river capture diverted large amounts of water from that valley into another river system; or (2) major climatic changes decreased the flow of water through the valley. The relationship between meander wavelength and river

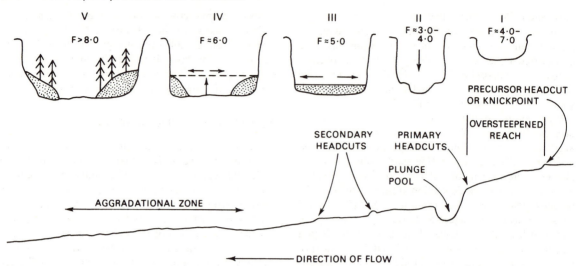

**Figure 12.35** Schematic longitudinal profile of an incising channel showing identifiable features; schematic cross-sectional profiles show the evolution of the reaches from type I to type V; typical values of width–depth ratio F are included; arrows indicate the relative importance and direction of the dominant processes.
*Source:* Schumm *et al.*, 1984, figure 3, p. 128.

discharge (see Figure 12.19) seems to be equally valid for valley meanders (Figure 12.36). In this case drainage area is used as a surrogate for discharge, and the scatter is very large, as one might expect, considering the influence of different types of bedrock on the valleys. Nevertheless, there is a significant relationship between discharge and valley meander dimensions. This should mean that the dimensions of valley meanders are related to paleodischarge; therefore, valley width and depth should also be related to discharge, and in Illinois it has been demonstrated that valley dimensions (e.g. floodplain width) are related to stream order which, in turn, is broadly related to discharge (Figure 12.37).

As we have seen, riffles and pools occur in alluvial rivers, and it seems that similar differential bedrock scour should produce highs and lows beneath alluvial deposits. Some of the preceding field observations suggest that this is true. Regular scour patterns of this type would be of interest to the economic and engineering geologist. In order to investigate valley morphology a series of experiments were performed which involved incision of straight and sinuous channels into cohesive materials in a large flume (Shepherd and Schumm, 1974). Contrary to expectations, during experimentation incision did not occur uniformly over the width of the channel floor. Rather, with water flowing bankfull in the straight trapezoidal channels, longitudinal lineations enlarged into predominant grooves (Figure 12.38). Although accompanying potholes and erosional ripples also grew in size, they were ultimately replaced by

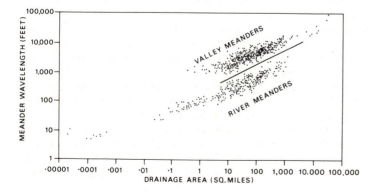

**Figure 12.36** Massed plots of wavelength against upstream drainage area in respect of valley meanders and river meanders. The largest magnitudes are taken from the rivers Dniestr and Don and the smallest from spoil tips; the majority of the data comes from rivers in the United States, England and eastern France.
*Source:* Dury, 1965, figure 6, p. 7.

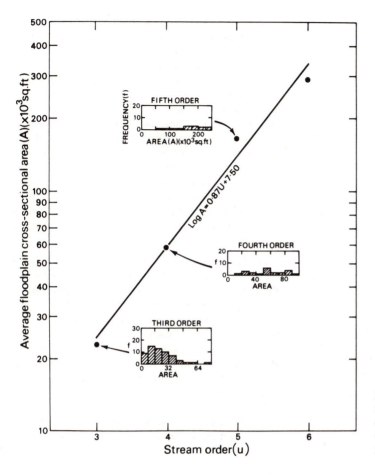

**Figure 12.37** Relationship between average floodplain cross-sectional area and stream order for the Big Muddy River, Illinois. *Source:* Bhowmik and Stall, 1979, figure 11, p. 26.

grooves. As incision progressed the discharge finally was conveyed within the banks of a narrow and deep inner channel, which was the result of the coalescence of grooves. Portions of the original channel bottom showing remnant erosional features remained above the water surface on each side of the inner channel. Of major significance is the fact that a narrow and deep inner channel finally conveyed the entire discharge, which initially flowed bankfull in a wider and shallower channel. Bedrock was scoured below the baselevel of the flume, as a result of the low width–depth ratio of the inner channel, which constrained the flow and caused erosional shear stresses to be effective even as gradient was decreased. The fact that scour occurred below baselevel in the experimental channels suggests the possibility that considerable depths of alluvium can exist upstream from local baselevels in rivers.

Some information is available that supports the conclusions reached from experimentation. For example, Bretz (1924) presents a detailed topographic map of a prominent inner channel in the basalt at Five Mile rapids at The Dalles of the Columbia River, and the topography there is remarkably similar to that which developed in the experiments. The fact that the Columbia River has eroded 115 ft (35 m) below sea level at a point 192 river-miles from the ocean provides a natural analogue for the erosion below baselevel in the experimental channels.

Inner channels also exist in many other river channels in nature. A cross-section taken at the location of Boulder Dam on the Colorado River shows an inner channel. Excavation of the dam site for the Prineville Dam in Oregon revealed a narrow and deep 'inner gorge', averaging 70 ft (21 m) in width with a depth as great as 60 ft, containing overhanging cliffs, potholes, and narrow and steep-sided interconnecting trenches. The Trinity dam site in California also has a prominent inner bedrock channel that had to be excavated before dam construction could proceed. Inner channels have long been observed by river engineers, but often they have been interpreted as low-flow channels, which need

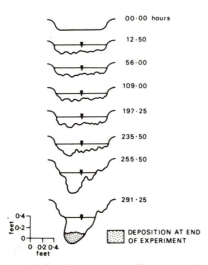

**Figure 12.38**   Series of transverse profiles measured as incision occurred at 34 ft upstream from baselevel in flume channel. Elapsed run time is shown in each section. Discharge was 0·14 cfs until 130 h, when it was increased to 0·20 cfs. Initial flume slope was 0·0048; it was increased to 0·0082 after 56 h and to 0·0175 after 109 h. After 214 h, erosion was still proceeding slowly, but then the coarse sand was introduced at the rate of 25 g/min. The sand-feed rate markedly influenced incision rates; sand feed was increased to 80g/min after 235 h. Stippled pattern (**h**) shows extent of deposition at end of experiment.
*Source:* Shepherd and Schumm, 1974, figure 3, p. 258.

not be the case. The existence of an inner channel on the bedrock floor of an incised valley could be a potential area for the localization and concentration of heavy minerals by hydraulic sorting. If the location and trend of an inner channel could be determined, in either paleofluvial deposits or Holocene valleys, the exploration for gold, diamonds and other placers would be facilitated.

Another result of negative movements of baselevel may be the production of *incised meanders*, although like many other features of polycyclic terrain, they may be produced by other means. Incision of meanders occurs when a river trenches its floodplain and incises into bedrock resistant enough to enclose the bends. Tabular rocks are particularly conducive to the preservation of incised meanders, which are of two general kinds – intrenched and ingrown (Figure 12.39A and B):

(1) Intrenched meanders result from symmetrical incision associated with rapid downcutting, often by the headward migration of a nickpoint after baselevel lowering. The incision must be fast enough to permit little concurrent lateral downstream migration of the meander bends and it is significant that well-developed intrenched meanders of the San Juan and Colorado Rivers occur where downstream migration is inhibited by the counter (upstream) dip of a large dome structure (Figure 12.40).

(2) Ingrown meanders exhibit asymmetrical trenching and result from slow incision during which migration of the meander bends occurs. This incision may be

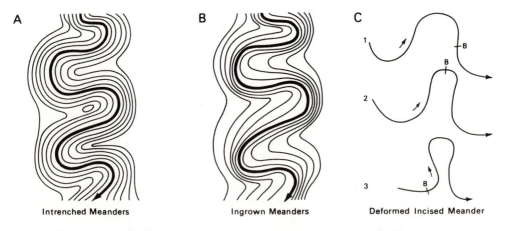

**Figure 12.39**   Incised meanders: A. intrenched meanders; B. ingrown meanders; C. deformed incised meanders; the downstream limb is fixed by bedrock and upstream the channel is in alluvium; **B** indicates point at which the channel enters bedrock as the deformation proceeds through stages 1–3.
*Sources:* Sparks, 1972, figure 9.11, p. 302, copyright © Longman, London; Schumm, 1977, figure 6-15, p. 200.

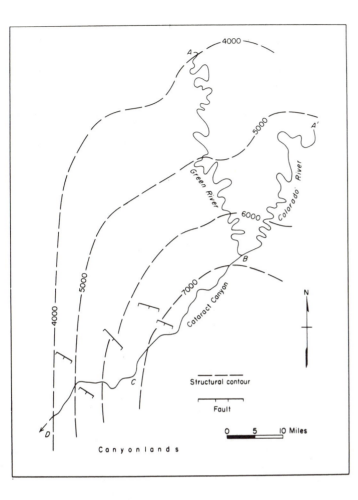

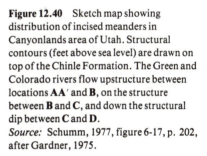

**Figure 12.40** Sketch map showing distribution of incised meanders in Canyonlands area of Utah. Structural contours (feet above sea level) are drawn on top of the Chinle Formation. The Green and Colorado rivers flow upstructure between locations **AA'** and **B**, on the structure between **B** and **C**, and down the structural dip between **C** and **D**.
*Source:* Schumm, 1977, figure 6-17, p. 202, after Gardner, 1975.

of polycyclic origin or due to slow uplift or regional tilting during a single cycle. It has been suggested that ingrown meanders tend to be associated with trenching rivers having abundant available bedload.

The downstream migration of intrenching meanders may result in their downstream limbs being arrested for a time by an outcrop of resistant rock leading to the development of deformed incisions characteristic of the San Juan River (Plate 16). Also, on occasions, meander curvature may increase during incision causing an incised meander loop to be cut off and left as a hanging abandoned incised meander. Instances of this type occur along the San Juan River and in the Lower Wye Valley of the Welsh borderlands. In the former, ancestral streambed gravel (15–40 ft; 5–12 m thick) indicates a river gradient similar to that of the present San Juan River.

During the experimental investigation of bedrock incision alluvial meanders developed in a sand cover placed over the simulated bedrock, but in every case when the slope of the flume was increased to induce incision, the alluvial meanders were destroyed and a relatively straight channel incised into bedrock. From these observations, it is difficult to accept the hypothesis that symmetrical incised meanders in nature inherited their patterns from an alluvial channel superimposed from a warped peneplain.

However, another way in which incised meanders form is by baselevel lowering or by essentially vertical uplift. Additional experiments (Gardner, 1975) showed that when a bedrock surface is horizontal, a lowering of baselevel will cause headward incision along the meander pattern. A nickpoint will migrate upstream following rejuvenation, and it will follow the existing pattern to form incised meanders in the underlying bedrock.

Of considerable interest is the fact that as headward erosion progressed along a given meander the downstream limb of the meander encountered bedrock first and became locked in position. The upstream limb of the meander continued to shift downstream as part of a normal downvalley sweep of the meander pattern, and the meander compressed. By the time the upstream limbs of the meander became fixed in bedrock a very deformed meander usually had developed (Figure 12.39C). Deformed meanders are common in the Colorado Plateau of the south-western United States where streams have incised deeply into massive sandstones. An excellent example is the Goosenecks of the San Juan River, which are difficult to explain in terms of normal

alluvial meander behaviour. However, when one considers the effect of rapid downcutting by the Colorado River which lowered the base-level of the tributary San Juan River, the explanation developed from the laboratory studies seems very reasonable. Similar deformed or compressed meanders are produced in alluvium when the shifting meander encounters resistant alluvium or bedrock. The upstream meanders compress down upon it, generating a reach of very high sinuosity. The Mississippi River has produced such meanders when in their downstream sweep meanders encounter clay plugs which are the fillings of old oxbow lakes.

# References

Ackers, P. and Charlton, F. G. (1970) 'The geometry of small meandering streams', *Proceedings of the Institute of Civil Engineers*, paper 7328S, 289–317.

Allen, J. R. L. (1974) 'Reaction, relaxation and lag in natural sedimentary systems; general principles, examples and lessons', *Earth Science Reviews*, vol. 10, 263–342.

Allen, J. R. L. (1976) 'Bedforms and unsteady processes: some concepts of classification and response illustrated by common one-way types', *Earth Surface Processes*, vol. 1, 361–74.

Betson, M. A. and Thomas, D. M. (1966) 'A definition of dominant discharge', *Bulletin of the International Association of Scientific Hydrology*, year 11, 76–80.

Bhowmik, N. G. and Stall, J. B. (1979) 'Hydraulic geometry and carry capacity of floodplains', *University of Illinois, Water Resources Center, Research Report* 145.

Bray, D. I. (1973) 'Regime relations for Alberta gravel-bed rivers', in *Fluvial Processes and Sedimentation*, Research Council of Canada, 440–52.

Bray, D. I. (1975) 'Representative discharges for several gravel-bed rivers in Alberta, Canada', *Journal of Hydrology*, vol. 27, 143–53.

Bretz, J. H. (1924) 'The Dalles type of river channel', *Journal of Geology*, vol. 24, 129–49.

Brice, J. C. (1975) 'Airphoto interpretation of the form and behavior of alluvial rivers', *Final Report for US Army Research Office*.

Brush, L. M. (1961) 'Drainage basins, channels, and flow characteristics of selected streams in central Pennsylvania', *US Geological Survey Professional Paper* 282-F, 145–81.

Colby, B. R. (1964) 'Discharge of sands and mean-velocity relationships in sand-bed streams', *US Geological Survey Professional Paper* 462-A.

Dury, G. H. (1964) 'Principles of underfit streams', *US Geological Survey Professional Paper* 452-A.

Dury, G. H. (1965) 'Theoretical implications of underfit streams', *US Geological Survey Professional Paper* 452-C.

Einstein, H. A. (1950) 'The bed-load function for sediment transportation in open channel flows', *US Department of*

*Agriculture Technical Bulletin* 1026.

Fisk, H. N. (1944) *Geological Investigation of the Alluvial Valley of the Lower Mississippi River*, Vicksburg, Miss., Mississippi River Commission.

Fisk, H. N. (1947) *Fine-Grained Alluvial Deposits and their Effects on Mississippi River Activity*, Vicksburg, Miss., US Corps of Engineers, Mississippi River Commission, Waterways Experiment Station, 2 vols.

Gardner, T. W. (1975) 'The history of part of the Colorado River and its tributaries: an experimental study', *Four Corners Geological Society Guidebook, 9th Field Conference, Canyonlands*, 87–95.

Hack, J. T. (1957) 'Studies of longitudinal stream profiles in Virginia and Maryland', *US Geological Survey Professional Paper* 294-B, 45–97.

Hickin, E. J. and Nanson, G. C. (1975) 'The character of channel migration on the Beatton River, N. E. British Columbia, Canada', *Bulletin of the Geological Society of America*, vol. 86, 487–94.

Jackson R. G. (1975) 'Velocity–bedform–texture patterns of meander bends in the lower Wabash River of Illinois and Indiana', *Bulletin of the Geological Society of America*, vol. 86, 1511–22.

Khan, H. R. (1971) 'Laboratory study of river morphology', unpublished PhD dissertation, Colorado State University.

Langbein, W. B. and Leopold, L. B. (1964) 'Quasi-equilibrium states in channel morphology', *American Journal of Science*, vol. 262, 782–94.

Langbein, W. B. and Leopold, L. B. (1966) 'River meanders – theory of minimum variance', *US Geological Survey Professional Paper* 422-H.

Langbein, W. B. and Leopold, L. B. (1968) 'River channel bars and dunes – theory of kinetic waves', *US Geological Survey Professional Paper* 422-L.

Laronne, J. B. and Carson, M. A. (1976) 'Interrelationships between bed morphology and bed-material transport for a small, gravel-bed channel', *Sedimentology*, vol. 23, 67–85.

Leopold, L. B. and Bull, W. B. (1979) 'Base level, aggradation, and grade', *Proceedings of the American Philosophical Society*, vol. 123, 168–202.

Leopold, L. B. and Maddock, T. (1953) 'The hydraulic geometry of stream channels and some physiographic implications', *US Geological Survey Professional Paper* 252.

Leopold, L. B. and Wolman, M. G. (1957) 'River channel patterns: braided, meandering and straight', *US Geological Survey Professional Paper* 282-B, 39–85.

Lewin, J. (1976) 'Initiation of bed forms and meanders in coarse-grained sediment', *Bulletin of the Geological Society of America*, vol. 87, 281–5.

Mollard, J. D. (1973) 'Airphoto interpretation of fluvial features', in *Fluvial Processes and Sedimentation*, Research Council of Canada, 341–80.

Nadler, C. T. and Schumm, S. A. (1981) 'Metamorphosis of South Platte and Arkansas Rivers, eastern Colorado', *Physical Geography*, vol. 2, 95–115.

Nevin, C. (1946) 'Competency of moving water to transport debris', *Bulletin of the Geological Society of America*, vol. 57, 651–74.

Richards, K. S. (1982) *Rivers: Form and Process in Alluvial Channels*, London, Methuen.

Schumm, S. A. (1960) 'The shape of alluvial channels in relation to sediment type', *US Geological Survey Professional Paper* 352-B, 17–30.

Schumm, S. A. (1963) 'Sinuosity of alluvial rivers on the Great Plains', *Bulletin of the Geological Society of America*, vol. 74, 1089–1100.

Schumm, S. A. (1977) *The Fluvial System*, New York, Wiley.

Schumm, S. A. (1981) 'Evolution and response of the fluvial system, sedimentological implications', *Society of Economic Paleontologists and Mineralogists Special Publication* 31, 19–29.

Schumm, S. A., Khan, H. R., Winkley, B. R. and Robbins, I. G. (1972) 'Variability of river patterns', *Nature (Physical Sciences)*, vol. 237, 75–6.

Schumm, S. A. and Lichty, R. W. (1963) 'Channel widening and flood-plain construction along Cimarron River in south-western Kansas', *US Geological Survey Professional Paper* 352-D, 71–88.

Schumm, S. A., Harvey, M. D. and Watson, C. C. (1984) *Incised Channels: Morphology, Dynamics and Control*, Littleton, Colo., Water Resources Publications.

Shen, H. W. and Schumm, S. A. (1981) 'Methods for assessment of stream-related hazards to highways and bridges', *Federal Highway Administration, Office of Research and Development Report* FHWA/RD-80/160.

Shepherd, R. G. and Schumm, S. A. (1974) 'Experimental study of river incision', *Bulletin of the Geological Society of America*, vol. 85, 257–68.

Simons, D. B. (1969) 'Open channel flow', in R. J. Chorley (ed.), *Water, Earth and Man*, London, Methuen, chapter 7.I, 297–318.

Sparks, B. W. (1972) *Geomorphology*, 2nd edn, London, Longman.

Sundborg, A. (1956) 'The River Klarälven, a study of fluvial processes', *Geografiska Annaler*, vol. 38, 127–316.

Vanoni, V. A. (1975) 'Sediment discharge formulas', in V. A. Vanoni (ed.), *Sedimentation Engineering*, New York, American Society of Civil Engineers, 190–229.

Williams, G. P. (1978) 'The case of the shrinking channels – the North Platte and Platte Rivers in Nebraska', *US Geological Survey Circular* 781.

Wolman, M. G. (1955), 'The natural channel of Brandywine Creek, Pennsylvania', *US Geological Survey Professional Paper* 271.

Yang, C. T. (1971), 'On river meanders', *Journal of Hydrology*, vol. 13, 231–53.

Yatsu, E. (1959), 'On the discontinuity of grainsize frequency distribution of fluvial deposits and its geomorphological significance', *Reports of the Faculty of Engineering*, Chuo University, G-1, no. 27.

# Thirteen  *Drainage basins*

Since the beginning of the nineteenth century three major geomorphic attitudes have provided bases for the spatial description and analysis of landforms. The first of these was naturally the geological framework of terrain which found its most complete expression in the physiographic regions of Fenneman (1914) during the period 1914–38. These regions were mainly based on considerations of structural geology but with subdivisions relying on certain gross morphometric attributes, notably relief and degree of dissection. The second spatial basis was that provided by the Davisian cycle and the associated denudation chronology, whereby regions of differing stage or cycle were believed to form identifiable units. In the 1930s Wooldridge (1932) was concerned to delimit such regions and to identify 'the physiographic atoms out of which the matter of regions is built', concluding that these atoms were the facets of flats and slopes forming the intersecting surfaces characteristic of polycyclic landscapes. These two approaches to geomorphic spatial analysis have variously resulted in more recent attempts to define morphometric regions as composed of 'component landscapes' and 'landscape patterns'.

## 13.1 The basin geomorphic unit

The third spatial basis for landform analysis, the erosional drainage basin, has been recognized as a viable process-response unit since the beginning of the last century. Playfair in 1802 spoke of the nice adjustment of a system of valleys communicating with the main trunk stream, and Gilbert in 1877 referred to the interdependence throughout this system leading to a dynamic equilibrium affecting all drainage lines and their flanking slopes (Plate 17). In 1899 Davis defined the 'river' as extending 'all over its basin and up to its very divides' such that 'ordinarily treated, the river is like the veins of a leaf; broadly viewed, it is like the entire leaf'. This approach was given its more powerful modern expression by the hydrophysical work of Robert E. Horton, who in 1945 described the morphometry of drainage basins, showed how morphometric features are interrelated and attempted to rationalize these features (notably drainage density) on the basis of hydrological processes. The work of Horton, extended brilliantly by Arthur N. Strahler (summarized 1964), established the erosional drainage basin as the basic geomorphic unit in many terrains because it appears to be:

(1) A limited, convenient, and usually clearly defined and unambiguous topographic unit, available in a nested hierarchy of scales on the basis of stream ordering. Unlike the geologist who must draw conclusions about the form and the characteristics of sedimentary units based on only limited exposures, the geomorphologist sees on the surface of the earth landforms displayed for complete description. Therefore, he can use standard statistical sampling techniques to obtain a quantitative description of the basin landforms. A drainage basin (watershed or catchment) is simply the area that gathers water from precipitation and delivers it to a larger stream, a lake, or ocean. It is an area limited by a drainage divide and occupied by a drainage network wherein the upstream drainage basin supplies water and sediment to the lower parts, reflecting the upstream geologic and hydrologic character of the watershed. Drainage basins, therefore, are the components of the continents and they are of all sizes. For example, the Mississippi River drainage basin is a hierarchy of drainage basins, the largest of which is, of course, the Mississippi–Ohio–Missouri river system, one of the largest river systems on earth, and the smallest is a single, tiny, unbranched badland tributary

located in the upper Missouri River basin.

(2) A physical process–response system open to a cascade of inputs and outputs. The inputs involve thermal energy from the sun, kinetic and potential energy from precipitation, potential energy from uplift and igneous activity and chemical energy released by weathering. The geomorphic outputs are mainly of water, sediment and dissolved material resulting in the maintenance or transformation of the topographic surface or near-surface.

If drainage basin and channel morphology are related to the geologic, climatic and hydrologic character of the basin, then it is necessary to describe the features quantitatively in order to investigate these relationships. For this reason, among others, numerous descriptors of basin morphology have been developed.

## 13.2 Morphometric analysis

Erosional terrain commonly represents a series of complex geometrical surfaces, particularly so at the medium scales with which the geomorphologist is concerned. The method employed in describing such surfaces is to select by sampling certain supposedly diagnostic geometric variables (e.g. heights, slope angles, stream lengths, basin areas, etc.) which are then used to generalize the features of landscape geometry which can be rationalized in terms of formative processes or history, or which are useful in some practical sense. The science of morphometry is concerned with the quantitative measurement and generalization of land surface geometry. The complexity of drainage basins derives from their being composed of divides, hillslopes (Chapter 11), valley heads, terraces, floodplains (Chapter 14) and channels (Chapter 12), each of which can be investigated separately. From the air, the most characteristic feature of a drainage basin is its drainage network, which is displayed in a variety of patterns and textures. On the ground the observer is more impressed with the characteristics of the valley-side slopes (Chapter 11). Nevertheless, the character of the drainage network is important because it can be used to interpret the geologic conditions responsible for certain patterns and, in addition, the texture of the pattern is controlled by, and in turn has an influence on, the hydrology of the drainage basin. In fact, the detailed study of drainage basin morphology was given its impetus by the work of a hydrologist, Robert E. Horton, whose 1945 paper has since become a classic. In this he demonstrated that the drainage basin and its network could be dissected and its components studied, on a stream-ordering basis, later modified by Strahler (1964).

The morphometric features of the erosional drainage

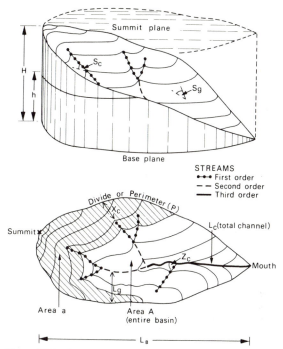

**Figure 13.1** Some morphometric attributes of the erosional drainage basin.
*Source:* Strahler, *Trans. Amer. Geophys. Union*, 1957.

basin have been classified by Strahler into linear, areal, and relief and gradient attributes (Figure 13.1). Clearly, the critical linear terrain elements are the perennial and ephemeral stream channels, commonly identified from detailed maps showing contour crenellations. As will appear later, there are problems both in identifying and interpreting the dynamic significance of stream channel networks, but the morphometry of an acceptable network can be described from two points of view:

(1) topological – by viewing the network as a 'directed graph' the interconnections of which are diagnostic of form;
(2) geometrical – by viewing the network as a geometrical form to be described in terms of lengths and orientations.

The distinction between the topology and geometry of stream networks is demonstrated in Figure 13.2A, and two simple types of topological ordering of a network are given in Figure 13.2B. Under the Strahler system a stream of a given order $(u + 1)$ is initiated at the junction of two streams of the next lower order $u$, such that for large samples orders are roughly expressive of stream basin areas and discharges. A difficulty with the Strahler ordering method is that it violates the distributive law in

A TOPOLOGICAL IDENTITY

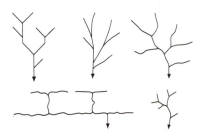

B ORDERING DIFFERENCE

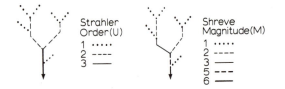

**Figure 13.2** Idealized stream networks showing
A. topological identity, and B. two different methods of
dividing or ordering the network.
*Source:* Shreve, 1966, figure 8, p. 28.

that the entry of a lower-order stream does not invariably increase the order of the main stream, a deficiency which is rectified by the use of Shreve's magnitude (*M*) ordering. One important feature of Strahler's ordering is that the number of streams of each order $N_u$ in a large basin or homogeneous region tends to approximate an inverse geometric series of which the first term is unity (see Figure 13.9A). The varying character of this series is expressed by the *bifurcation ratio* ($R_b = N_u/N_{u+1}$), which is higher for elongate channels fed directly by large numbers of first-order channels and lower for more intricately branching networks (see Figure 13.6). The bifurcation ratio is a dimensionless number varying only between about 3·0 and 5·0 for networks formed in homogeneous rocks, but exceeding 10 where pronounced structural control encourages the development of elongate narrow drainage basins. $R_b$ is rather insensitive to all but the most important structural controls, but the *orientation* of channel links (measured either in terms of entrance angles ($Z_c$: see Figure 13.1) or by absolute azimuth) is a much more sensitive expression of the effect of faults, joints, banding, bedding, etc. on the structure of stream networks. Figure 13.3 shows eight common types of drainage patterns expressive of homogeneous terrain (A), a steep regional dip (B), dipping or folded layers of sedimentary rock (C), right-angled faulting and

jointing (D), eroded domes or volcanoes (E), eroded domes or basins in layered rocks (F), hummocky surficial deposits or limestone solution (G) and complex metamorphic structures (H). Figure 13.4 compares the azimuths of stream channel directions and joint orientations in a dolomite locality of the Driftless Area of Wisconsin. It is important to recognize that $Z_c$ is not wholly controlled by structural trends, but is also inversely related to local relief and tends to increase as the order of the receiving stream segment increases.

The most important geometrical basin length measurements are (Figure 13.1):

$L_u$ the length of a stream segment of a given order. Histograms of values of $L_u$ for each order in a given basin or assemblage of homogeneous basins are characteristically right-skewed (log-normal) and mean values for each order ($\overline{L_u}$) significantly different.

$L_c$ the total length of the channel system within a basin.

$L_B$ the overall maximum basin length measured from the mouth.

$L_g$ the *length of overland flow*. This is the (map) distance from a point on a divide orthogonally (i.e.

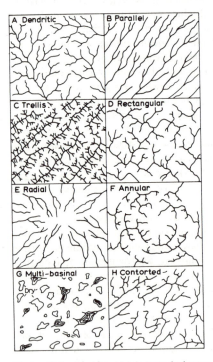

**Figure 13.3** Suggested basic-type drainage patterns, each of which can occur at a variety of scales.
*Source:* Howard, 1967, figure 1, p. 2248.

down the direction of maximum land slope) to the adjacent stream channel. The mean value of the length of overland flow $\overline{L_g}$ gives a measure of regional stream spacing and is approximately equal to the reciprocal of twice the drainage density ($\overline{L_g} \simeq 1/(2D)$).

$X_c$ the belt of no sheet erosion. The width (map) from the divide of the convex upper slope to the point where there is evidence of erosion by surface flow (e.g. rill or stream-channel heads).

$P$ the perimeter of the drainage basin.

The areal variables used to define basin morphometry are as follows:

$A_u$ the area of a drainage basin of a given order. Histograms of values of $A_u$ for each order are right-skewed (log-normal) and the mean values for each order $\overline{A_u}$ are significantly different.

$A$ the total area of a given drainage basin.

$D$ the *drainage density*, equal to $L_c/A$, which expresses the texture of fluvial dissection in terms of the average stream channel length per unit area. Values of $D$ vary widely, being about 3 (mile/sq. mile)* for chalk terrain, 4–5 for permeable sandstones, 20–30 for the metamorphic California Coast Ranges, 50–100 for the dryer areas of the American West, 200–400 for shale badlands and >1000 for unvegetated clay badlands (Figure 13.5).

$F$ the *stream frequency*, expressing the number of stream segments of all orders per unit area ($= \Sigma N_u/A$) (Figure 13.6). An allied measure is the dimensionless ratio $F/D^2$, which expresses the completeness with which the channel system fills the basin outline.

$R_c$ the *circularity ratio* ($= A/A_c$: where $A_c =$ the area of a circle having the same perimeter $P$ as the basin). Values for $R_c$ of 0·6–0·7 have been given for low-order basins in homogeneous shales and dolomites, and of 0·4–0·5 for basins in moderately steeply dipping quartzite in the Appalachian Ridge and Valley Province.

$R_e$ the *elongation ratio*, another measure of basin shape ($= d_A/L_B$: where $d_A =$ the diameter of a circle of area $A$). Values for $R_e$ have been obtained of 0·6 for areas of high relief and 1·0 for those of low relief.

Gradient measures include (Figure 13.1):

$S_g$ the maximum slope of the ground surface at a given point.

$\theta_{max}$ the maximum angle of a given valley-side slope

* Note the following equivalents: mile/sq. mile 4, 8, 20, 30, 50, 100, 200, 400, 1000 = km/km² 2, 5, 12, 19, 30, 60, 125, 250, 620. (See Table 13.1.)

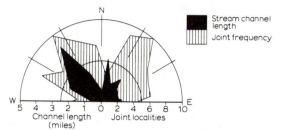

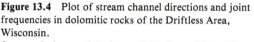

**Figure 13.4** Plot of stream channel directions and joint frequencies in dolomitic rocks of the Driftless Area, Wisconsin.
*Source:* Judson and Andrews, 1955, figure 2C, p. 330.

profile. Within a geologically uniform region samples of these angles exhibit normal distributions which cluster around characteristic mean (maximum) angles (Figure 13.7).

$S_c$ the slope of a stream channel at a point or averaged over a reach.

The relief of a basin may be simply described by (Figure 13.1):

$H$ the *relief*, expressing the elevation difference between high and low points. From this is derived the *relief ratio* ($R_h = H/L_B$) and the *ruggedness number* ($= HD$). The latter has been calculated at 0·06 for the Louisiana Coastal Plain and >1·0 for the clay badlands of South Dakota.

$\int$ the *hypsometric integral* (Figure 13.8) which is the percentage area under the dimensionless curve relating $h/H$ and $a/A$, and expresses the unconsumed volume of a drainage basin as a percentage of that delimited by the summit plane, base plane and perimeter (Figure 13.1). Where a particularly resistant geological outcrop maintains a portion of the summit plane during considerable erosion of the rest of the basin, $\int$ may reach low values (Figure 13.8D). However, in uniformly erodible material the continued erosion of the basin high point may stabilize $\int$ in a middle range of values between about 40 and 60 per cent (Figure 13.8B and C). For this reason, the value of $\int$ as a relative measure of erosion is limited to its higher values or to situations where the elevation of the original summit plane can be estimated.

Of the foregoing list of morphometric variables, it is important to draw attention to $H$, $D$ and $X_c$ which, in the common absence of pronounced basal slope concavities, together define the major diagnostic features of the geometry of fluvially eroded terrains.

Horton (1945) and others have studied the network components, and relations have been established

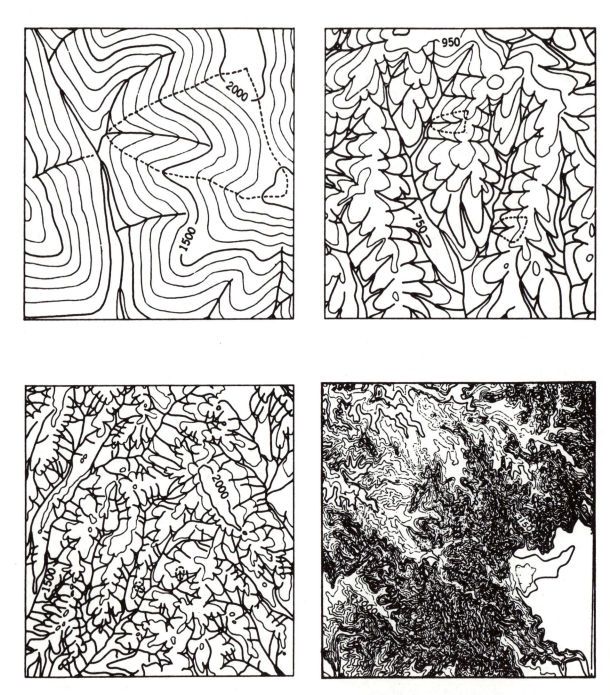

**Figure 13.5** Areas of 1 sq. mile each taken from the *US Geological Survey* topographic maps to illustrate natural range in drainage density: Top left – low drainage density (3–4 mile/sq. mile), Driftwood Quadrangle, Appalachian Plateau Province, Pennsylvania, Top right – medium drainage density (8–16 mile/sq. mile), Nashville Quadrangle, Interior Low Plateaus, Tennessee, Bottom left – high drainage density (20–30 mile/sq. mile), Little Tujunga Quadrangle, Coast Ranges, California, Bottom right – extremely high drainage density (200–400 + mile/sq. mile), Cuny Table West Quadrangle, Badlands, South Dakota.
*Source:* A. N. Strahler, *Introduction to Physical Geography*, 1965, figure 21.7, p. 310, copyright © John Wiley and Sons, by permission.

between stream order and the frequency or number of streams of each order and the lengths, gradients and drainage areas of streams of each order (Figure 13.9). A glance at Figure 13.2B shows that there are more first-order streams than any other order and that the first-order streams on the average are shorter and occupy smaller drainage basins. It has also been demonstrated that stream discharge increases systematically with order (Figure 13.10). These relations indicate, as should be

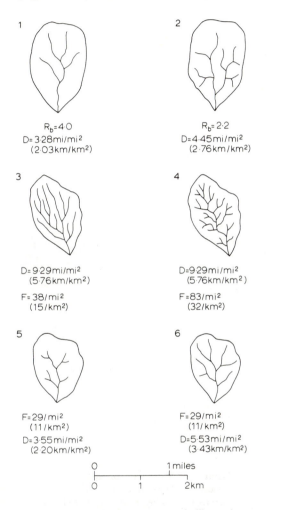

**1**

$R_b = 4.0$
D = 3.28 mi/mi²
(2.03 km/km²)

**2**

$R_b = 2.2$
D = 4.45 mi/mi²
(2.76 km/km²)

**3**

D = 9.29 mi/mi²
(5.76 km/km²)

F = 38/mi²
(15/km²)

**4**

D = 9.29 mi/mi²
(5.76 km/km²)

F = 83/mi²
(32/km²)

**5**

F = 29/mi²
(11/km²)

D = 3.55 mi/mi²
(2.20 km/km²)

**6**

F = 29/mi²
(11/km²)

D = 5.53 mi/mi²
(3.43 km/km²)

0        1 miles
0    1    2 km

**Figure 13.6** Idealized stream networks illustrating bifurcation ratio and the distinction between drainage density and stream frequency: 1 and 2, basins of somewhat similar drainage density but differing bifurcation ratio: 3 and 4, basins of identical drainage density but differing stream frequency: 5 and 6, basins of identical stream frequency but differing drainage density.
*Source:* Strahler, 1964, figures 4-II-4 and 4-II-16, pp. 44, 55, copyright © McGraw-Hill, New York.

expected, that the drainage network has developed in response to the erosive forces acting on the erodible materials that comprise the drainage basin. The result is a drainage pattern with characteristics that can be related to the erodibility of the material comprising the drainage basin as well as to climatic and hydrologic controls.

A problem frequently arises as to the accuracy of the required data which are obtained from maps, aerial photographs and measurements in the field. Fieldwork is often tedious and it can be expensive and time-consuming, but there can be no substitute for it. Drainage density is an important measure of the texture of the topography, and yet great care must be taken when measurements of drainage density are obtained from maps and photographs. For example, during the preparation of topographic maps cartographers rarely understand or appreciate the needs of the geomorphologist. Therefore, the drainage network that is shown by blue lines on topographic maps is not a total representation of that network. The blue lines on topographic maps in many cases designate streams that contained water at the time when the aerial photographs were taken. It is understandable, then, that depending on the time of year, the total length of blue lines on a topographic map will vary greatly, but for geomorphic purposes all drainage channels, wet or dry, must be measured or counted when drainage density, stream order, or stream frequency are determined.

A study of the *US Geological Survey* revealed that when four cartographic engineers were instructed to measure drainage density, they produced measured values of drainage density from 2.3 to 9.8 mile/sq. mile. When for the same area special maps of the area were prepared, it was found that the drainage density of perennial streams was 3.42 and the total drainage density was 10.31 (Figure 13.11). It is apparent that some cartographers were measuring only perennial streams or wet channels and others were attempting to measure not only the perennial, but also the dry ephemeral stream channels within the drainage basin. This study emphasizes the danger of attempting to use topographic maps alone for the study of drainage basin morphology. The ability to study drainage basins on maps and photographs depends to a large extent on the texture of the topography itself. Texture is simply the coarseness or fineness of the dissection of the drainage network. Drainage density varies from values of less than one to values in excess of a thousand mile/sq. mile in extremely fine-textured badland areas, where erosion is very rapid. The difficulty of comparing drainage density values for areas of different lithology is brought out clearly in Figure 13.12, which shows the change in drainage density values that are

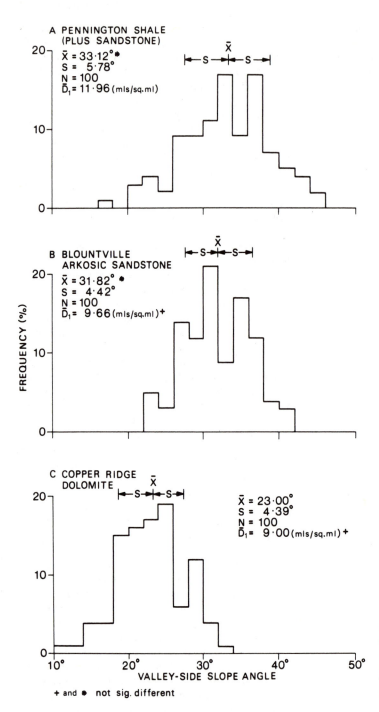

**Figure 13.7** Histograms of class frequencies of maximum valley-side slopes on three contrasting formations in the Ridge and Valley Province of the Appalachians, Virginia–Tennessee.
*Source:* Miller, 1953, figure 15.

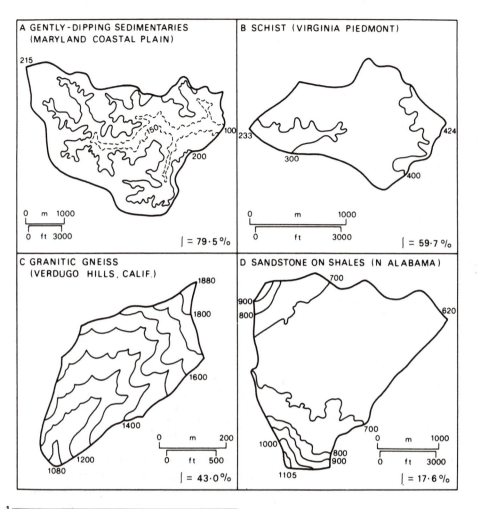

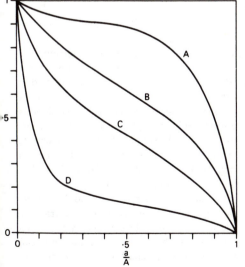

**Figure 13.8** The hypsometric integral: A.–D. four drainage basins showing heights in feet and hypsometric integrals; bottom, hypsometric curves for the four basins. *Source:* Strahler, 1952, figures 14, 15, 16, 17(1) and 18(1), pp. 1129–33.

obtained as photograph and map scale changes. Measurements of basins underlain by granite and sandstone show relatively minor, and perhaps insignificant, changes in drainage density as the photograph's scale changes from 1/10,000 to 1/70,000. Whereas the drainage basins formed on shale show a very significant decrease in drainage density. Hence comparisons of fine-textured topography should be made on photographs or maps of similar scale.

The preceding discussion was meant as a warning to those who use morphometric data. Remember that for some important descriptors, such as drainage density, subjectivity and differences of map and photograph scales may prevent a comparison of the results obtained by more than one investigator.

## 13.3 Morphometric controls

If the watershed is considered as a dynamic system with a history, then time can be important as the drainage basin passes through the various stages of its erosional evolution. However, for the moment, time is set aside and it is the character of the drainage basin in respect of the variables of relief, lithology, structure, climate and hydrology that concerns us.

*Relief* is an index of the potential energy available in the drainage basin and, of course, the greater the relief, the greater are the erosional forces acting on the basin. The clearest example of this is the rill experiment (see Figures 11.17 and 11.18) which shows the increase in drainage density, channel length and sediment produc-

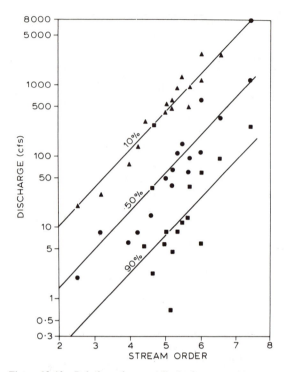

**Figure 13.10** Relation of stream discharge to stream order and frequency of occurrence (percentage of time a given discharge is exceeded). The symbols relate to the three frequency-of-occurrence lines.
*Source:* Stall and Fok, 1968, figure 13, p. 16.

tion from the higher relief surface. In addition to the greater drainage density, streams will be incised to greater depth in high-relief terrain and both channel gradients and valley-side slopes will be steeper (see Figure 11.5).

*Lithologic variations* are significant controls on morphology because they determine the erodibility of the surface materials and to a large extent determine the infiltration capacity of drainage basin materials. Figure 13.13 shows differences in drainage density as related to rock type with, of course, the resistant igneous rocks having the lowest drainage density and the weak shales a fine-textured topography of high *D*. Weak rocks in the Cheyenne River basin, Wyoming, show a considerable range of drainage density due more to difference in infiltration rates than to erodibility (Table 13.1). The shales produce large volumes of runoff from a given storm, whereas sandy formations do not, and drainage density is inversely related to infiltration. If a drainage network develops as a result of the interaction of eroding forces (i.e. runoff) and the resistance of soil and rock materials then, for a given equilibrium climatic and

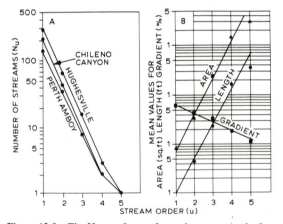

**Figure 13.9** The Horton Laws of morphometry: A. the Law of Stream Numbers for basins at Perth Amboy, New Jersey; Chileno Canyon, California; and Hughesville, Maryland.
B. The Law of Stream Lengths, Areas and Gradients for the basin at Perth Amboy, New Jersey.
*Source:* Schumm, 1956, figures 2 and 3, pp. 603, 604.

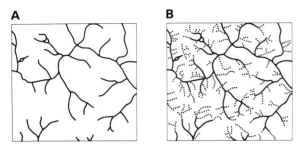

**Figure 13.11** Two maps of the drainage network of Beaver Run Creek area showing A. perennial streams and B. perennial and ephemeral streams.
*Source:* Schneider and Goodlett, 1962.

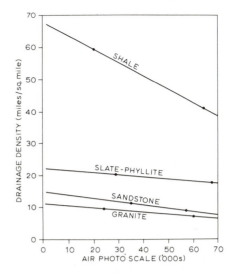

**Figure 13.12** Relation of measured drainage density to the scale of aerial photographs for four different rock types.
*Source:* Ray and Fischer, 1960, figure 1, p. 145.

vegetative situation, the texture of the pattern will vary significantly with lithology.

Most drainage networks are dendritic, having a branching treelike pattern, and it is interesting to speculate how drainage density may increase with time or due to changing erosional conditions. Obviously, existing channels lengthen, but much work suggests that significant differences in drainage density are due to the significantly larger number of channels that comprise the high drainage density patterns. Not only do the channels extend further towards the divide (see Figure 13.24), but many more tributaries fill the available space. Work by Melton (1958a, 1958b) indicates that the number of channel links increases as the square of drainage density (see Figure 13.23).

In addition to differences in texture alone, which reveal differences in rock type, the form of the drainage network itself is revealing of the underlying *geologic structure*. Figure 13.3 has already shown a number of drainage patterns associated with a variety of structural situations. These patterns can be equally clear on maps and air photographs, but often a pattern may occupy only a part of the area and close scrutiny is required to detect, for example, the annular component of a predominantly dendritic pattern.

Another significant control on both hydrology and drainage patterns is *climate* and its effects on vegetation and runoff. As demonstrated elsewhere, sediment yields are large from semi-arid regions, and not surprisingly drainage density is also greatest in semi-arid regions. This is very significant for interpreting the effects of climate change on the landscape during the Quaternary. The higher values in semi-arid regions are due to the protective influence of vegetation in humid regions and the lack of water to provide channels in arid regions. Thus the removal of vegetation will cause a major drainage density increase in subhumid and humid areas.

If a drainage basin and its network evolves with regard to climatic and lithologic controls, then there should exist clear relations between drainage basin morphology and its *hydrologic character*. For example, in small drainage basins where land use, geology and climate are similar,

**Table 13.1** Infiltration, sediment yield and drainage density related to lithologic units of the Cheyenne River basin.

| Stratigraphic units | Mean infiltration (in/h) | Sediment yield (acre-ft/sq. mile) | Drainage density (mile/sq. mile) | Drainage density (km/km²) |
|---|---|---|---|---|
| Wasatch Formation | 9·2 | 0·13 | 5·4 | 3·4 |
| Lance Formation | 5·0 | 0·5 | 7·1 | 4·4 |
| Fort Union Formation | 1·3 | 1·3 | 11·4 | 7·1 |
| Pierre Shale | 1·0 | 1·4 | 16·1 | 10·0 |
| White River group | 0·18 | 1·8 | 258 | 160·0 |

*Source:* Hadley and Schumm, 1961.

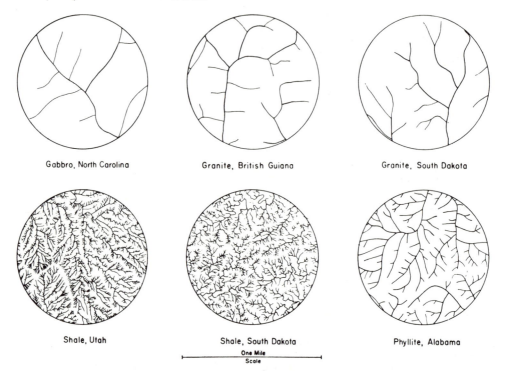

Gabbro, North Carolina

Granite, British Guiana

Granite, South Dakota

Shale, Utah

Shale, South Dakota

One Mile
Scale

Phyllite, Alabama

**Figure 13.13**  Drainage densities developed on four different rock types.
*Source:* Ray and Fischer, 1960, figure 2, p. 145.

highly significant relations can be established between runoff and sediment yield from these basins. In Figure 13.14 such relationships are presented for the Cheyenne River basin in Wyoming. The higher the drainage density, the more efficiently is the drainage basin drained and, therefore, the more rapid and the greater will be the quantity of water that leaves the basin. The greater the relief and the greater the drainage density, of course, the greater will be erosion and sediment yields. The relations shown in Figure 13.14 are very illuminating because they were developed for small drainage basins in an area of relatively uniform climate, vegetation and land use. A change in any of these variables throughout the region would have complicated these simple geomorphic hydrologic relations considerably.

Not only annual runoff, but also hydrograph characteristics are controlled by basin morphology. For example, Figure 13.15 shows the effect of increasing drainage density on a hydrograph generated by applying precipitation to the experimental watershed. A study by Carlston (1963) shows both the effect of basin morphology and geology on hydrology (Figure 13.16). He demonstrated that flood peaks increase significantly

with drainage density and the water applied to a high $D$ basin is evacuated quickly and efficiently. The result is that the low flow or baseflow is inversely related to drainage density because with a high drainage density little water infiltrates and is stored to supply the streams between storms. The differences again are related to infiltration rates.

Another important hydrologic characteristic of drainage basins that may be confusing is the inverse relation between sediment yield per unit area and drainage area (Figure 13.17). Obviously, the larger a drainage basin the more water and sediment is delivered from the basin, but on a unit area basis an increase of drainage area reduces sediment production. This apparent anomaly can be explained when the drainage basin is viewed in its entirety. Usually the upstream part of the basin near the drainage divide is the steepest, but downstream the valley widens, slopes become gentler, and the stream gradient and drainage density declines. Hence more sediment per unit area is produced in the headwaters, and the opportunity for sediment storage increases downstream. This will be better understood when the evolution of a basin through time is considered.

**Figure 13.14** Relations between A. mean annual runoff and drainage density, and B. mean annual sediment yield and relief ratio in respect of a small number of drainage basins.
*Source:* Hadley and Schumm, 1961, figures 31 and 33, pp. 174, 177.

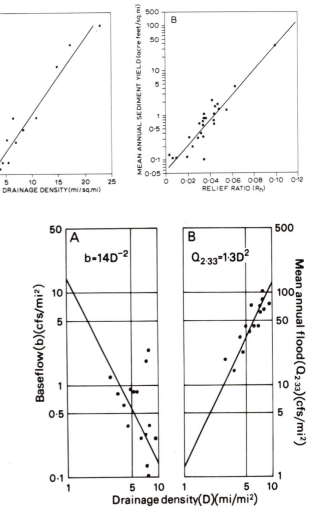

**Figure 13.16** The relations between drainage density and A. baseflow, and B. mean annual food, for fifteen locations in the eastern United States.
*Source:* Carlston, 1963, figure 3, p. 6.

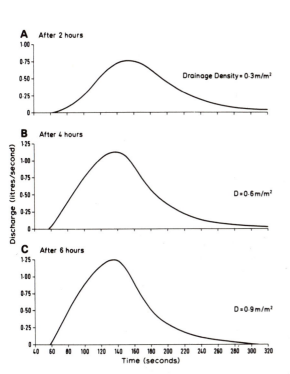

**Figure 13.15** The effect of increasing drainage density **D** on a hydrograph generated by applied rainfall to an experimental watershed after A. 2 h, B. 4 h and C. 6 h.
*Source:* Zimpfer, 1982.

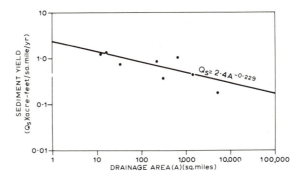

**Figure 13.17** Decrease of sediment yield per unit area with increase of drainage area for eight basins in the south-western United States.
*Source:* Strand, 1975.

## 13.4 Drainage basin evolution

As we have seen in Chapter 2, there are a number of strategies which are possible in order to try to visualize the long-term evolution of major landforms. Among these are the methods of denudation chronology, direct observation and measurement, mathematical simulation, the application of the ergodic hypothesis and simulation by physical experiments. The methods of denudation chronology are commonly limited in erosional drainage basins to dated surface, terrace, floodplain and channel sites. Direct observation and measurement is only possible in some categories of small basins on unresistant bedrock (e.g. badlands) or in

rapidly developing youthful basins. The spatial and temporal controls over erosion rates are so little understood that mathematical simulation is only in its infancy. Thus the most promising lines of research on drainage basin evolution are those relying on the ergodic assumption and on physical simulation.

### 13.4.1 The ergodic method

In the past the substitution of spatial variations for temporal ones has been a very popular geomorphic strategy to attack the problem presented by the long-term evolution of landforms (see Section 2.5.1). This is accomplished by selecting a sequence of landforms that appear

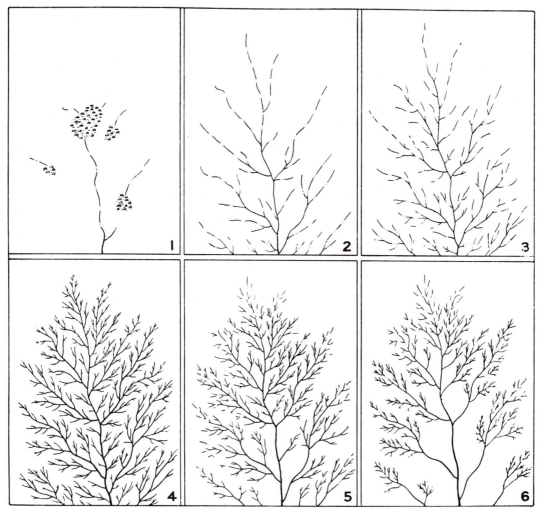

**Figure 13.18**   A proposed idealized development of drainage system through time showing: 1, initiation; 2, elongation: 3, elaboration: 4, maximum extension; and 5 and 6, integration.
*Source:* W. S. Glock, 'The development of drainage systems: a synoptic view', *Geographical Review*, vol. 21, 1931, figure 8, p. 481. Reprinted by permission of the American Geographical Society.

to show evolutionary development. A study by Glock (1931) of drainage pattern evolution is a good example (Figure 13.18). From his ergodic assumptions, Glock divided the evolutionary process into five stages: (1) initiation, the first development of a stream pattern on a pristine surface; (2) elongation, the growth of channels into the available area and the blocking out of the network; (3) elaboration, the filling in of the network by the addition of lower-order tributaries; (4) maximum extension, when drainage density is greatest and the drainage network completely fills the available space; and (5) integration, involving the loss of identity of the small order streams with progressive reduction of drainage density through time. Glock's scheme is reasonable and, in fact, at least a part of it is demonstrated by Ruhe's (1952) study of drainage patterns developed on four till sheets of different ages in Iowa (Figure 13.19), which suggests a rapid increase in drainage density and in

stream frequency through time to a maximum when drainage network growth ceases.

Drainage basins which appear to have evolved by progressive headward erosion into a terrace or tableland of known original form have proved to be the most fruitful ergodic devices in that such attributes as basin order, hypsometric integral and, on occasions, relief can be employed as surrogates for time. For example, comparisons of the relief of third-order, headward-eroding basins in Goulds Country, Tasmania, with bifurcation ratio (Figure 13.20) suggests that a decline in basin relief may be accompanied by a decrease in $R_b$ by the infill of first-order valleys and their blurring with adjoining hillsides. This seems to accord with Glock's stage of integration. Figure 11.25 has shown how the order of the basal stream channels might be used as an ergodic device for suggesting valley-side slope evolution where a channel network appears to be actively headcutting. A more

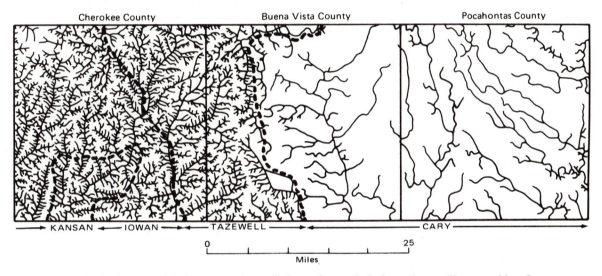

**Figure 13.19** The development of drainage networks on till sheets of successively decreasing age (Kansan – oldest; Cary – youngest) in north-west Iowa.
*Source:* R. V. Ruhe, *Amer. Journal Sci.*, 1952, figure 2, p. 51.

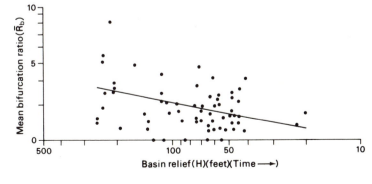

**Figure 13.20** The bifurcation ratio plotted against relief of third-order, headward-eroding basins in Goulds County, Tasmania.
*Source:* Abrahams, 1972, figure 4, p. 631.

elaborate analysis of ninety basins of the Rocksberg drainage system cut into a more than 160-m (500-ft) high retreating escarpment some 50 km north-west of Brisbane, Australia, employed stream order to predict sequential changes in four general slope characteristics (Figure 13.21). Progressive basin incision appears to be resulting in increasing mean valley-side slope length, an increase of mean maximum slope angle up to the fifth order and then decline (similar to the preceding Connecticut analysis), a sharpening of divides up to the fourth order and then rounding, and an inverse effect with regard to the lower slope concavity. Where a drainage network had cut back into the Perth Amboy clay badlands (possessing a known original terrace surface elevation), the hypsometric integrals of the basin hierarchy were employed as an ergodic device (Figure 13.22A) the value of which was assumed to be inversely related to the amount of elapsed erosional time. A plot of drainage density against hypsometric integral (i.e. inverse time) suggested that denudation was here accompanied by an initially rapid increase in drainage density which later decreased to an equilibrium value of about 1300 mile/sq. mile (800 km/km²) (Figure 13.22B).

A different type of inferential ergodic investigation into the possible manner of evolution of stream networks was based on the study of a large number of drainage basins in the western United States by Melton (1958a). It was assumed that basin forms must have evolved in some way during time and that therefore they must represent, at this time instant, a considerable range of ages, so that this large basin sample must include representatives of many stages of development. However, there were apparently none of the systematic variations which one might expect if absolute age is associated with variations in drainage density $D$. and it was assumed not altogether surprisingly, that the wide range of climatic, soil and geological environments represented must also have affected the range of network forms observed. As already shown, drainage density ($D = L_c/A$) is distinct from stream frequency ($F = \Sigma N_u/A$). For sample areas which are much larger than their component first-order basins, $D$ and $F$ showed no systematic variations with $A$, but data for the mature basins (i.e. in which smooth valley-side slopes continue up to the divides) showed a systematic relationship between $D$ and $F$, with a coefficient of correlation of $+0.97$ such that for all basins it was possible to propose that $F/D^2$ (the relative channel density) is a dimensionless constant (Figure 13.23). The relative channel density, therefore, appears to represent a basic law of behaviour of planimetric elements of maturely developed drainage basins and is a dimensionless measure of the completeness with which the channel net fills the basin outline for a given number of channel segments, and for any value of $A$. The function $F = 0.694D^2$ thus represents a drainage network growth model, which shows that as drainage density increases within a constant area, it does so by an accompanying increase in stream frequency.

### 13.4.2 Physical simulation

The other main way that drainage network evolution can be investigated is by experimental simulation studies, several of which have been carried out in the rainfall-erosion facility (REF) at Colorado State University. When precipitation was applied to a gently sloping surface composed of two intersecting planes which concentrated the runoff, a dendritic drainage pattern began to develop (Figure 13.24A). As more artificial precipitation was applied there was a rapid increase in drainage density and stream frequency before maximum extension of the drainage network was achieved (Figure 13.24B–F and Figure 13.25). For this experiment, drainage density remained relatively constant for a considerable time at maximum extension. However, as one views the drainage patterns, it is apparent that this constancy of drainage density is a result of a rather dynamic change within the drainage pattern itself (Figure 13.24E–F). For example, during this period of constant drainage density there is a loss of the lower-order tributaries in the centre of the drainage basin, whereas there is addition of first-order tributaries at the periphery of the drainage network. So

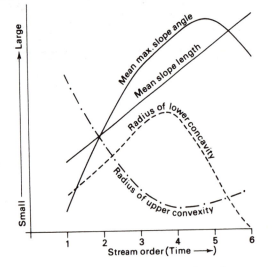

**Figure 13.21** Relations between stream order and mean valley-side slope length, mean maximum slope angle, radius of lower concavity of slope and radius of upper convexity. *Source:* Arnett and Conacher, 1973, figure 2, p. 239.

**A**

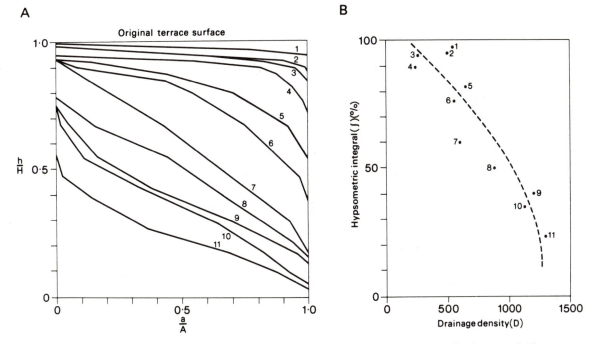

**B**

**Figure 13.22**  A. Hypsometric curves for eleven small drainage basins in the eroding Perth Amboy badlands terrace. B. The increase of drainage density (mile/sq. mile) with the progress of erosion through time (i.e. decrease of hypsometric integral) in respect of the eleven basins in A.
*Source:* Schumm, 1956, figures 19 and 20D, pp. 615 and 617.

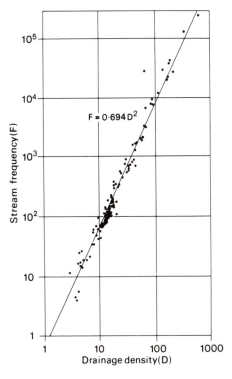

network growth is continuing at the margins of the network in those areas of steeper slope, whereas the older interior portions of the network enter Glock's phase of integration and channel loss (Figure 13.24 G–H). Erosion, of course, begins near the outlet of the basin and there is initially very high sediment production (Figure 13.26). However, the rate of sediment yield decreases rapidly from this peak through time. This is partly due to the fact that the rate of network growth is decreasing but, in addition, as the main channels enlarge and the valleys are widened there is an increasing opportunity for the storage of alluvium in the drainage basin. The growth of the experimental drainage network conforms to Glock's thesis, providing an interesting insight into the workings of a drainage basin in that it is very much like the expected result of Davis's theoretical cycle of erosion. Figure 13.24 shows the headward growth of a drainage network to maximum extension and then the beginning of abstraction on a surface unaffected by lithologic and structural differences, perhaps being

**Figure 13.23**  Relation between stream frequency and drainage density mile/sq. mile in respect of 156 basins in the south-western United States.
*Source:* Melton, 1958A, figure 1, p. 37.

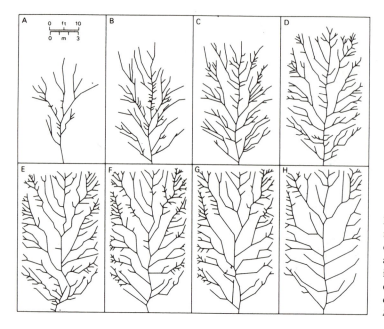

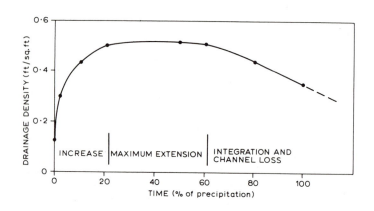

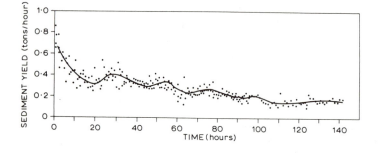

**Figure 13.24** Map of growth of a drainage network developed by spraying a model catchment of 3·2 per cent slope with artificial rainfall: A.–D. show a progressive increase, E. and F. show maximum extension and G. and H. integration and channel loss (see figure 13.25).
*Source:* Parker, 1977, pp. 54, 55.

**Figure 13.25** Drainage density change during the erosion of a large model catchment in the rainfall-erosion facility at Colorado State University; baselevel was stable during the experiment and the initial general slope of the catchment was 3·2 per cent. Time is expressed as a percentage of the total rainfall applied during the experiment.
*Source:* Parker, 1976.

**Figure 13.26** Sediment-yield variations during the erosion of an experimental drainage basin, following a 10-cm drop in baselevel. Note that the decrease in the initially high yield was punctuated by secondary sediment pulses at about 25, 55, 75 and 100 h suggestive of complex basin response. The fitted line is based on a moving mean of three points, with an average of ten points.
*Source:* Parker, 1976.

analogous to the pattern developments on an emerged coastal plain.

Another interesting aspect of drainage evolution is *stream capture*, the natural diversion of one stream by another. Usually one stream is either flowing at a lower level or is much more aggressive due to high discharge which permits it to erode into the adjacent drainage area (Figure 13.27). Numerous locations where capture is imminent or where it has probably occurred can be noted in Figure 13.24 (especially C–D and G–H). Capture thus appears to be a major ingredient of drainage pattern evolution. Capture is not only of geomorphic interest because, if the beheaded stream contains gold or diamonds, paleoplacers are preserved in the beheaded channel; and exploration for placers frequently involves working out the original drainage pattern that is now disrupted by capture.

## 13.5 Drainage basin response

The discussion of drainage basin controls and drainage network evolution reveals that there may be an orderly growth pattern reflecting the nature of the materials and the eroding forces applied. During the experimental studies, a drainage network filled the available space and developed a drainage density that at maximum extension

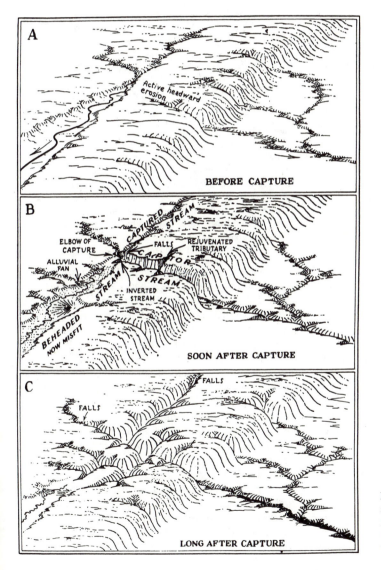

**Figure 13.27** Stages in the process of stream capture.
*Source:* A. K. Lobeck, *Geomorphology*, 1939, p. 198, copyright © McGraw-Hill, New York.

is as characteristic of that terrain as are the characteristic slopes defined earlier (Chapter 11). In fact, when the inverse of drainage density is calculated, this provides a measure of a unit drainage area required to maintain a unit length (i.e. 1 m) of stream channel. In badlands 1 m² is sufficient, but in coarser-textured topography much higher values are required. This number ideally should be a constant for any drainage basin, but except during an early stage of maximum drainage development, this is not the case. Nevertheless, the general existence of such a regular relationship shows the influence of the earth materials involved in network development. It also suggests explanations for the orderly drainage patterns observed on the earth's surface and for the variations of drainage pattern with lithology.

It is clear that a natural change, such as of climate or of baselevel lowering, may cause rejuvenation of the basin and an extension of the network with an increase of drainage density, runoff and sediment yield. Changes of drainage pattern would be similar to those shown in Figure 13.11. Man's activities such as lumbering, farming and the effects of overgrazing may have the same effects. Forest fires also induce serious erosion and drainage pattern expansion by destruction of the protective vegetative cover.

The preceding suggests that the evolutionary development of a drainage basin and the landscape will be complex. Any change in an external variable, such as climate, land use, or tectonics, will have its effect on the drainage pattern. Mosley's (1972) experiments on rill erosion, as described earlier (see Figure 11.17), illustrate how drainage density and sediment production change as slope increases.

An important external change is baselevel lowering, which is easily simulated experimentally in a model erosion facility (REF). In Figure 13.28A a dendritic drainage network has only partly filled the space available because relief is low and the network has reached maximum extension on a 0·75 per cent slope. During subsequent experimental runs (B–F) the baselevel was lowered, thereby increasing the relief from about 0·3 m to 1 m. The increase of drainage density following a drop in baselevel was greater than that with no change in baselevel. During other similar experiments the effect of baselevel change on stream frequency and sediment production were carefully monitored (Figure 13.29). The lowering of baselevel created a nickpoint in the stream profile which migrated upstream, successively rejuvenating tributaries as it did so. This can be viewed as a wave of rejuvenation that not only advanced up the main channel, but also up tributaries of decreasing order number. In each valley the passage of the nickpoint leaves the

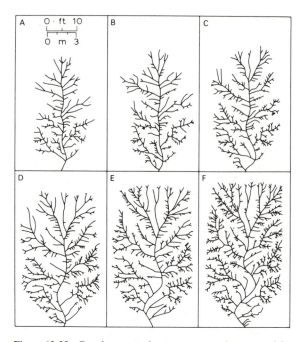

**Figure 13.28**   Development of a stream network on a model catchment of slope 0·75 per cent under artificial precipitation and with progressive drops in baselevel: A. network developed after 14 h of precipitation of an intensity of 2.61 in/h; B. network developed following a drop in baselevel at the outlet of 0·1 m (0·3 ft) and a further 13 h precipitation; C.–F. further networks developed after repeating the drop in baselevel four times, each followed by 13 h precipitation.
*Source:* Parker, 1977, pp. 52, 53.

incised previous floodplain preserved as a terrace. The rejuvenation of the drainage basin, therefore, not only has a significant influence on the drainage network, but also on the valleys and channels (see Chapter 14). In addition, the increased sediment production may significantly influence downstream areas, thereby causing aggradation. Strahler (1958) illustrates the effect of such a rejuvenation and drainage density transformation in an extreme situation where a major increase in drainage density induces valley aggradation (Figure 13.30). It is thus important to recognize that each component of the drainage basin is part of an evolving, dynamic system within which, as one morphologic component changes, others respond.

It would be wrong, however, to assume that the tendency towards this mutual adjustment of basin forms and processes in response to changing conditions is necessarily always rapid or synchronously accomplished throughout the basin. Most mesoscale landform assemblages, including the majority of higher-order drainage

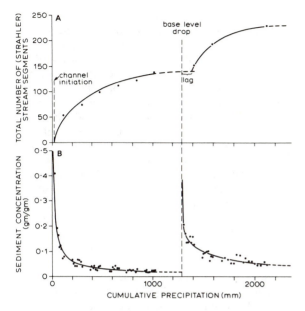

**Figure 13.29** Effect of cumulative precipitation and baselevel drop on A. total number of Strahler stream segments, and B. sediment concentration in an experimental stream network development. Precipitation is a surrogate for time and the exponential increase of stream segments (after a lag) and decrease of sediment concentration are noteworthy. *Source:* Schumm *et al.*, 1983, figure 3-8, after McLane, 1978.

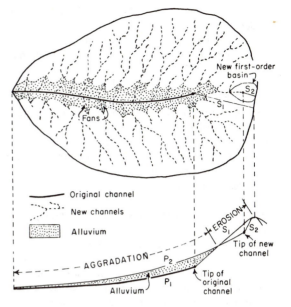

**Figure 13.30** Drainage basin transformation from low to high drainage density due to severe gullying and badland development on slopes accompanied by aggradation in the main valley axis. Profiles show that both the channel and slope gradients are increased by the transformation. *Source:* Strahler, 1958, figure 7, p. 297.

basins, represent a complex of variously linked process–response subsystems. The form and behaviour of each of these subsystems bears the current imprint of processes of differing antiquity and is being transformed at an individual rate under the dominant influence of processes of specific magnitudes.

Rates of transformation of individual landforms are determined by a tradeoff between their *sensitivity* and their *persistence* or *recovery* (see Figures 1.4 and 1.5). Sensitivity may be defined as the likelihood, extent and rapidity with which a given landform will change in response to a single unusual process event, or to a reinforcing sequence of such events (see Brunsden and Thornes, 1979, p. 476). Each such 'effective formative event' produces an effect which is immediately susceptible to recovery towards its original state. Different forms recover at different rates and their persistence depends partly on this recovery rate and partly on the reinforcing effects of recurring formative events (see Wolman and Gerson, 1978, p. 190). For example, infrequent, high-intensity rainfall may produce the following erosional and depositional changes in a drainage basin:

(1) Shallow slides on valley sides.

(2) Basal slope slips due to channel widening and undercutting.

(3) Increase in the width–depth ratio of the alluvial channel.

(4) Deposits on the floodplain of flood bars and of material slumped from the valley side.

Although comparative data are generally lacking, it is possible to make some estimates of the relations between these changes, with reference to the Exmoor, England, 150 + year flood of August 1952 (Anderson and Calver, 1977), the 100–200 + year flood of June 1972 on the Piedmont in Maryland resulting from Hurricane Agnes (Costa, 1974; Gupta and Fox, 1974) and the high-intensity flood of 1969 in Nelson County, Virginia (Williams and Guy, 1971). These tentative estimates include the following:

(a) In a region of reasonably high relief a 100–200 + year flood will mobilize sediment some 70 per cent of which may come from the hillslopes and 30 per cent from the channels.

(b) Shallow slides (12 m long and 0·5 m deep) developed on 20°–25° valley sides are still apparent after twenty years but would be expected to be obliterated long before the 100–200 + year recurrence interval. In Japan such obliteration may occur in under twenty-five years.

(c) Basal slope slips due to stream undercutting may be more persistent than (b), and some of these may show evidence of further development after the event, under events of lesser magnitude.

(d) The width of the active river channel appears to be well on the way to narrowing and recovery a mere year or two after the event, but decrease of scour depth may take longer.

(e) Depositional features on the floodplain appear quite persistent and may be the most persistent of the features mentioned.

The above considerations raise two important points. The first is what is meant by 'recovery'. For example, is a hillslope slump scar filled with fine alluvium 'recovered' – in the sense that a future event of a lesser magnitude than that which initially formed the scar would re-excavate it? The second is when a landscape is acted upon by an infrequent event of given magnitude and frequency – is it more or less susceptible to the future action of an event of similarly high magnitude?

In general, landforms in a given region may form a spectrum between mobile, fast-responding subsystems and slowly responding, insensitive areas. The 'fast' landforms, such as alluvial channels, have a high sensitivity to externally generated pulses; react quickly and relax to new system states with facility; are relatively sensitive to climatic variations and act as energy filters, removing the main impulse and passing on only minor changes to contiguous subsystems. These areas are morphologically complex because not only are they subject to rapid change and therefore exhibit transient forms, but they are also capable of rapid restoration and achievement of new stable states. The insensitive landforms, such as interfluves, occur where the ratio of stress to resistance is low. They are passive, insensitive to external effects (such as climate) and therefore change but slowly (mostly verbatim from Brunsden and Thornes, 1979, p. 479).

For example, a well-vegetated humid drainage basin of some tens of square kilometres in area may exhibit landforms of widely differing antiquity. The hydraulic geometry of the alluvial channel may be a reflection of the last bankfull event ($10^0$ years ago); the abandoned floodplain features of meander scrolls occurring with a frequency of tens or hundreds of years ($10^1$–$10^2$ years); and the cut and fill of valley heads and the cycles of trenching of lower slope concavities may be accomplished within hundreds of years ($10^2$ years). At the other extreme, the elevation of major divides and the general form of the upper convexity of slopes may reflect uplift events which took place hundreds of thousands or even millions of years ago ($10^5$–$10^6$ years) or rates and patterns of surficial creep which were dominant within a period of thousands or tens of thousands of years ago ($10^3$–$10^4$ years). Intermediate between these extremes lie the extensive and topographically dominant valley-side slopes where the complexities of the association of features of differing age reach a maximum. Gullies and slump scars resulting from storms of a recurrence interval of a hundred years or more ($10^2$ years) may persist before infilling or, indeed, may form permanent features of the drainage network, whereas the general form of the whole slope is geared to the hundreds of thousands or millions of years during which the associated downcutting of the basal stream channels and the divides occurs ($10^5$–$10^6$ years). It is this degree of complexity which presents the biggest challenge for the geomorphologist who wishes to study drainage basin evolutional response.

There are two main ways by which it is possible to gain some information regarding the nature of such a response. The first is to measure the comparative rates and directions of change of associated landforms, and the second is to try to establish correlations between the features of spatially linked landforms which are assumed to have been influenced by the same set of processes.

The non-steady states existing between the rate at which material is eroded from upland slopes and that at which it is evacuated from the lower floodplains is an important aspect of basin measurement. Sediment appears to move episodically through many larger drainage basins, due to the existence of internal and external thresholds leading to a complex response. The present average upland erosion rate in ten catchments in the southern Appalachians of 950 mm/1000 years compares with sediment yields to the sea of only 53 mm/1000 years (delivery ratio 6 per cent), and the delivery ratio for the southern Piedmont is only 4·7 per cent (Trimble, 1975, 1977). This discrepancy is due to sediment storage in floodplains, channels and reservoirs, and it has been demonstrated by studying the sediments in the valleys of northern Mississippi which contain recent fallout cesium-137 that present sedimentation rates there are as much as 9–65 mm/year (Ritchie *et al.*, 1975). Within small drainage basins in the Colorado Rockies measured denudation rates for the convex divides (<0·14 mm/year) exceed average basin sediment export rates (<0·093 mm/year) (Caine, 1974). For one of these basins, Williams Fork, San Juan Mountains (0·98 km²), an attempt has been made to state measured rates of interfluve and channel erosion in comparable terms, employing estimates of work (in watts per square kilometre) (Table 13.2).

**Table 13.2**  Erosional Work (watt/km²)

| Process | | Williams Fork, San Juan Mountains | Kärkevagge, Sweden | Two O'Clock Creek, Canadian Rockies |
|---|---|---|---|---|
| Slopes and divides | Surface water erosion | 0·147* | small | |
| | Soil creep and solifluction | 0·109 | 0·166 | |
| | Earth slides and mudflows | 0·362† | 1·997 | |
| | Rockfall | 0·133 | 0·405 | |
| | Avalanches | 0·001 | 0·453 | |
| Channels | Dissolved load | 0·104 | 2·828 | 20·104 |
| | Suspended load | small? | ? | 587·200‡ |
| | Bedload | small? | ? | 2·970‡ |

\* Mainly summer rainbeat; however, total annual rainbeat generates 96 watt/km²!
† Mainly due to one eighteen-year recurrence interval mudflow.
‡ Eighty-seven per cent of suspended load and 65 per cent of bedload removed by a snowmelt flood during the first few days in June.
*Source:* Caine, 1976, table 2, p. 139.

The table shows some comparability with the periglacial conditions of the Kärkevagge area of northern Sweden, but also a great disparity between measured geomorphic work done and the total receipt of radiant solar energy by the Williams Fork basin. It was estimated that even if only 5 per cent of the total basin heat budget was involved in exchanges with the soil in the form of geomorphic work, this would still be seven or eight orders of magnitude greater than that producing the observed movement of debris within the basin! However, the possible episodic movement of slope and stream debris may have caused the effects of certain processes to be grossly underestimated. Two O'Clock Creek in the Canadian Rockies (9·1 km²) is, by contrast, yielding suspended load at a rate equivalent to an average denudation of 0·480 mm/year over the whole basin, and is probably deepening (McPherson, 1971).

Correlations of landform features have commonly involved sets of valley-side slope properties and those of basal streams. Where streams are actively deepening or removing material from a slope base, the form of the valley-side slope is dominantly associated with the rate of debris removal by the stream, as has been demonstrated by studies of valley-side slope asymmetry. This is further supported by the use of stream channel gradients $S_c$ as a surrogate for stream activity such that correlations between $S_c$ and valley-side slope angles $S_g$ are held to express the above interdependence. For a given region, significant relations between mean values of $S_c$ and $S_g$ have been held to imply some adjustment between rates of debris transport in the slope and channel systems. Figure 13.31 suggests such a stream/slope adjustment for

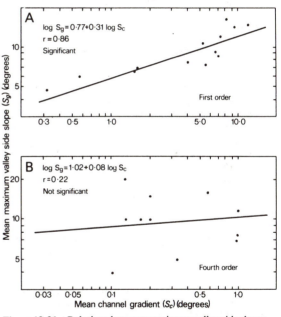

**Figure 13.31**  Relations between maximum valley-side slopes and basal channel gradients in respect of first- and fourth-order streams in the River Yeo catchment, Devon, England; data from 1:25,000 maps.
*Source:* Richards, 1977, figure 2, p. 89, copyright © John Wiley and Sons, by permission.

first-order basins which does not exist for fourth-order basins where it is probable that downcutting is slower, that debris storage in floodplains is taking place and that slope basal undercutting is occurring only intermittently.

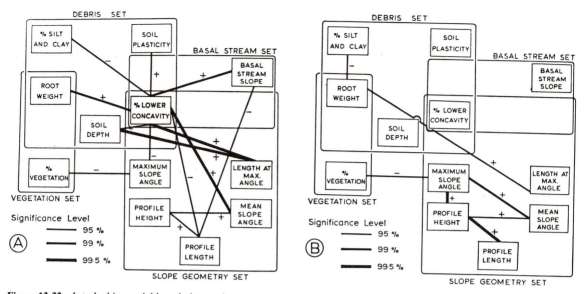

**Figure 13.32** Interlocking variables relating to slope geometry, debris, vegetation and basal stream activity for slopes on the Charmouthian Limestone, Plateau de Bassigny, northern France: A. where the stream is adjacent to the slope base; B. where the stream is away from the slope base. Significant correlation links are shown, with thicknesses indicative of significance levels. Direct relationships are shown by a plus sign and indirect ones by a minus sign.
*Source:* Chorley and Kennedy, 1971, figure 6.7, p. 211, after Kennedy.

The ability of active basal stream action to produce well-correlated sets of geomorphic variables (Melton, 1958b) can be illustrated by correlation structures derived by Kennedy in respect of slope localities with and without basal streams in the region of Charmouthian Limestone, Plateau de Bassigny, northern France (Figure 13.32). Of course, to obtain a universally adjusted steady state surface, rather than scattered adjustments of individual slopes with basal stream segments, would require an unreasonably long, stable period of denudation.

# References

Abrahams, A. D. (1972) 'Factor analysis of drainage properties: evidence for stream abstraction accompanying the degradation of relief', *Water Resources Research*, vol. 8, 624–33.

Ahnert, F. (1976) 'Brief description of a comprehensive three-dimensional process–response model of landform development', *Zeitschrift für Geomorphologie*, Supplement 25, 29–49.

Anderson, M. A. and Calver, A. (1977) 'On the persistence of landscape features formed by a large flood', *Transactions of the Institute of British Geographers*, n.s., vol. 2, 243–54.

Arnett, R. R. and Conacher, A. J. (1973) 'Drainage basin expansion and the nine unit landsurface model', *The Australian Geographer*, vol. 12, 237–49.

Brunsden, D. and Thornes, J. B. (1979) 'Landscape sensitivity and change', *Transactions of the Institute of British Geographers*, vol. 4, 463–84.

Caine, N. (1974) 'The geomorphic processes of the Alpine environment', in J. D. Ives and R. G. Barry (eds), *Arctic and Alpine Environments*, London, Methuen, 721–48.

Caine, N. (1976) 'A uniform measure of subaerial erosion', *Bulletin of the Geological Society of America*, vol. 87, 137–40.

Carlston, C. W. (1963) 'Drainage density and stream flow', *US Geological Survey Professional Paper* 422-C.

Carter, C. S. and Chorley, R. J. (1961) 'Early slope development in an expanding stream system', *Geological Magazine*, vol. 98, 117–30.

Chorley, R. J. and Kennedy, B. A. (1971) *Physical Geography: A Systems Approach*, London, Prentice-Hall.

Costa, J. E. (1974) 'Response and recovery of a Piedmont watershed from tropical storm Agnes, June 1972', *Water Resources Research*, vol. 10, 106–12.

Davis, W. M. (1899) 'The geographical cycle', *Geographical Journal*, vol. 14, 481–504.

Fenneman, N. M. (1914) 'Physiographic boundaries within the

United States', *Annals of the Association of American Geographers*, vol. 4, 84–134.

Gilbert, G. K. (1877) *Report on the Geology of the Henry Mountains*, Washington, DC, US Department of the Interior.

Glock, W. S. (1931) 'The development of drainage systems: a synoptic view', *Geographical Review*, vol. 21, 475–82.

Gupta, A. and Fox, H. (1974) 'Effects of high-magnitude flow on channel form: A case study in Maryland Piedmont', *Water Resources Research*, vol. 10, 499–509.

Hadley, R. F. and Schumm, S. A. (1961) 'Sediment sources and drainage-basin characteristics in upper Cheyenne River basin', *US Geological Survey Water Supply Paper* 1531-B, 137–96.

Horton, R. E. (1945) 'Erosional development of streams and their drainage basins: hydrophysical approach to quantitative morphology', *Bulletin of the Geological Society of America*, vol. 56, 275–370.

Howard, A. D. (1967) 'Drainage analysis in geologic interpretation: a summation', *Bulletin of the American Association of Petroleum Geologists*, vol. 51, 2246–59.

Judson, S. and Andrews, G. W. (1955) 'Pattern and form of some valleys in the Driftless Area, Wisconsin', *Journal of Geology*, vol. 63, 328–36.

Lobeck, A. K. (1939) *Geomorphology*, New York, McGraw-Hill.

Luke, J. C. (1972) 'Mathematical models for landform evolution', *Journal of Geophysical Research*, vol. 77, 2460–4.

McLane, C. F., III (1978) 'Channel network growth: an experimental study', unpublished MS thesis, Colorado State University.

McPherson, H. J. (1971) 'Dissolved, suspended and bed load movement patterns in Two O'Clock Creek, Rocky Mountains, Canada, Summer 1969', *Journal of Hydrology*, vol. 12, 221–33.

McPherson, H. J. (1975) 'Sediment yields from intermediate-sized stream basins in southern Alberta', *Journal of Hydrology*, vol. 25, 243–57.

Melton, M. A. (1958a) 'Geometric properties of mature drainage systems and their representation in an $E_4$ phase space', *Journal of Geology*, vol. 66, 35–56.

Melton, M. A. (1958b) 'Correlation structure of morphometric properties of drainage systems and their controlling agents', *Journal of Geology*, vol. 66, 442–60.

Miller, V. C. (1953) 'A quantitative geomorphic study of drainage basin characteristics in the Clinch Mountain area, Virginia and Tennessee', Columbia University, Department of Geology, Technical Report No. 3, Contract N6 ONR 271–30.

Mosley, M. P. (1972) 'An experimental study of rill erosion', unpublished MS thesis, Colorado State University.

Parker, R. S. (1976) 'Experimental study of drainage system evolution', unpublished report, Colorado State University.

Parker, R. S. (1977) 'Experimental study of drainage basin evolution and its hydrologic implications', *Colorado State University Hydrology Paper* 90.

Peltier, L. C. (1962) 'Area sampling for terrain analysis', *Professional Geographer*, vol. 14, 24–8.

Playfair, J. (1802) *Illustrations of the Huttonian Theory of the Earth*, Edinburgh, William Creech.

Ray, R. G. and Fischer, W. A. (1960) 'Quantitative photography – a geologic research tool', *Photogrammetric Engineering*, vol. 25, 143–50.

Richards, K. S. (1977) 'Slope form and basal stream relationships', *Earth Surface Processes*, vol. 2, 87–95.

Ritchie, J. C., Hawks, P. H. and McHenry, J. R. (1975) 'Deposition rates in valleys using Fallout Cesium-137', *Bulletin of the Geological Society of America*, vol. 86, 1128–30.

Ruhe, R. V. (1975) *Geomorphology*, Boston, Mass., Houghton Mifflin.

Schneider, W. J. and Goodlett, J. C. (1962) 'Portrayal of drainage and vegetation on topographic maps', *US Geological Survey, Water Resources Division, Open-File Report*.

Schumm, S. A. (1956) 'The evolution of drainage systems and slopes in badlands at Perth Amboy, New Jersey', *Bulletin of the Geological Society of America*, vol. 67, 597–646.

Schumm, S. A. (1977) *The Fluvial System*, New York, Wiley-Interscience.

Schumm, S. A., Harvey, M. D. and Watson, C. C. (1983) *Incised Channels*, Littleton, Colo., Water Resources Publications.

Shreve, R. L. (1966) 'Statistical law of stream numbers', *Journal of Geology*, vol. 74, 17–37.

Smith, T. R. and Bretherton, F. P. (1972) 'Stability and conservation of mass in drainage basin evolution', *Water Resources Research*, vol. 8, 1506–29.

Stall, J. B. and Fok, Yu-Si (1968) 'Hydraulic geometry of Illinois streams', *University of Illinois, Water Resources Center, Research Report* 15.

Strahler, A. N. (1952) 'Hypsometric (area–altitude) analysis of erosional topography', *Bulletin of the Geological Society of America*, vol. 63, 1117–42.

Strahler, A. N. (1958) 'Dimensional analysis applied to fluvially eroded landforms', *Bulletin of the Geological Society of America*, vol. 69, 279–300.

Strahler, A. N. (1964) 'Quantitative geomorphology of drainage basins and channel networks', in V. T. Chow (ed.), *Handbook of Applied Hydrology*, New York, McGraw-Hill, section 4-II.

Strand, R. I. (1975) 'Bureau of Reclamation procedures for predicting sediment yield', *Agricultural Research Service ARS-S-40*, 10–15.

Trimble, S. W. (1975) 'Denudation studies: can we assume stream steady state?', *Science*, vol. 188, 1207–8.

Trimble, S. W. (1977) 'The fallacy of stream equilibrium in contemporary denudation studies', *American Journal of Science*, vol. 277, 876–87.

Williams, G. P. and Guy, H. P. (1971) 'Debris avalanches: a geomorphic hazard', in D. R. Coates (ed.), *Environmental Geomorphology*, Binghamton, State University of New York, Publications in Geomorphology, 25–48.

Wolman, M. G. and Gerson, R. (1978) 'Relative scales of time and effectiveness of climate in watershed geomorphology', *Earth Surface Processes*, vol. 3, 189–208.

Wooldridge, S. W. (1932) 'The cycle of erosion and the representation of relief', *Scottish Geographical Magazine*, vol. 48, 30–6.

Zimpfer, G. L. (1982) 'Hydrology and geomorphology of an experimental drainage basin', unpublished PhD dissertation, Colorado State University.

# Fourteen *Fluvial depositional landforms*

In topographically low parts of the landscape (e.g. valleys, piedmonts, coastal plains etc.) sediments are deposited. Many of the landforms discussed previously have been dominantly erosional in origin, and it is now appropriate to treat fluvial depositional landforms, such as valley fills, alluvial fans and deltas. Coastal, aeolian and glacial depositional landforms will be treated in the next three chapters.

Fluvial sediments are laid down, and depositional landforms created, at three dominantly or partially sub-aerial locations; topographic discontinuities, valleys and the margins of water bodies:

(1) Topographic discontinuities may be created by faulting, tectonic movements, glacial overdeepening, marine abrasion, etc. These result, under suitable conditions, in the deposition of such features as alluvial fans, screes and talus slopes.

(2) Valley fills reflect a complex set of alluvial processes involved in the production of floodplains and terraces.

(3) Water margin deposits are the result of sedimentation in standing bodies of water where the velocities of transporting currents are decreased. Deltas and beach deposits are the major landform categories here, the second of these being treated in Chapter 15.

It is important to note that all dominantly depositional landforms show evidence of a complex history involving erosional as well as depositional episodes.

The major characteristic of those river channels that operate near base level is that they tend to be unconfined for the most part, i.e. the channels are able to shift position over rather wide areas on deltas and on coastal and alluvial plains, whereas upstream channels are confined within their valleys, and their lateral movement is relatively restricted. Hence the possibility for significant migration by avulsion (see Section 14.3.3) is present. In addition, the downstream effects of changing sea level exercise a very important role in determining the morphology of these streams and the character of the sediments that are delivered to a depositional basin.

These fluvial deposits are of considerable interest because of their economic importance. They normally are sources of groundwater and a knowledge of permeability variation is essential but, in addition, they may contain heavy mineral deposits, placers, diamonds and gold. Paleochannels and paleovalley fills not only contain placer deposits, but also they provide conduits for the migration of uranium-bearing solutions and petroleum. Ancient delta and beach deposits are frequently sources of petroleum and natural gas.

## 14.1 Alluvial fans

Alluvial fans occur in a variety of environments, particularly in arid and semi-arid or seasonally dry regions, where there is a large sediment supply to a point where accumulation can occur. Such locations are particularly common along faulted or tectonic mountain fronts. Depending on the magnitude and persistence of the fluvial processes, the radii of alluvial fans can vary from a few hundred metres to tens of kilometres. Smaller fans commonly have gradients of 3°–6° or less, steepening to some 10° near the apex where they may grade into marginal screes lying at angles approaching 30°. The gross form of a fan depends on the magnitude of the sediment supply from the contributing drainage basin and the climatic controls over the local hydrological characteristics.

Semi-arid and arid fans are of interest because of their propensity for trenching and, therefore, fan stability is of real concern in the selection of routes for pipelines, powerlines and roads, but it is unlikely that concentrations of heavy minerals will develop in the poorly sorted modern fans. All alluvial fans represent potentially important aquifers.

### 14.1.1 Fan structure

The upper part of the fan commonly exhibits channels incised 1–10 m, passing outward into braided patterns along with a regular decline in debris calibre which imparts to the fan a profile which is that of a concave-up, negative exponential curve. Despite the tendency for debris to pass outward from boulder beds to conglomerates to pebbly sandstones to siltstones, alluvial fan debris is poorly sorted due to the torrential deposition, especially near the fan head. The primary fan deposits are poorly sorted, immature, coarse-grained clastics,

becoming better sorted and finer-grained away from the source, and commonly exhibiting a complex of cross-bedding and flow structures, together with occasional graded bedding and finer lake deposits, the latter especially towards the base of the deposit (Figure 14.1). The extent and pattern of the coarser facies is governed by a balance between basin size and relief, vegetation cover and hydrology, and the tectonic behaviour of the mountain mass.

Normally the coarsest sediments will be found at the fanhead. However, when fanhead trenching occurs, there is reworking and flushing of the sediment further downstream. Sometimes this results in a slight increase in the size of sediment along the fan radius (Denny, 1965). Shifting patterns of deposition on an alluvial fan will result in considerable sediment variability. Lustig (1965) shows the distribution of tractive force on the Antelope Springs fan, and the figure permits an appreciation of

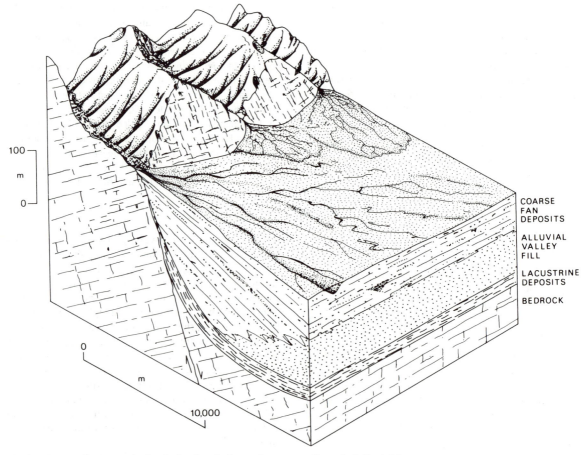

**Figure 14.1**   A diagrammatic sketch showing the internal structure of a typical alluvial fan.
*Source:* Bull, 1977, figure 10, p. 241.

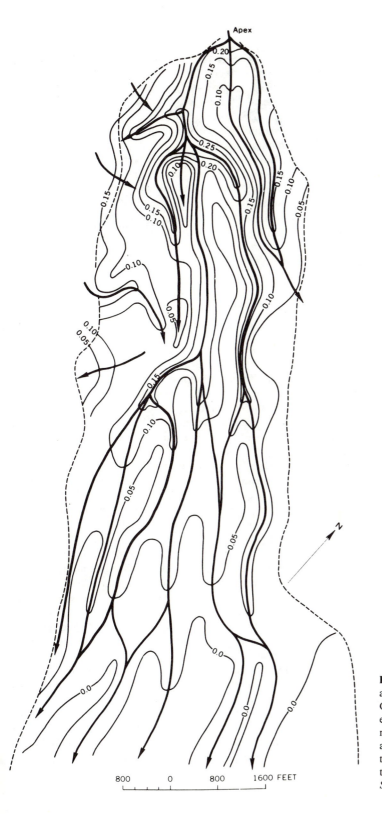

**Figure 14.2** Map of Antelope Springs alluvial fan, Deep Springs Valley, California, showing isopleths related to estimated tractive force (i.e. product of maximum ground slope and particle size) and arrows indicating inferred sediment transport paths; the broken lines represent the approximate fan boundary.
*Source:* Lustig, 1965, figure 113, p. 169.

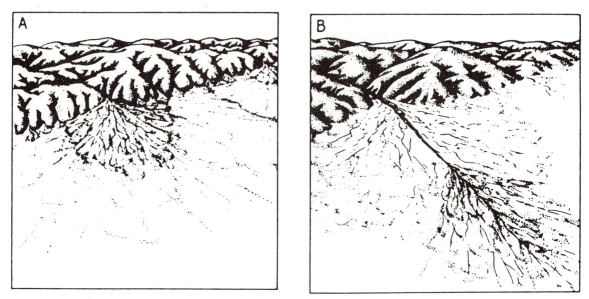

**Figure 14.3** Diverse morphology of alluvial fans: A. area of deposition at fan head: B. fan head trench with deposition at fan toe.
*Source:* Bull, 1968.

how sediments are distributed over a fan surface at one stage in its growth (Figure 14.2). The variability of sediment size on the fan surface is also encountered at depth; however, the overall vertical change in sediment size should be from finest at the base to progressively coarser as the fan grows up and out from the canyon mouth (Ryder, 1971), but then grain size should decrease as erosion reduces the source area. A fan deposited by perennial streamflow, such as the Kosi fan in India (see Figure 14.8) should show better-defined stratification and less sediment size variability.

The above distinction between the smaller, 'drier' fans associated with arid and semi-arid ephemeral streamflow and the larger, 'wetter' fans resulting from seasonally dry tropical streamflow is an important one.

### 14.1.2 Dry fans

Bull (1964) has graphically summarized his observations on modern fans (Figure 14.3), and he has identified two situations. The first is when deposition is near the mountain front and the fan surface is undissected, and the second is when deposition is at the toe of the fan and water and sediment move to this location through a trench. A logical explanation for these differences is climate change, tectonism and even land use. However, if the concept of geomorphic thresholds is applicable to fans, then the progressive growth of a fan may be inter-

rupted by periods of instability, trenching and reworking of fanhead deposits. A related model was proposed by Eckis (1928), who suggested that trenching of fans will take place as sediment yields from the source area decrease through time during the normal progress of the erosion cycle.

All of the above are acceptable explanations of fan differences, but none can be applied indiscriminately to all fans. Modern alluvial fans have been studied sufficiently, so that quantitative relations have been developed between fan size and slope and the size, relief and geology of the source area. Anstey (1965) made a detailed comparison of fans in the Great Basin of the western United States, Baluchistan and Pakistan, and he demonstrates, from a sample of almost 2000 fans, that the greatest number of modern fans have radii between about 1 and 5 miles (1·6–8 km) (Figure 14.4). Considerable scatter exists but considering the range of geologic and climatic conditions, this is not surprising, and the scatter emphasizes that it is not possible to understand the fan without an understanding of the source area. The same figure also shows an inverse relationship between fan size and its mean gradient.

Fan size should be significantly related to size of the sediment source area. An example of this type of relationship is provided by Bull (1964) from the Fresno area, California (Figure 14.5). He shows that a good relationship exists between fan area and drainage basin

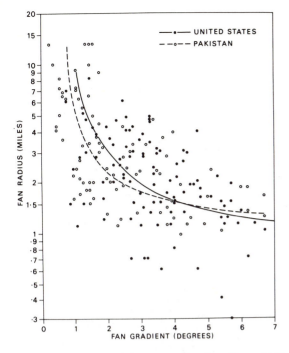

**Figure 14.4** Relationship between radii and gradients of 100 random samples each from the United States and Pakistan. *Source:* Anstey, 1965.

area, and that geology influences the relation, i.e. the intercept of the regression lines varies depending on lithology. For a given drainage basin area, fans related to basins underlain primarily by sandstone have a smaller area than fans associated with drainage basins underlain by mudstone and shale. This relationship is reasonable, as the highly erodible mudstone and shale drainage basins will produce more sediment, and the alluvial fans associated with them will be larger. An inverse relation between drainage basin area and fan slope was also developed (Figure 14.5).

A detailed morphological analysis of alluvial fans has been made by Ryder (1971) for fans located in five valleys in south-west British Columbia. These fans are paraglacial, i.e. they formed during deglaciation, and their development was dependent upon the temporary abundance of glacial debris. They are trenched because of the relatively rapid depletion of the erodible glacial debris and the subsequent reduction of sediment loads from the drainage basins. This is an example of the Eckistype trenching. Ryder found a relationship between drainage basin area, fan area and gradient similar to that of Figure 14.5, and she also found that the greater the ruggedness or average slope of the basin, the greater is the fan gradient (Figure 14.6).

The relations between fan slope, fan area and drainage

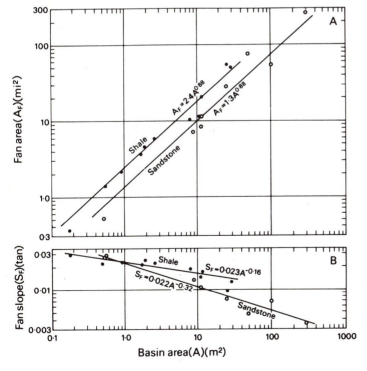

**Figure 14.5** Relations between basin areas supplying alluvial fans in the vicinity of Fresno, California, and A. fan areas $A_F$ and B. fan slopes $S_F$ for localities underlain by sandstone and shale. *Source:* Bull, 1964, figure 54, p. 95.

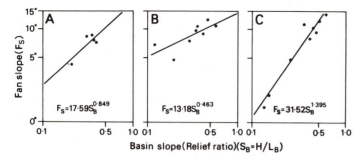

$F_S=17.59S_B^{0.849}$   $F_S=13.18S_B^{0.463}$   $F_S=31.52S_B^{1.395}$

**Figure 14.6** Graphs showing relationships between mean slope of alluvial fans $F_S$ and mean slopes of contributing drainage basins $S_B$ for three valleys in southern British Columbia: A. Fraser West, B. Thompson and C. Similkameen.
*Source:* Ryder, 1971, figure 6, p. 1261.

basin characteristics are not surprising. What is surprising is the variation in fan surface morphology within a given region, i.e. the presence or absence of fan-head trenches. A series of fan maps prepared by Hunt and Mabey (1966) shows the great differences in fan morphology in Death Valley, California (Figure 14.7). These differences have been attributed to tilting and to

the obvious tectonic activity in the Death Valley area. The patterns of recent gravel distribution on the Trail Canyon and Johnson Canyon fans imply a considerable range in fan character. The Trail Canyon map suggests that water has spread widely over the fan surface, and that recent gravels are distributed from the head to the toe of the fan in a number of distributary channels. In

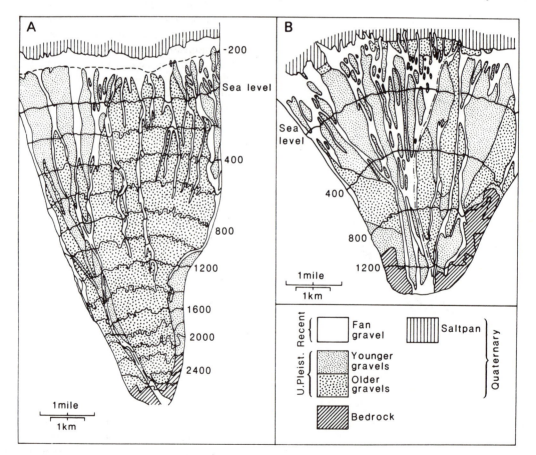

**Figure 14.7** Morphology and geology of A. Johnson Canyon and B. Trail Canyon alluvial fans, Death Valley, California; heights are in feet.
*Source:* Hunt and Mabey, 1966, plate 2.

contrast, the Johnson Canyon fan shows one main channel leading from the canyon mouth to the toe of the fan. Minor distributaries branch from this channel, but over the remaining fan area there are smaller channels that seem to have formed independently of the main channel, as a result of local precipitation and drainage. The fans appear to form a series that resembles that of Figure 14.3. In one case (Trail Canyon) the modern sediments are being distributed widely over the fan. At the other extreme, the sediments are being delivered to the distal end of the fan, with perhaps only minor deposition and minor reworking of the sediments at the head of the fan (Johnson Canyon).

According the Bull (1964), the overall longitudinal profile of a fan is concave. However, this concavity may be formed by straight segments. He has described segmented fans from the Fresno area and concluded that these are of two types one of which is due to intermittent uplift of the mountains, which produces large quantities of sediment and builds a steep fanhead. Through time the fanhead steepens, and each major episode of uplift is associated with a steeper segment of the fanhead, as deposition occurs near the fanhead. The other extreme is due to trenching and the building out of the lower, flatter segment of recent alluvial material at the toe of the fan. This is the reverse of the preceding situation in that the most recent material is being deposited further down the fan on gentler slopes. In any case the longitudinal profile of fans can be complex, and depending on the nature of the profile and the events that caused it, the character of the sediment along a given radius may show considerable variation.

### 14.1.3 Wet fans

The fans discussed above are in dry regions, and the streams that occupy their surfaces are ephemeral. For this reason, the fans are relatively small. However, the perennial Kosi River in India has built a large fan which has been described by Gole and Chitale (1966), and it provides an interesting contrast to the fans previously discussed. The Kosi River (drainage from the high Himalayas) delivers a tremendous sediment load to the piedmont area. For example, the Kosi River produces 3·4 acre-feet of sediment per square mile of drainage area, in contrast to 2·8 acre-feet of sediment per square mile for the Colorado River at the Hoover Dam. The average concentration of sediment in the Kosi River (0·2 per cent) is less than that for the Colorado River (0·9 per cent). This great difference in concentration is due to the fact that the average annual rainfall in the Kosi River catchment is in excess of 60 in (1500 mm) per year, and for the Colorado River at Hoover Dam, 10 in per year. The Kosi River is in the process of building a large fan-shaped piedmont deposit. During the period 1736–1963 the river shifted 70 miles from east to west, and in this process 3500 square miles of land were laid waste as a result of sand deposition and of bank erosion.

At the head of the fan, at Chatra (Figure 14.8A), the channel consists of boulders and cobbles and pebbles. These coarser materials rarely are transported downstream for more than 9·5 miles from Chatra, where the coarse material practically disappears, and the bed material is composed of medium- and fine-grained sediments. The river widens downstream of Chatra, and braiding is characteristic from Belka downstream, where interlacing channels are spread over a width of approximately 4 miles. From the head of the fan, the gradient decreases from 5 ft to about 1 ft per mile. At the lower gradients the river divides into several channels occupying a width of as much as 10 miles.

The progressive shift of the Kosi River to the west is shown in Figure 14.8A. The fact that this is a wet, alluvial fan of considerable size explains the difference between it and the arid fans discussed above. The almost random distribution of patterns on the arid fan is replaced in this case by a progressively shifting channel and an apparently rather orderly development. Comparison of the position of the contours on Figure 14.8B shows that between 1936 and 1957 the river moved westward into a low area on the westernmost part of the fan, which it built up to a considerable extent. This movement of the river to the west could have been anticipated by the position of the 1936 contours, as shown on Figure 14.8B. The shaded areas represent the additions of sediment in this area, and the contours have straightened due to deposition. If this progressive shift of the Kosi River to the west is not due to tilting of the piedmont area to the west, then the possibility that the river will soon begin a migration to the east is worthy of consideration. However, it is equally possible that if there is a low area on the east side of the fan at present, the channel might shift to that position by avulsion (Section 14.3.3).

### 14.1.4 Depositional belts

Where alluvial fans are large enough or sufficiently closely spaced, they may coalesce laterally to produce a depositional belt along the piedmont zone. In arid regions dry fans may unite to form a bajada surface along a mountain front (see Figure 14.1), whereas in wetter areas huge, gently sloping alluvial aprons may result. The bajadas flanking Death Valley and the

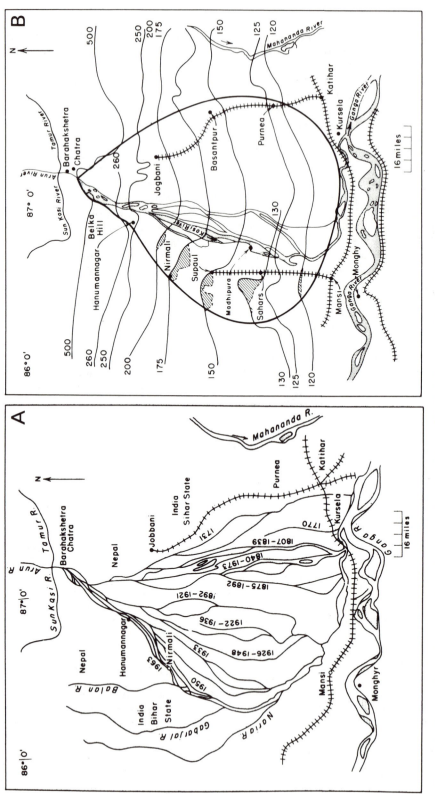

**Figure 14.8** The Kosi River alluvial fan, India: A. courses of the Kosi River, 1731–1963; B. extent of the alluvial fan; full contours show fan configuration in 1936, broken contours the changes by 1957, shaded areas the deposition on the west side of the fan in 1936–57.

*Source:* Schumm, 1977, figures 7-5 and 7-6, pp. 253, 254, after Gole and Chitale, 1966, figures 2 and 3, 116, 117.

extensive alluvial slope lying between the Himalayas and the River Ganges are examples of these.

## 14.2 Valley fills

Most valleys contain a valley-fill alluvial deposit which has been eroded to form terraces and a floodplain on which the present river flows. The sediments that fill the valley normally 'fine upward', with the coarsest sediments deposited on the bedrock valley floor and finer materials deposited progressively above it. This means, of course, that as the filling has occurred the morphology of the channel has changed, perhaps from braided to meandering, as in the Mississippi Valley. Much also depends on the amount and nature of the sediment delivered to the valley from upstream.

Braided channels usually occur on higher gradients, at greater stream discharges, with coarser debris and with less vegetation stabilization than do meandering channels. Individual channels in the braided pattern have low sinuosity, but they are constantly subdividing and shifting their position as the result of the building of

longitudinal sand bars with horizontal bedding and of transverse bars with planar cross-bedding. The resulting facies present a complex aggradational association of channel lag deposits, overbank finer material and cross-bedded and rippled sand-bar material (Figure 14.9). *Sandurs* represent an extreme case of valley fill due to braiding associated with the large seasonal discharge variations of glacially fed rivers or the bursting of glacial lakes. For example, the Skeiðarársandur in Iceland is some 1000 km² in area and is fed by the bursting of a glacial lake about once every ten years, often associated with volcanic eruptions (Thorarinsson, 1953). Sandurs are gently sloping aggradation plains made up of a chaotic complex of raised lozenge-shaped bars, small plateaus and terracelike forms produced from segregated gravel sheets and separated by a maze of seasonally-abandoned channels (Krigström, 1962). The rivers are wide and braided, exhibiting pools and riffles, and transport mainly coarse bedload some 75 per cent of which may be moved in only 8 per cent of the total *runoff time* which occurs during short summer periods. The longitudinal stream profiles tend to be steep and

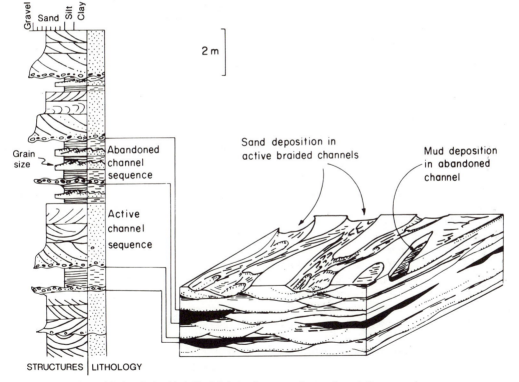

**Figure 14.9** Physiography and facies of a braided alluvial channel system, where sedimentation occurs in a rapidly shifting complex of channels; key is given in Figure 14.10.
*Source:* R. C. Selley, *An Introduction to Sedimentology*, 1976, figure 101, p. 265, copyright © Academic Press, London.

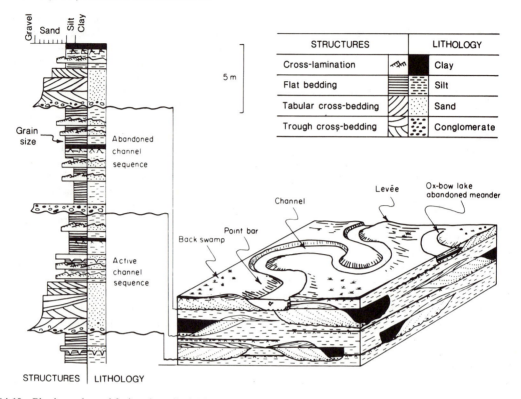

**Figure 14.10**　Physiography and facies of an alluvial floodplain produced by meandering channels. *Source:* Selley, 1976, figure 99, p. 261, after Visher.

parabolic in form (Church, 1972) (Plate 18).

In contrast to the above, the deposits of aggrading and laterally shifting meandering rivers form alluvial fills the facies of which occur in characteristic sequences depending on whether deposition was associated with active channel or backswamp conditions (Figure 14.10).

The actual process of valley filling may take place in three ways, depending on the nature of the controls that are causing deposition. First, there can be a change in baselevel that will reduce gradient in the lower end of the valley. This downstream control will initiate a progressive back-filling of the valley. Second, if there is uplift in the source area or a change in climate such that a great increase in sediment production results, aggradation will also result. The effect of this upstream control, in contrast to the back-filling of the valley, will be a progressive downfilling of the valley, as the coarse sediment load progressively moves down through the system. Third, there may be a uniform filling of the valley, from bottom up, by a great increase in production of sediment from numerous tributaries such that a uniform raising of the level of the valley floor occurs

without any significant change in slope.

The post-glacial rise in sea level has resulted in the lower parts of most valleys in coastal locations having been aggraded and in the case of large rivers this aggradation may have extended hundreds of kilometres inland. Commonly the character of the valley fill and the buried valley features are poorly known, except where engineering excavations or deep borings have been made. Nor is the present position of the alluviating river a necessary guide to subsurface features in that the position of the river is conditioned by many influences, for example, by regional tilting or by the direction from which the tributaries are supplying the maximum of sediment. Figure 14.11 shows the thickness of Mississippi Valley alluvium in the vicinity of East St Louis, Illinois. The rather irregular bedrock floor is deepest near the confluence of the Mississippi and Missouri rivers (showing that irregularities in the longitudinal profiles of rivers are often associated with action at tributary junctions); the modern river is not located over the thickest eastern part of the valley fill, but over the thinnest western part (being even on bedrock in places).

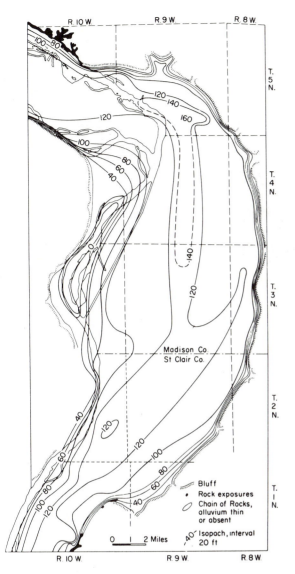

**Figure 14.11** Thickness of the Mississippi Valley alluvium, East St Louis, Illinois.
*Source:* Bergstrom and Walker, 1956, figure 3, p. 27.

### 14.2.1 Floodplains

A floodplain is the alluvial surface adjacent to a channel that is frequently inundated (Figure 14.12). Except in underfit rivers, floodplain size is generally related to the discharge of the river. For example, floodplain widths for three major Welsh rivers vary between only 250 and 1100 m, whereas the active floodplain width of the lower Mississippi is 16 km (10 miles), lying within a terraced valley floor varying between 40 km and 200 km (25–125

miles) wide. In situations of aggradation and levee-building the concept of bankfull discharge as a simple discriminator between channel-forming and floodplain-forming processes is not easy to support, and it is especially difficult to generalize regarding the magnitude and frequency of floods which inundate the floodplain. For some time, it was assumed that bankfull discharge has a general recurrence interval similar to that of the mean annual flood (2·33 years) or the most likely flood (1·6 years). However, a study of thirty-six active floodplains in the United States has shown bankfull recurrence intervals to vary between 1·01 and 32 years, and although the modal interval is 1·5 years, only about one-third of the floodplains conform to this (Figure 14.13). It was found impossible to find a simple relationship between the recurrence interval of discharges and the degree of floodplain inundation between different areas in Wales. The case of the Mississippi is naturally much more complex and, before canalization, some channels near the mouth flooded five or six times a year. Further upriver in Wisconsin the 1952 flood (RI = 40 years) inundated only 56 per cent of the local Mississippi floodplain and even the 1965 flood (RI = approx. 175 years) covered only 62 per cent.

Table 14.1 gives a classification of valley sediments and suggests the three major ways by which floodplains may be formed: by vertical accretion, lateral accretion, and island formation and channel abandonment (Figure 14.14A–C).

Meandering channels shift laterally reworking existing sediments and building up newer point bar, channel and levee deposits (see Figure 14.12). The floodplain thus becomes in large part a palimpsest of such deposits which contribute very significantly to its internal structure (see Figure 14.10) and impart to it most of its relief. Abandoned point bars, successive bars built up within developing meander loops, sometimes having been trenched to produce chutes and flanking chute bars (Figure 14.15), are dominant in producing the ridge and swale topography so characteristic of most floodplains. The highest elevation on the floodplain normally occurs on abandoned levees and point-bar ridges, and the magnitude of the relief is naturally related to the associated historical discharges. Small Welsh floodplains commonly have a maximum relief of 1·7–3·3 m; whereas that for the active Lower Mississippi floodplain is as much as 20 m. There are also great variations in measured maximum rates of lateral channel shifting (Table 14.2).

Vertical accretion of floodplains results from overbank flow which produces crevasse splays and layers of silt, clay and organic deposits in the backswamp basins

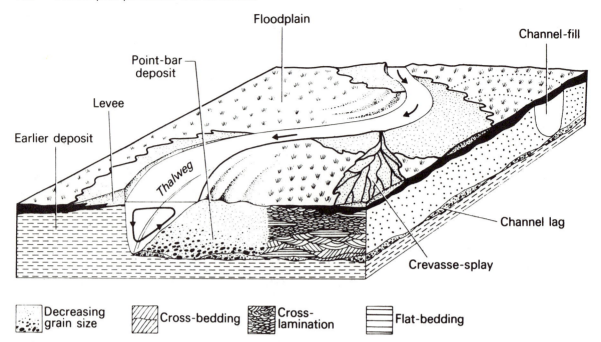

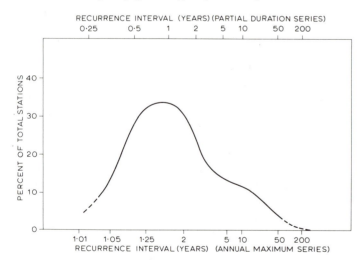

**Figure 14.12** The classical point bar model for a meandering river; a crevasse is a break in the levee.
*Source:* Collinson, 1978, figure 3. 24, p. 33, after Allen, 1970.

**Figure 14.13** Frequency distribution of recurrence intervals for bankfull discharge on thirty-six active floodplains in the United States.
*Source:* Williams, 1978, figure 2, p. 1151.

(see Figures 14.10 and 14.12). Using aerial survey techniques to track the movement of floats, Russian hydrologists have investigated the dynamics of the inundation of the floodplain of the Oka River, some 300 km east of Moscow (Figure 14.16). Here the left-bank floodplain is 3–8 m above low-water level, is underlain by 12–17 m of floodplain deposits and is flanked by a terrace 8–12 m above the floodplain. Figure 14.16 shows two stages of inundation by a flood of April 1968: (1) just above bankfull, mean water elevation about 80·17 m and 11 per cent of the floodplain inundated by flows from both upstream and downstream; (2) mean water elevation 81·16 m, 89 per cent of the floodplain (including six distinct basins) inundated by flood flow from upstream, this discharge occurring some nine years out of every ten. The frequency of floodplain inundation

**Table 14.1**   Classification of valley sediments

| *Place of Deposition* (1) | *Name* (2) | *Characteristics* (3) |
|---|---|---|
| Channel | Transitory channel deposits | Primarily bedload temporarily at rest; part may be preserved in more durable channel fills or lateral accretions |
| | Lag deposits | Segregations of larger or heavier particles, more persistent than transitory channel deposits, and including heavy mineral placers |
| | Channel fills | Accumulations in abandoned or aggrading channel segments, ranging from relatively coarse bedload to plugs of clay and organic muds filling abandoned meanders |
| Channel margin | Lateral accretion deposits | Point and marginal bars which may be preserved by channel shifting and added to overbank floodplain by vertical accretion deposits at top; point-bar sands and silts are commonly trough cross-bedded and usually form the thickest members of the active channel sequence (see Figure 14.10) |
| Overbank flood plain | Vertical accretion deposits | Fine-grained sediment deposited from suspended load of overbank floodwater, including natural levee and backland (backswamp) deposits; levee deposits are usually horizontally bedded and rippled fine sands, grading laterally and vertically into point-bar deposits. Backswamp deposits are mainly silts, clays and peats |
| | Splays | Local accumulations of bedload materials, spread from channels on to adjacent floodplains; splays are cross-bedded sands spreading across the inner floodplain from crevasse breaches |
| Valley margin | Colluvium | Deposits derived chiefly from unconcentrated slope wash and soil creep on adjacent valley sides |
| | Mass movement deposits | Earthflow, debris avalanche and landslide deposits commonly intermix with marginal colluvium; mudflows usually follow channels but also spill overbank |

*Source:* Happ, 1975, table 2.25, p. 287.

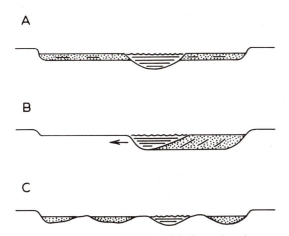

**Figure 14.14**   Three types of floodplain formation: A. vertical accretion: B. lateral accretion; C. island formation and side-channel abandonment.

is not only a matter of simple discharge frequency, but also of changing channel capacity. Paradoxically, a floodplain may be extended when a decrease of flood peak magnitude or frequency causes a reduction in channel

**Table 14.2**   Variation in maximum rates of lateral channel shifting

| *River* | *Rate of channel shifting* | |
|---|---|---|
| | *(m/year)* | *ft/year* |
| Kosi, northern Bihar, India | 750 | 2460 |
| Colorado, Needles, California | 244 | 800 |
| Ramganga, Shahabad, India | 80 | 264 |
| Missouri, Peru, Nebraska | 76 | 250 |
| Mississippi, Rosedale, Mississippi | 48 | 158 |
| Yukon, Holy Cross, Alaska | 37 | 120 |

*Source:* Wolman and Leopold, 1957, table 4, p. 97.

size. For example, such a reduction of peak discharge caused the formation of a new floodplain by vertical accretion in the Cimarron River Valley, Kansas, because vegetation moved in to colonize the area adjacent to the low-water channel. Subsequent floods deposited sediment on these surfaces and they rapidly increased in height above the low-water channel to extend the floodplain. However, in general rates of vertical floodplain accretion appear to be less than those due to horizontal

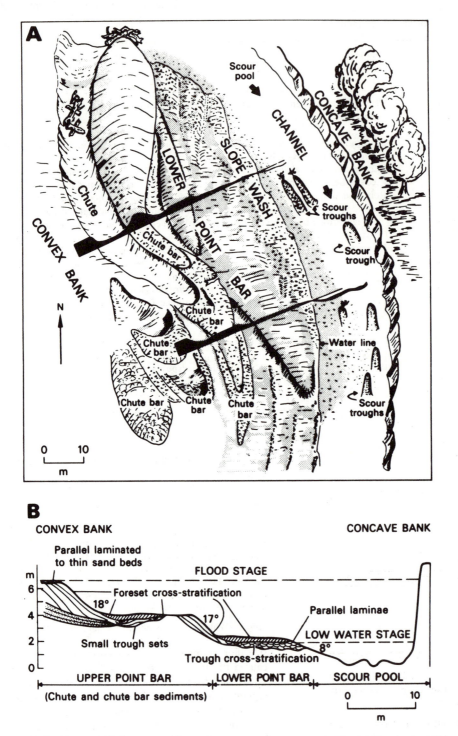

**Figure 14.15** Topographic features and internal structures of a coarse-grained point bar; A. plan; B. cross section.

*Source:* Collinson, 1978, figure 3.26, p. 34, after McGowen and Garner, 1970.

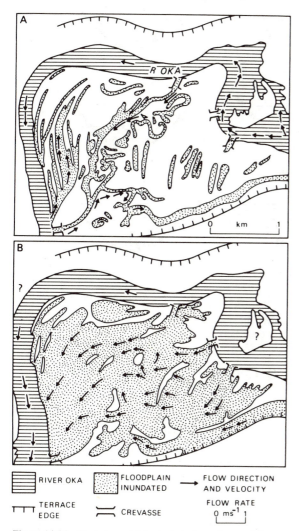

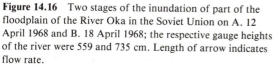

**Figure 14.16** Two stages of the inundation of part of the floodplain of the River Oka in the Soviet Union on A. 12 April 1968 and B. 18 April 1968; the respective gauge heights of the river were 559 and 735 cm. Length of arrow indicates flow rate.
*Source:* Popov and Gavrin, 1970, figures 4 and 6, pp. 418, 419.

accretion because of the comparatively high floodplain flow velocities (0·3–1·2 m/s or 1–4 ft/s) (see Figure 14.16) and the low concentrations of suspended sediment at high discharges. It was estimated that the 100 + year flood of the Ohio River in 1937 deposited on average only 0·002 m (0·008 ft) of floodplain alluvium.

Island formation and channel abandonment are commonly associated with the braiding which accompanies widely fluctuating discharges operating on coarse

channel debris (see Figure 14.9). An extreme case of this is that of the proglacial sandurs. Braided rivers, such as the North and South Platte in Colorado and Nebraska, have developed floodplains as a result of irrigation diversions and flow regulation by upstream dams. The first stage in the process was the development of islands as sand bars were colonized by vegetation. Then, as discharge decreased, the side channels on the landward side of the islands became progressively filled with sediment until they were abandoned and the island was incorporated into the floodplain. Until overbank deposition has smoothed the floodplain surface it will retain the hummocky appearance of a braided river.

### 14.2.2 River terraces

River terraces may occasionally be produced by the incision of bedrock surfaces but they are most usually the remnants of floodplains which have been trenched by rivers. Terraces border most valleys, and where the valley has been filled with alluvium, it is logical to assume that buried terraces lie beneath the floor of the valley in a wide variety of possible configurations (Figure 14.17). Floodplain trenching is produced either by negative baselevel movements (see Chapter 2) or by complex climatic changes leading to the provision of less sediment or more water. The former causes produce headward surges of rejuvenation leading to composite polycyclic stream segments, graded to different baselevels, separated by nickpoints and correlated downstream with river terrace remnants which may be paired on opposite sides of the valley (Figure 14.18A). Of course, these polycyclic terraces are commonly complicated by the occurrence of subsidiary terraces caused by the complex response of drainage basins to renewed erosion referred to earlier. Other non-cyclic terraces can be produced by a variety of mechanisms and, for example, unpaired terraces can be generated by the slow lowering of a broad floodplain by a widely swinging meander belt which produces terrace remnants which converge downstream (Figure 14. 18B), rather than being more or less parallel as in the case of polycyclic terraces. In reality, most rivers have complex terrace sequences, as well as nickpoints resulting from polycyclic, geological, tributary entrance, or changed bedload/discharge causes.

Davisian denudation chronology is based on the assumption that most river terraces, like marine terraces, are *thalassostatic* (Greek: *thalassa* – 'the sea'), i.e. were formed with respect to a given sea or baselevel. However, it is now clear that floodplain aggradation and terrace trenching can be produced by a variety of combinations of stream load, vegetation, erosion rates and discharges,

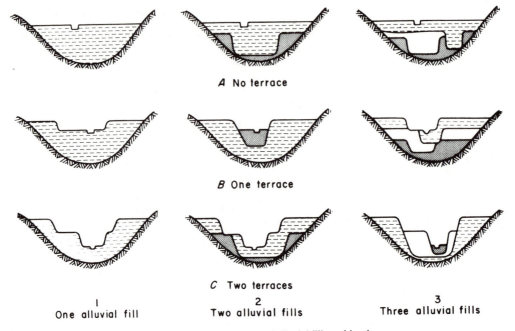

A No terrace

B One terrace

C Two terraces

| 1 | 2 | 3 |
| One alluvial fill | Two alluvial fills | Three alluvial fills |

**Figure 14.17** Valley cross sections showing some possible terrace and alluvial fill combinations.
*Source:* Schumm, 1977, figure 6-22, p. 211, after Leopold and Miller, U.S. Geol. Surv.

## A  Paired, poly-cyclic terraces

## B  Unpaired, non-cyclic terraces

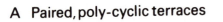
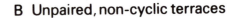
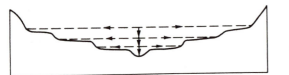
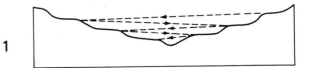
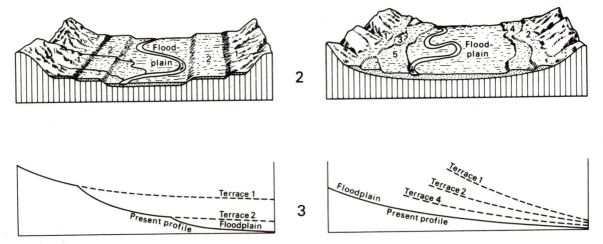

**Figure 14.18**  A. Paired, polycyclic terraces and B. unpaired non-cyclic terraces: 1, formation (arrows show vertical and horizontal shift of river; 2, three-dimensional view; 3, longitudinal profiles.
*Sources:* Sparks, 1972, figures 9.5, 9.7 and 9.8, pp. 296, 297, copyright © Longman, London; Thornbury, 1969, figure 6.9, p. 157.

especially during the glacial, periglacial and interglacial fluctuations which have characterized the past 2 million years in the middle and higher latitudes (Figure 14.19). This being so, it is possible to envisage complex denudation chronologies of valley cut and fill resulting from climatic fluctuations in which the effects of changing baselevel may be of minor importance or, indeed, absent altogether. Load/discharge variations in streams bear a complex relationship to climatic, vegetational and human factors commonly controlled by threshold effects, but their results are valley cut and fill terrace sequences which present a chronology not capable of precise correlation with baselevel changes. The case for Davisian treatment of river terraces on thalassostatic assumptions holds best for the lower parts of river valleys near present sea level.

Geomorphologists have always been intrigued by terraces, in particular, their correlation from valley to valley and from region to region. If terraces are climatically induced, being formed by major climatic changes, then it is logical to assume that because these climatic changes are regional then the terraces found flanking the modern floodplain can be correlated regionally, and perhaps continentally. This reasoning is sound but, of course, the climatic and geomorphic history of the globe is much more complicated than it has been assumed to be by this simple model of terrace formation. Terraces are, in fact, dependent on the behaviour of a river at a given location and rivers can respond in different ways to similar climatic changes as sediment loads change. For example, correlation of Mississippi Valley terraces in the lower valley with those in the upper valley was not possible because when sea level was lower along the Gulf Coast glaciation was active to the north. Near sea level rivers would be entrenching, whereas rivers to the north would be aggrading as a result of the high sediment load

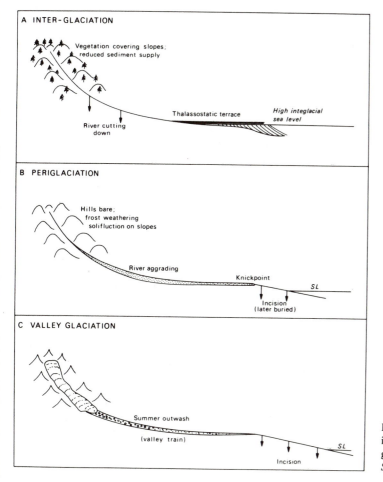

Figure 14.19 River behaviour during A. interglacial, B. periglacial and C. valley glacial times.
*Source:* Clayton, 1977, figure 12.1, p. 157.

delivered by glacial meltwater. Therefore, terraces in the upper valleys are out phase with terraces in the lower valley.

The importance of identifying the separate fills and terraces in a valley is, of course, that each represents a change in the behaviour of the river system. In addition, terrace surfaces may represent periods of sediment reworking and, therefore, they may contain heavy mineral concentrations. Table 20.3 shows a response of rivers to climatic change depending on original climatic conditions and geographic location. Obviously, rivers draining to the ocean, when sea levels were lower, incised; whereas those draining to closed basins or to pluvial lakes aggraded during the same time. Climatic controls are important, but even if one can assume a

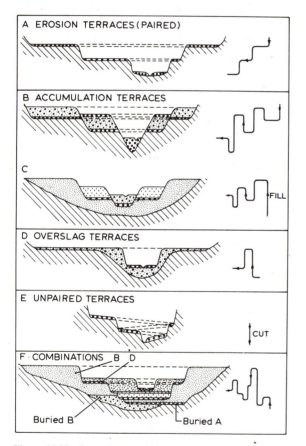

**Figure 14.20**   Several types of river terraces: the diagrams to the right indicate schematically the vertical (cut and fill) and horizontal movement of the river during the successive phases of terrace formation; for example, overslag terraces (Dutch: *overslaugh* – 'pass over') are formed by a simple succession of fill and partial cut.
*Source:* Zonneveld, 1975, figure 1, p. 278.

worldwide increase in precipitation and a decrease in temperature during glaciation, the rivers behaved differently, depending on the climatic regions in which they were located (see Table 20.3). In fact, climatic changes were not uniform over the entire world, some areas were cooler and drier, whereas other areas were cooler and wetter.

Figure 14.20 shows a series of idealized situations depicting a river valley with a series of terraced alluvial fill deposits, the diagrams to the right of each cross section showing the behaviour of the river during the period of time that it took to form these terraces. The terraces often can be correlated over large areas, but the valley fill and the terraces may be composed of a number of deposits of different ages that reflect the erosional and depositional history of a valley. As Figure 14.20 demonstrates, valleys with similar morphologic features may have very different histories. Obviously, the investigator will not see this evidence unless there has been trenching of the deposit. A typical explanation for all of the details of these valleys and valley fills has been that they reflect a climatic change, and if a valley fill or terrace does not correlate with a known Pleistocene climatic change, then they are attributed to climatic fluctuations.

The effects of such fluctuations are best marked in semi-arid regions where even small climatic changes in the past have been accompanied by marked variations in vegetation and sediment supply. An example of this is found in the Arroyo de los Frijoles, New Mexico (Figure 14.21). Here the Coyote Terrace deposits contain charcoal dated by [14]C methods at 2800 ± 250 years BP and possess Pre-Columbian pottery fragments (*c*.AD1100;900 BP) in their surficial layers. This period of alluviation appears to have been followed by one of ephemeral stream (arroyo) trenching until about AD1400 (600 BP), also due to climatic fluctuations, and succeeded by the alluviation of the Low Terrace deposits until historically documented arroyo-trenching set in about 1865 throughout much of the south-western United States. Unlike previous events, this arroyo-cutting, which lasted until it abruptly terminated about AD1915, appears to have been of complex origin involving not only subtle climatic fluctuations, but also human and agricultural impacts on surface erosion. Since the First World War a similarly complex set of climatic and human controls has resulted in the alluviation of the present stream bed (Leopold *et al.*, 1966).

A more complex denudation chronology has been worked out for wadis in Tripolitania (Figure 14.22). Initial valleys (A) were alluviated with Older Fill, with resistant calcareous crusts, possibly associated with climatic changes accompanying the last glaciation (B).

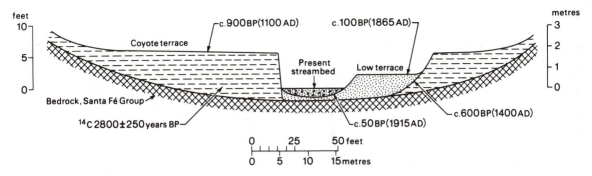

**Figure 14.21**  Terraces in the Arroyo de los Frijoles, New Mexico.
*Source:* Leopold *et al.*, 1966, figure 141, p. 198.

Between Neolithic and Roman times climatic changes promoted shallow wadi (arroyo) cutting (C), and after AD 69 the Romans built dams across many of these wadis (D). Increasing human occupation makes it difficult to assign uniquely climatic mechanisms to post-Roman events, but it seems clear that renewed wadi-cutting in late classical times breached the calcareous crusts, allowing deep trenching (E) which was followed by up to 10 m (30 ft) of Younger Fill in the Middle Ages (F), to be trenched, in turn, by modern wadis (G) (Vita Finzi, 1969).

A final point to be made here, however, is that many unpaired terraces (see Figures 14.18B and 14.20E) are the result of the manner in which a river incises, rather than due to any external influence. The filling of the valleys of the Yuba River, California, by hydraulic mining debris up to 1890 (Figure 14.23A) provides an excellent example of this process. When hydraulic mining was prohibited by law, the reduction of sediment loads permitted incision of the deposits, and during the past 100 years multiple terraces and a new floodplain formed. The channel was unable to incise progressively through the thick alluvial deposit, and it paused several times during downcutting until it established a new longitudinal profile. Eventually the multiple terraces (Figure 14.23B) will be destroyed, and only the upper alluvial surface will remain as evidence of the period of valley filling. This episodic behaviour of the Sierra Nevada streams is similar to the episodic erosion observed experimentally. Many details of a depositional landform, such as multiple unpaired terraces and alluvial deposits, that cannot be explained by known climatic fluctuation are understandable when the complex response of a fluvial system to change is appreciated.

## 14.3  Deltas

Deltas are formed where sediment-laden rivers flow into standing bodies of water. The geomorphic characteristics

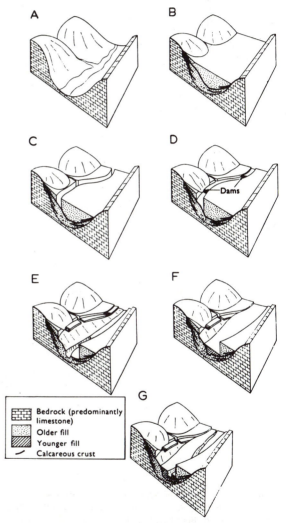

**Figure 14.22**  Schematic block diagrams showing the history of a Tripolitanian wadi; the height of the block is approximately 40 m.
*Source:* Vita-Finzi, 1969, figure 3, p. 10.

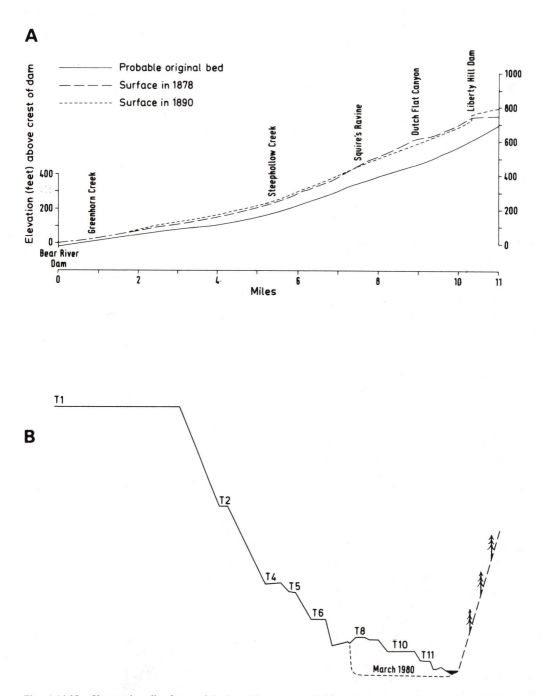

**Figure 14.23** Changes in valley forms of the Bear River system, California, due to the effects and cessation of hydraulic mining: A. longitudinal profiles of the Bear River; B. multiple terraces of Greenhorn Creek, a lower tributary of Bear River, at this point incised 21 m (69 ft).
*Source:* Wildman, 1981, figure 18, p. 54.

of deltas are determined by a number of groups of factors, including:

(1) The relative densities of the river water and the standing water (Bates, 1953).

(2) The river hydrology, including the magnitude and variations of the discharge.

(3) The amount and calibre of the river load, especially the proportion of bedload to suspended load.

(4) The intensity of the coastal processes, particularly of wave action with associated longshore currents and of tidal scour.

(5) The geometry of the coast, including its plan view and seaward slope.

(6) The tectonic stability of the shoreline, particularly in the vertical sense.

(7) The climate, in so far as it exerts additional influences over such matters as the amount and type of vegetational cover and the growth of marine organisms.

### 14.3.1 Delta morphology

Except where very high concentrations of fine suspended sediment are carried by the debouching river, the dynamics of water mixing at a river mouth are strongly influenced by the relative densities of the inflowing and standing waters. Where the inflow is of equal density, it is termed homopycnal (Greek: *homo* – 'same'; *pycno* – 'dense'), of greater density hyperpycnal (Greek: *hyper* – 'more') and of less density hypopycnal (Greek: *hypo* – 'less') (Figure 14.24). Homopycnal flow (Figure 14.24A) naturally occurs where rivers flow into freshwater lakes, and the simplest deltas develop where small rivers carrying a substantial proportion of bedload flow into shallow lakes devoid of tidal or of significant wave forces. Here the bedload settles immediately and the suspended load quite rapidly to produce a delta approximating a semi-circular form and having the hydrodynamic and sedimentary structure shown in Figure 14.25, the latter dominated by topset, foreset and bottomset bedding. The relative abundance of the river bedload and suspended load influences both the geometry of the foreset beds and the relative thicknesses of the topset, foreset and bottomset beds. Hyperpycnal flow (Figure 14.24B) is naturally rare but does occur when the river water is carrying large proportions of fine suspended sediment, for example, at the snout of a tide water glacier (see Figure 19.18), or where submarine slumps produce turbidity currents. Such flows spread graded beds of fine sediment far out into the basin of deposition. Hypopycnal flow (Figure 14.24C) is the most common accompaniment of delta-building when fresh-

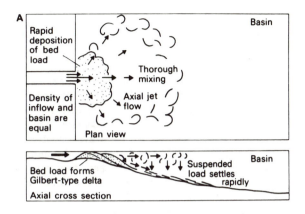

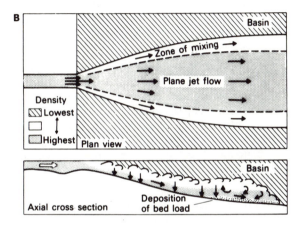

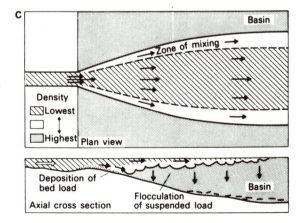

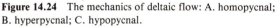

**Figure 14.24** The mechanics of deltaic flow: A. homopycnal; B. hyperpycnal; C. hypopycnal.
*Source:* Elliott, 1978, figure 6.17, p. 111, after Bates and Fisher.

water rivers flow into salt water. Here the coarse bedload is deposited as a shoal near the mouth and the coarser suspended sediment as levees (submarine in the first instance) flanking the currents into which the flow is sub-divided by the shoal. The finer suspended sediment is carried seaward in the surface plane jet which spreads out over the denser sea water and produces the extensive, gently sloping submarine delta front. Such deltas build into deeper water, in the absence of strong waves or tides, around an extending framework of coarser channel, shoal and levee deposits over offshore clays and interdistributary silty clays and peats. As progradation continues the latter deposits, in places reinforced by crevasse-splay sands, are covered by topset floodplain and foreset silts and fine sands. The constant generation and shift of distributaries during deposition, as well as possible major changes in delta location by avulsion (see Chapter 14.3.3), may produce a complex geometry of delta facies (see Figure 14.32), involving nevertheless the following three basic sedimentary units:

Topset beds – floodplain sands and silts; marsh organic deposits; platform sands.

Foreset beds – cross-bedded coarser sands and silts of the delta slope, together with channel fingers grading laterally into clays, marsh deposits and sand splays.

Bottomset beds – offshore clays and toeslope turbidity silt deposits.

Deltas may be classified according to their plan-view geometry as either high-destructive or high-constructive deltas (Figure 14.26). The differences result from a complex interaction of the previously noted variables (e.g. energy of wave attack, amount and size of sediment load, vegetation effects):

High-destructive deltas are dominated by the removal of river debris by wave and tidal currents such that distributaries do not become blocked and tend to remain stable in position:

Wave-influenced deltas are flattened in plan because of the rapid removal of debris by strong longshore wave currents (e.g. the São Francisco River of Brazil) (Figure 14.26A).

Tide-influenced deltas have funnel-shaped distributaries which are kept open and straight by strong tidal scour and whose stability may be assisted in the tropics by the growth of mangrove vegetation (e.g. the Ganges, Niger and Mekong rivers) (Figure 14.26B).

High-constructive deltas are dominated by a large debris supply from the river:

Lobate (fan) deltas form where much of the river debris is coarse, giving a compact form over which the distributaries constantly change position, as on an alluvial fan (e.g. the Nile and Rhône rivers) (Figure 14.26C).

Elongate (bird's-foot) deltas occur when a river is delivering large quantities of fine sediment, and the delta builds out around constantly shifting distributaries flanked by levees little affected by wave action (e.g. modern Mississippi River) (Figure 14.26D).

Of course, the plan-form of a delta may be constrained by the broad geometry of the coast, as when a river is delivering sediment into an elongate trough or drowned valley to form an estuarine delta, such as the Colorado River delta.

Silvester and de la Cruz (1970) attempted a quantitative investigation of delta morphology, after collecting information on fifty-three deltas throughout the world, for which information on morphology, climate and hydrology was available. They studied several delta characteristics, such as delta length, area, maximum width, ratios between maximum width and the number of distributary mouths, and between maximum width of the delta and the number of distributaries. They found

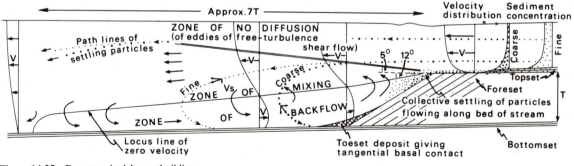

**Figure 14.25** Processes in delta outbuilding.
*Source:* Jopling, 1963, figure 2, p. 118.

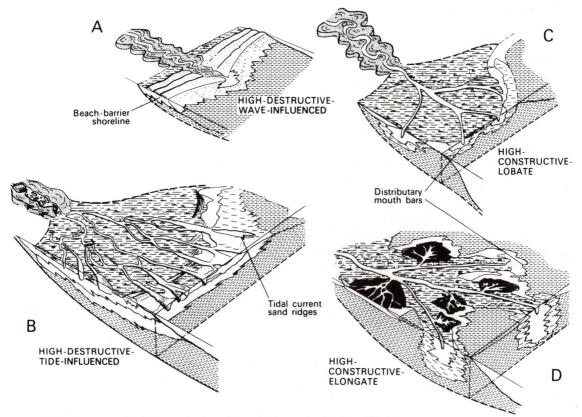

**Figure 14.26**  Two types of high-destructive A. and B. and high-constructive C. and D. deltas.
*Source:* Elliott, 1978, figure 6.3, p. 101, after Fisher *et al.*, 1969.

that both delta length and delta area are directly correlated with water discharge. However, there is an
inverse relationship between delta length and area and
river slope and the continental shelf slope. The greater
the water discharge, the larger is the drainage basin and,
therefore, the greater the sediment production, and a
large delta results – but the larger the discharge, the
flatter is the slope of the stream. The flatter the gradient
of the continental shelf, the less the water depth and,
therefore, the more rapidly the delta grows outward from
the shoreline.

Temperature plus swell duration and swell direction
are important factors determining delta width. The
greater the incidence of swell, the more energy is available for spreading the deltaic material along the coast,
and the spreading is enhanced by the angle at which the
waves intersect the coast. Vegetation is also important.
The more vegetation, the less likely is the sediment to
move out to sea, and the wider is the delta.

There is positive correlation between delta slope and
river slope. River slope will reflect discharge and the

nature of the sediment load, and the delta will also
respond to this control in the same way as the river
channel. Concerning shape, the greater the river slope,
the higher the river velocity and hence a more elongate
delta forms.

The most significant variables that influence delta size
are discharge, sediment production and river gradient, as
well as temperature, which is reflected in vegetative
growth. The results are not surprising and resemble those
developed for alluvial fans. The larger the drainage
basin, the larger the alluvial fan, and the larger the delta.
The variables that influence sediment production from a
watershed, and all the factors previously discussed as
controlling drainage basin morphology are also important in influencing delta growth.

An increase or decrease of sediment delivered to a delta
can be the result of several factors. A sediment-load
increase should result in a rapid prograding of the delta,
aggradation of the stream channels and blockage of
distributaries. A reduction in sediment delivering to a
delta should cause channel incision.

### 14.3.2 Experimental study of delta morphology

An experimental study of fan delta growth by Chang (1967) provides some valuable information on delta response to fluctuations of sediment yield and discharge. Figure 14.27 is a map showing the progressive development of one of Chang's deltas through time. Figure 14.28 shows the rate of delta growth through time, for his runs 15, 16 and 17. The rate of growth was least during run 15, which had the smallest amount of sediment fed into, and transported through, the channel. The sediment transported through the channel was identical for runs 16 and 17, but the depth of water was shallower for run 17; therefore, there is a much more rapid increase in area through time. During run 17a the sand load was reduced to zero. The channel degraded and a very rapid increase in delta area resulted, as sediment was flushed out of the main channel and from the delta head. At the same time a dramatic change in the shape of the delta occurred. Figure 14.29 shows a marked elongation of the delta, as the upper part of the delta was eroded, and new lobes were built at the toe of the delta. Note, however, that the

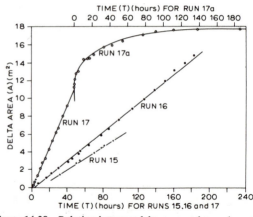

**Figure 14.28** Relation between delta area and experimental run time for runs 15, 16 and 17.
*Source:* Schumm, 1977, figure 8-8, p. 311, after Chang, 1967.

increase was short-lived and after rapid growth for about 10 h, growth decreased.

Figure 14.30 shows a somewhat similar relationship, when during run 20a, the amount of sediment fed to the

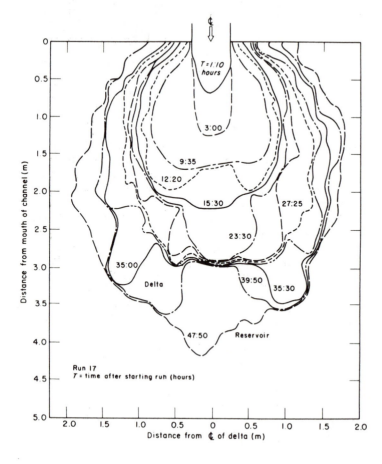

**Figure 14.27** Plan view of an experimental delta showing its progressive expansion during a period of up to 47 h and 50 min after the start of the experiment; centre line = long axis of delta.
*Source:* Schumm, 1977, figure 8-7, p. 310, after Chang, 1967.

channel was doubled, and an obvious increase in the rate of delta growth is apparent. It seems therefore, that the results of both a reduction in sediment load and an increase in sediment load, at least initially, had the same effects on increase in delta area with time. However, there were major differences in the shape of the delta. A marked elongation of the delta during channel and delta-head degradation occurred because of reduction of sediment load (run 17). A much higher width–length ratio of the delta developed with an increase of sediment loads (run 20).

A decrease in the rate of delta growth occurred with aggradation, whereas degradation increased the rate of delta growth. Aggradation stored sediment near the delta head as during alluvial fan growth, and as in the alluvial fan situation, when trenching occurred, sediment was moved to the distal part of the delta (Figure 14.29).

Chang (1967) has demonstrated that changes of

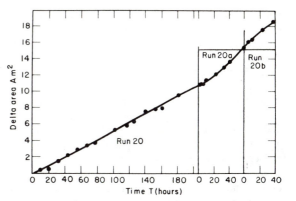

**Figure 14.30** Rate of experimental delta area growth with uniform sediment delivery (run 20), an increase of sediment load (run 20a) and a reduction of sediment load (run 20b). *Source:* Schumm, 1977, figure 8-10, p. 313, after Chang, 1967.

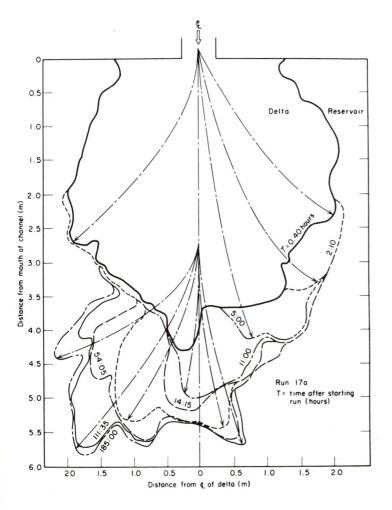

**Figure 14.29** Changes in the experimental delta shown in Figure 14.27 during a further period of 185 h (run 17a) after reduction of the injected sediment load at the delta head to zero.
*Source:* Schumm, 1977, figure 8-9, p. 312, after Chang, 1967.

sediment production from the source area will cause changes of delta morphology. Similar changes will result from changes of baselevel. Both of these phenomena will have an effect not only on the delta itself, but on the channels upstream. The response of these channels will depend on their nature. A bedload channel that is straight and transporting relatively coarse sediments to a fan delta or to an essentially lobate delta, will respond by degradation. The channel will scour, bank instabilities will develop and very large quantities of sediment will be delivered to the delta as the gradient of the channel steepens. On the other hand, a finer mixed-load channel or suspended-load channel, that is relatively sinuous, might respond to the need for an increased gradient by a major change in pattern. It was demonstrated earlier that river patterns are related to the surface on which they flow, and if a major increase in gradient is required due either to a lowering of baselevel or to scour on a delta, a sinuous channel could compensate and increase its gradient by reducing its sinuosity.

### 14.3.3 Avulsion

As noted above, it is characteristic of the coastal plain situation that rivers are not confined in valleys, and they shift position. One of the most dramatic examples of this is the avulsion and shift of the mouth of the Yellow River 650 km to the north of its former position in 1851. This type of avulsion generally is in response to two factors: (1) channel aggradation due to progressive extension of the delta into the sea (the increased length of the stream requiring aggradation upstream), and (2) the fact that in many cases there is usually a shorter route to the sea that the river may take.

This type of event has occurred many times on the Mississippi River delta and, as a result, the Mississippi delta is a complex deposit of several delta lobes each reflecting a different position of the lower Mississippi River (Figure 14.31). At present, in fact, without control structures built by the Corps of Engineers, it is likely that the Mississippi River would now follow a shorter course to the Gulf of Mexico, through the Atchafalaya River channel, which is approximately a 160 km shorter route. Such a dramatic shortening of its course would have the effect of a major baselevel lowering which, in turn, would result in scour of the channel, rejuvenation of tributaries, a greatly increased sediment production and movement of large quantities of alluvium to the new delta lobe.

The most detailed study of a river that has undergone periodic avulsion on a grand scale has been carried out on the Mississippi River, by Fisk and his colleagues (1944, 1952), using vast quantities of information provided by the US Army Corps of Engineers, which provides a fairly clear picture of the response of the Mississippi River to delta growth through time. Two types of avulsion have taken place, in one of which a major diversion of the river

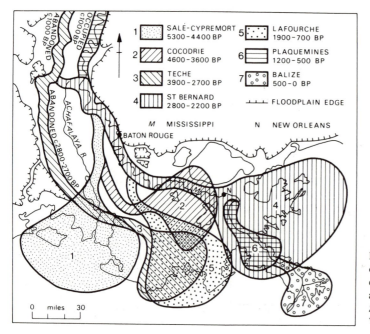

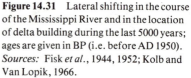

**Figure 14.31** Lateral shifting in the course of the Mississippi River and in the location of delta building during the last 5000 years; ages are given in BP (i.e. before AD 1950). *Sources:* Fisk *et al.*, 1944, 1952; Kolb and Van Lopik, 1966.

**A INITIAL PROGRADATION**

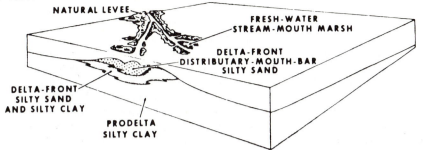

NATURAL LEVEE

FRESH-WATER
STREAM-MOUTH MARSH

DELTA-FRONT
DISTRIBUTARY-MOUTH-BAR
SILTY SAND

DELTA-FRONT
SILTY SAND
AND SILTY CLAY

PRODELTA
SILTY CLAY

**B ENLARGEMENT BY FURTHER PROGRADATION**

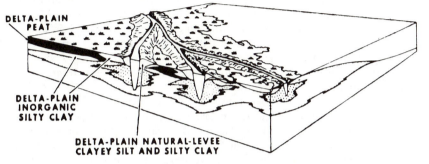

DELTA-PLAIN
PEAT

DELTA-PLAIN
INORGANIC
SILTY CLAY

DELTA-PLAIN NATURAL-LEVEE
CLAYEY SILT AND SILTY CLAY

**C DISTRIBUTARY ABANDONMENT AND TRANSGRESSION**

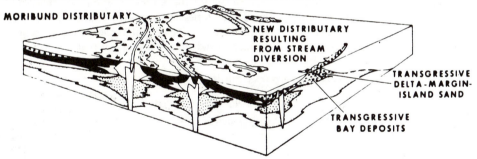

MORIBUND DISTRIBUTARY

NEW DISTRIBUTARY
RESULTING
FROM STREAM
DIVERSION

TRANSGRESSIVE
DELTA-MARGIN-
ISLAND SAND

TRANSGRESSIVE
BAY DEPOSITS

**D REPETITION OF CYCLE**

REOCCUPATION OF OLD
DISTRIBUTARY COURSE

CREVASSE
DEPOSIT

**Figure 14.32** Development of delta sequence in the Mississippi delta.
*Source:* D. E. Frazier, *Trans. Gulf Coast Assoc. of Geol. Societies,* 1967.

hundreds of kilometres from the sea results in it forming a new delta, and the other when the avulsion occurs on the delta itself to form a new delta lobe. When the river shifted from construction of the Teche delta to the beginning of the construction of the St Bernard and LaFourche deltas (Figure 14.31), the Mississippi River, at a position approximately 240 kilometres from the sea, shifted from the west to the east side of its valley. This abruptly shortened the course of the river. The river remained in this position, with progressive growth of the LaFourche delta. When the LaFourche delta had grown to essentially its maximum extent, the river was approximately 2146 kilometres long. There was then a shift in the position of the river upstream to its present position along the east bluff below Vicksburg, and the LaFourche delta was abandoned. Both of these events dramatically shortened the course of the river.

No indication has been found that any permanent diversion of the Mississippi River has been brought about by crevassing or overtopping of banks during floods, such as occurred on the Yellow River. Instead abandonment of Mississippi River courses was accomplished through gradual and progressive enlargement of the diversion channel over time periods of a century or more, as demonstrated by ages of the deltas (Figure 14.31). There is a considerable time overlap among these deltas, and this supports Fisk's conclusion that the diversion of the Mississippi River from one course to another requires periods of at least 100 years. Fisk concludes that the course changes of the Mississippi are fundamentally

different from the rapid diversions of rivers like the Yellow River, because the Mississippi channel is relatively deep and is able to transport its sediment load without difficulty.

Figure 14.32 shows the growth of a bird's-foot delta. Unlike the fan delta which builds a well-defined fanfront into the water, the bird's-foot advances its distributary channels. As they lengthen they will be abandoned and new distributaries form. Growth is further complicated by subsidence of the distributary channels into the pro-delta sediments which are fine. The delta advances tentatively, with many episodes of channel abandonment and shift above sea level, and by a complex process of submarine slumps and slides trenched by mudflow gullies below sea level (Figure 14.33). The present Mississippi delta is prograding on a slope of 0·2°–1·5° in a water depth of 5–100 m by flows and slumps which may move given sediment up to 1·5 km per year.

The ultimate destination of fluvial transport is the deep marine environment, where river sediments move down canyons in the continental shelf (see Chapter 15) and accumulate at their mouths as deep-sea fans (Figure 14.34). These fans are larger and of lower gradient (i.e. less than 1°) than wet alluvial fans, but are somewhat similar in morphology and evolution. Upper fans are commonly trenched by turbidity currents, bordered by natural levees, whereas the lower fans are characterized by distributaries. Fan growth appears to take place through the development of a series of smaller suprafans formed by individual turbidity currents.

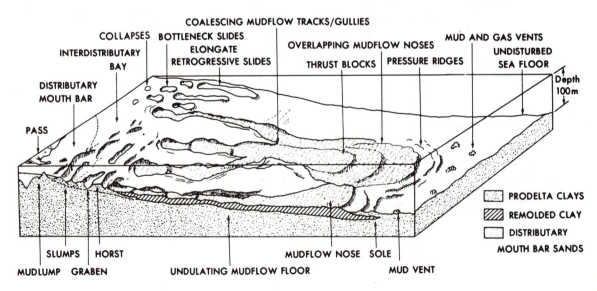

**Figure 14.33** Schematic diagram of the subaqueous morphology of part of the Mississippi delta front slope.
*Source:* Prior and Coleman, 1978, figure 1, p. 43.

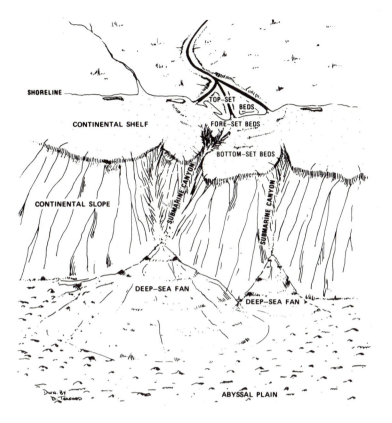

**Figure 14.34** Schematic diagram illustrating the components of a post-glacial (Holocene) subaerial delta–submarine canyon–deep-sea fan system of a continental margin.
*Source:* Moore and Asquith, 1971, figure 1, p. 2564.

# References

Allen, J. R. L. (1970) 'A quantitative model of grain size and sedimentary structures in lateral deposits', *Geological Journal*, vol. 7, 129–46.

Anstey, R. L. (1965) 'Physical characteristics of alluvial fans', *US Army Natick Laboratory Technical Report* ES-20.

Bates, C. C. (1953) 'Rational theory of delta formation', *Bulletin of the American Association of Petroleum Geologists*, vol. 37, 2119–62.

Bergstrom, R. E. and Walker, T. R. (1956) 'Groundwater geology of the East St Louis area, Illinois', *Illinois State Geological Survey Report of Investigations* 191.

Blatt H., Middleton, G. and Murray, R. (1972) *Origin of Sedimentary Rocks*, Englewood Cliffs, NJ, Prentice-Hall.

Brice, J. C. (1975) 'Airphoto interpretation of the form and behavior of alluvial rivers', *Final Report for the US Army Research Office*.

Bull, W. B. (1964) 'Geomorphology of segmented alluvial fans in western Fresno County, California', *US Geological Survey Professional Paper* 352-E, 89–128.

Bull, W. B. (1968) 'Alluvial fans', *Journal of Geological Education*, vol. 16, 101–6.

Bull, W. B. (1977) 'The alluvial-fan environment', *Progress in Physical Geography*, vol. 1, 222–70.

Cain, J. M. and Beatty, M. T. (1968) 'The use of soil maps in the delineation of flood plains', *Water Resources Research*, vol. 4, 173–82.

Chang, Hai-Yang (1967) 'Hydraulics of rivers and deltas', unpublished PhD dissertation, Colorado State University.

Church, M. (1972) 'Baffin Island sandurs: a study of arctic fluvial processes', *Geological Survey of Canada, Bulletin* 216.

Clayton, K. M. (1977) 'River terraces', in F. W. Shotton (ed.), *British Quaternary Studies*, Oxford University Press, chapter 12, 152–67.

Collinson, J. D. (1978) 'Alluvial sediments', in H. G. Reading (ed.), *Sedimentary Environments and Facies*, New York, Elsevier, chapter 3, 15–60.

Denny, C. S. (1965) 'Alluvial fans in the Death Valley region, California and Nevada', *US Geological Survey Professional Paper* 466.

Eckis, R. (1928) 'Alluvial fans of the Cucamonga district, southern California', *Journal of Geology*, vol. 36, 225–47.

Elliott, T. (1978) 'Deltas', in H. G. Reading (ed.), *Sedimentary Environments and Facies*, New York, Elsevier, chapter 6, 97–142.

Fisher, W. L., Brown, L. F., Scott, A. J. and McGowen, J. H. (1969) 'Delta systems in the exploration for oil and gas', Bureau of Economic Geology, University of Texas.

Fisk, H. N., *et al.* (1944) 'Geological investigation, of the alluvial valley of the lower Mississippi River', Vicksburg, Miss., US Army Corps of Engineers, Mississippi River Commission.

Fisk, H. N. *et al.* (1952) 'Geological investigation of the Atchafalaya Basin and the problem of Mississippi River diversion', US Army Corps of Engineers, Waterways Experiment Station, 2 vols.

Gole, C. V. and Chitale, S. V. (1966) 'Inland delta building activity of Kosi River', *American Society of Civil Engineers, Journal of the Hydraulics Division* HY-2, 111–26.

Happ, S. C. (1975) 'Genetic classification of valley sediment deposits', in V. A. Vanoni (ed.), *Sedimentation Engineering*, New York, American Society of Civil Engineers, 286–92.

Hunt, C. B. and Mabey, D. R. (1966) 'Stratigraphy and structure, Death Valley, California', *US Geological Survey Professional Paper* 494-A.

Kolb, C. R. and Van Lopik, R. R. (1966) 'Depositional environments of the Mississippi River deltaic plain southeastern Louisiana', in M. L. Shirley and J. A. Ragsdale (eds), *Deltas and their Geologic Framework*, Houston Geological Society, 17–61.

Krigström, A. (1962) 'Geomorphological studies of sandur plains and their braided rivers in Iceland', *Geografiska Annaler*, vol. 54, 328–46.

Jopling, A. V. (1963) 'Hydraulic studies on the origin of bedding', *Sedimentology*, vol. 2, 115–21.

LeBlanc, R. J. (1972) 'Geometry of sandstone reservoir bodies', *American Association of Petroleum Geologists Memoir* 18, 133–89.

Leopold, L. B. (1976) 'Reversal of erosion cycle and climatic change', *Quaternary Research*, vol. 6, 557–62.

Leopold, L. B., Emmett, W. W. and Myrick, R. M. (1966) 'Channel and hillslope processes in a semiarid area, New Mexico', *US Geological Survey Professional Paper* 352-G, 193–253.

Lewin, J. and Manton, M. M. M. (1975) 'Welsh floodplain studies: the nature of floodplain geometry', *Journal of Hydrology*, vol. 25, 37–50.

Lustig, L. K. (1965) 'Clastic sedimentation in Deep Springs Valley, California', *US Geological Survey Professional Paper* 352-F, 131–92.

McGowen, J. H. (1970) 'Gum Hollow fan delta, Nueces Bay, Texas', *Bureau of Economic Geology, University of Texas, Report Investigation* 69.

McGowen, J. H. and Garner, L. E. (1970) 'Physiographic features and stratification types of coarse-grained point bars: modern and ancient examples', *Sedimentology*, vol. 14, 77–111.

Moore, G. T. and Asquith, D. O. (1971) 'Delta – term and concept', *Bulletin of the Geological Society of America*, vol. 83, 2563–8.

Pettijohn, F. J., Potter, P. E. and Siever, R. (1972) *Sand and Sandstone*, New York, Springer-Verlag.

Popov, I. V. and Gavrin, Yu. S. (1970) 'Use of aerial photography in evaluating the flooding and emptying of river flood plains and the development of flood-plain currents', *Soviet Hydrology: Selected Papers*, no. 5, 413–25.

Prior, D. B. and Coleman, J. M. (1978) 'Submarine landslides on the Mississippi delta-front slope', *Geoscience and Man*, vol. 19, 41–53.

Rachochi, A. (1981) *Alluvial Fans: An Attempt at an Empirical Approach*, Chichester, Wiley.

Ryder, J. M. (1971) 'Some aspects of the morphometry of paraglacial alluvial fans in south-central British Columbia', *Canadian Journal of Earth Science*, vol. 8, 1252–64.

Schumm, S. A. (1965) 'Quaternary paleohydrology', in H. E. Wright and D. G. Frey (eds), *The Quaternary of the United States*, Princeton, NJ, Princeton University Press, 783–94.

Schumm, S. A. (1977) *The Fluvial System*, New York, Wiley.

Selley, R. C. (1976) *An Introduction to Sedimentology*, London, Academic Press.

Silvester, R. and de la Cruz, D. de R. (1970) 'Pattern forming processes in deltas', *Journal of the Waterways and Harbors Division, American Society of Civil Engineers*, vol. 96, 201–17.

Thorarinsson. S. (1953) 'Some new aspects of the Grimsvötn problem', *Journal of Glaciology*, vol. 2, 267–74.

Thornbury, W. D. (1969) *Principles of Geomorphology*, 2nd edn, New York, Wiley.

Vita-Finzi, C. (1969) *The Mediterranean Valleys*, Cambridge University Press.

Wildman, N. A. (1981) 'Episodic removal of hydraulic-mining debris, Yuba and Bear River basins, California', unpublished MS thesis, Colorado State University, Fort Collins.

Williams, G. P. (1978) 'Bank-full discharge of rivers', *Water Resources Research*, vol. 14, 1141–58.

Wolman, M. G. and Leopold, L. B. (1957) 'River flood plains: some observations on their formation', *US Geological Survey Professional Paper* 282-C, 87–109.

Zonneveld, J. I. S. (1975) 'River terraces and Quaternary chronology in the Netherlands', *Geologie en Mijnbouw*, vol. 19, 277–85.

# Fifteen  *Coastal geomorphology*

Coastal geomorphology is concerned with the processes and landforms, both subaerial and submarine, within the littoral zone together with any associated cliff features, sand dunes, or organic deposits lying to the landward. The major divisions of the littoral zone are presented in Figure 15.1 and have been defined by Komar and others as follows (see Komar, 1976, pp. 12–14):

*Offshore* The shallowing water seaward of the breaker zone swept by sinusoidal wind waves (see Section 15.1) which are transformed shoreward into solitary waves.

*Breaker zone* The belt of variable width in which incoming waves become unstable and break.

Longshore bar  A low ridge, or series of ridges, aligned parallel to the shore, usually composed of sand, which are formed within the high-water breaker zone and may be exposed at low tide.

*Surf zone* The zone between the breaker and swash zones swept by turbulent borelike translation waves following the initial wave breaking. Together with the breaker zone, this forms the location of longshore currents.

Longshore trough An elongated depression on the shoreward side of the longshore bar.

*Swash zone* A normally more steeply sloping belt exposed to swash and backwash lying between the upper limit of high-tide wave swash and the lower limit of low-tide backwash.

Beach slope  The general sloping surface shoreward of the breaker zone.

Beach face The part of the beach profile usually exposed to wave swash.

*Backshore* The part of the beach above the level of normal high-tide wave swash.

Berm  A low-gradient surface built by sediment deposited by receding waves. The berm crest is its outer edge

and multiple berms may be separated by beach scarps which are normally less than 1 m high.

The clastic material composing marine beaches presents a wide range of composition and size. Although most commonly quartzose, some tropical coasts have coral sand and other volcanic coasts with rapid erosion may have beach material predominantly of ferromagnesian minerals. The coarsest sand material appears in the two zones of maximum turbulence or high energy where larger and smaller waves break (i.e. the breaker and transition zones: Figure 15.1), and it is here that sorting is poorest. The coarsest material of all (i.e. gravel, cobbles and boulders) is commonly found in the backshore zone where it has been thrown by storm waves. This coarser material is subjected to considerable abrasion by high-energy wave action and, for example, it has been estimated that a 10-cm cube of limestone would become well rounded after only about one week under the action of 0·5-m-high waves. Beach cobbles are well rounded but are commonly of low sphericity due to wave sliding action, compared with the low roundness and high sphericity characteristic of many stream cobbles. Each littoral zone possesses its own distinctive sedimentary structures but the most important of these are the cross-laminated sands dipping seawards at low angles characteristic of both the foreshore and the longshore bar.

## 15.1 Sea level, waves and currents

The height of sea level, both absolute and relative to that of the adjacent coast, is constant neither in space nor time. It is thus not possible to separate logically the causes and consequences of short-term changes (i.e. those having a prevalence of less than $10^2$ years) from those occurring in the longer term (see Section 15.5.1). Table 15.1 lists the major short-term sea-level variations.

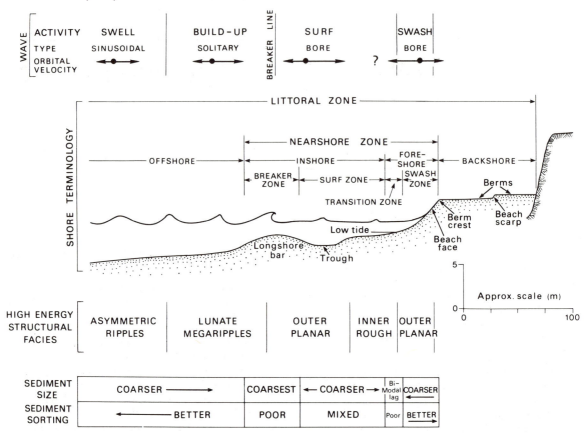

**Figure 15.1**  The major divisions of the littoral zone.
*Sources:* Komar, 1976, figure 2-1, p. 12; Ingle, 1966, figure 116, p. 181.

Wave characteristics are described later. A tsunami, produced by an earthquake shock, is characterized by waves in the open sea of wavelength 150–250 km or more, and velocities of up to 800 km/h – these figures compare with wind-generated wavelengths of < 300 m and velocities of < 90 km/h. Although deep-water tsunami waveheights are only 3–5 m, they have huge energy and greatly increase their height in shoaling water such that an open-sea wave of height 5 m and velocity 600 km/h may suddenly become one 30 m high travelling at 50 km/h when it breaks on the shore (Hindley, 1978). Other sea-level rises occur in certain locations with greater frequency as a result of onshore winds associated with atmospheric low-pressure centres acting in relatively

**Table 15.1**  Short-term sea-level variations

| Cause | Period | Prevalence (Frequency) | Amplitude |
|---|---|---|---|
| Waves | seconds | virtually continuous | up to 20 m |
| Tsunamis | minutes to hours | occasional | up to 10 m |
| Tides | $12\frac{1}{2}$ h | daily | up to 20 m |
| Storm tides and surges | days to years | occasional | 1 to several metres |
| Annual tides | 1 year | yearly | up to 1 m |

*Source:* Fairbridge, 1966, table 1, p. 477.

enclosed shallow seas like the Gulf of Mexico and the North Sea. A fall in atmospheric pressure leads to a rise in sea level, and this can be as much as 30 cm in the case of a pressure fall of 34 mb. This effect, combined with onshore winds over shallow water, results in sea-level rises of 3–5 m along the Gulf Coast associated with hurricane conditions. Abnormal weather conditions can also generate long sea waves which cause short-term high water in coastal areas, as in the case of the North Sea surges. In 1953 a deep low-pressure centre to the north of Scotland moved east causing south-west winds over the North Sea to veer suddenly north, to strengthen locally up to 280 km/h causing a surge peak to move south along the east coast of Britain during the next 12 h. This coincided with high tide in places, causing sea level to rise 2·4 m above predicted high tide and extreme wave swash to reach 6 m above this (King, 1972). Of course, sea-level changes at a scale more frequent and regular than those produced by meteorological effects are the semi-diurnal and fortnightly tidal effects. The periodicity of high tides is about 12 h 25 min (i.e. one-half lunar day), and this is superimposed on a two-week cycle of high (spring) and lower (neap) tides when the earth, moon and sun are in and out of phase. Spring and neap high tides are, on average, 20 per cent higher and lower than the average high-tide level, respectively. Tidal ranges depend on the size, configuration and bottom topography of the marine basin. The maximum range in the central Pacific is about 50 cm; in the restricted and shallow North Sea 4 m + ; in the funnel-shaped Severn estuary 7 m + ; and in the Bay of Fundy 15 m. Spring tides along the Texas coast rise 1 m above mean sea level and even in the land-locked Lake Michigan the tidal rise is 10 cm. Almost all coastal locations experience annual tides, or fluctuations in sea level of 20–30 cm or more; and although their causes are not known for certain, they are due to a complex interaction of:

(1) sea density – this is controlled mainly by temperature and to a much lesser extent by salinity; the lower the temperature and the greater the salinity, the higher the density and the lower is sea level; it is partly for reasons of density that the eastern Pacific is 30–50 cm higher than the Atlantic, and that sea level off Nova Scotia is some 35 cm higher than off Florida;
(2) atmospheric pressure – high atmospheric pressures give lower local sea levels;
(3) speed of ocean currents – these speeds vary with total global energy budgets and there may be as much as 18 cm difference in sea levels on opposite sides of a curving, fast-flowing current;
(4) water locked up in winter snow in the northern hemisphere;

(5) seasonal piling up of water along windward coasts – e.g. South and East Asia during the summer monsoon.

Sea-level changes off California illustrate the effect of these variables. Here sea level varies some 25 cm or so, being low in March–April and high in August–September, mainly as a result of actively upwelling cold water during spring. It has been estimated that the sea-level rise of more than 6 cm measured off southern California during the period 1958–69 could be attributed to: 3·7 cm due to an increase of 1°C in water temperature; 0·6 cm due to atmospheric pressure change; 1 cm due to changes in the dynamics of ocean currents; and 1 cm due to world glacier melting. The effect of glaciers on eustatic changes will be treated in Section 19.4.

Waves are described in terms of their height $H$, wavelength $L$, steepness or slope $H/L$, ratio of their height to water depth $H/h$, period $T$ (in seconds), velocity $C$ (phase velocity), energy density $E$ (per unit width per wavelength) and energy flux or wave power $P$ (see Shepard, 1963; King, 1972; Komar, 1976; Davis, 1978). Equations for energy density and energy flux are as follows:

$$E = \rho g H^2/8$$
$$\text{and } P = E C n$$

where: $g$ = the acceleration of gravity;
$\rho$ = water density;
and $n$ = a constant, depending on water depth (deep water $n = \frac{1}{2}$; shallow water $n = 1$).

Of course, waveheights vary considerably in a wave sequence (spectrum) over a period of time and their group characteristics are described by the following statistics:

$H_{1/3}$ = the significant waveheight; being the mean height of the highest third of the waves;
$H_{1/10}$ = the mean height of the highest 10 per cent of the waves;
$H_m$ = the height of the single maximum wave;
$\overline{H}$ = the mean of all wave heights.

Expectable relations have been established between these statistics, namely:

$$\overline{H}/H_{1/3} = 0·64; \quad H_{1/10}/H_{1/3} = 1·27 \text{ and}$$
$$H_m/H_{1/3} = 1·53–1·85.$$

Most waves are generated by winds of speed $W$ blowing over a fetch distance of $F$, and deep-water relations have been predicted (Table 15.2). For short fetches ($F < 32$ km or 20 miles), $H$ increases directly with wind speed $W$ and wave period $T$ increases as $\sqrt{W}$; on small lakes wavelengths $L$ are usually less than 7 m (20 ft) and periods less than 2 s; with long oceanic fetches, wavelengths exceed

**Table 15.2**   Deep-water wave relations

| Wind speed (W) (m/s) | Fetch distance (F) (km) | $H_{1/3}$ (m) | Wave period (T) (s) |
|---|---|---|---|
| 10 | 200 | 2·1–2·4 | 7·0–8·0 |
| 10 | 1000 | 2·4–2·7 | 8·0–11·0 |
| 20 | 200 | 4·3–5·2 | 8·5–10·0 |
| 20 | 1000 | 8·9–11·0 | 15·0–16·0 |
| 30 | 200 | 7·9–8·2 | 10·0–12·0 |
| 30 | 1000 | 15·0 | 15·0–19·0 |

*Source:* Komar, 1976, table 4–2, p. 93; after Darbyshire and Draper and Longuet-Higgins.

700 m (2000 ft) and periods exceed 20 s.

Wave movement can be viewed in terms of deep-water $(h_\infty > L_\infty/4)$,* shallow-water $(h_s < L_\infty/20)$ and breaking wave $(h_b$: where $H_b/h_b \simeq 0.8–1.25)$ water-depth conditions, respectively.* Deep-water wind-generated (Airy) waves involve a sinusoidal motion (Figure 15.2A) with individual particles describing circular orbits, having a maximum orbital velocity of $u_m$ ($= \pi H_\infty/T_\infty$). Such waves are of long wavelength and small steepness, with phase velocities expressed by the following relationships:

$$C_\infty = \sqrt{[gL_\infty/(2\pi)]} = 1.56\ T_\infty\ \text{m/s}.$$

* The subscripts $\infty$, $s$ and $b$ refer to deep water, shallow water and breaking wave conditions, respectively.

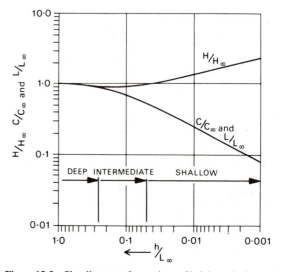

**Figure 15.3**   Shoaling transformations of height, velocity and wavelength of deep-water sinusoidal waves.
*Source:* Komar, 1976, figure 3-5, p. 44, copyright Prentice-Hall Inc.

As the deep-water waves move into shoaling water changes in waveform and motion occur. At depths of approximately $\frac{1}{2}$ $L_\infty$ the waves noticeably 'touch bottom' and begin to decrease velocity and wavelength; whereas when the depth $h_s$ becomes less than one-twentieth of the deep-water wavelength ($L_\infty$), the wave-

**A**   Deep-Water Sinusoidal Waves

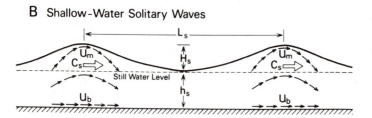

**B**   Shallow-Water Solitary Waves

**Figure 15.2**   Geometry and water-particle movements in respect of A. deep-water sinusoidal waves, and B. shallow-water solitary waves: **C** = velocity of waveform; **U** = orbital velocity of water particles; **H** = waveheight; **h** = water depth; **L** = wavelength. Subscripts: $\infty$ = deep water; **s** = shallow water; **m** = maximum; **b** = bottom.

height increases markedly (Figure 15.3). Particle orbits then become more elliptical and each wave tends to act independently in a solitary manner (Figure 15.2B), where $C_s = \sqrt{\{g(h_s + H_s)\}}$ with a net shoreward particle orbital velocity being at a maximum at the wave crest $(u_m = \frac{1}{2} \cdot \frac{H_s}{h_s} \cdot C_s)$. When $H/h$ exceeds $0.8$, the waves steepen sufficiently so as to become unstable and break, but particularly steep deep-sea waves may break when $H/h$ is as little as $0.6$. Figure 15.4 shows that varying combinations of two ratios of deep-water and breaking wave conditions (including beach slope – tan $\beta$) determine the three types of breakers commonly encountered:

(1) Spilling breakers – where water spills down the steepened shoreward wave limb; especially associated with steep deep-water waves and onshore winds.

(2) Plunging breakers – the conventional breakers, where the shoreward face becomes vertical and plunges shoreward; especially associated with long, low deep-water swells and offshore winds,

(3) Surging breakers – where the base of the wave surges up the beach face and the crest collapses and dis-

appears; especially associated with steep beach slopes.

At the break point maximum orbital velocities are slightly less than those in the shallow water ($u_m = \frac{1}{3} \cdot \frac{H_b}{h_b} \cdot C_b$) and the breaking wave is bodily propelled up the surf and swash zone as a bore which retreats as a backwash.

Wave energy is consumed by reflection from the beach, by frictional drag on the bottom (i.e. debris-transporting forces) and by the generation of longshore currents parallel to the shore in the surf zone. Longshore currents are generated by differences of waveheight along the beach at a given time or by an oblique wave approach (angle of incidence – $\alpha$) (Figure 15.5). Longshore currents would continue to increase their mean velocity ($\overline{V}_e$) and discharge ($\overline{q}_e$) with distance along the beach unless relieved by rip currents – strong, narrow currents flowing seaward from the surf zone. It is thus possible to define these variables theoretically in terms of shoreward wave discharge, per unit length of breaking wave crest, per unit time ($q = \frac{\pi}{4} \cdot \frac{H_b^2}{T_b}$) and average spacing of rip currents $\ell$:

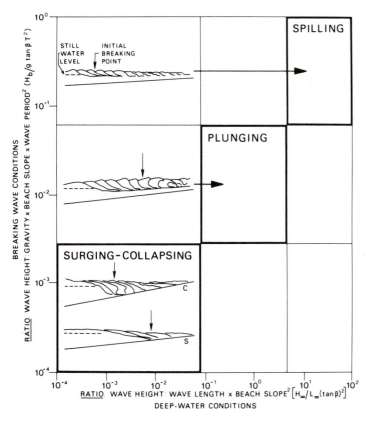

**Figure 15.4** Relationship between breaking wave and deep-water conditions which controls the occurrence of spilling, plunging, or surging–collapsing waves. In general, steep deep-water waves (i.e. $H_\infty/L_\infty$ being large) breaking on a gentle shore tend to spill, flatter waves breaking on a steeper shore collapse, whereas the more common intermediate waves are plunging.
*Source:* Komar, 1976, figures 4–18, and 4–19, pp. 107, 108, after Galvin.

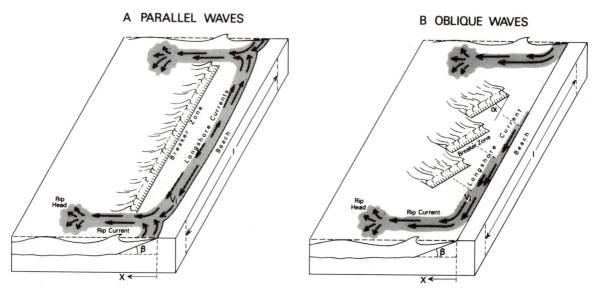

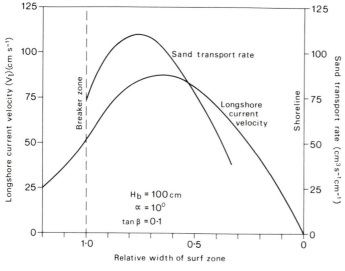

**Figure 15.5** Schematic drawings showing the generation of long shore and rip currents by waves A. parallel and B. oblique to the beach; longshore current velocity increases along the beach until it flows seaward as a rip current.
*Source:* F. P. Shepard, *Submarine Geology*, 1963, 2nd edn, figure 41, p. 76, copyright © 1948, 1963 by Francis P. Shepard, reproduced by permission of Harper & Row, Publishers, Inc.

**Figure 15.6** Theoretical distribution of longshore sand transport and longshore current velocity across the surf zone, assuming a breaking waveheight of 100 cm, an angle of incidence of 10° and a tangent of beach slope of 0·1.
*Source:* P. D. Komar, *Beach Processes and Sedimentation*, 1976, figure 8–6, p. 213, copyright © Prentice-Hall Inc.

$$\overline{V}_\ell = \frac{q\ell}{h_b^2} \cdot \tan\beta \cdot \sin\alpha \cdot \cos\alpha \text{ (empirically equal to } 2\cdot7\,u_m \cdot \sin\alpha \cdot \cos\alpha)$$

$$\overline{q}_\ell = q\ell \cdot \sin\alpha \cdot \cos\alpha/2.$$

Figure 15.6 gives theoretical values of $V_e$ (centimetres per second) across a surf zone for values of $H_b = 100$ cm, $\alpha = 10°$ and $\tan\beta = 0\cdot1$.

## 15.2 Beach processes and profiles

As solitary waves move shoreward across the shoaling offshore zone, where the ratio $h_s/L_s$ is small, they disturb the particles lying on the bottom. This disturbance is strongly controlled by waveheight and particle diameter (Figure 15.7) such that for normal waveheights of 1–2 m coarse silt can be disturbed at twice the depth as can medium sand. Once disturbed, the net movement of bottom particles depends on the balance between the forces of net onshore wave drag $F_D$, downslope offshore gravity $F_g \sin\beta$ and frictional resistance $F_r$ (Figure 15.8). Solitary waves induce a near-bottom mean mass trans-

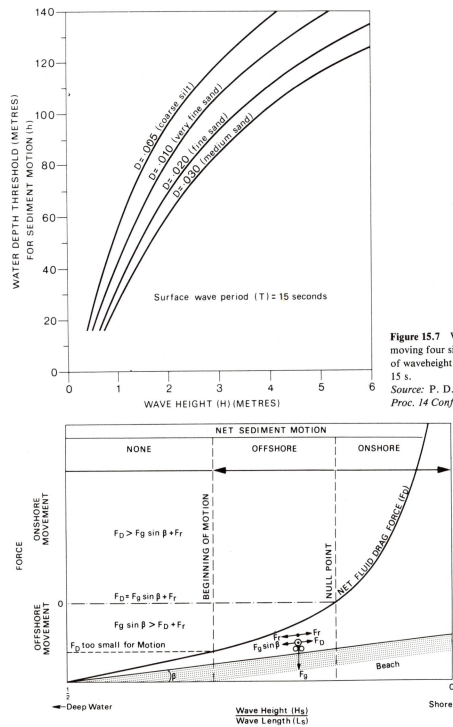

**Figure 15.7** Water depth threshold for moving four sizes of material as a function of waveheight, assuming a wave period of 15 s.
*Source:* P. D. Komar and M. C. Miller, *Proc. 14 Conf. Coastal Eng.*, 1975.

**Figure 15.8** Forces operating on near-shore sedimentary particles.
*Source:* Pettijohn *et al.*, 1972, figure 9-20, p. 356, after Johnson and Eagleson.

port velocity $\bar{u}$ in the direction of their travel, where $h_s/L_s$ is small such that:

$$\bar{u} = \frac{5}{4} \cdot \frac{u_b^2}{C_s} = \frac{5}{4} [H_s/(2h_s)]^2/C_s$$

where: $u_b$ = maximum horizontal component of the orbital velocity near the bottom (see Figure 15.2B);

$C_s$ = shallow-water wave phase velocity;
$H_s$ = waveheight;
$h_s$ = still water depth.

The mean mass transport velocity induces the net onshore force which operates on the disturbed bed particles.

Where: $F_D = F_g \cdot \sin \beta + F_r$

particles of a given size and settling velocity oscillate to and fro in dynamic equilibrium at the so-called null-point, to the seaward of which there is a slow offshore movement mainly by bed traction and to landward a more active and complex onshore movement by bed traction, suspension and mixed traction/suspension (Figure 15.8). The positions of null-points for a given set of waves can thus be defined for each particle size such that at any null-point coarser particles are moving onshore and finer particles offshore. Thus mixtures of different populations of sediment sizes may occur at each given location. As the water becomes shallower onshore the value of $F_D$ increases dramatically and larger and larger particles are capable of onshore movement. Within the breaker zone material of many sizes tends to accumulate and coarse, poorly sorted sediments are found (see Figure 15.1), but just to the seaward of this the

maximum rate of onshore particle movement occurs (Figure 15.9). Landward of the breaker and surf zones, as will be seen later, debris movement is more irregular with steep or less steep waves causing backwash or swash transport, respectively, to dominate.

Cross-bedded longshore bars may develop in the breaker zone. They are particularly associated with steep ($H_s/L_s > 0.012$), plunging breakers operating on low-gradient beach slopes when onshore transport outside the breaker zone converges on offshore movement of material eroded from the surf zone and foreshore. Multiple longshore bars may result from:

(1) different populations of wave sizes producing a series of breaker zones, even on tideless coasts;
(2) the reforming of waves inside the main breaker zone and their secondary breaking;
(3) a range of tidal levels – although a large tidal range usually produces a broad coastal terrace with a single longshore bar.

Landward of the longshore bar there is commonly a parallel trench or runnel, often associated with a pronounced longshore current. To the shoreward of the breaker zone a breaking wave projects a shallow bore up the beach, moving coarser material as bedload and short-lived clouds of suspended sediment. Material is moved seaward here by return gravity currents and, locally, by rip currents which erode material from nearshore and deposit it as offshore submarine fans out to depths of some 5 m.

The swash zone is dominated by a cycle of shallow, high-velocity onshore flow followed by a possibly shallower and lower-velocity backwash under gravity (Figure 15.10). The ratio of the time to uprush $t$ and the breaking

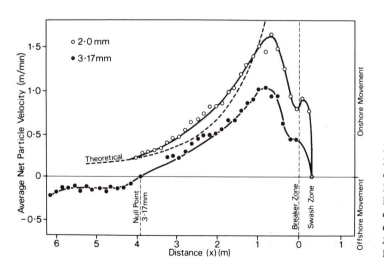

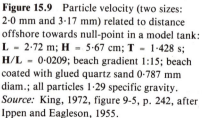

**Figure 15.9** Particle velocity (two sizes: 2·0 mm and 3·17 mm) related to distance offshore towards null-point in a model tank: **L** = 2·72 m; **H** = 5·67 cm; **T** = 1·428 s; **H/L** = 0·0209; beach gradient 1:15; beach coated with glued quartz sand 0·787 mm diam.; all particles 1·29 specific gravity.
*Source:* King, 1972, figure 9-5, p. 242, after Ippen and Eagleson, 1955.

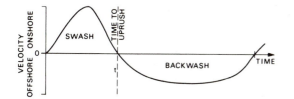

**Figure 15.10** Asymmetrical wave velocity in the near-shore zone.
*Source:* Clark, 1979, figure 11.12, p. 369.

wave period $T$ is expressed as a phase difference $t/T$ such that when:

$t/T < 0.3$, swash and backwash occur before the next wave breaks;

$0.3 < t/T < 1.0$, the swash–backwash cycle is not completed and backwash interferes with succeeding swash;

$t/T > 1.0$, there are continuous surf conditions.

Backwash is commonly assumed to be depleted by beach percolation, but for beach material smaller than 1 mm permeability is virtually nil within the time of a wave period. Assymmetry of swash and backwash is most marked either when the beach is composed of coarse, very permeable material or during rising-tide conditions before the beach material is saturated, producing dominant swash and beach face aggradation and steepening. Within the swash material is moved as bedload, but fine material may be thrown into suspension. Indeed, one of the most important aspects of breaking waves is that they may put material into suspension to be moved by other processes. The swash bore may disturb beach sand to a depth of as much as 5 cm and the net movement of these suspended particles depends on the ratio of their fall (settling) time $f$ to the wave period $T$ such that when $f/T$ is less than 0.5 there is onshore suspended sediment movement, and when $f/T$ is greater than 0.5 there is offshore movement and erosion of these finer fractions.

The steepness of the beach face slope is mainly controlled by two factors (Figure 15.11):

(a) Mean particle diameter. This is the most important influence which may explain up to 70 per cent of the observed variance of beach slopes. Calibre of beach material and beach slopes are directly related.

(b) Wave energy. High-energy, steeper waves tend to transport beach material offshore, resulting in less steep beach face slopes. The influence of wave action on beach slope is complex, however, in that high energy can result both from steep and high waves.

Swash and backwash under differing wave energy condi-

tions produce a cycle of sand structures in which symmetrical, lower-energy ripples alternate with high-energy asymmetrical ripples and antidunes.

One might suppose that, landward of the breaker and surf zones, beach aggradation and steepening by a dominant swash will lead inevitably to an increase in the erosive power of the backwash by a process of negative feedback until an equilibrium beach slope is generated. Although this may occur during short periods of hours or days, in the longer term a beach profile is normally in a constant state of change. The most common cycle of change is that associated with the alternation of swell and storm profiles (Figure 15.12) (Komar, 1976):

Swell profiles – these are associated with waves of greater wavelength and low steepness which tend to move debris onshore at all depths, particularly during the summer in the higher latitudes. Swell waves produce net beach aggradation, a wide berm, a smooth profile, a steep beach face slope and no longshore bars. Swell profiles are also promoted by larger debris sizes, which because of their rapid settling time after wave disturbance, tend to be preferentially moved onshore; whereas finer material may move offshore and assist the production of storm profiles.

Storm profiles – these are associated with steep waves of short wavelength which tend to move material onshore, particularly the finer material which is maintained in suspension. Storm waves are more characteristic of winter conditions, they erode the foreshore, remove the berm, decrease the foreshore slope, and produce offshore bars and troughs under accentuated longshore drift conditions.

Cycles of beach changes which disturb equilibrium may thus be biannual, where the volume of mobile debris remains constant (see Figure 2.18, where the shoreline at Carmel, California was aggraded during April–September 1946 and eroded during the succeeding October–February). There are also occasional disturbances by storm events and associated with tidal fluctuations. For example, spring tides may remove sand deposited by neap tides.

## 15.3 Shoreline processes and depositional forms

### 15.3.1 Longshore movement

As has been shown in Figure 15.5, the nearshore circulation of water is composed of a series of narrow, high-velocity rip currents directed seaward from the breaker and surf zones, which may erode channels in the beach face connected with shallow submarine deltas. These

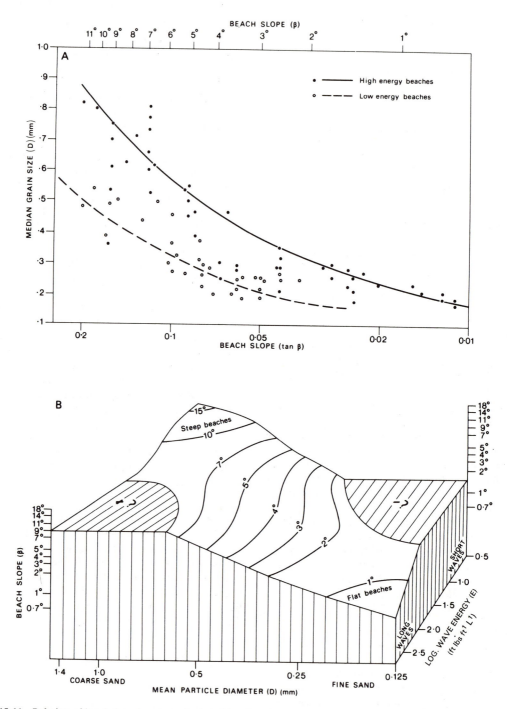

**Figure 15.11** Relation of beach face slope to grain size of beach sediments at mid-tide level and wave energy. *Sources:* W. H. Bascomb, *Trans. Amer. Geophysical U.*, 1951, figure 2, p. 867; King, 1972, figure 12-18A, p. 332.

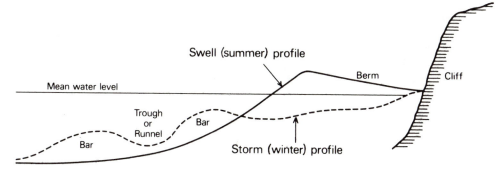

**Figure 15.12**   The storm beach profile with bars v. profile with pronounced berm that occurs under swell wave conditions.
*Source:* P. D. Komar, *Beach Processes and Sedimentation*, 1976, figure 11-1, p. 289, copyright © Prentice-Hall Inc.

currents are fed by the longshore movement of water generated by water level differences associated with the refracted wave approach oblique to the shore. The component of wave energy directed parallel to the shore $E_\ell$ has been theoretically calculated as:

$$E_\ell = E_\infty \sin\alpha_\infty \cos\alpha_\infty \, (L_b/L_\infty)$$

and empirically estimated (where $E_\ell$ is given in $10^3$ ft-lb/min/ft of beach) as:

$$E_\ell = 0 \cdot 0625 \; wH_b^2(L_b/T)\sin (2\alpha_b).$$

where:
$E_\infty$   =   deep-water wave energy;
$\alpha_\infty$   =   angle of wave incidence in deep water;
$\alpha_b$   =   angle of wave incidence in breaker zone;
$L_\infty$   =   deep-water wavelength;
$L_b$   =   breaker zone wavelength;
$H_b$   =   waveheight in breaker zone;
$T$   =   wave period;
$w$   =   weight (lb) of 1 ft³ of sea water ($\simeq$ 64 lb).

$E_\ell$ is at a theoretical maximum, in respect of $E_\infty$, when $\alpha_\infty$ is 47° due to wave crest spreading by refraction, rather than at the expectable 45° (Shepard, 1963; Komar, 1976).

   The longshore movement of debris takes place. by rather different processes within the breaker, surf and swash zones. On gently sloping shores the zones are distinct with longshore transport within and just landward of the breaker zone dominant. On steeper shores underlain by coarser material the three zones become fused into one narrow zone of complex longshore debris transport.

**(1) Breaker zone transport.** It is commonly believed that the most effective longshore sediment transport occurs

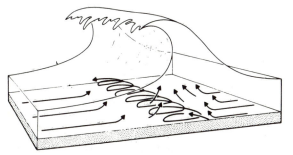

**Figure 15.13**   Sediment transport associated with a breaking wave. Coarse sediment moves as bedload in a series of elliptical paths parallel to the coast, while finer sediment is suspended.
*Source:* Ingle, 1966, figure 46, p. 53.

when breaking waves have put most sand into suspension to be then moved by a longshore current as a 'river of sand' in a belt landward of the breaker line (Figure 15.3). This view is supported by the theoretical calculation that the most effective longshore sediment movement does not occur where the velocity of the current is greatest but closer to the breaker zone (see Figure 15.6). However, some authorities believe that too much emphasis has been placed on the movement of sand in suspension by longshore currents, as distinct from transport at or near the bed. While transport in suspension has been suggested as dominant in the transition zone (see Figure 15.1), in the breaker and surf zones sand may only remain in suspension for 4–5 s at a time and most waves break on a cushion of water which prevents the sand bed from being effectively churned up more frequently than once every few minutes. Indeed, it is believed by some workers that 'bedload' longshore transport may account

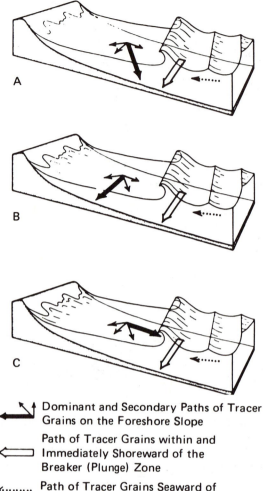

Dominant and Secondary Paths of Tracer Grains on the Foreshore Slope

Path of Tracer Grains within and Immediately Shoreward of the Breaker (Plunge) Zone

Path of Tracer Grains Seaward of the Breaker Zone

**Figure 15.14** Predominant paths of tracer grain movement: A. longshore current velocity and wave motion exert equal influences; B. longshore current dominates; C. wave motion dominates.
*Source:* Ingle, 1966, figure 49, p. 59.

for some 86 per cent of all sand moved parallel to the shore.
(2) Surf zone transport. Shoreward of the longshore stream of sediment transport associated with breaker action a number of vectors act to produce resultant particle movement in the surf zone (Figure 15.14). The predominant movement is determined by the balance between wave velocity, acting more or less in an offshore direction, and the velocity of the longshore current. A less important contributory vector, which becomes dominant in the swash zone, is the velocity directed in

respect of the angle of incidence of surf uprush. Figure 15.14 suggests directions of dominant grain movement in the swash zone when longshore current and wave velocities are balanced and unequal. In general, the longshore current direction dominates if its velocity is in excess of 2 ft/s (0·6 m/s), whereas if its velocity is less than 1 ft/s (0·3 m/s) wave direction dominates.
(3) Swash zone transport. Swash directed obliquely to the shore may produce longshore debris transport by *beach drifting* – a sawtooth movement effected by alternating swash and backwash action (Clark, 1979). Where the shoreward component of swash velocity is equal to the offshore component of backwash, material tends to remain at the same elevation on the foreshore and equilibrium of beach profile occurs (Figure 15.15). Where the former dominates, material moves obliquely up the beach, which aggrades, and where the latter dominates there is net seaward movement and consequent erosion. Experimental results using sand tracers suggest that swash velocity is more or less independent of foreshore slope $\beta$ and that even backwash velocity is more influenced by discharge (controlled by beach permeability) than by slope. Where the swash is directed normal to the beach, small, crescentic seaward projections or mounds of coarser sediment (*beach cusps*) are formed.

Estimates of total longshore sediment transport are made by a number of means, none of which is free from error or ambiguity (King, 1972; Komar 1976).
(1) Measured rates of accretion behind breakwaters and headlands. However, these do not trap all the moving material, some of which by-passes the obstruction or is lost out to sea by offshore transport or rip currents. In addition, the value of such estimates has been shown to

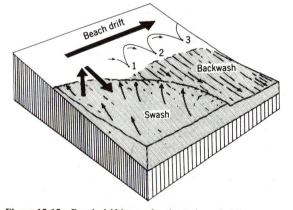

**Figure 15.15** Beach drifting under the action of oblique swash where the longshore current dominates. The path of an individual particle is shown (1–3).
*Source:* A. N. Strahler, *Introduction to Physical Geography*, 1965, figure 23.5, p. 335, copyright © John Wiley and Sons.

vary significantly with different wave energies and directions and, indeed, breakwaters themselves appear to interfere with the whole regime of beach debris movement.

(2) Computations based on wave energy estimates. For example, Bagnold (1963) has given the following equation expressing the total immersed weight of suspended sediment transported past a section of beach in a unit time ($I_\ell$: energy flux, dyne/s):

$$I_\ell = K' (ECn)(\cos\alpha)\overline{V_\ell}/u_b$$

where: ($ECn$)  = wave energy flux in breaker zone (see Section 15.1);

$K'$  = a dimensionless coefficient (experimentally estimated as approximately 0·28);

$\alpha$  = angle of wave incidence in the breaker zone (see Figure 15.5);

$\overline{V_\ell}$  = mean velocity of longshore current (see Section 15.1);

$u_b$  = mean wave orbital velocity near bottom just before wave breaks (see Figure 15.2B).

(3) Estimates from use of tracers or 'tagged debris', such as painted or radioactive pebbles and sand, have been used. The relatively short radioactive decay times (phosphorus-32, halflife 14·6 days, traceable for several months; barium 140-lanthanum 140, halflife 12 days, traceable for up to 7 weeks) restrict observations, but the major difficulty is that the marked material is only a very small proportion of that moving on the beach, making recovery and interpretation difficult. Burial at depths greater than 23 cm makes the material difficult to trace. Empirical work suggests that mean grain velocities $\overline{V_g}$ are about one-sixth of the mean longshore current velocity $\overline{V_\ell}$ and that there is a simple linear relation between the former and the net longshore energy flux $E_\ell$ (Figure 15.16).

(4) Sediment traps present problems in that wave action is destructive of equipment; the latter may itself affect movement; the zone of wave action moves up and down the beach with the tides; the directions of debris movement are both along shore and at right angles to it; different wind directions can completely reverse the direction of longshore transport; and rates of mass transport depend both on the very variable wave energy, as well as on the sediment characteristics. Bedload traps have been tried on Florida sand beaches, indicating 10–40 per cent of movement by bed transport, but this method is not thought to be satisfactory.

(5) Measured rates of 'dilution' of heavy minerals (e.g. hornblende, augite, zircon, etc.) naturally occurring in

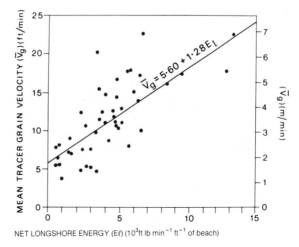

**Figure 15.16** Correlation between a value representing the intensity of net longshore energy and average tracer grain velocity; correlation coefficient = 0·73; empirically derived from Californian beaches, assuming $E\ell = 0·0625\ wH_b^2 (L_b/T)\sin 2\alpha_b$.

*Source:* US Army Beach Erosion Board Tech. Memo. 68, 1956.

beach sands, moving less readily and hence tending to become concentrated. Naturally such observations are very difficult to interpret.

(6) Model studies are also ambiguous but tend to suggest maximum longshore sediment transport at $\alpha = 30°$ and that specific gravity of debris is a very important control, with pumice moving at a rate of some fourteen times that of quartz for a similar grain size.

(7) Estimates of sediment budgets for major coastal compartments separated by headlands or rip currents, and within which there may be equilibrium or disequilibrium of debris input and output.

input — longshore transport in; river sediments; cliff erosion; onshore transport; organic deposits; wind transport in; artificial beach nourishment;

output – longshore transport out; solution and abrasion loss; offshore transport; deposition in submarine canyons; wind transport out; artificial obstruction.

Southern California has five major coastal sediment budget cells operating as self-contained entities. The following longshore sand transport rates ($10^3$ m³/year) have been estimated for certain important beaches; Sandy Hook, New Jersey, 377; Palm Beach, Florida, 115–72; Galveston, Texas, 335; Santa Barbara, California, 214; Port Said, Egypt, 696; Durban, South Africa, 293; Madras, India, 566.

### 15.3.2 Shoreline configuration

The longshore movement of coarser beach material affects the geographical configuration of the shoreline in two major respects. First, the building of extended spits tends to straighten the shoreline by blocking embayments which become sediment traps and fill. Second, erosion at one end of a beach segment and sedimentation at the other causes the shoreline to align itself at right angles to the dominant wave attack. The direction of dominant wave attack may be determined by the long swell waves of maximum fetch, the resultant wind direction, infrequent storm waves driven by winds in excess of 20 km/h

or by the refraction of the wave fronts by the submarine topography. The last-named influence involves an important negative feedback effect in so far as the alignment of the shoreline influences the offshore submarine topography, which influences wave refraction which, in turn, influences shoreline alignment. The concept of an equilibrium beach has thus been extended by Komar to include both plan curvature and profile aspects which may be 'so adjusted so that over a period of years or decades the waves breaking on it provide just the energy needed to transport the sediments supplied to the beach'.

Spits are constructed by dominant wave action in the presence of abundant clastic debris and as they exten

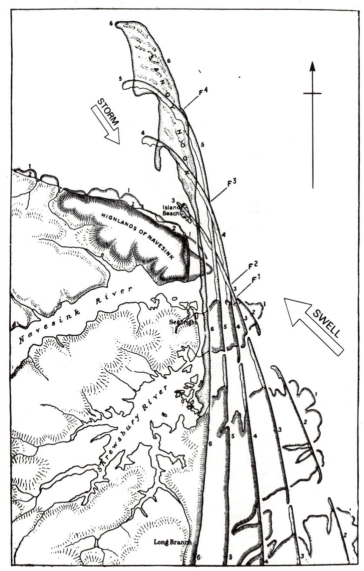

**Figure 15.17**   The development of Sandy Hook spit, New Jersey, from the assumed original coastline (1) through a sequence of recession and spit building (2–5) to the present spit configuration (6).
*Source:* Johnson, 1919, figure 57, p. 296.

into deeper water they are attacked by concentrated refracted wave action and by waves generated by infrequent storm events. Concentrated wave action at spit ends, together with that related to secondary storm directions is shown by the recurving of Sandy Hook, New Jersey (Figure 15.17). Spit growth may take place in a number of ways, depending on debris supply and on the relative importance of wave action from different directions. Sandy Hook is an example of northward spit prograding (outward building) and extension associated with erosion, cliff recession and debris production in the south; with the point of division (fulcrum) between the two zones shifting progressively north ($F^1$–$F^4$). Spurn Head, on the east coast of England, provides a contrasting and more complex development sequence (De Boer, 1964). Historical records suggest that a recession of the weak till and alluvial shoreline of more than 2 km has occurred since about AD 1100 under the action of dominant north-easterly waves, and that this has led to three cycles of spit building towards the south-west. Each cycle, lasting about 250 years (Figure 15.18A) appears to have consisted of spit extension, narrowing of the neck, breaching, rapid destruction of the isolated island tip and the beginning of a new cycle of spit extension – all superimposed on a general westward displacement of the shoreline (Figure 15.18B) (Plate 19).

Cuspate forelands, such as Cape Canaveral, Florida, and Dungeness, in southern England, are accumulations of coastal debris of triangular form, contrasting with that of spits. They were originally thought to be due either to the junction of two large opposed coastal eddies or to wavefront convergence caused by a submerged offshore obstruction. Detailed work on Dungeness by Lewis (1932), however, suggests that their development may be closely allied to that of spits and that changes of sea level, sources of debris supply and wave directions may be important. In Roman times when sea level was some 1·7 m lower than at present, the present area of Romney Marsh was probably a shallow cliff-fringed bay, perhaps more or less isolated from the open sea by a spit (Figure 15.19). It is possible that the following rise of sea level increased the longshore debris supply from the southwest which led to the building of a series of spits (AB, AC, AE, AFG). This phase may have been terminated by a breach in the spit at the southern end near Fairlight, which cut off the debris supply and caused the spit to realign at right angles to the dominant south-south-west swell waves and to retreat northward (HK, MP, BP). As the spit built eastward into deeper water it came under the action of infrequent east-north-easterly storm waves from upchannel and recurved (PNO, PQR), and at the same time the depletion of debris in the north causing

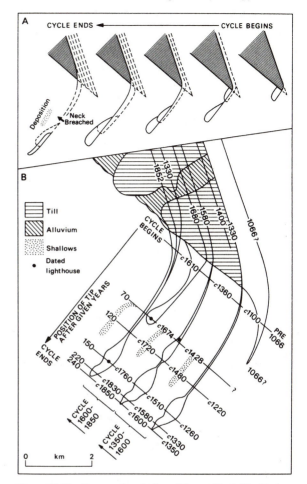

**Figure 15.18** Proposed evolution of Spurn Head, England: A. schematic cycle; B. actual development through at least two cycles.
*Source:* de Boer, 1964, figures 2 and 4, pp. 74, 81.

erosion between New Romney and Hythe, which was checked in 1803–4 by the building of the Dymchurch Wall. The northward movement of the southerly spit-front was succeeded after 1538 by the building of a prograded sequence of smaller spits to the east of Camber Castle, which was constructed in that year on the then coastline.

The study of the temporal evolution of spits is made difficult because of a paucity of means by which spit segments may be precisely dated and because of ambiguities present in their sedimentary petrology. We have already seen how short-term surveys can throw light on spit evolution in the case of Scolt Head Island, Norfolk, England (see Figure 2.13), but such approaches are obviously limited in time. Attempts to examine the

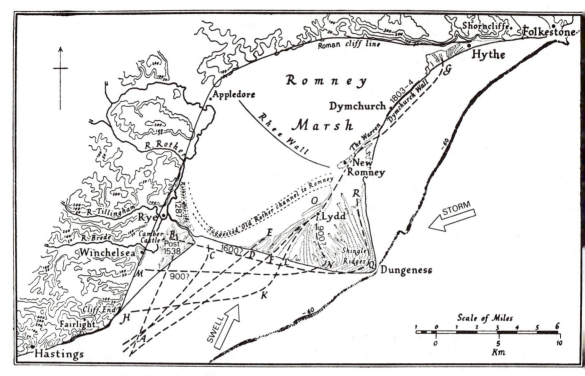

**Figure 15.19**   The evolution of Dungeness, Kent, England.
*Source:* Lewis, 1932, figure 1, p. 310; reproduced in Steers, 1964, figure 71, p. 326.

mechanics of spit-building by observations of the distribution and movement of sedimentary particles have, unfortunately, often raised as many problems as they have solved:

(1)  Most spits have been built several metres above present mean sea level. To what extent is this due to the constructional action associated with storm waves and high tides, or have sea-level changes been a natural adjunct of spit-building?

(2)  Material forming a spit is usually well sorted at any given location and debris calibre decreases systematically from the shoreward end to the tip of the spit.

(3)  The longshore decrease of calibre seems to occur in too short a distance for it to be attributable to abrasion action alone. To what extent are the observed systematic variations of debris calibre due to sorting action, as distinct from abrasion?

(4)  Tagged debris on spits is seldom observed to move simply towards the spit end. On many spits there appears to be no net longshore movement, while on others significant movement occurs, but in different directions depending on the state of wind and tide.

(5)  The most systematic debris sorting along spits

appears to occur where longshore movement is at a minimum and the spit has aligned itself normal to the dominant wave attack, so that debris movement is mainly normal to the beach line.

The above ambiguities are especially important in respect of Chesil Beach, Dorset, southern England, a 28·8-km long shingle ridge lying at, or close to, the present shore and tied to it at both ends (Figure 15.20). The spit reaches its maximum elevation of more than 13 m above mean high water at its south-eastern end and the debris is very well sorted at each location, decreasing in size towards the north-west. This debris does not appear to have been derived from the rocks of the Isle of Portland; the size distribution appears to have remained constant for at least fifty years; there has been no observed net transport towards the north-west (except perhaps for some of the smallest particles by the south-south-westerly winds of secondary fetch); debris movement is mostly normal to the shoreline; the longshore size grading does not occur offshore and it is unlikely that the observed decrease in grain size could have occurred by longshore abrasion. It must be concluded that the present features of the spit are those of a relatively closed sedimentary system oriented at right angles to the dominant south-westerly swell.

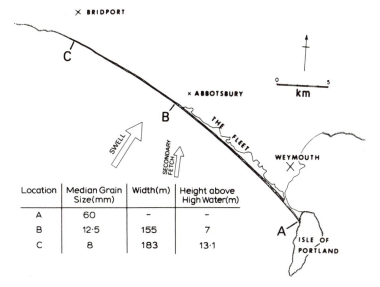

| Location | Median Grain Size(mm) | Width(m) | Height above High Water(m) |
|----------|------------------------|----------|-----------------------------|
| A | 60 | – | – |
| B | 12·5 | 155 | 7 |
| C | 8 | 183 | 13·1 |

**Figure 15.20** Longshore sorting of beach sediments along Chesil Beach, Dorset, England. The included table gives the grain size, width and height of the ridge which is composed of shingle at **A** and pea-sized gravel and sand at **C**.
*Source:* Komar, 1976, figure 13-6, p. 352.

waves, differences of which along shore (i.e. higher energy in the south-east) explain differences in the magnitude and debris characteristics of the spit. However, Chesil Beach has probably had a complex history, perhaps originating as an offshore bar (barrier island) during the Neolithic rise of sea level and having been subjected to more vigorous longshore debris movement at an earlier stage of its history. Chesil Beach is an example of the complex relationship which exists between large spits and barrier islands.

### 15.3.3 Barrier islands

Barrier coasts make up some 13 per cent of the present world's coastlines, with notable absences in China and south-east Asia, and are characterized by major longshore bars (barrier beaches) having lengths of up to 100 km. They are especially formed on depositional coasts of low gradient and abundant supplies of loose sediment. They also appear to be favoured by steeply plunging waves generated by swell of low gradient, but they also occur with storm waves in places. A small tidal range appears to be a favouring, but not essential, factor. Barrier beaches are frequently breached by wave attack to produce inlets landward of which submarine 'washover' fans are formed in the backing lagoons. Excellent examples of barrier islands occur along a stretch of more than 300 km of the coast of the Carolinas (Figure 15.21). They differ from spits in not necessarily being linked to the shore. Barrier islands are composed of cross-bedded, well-sorted sands and pebbles capped with wind-blown sand. Seaward these sands grade into shoreface silts and fine sand and then into the current-rippled marine muds

of the shelf. Landward there is a facies sequence of lagoonal and tidal flat silts and clays, marsh peats and organic clays, with perhaps floodplain deposits.

There have been four main suggestions as to the circumstances necessary for the formation of barrier islands, and these are also associated with some features of spit development (Schwartz, 1973):

(1) Formation during stationary sea level by the build-up of longshore bars by storm swash and blown sand accumulation in the presence of newly colonizing vegetation. Although these features are common in barrier islands (Figure 15.22A), it is now generally agreed that longshore bars cannot build up above mean still water level. This is supported by the fact that no longshore bars are at present observed as being converted into barrier beaches and also because the lagoonal sediments and organisms behind barrier islands show no signs of having been linked to open-sea circulation.

(2) Formation as the result of a fall of sea level exposing a spit or longshore bar. However, the recent history of sea level has been dominated by a postglacial rise and, in addition, barrier beaches commonly occur along tectonically subsiding coasts (e.g. Gulf of Mexico).

(3) Formation associated with a sea-level rise isolating a storm beach or dune ridge by drowning the back beach to form a lagoon (Figure 15.22B). In the presence of a supply of marine and aeolian sand the barrier island may continue to grow and to exhibit the observed sedimentary features (Figure 15.22A). This theory is also supported by Holocene and more recent

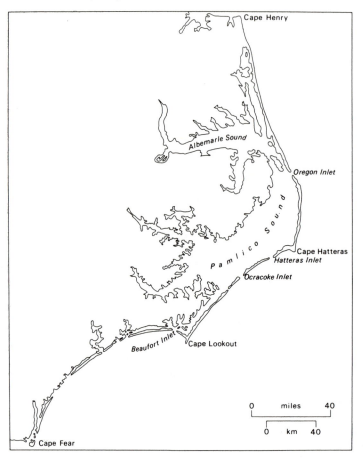

**Figure 15.21** Barrier islands between Cape Henry and Cape Fear, east coast of the United States.

rises of world sea level. In the absence of adequate debris supplies, however, the newly formed barrier island may be eroded by wave action, or driven back on to the shore, as perhaps has been the case with Chesil Beach (see Section 15.3.2).

(4) Formation by build-up of spits by storm wave action and their subsequent breaching by wave attack (Figure 15.22C). Such evolution is possible in certain cases but has probably not been widespread.

## 15.4 Erosional coasts

### 15.4.1 Processes of coastal bedrock erosion

The erosion of bedrock coasts is brought about by the combined influences of three sets of interrelated processes – mechanical wave action, weathering and biological erosion – to which must be added the subaerial processes of mass movements and of fluvial, glacial and aeolian action.

Mechanical wave action operates particularly effectively where long steep waves attack a shoreline, where

**Figure 15.22** Barrier islands. A. Cross section of Galveston Island, Texas, showing the vertical sequence of facies and the rate of outbuilding of the beach and marine sands; B. formation of a barrier island by submergence: 1, a beach or dune ridge forms adjacent to the shoreline; 2, submergence floods the area landward of the ridge forming a barrier island and lagoon; C. idealized formation of an elongated spit which might be breached to form a barrier island. Dunes (1) and shallow sediments (2) formed along the original unprotected shore, to be followed in sequence by the spit initiation and prograding (3–8) and the lagoon sediments.
*Sources:* Komar, 1976, figure 13-23, p. 395, after H. A. Bernard *et al.*, *Geol. Soc. Amer.*; Schwartz, 1973, figure 1, p. 286; after J. H. Hoyt, *Bull. Geol. Soc. Amer.*, 1967, figure 5, p. 1130 and 1968, figure 1, p. 1429.

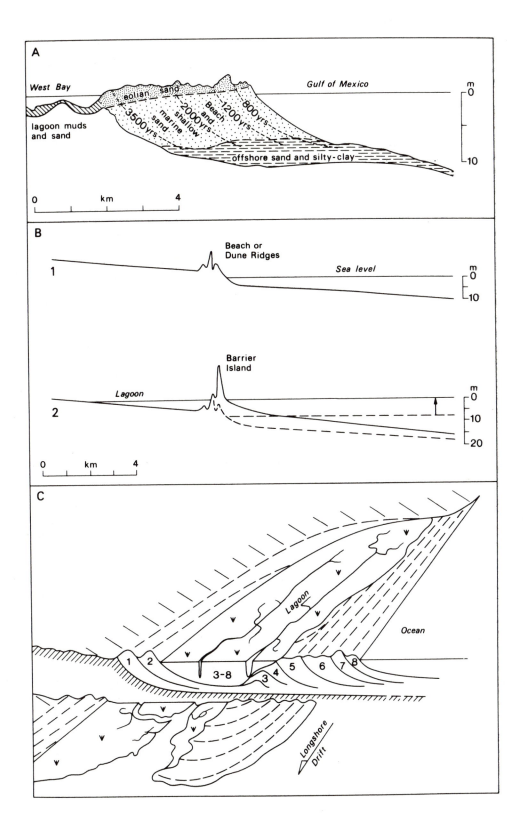

A

West Bay
Gulf of Mexico
m
0

eolian sand
800 yrs
1200 yrs
2000 yrs
Beach and shallow marine sand
3500 yrs

lagoon muds and sand

offshore sand and silty-clay
10

0          km          4

B

1

Beach or Dune Ridges
Sea level
m
0
10

2

Barrier Island
Lagoon
m
0
10
20

0          km          4

C

Lagoon

Ocean

1  2
3-8
3  4  5  6  7  8

Longshore Drift

there is not sufficient beach debris to inhibit their action and where the coastal configuration concentrates wave attack around exposed headlands. Wave processes involve wave shock, wave hammer, air compression, hydrostatic pressure and abrasion. The first four of these operate mainly in the breaker and surf zones and are especially effective in the storm wave environments of the northern hemisphere and the swell wave environments of the southern hemisphere. If waves break at a cliff face, air trapped between the wave and rock may achieve high pressures, where (see Figure 12.4B):

$$\text{Pressure} = 1 \cdot 31 \, (C_s + u_m)^2 \, \rho/(2g).$$

Waves of height 3·05 m and length 45·8 m have been observed by dynamometers to exert pressures of 0·608 kg/cm$^2$ (1241 lb/ft$^2$), although much of the compression may be relieved by air escaping upward through the wavetop. The effect of this hydraulic wave action is mainly to cause the movement of previously separated particles (e.g. by jointing, weathering, etc.) by the process of quarrying, which is naturally most effective where bedrock structural planes dip seaward. Abrasion takes place where wave action operates on loose debris to wear away the cliff face by impact, mainly during winter storms, where debris is comminuted and where the shore platform is lowered by the action of loose debris moving across it. Abrasion is especially active at the back of the shore platform often leading to cliff failure by mass movements in softer rocks. On the shore platform abrasion is enhanced by high-energy waves, by an abundance (but not excess) of mobile debris and by steep gradients which favour backwash and inhibit the clogging of their inner edges by debris (King, 1972).

Marine weathering operates most effectively around high-water level and up to about 0·3 m above it by the processes of hydration, salt crystallization, swelling, loosening, frost action and solution. Water-level weathering is most active in homogeneous, fine-grained, flat-lying permeable sedimentary rocks, but it mainly acts to prepare debris for subsequent removal by mechanical wave action. Coastal frost weathering may be an important high-latitude mechanism of rock break-up, particularly where cliffs are supplied with fresh water from permafrost and snow, and some authorities have ascribed the Strandflat, a platform near sea level up to 60 km wide off Norway, Spitsbergen, Iceland and Greenland to the initial action of frost weathering, followed by wave action and glaciation. It has also been suggested that some basic minerals and orthoclase may be more than ten times more soluble in salt water than in fresh, but most attention on solution by sea water has naturally been directed to limestone coasts. Theoretically

surface sea water should be fully saturated with $CO_2$ and suggestions have been made as to the prime importance of water-level solution by fresh groundwater from the land. However, solution notches have been inferred in the tropics at locations where groundwater supplies are limited due to low rainfall and small catchment areas. The significance of water-level weathering of limestones along tropical coasts may be more apparent than real due to the relative abundance of coastal coralline limestone in the tropics (see Section 15.6), or an actual result of enhanced solution in shallow tropical waters.

Biological erosion takes place by the leaching, rasping and boring action of burrowing and browsing organisms, such as molluscs, sponges and barnacles. Although burrowing clams (e.g. *Pholadidae*) have been observed to lower a marine surface on the English Chalk by some 2·3 cm/year, most biological erosion appears to occur in shallow tropical seas where, for example, bluish-green algae cause limestone solution around and just above mean high-water level. Shallow warm seas may become mild solvents of limestone at night when marine plant photosynthesis ceases causing complex $CO_2$ exchanges.

### 15.4.2 Cliffs

Steep cliffs occur where a post-glacial rise of sea level has drowned steeply sloping terrain or where strong wave attack is taking place on easily quarried rock which exhibits low rates of subaerial weathering and mass movements (e.g. limestones, chalk, horizontally bedded resistant sandstones, massively jointed igneous and metamorphic rocks). It is possible to view the morphology of cliffs in terms of two major influences: on the one hand the effect of bedrock lithology and structure, and on the other the balance which exists (or which may have existed in the past) between marine and subaerial erosional processes. Figure 15.23 illustrates these influences on (A and B) resistant, (C) weak and (D and E) composite structures; while (F) shows a complex coast suggesting interactions through time of structural controls, marine action, subaerial processes and changes of sea level. It is thus clear that, besides influencing the coastal shoreline geometry, variations of lithology and geological structure exercise a profound control over cliff form, especially where marine erosional processes predominate over subaerial ones. If wave removal of beach debris becomes less effective and basal accumulation begins, subaerial erosional processes may begin to influence cliff form, although operating at very different speeds depending on rock resistance. An obvious lithological influence on cliff form is exercised by angle of repose, varying between 90° and some 13°, but this finds

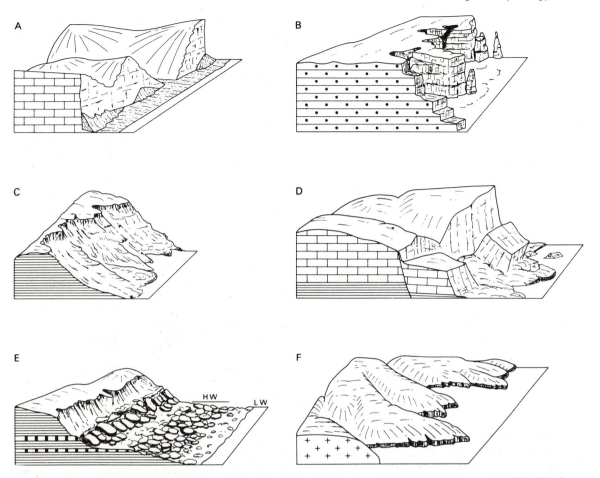

**Figure 15.23** Classification of cliffs. Resistant cliffs of A. chalk and B. horizontally bedded sandstone; weak cliffs of C. clays and shales; composite cliffs of D. chalk overlying clay and E. interbedded sandstones and shales; complex coast F. suggesting interactions through time of geological controls, marine v. subaerial processes, and change of sea level.
*Source:* Guilcher, 1958, figure 4, p. 72.

morphological expression with regard both to cliff height and rate of basal debris removal by marine processes. Figure 15.24 suggests some variations in cliff form depending on the relationship between marine and subaerial processes. It is important to recognize, however, that geological structures may exercise a strong influence over subaerial processes, especially where there is a seaward structural dip. For example, at Portuguese Bend, Palos Verdes Hills, near Los Angeles, a seaward dip is assisting a mass of area $4 \times 10^6$ m² to slip seaward at a rate of 1–3 cm/day, and near Cromer, Norfolk, similar slipping is taking place assisted by water draining seaward in sands dipping at 20°. With the obvious differences between subaerial processes in different climatic regions in mind, Davies (1980) has suggested the following possible climatic contrasts between cliff forms:

(1) Tropical cliffs – protected by coral reefs and much vegetation, they generally retreat slowly and present low angles.

(2) Arid desert cliffs – the lack of abundant debris brought down by rivers assists wave attack and the production of cliffs.

(3) Temperate cliffs – high wave energies in the westerlies can produce steep cliffs.

(4) High latitude cliffs – these tend to have low angles because of inhibited wave energy (e.g. by sea ice) and due to strong periglacial cliff slope processes.

The form of cliffs in resistant rock is dominated by the attitudes of the planes of weakness represented by the bedding and joint sets. Where beds are horizontal or

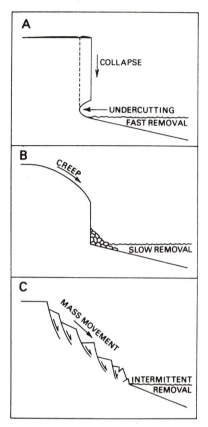

**Figure 15.24** Relative effects of marine and subaerial processes on cliff morphology: A. strong marine, weak subaerial; B. moderate marine, moderate subaerial; C. strong mass movements, intermittent marine action.
*Source:* Davies, 1980, figure 49, p. 76, after Hutchinson.

slightly dipping inland and cut by near-vertical joints, blocks are removed with difficulty and the cliff profile is steep or vertical as with resistant Devonian sandstones at Duncansby Head, Caithness, Scotland (Figure 15.23B), and those 300–400 m high at the Dingle Peninsula, Eire. It is interesting that vertical bedding and near-horizontal jointing produce similar steep cliff forms, as at Bats Head, Durdle Cove, Dorset (Figure 15.25C). Where beds or joints dip more steeply towards the sea the cliff-form is controlled by this dip, as with the 65°–70° seaward-dipping beds in the foreground chalk cliffs in Figure 15.25C. The forms of cliffs in igneous and metamorphic rocks are dominated by jointing, so explaining the contrasts between the strongly jointed granite cliffs of Land's End, Cornwall (Figure 15.25A) and the less-jointed serpentines of Mullion Cove, Cornwall (Figure 15.25B).

The form of slopes developed in weak rocks, notably clays and glacial drift, has been discussed in Chapters 10 and 11. Normal and overconsolidated clays actively subjected to basal wave attack tend to retreat rapidly by single rotational slips (see Figure 15.26A), as in the case of the glacial deposits of Holderness and Cape Cod (Highland Light), the fissured London Clay at Warden Point (north Kent; 50 m high) and the Oxford Clay at Furzy Cliff, Dorset (Figure 15.26C). Where active marine basal removal ceases, the slopes continue to fail in shallow rotational slips (see Figure 15.26B) (with terraces 9–12 m wide separated by scarps some 1.5 m high) giving angles of slope less than 13°, approaching complete stability at about 8°. An example of such development is at Barton, Hampshire (Figure 15.26D). Where fissures allow water to enter overconsolidated clays or poorly drained glacial tills containing fine sands or silts, high porewater pressures may be generated and mudslides produced on slopes, even as low as less than 5°. Such slides are common along the 200-m-high clay, shale, marl and interbedded limestone cliffs at Black Ven, Dorset, and at nearby Fairy Dell mudslides in the sandy facies of the Gault overlying clays and marls have been observed to move at 25 m/year for three years on a 9° slope (May, 1977). Glacial drift cliffs of north Norfolk suffer 83 per cent of their erosion by landslides due to wave undercutting; 8 per cent by wind erosion; 7 per cent by fluvial erosion; and only 2 per cent by mudflows.

Most existing cliffs are being currently attacked by waves which have risen to their present post-glacial level during only the last few thousand years, 5000 at most (see Section 15.5). While rapidly retreating cliffs owe their forms to a combination of structural control and wave attack, many cliffs of highly-resistant material were probably formed at much earlier periods and their bases have been reoccupied by the recent rise in sea level. This long-continued development means that many of them cannot be understood without reference to the subaerial processes which have helped to shape them, particularly during the period of low sea levels during the glaciations which preceded the recent rise of sea level. A good example of such cliffs are the hog's back (plunging or slope-over-wall) cliffs in the Devonian sandstones of North Devon. These may decline from levels of more than 300 m (1000 ft) down steep subaerial slopes to basal cliffs 100 m high or less. Although this form is undoubtedly assisted by a landward dip of the bedding, it is probably the case that Pleistocene periglacial processes were active here for long periods and were responsible for these forms.

The rate of cliff retreat is very variable both in space and time (Table 15.3), being dependent on:

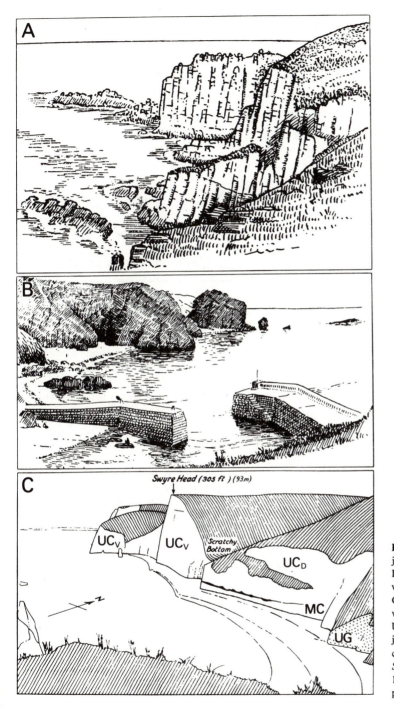

**Figure 15.25** Resistant cliffs with various jointing effects: A. granite; vertical jointing, Land's End, Cornwall; B. serpentine; some vertical jointing, Mullion Cove, Cornwall; C. Upper Chalk UC$_V$ – Upper chalk with vertical bedding and joints dipping 0°–20°S. UC$_D$ – Upper chalk dipping 65°–70°S and joints dipping 65°–75°N. MC – Middle chalk. UG – Upper greensand.
*Sources:* Trueman, 1938, figures 109 and 110, pp. 313, 317; House, 1958, figure 5, p. 9.

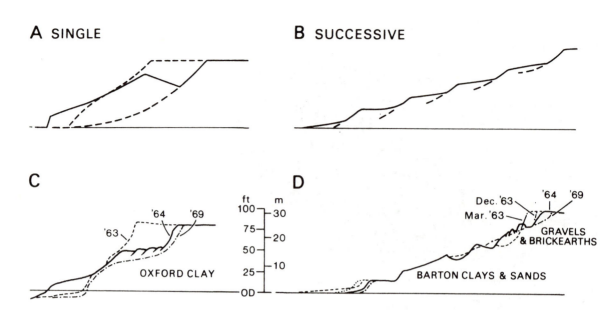

**Figure 15.26** The failure of clay cliffs: A. single rotational slips: B. successive rotational slips: C. single slip, Furzy Cliff, Dorset, 29 January 1964 and subsequent erosion: D. successive rotational slips, Barton, near Bournemouth, Hampshire. *Source:* May, 1977, figure 72, p. 219.

**Table 15.3**   Rates of cliff retreat

| Location | Rock | Cliff height(m) | Retreat (m/year) | Remarks |
|---|---|---|---|---|
| Warden Point, Kent | London Clay | 50 | 3·0 | |
| North Kent | London Clay | | 0·81 | Average |
| Dunwich, Suffolk | Glacial drift | | >4·0 | For more than 100 years |
| Suffolk Cliffs | Glacial drift | 15 | 0·8 | Average 1880–1950 |
| North Norfolk | Glacial drift | 30 | 0·9 | Average 1880–1967; 30 km stretch |
| Holderness, England | Glacial drift | 20 | 1·2–1·75 | Average 1852–1952; 60 km stretch |
| Cape Cod | Glacial drift | | 0·3–1·0 | |
| New Jersey | Sand, clay, gravel | | 1·8 | |
| Louisiana | Sands, clays | | 8·0–38·0 | Coastal islands |
| Ishikawa, Japan | | | 1·0 | For 1800 years |
| Oregon | Miocene sandstones | | 0·6 | Since 1880 |
| Sussex, England | Chalk | | 0·5 | |
| English Channel coast | Chalk | | 0·29 | Average |
| North-east England coast | Shale and sandstone | | 0·04 | |

(1) Rock lithology and structure – cliffs of resistant, massively jointed and impermeable rocks may remain little changed for long periods; those of south-west England may have been little altered for tens of thousands of years (rates less than 0·01 m/year). Greater frequency of joints and bedding planes tends to increase the rate and to assist the formation of stacks, arches and caves. Recession is usually greatest in uncemented rocks (volcanic ash at Krakatoa was cut back 30 m/year between 1883 and 1928).

(2) Susceptibility to chemical weathering, mass movements and subaerial processes.

(3) Height of the cliffs, which controls the amount of basal debris which must be removed before renewed cliff-face erosion can occur.

(4) Orientation of the coast, by its control over wave energy.

(5) Wave energy – in the northern hemisphere winter storm conditions normally produce the greatest rates of cliff retreat.

(6) Height of sea level – high spring tides, storm surges and frequent tidal inundation of the cliff base are most conducive to rapid cliff retreat. Parts of the 13 m-high glacial drift cliffs of Suffolk retreated 13 m in one day during the 1953 storm surge (a 20–200-year recurrence interval event). It is possible that much present coastal erosion is associated with the contemporary sea-level rise of some 0·15 cm/year.

(7) The offshore topography, especially the width of the shore platform (see Section 15.4.3).

### 15.4.3 Shore platforms

The development of shore platforms is clearly related to cliff recession, and shore platforms occur where cliff retreat has been pronounced and the seaward or longshore removal of debris has been efficient. Transient features, such as arches, stacks and marine caves, are those which normally accompany rapid platform expansion under strong differential structural controls. Quite apart from other influences, geological factors exert a strong control over the form of shore platforms. The widest platforms tend to be associated with the least-resistant rocks, which are thinly bedded and densely jointed, flat-lying and with the strike parallel to the shoreline. As the dip increases the platform width decreases, until with very steep dip the widest platforms are associated with structures striking at right angles to the shoreline. Resistant rocks tend to give narrower and steeper platforms with high mean elevations. In weaker rocks the platform–cliff junction tends to occur at about mean high-water spring tides (MHWS), whereas that in more

resistant rocks tends to reflect geological controls. Platforms backed by high, slowly retreating cliffs may have a lower platform–cliff junction because of the greater time available at any point for platform downwearing, reported rates of which range between $1 \times 10^{-4}$ and $3·5 \times 10^{-2}$ m/year.

Marine processes naturally exert important controls over the form of shore platforms. Quarrying and abrasion by large, high-energy storm waves may produce wide platforms, but it should be noted that platform width is not a simple function of wave energy – for example, the c. 25 ft (7·5 m) raised platform of the west coast of Scotland occurs at many locations not exposed to the open sea. Abrasion platforms tend to slope from HW to LW marks and, therefore, their gradients are directly related to the tidal range, although the greatest platform width tends to occur at the levels at which tide levels are most frequent (Figure 15.27) (Trenhaile, 1980). The height of the tide clearly influences the elevation of the cliff–platform junction, which in southern Britain averages about 2 m higher at bayheads than at headlands. Water-level weathering, especially near HW, favours subhorizontal platform development, particularly in permeable, horizontally bedded rocks in tropical seas (e.g. Hawaii, Senegal, Australia, Madagascar, Peru) where the tidal range is low. Biological erosion is an important contributory factor along tropical limestone coasts where wave energy is low.

The number of factors influencing shore platform formation make it difficult to generalize regarding platform morphology, but among others three types deserve comment (Figure 15.28) (Wright, 1967):

(1) Inclined plane. Characteristic of northern

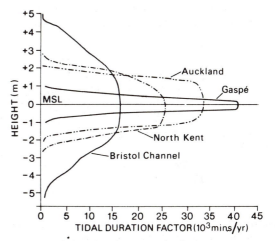

**Figure 15.27** Tidal duration curves for four areas.
*Source:* Trenhaile, 1980, figure 4, p. 14.

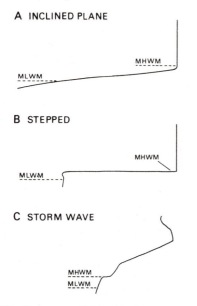

**Figure 15.28**   Three types of shore platform.
*Source:* Wright, 1967, figure 3, p. 44.

hemisphere, west coast, high-energy coasts. Some 35 per cent of southern Britain is fringed by such platforms (Figure 15.29) and east of Torquay they have a mean width of 100 m, a maximum width of 270 m and slope of between 1° and 4°. The cliff–platform junction is usually at MHWS or up to 1 m above highest tide level, and the greater the tidal range, the wider and steeper the platform.

(2) Stepped platforms. These have very low slopes, are located near MHW mark and are terminated seaward by a cliff (often 3 m high) or ramp. These tend to be associated with tropical water-level weathering, biological action and small tidal ranges. They are unknown in southern England, but occur in the North Island of New Zealand, south-east Australia, southern California (average platform gradient 1°), Gaspé and the tropics.

(3) Storm wave platforms. These are also unknown in southern England, but are common in the tropics. They may be caused by very infrequent storm wave conditions separated by a long period of low-energy wave attack.

One must be cautioned against the simple relation of platform morphology to present-day wave attack as it is clear that, on resistant coasts especially, most platforms may have been reoccupied by a recent sea-level rise and their formation may not have been associated with the present still-stand. This is particularly true of platforms west of Lyme Regis, Dorset (Figure 15.30), where two or three levels may be found at any one location, some well above MHWS level.

The lateral extent of shore platform-cutting is related to the depth below sea level of effective marine abrasion. Earlier workers assumed that such abrasion was effective down to a wavebase 183 m below sea level and that extensive marine surfaces could, therefore, be cut by the sea (Figure 15.31A). Recent work on California shore platform processes, however, suggests that little erosion occurs below about 10 m and that the minimum equilibrium slope of shore platforms is about 1:100 (i.e. 0·57°). Extreme maximum platform widths are probably 1000 m with a small tidal range and 1500 m with a tidal range of 5 m, although attainable widths are probably not much more than half these amounts. Southern California shore platforms average 484 m wide at a depth of 9·15 m, with an average slope of 1·53 (i.e. 1°). This work suggests that long-continued marine erosion and planation requires a continual, gradual rise of sea level (Figure

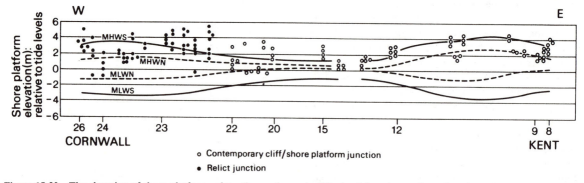

**Figure 15.30**   The elevation of shore platforms along the south coast of England (numbers refer to locations shown in Figure 15.29) relative to tide levels: **MHWS** – mean high-water spring tides; **MHWN** – mean high-water neap tides; **MLWN** – mean low-water neap tides; **MLWS** – mean low-water spring tides.
*Source:* Wright, 1970, figure 2, p. 348, copyright © Elsevier, Amsterdam.

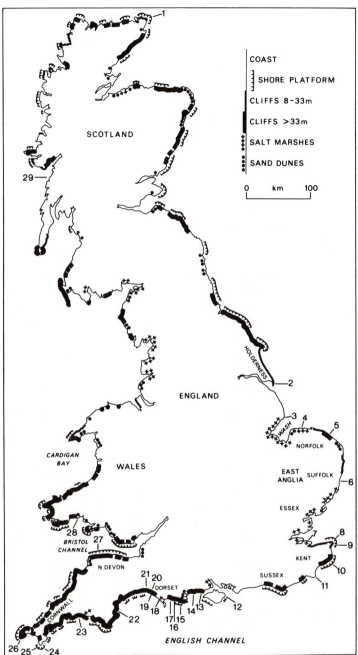

**Figure 15.29** A classification of some
mainland British coastal types, together with
a numbered key of the following locations:
1. Dunscansby Head
2. Spurn Head
3. Gibraltar Point
4. Scolt Head Island
5. Cromer
6. Dunwich
7. Warden Point
8. Grenham Bay
9. Ramsgate
10. Folkestone Warren
11. Dungeness
12. Bembridge
13. Hurst Castle Spit
14. Barton
15. Kimmeridge
16. Durdle Cove
17. Furzy Cliff
18. Chesil Beach
19. Fairy Dell
20. Lyme Regis
21. Black Ven
22. Torquay
23. Looe
24. Lizard
25. Mullion Cove
26. Land's End
27. Lynmouth
28. Laugharne Burrows
29. Oban
*Sources:* Partly after *Atlas of Britain*,
copyright © Oxford University Press, 1963;
Wright, 1967, figure 1, p. 38.

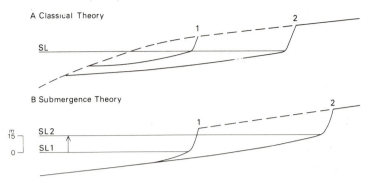

**Figure 15.31**   Theories of shore platform-cutting.
*Source:* Bradley, 1958, figures 3 and 6, pp. 973, 969.

15.31B), such as must have been involved in cutting the 1·6-km (1-mile) wide platform at Santa Cruz, California, which extends out to a depth of more than 21 m (70 ft). It thus seems unlikely that the cutting of extensive marine erosion surfaces, such as the possible 180–200 m (590–650 ft) Plio-Pleistocene surface in southern Britain (see Section 2.3), could have occurred in the absence of a significant marine transgression.

## 15.5   Sea-level variations

From the foregoing it is clear that movements of sea level have been important factors in the formation of such existing coastal features as cliffs, barrier islands and coastal platforms; in developing submarine features around our coasts; and in the formation of high-level erosional and depositional surfaces. The history of sea-level variations is not easy to determine, partly because of the difficulties of defining mean sea level (see Section 15.1), partly because of problems in interpreting the possible evidence for old sea levels and partly again because of the complications introduced by local tectonic and isostatic vertical movements of the continental margins. For example, although during the period 1850–1950 there is estimated to have been a worldwide sea-level rise of 12 cm (1·2 mm/year), this rate was very variable both in time and space. Much more rapid rates occurred during the periods 1875–77 and 1946–50, and the following data show considerable recent variations in the rate of rise around the coasts of the United States (Table 15.4). Clearly, there have been local differential tectonic movements in American coastal areas (see

**Table 15.4**   Mean observed rise of sea level in the United States (mm/year) during 1940–60

| | |
|---|---|
| East coast | 2·2 (range 0·91 to 4·53 at different stations) |
| West coast | 0·6 (range – 1·52 to + 2·13) |
| Gulf coast | 3·7 (range 0·61 to 9·15) |

Section 3.4.2). Such observations complicate the task of generalizing worldwide, eustatic changes of sea level.

### 15.5.1   Eustatic changes

Eustatic (i.e. worldwide) changes of average sea level are due to four mechanisms of which the last two are of much lesser importance:

(1) tectonic causes, associated with changing the capacity of the ocean basins;
(2) glacial causes, concerned with the amount of water locked up in ice sheets and glaciers;
(3) sedimentary causes, to do with the slow sedimentary filling of the ocean basins;
(4) ocean density causes, which change the volume of sea water.

Some authorities have suggested that the drop in average ocean temperatures of 5°C during the Pleistocene, estimated from $O^{18}$ isotopes of ocean sediments, could have led to a drop of sea level of as much as 10 m. It is also clear that the glacial and sedimentary mechanisms have an added isostatic effect to do with the loading of the crust, some of the implications of which are discussed in Section 19.4.

Tectonic eustatism has long been of interest to geologists, who have been concerned to establish causal and temporal links between orogenic activity and high sea levels. Recently suggestions have been made which may provide a linking mechanism in that the elevation of the ocean floor is believed to be greatest at times of maximum rates of sea floor spreading, and thus ocean capacity least and sea-level highest. A slowdown or cessation of crustal spreading might cause a drop in sea level of the order of 200 m in a period of $10^3$–$10^7$ years. Geological evidence points to generally high sea levels during the Miocene (22,500,000–5,500,000 BP) when crustal spreading was rapid, after which sea level fell slowly during a period of slow crustal spreading.

Superimposed on any changes of sea level attributable to other causes are the important glacial eustatic swings of sea level during the Pleistocene, on the subject of which there is considerable controversy, not least because the recent realization of the antiquity of the Antarctic glaciation has led some experts to suggest that at the beginning of the Pleistocene sea level may have been close to its present elevation and may have been much lower than at present many times in the past. However, a 'classic' eustatic view is represented in Figure 15.32, which suggests a series of successively lower sea levels corresponding with four major Pleistocene glaciations superimposed on an overall tectonic-eustatic fall. In this scheme the last glaciation was the first to carry sea level to more than 130 m below its present position about 18,000 years BP. There is some doubt as to whether the overall eustatic fall during the Pleistocene was more imaginary than real, in view of the possibility of an already low Pre-Pleistocene sea level. Not only this, there may have been as many as seventeen Pleistocene glaciations! Evidence of lower sea levels is believed to exist off the Australian coast ( – 175 m to – 200 m) and on northern Pacific seamounts ( – 220 m to – 250 m), but these features may have suffered additional tectonic depression.

The most recent Post-Wisconsin (Holocene) rise in sea level took place at a quite rapid rate (1 m/100 year) between 18,000 and 8000 years BP, but thereafter at a decreased and controversial rate (Figure 15.33). The last 10,000 years have witnessed considerable climatic oscillations including some nine cold phases of glacier advance in Europe, the last two of which were the Medieval Advance (AD 1200–1400) and the Little Ice Age (AD 1550–1800). Ice sheets appear to be very sensitive to world climatic changes such that a change of 1 degC in mean atmospheric temperature might be expected to cause a 50 mm change of sea level over a fifty-year period.

### 15.5.2 Submergence features

Most near-contemporary coastal features are dominated by the Holocene sea-level rise during the past 18,000 years or so. As we have seen, it is now suggested that even features such as barrier beaches, previously considered as evidence of coastal emergence, are now viewed as having been produced by a rise in sea level. On the other hand, drowned valleys have long attracted the attention of coastal geomorphologists. Drowned valleys (rias) are widespread, for example, in Cornwall, Brittany, north-west Spain, Virginia and Maryland; those lying at right angles to the coast (e.g. in south-west Eire) contrasting with those produced by the inundation of longitudinal fold valleys parallel to the main coastal trend (e.g. Dalmatia). Drowned glacial valleys (fjords) abound in Norway (see Section 19.2.2), the South Island of New Zealand, Chile, British Columbia, Vancouver Island and elsewhere. Some of these are very deep and their form cannot be completely explained satisfactorily by a eustatic rise of sea level. The Sogne fjord of Norway is some 1300 m deep, shallowing to a characteristic seaward threshold of 150 m. It is clear that most of this overdeepening must be due to deep glacier scour produced by thick, fast-flowing ice in zones favouring glacial quarrying. Localized quarrying may well have been assisted by:

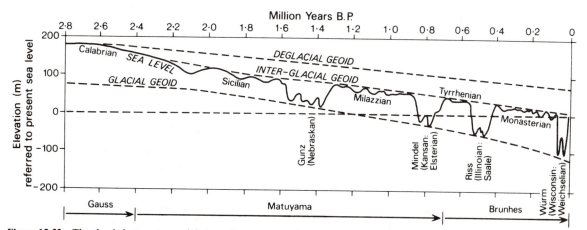

**Figure 15.32** The classic but controversial view of movements of sea level since the beginning of the Pleistocene, suggesting four main glacial periods (Nebraskan–Wisconsin) and the five elevated marine beaches of the western Mediterranean (Calabrian 180 m; Sicilian 80–100 m; Milazzian 50–60 m; Tyrrhenian 32–45 m; Monasterian 7.5–18 m).
*Sources:* Holmes, 1965, figure 527, pp. 702–3; Fairbridge, 1966, figure 6, p. 484, Fairbridge, 1971, figure 3, p. 102; Frakes, 1979.

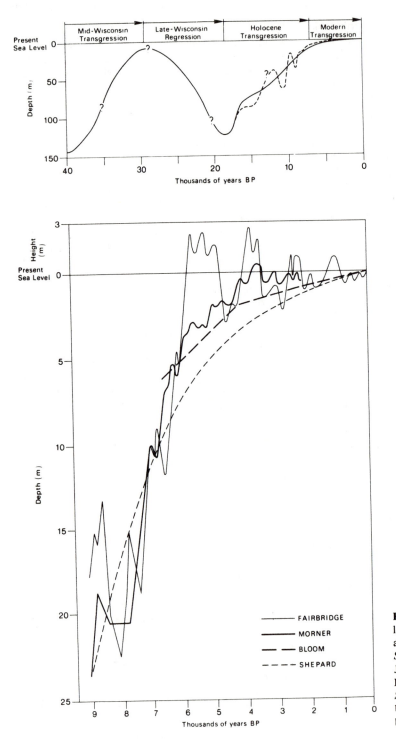

**Figure 15.33** Suggested movements of sea level (above) during the last 40,000 years, and (below) during the last 9000 years. *Sources:* Mörner, 1972, figure 5, p. 589; J. R. Curray in H. E. Wright and D. G. Frey (eds), *The Quaternary of the United States*, copyright © 1965 by Princeton University Press, figure 2, p. 725, reprinted by permission of Princeton University Press.

(1) Pre-glacial and interglacial frost shattering.

(2) Selective glacial erosion caused, for example, by topographic confinement of an ice stream along the line of a pre-existing river valley.

(3) The fact that glaciers entering the sea continue to exert the same pressure on their beds as subaerial glaciers of the same thickness until the process of floatation actually begins.

A more speculative example of coastal submergence is the continental shelf itself (see Section 5.1.1), the average width of which is 74 km (46 miles), the average depth of its flattest portion 76 m (210 ft) and its average slope 0°07′. However, in respect of the United States the morphology of the continental shelf is very varied with its outer edge varying between 32 m and 220 m (90–600 ft) below sea level. Hills of relief greater than 22 m (60 ft) are found on 60 per cent of the world's continental shelves and depressions and valleys deeper than 22 m on 35 per cent of them. It is clear that continental shelves have had varied and complex histories, and although many of their outer edges may be related to low-level wave abrasion or delta-building, many wide shelves may be due to depositional fill or coastal downwarping unconnected with previous low sea levels.

The greatest contributors to the topographic irregularity of the continental slopes and shelves are submarine canyons (Shepard, 1963). They occur at a variety of locations; many appear to be unrelated to major subaerial river basins, whereas others occur off major river mouths (e.g. Hudson and Congo) or seaward of submarine deltas (e.g. Niger and Ganges). Some canyons are cut in resistant bedrock, others in unconsolidated sediments; some extend to considerable depths and some appear to have been cut back into the continental shelf; some have steep or precipitous walls; many bifurcate-like river valleys; and some may have existed for millions of years. Three major submarine canyons off the United States coast exhibit some of the above features:

Monterey Canyon, southern California (Figure 15.34) – possibly the largest submarine canyon in the world, extending from less than 100 m below sea level down to depths of more than 2000 m.

La Jolla Canyon, southern California – possessing steep walls 280–380 m (780–1080 ft) high in places, extending down to some 1300 m (3600 ft) and being partly cut in bedrock down to 650 m (1800 ft).

Hudson Canyon, north-eastern United States – in line with the outlet of the Hudson River, having a high bifurcation ratio with many small tributaries but no large ones, cut in sediments and bedrock with steep walls (precipitous in places) and extending down to more than 2000 m below sea level.

Two major possible causes have been suggested for the origin of these canyons:

(a) Subaerial erosion. This is supported by their dendritic form and their incision into bedrock. However, few canyons are aligned with major subaerial rivers and it is not possible to postulate the falls of sea level required to form any more than small upper parts of the canyons by subaerial river action.

(b) Turbidity currents. These high-density submarine flows, moving at speeds of up to 2·2 m/s (80 km/h; 50 m.p.h.), are capable of sediment movement and erosion on low slopes, could produce the graded sand bedding observed on the floors of the canyons and could account for the branching canyon forms. However, they do not appear to produce experimentally the V-shaped canyon forms and their ability to erode lithified bedrock is in question.

Weighing the evidence for and against these two possible mechanisms, the general view inclines towards the importance of turbidity currents as submarine canyon formers, allowing for some river erosion on the continental margins at times of low sea levels.

### 15.5.3 Emergence features

Elevated shoreline features are widespread inland of the present world's coastlines, but because they are generally of greater antiquity than those of more recent submergence, their age and even existence is sometimes unclear. The denudational evidence for higher sea levels has been treated in Chapter 2, for example, that of the Plio-Pleistocene surface at about 180–200 m (590–650 ft) in southeast England, below which there appear to be other possible later marine benches at 147 m (475 ft) and 25–40 m (80–130 ft). It is clear, however, that the elevations of former beach levels are not simple indicators of past eustatic movements alone, including glacial eustatism (see Section 19.4). This is mainly due to, first, differential glacial isostatic rebound following ice unloading (see Section 19.4), and, second, to differential tectonic uplift. Glacial isostatic rebound has displaced the main post-glacial 'twenty-five foot beach' (7·5 m) of western Scotland, so that it lies at 5·5 m in western Mull and 13·7 m east of Oban; whereas tectonic uplift has contributed to the existence of some seventeen marine benches in Santa Barbara County, southern California, extending up to more than 300 m and to thirteen benches near San Pedro, the highest located at more than 400 m. It has been proposed that a flight of elevated marine benches in the western Mediterranean (mainly in Algeria, southern Italy and Provence) are indicative of higher eustatic

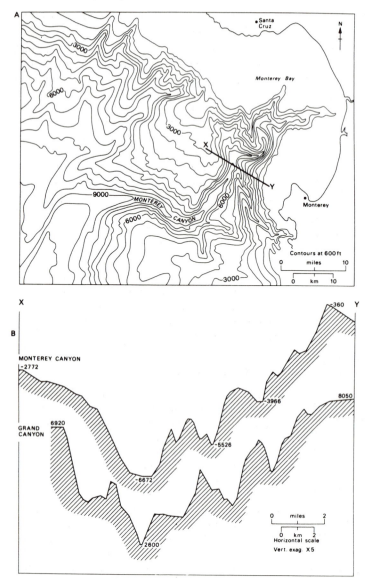

**Figure 15.34**   Monterey submarine canyon, California: A. map with submarine contours at 600-ft (183 m) intervals;' B. cross section along **XY** compared with a section of the Grand Canyon, Arizona.
*Source:* F. P. Shepard, *Submarine Geology*, 1963, 2nd edn, figure 150, p. 322, copyright © 1948, 1963 by Francis P. Shepard, reproduced by permission of Harper & Row, Publishers, Inc.

levels and that they may be correlated with terraces of the River Thames and with interglacial sea levels (see Figure 15.32). Recent work on the Quaternary shorelines of the Mediterranean, however, indicates that although they are horizontal, they are fragmentary and show little preference for any particular levels, probably having suffered differential vertical displacement by earth movements. Other marine-level correlations are even more speculative. The two major marine terraces of the east coast of the United States, the Suffolk (6–9 m; 20–30 ft) which appears discontinuously from New Jersey to the eastern Gulf Coast and the Surry (28–31 m; 90–100 ft) in

South Carolina and Virginia, are so uncertainly dated that even their relative ages are obscure.

### 15.5.4 Coastal classification

The importance of sea-level movements to the development of shore features has naturally led to their being used as one of the major criteria whereby coasts have been classified. These criteria include:

(1) relative sea-level movement;
(2) the effect of marine processes;

(3) relative shoreline movement;

(4) type of marine process;

(5) the composition and form of the coast.

It should be noted that the first four of these criteria are basically genetic in character, while only the last is generic (i.e. based on observable, objective features). The following are classificatory schemes by various authors which adhere generally to each of the following five criteria:

I D. W. Johnson (1919)

(1) Submergent coasts – ria, fjord ⎫
(2) Emergent coasts – barrier island ⎬ each subject to a youth–old-age cycle of development ⎭

(3) Neutral coasts – (e.g. delta, volcanic, fault, coral, etc.).

(4) Compound coasts – combinations of points 1–3 (strictly, of course, all coasts fall into this category).

II F. P. Shepard (1963)

(1) Primary coasts – unmodified by coastal processes.

    (a) Land erosion coasts – drowned coasts (ria, fjord).

    (b) Subaerial deposition coasts – fluvial, glacial, wind and landslide.

    (c) Volcanic coasts.

    (d) Coasts shaped by diastrophic movements (folds, faults, etc.).

(2) Secondary coasts – modified by coastal processes.

    (a) Wave erosion coasts – straightened or made more irregular.

    (b) Marine deposition coasts – barrier beaches, cuspate forelands, spits, etc.

    (c) Coasts built of organisms – corals, mangroves, salt marshes.

III H. Valentin (1952)

(1) Coasts that have advanced.

    (a) Emerged sea floors.

    (b) Prograded coasts – deltas, barrier islands, corals, mangroves, etc.

(2) Coasts that have retreated.

    (a) Submerged coasts – rias, fjords.

    (b) Retrograded coasts – cliffed coasts.

IV J. L. Davies (1964)

(1) Storm wave environments – high-latitude coasts with strong onshore winds, especially in winter.

(2) Swell wave environments – lower latitude coasts.

    (a) West coast – stronger swell waves.

    (b) East coast – weaker swell waves.

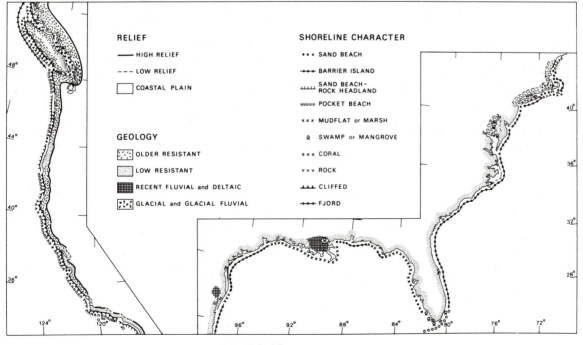

**Figure 15.35** Classification of coastal features of the United States. *Source:* Dolan *et al.*, 1972.

(3) Low-energy coasts – e.g. protected by sea ice or in enclosed seas.

V  R. Dolan *et al.* (1972)

A composite classification based on coastal relief, coastal geology and shoreline character (Figure 15.35).

## 15.6  Organic coasts

Along extensive stretches of the world's coastlines the configuration of the shore and its evolution is dominantly controlled by the growth and ecology of organisms. Among these, corals, salt marsh vegetation and mangroves are the most important.

### 15.6.1  Coral reefs

Coral reefs are organic structures built in shallow tropical marine waters by corals, bryozoa, algae and sponges. Reefs are composed of three main sedimentary facies (Figure 15.36):

reef facies – containing the main active seaward coralline growth, having an initial porosity of 25–50 per cent and making up some 30 per cent of present reef structures;

forereef facies – coralline screes and breccias dipping seaward off the reef front at angles of up to 30° and subject to secondary organic growths;

backreef facies – horizontally bedded carbonates, carbonate muds, evaporites (formed in drier tropical marine locations) and reef patches.

Coral reefs are of geomorphic significance in that they:

(1) form large existing topographic structures;

(2) exist in a variety of configurations along tropical coastlines and in shallow water;

(3) were responsible for much limestone in the geological record;

(4) fossil reefs occur in the geological record virtually intact and may form significant contemporary topographic features. An obvious example is the Permian reef complex of New Mexico–west Texas where the front of the Guadalupe Mountains is composed of 1300 m of forereef deposits;

(5) reef formation casts light both on the history of ocean basins and on sea-level changes.

Corals grow within a relatively narrow ecological range. They grow mostly in water shallower than 25 m, with an optimum of less than 10 m, but some species grow at depths as much as 165 m. Optimum water temperatures are 25°C–29°C, with winter temperatures not falling below 18°C. Corals need saline water of 27–40 ‰ salinity but do grow with salinities of up to 48 ‰ in the Persian Gulf. They do not favour muddy water and are, therefore, not found in association with deltas or areas of turbidity. Coral reefs are very characteristic of mud-free tropical coasts between latitudes of about 30°N and S.

There are four major configurations of large-scale coral reefs, three of which are shown in Figure 15.37: fringing reefs, barrier reefs, atolls and table reefs (Stoddart, 1969, 1973). In addition, there are minor forms composed of ring-shaped reefs on banks or shallows and reef knolls or patch reefs growing in lagoons. Whereas fringing reefs are connected with the land, barrier reefs are separated from it by a lagoon which may be up to several kilometres wide. Most barrier reefs are 300–1000 m wide but the Great Barrier Reef which parallels the Queensland coast for some 1950 km is a much larger-scale feature. It is composed of some 2500 separate reefs, few of which exceed a few square kilometres in area, backed by a lagoonal sea tens of kilometres wide. Atolls, like barrier reefs, rise from deep water and enclose a lagoon within an irregular elliptical form the diameters of which measure tens of kilometres. No land is enclosed within the atoll ring and the lagoon

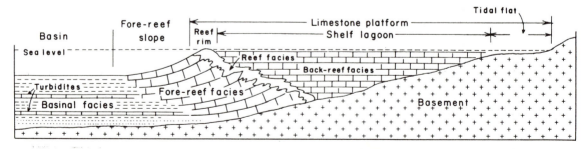

**Figure 15.36**  Cross section of typical reef facies: backreef – Carbonates, evaporites, passing landward into carbonate muds; reef – biothermal reef structures; forereef – submarine coralline screes dipping up to 30°, bound together by secondary organic growth; basinal – calcareous sands, turbidites (flows) and muds.
*Source:* F. J. Pettijohn, *Sedimentary Rocks*, 1975, 3rd edn, figure 10-2, p. 320, copyright © 1949, 1957 by Harper & Row, Publishers, Inc., copyright © 1975 by Francis J. Pettijohn, reproduced by permission of Harper & Row, Publishers, Inc., after Playfair.

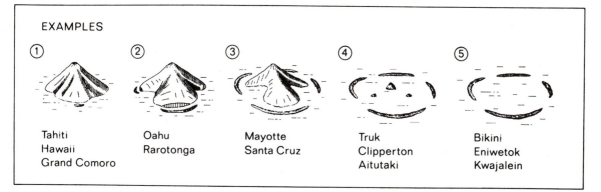

**Figure 15.37** Examples of fringing (1 and 2), barrier (3 and 4) and atoll (5) reefs. These may be regarded as exemplifying the sequence of reef forms associated with upward reef growth around a subsiding volcanic island: Tahiti, Society Islands, central Pacific; Hawaii, Hawaiian Islands, central Pacific; Grand Comoro, Comores Islands, NW of Malagasy, Indian Ocean; Oahu, Hawaiian Islands, central Pacific, Rarotonga, Cook Islands, central Pacific; Mayotte, Comores Islands, NW of Malagasy, Indian Ocean; Santa Cruz, SE of Solomon Islands, western Pacific; Truk, Caroline Islands, western Pacific; Clipperton, eastern Pacific; Aitutaki, Cook Islands, central Pacific; Bikini, Marshall Islands, western Pacific; Eniwetok, Marshall Islands, western Pacific; Kwajelein, Marshall Islands, western Pacific.
*Source:* Stoddart, 1971, figure p. 610. Reproduction from *The Geographical Magazine,* London.

floors (sediment covered with reef knolls) average some 40 m deep, but vary in depth between near sea level and 80 m. The outer faces of the reefs and forereefs of atolls may be steep or even vertical down to depths of approximately 50 m and have angles of as much as 45° down to depths of 200 m, or even 500 m in places. It is interesting that, of the 330 or so known atolls, only about ten lie outside the Indo-Pacific Zone in the Atlantic. Table reefs rise from the sea floor as a shallow bank being capped with reef growths which are not atoll-shaped – indeed, the evolutionary relationships of table reefs to other reef forms is not clear. One of the largest table reefs is the Bahama Platform covering 96,000 km² of shallows and surrounded by submarine slopes having angles exceeding 40° in places and descending hundreds of metres to the ocean floor.

We may turn now to the more detailed geomorphology of atoll reefs, as analysed by Stoddart (1969) (Figure 15.38), who recognizes the following zones, not all of which may be present at any one locality:

(a) A steep seaward slope below 18 m.

(b) A '10-fathom' terrace at 14·5–18-m depth of unknown origin.

(c) The reef front of steep submarine scree slopes dissected by chutes caused by slumps and presenting a groove and spur pattern in plan view.

(d) An algal ridge built up in the wavesplash zone on windward coasts. These ridges are rare in the Caribbean but common in the Indo-Pacific region. The ridge is broken by surge channels.

(e) A moat and boulder zone.

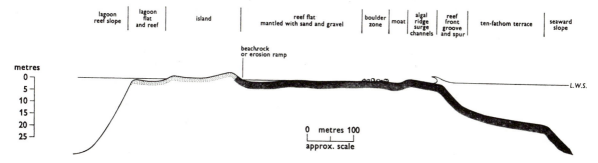

**Figure 15.38** Schematic section across a windward Indo-Pacific atoll reef showing possible geomorphic zones, all of which may not be present at any one locality. Reef flats may be lower in the absence of islands.
*Source:* Stoddart, 1969, figure 3, p. 454.

(f) A reef flat commonly 300–500 m wide, floored with dead corals, sand and gravel.

(g) A sandy island of limited size, which is much more likely to exist where protected by a pronounced algal ridge.

(h) The lagoon flat.

(i) The lagoon reef slope.

(j) The lagoon floor with reef knolls.

The origin of, and relationships between, coral-reef types has centred upon a debate between the supporters of Charles Darwin's subsidence theory (especially W. M. Davis) and the glacial control theory of R. A. Daly. Darwin believed that fringing, barrier and atoll reefs form a sequential progression as upward reef growth occurs around a subsiding volcanic island (see Figure 15.37). Daly proposed that the lower sea level and colder water accompanying the Pleistocene glacial periods would have led to coral destruction and the development of broad wave-cut platforms of dead coral, of which the existing lagoon floors are remaining evidence. The subsequent post-glacial rises of sea level would, Daly believed, lead to coral-reef growth on the outer edges of these bevelled platforms to produce modern barrier and atoll reefs.

Stoddart (1969, 1973) has marshalled the following factual evidence bearing on this debate:

(i) Deep borings on Pacific atolls have shown shallow-water coral rock, dating back to the earliest Tertiary times, to form continuous sequences hundreds of metres thick and lying on olivine basalts which are assumed to be original sea floor material. Borings on Eniwetok (Marshall Islands, 1951–2) showed a sequence of reef limestones floored by Eocene limestone lying on basalt at a depth of 1411 m below sea level. Borings on Mururoa (Tuamotus, 1964–5) in the south-east Pacific showed coral limestone on basalt at 577 m below sea level, at a location where the sea floor can be assumed to be about 65 million years old (i.e. earliest Tertiary). These islands appear to have subsided at average rates of some 2·3 cm/1000 year for long periods of time.

(ii) During the Pleistocene sea surface temperatures probably remained high enough for corals to survive and grow over most of the tropical oceans.

(iii) Recent measurements of maximum erosion rates on exposed coral reefs (i.e. horizontal wave erosion 1–7 mm/year and vertical rainwater solution 0·1–0·5 mm/year) implies that it would have taken 0·5–1 million years to bevel a reef by wave action. This is clearly too slow for extensive wave-bevelled surfaces to have been cut during the possibly transitory low sea levels of the Pleistocene.

(iv) There is not the coherent existence of the elevated reef terraces which might be expected to have resulted from the high interglacial sea levels.

(v) Lagoon floors show more relief than was hitherto believed and do not provide conclusive evidence for extensive marine bevelling.

In short, all modern evidence combines to support the subsidence theory of reef formation, with living coral (i.e. unlithified) able to grow rapidly upward at rates of some 1·8 m/1000 year easily keeping pace with sea-level rises. The subsidence theory also accords well with modern ideas of sea floor subsidence away from the oceanic axes of sea floor spreading (see Section 5.2.2). This model of Stoddart envisages reef subsidence to occur in the Pacific at rates of approximately 0·02 m/1000 year as the sea floor moves westward from the axis of spreading of the East Pacific Ridge at a rate of about 100 m/1000 year into deeper water. The role of Pleistocene sea-level changes has been mainly relegated to the production of some karst-type erosion of exposed reefs during low glacial sea levels.

### 15.6.2 Salt marshes and mangrove coasts

Salt marshes are characterized by salt-tolerant (halophytic) grassland and dwarf brushwood growing on, and trapping, alluvial sediments in tidal situations of low wave energy along protected stretches of extratropical coasts (Beeftink, 1977; Frey and Basan, 1978). Such locations are mainly lagoonal, behind barrier beaches (e.g. the coasts of South Carolina and Georgia), behind spits (e.g. the north Norfolk coast, eastern England) and in estuaries (e.g. the Wash, eastern England), where salinity is between > 38 and 5 ‰. At lower salinities salt marshes are replaced by reeds and rush marsh. Salt marsh development commonly begins by the deposition of mud on a hummocky sand surface upon which algae and plants like eelgrass become established near high-water mark and begin to trap increasing amounts of sediment aided by seaweed. Later a more dense halophytic vegetation succeeds (e.g. *Spartina townsendei* in north-western Europe and *Spartina alterniflora* along the east coast of the United States). This vegetation, and its successors (e.g. *Salicornia* and *Juncus* in the eastern United States), may ultimately form a dense mat some 15 cm high, decreasing wave heights by up to 70 per cent and wave energy by 90 per cent. The accompanying increase of sediment-trapping leads to the upward and outward building of the sedimentary hummocks restricting the lower tidal flows to a branching network of highly sinuous channels (creeks). The patterns and density of these creeks are diverse and are believed to represent some form of equilibrium between the cutting and filling action of the two-way tidal flows and the growth of the marsh surface. Rates of marsh upbuilding may reach a maximum (e.g. 1 cm/year at Scolt Head Island, Norfolk) on the lower vegetated marsh surfaces and along the leveelike banks of the major stabilized creeks where a particularly dense vegetation is favoured (e.g. *Halimione portulacoides* in southern England). The dynamics of salt marsh sedimentation may be quite complex. Clearly, rates of marsh sedimentation are controlled not only by sediment availability and vegetation trapping, but also by the relative flood and ebb–flow velocities and the magnitude and frequency of tidal levels. Work by Bayliss-Smith *et al.* (1979) in Norfolk, England, suggests flood-tide velocities in the larger creeks (20 m wide and 2 m deep) of 0·2 m/s for neap, 0·5 m/s for spring and 0·75 m/s for storm surge tides; together with three possible sedimentary regimes separated by two thresholds:

(3) *Deep marsh flow* – flood tide pulses stronger than ebb tide.
Overmarsh threshold: marsh covered by more than 0·2 m of water.
(2) *Shallow marsh flow* – ebb tide pulse dominant.
Marshfull threshold: creeks filled to bankfull.
(1) *Creek flow* – both flood and ebb tide flows of low velocity.

Such variations may go some way to explain characteristic differences in the levels of upper and lower parts of individual salt marshes (as much as 1 m in Norfolk), to which must be added differential rates of sediment trapping, tidal magnitude and frequency, and location with respect to major creeks. Ultimately, however, the decrease in frequency of tidal inundations with increasing marsh elevation reduces the rate of upward accretion. Most salt marshes are only of the order of 1–2 km wide but they may be very extensive laterally; fringing most of the northern coasts of Canada, comprising up to 80 per cent of the Atlantic and Gulf coasts of the United States and occupying substantial pockets along the Nova Scotia, New England and Pacific coasts.

Mangrove swamps (Scholl, 1968) tend to occupy similar sedimentary environments between 30°N and S to those of salt marshes in higher latitudes. They are tidally submerged tropical coastal woodlands with low trees and shrubs, characterized by an entanglement of arching prop roots entrapping organic-rich sandy or clayey mud, forming prograding tidal flats intersected by meandering tidal channels and branching tidal creeks. Mangrove growth is favoured by:

(a) A high tidal range, where water is less than 1 m deep at low tide and the whole depositional surface is completely submerged at extreme high tides.
(b) Low coastal relief.
(c) The presence of saline or, preferably, brackish water. Mangroves tend to be absent near the mouths of active present deltas where the tidal range is small (e.g. the Mississippi) but present where the tidal range is great (e.g. the 'sundarbans' of the Ganges delta). They grow especially well in tidal estuaries, where they may extend 30–60 km inland, and occur along delta shorelines which have been abandoned by the process of avulsion (e.g. parts of the Orinoco delta).
(d) A large supply of fine sediment.
(e) Low wave energy.

Mangroves form distinctive coastlines up to 20 km in width in such areas as the Philippines, south-west Florida, the Pacific coast of Colombia, the west coast of Malaysia, the north coast of Borneo and the Ord River delta of north-western Australia.

## 15.7 Coastal management

It is clear that many coasts represent dynamic process-response systems, particularly those involving mobile beaches, sand dunes and vulnerable organisms. Sandy barrier islands, such as those along the Atlantic coast of the United States (see Figure 15.21), are particularly in need of careful management in that both the coastal forms and the material of which they are constructed are subjected to constant change and human prevention of change in one place serves only to accelerate it in another. In particular, jetties, groynes and sea walls are designed to inhibit local sand movement, but this leads to beaches further along the coast being starved of sand and eroded.

The natural mobility of barrier beaches in the United States makes them particularly vulnerable both to human interference, which is common due to their environmental attractiveness, and to natural erosional processes accentuated by the occurrence of hurricanes and by the recent rise of sea level (6–7 in. or 150–180 mm in the past fifty years). A proposed protection scheme for Miami Beach (property value $235·4 million per mile) has been costed at $3·8 million per mile, but the beach material would need to be replenished every three to four years. Similar schemes for the New Jersey beaches would involve facilities for sand 'by-passing'. Such by-passing can either be mechanized (i.e. the pumping of sand and water in pipes or the dredging and transporting of sand) or 'natural' as permitted by the special design of groynes and jetties.

Problems of beach mobility are particularly acute at inlets used by shipping, for example the Oregon Inlet, North Carolina (see Figure 15.21). To keep this open and stable, while not starving the coast further south of sand, requires either constant dredging (at a high cost, particu- larly if the sand is to be transported further south), or permanent jetties with a mechanized by-pass scheme (somewhat less costly but environmentally undesirable), or a more speculative scheme with a floating jetty, more natural by-passing and some limited dredging.

## References

Arber, M. A. (1949) 'Cliff profiles of Devon and Cornwall', *Geographical Journal*, vol. 114, 191-7.

Bagnold, R. A. (1963) 'Mechanics of marine sedimentation', in M.N. Hill (ed.), *The Sea*, New York, Interscience, vol. 3, 507-28.

Bayliss-Smith, T. P., Healey, R., Lailey, R., Spencer, T. and Stoddart, D. R. (1979) 'Tidal flows in salt marsh creeks', *Estuarine and Coastal Marine Science*, vol. 9, 235-55.

Beeftink, W. G. (1977) 'Salt-marshes', in R. S. K. Barnes (ed.), *The Coastline*, London, Wiley, 93-121.

Bloom, A. L. (1978) *Geomorphology; A Systematic Analysis of Late Cenozoic Landforms*, Englewood Cliffs, NJ, Prentice-Hall.

Bradley, W. C. (1958) 'Submarine abrasion and wave-cut plat- forms', *Bulletin of the Geological Society of America*, vol. 69, 967-74.

Brunsden, D. W. (1979) 'Mass movements', in C. Embleton and J. Thornes (eds), *Process in Geomorphology*, London, Arnold, chapter 5, 130-86.

Cambers, G. (1976) 'Temporal scales in coastal erosion systems', *Transactions of the Institute of British Geogra- phers*, vol. 1, 246-56.

Clark, M. W. (1979) 'Marine processes', in C. Embleton and J. Thornes (eds), *Process in Geomorphology*, London, Arnold, chapter 11, 352-77.

Davies, J. L. (1964) 'A morphogenetic approach to world shorelines', *Zeitschrift für Geomorphologie*, vol. 8, 127-42.

Davies, J. L. (1980) *Geographical Variation in Coastal Devel- opment*, 2nd edn, London, Longman.

Davis, R. A. (1978) 'Beach and nearshore zone', in R. A. Davis (ed.), *Coastal Sedimentary Environments*, New York, Springer-Verlag, chapter 5.

De Boer, G. (1964) 'Spurn Head: its history and evolution', *Transactions of the Institute of British Geographers*, no. 34, 71-89.

Dolan, R., Hayden, B., Hornberger, G., Zieman, J. and Vincent, M. (1972) 'Classification of the coastal environ- ments of the world. Part I, The Americas', *University of Virginia, Department of Environmental Sciences, Technical Report 1.*

Elliott, T. (1978) 'Clastic shorelines', in H. G. Reading (ed.), *Sedimentary Environments and Facies*, New York, Elsevier, chapter 7, 143-77.

Fairbridge, R. W. (ed.) (1966) *The Encyclopedia of Oceano- graphy*, New York, Reinhold.

Fairbridge, R. W. (1971) 'Quaternary shoreline problems at INQUA', *Quaternaria*, vol. 15, 1-17.

Frey, R. W. and Basan, P. B. (1978) 'Coastal salt marshes', in R. A. Davis (ed.), *Coastal Sedimentary Environments*, New York, Springer-Verlag, 101-69.

Guilcher, A. (1958) *Coastal and Submarine Morphology*, London, Methuen.

Hey, R. W. (1978) 'Horizontal Quaternary shorelines of the Mediterranean', *Quaternary Research*, vol. 10, 197-203.

Hindley, K. (1978) 'Beware the big wave', *New Scientist* (9 February), 346-7.

Holmes, A. (1965) *Principles of Physical Geology*, London, Nelson.

House, M. R. (1958) 'The Dorset coast from Poole to Chesil Beach', *Geologists Association Guides*, 22.

Ingle, J. C. (1966) *The Movement of Beach Sand*, Develop- ments in Sedimentology, Amsterdam, Elsevier.

Johnson, D. W. (1919) *Shore Processes and Shoreline Develop- ment*, New York, Wiley.

King, C. A. M. (1972) *Beaches and Coasts*, 2nd edn, London, Arnold.

Komar, P. D. (1976) *Beach Processes and Sedimentation*, Englewood Cliffs, NJ, Prentice-Hall.

Lewis, W. V. (1932) 'The formation of Dungeness foreland', *Geographical Journal*, vol. 80, 309-24.

May, V. D. (1977) 'Earth cliffs', in R. S. K. Barnes (ed.), *The Coastline*, London, Wiley, 215-35.

Mörner, N-A. (1972) 'Isostasy, eustacy and crustal sensitivity', *Tellus*, vol. 24, 586-92.

Pettijohn, F. J. (1975) *Sedimentary Rocks*, 3rd edn, New York, Harper & Row.

Pettijohn, F. J., Potter, P. E. and Siever, R. (1972) *Sand and Sandstone*, New York, Springer-Verlag.

Scholl, D. W. (1968) 'Mangrove swamps: geology and sedimen- tation', in R. W. Fairbridge (ed.), *The Encyclopedia of Geo- morphology*, New York, Reinhold, 683-8.

Schwartz, M. L. (ed.) (1973) *Barrier Islands*, Benchmark Papers in Geology, Stroudsburg, Pa, Dowden, Hutchinson & Ross.

Selley, R. C. (1976) *An Introduction to Sedimentology*, London, Academic Press.

Sellwood, B. W. (1978) 'Shallow-water carbonate environ- ments', in H. G. Reading (ed.), *Sedimentary Environments and Facies*, New York, Elsevier, 259-313.

Shepard, F. P. (1963) *Submarine Geology*, 2nd edn, New York, Harper & Row.

Small, R. J. (1978) *The Study of Landforms*, 2nd edn, Cambridge University Press.

Sokal, R. R. (1974) 'Classification: purposes, principles, progress, prospects', *Science*, vol. 185, 1115–23.

Sparks, B. W. (1949) 'The denudation chronology of the dip slope of the South Downs', *Proceedings of the Geologists Association of London*, vol. 60, 165–215.

Sparks, B. W. (1972) *Geomorphology*, 2nd edn, London, Longman.

Steers, J. A. (1964) *The Coastline of England and Wales*, 2nd edn, Cambridge University Press.

Stoddart, D. R. (1969) 'Ecology and morphology of recent coral reefs', *Biological Reviews*, vol. 44, 433–98.

Stoddart, D. R. (1971) 'Variations on a coral theme', *Geographical Magazine*, vol. LXIII, 610–15.

Stoddart, D. R. (1973) 'Coral reefs: the last two million years', *Geography*, vol. 58, 313–23.

Strahler, A. N. (1965) *Introduction to Physical Geography*, New York, Wiley.

Trenhaile, A. S. (1980) 'Shore platforms: a neglected coastal feature', *Progress in Physical Geography*, vol. 4, 1–23.

Trueman, A. E. (1938) *The Scenery of England and Wales*, London, Gollancz.

Valentine, H. (1952) *Die Küsten der Erde*, Gotha, Justus Perthes.

Wood, R. M. (1980) 'The fight between land and sea', *New Scientist*, vol. 87, 512–15.

Wright, H. E. and Frey, D. G. (eds.) (1965) *The Quaternary of the United States*, Princeton, NJ, Princeton University Press.

Wright, L. W. (1967) 'Some characteristics of the shore platforms of the English Channel coast and the northern part of North Island, New Zealand', *Zeitschrift für Geomorphologie*, N.F., vol. 11, 36–46.

Wright, L. W. (1970) 'Variation in level of the cliff/shore platform junction along the south coast of Great Britain', *Marine Geology*, vol. 9, 347–53.

# Sixteen  *Aeolian processes and landforms*

Aeolian processes require special conditions for their operation. First, it is necessary that there be an atmosphere. Without this there can be no wind action and on the moon, for example, where there is no atmosphere, aeolian forms are unknown. However, on Mars, where the atmospheric pressure is only 1–2 per cent of that on earth, sand dunes have been identified and, in fact, it has been possible to draw some conclusions about the circulation of the Martian atmosphere based on the orientation of these dunes. The Martian atmosphere is composed primarily of carbon dioxide, which is a dense gas, and so – although it forms part of a relatively sparse total atmosphere – it can be effective in moving sediment near the surface. In fact, when the first Mariner spacecraft reached Mars, a severe duststorm had shrouded the planet in a yellow curtain of dust, and it was impossible to see the Martian surface for at least a month. The second condition is that vegetation be sparse. The wind cannot be effective unless it can attack the ground surface, and if the surface is vegetated not only is the soil held by the roots of the plants, but the portion of the plant above the surface creates a roughness which decreases wind velocity.

## 16.1  Aeolian environments

Some 36 per cent of the world's land surface is classed as dry savanna, semi-arid and arid; 19 per cent is arid and largely devoid of vegetation, and of this one-quarter to one-third is covered with mobile sand (Cooke and Warren, 1973). Wilson (1970) has shown that virtually all this mobile sand is contained in individual 'sand seas' or *ergs* (Arabic: *erg* – 'region of shifting sand') larger than 125 km$^2$, and fully 85 per cent of it in ergs greater than 32,000 km$^2$. The largest erg is the Rub al Khali in Arabia (560,000 km$^2$) and the modal erg size is about

188,000 km$^2$. Most sandy deserts are associated with the subtropical high-pressure cells and are located between latitudes 10° and 33°, the major exceptions being the Asiatic interior regions of Turkistan and the Gobi which lie far from maritime influences and experience high atmospheric pressure in winter. Some sandy deserts lie at quite high elevations (e.g. parts of Chile and Turkistan) and may be very cold in winter. The present areas of moving sand lie generally within the 150 mm isohyet and occur in a variety of tectonic environments, from the stable craton of the Sahara to the faulted basins of the south-western United States. In moister areas there are vegetated, stabilized tracts of ancient dunes indicative of climatic change (e.g. Nebraska Sand Hills, a broad belt south of the Sahara, and Botswana) (Figure 16.1). Desert sand derives from ephemeral stream channels and other fluvial deposits, coastal deposits, earlier dunes and from the weathering of sandstones and other siliceous granular rocks.

Aeolian processes can be treated as any other sediment transport situation in that there is erosion, transportation and deposition. Sand, of course, is a common aeolian-transported sediment but a variety of other particulate materials can be transported in the atmosphere: silt, clay, pollen, salts, snow and ice, ash from fires and, of course, volcanic ash. It has been estimated that 500 million tons of dust is produced by soil erosion each year.

## 16.2  Aeolian sand movement

Before discussing the landforms produced by wind action, it is necessary to understand aeolian erosion and transportation of sediment. Some of the most significant work in this regard was performed by Major Ralph Bagnold (1941, 1953), a British engineering officer stationed in Egypt during the 1930s. During that time he

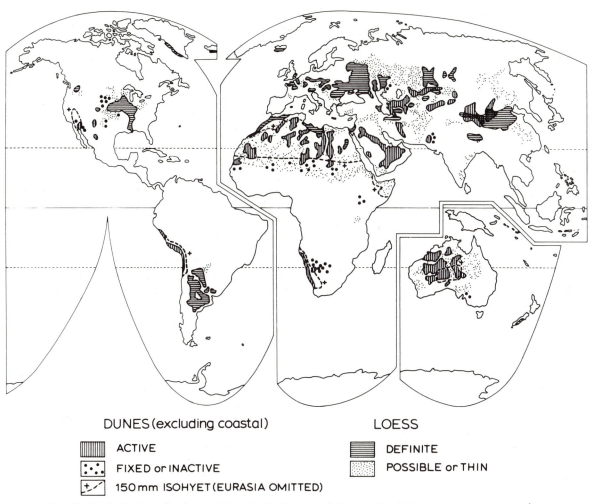

DUNES (excluding coastal)     LOESS

|||||| ACTIVE
∴∴∴ FIXED or INACTIVE
+ ⁄ 150 mm ISOHYET (EURASIA OMITTED)

═══ DEFINITE
∴∴ POSSIBLE or THIN

**Figure 16.1** World distribution of active sand dunes (excluding coastal), inactive (fossil) dunes, present rainfall (excluding Eurasia) and loess.
*Sources:* Snead, 1972, pp. 134–5, 138–9; Cooke and Warren, 1973, figure 4.1, p. 230; Smalley and Vita-Finzi, 1968, figure 1, p. 768, basic map, copyright © John Wiley and Sons.

used his leaves to explore the desert to the west of the Nile River using a Ford car and rolls of chickenwire to help him traverse loose sand areas. He travelled widely in the desert, and he studied sand movement and the migration of sand dunes. Bagnold's knowledge of desert conditions was very valuable to the Allied forces during the Second World War.

Material which is moved near to the ground by the action of the wind occurs within the narrow range of sizes of approximately 0·1–1·0 mm (i.e. very fine to coarse sand). Sizes larger than 1 mm require very infrequent winds of high velocity to move them and material smaller than some 0·1 mm is either composed of cohesive clay which is difficult to move or of quartz silt which is commonly transported with ease over larger distances and at higher elevations as loess (see Section 16.6).

Sand moves in the wind by two closely related processes – saltation (Latin: *saltare* – 'to leap') and surface creep, of which the former performs about four-fifths of the total transport and the latter one-fifth (Bagnold, 1941; Warren, 1979). Saltating grains move by being propelled into the near-surface moving air layer by the combined action of aerodynamic lift (due to velocity increase with height) and the impact of other saltating grains returning to the surface. The trajectory of saltation rises almost vertically from the surface and then bends downwind to reach the surface again at an angle of 6°–12°, generally rising less than a metre from the

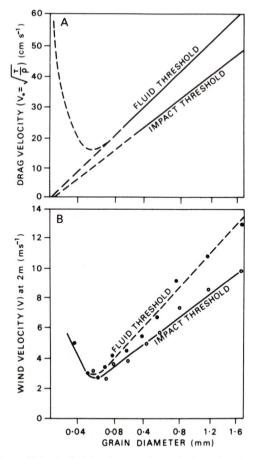

**Figure 16.2**  A. Relation between drag velocity and grain diameter showing fluid and impact thresholds. B. relation between quartz particle size and the threshold velocity for movement at a height of 2 m; the diagram has been constructed with data from Chepil and Hsu. For particles finer than 0·06 mm in diameter, the threshold velocity for movement increases as size decreases.

*Sources:* Bagnold, 1941, figure 28, p. 88; Warren, 1979, figure 10·2, p. 327.

surface, Surface creep, a characteristic of grains which are too large to saltate at a given wind velocity, is generated by a combination of surface wind shear and, particularly, by the impact of the near-surface cloud of smaller saltating grains. Figure 16.2A shows the result of an experimental study by Bagnold in which he established the critical wind velocities required to set in motion particles of different size. There are two curves on this figure, one labelled the fluid threshold and the other the impact threshold. Bagnold found that, as wind drag velocity increased, there would be no sediment movement until a critical wind drag velocity was achieved

when sediment particles would be set in motion. This he called the fluid threshold, which marked the beginning of motion of particles under the influence of wind alone. However, once sediment was moving, the impact of the moving grains started movement of larger grains. It is thus obvious that the 'fluid' shear of the wind alone requires a higher velocity to move sand of a given size than that associated with the 'impact' of already saltating grains. After the fluid threshold is crossed and aeolian transport is taking place, wind velocities can decrease to the line labelled impact threshold before sediment motion ceases. In figure 16.2B more recent threshold data have been related to actual wind velocities, measured at 2 m above the surface, showing that approximately 16 km/h (4·44 m/s) is the 2-m threshold velocity for most desert sands.

The key factor in the movement of sediments by wind, as by water, is the shear force exerted by the transporting medium in the river bed by water and on the ground by wind. This shear or tractive force is expressed for water as:

$$\tau = \gamma \, d \, s$$

or as

$$\tau = \gamma V^2$$

because in the Chézy or DuBoys equation (see Chapter 12) the product of depth and slope is proportional to $V^2$. Therefore, for air the equation for shear or drag is:

$$\tau = \rho V_*^2 \text{ such that } V_* = \sqrt{\frac{\tau}{\rho}}$$

where: $V_*$ = drag velocity (cm/s);
$\tau$ = mean drag per unit surface area;
$\rho$ = air density.

$V_*$ is also related to the rate of change in wind velocity above the ground.

Wind velocity near the ground depends largely upon the roughness of the ground surface, and velocity decreases to zero at the surface (Figure 16.3A). The roughness of the ground surface, whether it be due to vegetation or to the irregularity of the surface itself, has a significant effect on wind velocity. The logarithmic plots of Figure 16.3B intercept the Y-axis at some value greater than zero indicating that wind velocity is, in fact, zero at some small height above the ground surface. This height is approximately one-thirtieth of the mean grain diameter on the surface; i.e. if the average diameter of particles on the ground surface is 30 mm, then wind velocity is zero at 1 mm above the ground surface. This is important because the larger particles prevent the movement of the smaller particles by protecting them from the wind. Therefore, a rough surface or a surface covered with an armour of scattered boulders or cobbles is very effective in reducing aeolian erosion. Figure 16.4 shows the aeolian transport of various sediment sizes with height above the surface, as collected in vertical sand traps in Coachella Valley, California, during 146 days prior to 11 December 1953 (Sharp, 1964). This indicates the small range of sizes being moved, the predominance of fine and medium sand and that most of the transport takes place quite close to the ground surface.

The rate of sand transport by the wind is generally proportional to the cube of the wind velocity in excess of its threshold value $V_t$. In Figure 16.5 Warren (1979) shows a theoretical relation between an assumed frequency distribution of annual wind velocities $V - V_t$ and the discharge of sand $Q$ (in tonnes per metre width per annum). $Z$ is the

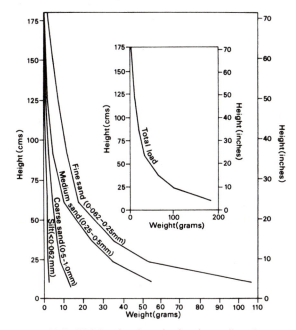

**Figure 16.4**  Weight of various size fractions collected at different heights up to 175 cm moved by wind in the Coachella Valley, California, during 146 days prior to 11 December 1953.
*Source:* Sharp, 1964, figure 4b, p. 790.

height (in metres) at which wind speed measurements were made. This supports the view that most sand movement occurs during fresh-gale force winds.

## 16.3  Wind abrasion

The effectiveness of the action of particle-laden air as an erosional agent in the desert landscape was a controversial topic for many years. In the extreme case it was suggested that the major structural valleys or basins in western United States were formed by deflation or wind erosion. This is unreasonable, but rounded, pitted and hollowed cliff faces frequently have been attributed to wind erosion. The present consensus is that wind action is of relatively minor significance in the forming of erosional landforms, and of limited importance in near-surface abrasion.

The curves of Figures 16.3 and 16.4 provide sufficient information for an evaluation of wind abrasion. Figure 16.3 shows that wind velocities increase rapidly above the ground surface; and the higher the wind velocity, the greater will be the abrasive force of the particles that are transported by the wind. However, Figure 16.4 indicates that the greatest quantity of sediment is transported near the ground surface, and obviously the quantity of

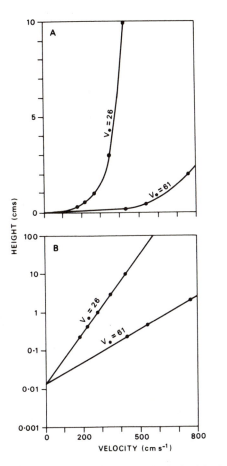

**Figure 16.3**  Variations of wind velocity with height plotted A. arithmetically and B. logarithmically for two different drag velocities.
*Source:* Bagnold, 1941, figure 15A and B, pp. 48, 49.

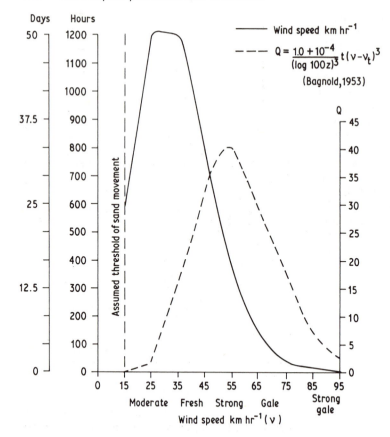

**Figure 16.5** The relation between wind velocity and sand movement; the wind data are based on a hypothetical, but realistic, Poisson distribution through a yearly cycle; **Q** refers to the rate of sand movement in tonne/m-width$^{-1}$ year$^{-1}$; $t$ is the period of time that winds of velocity $v$ blow during the year and $v_t$ is the threshold of movement; neither 'moderate' nor 'gale'-force winds transport nearly as much as 'strong' winds blowing for no more than about thirty days in the year.
*Source:* Warren, 1979, figure 10.8, p. 335.

material moved is important in determining the extent of abrasion. However, near the ground surface the velocity of the wind and the velocity of particles transported by the wind will be low. Hence there is a threshold velocity above which abrasion begins. The gentle rolling of sand grains against a rock surface presumably is not sufficient to cause abrasion of that surface. Therefore, it is necessary to combine the curves of Figures 16.3 and 16.4 to determine at what height above the ground abrasion is most efficient. Consideration of the combined two figures suggests that wind abrasion is a maximum at a short distance above the ground, where wind velocity and sediment movement are moderate. At the ground sediment transport is high but velocity of transport is low. At a height velocity is high but the quantity of sand available for abrasion is small.

Studies of wind abrasion were carried out in the Mohave Desert during an eleven-year period (1952–63) by Robert Sharp (1964, 1980). He installed 113-cm-high lucite rods and measured abrasion on the windward side of these rods. Maximum abrasion occurred at a height of 23 cm above the ground surface (Figure 16.6), con-

firming the theoretical relationships. However, an additional five years of data (1964–9) showed that maximum abrasion occurred at a height of only about 10 to 15 cm above the ground, which was attributed to an increased supply of larger sediment particles provided upwind by a flood.

One of the most interesting features associated with wind erosion are *ventifacts*, faceted rocks abraded by long periods of wind action. As part of the same study of wind abrasion, Sharp placed a number of objects on the desert surface that he presumed would be abraded by the wind. During sixteen years it was found that the softer materials were, indeed, being faceted by wind action, the development of multiple facets on ventifacts being the most intriguing aspect of this study. The blocks were faceted by the abrasion; but, in addition, there was scour on the upwind side and along the flanks of the block, so that it was undermined and toppled forward into the wind. This toppling exposed another portion of the rock surface to wind abrasion and a multifaceted ventifact was formed. New facets also resulted from the rotation of a brick by the wind.

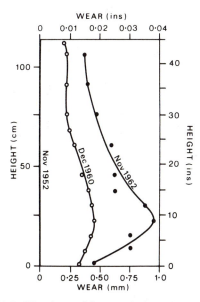

**Figure 16.6** Wear by sandblast on a lucite rod, anchored in concrete, 4 ft (1·2 m) high and 1·125 in (29 mm) in diameter (hardness 2·5 on Moh's scale) between November 1952 and November 1962 in the Mohave Desert; measurements on upwind face (facing N75°W).
*Source:* Sharp, 1964, figure 10, p. 798.

Obviously, the ventifact must be large enough to escape transport by wind. If the desert surface is composed of a poorly sorted sediment, the fine particles will be removed by deflation leaving a concentration or armour of the immobile particles. These form a monogranular layer termed *desert pavement*. This mosaic of pebbles is a very effective deterrent to wind erosion.

In addition to small abrasion features (ventifacts), the breaking of a protective pavement or vegetational cover may permit wind erosion to remove the underlying sediments and a depression is formed. These shallow deflation basins are common on the Great Plains where they were erroneously attributed to the wallowing of buffalo.

Large-scale streamlined hills have been recognized, and these *yardangs* may range in size from a metre to a kilometre (McCauley *et al.*, 1977). Yardangs of all sizes can be recognized by their parallelism and similarity to inverted ship hulls (Plate 20), which offer the least resistance to wind. They are formed, however, more by deflation (i.e. the *removal* of fine material by the wind) than abrasion (i.e. the *wearing away* of bedrock by wind action). Weathered material is removed by wind, and weakly resistant rocks are eroded. In addition, the detached particles may aid in the abrasion and deepening of adjacent troughs. The recent interest in yardangs has

resulted from observation of similar streamlined features on Mars.

## 16.4 Aeolian bedforms

According to Bagnold, a dune forms where a patch of sand accumulates, perhaps in the lee of a roughness element (rock, vegetation). This patch of sand acts as a trap for other saltating grains because on impact they will not rebound, and so the patch grows to form a dune.

McKee (1979), as a result of his global study of sand seas, based a classification of sand dunes not only on their morphology or ground pattern, but also on their internal structure, as determined by the number of slipfaces that is characteristic of the dune type (Table 16.1 and Figures 16.7 and 16.8). A slipface is the steep face on the lee side of a dune that is usually at the angle of repose of dry sand (30°–34°). A single slipface will produce one set of cross stratification dipping downwind. Two slipfaces develop as a result of two dominant wind directions, and a more complex internal dune structure is produced. Although the number of basic dune forms is limited, a number of combinations of the basic types can occur. For example, compound dunes are composed of two or more dunes of the same type, and complex dunes are composed of two or more different types of dunes (Plate 21).

Another general classification of aeolian bedforms, developed by Wilson (1970, 1972), Cooke and Warren (1973, 1976) and others, is based on a hierarchy of sizes and a variety of forms controlled by wind strength and direction, airflow dynamics, and sand size and supply. The existence of this hierarchy is indicated by the discontinuities in the frequency scatter of wavelengths (or spacing) for ripples (measured in the field in Algeria) and for dunes and draas (North African: draa – 'large sandhill') (measured from air photographs of Algeria, Arabia, Australia, Mali, Niger, Mauritania, Chad and the United States). This hierarchy shows four discrete modal wavelengths for ripples, dunes (2) and draas, each having associated height ranges and each represented by both transverse and longitudinal forms (Figure 16.9 and Table 16.2). Grain sizes are of surface material, and these suggest that although there may be considerable wavelength overlap between large ripples formed in coarse sand and small dunes in fine sand, grain size exercises a direct control over the wavelength of aeolian bedforms at all scales. The hierarchy of bedform sizes is also supported by certain high correlations which have been obtained between morphometric measures within given dune types shown in Table 16.1 (note that these values have a limitation imposed by the resolution of

**Table 16.1**  Basic dune types, structure and morphometry

| Name† | Form | Number of slipfaces | Morphometry*: mean and range (km) | | | | Comments† |
|---|---|---|---|---|---|---|---|
| | | | Mean length ($\overline{L}$) | Mean width ($\overline{W}$) | Mean wave length ($\lambda$) | Mean diameter ($\overline{D}$) | |
| Sheet | Sheetlike with broad, flat surface showing no dune forms | None | | | | | Too little relief to be termed 'dunes' |
| Stringer | Thin, elongate strip on bedrock | None | | | | | |
| Dome | Circular or elliptical mound | None or as barchan | | | 2·98 (0·8–5·4) | 1·28 (0·6–2·0) | Modified barchans? |
| Barchan | Crescent in plan view | 1 | 0·56 (0·03–1·5) | 0·90 (0·11–1·66) | 0·68 (0·11–1·1) | | (small) |
| Barchanoid | Row of connected crescents in plan view | 1 | | | | | Sand supply → One dominant wind direction |
| Transverse | Asymmetrical ridge | 1 | 1·27 (0·3–3·4) | 2·11 (0·5–9·6) | 1·90 (0·4–5·5) | | (large) (aklé dune fields) |
| Blowout | Circular rim of depression | 1 or more | | | | | Deflation dunes controlled by vegetation |
| Parabolic | U-shape in plan view | 1 or more | | | | | |
| Linear or longitudinal (seif) | Symmetrical ridge | 2 | 18·14 (3·65–26·0) | 0·24 (0·04–0·38) | 0·81 (0·15–1·41) | | Bimodal dominant wind directions (large, fossil(?) dunes termed 'draas') |

**Table 16.1** — *continued.*

| Name[†] | Form | Number of slipfaces | Morphometry*: mean and range (km) | | | Comments[†] |
|---|---|---|---|---|---|---|
| | | | Mean length $(\overline{L})$ | Mean width $(\overline{W})$ | Mean wave length $(\overline{\lambda})$ | Mean diameter $(\overline{D})$ | |

| Name[†] | Form | Number of slipfaces | Mean length $(\overline{L})$ | Mean width $(\overline{W})$ | Mean wave length $(\overline{\lambda})$ | Mean diameter $(\overline{D})$ | Comments[†] |
|---|---|---|---|---|---|---|---|
| Reversing | Asymmetrical ridge | 2 | | | | | Intermediate between star dune and transverse ridge |
| Star | Central peak with 3 or more arms | 3 or more | | | 1·76 (0·1–6·7) | 0·86 (0·2–3·0) | Tend to grow vertically (large ones termed 'rhourds') |

†See definitions in Figures 16.7 and 16.19.
*In sand seas observed by remote sensing (from Breed and Grow, 1979) (see Figure 16.8).
*Sources:* McKee, 1979, table 1, p. 10; Breed and Grow, 1979.

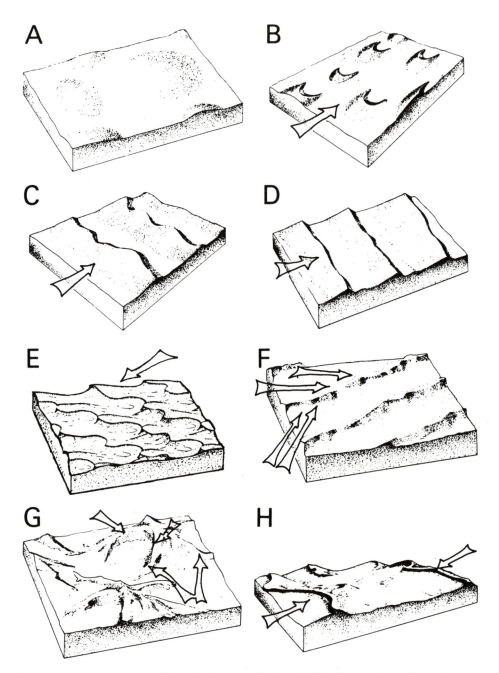

**Figure 16.7** Dune types; arrows show prevailing, dominant, or effective wind directions: A. dome dunes; B. barchan dunes; C. barchanoid ridges; D. transverse dunes; E. aklé dunes; F. longitudinal dunes; G. star dunes; H. reversing dunes.
*Sources:* A–D and F–H from McKee, 1979b figures 7, 3, 4, 5, 10, 11 and 12, pp. 11–13; E. modified after Cooke and Warren, 1973, figures 4, 27h, p. 288.

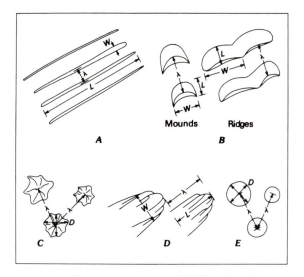

**Figure 16.8** Measures for defining the plan form of sand dunes from satellite imagery and aerial photographs; involving length **L**, width **W**, diameter **D** and wavelength λ. A. linear dunes; B. crescentic dunes; mounds are either single barchans or compound barchans presenting a simple form; ridges are barchanoid ridges or form aklé dune fields; C. star dunes; D. compound parabolic dunes; E. dome-shaped dunes. *Source:* Breed and Grow, 1979, figure 166, p. 258.

LANDSAT images, compared with the smaller features observed from air photographs and on the ground). For example, good positive linear correlations have been shown to obtain between remotely sensed measures of mean length $\overline{L}$ width $\overline{W}$ and wavelength λ in respect of barchanoid and transverse (aklé) dunes.

Aeolian bedforms occur due to shear at the air–sand interface, and are sustained and moved by the orderly, if intermittent, passage of sand grains across the surface of the forms. *Ripples* are small-scale features (see Figure 16.9 and Table 16.2) formed mainly by saltation impact, having wavelengths related to the length of flight of sal-

tating paths, or by aerodynamic instability. Impact ripples are transverse but others may be longitudinal, but they are superimposed on the whole range of larger aeolian depositional forms – stringers, sheets and true dunes. Despite some size overlap mentioned earlier, there is commonly a distinct scale break between ripples (wavelengths generally 1–300 cm) and true dunes (wavelengths > 20 m).

*Transverse dunes* (Figure 16.7D) appear to have an origin related to the initiation of oscillatory aerodynamic drag (perhaps generated by an obstruction or by a low-level atmospheric inversion layer and reinforced by positive feedback as the transverse forms themselves develop), causing waves to form in the surface airflow (Figure 16.10). These long ridges develop at right angles to the dominant wind direction (a narrow unimodal wind regime: see Figure 16.15A) or where the large size of the surface sand particles only allows the surface material to be moved by the stronger winds of preferred orientation. They migrate downwind by repeated additions to the steep (i.e. about 33°) lee-side slipface. Transverse dunes may become sinuous by the development of alternating barchanoid (concave downwind) and linguoid (convex downwind) forms due to the setting up of horizontal vortex eddies parallel to the dominant wind direction either under higher-velocity conditions or where the general airflow is disturbed by thermal eddies (Figure 16.11). The rate of movement of sections of transverse dunes varies with their size, the rate of sand supply, the geometry of the horizontal eddies and the strength of the dominant winds. Unidirectional winds with a large sand supply normally result in a complex pattern of sinuous transverse dunes producing an *aklé* dune field (Figure 16.7E) (French: *écaille* – 'fish scale'). If the sand supply becomes more limited downwind, transverse dunes may pass into *barchanoid ridges* (Figure 16.7C) and then into isolated, crescent-shaped *barchans* (Turkish: *barkan* – 'sandhill') (Figures 16.7B and 16.12). Barchans may move over less mobile surfaces at velocities of up to 50 m

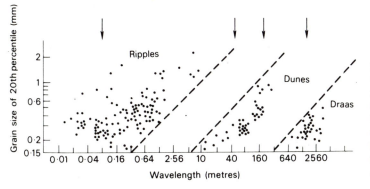

**Figure 16.9** Relationship of grain size to wavelength for aeolian bedforms, selected from the whole world; the 20th percentile is the mean diameter for which 20 per cent of the sand is finer; arrows indicate the modes of the frequency distributions of ripples (8 cm), dunes (40 m and approx. 200 m) and draas (approx. 1500 m). *Source:* Wilson, 1972, figure 1b, p. 668.

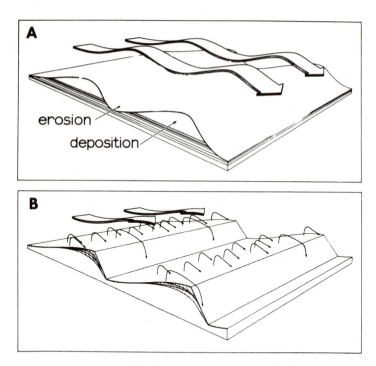

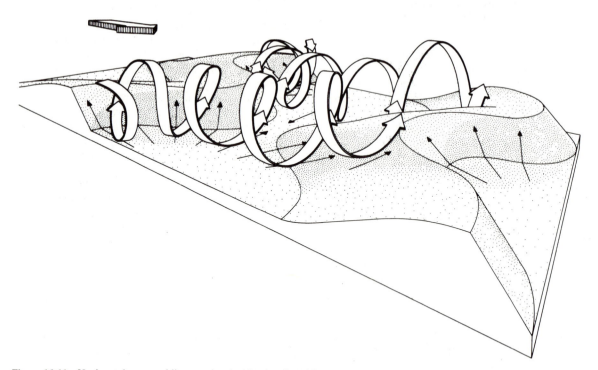

**Figure 16.10** The hypothetical first two stages of transverse dune formation: A. wavelike pattern in the wind erodes and deposits alternate ridges and hollows; B. slipfaces develop.
*Source:* Warren, 1979, figure 10.10, p. 337.

**Figure 16.11** Horizontal vortex eddies associated with a barchanoid transverse dune which propagate downwind distortions in the subsequent ridge.
*Source:* Warren, 1979, figure 10.12, p. 340.

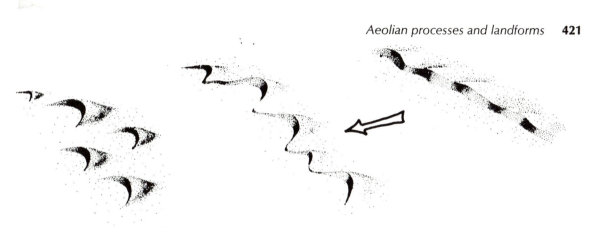

**Figure 16.12** Sequence of dune types with a narrow unimodal wind regime and (from right to left) a diminishing sand supply; from right to left, a transverse dune, a barchanoid ridge and a group of barchans are shown.
*Source:* McKee, 1979b, figure 6, p. 12.

per year with strong unidirectional winds depending on slipface height (i.e. barchan volume) and wind characteristics (Figure 16.13). Isolated, stationary barchans may become remoulded into *dome dunes* (Figure 16.7A).

**Table 16.2**  The hierarchy of aeolian bedforms

| Name | Modal value of wavelength | Range of wavelength | Range of height |
|---|---|---|---|
| Ripples | 8 cm | 1–300 cm | 0·001–20 cm |
| Dunes (1) | 40 m | | |
| | | 20–300 m | 1–30 m |
| Dunes (2) | 200 m | | |
| Draas | 1500 m | 1–3 km | 20–200 + m |

*Source:* Wilson, 1972, p. 668.

The formation of *longitudinal dunes* (Figure 16.7F) is more difficult to explain than that of transverse dunes. Longitudinal dunes compose some 30 per cent of all wind-deposited surfaces, varying from 85 per cent in the south-west Kalahari to virtual absence in the Saharan erg oriental. One type of longitudinal dune (long linear dune) is relatively straight and of considerable length, as in the Simpson Desert of Australia and the Kalahari – in the former being on average 10–35 m high, 150–300 m apart and 20–25 km long. These dunes form parallel to the resultant wind direction, or to the dominant (higher-velocity) direction, associated with a wide unimodal range of wind directions or with acute bimodal wind directions (Figure 16.15B and C). An important mechanism is produced by large-scale adjacent, counterrotating helical eddies (horizontal roll vortices) (Figure 16.14) which propel the sand grains up the sides (slipfaces) of the dune in a downwind direction. A more common type of longitudinal dune is the *seif dune* (Arabic *seif* – 'sword') of which the best examples are in Arabia and the

Sahara. Seif dunes are only a few kilometres long, have somewhat sinuous crests, are rounded upwind and pointed downwind. Their side slopes are some 20° towards the base but increase to the repose angle (about 33°) at their sharp crests. Although, as we have seen, it is not possible to demonstrate complete correlations between dune type and wind regime, it seems clear that seif dunes are associated with obtuse bimodal regimes (Figure 16.15D) generated by the alternation of two oblique seasonally dominant wind directions, which also produce horizontal eddies of limited extent (see Figure 16.14). For example, the wind regime shown for the Sinai (Figure 16.15D) reflects the broadly south-west winds of winter and the north-north-west in summer. It has also been suggested that seif dunes may be characteristic of finer sands which are susceptible to movement by a wider range of medium as well as strong wind velocities.

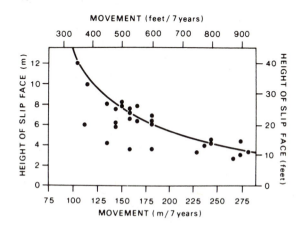

**Figure 16.13**  Plot of distance moved against height of dune slipface for twenty-seven barchans measured during 1956–63.
*Source:* Long and Sharp, 1964, figure 3, p. 154.

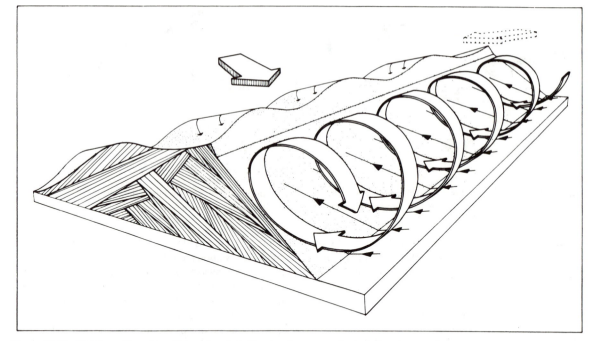

**Figure 16.14**   The formation of a seif dune by winds blowing from two principal directions; the slipface and the shallower beds dip in two directions; the lateral eddies are sand-transporting up the slipfaces at low levels and relatively sand-free at higher levels.
*Source:* Warren, 1979, figure 10.15, p. 342.

Dunes of multiple slipfaces are the result of winds from several directions (Figure 16.15E); they commonly have a high central peak and three or more arms extending radially. Such forms are called *star dunes* (Figure 16.7G), and they grow vertically, rather than migrating laterally. Intermediate in character between the star dune and the transverse ridge is the *reversing dune* (Figure 16.7H). It characteristically forms where winds from nearly opposite directions are balanced with respect to strength and duration. Such dunes may have the general form of transverse ridges but a second slipface, oriented in a direction nearly opposite that of the primary slipface, develops periodically.

*Draas* are the largest and most enigmatic aeolian bedforms (Figure 16.16). These are very large whaleback dunes, usually longitudinal, with extreme heights reaching 400 m and wavelengths of 1000–3000 m. They are characteristic of parts of the Sahara, are present in the Namib Desert of south-west Africa but are generally absent, for example, in the Australian deserts. The huge volumes of sand involved imply that draas take long periods to form (perhaps $10^4$–$10^6$ years), move and develop very slowly, and may thus reflect the wind regimes of earlier periods. However, the persistence of the subtropical high-pressure locations, even during

much of the Pleistocene, implies a minimum of difference between present and past dominant wind directions, especially in the Sahara. There are two groups of related theories which have been applied to the development of draas (Cooke and Warren, 1973):

(1) Their common orientation with respect to present dominant wind directions implies that draas and associated large features may be generated and self-sustained by mesoscale atmospheric circulations (e.g. convective cells, lee waves, large and persistent horizontal roll vortices, etc.). Draa ridges are commonly interrupted by large nodal star dunes (*rhourds*) rising to 200 m in places where there are assumed to be complex interactions between persistent convective cells.

(2) Most draas structurally appear to be either *compound dunes* (consisting of several dunes of the same type combined by overlapping or superimposition) (Figure 16.17) or, most commonly, *complex dunes* (resulting from the coalescence or growing together of two or more basic types) (Figure 16.18). Draas are usually large (compound?) longitudinal dunes made complex by the superimposition of chains of star dunes (Figure 16.18A). In the Namib Desert,

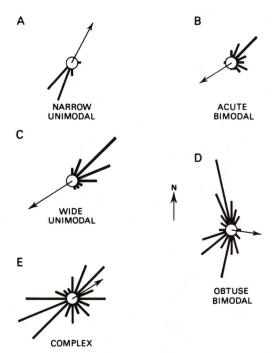

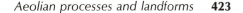

**Figure 16.15** High-energy wind regimes associated with four basic dune types. Roses depict the annual distribution of effective, sand-moving winds; the length of the bars represent calculated potential sand drift (based on wind velocities above an assumed threshold and length of time of occurrence); and the bars extend in an upwind direction: A. barchanoid dunes, Pelican Point, south-west Africa; B. long linear dunes, Fort Gouraud, Mauritania; C. long linear dunes, Blima, Niger; D. seif dunes, south of El Arish, Sinai; E. complex star dunes, Ghudāmis, Libya.
*Sources:* Fryberger and Dean, 1979, figure 97, p. 151; Lancaster, 1982, figure 6B, p. 488.

for example, chains of star dunes superimposed on linear features form draa ridges 50–150 m high and 600–1000 m wide.

Of course, most compound and complex dunes are on a much smaller scale, but the latter are usually the larger. Another class of sand dune includes those which are developing in conjunction with some vegetative growth, usually in areas of increasing precipitation. These dunes that are controlled by partial stabilization by vegetation are largely features of deflation, where the sand supply is limited, having slipfaces sloping in one or many directions. The *blowout dune* (Figure 16.19A) is normally a circular bowl with a dune on the downwind side, whereas the *parabolic dune* (Figure 16.19B) develops a U-shape with a convex nose which migrates downwind. Such dunes have been described as occurring in the Navajo Country of Utah and Arizona, and Hack (1941) has suggested a general model for the development of dune types there based on the three variables of wind strength, sand supply and vegetation cover (Figure 16.20).

For all dune types – simple, compound and complex – an almost infinite number of varieties occur which represent transitions from one basic type to another and result from differences in the direction and strength of wind, the amount of available sand, the physical obstructions and other factors that control dune types. Some varieties are barchans with one horn greatly extended, linear dunes with many branches diverging from one end suggesting the name feather dunes, and parabolic dunes that are V-shaped rather than U-shaped.

Unlike many other landforms, few evolutionary stages in dune development have been identified. They are too dynamic to be classified except by their present form

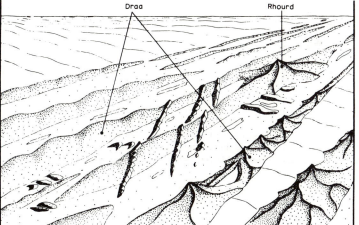

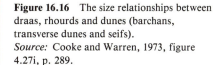

**Figure 16.16** The size relationships between draas, rhourds and dunes (barchans, transverse dunes and seifs).
*Source:* Cooke and Warren, 1973, figure 4.27i, p. 289.

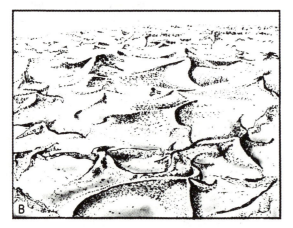

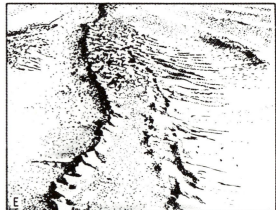

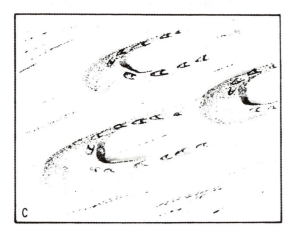

**Figure 16.17** Compound dunes: A. barchanoid ridges coalescing; B. star dunes coalescing; C. small barchans on large barchan; D. parabolic dunes within a large parabolic dune; E. linear dunes on large linear ridge.
*Source:* McKee, 1979b, figure 13, p. 14.

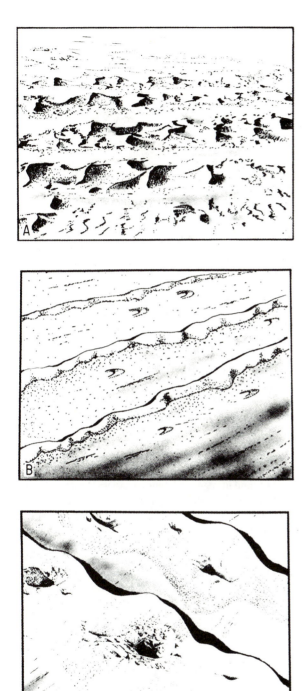

**Figure 16.18** Complex dunes: A. star dunes on linear dune; B. linear dunes with barchans in interdunes; C. blowouts on transverse dunes.
*Source:* McKee, 1979b, figure 14, p. 15.

and internal structure. However, owing to climate change dune fields can be stabilized by vegetation, and the Sand Hills of Nebraska are such a stabilized sand sea. Here the individual dunes no longer retain the sharpness of active dunes and support a lush growth of grass (Warren, 1976). A major problem, of course, is that climatic fluctuations, overgrazing, or weakening of the protective vegetation by any means will permit reactivation of the dunes. This phenomenon of *desertification* is a major problem along the southern border of the Sahara.

## 16.5 Coastal sand dunes

Coastal sand dunes form in the back beach areas where there is a large sand supply of suitable calibre, winds strong and persistent enough to move it and a suitable place for it to accumulate (Goldsmith, 1978). Large sand supplies are commonly associated with a large tidal range, which exposes an extensive sand area at low tide and results in storm overwash of sand at higher levels during high water. Both wind strength and onshore direction are also important, and it is possible that persistent winds of moderate velocity may be more effective movers of large quantities of sand than infrequent storm winds. Vegetation also plays an important role in some dune sand accumulation. Coastal dunes are widespread along humid temperate and arid tropical coasts, but are much less common on humid tropical coasts. Suggested reasons for this are the tendency of humid tropical weathering to produce silts rather than sands, an excess of vegetation near the shore, the inhibition of sand movement in the damp climate and the generally low mean wind velocities.

The classification of coastal dunes naturally bears a relationship to general dune classifications, but three main types of coastal dunes have been identified by King (1972).

(1) Transverse dune ridges. These are unvegetated sinuous ridges up to 1 km long and as much as 30–50 m high or more. They are asymmetrical with a well-defined steep (30°–34°) landward face and migrate inland at rates of up to 30 m/year. Transverse dunes occur with large and constant supplies of sand under coastal conditions where vegetational fixing is ineffective. Inland, or where the sand supply decreases, the transverse dunes may pass into smaller barchans. In the United States such dune types occur particularly at:

Coos Bay, Oregon, along a 60-km stretch of coast where lines of dunes of 34° landward slope and 3°–12° seaward slope are moving inland under the action of north and north-west summer winds inundating forests. The absence of salt-resistant vegetation

**A**

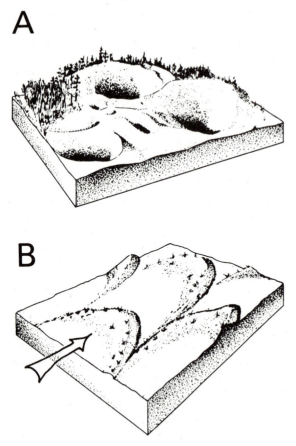

**B**

**Figure 16.19** Dunes formed under partial-vegetation cover:
A. blowout dunes; B. parabolic dunes; arrow shows
prevailing wind direction.
*Source:* McKee, 1979b, figures 8 and 9, p. 12.

(especially of marram grass which was only artificially
introduced in 1910) has been an important factor in the
development of this coastal dune field.

Southern end of Lake Michigan, Indiana, where
dunes of wind-eroded glacial drift material cover some
400 km² and reach 50–70 m high.

Provincetown, Cape Cod, where four lines of dunes
parallel the coast reaching 30 m high and covering
26 km².

In the classic coastal dune field of south-west Peru,
where rainfall and vegetation are almost entirely lacking,
barchans 3·5 m high and 37 m wide are advancing north
at 10–30 m/year across an uplifted coastal shelf.

(2) Vegetated dunes. These are the most common, if not
the most spectacular, of coastal sand dunes. They form
ridges with flat or undulating upper surfaces which are

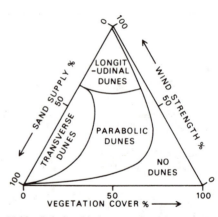

**Figure 16.20** Relationship between dune forms, effectiveness
of vegetation cover, supply of sand available to the wind, and
strength of wind in the Navajo Country, south-western United
States; it is assumed that the wind blows almost constantly
from one direction.
*Source:* J. T. Hack, 'Dunes of the western Navajo Country',
*Geographical Review*, vol. 31, 1941, figure 19, p. 260.
Reprinted by permission of the American Geographical
Society.

continuous but irregular and punctuated by blowouts
where the vegetation cover has been breached. Each dune
line is composed of a complex cross-bedding at a wide
variety of angles, and there is commonly a series of such
lines not necessarily parallel to each other or to the
shoreline. For example, at Laugharne Burrows, South
Wales (see Figure 2.12), a series of ridges between
Pendine and Laugharne reach 16 m high. The accumulation of sand in vegetated dunes is assisted by colonization
by long-rooted, salt-resistant plants. In Britain the first
colonizer of sand and shingle near the limit of high spring
tides is sea couch grass (*Agropyron junciforma*), which is
followed by marram grass (*Ammophila arenaria*), the
most important dune-fixing plant possessing a complex
of long, bifurcating roots. As a primitive soil profile
begins to develop sea buckthorn (*Hippophae rham-
noides*) and elder (*Sambucus nigra*) move in. On Cape
Cod marram grass and saltwort (*Salsola kali*) are
important. From the first colonization vegetation is
capable of causing sand stabilization and rapid dune
growth. At Gibraltar Point, Lincolnshire, England, the
4–7-m-high dunes have been observed to grow at a rate of
1·5 m in six years and growths of up to 1 m/year have
been observed in other areas during the first three years
of vegetation colonization (Steers, 1964; King, 1972).

(3) Parabolic dunes are isolated features up to 5–10 m
high and as much as 1–2 km long flanking the depressions caused by local vegetation destruction and
blowout.

## 16.6 Loess

When quantities of quartz silt and dust (<0·05 mm diam.) are produced, the particles are commonly susceptible to wind entrainment because of the considerable roughness of the generating surfaces and the consequent near-surface microturbulence (Péwé, 1981; Smalley and Vita-Finzi, 1968). Once airborne, fine particles may be transported considerable distances, as is shown by Von Karman's calculations given in Table 16.3. The very sharp difference in aerodynamic properties of particles of 0·1 mm and 0·01 mm is of great significance, with maximum flight times for the smaller particles being up to eight-plus orders of magnitude greater and maximum ranges two to three orders greater. The most important wind-blown deposit is a buff-coloured, non-indurated, well-sorted quartz silt (mainly 0·05–0·02 mm or less) which occurs with a small clay fraction and possibly as much as 40 per cent carbonate minerals. This material is termed loess (German: *Löss* – originally the name for fine loam in north Germany). Loess occurs, sometimes far from its original source, as extensive sheets in places hundreds of metres thick, as shallow and discontinuous patches, or as very thin extensive blankets. Once deposited, its small diameter makes it very resistant to further wind erosion, and it rarely forms dunes. Much loess accumulation has occurred in steppe and grassland environments, and the buried vertical root structures, replaced by finer material or calcium carbonate, have helped to produce the vertical cleavage so characteristic of some loesses – other possible factors being the formation of tension joints by superincumbent pressure and the electrostatic forces generated in the small particles during flight. Although resistant to wind erosion, loess is readily eroded by running water and it is noteworthy that the cultivated loess hills of eastern Iowa

provide the highest sediment yields in the United States. Rapid erosion of thick loess may produce vertical cliffs and bluffs in places.

As Smalley and Vita-Finzi (1968) have stressed, there appear to be two types of loess – the first a 'true' loess derived from glacial action and later wind transported; the second a 'desert' loess of much more questionable origin:

(1) The glacial loess derived from glacially ground quartz particles, was transported by fluvioglacial action to outwash plains in the middle and high latitudes during the Pleistocene and then moved by strong winds blowing outward from the icecaps to produce extensive deposits in China, Central Asia, Northern Europe, North America and Argentina. Figure 16.21 shows the most extensive loess area in northern China derived from the Pleistocene ice sheets in western and south-west China; and Figure 16.22 shows the loess of North America, much of which has been fluvially reworked by the Mississippi and its tributaries. Although fine material may remain in air suspension for long periods, much loess was deposited quite close to the braided outwash plains which were its source, and the thickness of the loess deposits generally decreases logarithmically downwind.

(2) Desert loess does not occur in sandy deserts but as thin patches near its margins or as thin sheets in distant locations. Péwé has estimated that in central Arizona an average of 3·5 dust storms per year deposit some 54 gm/m$^2$/year of fine material. Considerable controversy surrounds its origin, particularly as some of it (e.g. in Tripolitania) consists of very fine sand. There is some suggestion that aeolian abrasion of sand may produce silt-sized particles, but the experimental evidence is conflicting. It is clear that the silt originates in

**Table 16.3**  Time of flight, range and height of rise of particles moving in moderately strong winds (15 m/s)

| Diameter (mm) | Fall Velocity (cm/s) | Flight time (Max.) | Range (Max. transportation distance) | Height (Max.) | Height (Mean) | % moved by saltation |
|---|---|---|---|---|---|---|
| 0·001(clay) | 0.00824 | 9–90 years | 4–40 × 10$^6$ km | 6·1–61 km | | 0 |
| 0·01 (silt) | 0·824 | 8–80 years | 4–40 × 10$^2$ km | 61–610 km | | 0 |
| 0·1 (very fine sand) | 82·4 | 0·3–3 s | 46–460 m | 0·61–6·1 m | | ? |
| 0·15–0·25 (fine sand) | 109·0–156·0 | very variable | | 2 m on boulders; 9 cm on sand | 63 cm on boulders (90% < 87 cm); | 84 |
| 0·25–0·83 (medium–coarse sand) | 156·0–218·0 | | | | 90% < 31 cm on sand | 75 |

*Source:*  Cooke and Warren, 1973, after Malina and Committee on Sedimentation.

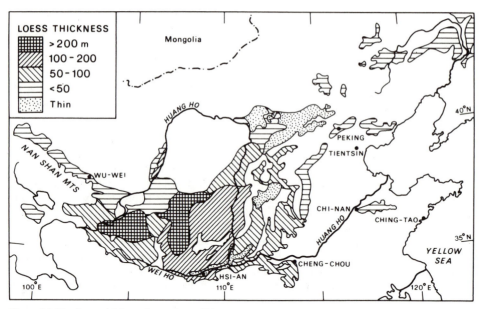

**Figure 16.21**  Loess thickness in northern China.
*Source:* Wang Yong-yan and Zhang Zong-hu, 1980.

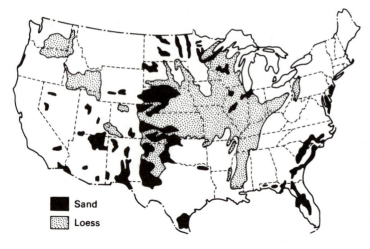

**Figure 16.22**  Wind-blown sediments forming surficial sand and loess deposits in the United States: this includes coastal dunes, fossil dunes and active dunes.
*Source:* Ruhe, 1975, figure 8.2, p. 153, after J. P. Thorp and H. T. U. Smith, *Map of Pleistocene Eolian Deposits of the U.S.*, 1952, Geol. Soc. Amer.

arid and semi-arid regions from bedrock weathering (including salt weathering); possible sandblast action; from floodplain, playa and alluvial fan deposits; and locally from the breakdown of chert formed from quartz silt. In present dust storms clouds of silt have been observed to rise to 2500 m and travel at speeds of up to 200 m/s along a front of as much a 600 km. Modal diameters of such windblown material moved west from the Sahara have been observed as 0·02 mm at 1000 km, 0·02–0·008 mm at 2000 km and 0·002 mm at 5000 km. Some silt from Central Asia is presently observed to move as much as 5000 km, particularly during the winter

drying period, and it is a fairly common occurrence for dust from 1500 km to the west to fall on Peking. In some locations of Central Asia the close juxtaposition of glacial and desert loess adds confusion to the question of origin.

## 16.7  Snow drifting

The processes involved in, and the forms resulting from, snow drifting exhibit both resemblances to and differences from those associated with sand drifting. As Ward (1981) has shown, the intensity and balance of processes occurring when snow is falling differ from those when it is not, although the role of repeated snowfalls is important in initiating all drifting. The transport of deposited (i.e. non-falling) snow takes place by surface creep, saltation and turbulent suspension:

(1) Creep may affect the top millimetre or two of fresh snow, producing ripples, but its total contribution to drifting is probably small.

(2) Saltation is essential for significant drifting with particles moving in a layer from a few centimetres to 1 m above the surface.

(3) Turbulent suspension, the movement of snow as an aerosol, is the most important mechanism of snow transport. There is some controversy as to whether the thickness of the major turbulent layer is tens of centimetres or tens of metres. In the bottom 10 cm or so turbulent suspension and saltation may occur together and, for example, at wind speeds of 10 m/s up to 90 per cent of snow movement may be in this layer, largely by turbulent suspension.

There has been a variety of suggested relationships between the mass transport of drifting snow and wind

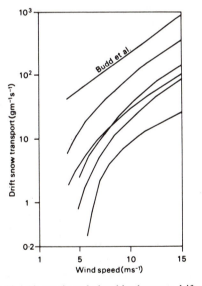

**Figure 16.23** Alternative relationships between drift snow transport and wind speed which have been suggested by various authorities.
*Source:* Ward, 1981, figure 2-13, p. 57, after Takeuchi.

velocity (Figure 16.23) both because of difficulties of measurement and because of the possible range of physical changes undergone by blowing snow. The latter include frictional melting; hardening and entrainment inhibition of snow surfaces at higher wind speeds; the increase of surface snow density with wind speed; differences of surface hardness due to normal melting processes; the variation of the threshold separating saltation and suspension transport; and the important effect of *sublimation* – the direct evaporation of ice crystals. It has been estimated that snow particles of 0·1, 1·5 and 2 mm diameter will sublimate completely after moving in air through some 450 m, 900 m, and 1400 m, respectively, and wind-blown snow in Wyoming has been shown to lose up to 25 per cent of its mass through sublimation after moving 500 m and some 80 per cent after 3000 m. Another factor affecting snow transport is the rate of snow supply, and it requires at least a length of 100 m of exposed snow surface to initiate a steady state of downwind drift movement. The transport curve of Budd *et al.* (1966) (see Figure 16.23) implies that most drift transport occurs during a few high-magnitude wind storm events, but it has been shown empirically that significant drifting occurs at lower windspeeds with, for example, 54–69 per cent of the total annual drift accumulation in some small Colorado catchments occurring during five weeks with windspeeds not exceeding 9 m/s.

Drifting during a snowstorm is naturally greater than when only deposited snow is involved, although the two may well occur together. However, heavy snowfall intensities may cause a saturation of the air–snow mixture and a limiting maximum of drift transport to occur at only moderate windspeeds.

Snow dunes may result from wind drifting and these are broadly divided into two types – the smaller transient snow dunes and the larger, more permanent snow and ice features. The most characteristic transient dunes, common in Greenland and Antarctica, are the *sastrugi*. As described by Whillans (1975), these are longitudinal dunes of hard-packed snow, reaching heights of 2 m, formed both in the lee of obstacles and in broad zones, their dimensions being partly related to wind velocity. More permanent dunes are formed downwind of larger topographic features (i.e. with dimensions greater than 3 km or so in the downwind direction), which generate lee waves having a wavelength of some 6 km or more which produce large transverse snow and ice ridges.

## References

Bagnold, R. A. (1941) *The Physics of Blown Sand and Desert Dunes*, London, Methuen.

Bagnold, R. A. (1953) 'Forme des dunes de sable et régime des vents', in *'Actions éoliennes', Centre Nat. Rech. Sci., Paris, Coll. Int.*, vol. 35, 29–32.

Baker, V. R. (1981) 'The geomorphology of Mars', *Progress in Physical Geography*, vol. 5, 473–513.

Breed, C. S. and Grow, T. (1979) 'Morphology and distribution of dunes in sand seas observed by remote sensing', *US Geological Survey Professional Paper* 1052, 253–302.

Budd, W. F., Dingle, W. R. J. and Radok, U. (1966) 'The Byrd snow drift project: outline and basic results', in M. J. Rubin (ed.), *Studies in Antarctic Meteorology, American Geophysical Union, Antarctic Research Series*, vol. 9, 71–134.

Collinson, J. D. (1978) 'Deserts', in H. G. Reading (ed.), *Sedimentary Environments and Facies*, New York, Elsevier, chapter 5, 80–96.

Cooke, R. U. and Warren, A. (1973) *Geomorphology in Deserts*, London, Batsford.

Fryberger, S. G. and Dean, G. (1979) 'Dune forms and wind regime', *US Geological Survey Professional Paper* 1052, 137–69.

Goldsmith, V. (1978) 'Coastal dunes', in R. A. Davis (ed.), *Coastal Sedimentary Environments*, New York, Springer-Verlag, chapter 4, 171–235.

Hack, J. T. (1941) 'Dunes of the western Navajo Country', *Geographical Review*, vol. 31, 240–63.

Horikawa, K. and Shen, H. W. (1960) 'Sand movement by wind action', *US Army Corps of Engineers, Beach Erosion Board Technical Memorandum* 119.

Jackson, M. L. *et al.* (1973) 'Global dustfall during the Quaternary as related to environments', *Soil Science*, vol. 116, 135–45.

King, C. A. M. (1972) *Beaches and Coasts*, 2nd edn, London, Arnold.

Lancaster, N. (1982) 'Linear dunes', *Progress in Physical Geography*, vol. 6, 475–504.

Long, J. T. and Sharp, R. P. (1964) 'Barchan-dune movement in Imperial Valley, California', *Bulletin of the Geological Society of America*, vol. 75, 149–56.

Mabbut, J. A. (1977) *Desert Landforms*, Cambridge, Mass., MIT Press.

McCauley, J. F., Grolier, M. J. and Breed, C. S. (1977) 'Yardangs', in D. O. Doehring (ed.), *Geomorphology in Arid Regions*, Binghamton, NY, Proceedings 8th Geomorphology Symposium, 233–69.

McKee, E. D. (ed.) (1979a) 'A study of global sand seas', *US Geological Survey Professional Paper* 1052.

McKee, E. D. (1979b) 'Introduction to a study of global sand seas', *US Geological Survey Professional Paper* 1052, 1–19.

Péwé, T. L. (1981) 'Desert dust: origin, characteristics, and effect on man', *US Geological Survey Special Paper* 186.

Ruhe, R. V. (1975) *Geomorphology*, Boston, Mass., Houghton Mifflin.

Sharp, R. P. (1964) 'Wind-driven sand in Coachella Valley Calif.', *Bulletin of the Geological Society of America*, vol. 75, 785–804.

Sharp, R. P. (1966) 'Kelso Dunes, Mojave Desert, Calif.', *Bulletin of the Geological Society of America*, vol. 77, 1045–74.

Sharp, R. P. (1980) 'Wind-driven sand in Coachella Valley, Calif., further data', *Bulletin of the Geological Society of America*, vol. 91, 724–30.

Smalley, I. J. and Vita-Finzi, C. (1968) 'The formation of fine particles in sandy deserts and the nature of "desert" loess', *Journal of Sedimentary Petrology*, vol. 38, 766–74.

Snead, R. E. (1972) *Atlas of World Physical Features*, New York, Wiley.

Steers, J. A. (1964) *The Coastline of England and Wales*, 2nd edn, Cambridge University Press.

Wang Yong-yan and Zhang Zong-hu (eds) (1980) *Loess in China*, Shaanxi People's Art Publishing House.

Ward, R. G. (1981) 'Snow avalanches in Scotland', unpublished Ph D dissertation, University of Aberdeen, Department of Geography.

Warren, A. (1976) 'Morphology and sediments of the Nebraska Sand Hills in relation to Pleistocene winds and the development of aeolian bedforms', *Journal of Geology*, vol. 84, 685–700.

Warren, A. (1979) 'Aeolian processes', in C. Embleton and J. Thornes (eds), *Process in Geomorphology*, London, Arnold, 325–51.

Whillans, I. M. (1975) 'Effect of inversion winds on topographic detail and mass balance on inland ice sheets', *Journal of Glaciology*, vol. 14, 85–90.

Wilson, I. G. (1970) 'The external morphology of wind-laid sand deposits', unpublished Ph D dissertation, University of Reading.

Wilson, I. G. (1972) 'Universal discontinuities in bedforms produced by the wind', *Journal of Sedimentary Petrology*, vol. 42, 667–9.

Woodruff, W. J., Chepil, W. J. and Zingg, A. W. (1975) 'Wind erosion and transportation', *American Society of Civil Engineers, Manual on Engineering Practice* 54, 230–45.

# Seventeen  *The glacier sedimentary system*

## 17.1 Glaciers

Approximately 10 per cent of the earth's land area is too cold for normal river activity and instead water exists in its frozen form as glaciers. Indeed, for most of the last 2–3 million years glaciers have extended over almost one-third of the earth's land surface. Ice is much less easily deformed under the influence of gravity than water and there are several consequences which follow from this. First, as the ice moves across the surface of a continent, the interaction of a deforming solid with the underlying topography produces a distinctive geomorphological process/landform system. A second consequence is that ice can accumulate to great thicknesses – over 4 km in the case of the Antarctic ice sheet. This means that glaciers represent an important store in the earth's hydrological system and, indeed, almost three-quarters of the earth's fresh water is locked up in the polar ice sheets. A third consequence is that the response time of a particular glacier system is slow and, for example, the effects of an exceptionally heavy snowfall may not make themselves felt at a glacier snout for many years and, indeed, may often be suppressed altogether. A fourth consequence is that the flow process has a distinct morphological form. Thus there is a field of geomorphological study which views a glacier as a direct morphological expression of the sedimentary and metamorphic processes associated with ice accumulation and deformation. Finally, the great thickness of many glaciers implies that the crucial subglacial ice–rock interface is difficult to reach except round glacier margins or at the bottom of boreholes. This has meant that, compared to other geomorphological fields, understanding of subglacial processes depends more on theoretical models, laboratory experiments and inferences from landforms than on direct measurement of processes in the field.

A glacier can be considered as a sedimentary system that is involved with the accumulation, transfer and deposition of mass (snow, ice, water, rock debris) in response to additions and losses of mass and energy. This chapter focuses on the way the system operates and the processes by which it creates landforms. Chapter 19 contains the discussion of landscapes resulting from this modification, and a consideration of the glacial morphogenetic regions is presented in Chapter 18.

It is helpful to recognize three main types of glacier on the basis of the relative importance of ice supply and the nature of the topography (Figure 17.1). *Ice sheets* or *icecaps* form broad domes which submerge the underlying topography. Ice radiates out from the centre as a sheet, although near the periphery it may be confined and flow through uplands in outlet glaciers. The difference between an ice sheet and an icecap is conventionally accepted to be one of scale, with icecaps generally less than around 50,000 km² in area. The Antarctic provides a fine example of a large ice sheet, with two components, one in west Antarctica over 2000 m high and one in east Antarctica nearly 4000 m high (Figure 17.2). The latter is some 5000 km across. Smaller icecaps occur in Arctic Canada, Iceland and Norway and may measure only a few tens of kilometres across. An *alpine* or *valley glacier* is one which is severely controlled by topography in that it flows in a mountain valley and is overlooked by rock valley walls. Unlike an ice sheet or icecap, glacier flow is strongly influenced by topography. Such glaciers are characteristic of steep mountains, both in the polar regions and in mountain ranges worldwide. An *ice shelf* is in essence a floating ice sheet or icecap. It is loosely controlled by topography in that one side is bounded by a coastal embayment. The main feature is that, unlike other glaciers, there is no friction with the bed and the

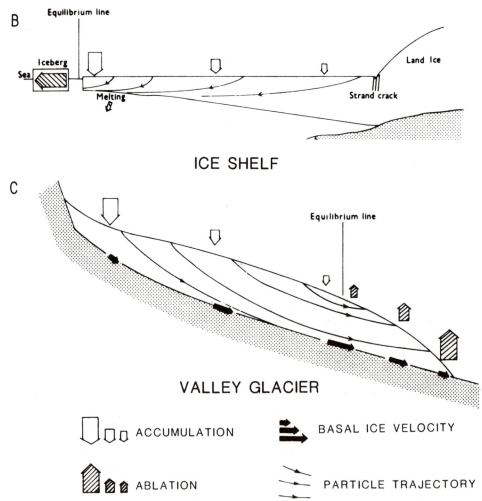

**Figure 17.1** Models of A. an ice sheet, B. an ice shelf and C. a valley glacier showing distribution of snow input and output, together with related flow characteristics. Basal slipping is assumed to occur in models A. and C. and is a maximum in the vicinity of the equilibrium line.
*Source:* Sugden and John, 1976, figure 4.8, p. 63.

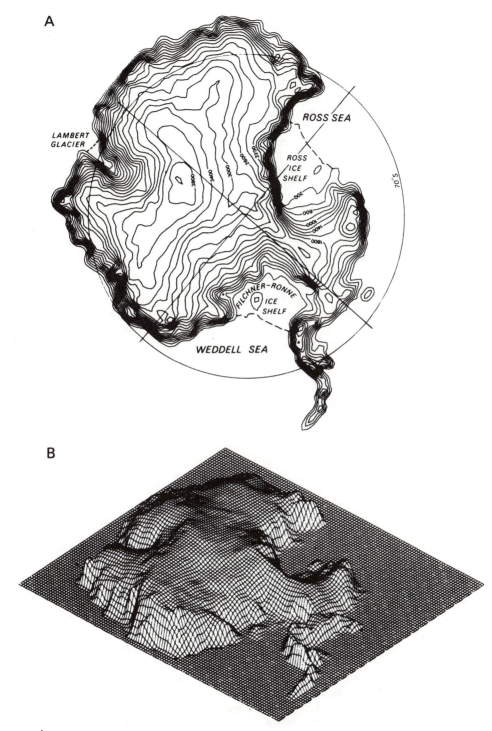

**Figure 17.2** The Antarctic ice sheet: A. surface contours (in metres); note that the lowest contour is 200 m and that the coastline (except for the edges of the major ice shelves) is not shown; B. three-dimensional isometric view of the ice surface.

*Source:* Drewry *et al.*, 1982, figures 2 and 3, p. 86.

ice can spread freely. The largest examples occur in Antarctica, namely the Ross and Ronne-Filchner ice shelves (Figure 17.2). The former is larger than France and yet moves up and down with each tide.

As noted above, a glacier can profitably be viewed as a sedimentary system. Obviously, the ice itself is the crucial factor and the input (or *accumulation*) processes include all ways in which mass is added to the glacier. Snowfall is the most important factor, but avalanching, basal freezing and the growth of superimposed ice are also involved. The latter two processes occur when water (rain or meltwater) freezes as it comes into contact with ice at temperatures below the melting point. Output (or *ablation*) includes all the ways in which mass is lost. Surface melting is the most important process; but internal and basal melting, evaporation, sublimation, deflation and calving are often locally important. The difference between accumulation and ablation for a whole glacier over one year is called the *net balance*.

There is a variation in the net balance from place to place on a glacier and as a result two subsystems can be recognized. The accumulation subsystem includes all those areas where there is an excess of accumulation over ablation in a year, while the ablation subsystem describes those areas where ablation exceeds accumulation in a year. The boundary between the two is the *equilibrium line* and occurs where the annual ablation equals accumulation (Figure 17.3).

The spatial variation of the net balance over a glacier is the basic driving-force of the system. On most glaciers there is a tendency for net ablation to be greatest at the snout, falling to zero at the equilibrium line. As one proceeds above the equilibrium line net accumulation tends to rise. As shown in Figure 17.3, this tends to remove a wedge of snow and ice from below the equilibrium line and add on a wedge of snow and/or ice above it. If the glacier is to remain in equilibrium, then it must restore the profile by transferring ice from the accumulation zone above the equilibrium line to the ablation zone below. In effect the gradient of wedges represents what is termed the *net balance gradient* which is expressed as the increase in net balance in millimetres per metre of altitude. The significance of this gradient is that it is a measure of the activity of a glacier. The steeper the net balance gradient is, the faster the glacier must flow to restore equilibrium. Conversely, the lower the gradient, the less flow is needed to restore equilibrium.

The net balance gradient is correlated with the magnitude of the snowfall on to a glacier and is higher where snowfall is high. This is why glaciers tend to be more active in humid areas than in dry areas. There is a latitudinal decline in activity from temperate to polar latitudes

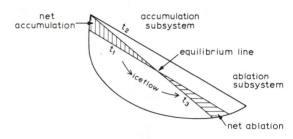

**Figure 17.3**   Cross-sectional patterns of accumulation and ablation on an idealized glacier, which provide the basic driving-forces for glacier flow: $t_1$ = glacier surface at the end of a summer; $t_2$ = surface at the end of succeeding winter; $t_3$ = surface at end of following summer; glacier flow restores the glacier surface profile.

while, superimposed on this general tendency, there is a decrease in activity from maritime to continental climates.

The output of a glacier is intimately linked to the input in that in most land-based situations the snow and ice has to be carried to an area with sufficiently mild a climate to melt it. The greater the input of snow into the accumulation zone, the warmer the climate of the ablation area needs to be to melt it. This is why glaciers in maritime situations, for example, Franz Josef Glacier in New Zealand and many in southern Chile, are able to extend their snouts into environments sufficiently warm to support temperate forests. In contrast, in continental climates, such as northern Canada, the glaciers end in cold Arctic tundra; here the summer warmth necessary for ablation is negligible but so also is the quantity of ice to be removed.

The above arguments apply to active flowing glaciers where the input of snow is sufficient for the maintenance of flow under the influence of gravity. In some situations, particularly in dry continental areas (e.g. arctic Canada), the snow supply may be insufficient to maintain glacier flow and there may be a stagnant glacier with low surface profile whose altitude at any point depends directly on the net mass balance *at that point*.

## 17.2 Glacier ice

Snow in the accumulation zone undergoes a series of changes which transform it to glacier ice. The term *firn* is generally applied to snow which has survived a summer melt season and has begun this transformation. It consists of loosely consolidated, randomly oriented ice crystals with interconnecting air passages and a density generally greater than $0.4$ mg/m$^3$. Transformation involves the regrowth of ice crystals and the elimination of air passages. When consolidation has proceeded sufficiently to isolate the air into separate bubbles, the firn becomes *glacier ice*. This change to ice takes place at densities of between 0·8 and 0·85 kg/m³. At higher temperatures, and especially in the presence of water, transformation to ice can take place within one year. In cold subzero environments, such as on the Antarctic ice sheet, transformation may take several thousand years.

Once formed, the behaviour of ice is closely related to its temperature, and there are two quite distinct situations. In one the ice is below the pressure melting point and known as *cold* or *polar* ice; in the other the ice is sufficiently close to its pressure melting point to contain water and is known as *warm* or *temperate* ice. The term *pressure melting point* is used because the temperature at which water freezes diminishes under additional pressure by a rate of approximately 1°C for every 140 bars (1 bar = 100 kPa).

Cold ice forms in two main situations. The first is where the firn accumulates in environments so cold that there is little or no surface melting in summer. Such a situation occurs over much of the Antarctic ice sheet, the northern Greenland ice sheet and at higher elevations in mountains like the European Alps. Where there is no surface melting, the firn accumulates at a temperature close to the mean annual temperature. Thus in Antarctica the surface temperature ranges between −30° and −60°C. As with most sedimentary deposits on the earth's surface, the temperature increases with depth mainly due to the geothermal heat flux, but also due to the effect of internal deformation of the ice by glacier flow (Figure 17.4). The second situation where cold ice forms is related to the cooling of surface layers of a glacier by winter cold. Cold ice of this type occurs on the surface of all glaciers in winter. In certain situations, especially in the ablation zone of glaciers, a greater thickness of ice is cooled in winter than can be warmed in summer. Thus a cold subsurface layer can persist and extend sufficiently deep to freeze the toe of a glacier on to the bottom. As will be seen later, this latter situation can have a profound effect on erosional and depositional processes.

*Warm ice* forms whenever there is sufficient heat to raise ice temperatures to the pressure melting point. At the surface this commonly occurs when firn experiences summer melting. Under such circumstances surface meltwater sinks into the firn and freezes if it comes into contact with snow or ice below the pressure melting point. Each gram of meltwater which freezes releases enough latent heat to raise the temperature of 160 g of ice by 1°C. This is so efficient a heating mechanism that on any glacier where firn is subjected to significant summer melting, there is usually enough heat to bring the firn, or ice, to the melting point during its transformation to ice. In such situations the whole glacier may consist of ice at pressure melting point. In such a case there is a modest *decrease* of temperature with depth in association with increased pressure (Figure 17.4). Warm ice can also occur at the bottom of glaciers whose surface layers consist of cold ice. Whether or not this threshold is reached depends on a number of variables (Table 17.1), but the most favourable situations are beneath thick ice sheets or swiftly moving glaciers. The various combinations of cold and warm ice can be crystallized into three main situations, namely (a) glaciers consisting wholly of cold ice; (b) glaciers consisting wholly of warm ice; and (c) glaciers consisting of cold ice underlain by warm ice (Figure 17.4).

The temperature of glacier ice is of crucial importance for glacial geomorphology. If the basal ice is at the pressure melting point, then the negative temperature gradient prevents geothermal heat from being conducted upwards into the ice. Instead the geothermal heat melts a basal layer of ice, with an average thickness of 6 mm. The implications of this layer of water on meltwater

**Table 17.1** Variables which affect basal ice temperatures and the direction of change

| Variables | Effect of an increase on basal temperature |
|---|---|
| Ice thickness | + |
| Ice surface temperature | + |
| Surface accumulation rate (which incorporates cold snow) | − |
| Surface warming rate (as a result of the flow into lower warmer environments) | − |
| Geothermal heat | + |
| Frictional heat (related to ice velocity) | + |

*Source:* Budd *et al.*, 1970.

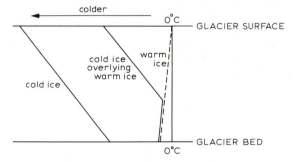

**Figure 17.4** Idealized temperature profiles illustrating three characteristic thermal situations; in practice, glacier flow and snow accumulation modify the temperature gradients.

processes, glacier sliding, deposition and erosion are profound.

An important point to stress is that the temperature of the basal ice can vary spatially in response to the varying interplay of the factors listed in Table 17.1. Possible variations in the basal temperatures beneath a reconstruction of the Laurentide ice sheet are shown in Figure 17.5. Here a central zone of ice at the pressure melting point with associated basal melting is succeeded towards the periphery by a cold zone, another warm melting zone and in certain sectors another cold zone. At the smaller scale of an individual glacier or icecap it is common to find situations, especially in subpolar areas, where the accumulation zone is underlain by warm basal ice inherited from the accumulation of firn at the pressure melting point, while parts of the ablation zone and particularly the margin comprise ice frozen to the bed due to excess winter cooling.

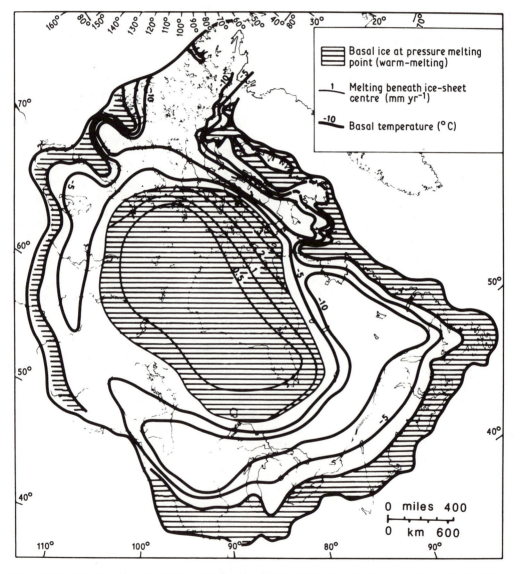

**Figure 17.5** Calculated basal temperatures and zones of basal melting beneath the Laurentide ice sheet at its maximum. *Source:* Sugden, 1977, figure 10, p. 37. Reproduced with permission of the Regents of the University of Colorado from *Arctic and Alpine Research*, vol. 9.

## 17.3 Glacier flow

A glacier flows because it deforms in response to stress set up in the ice mass by the force of gravity. Any point within the glacier is subjected to stress as a result of the weight of the overlying ice. This stress can be envisaged as having two components – hydrostatic pressure and shear stresses. The hydrostatic pressure which is related to the weight of the overlying ice is the same in all directions. It is the shear stresses that cause particles to slip past one another and they are related to both the weight of the overlying ice and the surface slope of the glacier. The shear stress at a point can be calculated from the equation:

$$\tau = \rho g h \sin\alpha$$

where: $\tau$ is the shear stress
$\rho$ is the density of ice
$g$ is the acceleration of gravity ⎫
$h$ is the thickness of the glacier ⎬ weight
$\alpha$ is the slope of the upper glacier ⎭
surface.

The important conclusion is that shear stresses vary according to the thickness of the glacier and the surface slope, with high values formed by thick ice or a steep surface slope. In practice, ice deforms under relatively low shear stresses and calculations for a wide variety of glaciers suggest that shear stresses generally vary from 0 beneath an ice divide with horizontal surface to 1·5 bar (150 kPa) on steep glaciers.

Glacier flow consists of three groups of processes which are conveniently described under the headings internal deformation, basal sliding and bed deformation (Figure 17.6).

### 17.3.1 Internal deformation

The fundamental mechanism of internal deformation is

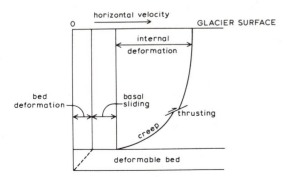

**Figure 17.6** The component processes leading to glacier flow.

creep, whereby there is mutual displacement of ice crystals relative to one another. The rate of deformation or *strain rate* of ice has been studied in the laboratory and the results form the basis of what is termed Glen's Law (Glen, 1955). The behaviour of ice under stress can be approximated by the power function $\dot{\epsilon} = A\tau^n$, where $\dot{\epsilon}$ represents the strain rate, $A$ is a constant approximating to the temperature of ice; $\tau$ is the effective shear stress and $n$ is an exponent with a mean value of 3. The significance of this relationship is that it demonstrates that the rate of deformation is highly sensitive to changes in the shear stress, for example, a doubling of the shear stress will increase the rate of deformation eightfold. Also the rate of deformation is related to ice temperature, though less sensitively, and for example declines fivefold with a temperature reduction from $-10°C$ to $-25°C$.

Application of Glen's Law gives important insights into glacier behaviour. It explains why most deformation takes place in the basal layers of a glacier where shear stresses are highest. It explains how there is movement within cold-based glaciers which are unable to slide. Also it explains the apparent weakness of ice in that it is unable to withstand high shear stresses without deforming. In several field tests Glen's Law seems at face-value to underestimate the softness of basal ice which appears to deform more easily than predicted. It is likely that the higher than expected plasticity reflects factors such as crystal orientation, longitudinal compression, deformation history and debris content, all of which are active areas of current research.

Larger-scale mechanisms of internal deformation are folding and thrusting. Sections through folds can often be seen in marginal ice cliffs and also picked out in patterns on glacier surfaces. They reflect differential flow down the length of a glacier induced either by variation in ice discharge or in the drag induced by the glacier bed. Under certain conditions creep cannot adjust sufficiently to the stresses within the ice and faults occur. In zones of longitudinal tension, as in ice falls, a succession of rotational slips may occur and account for a considerable proportion of glacier movement. In zones of compression overthrusting may occur. Such thrust planes can be observed in glacier cliffs or in the dislocation of such features as englacial meltwater tunnels or crevasses.

Perspective on the longitudinal distribution of such zones of longitudinal tension and compression is given by Nye (1952) (Figure 17.7). Compressive flow describes a situation where the longitudinal stress is compressive throughout the depth of the glacier and is marked by a reduction in forward glacier velocity. Extending flow describes a situation where the longitudinal stress is more

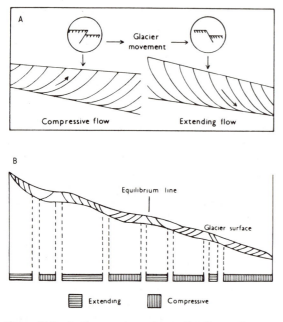

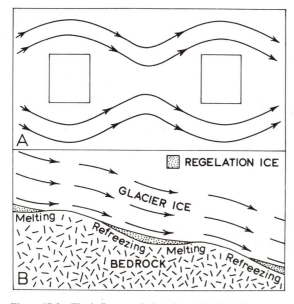

**Figure 17.7** A. Compressive and extending flow and associated sliplines; B. distribution of compressive and extending flow in an idealized glacier.
*Source:* Nye, 1952, figure 9, p. 91.

**Figure 17.8** The influence of obstacles on iceflow: A. mechanism of enhanced basal creep viewed in plan; high stresses on the upstream side of an obstacle cause high rates of deformation and allow basal ice to by-pass the obstacle; B. pressure melting mechanism; melting takes place on upstream side of the obstacle and regelation on the downstream.

tensile than the overburden pressure and velocity increases down glacier. In the case of compressive flow slip planes and differential creep produce an upward flow of ice, while in the case of extending flow the movement is downward. At the scale of a glacier as a whole compressive flow tends to occur where the ice thickness decreases downglacier – i.e. in the ablation zone – and extending flow where the thicknesses tend to increase downglacier – i.e. the accumulation zone. At a smaller scale compressive and extending flow are also related to the underlying topography. Extending flow occurs on bedrock slopes which steepen downglacier, while compressive flow occurs on bedrock which becomes shallower downglacier, for example, behind a rock step. Another important situation affecting longitudinal flow is any change in the basal temperature regime. If the margin of a glacier is frozen to its bed while the interior is warm-based, the reduction in sliding velocity accentuates the upward flow associated with compressive flow near the margin. The compressive flow which tends to occur in the ablation area of a glacier and particularly near its snout is an important way of raising basal debris to the surface of the glacier, with important results that are detailed on p. 457.

### 17.3.2 Sliding and bed deformation

The processes by which a glacier slides over its bed are of obvious interest to geomorphology. Two main processes were postulated by Weertman (1957) on theoretical grounds and have been largely substantiated by field studies. *Enhanced basal creep* occurs within the lower layers of a glacier and can occur in ice at any temperature (Figure 17.8A). *Regelation slip* describes a process at the actual ice–rock interface. It occurs only when the basal ice is close to the pressure melting point (Figure 17.8B). An important implication is that cold-based glaciers cannot slip at the ice–rock interface. Instead all movement must take place within the lower layers of the glacier.

Enhanced basal creep is an extension of Glen's flow law in that it explains how creep operates to circumvent irregularities in the bed. If an obstacle such as a bedrock bump or a boulder lodged on the glacier bed protrudes into the bottom of the moving glacier, then the stresses on the upstream side increase and this, in turn, increases the rate of deformation of the ice. This allows flow of ice around the obstacle.

The average rate of deformation $\langle \dot{\varepsilon} \rangle$ is related to the size of the obstacle and, assuming that the obstacle is cubic with side $\ell$ and area $d^2$, can be calculated from:

$$\langle \dot{\varepsilon} \rangle = B \left( \frac{\langle \tau \rangle}{2} R^2 \right)^n$$

where $B$ and $n$ are constants with $n = 3$;
   $\langle \tau \rangle$ is the average drag between a glacier and its bed;
   $R^2$ is an index of roughness related to obstacle size.

The important conclusion to arise is that the rate of deformation increases with obstacle size. The process is least effective at obstacle sizes less than around 10 cm across.

*Regelation slip* occurs when ice at the pressure melting point moves across a series of bumps. Ice melts under the influence of high pressures on the upstream side of the obstacle. The meltwater produced then flows round the bump to the downstream side where the pressure is lower and it refreezes to form regelation ice. The process is most effective when the latent heat released by freezing can be transferred from the downstream side of the obstacle through the rock and assists in further melting on the upstream side. Such a transfer occurs most efficiently on small obstacles generally less than 10 cm across. The rate of regelation slip ($V_{pm}$) can be calculated for an idealized glacier bed from the following equation:

$$V_{pm} = \frac{CK\langle \tau \rangle R^2}{L\rho_i \ell}$$

where: $C$ = constant relating pressure and the resultant lowering of the freezing point;
   $K$ = thermal conductivity of the obstacle;
   $L$ = latent heat of the fusion of ice;
   $\rho_i$ = density of ice;
   $\ell$ = length of the obstacle (cubic in shape).

Confirmation of regelation slip first came from observations beneath Blue Glacier, in Washington State, which revealed a 30-mm basal layer of clear ice with few bubbles over hollows, but which declined in thickness towards the top of bumps (Kamb and LaChapelle, 1964). Subsequently such layers have been recognized widely and are generally less than 0·5–1 m thick beneath ice at the pressure melting point. The isotopic and chemical composition, the crystal characteristics and debris inclusions all point to the process of refreezing at the base (Souchez *et al.*, 1978; Jouzel and Souchez, 1982).

An important implication follows from the recognition of two sliding mechanisms, one operating best to circumvent small obstacles and the other large obstacles. In between there is a critical obstacle size where neither mechanism is effective. This size is the range of 10–15 cm and offers the greatest resistance to sliding.

The operation of the two basic sliding mechanisms is greatly influenced by the presence of water beneath a glacier. A film of water only a few millimetres thick is sufficient to reduce the friction between ice and rock and is sufficient to double sliding velocities. In such a case the decreased number of prominences holding the glacier back engender higher stresses in the ice, thus enhancing the operation of the two sliding mechanisms. Lliboutry (1965) has stressed that the water pressure at the base of a warm glacier can be regarded as independent of the overlying weight of ice and instead depends on such factors as melting rates and the head of water. Where water pressures in cavities are sufficiently high, they may grow. A positive feedback mechanism may then come into play with growing water-filled cavities leading to higher sliding velocities which then lead to still larger cavities. Apparent evidence of such a process has been observed on the Unteraargletscher in the European Alps where a sharp increase in glacier velocity coincided with it being raised bodily by 0·6 m, presumably by the increasing thickness of a basal water layer (Iken *et al.*, 1979).

Field observations over many decades have suggested that glacier sliding is not a continuous process, but takes place in a jerky, stick–slip fashion. Slip events of up to 20–30 mm may be interspersed by periods of hours when no basal movement takes place. Robin (1976) has suggested an elegant explanation of such a process – the heat-pump effect. The high ice pressure on the upstream side of a bump causes water to be squeezed out of the high-pressure zone along the pressure gradient. When the pressure is reduced on the top and/or lee side of the bump, the water is no longer available within the ice to refreeze and maintain the lee temperature at the pressure melting point. The effect will be for cold patches of ice to exist on the top and in the lee of bumps. It is tempting to regard the stick phases as representing periods when the ice is frozen to the rock bed. For a time the ice will resist the stresses created by the movement of the glacier as a whole, but eventually failure will take place to cause a slip phase. In this scenario the bottom of a 'warm' glacier may in reality consist of warm ice interspersed by islands of small cold patches in the lee of zones of pressure melting.

A new mechanism accounting for the forward motion of glaciers emerged in the 1970s. It is the deformation of unfrozen sediments beneath glaciers (Haeberli, 1981). Boulton (1979) measured the deformation of saturated till beneath the edge of an Icelandic glacier and noted that in this case it accounted for 90 per cent of the forward movement (Figure 17.9). The effect has been likened to a glacier on 'rollerskates' and occurs when high porewater pressures within the till reduce the resistance to shear stress sufficiently for it to deform beneath the weight of the glacier. Such a process is confined to warm-based

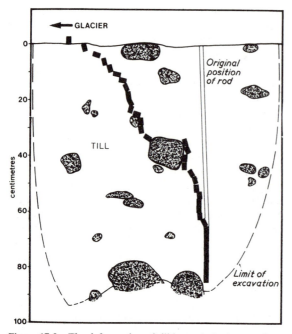

**Figure 17.9**   The deformation of till beneath the ice margin of the Breiðamerkurjökull, Iceland, illustrated by the displacement of separated segments of a rod inserted vertically 244 h previously. The till discharge here is 1122 cm³/cm in 244 h, and the effect of boulders on till deformation is clearly shown.
*Source:* Boulton, 1979, figure 7, p. 29.

glaciers moving over deformable sediments. Although the process is known to be locally important, Boulton and Jones (1979) have suggested that it could have played a major role in the marginal zones of the former European and North American ice sheets underlain by till.

### 17.3.3 Velocities

Most glaciers have velocities in the range of 3–300 m/year for most of their length, but the velocity can reach 1–2 km/year in steep ice falls. Movement at the lower velocities is largely by internal deformation. For example, the sluggish cold-based Meserve Glacier in Victoria Land, Antarctica, has a velocity of 3–4 m/year at the equilibrium line. Higher velocities reflect the component added by basal sliding which can contribute up to 90 per cent of the total. Steep, active glaciers like Franz Josef Glacier in New Zealand have velocities over 300 m/year, most of which is due to sliding.

A few glaciers flow at speeds an order of magnitude higher than these figures. For example, Jakobshavn, Isbrae, one of the largest outlet glaciers in Greenland,

flows at 7–12 km/year. Also there are a number of glaciers which experience periodic surges in which a wave of ice moves downglacier at velocities of 4–7 km/year. In these cases the surging ice velocities may be 10–100 times greater than the previous velocity. Further, the jump from normal to surging velocities takes place very suddenly.

One possible explanation of these different sets of flow velocity is that there are two modes of basal sliding, one 'normal' and one 'fast'. Budd (1975) has related the discharge of glaciers to glacier velocity and suggested that large outlet glaciers flowing from major ice sheets have sufficient ice supply to maintain the fast mode of sliding. 'Normal' glaciers have only enough ice to maintain normal flow. Periodically surging glaciers occupy an intermediate position and may have enough ice supply for them to be able to cross the threshold from 'normal' to 'fast' flow. But then they cannot provide enough ice to maintain the fast mode of flow and fall back to a 'normal' mode. Hutter (1982) has suggested that the two modes of flow may be related to the presence or absence of cavities beneath the glacier, and he has looked at the problem in terms of catastrophe theory, with sliding related to basal shear stress and the thickness of the basal water layer (Figure 17.10). At low water thicknesses the sliding rate increases regularly with shear stress. At

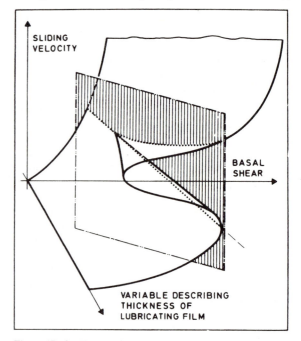

**Figure 17.10**   Two modes of glacier sliding viewed in terms of catastrophe theory; see text, for description.
*Source:* Hutter, 1982, figure 2, p. 31.

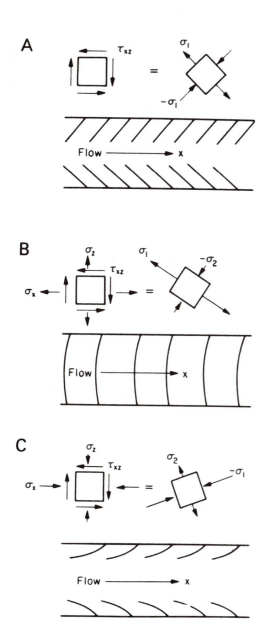

**Figure 17.11** Crevasse patterns in a valley glacier: A. effect of shear stress exerted by valley walls only (chevron); B. shear stress and extending flow (arcuate); C. shear stress and compressive flow (splaying).
*Source:* Paterson, 1981, figure 11.4, p. 227, after Nye, 1952.

critical water thicknesses sliding velocity increases with shear stress until suddenly it jumps to a much higher value. Conversely, when starting at high shear stresses, sliding velocity decreases with decreasing shear stress until suddenly it drops to a much lower value, which corresponds to a sudden slowing of the ice. This is an interesting example of a complex threshold which serves to separate two distinct process–response systems.

### 17.3.4 Ice surface features

There are several surface features which result from the flow of glaciers. The first group are crevasses, more or less vertical cracks, which reflect tensional stresses in the surface of the glacier introduced by friction with the valley side, as well as extending and compressive flow along the length of the glacier. Characteristic crevasse patterns are shown in Figure 17.11 (Plate 22). The arcuate transverse crevasses which are common in ice falls are a type of rotational slip discussed in Chapter 10.

Once formed, a crevasse is modified by glacier flow. As it moves downglacier the initial crevasse is distorted by the differential flow between the glacier centre and margin. A marginal chevron crevasse will tend to pivot on the wall and rotate downglacier while both splaying and transverse crevasses tend to be straightened. The life of an open crevasse is limited and it soon closes as it moves to an area of less tension and its place is taken by another more suitably oriented crevasse. Once closed, it is represented by a thin blue vein of ice.

The surfaces of the ablation areas of glaciers are commonly marked by foliation, a banding in the ice commonly consisting of thin, blue ice layers interspersed by thicker layers of white bubbly ice. These structures

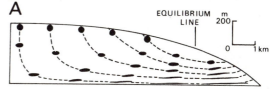

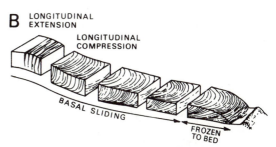

**Figure 17.12** Deformation within flowing ice: A. schematic illustration of cumulative strain at selected points along particle paths; the initial circles become deformed into ellipses during flow; B. schematic block diagram illustrating the transition from crevasse traces to foliation and shear planes in a valley glacier.
*Sources:* Hooke and Hudelston, 1978, figure 3B, p. 290; Hambrey and Müller, 1978, figure 14, p. 61.

reflect inhomogeneities in the ice which have been deformed by the normal processes of creep (Figure 17.12A). The bands may be modified sedimentary structures, reflecting the accumulation of firn in the accumulation zone, or modified crevasse traces. Figure 17.12B shows how an originally vertical transverse crevasse is first closed and then tilted forwards and bent in response to glacier flow. Eventually a succession of such crevasse traces resemble nested spoons with the strike of the foliation lying approximately parallel to the glacier margin and dipping most steeply at the margin. These structures are important in that they may form lines of weakness in the glacier which can be exploited by compressional thrusting, especially if the margin is frozen to the bed.

## 17.4 Rock debris in glaciers

The geomorphological effects of glaciers depend on their ability to pick up, transport and deposit rock debris. Figure 17.13 shows some idealized rock debris paths within a glacier. Category A describes debris which falls into a glacier and is transported without touching the glacier bottom. If the rock debris is derived from above the equilibrium line, it is buried beneath successive layers of accumulating snow as it moves downglacier. Below the equilibrium line it moves towards the glacier surface as ablation removes the overlying ice. If the rock debris falls on to the glacier surface below the equilibrium line, it will stay on the glacier surface until it reaches the snout. The material is derived from rock walls overlooking the glacier and thus is characteristic of valley glaciers.

Category B describes rock material which has experienced some transport at the bottom of the glacier. The material may be derived from the bed by subglacial erosion or indirectly from debris which falls on to the glacier surface and makes its way to the bed via crevasses, meltwater tunnels and downward flow associated with longitudinal extension or bottom melting. Once on the bottom, it may remain there until it is deposited on the glacier bed or at the glacier snout. Some, however, may find its way to the glacier surface due to upward compressive flow, especially in the region of the snout.

The two categories of debris are fundamentally different. Material which has remained in a surface or englacial transport position throughout consists of talus and other rockfall material which has escaped modification by glacial movement. Rock fragments tend to be angular and coarse and the fines unimportant, as in normal talus. Material which has touched the bottom is quite different. It has experienced crushing and abrasion. The most important modifications are the greater rounding of clasts and the higher proportion of fine matrix (Figure 17.14). The differences between the two types of material are often striking visually. In South Georgia valley glaciers are dominated by steep mountains of greywacke and discharge reddish, weathered fragments on to the glaciers. Above the equilibrium line this material is buried by snow, but reappears on the glacier surface near the snout with the reddish hue intact. In contrast, a few zones of basal material near the glacier snouts are distinctive because of their unweathered grey colour (Birnie, 1978). As will be seen, the different sedimentary characteristics of the two types of debris produce quite different landforms.

## 17.5 Processes affecting debris at the glacier sole

The processes operating on rock debris at the bottom of a glacier are difficult to observe directly. Partly as a result, understanding leans heavily on theories which are as yet only loosely constrained by field observations. There are two important issues to discuss – namely, entrainment of basal rock debris, and the forces associated with basal rock debris in motion.

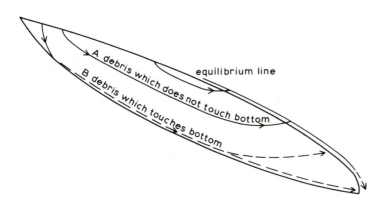

**Figure 17.13**  Rock debris transport paths in a glacier.

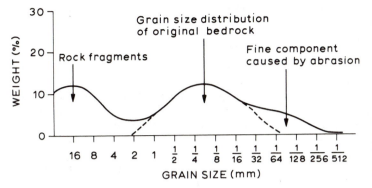

**Figure 17.14** Characteristic grain-size distribution of fine debris which has experienced transport at the bottom of a glacier.
*Source:* Haldorsen, 1981, figure 5, p. 97.

### 17.5.1 Entrainment

Two main mechanisms allow rock debris to be incorporated in the bottom of a glacier – plastic flow and freezing on. Plastic flow describes the situation where ice envelops a rock particle and exerts sufficient tractive force to overcome the friction between it and the bed. If this situation is envisaged from the point of view of the 'obstacle', then following Weertman's predictions, one would expect the drag of the overriding glacier to be most effective on particles approximating to the 'critical obstacle size' of around 10–15 cm.

The second major entrainment mechanism is freezing-on. Good evidence of such a process comes from the association of debris with zones of basal regelation ice beneath warm-based glaciers. Commonly the debris is arranged in roughly horizontal bands separated by clear ice bands. Many bands are only a few millimetres in thickness, while others may be a few centimetres across. Debris contents frequently range from 5–55 per cent of the total volume by weight. The debris is thought to freeze on to the ice in the immediate lee of depressions as a result of pressure-induced regelation (Figure 17.15).

Bands of clear ice presumably freeze on in the absence of debris. Such a process is clearly related to pressure differences associated with small bumps beneath a glacier, and the debris melts out and refreezes repeatedly as it passes from bump to bump.

The pressure-melting mechanism is attractive for thin basal debris layers but does not explain the great thickness of debris-bearing regelation ice exposed in the snouts of some glaciers or the wholesale freezing on of basal sediment evidenced by the inherited sedimentary structures within some thicker debris bands. Thus it is necessary to envisage freezing-on at a larger scale. One possible mechanism is illustrated in Figure 17.15, which shows the effect of successive freezing-on of ice on to the bottom of a glacier as it moves from a warm-based to cold-based regime. Such a situation is common in glaciers which are cold-based at their margin and warm-based further upstream.

Another situation allowing successive freezing-on is related to Robin's heat-pump effect. If some meltwater expelled by pressure melting on the upstream side of a bump is lost, for example, into a subglacial meltwater stream, then a cold patch may exist in the lee of the

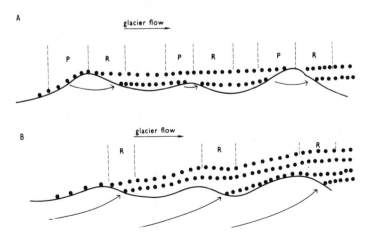

**Figure 17.15** Debris incorporated by refreezing **R** in the lee of bumps: A. beneath a warm-based glacier the layers tend to be destroyed by pressure melting **P** associated with comparable bumps and remain thin; B. beneath a cold glacier successive freezing-on can build up considerable thicknesses.
*Source:* Sugden and John, 1976, figure 8.11, p. 162, after Boulton, 1972a, figure 5, p. 10.

bump. The glacier bed which freezes on may, if sufficiently weak, be torn off by further basal sliding, perhaps during a stick–slip phase. If the meltwater is lost from several successive bumps crossed by the glacier, then there could be successive freezing-on.

Whereas Robin's heat-pump is a spatial process related to individual bumps, Röthlisberger and Iken (1981) pointed out that the process could operate over time. They suggested that increasing basal water pressures in cavities can have the effect of lifting the ice and reducing ice pressures where the glacier remains in touch with the bed. This, in turn, could allow bottom freezing at the points of contact.

In certain situations enormous erratics of the order of hundreds or thousands of metres across may be entrained in glaciers. One such erratic in Germany is $2 \times 4$ km in area and 120 m thick. Boulton (1972a) suggested that such erratics may be moved by cold-based glaciers if the $0°C$ isotherm is close to the glacier bed. If the friction at the frozen ice–rock interface exceeds that on a lower unfrozen plane of weakness in the rock, then the whole block may be dragged along. This is freezing-on with a vengeance!

Once attached to the glacier sole, there are various ways in which basal debris-bearing ice can be thickened. Folding and thrusting have been observed and are two mechanisms that are particularly important in the zone of compressive flow near ice margins. It is common to find a glacier overriding largely stagnant ice at the ice margin along a thrust plane carrying debris in the over-riding ice. In time such a process can lead to progressive stacking of debris-bearing ice. The process seems to be common where compressive flow is especially favoured, for example, when a warm-based glacier has a frozen margin and the ice moves from a sliding to non-sliding mode of flow, or when a wave of surging glacier ice ploughs into the stagnant lower reaches of a glacier (Sharp, 1982).

### 17.5.2 Rock particles in traction at the bed

Once entrained, debris becomes an abrasive which is an extremely effective agent of erosion and deposition. It is helpful to consider two theories concerning the forces associated with rock particles in traction beneath a glacier.

Boulton (1974) used an approach common in slope mechanics and assumed that the contact pressure on a rock particle against the glacier bed is related to the weight of overlying ice. These contact pressures are reduced if there is a basal layer of water with significant water pressure. In such a case the contact pressure becomes an effective pressure:

$$N = \rho_i g h - W_p$$

where: $\rho_i g h$ is the weight of overlying ice; and $W_p$ is the basal water pressure.

If the rock particle is in movement, then the drag of the particle against the bed $F$ takes into consideration the roughness of the contact and its surface area, so that:

$$F = (\rho_i g h - W_p) S_i \mu$$

where $S_i$ is the surface area of the particle, and $\mu$ the coefficient of friction.

In this analysis the force required to overcome drag is estimated from Weertman's analysis of basal sliding, and Boulton arrives at the important conclusion that the force varies with particle size (and shape). Whereas enhanced basal creep allows ice to flow round large particles and regelation slip round small particles, there is an intermediate critical size of the order of 10–15 cm which ice is unable to by-pass efficiently. These intermediate-sized particles will, therefore, tend to be transported faster than larger and smaller particles. This means that for particular combinations of ice velocity and effective normal pressure, some particles may be moving in traction, while others will have ceased to move altogether and become 'lodged'.

If we equate erosion with particles moving in traction and subglacial deposition with their lack of movement, then it can be seen that the theory has important implications for glacial geomorphology. In particular, one can highlight the following:

(1) Ice thickness has an important influence on patterns of erosion and deposition. Erosion will increase with increasing ice thickness up to a point, but then increasing frictional drag will cause lodgement.

(2) The permeability of the bed is important in influencing basal water presures. One would expect increasing effective normal stress over permeable rocks. In most cases this would favour lodgement.

(3) There is a crude sorting mechanism of the debris in traction which will influence the characteristics of glacial deposits.

Hallet (1979, 1981) used an alternative approach in which the rock particles are envisaged as essentially floating in the ice. They are buoyed up by the hydrostatic pressure in the ice which envelops them. In this analysis the key factors affecting the pressures between rock particle and bed are related to (1) ice flow towards the ice bed associated with bottom melting or ice deformation, and (2) the weight of the particle over and above the ice that it displaces. Generally the effect of bottom melting is the most important factor controlling the contact pressures for particles less than 20 cm in size, while the buoyant weight of particles is also important for larger particles. As in Boulton's model, the average shear stress $\tau$ must be balanced by the basal drag due to bed irregularities and rock fragments dragged against the bed, and can be expressed as:

$$\tau = \xi\eta U + \mu c\overline{F}$$

where   $\xi$ is a proportionality factor dependent on bed roughness;

 $\eta$ is the effective viscosity of basal ice;

 $U$ is the sliding velocity;

 $\mu$ is the coefficient of rock to rock friction;

 $c$ is the areal concentration of rock fragments in contact with the bed;

 $\overline{F}$ is the average particle–bed contact force.

The main implications for glacial geomorphology arising from Hallet's model are:

(a) abrasion is highest where basal melting is greatest;

(b) glacier thickness is irrelevant in influencing abrasion and lodgement;

(c) lodgement of particles takes place at low velocities roughly equivalent to the rate of melting – i.e. a matter of a few centimetres per year. This is in complete contrast to Boulton's prediction which envisages lodgement under velocities up to 100 m/year.

(d) the lodgement of particles is independent of particle size, again a contrast with Boulton's prediction.

At first sight it seems strange that two theories with such widely differing assumptions and predictions can survive side by side. But it may be that each model represents different but equally valid subglacial conditions. Boulton's model will apply if particles are not surrounded completely by ice and may, for instance, be valid when the glacier has a distinct layer of basal debris. In cases where particles are thinly distributed in the basal ice Hallet's model may be the more appropriate. This is a clear case where field geomorphological observation can be usefully designed to test the links between process and form.

## 17.6 Erosion by glaciers

As ice and its contained debris flow across the landscape the ground surface is modified, to produce characteristic landforms of glacial erosion and deposition. The fundamental landforms of glacial erosion are relatively few.

One can recognize *bumps* and *depressions* and these may be subdivided into streamlined and partly streamlined forms (Table 17.2).

*Whale-back* forms occur at a scale of tens to a few hundred metres and consist of bumps smoothed by ice on all sides. Sometimes their sides may slope at angles of 40°. *Rock drumlins* are bigger forms which are more streamlined in the direction of iceflow, but again they are smooth on all sides. They tend to have length and breadth ratios of 4:1, while heights vary from 5 m to 50 m.

**Table 17.2**  Landforms of glacial erosion

| Bumps | streamlined | whale-back, rock drumlin |
|---|---|---|
| | part streamlined | roche moutonnée |
| Depressions | streamlined | groove |
| | part streamlined | rock basin |

*Roches moutonnées* comprise a family of partly streamlined forms generally regarded as the hallmark of glacial erosion. They comprise asymmetric hillocks or hills with one side smoothly moulded and the other side steepened and often craggy. They occur at a wide range of scales. At the scale of 1 m it is common to find a convex, ice-smoothed surface sharply truncated by individual joint planes. Large roches moutonnées may measure several hundred metres across and their dimensions are closely related to lines of structural weakness. On their backs are smaller roches moutonnées measuring tens of metres across. At the largest scale complete hills have been steepened on one side. Perhaps the best-known examples were described by Jahns (1943) in New England. Here the tops of granite hills 300–1300 m in length and 30–50 m high have been moulded by ice. By examining the pre-glacial sheeting structures in the granite, Jahns deduced that up to 33 m of granite had been removed from the lee side of the hills by ice, compared with 3–4 m from the smoothed upstream sides.

*Grooves* comprise a family of medium-scale streamlined depressions. In the Mackenzie Valley, Canada, individual grooves may be 12 km long, 100 m wide and 30 m deep (Smith, 1948). Usually, however, they are smaller by an order of magnitude. Sometimes grooves are straight and excavated indiscriminately across the underlying bedrock structure. In other situations, especially at smaller scales, grooves may be sinuous with soft flowing outlines. The lip of a groove may be overhanging and in places there may be lunate scars interrupting its form.

Partly streamlined depressions include a little-studied but extremely common form of glacial erosion – the *rock basin*. These may vary from a few metres to many hundreds of kilometres in size. The sides of rock basins bear signs of smoothing and/or roughening by ice and their main characteristic is that they have been overdeepened.

**Table 17.3**    Tentative list of variables affecting the process of glacial abrasion

| Fundamental requirements for abrasion | 1 Basal debris |
| | 2 Sliding of basal ice |
| | 3 Movement of debris towards bedrock |
| Other factors affecting rate and type of abrasion | 4 Relative hardness of rock particles and bedrock |
| | 5 Particle shape and size |
| | 6 Debris concentration (Hallet) |
| | 7 Effective normal stress ($\rho gh-Wp$) (Boulton) |
| | 8 Efficient removal of rock flour by meltwater |

For a long time the basic processes of erosion have been described as *abrasion*, a process of rock wear associated with rock sliding on rock, and *plucking*, a process whereby discrete blocks of bedrock are removed. Although these two processes involve several different mechanisms, they remain a convenient framework for discussion.

Abrasion of a glacier bed requires three fundamental conditions before it can occur (Table 17.3). There must be basal debris, it must be sliding across the rock and it must be pressed against the bedrock. The rate of abrasion will be influenced by these and the other factors listed in the table. According to Hallet (1979), the most critical of these factors are the flux of particles in contact with the bed, the force with which the particles press against the bed and the relative hardness of the bed and abrasive particles. The average thickness of rock removed by abrasion per unit time $\overset{\circ}{A}$ can be expressed as:

$$\overset{\circ}{A} = \alpha C_r V_p F^n$$

where $\alpha$ and $n$ are empirical constants depending on the
hardnesses of the striator and the bed being striated;

$C_r$ is a measure of rock concentration;
$V_p$ is the particle velocity;
$F$ is the effective force pressing the particles against the bed.

Plucking is the result of rock failure along joint planes. Two important mechanisms seem relevant. One is wedging by the pressure of overriding rock particles. In a way this can be regarded as the same process as abrasion,

except that the rock fails in a different manner. Instead of small-scale fracturing, the fractures are larger and can follow joints. The contrast between the two modes of fracture might simply reflect the contrast between high confining ice pressures, associated with abrasion, and low confining ice pressures, associated with plucking. A second mechanism is freezing-on. If the ice base freezes to a rock surface, it may apply sufficient drag to pull off the rock fragment. Such a process occurs at a small scale associated with regelation in the lee of bedrock bumps. More important, it is likely to occur at a larger scale when associated with rising basal water pressures in basal cavities; in such situations the high water pressures support more of the weight of the glacier and this allows freezing in intermediate areas where the ice is in contact with the bed (Röthlisberger and Iken, 1981). The increased basal slip associated with such changes offers a powerful means of removing the rock particles (Figure 17.16). When considering the forces required to remove blocks by plucking, it is likely that high water pressures in the bedrock will contribute to the weakening of the rock – as occurs in rock failure beyond the limits of glaciers. Addison (1981) has stressed the importance of this factor in the failure of rock masses in Snowdonia, North Wales.

Armed with these general considerations, it is possible to predict the morphology of subglacial bumps eroded by glaciers. Figure 17.17 is an idealized example. Abrasion is dominant on the upstream side of the bump. Here pressure melting takes place and provides the high contact forces required for abrasion. The ice diverges round the bump as envisaged in Weertman's enhanced basal creep, and this thinning also has the effect of bringing debris into contact with the rock. The abrasion is marked by scratches or striations at a variety of scales which reflect the divergent movement of ice around the bump. The striations are likely to be deeper on the upstream side of the bump than on the sides of the bump because high rates of pressure melting increase the

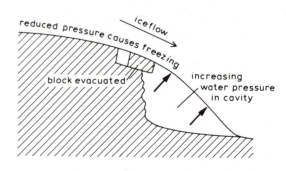

**Figure 17.16**    Block removal as a result of increasing water pressure in a cavity.

contact forces. On the other hand, there may be a higher density of scratches on the sides because of the diversion of the thin debris-bearing basal ice around the bump rather than over it. On the top of the bump one might predict that stick–slip glacier sliding associated with the presence of a cold patch of ice might produce discrete fractures associated with periodic slip phases. Such a location would be the obvious place for crescentic friction cracks, long-observed characteristics of glacial abrasion. Streamlining of the upstream side of the bump will respond to differential pressures and sliding velocities over the obstruction tempered in the real world by the effects of rock structure.

The lee side of the bump is likely to experience different processes. The ice pressures will be less than on the up-ice side and in some circumstances the ice may be separated from the rock by an air-filled or water-filled cavity. Plucking by wedging will reflect the pressure differences across the obstacle. It will be at a maximum at the top of the leeside where the pressure difference is greatest, especially if there is a cavity. Plucking by regelation will occur in the lee of the smallest obstacles. If the ice presses down on the lee side of the obstacle, there could be light striae swinging round the obstacle. Such striae would be more likely in the case of Boulton's theory of abrasion than in Hallet's theory.

There are surprisingly few field observations to test the above predictions. Yet those that do exist agree well. Measurements of the pressure on either side of artificial bumps beneath 160 m of ice under an ice fall on Bondisbreen, Norway, reveal that differential pressures of 30 bar existed across the obstacle (Hagen *et al.*, 1983). Furthermore, one boulder of about 0·3 m³ moved across one of the sensors on top of the bump and recorded a pressure of 90 bar before breaking the sensor. These pressures are far in excess of the 16 bar overburden pressure. One interesting feature was that a low pressure area moved downglacier in front of the boulder. This is assumed to represent a cavity formed in the lee of the boulder which was being retarded by friction with the base. The implication is that there may be locally very high pressure gradients in the underlying rock, facilitating its fracture.

Observations on striations and friction cracks agree with the predictions above. Rastas and Seppälä (1981) measured several hundred roches moutonnées in Finland. They found that striations were diverted round the upstream flanks of roches moutonnées, that friction cracks were confined to the upper surface of the upstream side and that plucking was confined to the downstream side (Figure 17.17). In Skye Smith (1982) found that striations are deeper on the top of roches

moutonnées than on the sides.

The morphology of roches moutonnées is closely influenced by bedrock structure. Rastas and Seppälä (1981) note that in Finland the dimensions of individual rock bosses reflect dominant joint spacing in the underlying bedrock. In west Greenland and Scotland rock structures have been found to influence the morphology of both the upstream streamlined flank and the plucked lee side in predictable ways (Gordon, 1981) (Figure 17.18).

One can summarize this discussion of the morphology of bumps sculpted by glacial erosion by suggesting that there is likely to be a continuum of forms. At one extreme there are bumps where the differential pressures between the upstream and lee side are minimal. In such cases one can predict a streamlined form moulded mainly by abrasion on all sides, such as whale-backs and rock drumlins. At the other extreme are bumps where the differential is great and, as a result, there is asymmetry between abrasion on the upstream side and plucking on the lee side. These are the classic roches moutonnées. Whale-backs are likely to occur where the ice is thick, slowly moving and where bedrock structures favour small bumps. Asymmetric roches moutonnées are more likely to occur where the ice is thinner and fast-moving, and where bedrock structures form high amplitude bumps.

The morphology of grooves seems adequately explained by the action of abrasion. Given an irregular bed, the thin basal debris layer will tend to be concentrated in between bumps. The accelerated flow and higher flux of rock debris will favour abrasion. The deeper the groove, the greater this channelling effect of debris will be. The overhang apparent in some areas of grooving may be explained by the greater debris concentration at the bottom of the ice, allowing greater rates of abrasion at the bottom of the groove. Lunate marks associated with the grooves are explained by enhanced basal creep in front of obstacles. Upstream of rock knobs a scour mark forms and is analogous to wind scour upstream of obstacle, presumably reflecting locally high ice velocities associated with diversion. Scour marks not related to bedrock obstacles may reflect diversion round a temporarily lodged boulder. In most cases small-scale groove morphology can be related to bedrock structure, notably differential rock hardness and jointing.

Irregular rock basins are not straightforward to explain, in spite of their ubiquity. It is known that their locations conform closely to structure. Rudberg (1973) has shown that in Sweden fracture frequences are denser beneath depressions than in rock on either side, and the association of lakes with joint directions has long been recorded. On a larger scale lake basins are preferentially

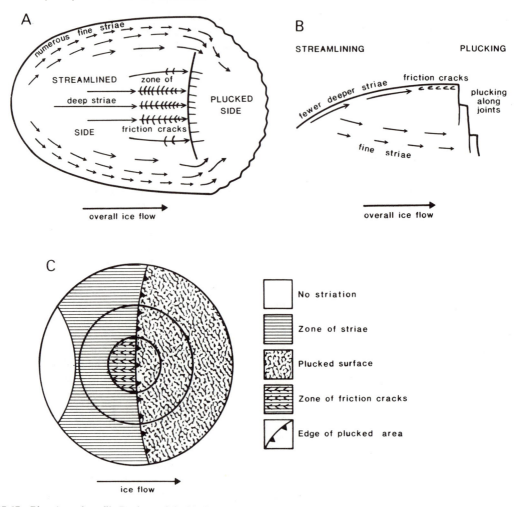

**Figure 17.17**   Plan A. and profile B. views of the ideal roches moutonnées; C. stereographic model of the distribution of different abrasion features on roches moutonnées in Finland. *Source of C:* Rastas and Seppälä, 1981, figure 8, p. 162.

associated with softer rocks in areas of lithological differences. Thus in part of the eastern Canadian shield 60 per cent of the area underlain by schist is covered by lake basins, but only 1 per cent of the gabbro (Brochu, 1954). In Minnesota lakes selectively form over the sites of weak rocks (Zumberge, 1952).

It is tempting to suggest that depressions form over the site of fracture planes in the rock because it is here that the normal process of abrasion can be supplemented by plucking. In such situations plucking is favoured by the weaker than average rock and the higher than average concentrations of debris which will allow wedging. Freezing on as a plucking mechanism is, however, unlikely to be important. Depressions over the sites of

weak rocks result from differential rates of abrasion. Hard rock 'tools' flowing over weak rock can accomplish rapid abrasion and form basins. Conversely, till derived from weak rock areas cannot abrade harder rock. Thus the greater the differential hardness of adjacent lithologies, the greater the effect of differential overdeepening will be.

## 17.7   Deposition by glaciers

The debris obtained by glacial erosion of the bed or from rockfall from surrounding rock walls can have several destinations. It may be carried to the glacier margin, deposited on the glacier bed or lost into icebergs or sub-

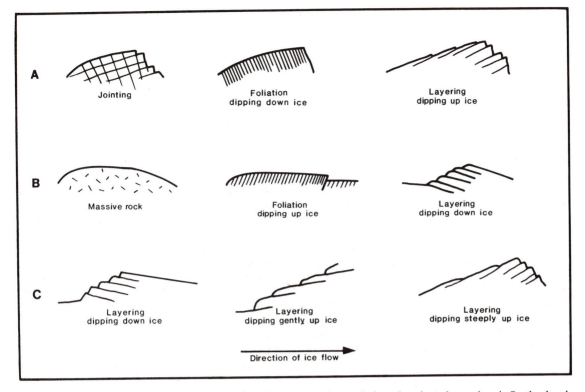

**Figure 17.18** Some relationships between structure and roche moutonnée morphology, based on observations in Scotland and Greenland: A. plucked lee-side slopes: B. abraded slopes; C. abraded and plucked stoss slopes; in all cases the ice moved from left to right.
*Source:* Gordon, 1981, figure 6, p. 60.

glacial meltwater flow. This section, first, discusses the main processes involved in deposition directly by a glacier, and, second, the main landforms which result.

Debris deposited directly by a glacier is known as till. Table 17.4 shows a classification based on the processes of debris release and the position of the debris during deposition. The two important domains are subglacial and supraglacial. Following Lawson (1981a), the processes are subdivided into those which are primary and influence the nature of the sediment directly and those which are closely related secondary processes, which modify the sediment.

*Meltout* till forms by the direct release of debris from a body of stagnant, debris-rich ice by melting of the interstitial ice. When it occurs subglacially, the debris is let down on to the subglacial surface without modification. Thus it retains a fabric inherited from transport in the ice with preferentially oriented pebbles. The deposit can consist of structureless pebbly, sandy silt, discontinuous laminae and bands of sediment which are frequently draped over large clasts (Figure 17.19). These charac-

teristics are the direct result of the melting out of debris-bearing regelation ice. Meltout also occurs at the surface, but here secondary processes of flow and slumping may affect the sediment.

In cold arid environments, such as Victoria Land in Antarctica, the interstitial ice may be lost by *sublimation* rather than melting (Shaw, 1977). In view of the lack of meltwater the deposit may retain textural and structural characteristics inherited from glacial transport.

The third primary depositional process is subglacial *lodgement*. Particles lodge when the frictional resistance with the bed exceeds the drag of the overlying ice. Where the tills are deposited beneath the weight of overlying ice, they tend to be characterized by a high degree of compaction, high shear strength and low porosity. Foliation slickensides (striations on slip planes), fine lamination and horizontal joints may reflect the effect of the overriding ice (Figure 17.20). The fabric of the till usually shows a preferred orientation with elongated pebbles parallel to the direction of iceflow. Occasionally large boulders may have been moulded into minor roche

**Table 17.4**   Classification of till based on processes of debris release and position in relation to the glacier

|  | *Primary* | *Secondary* |  |
|---|---|---|---|
| Subglacial | meltout<br>sublimation<br>lodgement | deformation | settling<br>through<br>standing water |
| Supraglacial | meltout<br>sublimation | sediment flow<br>gravitational slumping | |

*Source*: Lawson, 1981.

moutonnée shapes as a result of overriding during lodgement (Sharp, 1982a).

The subglacial conditions favouring lodgement are not clear largely because the models of Boulton and Hallet yield contrasting predictions. Both theories agree that the friction provided by a rough bed helps to arrest debris and this favours lodgement in hollows and against irregularities. Boulton's analysis suggests that lodgement is favoured by high effective pressures. Below most glaciers these are high when water pressures are low, and one would expect lodgement to occur most readily on permeable rocks. Further, Boulton envisages lodgement affected by particle size with large and small particles

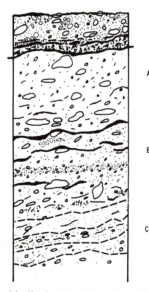

**Figure 17.19**   Idealized cross section of typical features of meltout till at Matanuska Glacier: A. structureless pebbly sand silt; B. discontinuous laminae, stratified lenses and pods of texturally distinct sediment in massive pebbly silt; C. bands of texturally, compositionally, or colour-contrasted sediment. Laminae and layers may appear draped over large clasts; pebbles are preferentially oriented throughout the deposit. *Source:* Lawson, 1981b, figure 3, p. 79.

lodging first and intermediate sizes last. Large boulders, once lodged, increase the roughness of the bed and impede further boulders, thus accelerating the process of lodgement and causing clusters of coarse boulders within the deposit. Hallet's theory, on the other hand, predicts that effective normal pressures and size sorting are unimportant. Instead low forward ice velocities of the order of a few centimetres per year are critical. Lodgement affects all particles regardless of size. Once lodged, the increase of bed roughness encourages more lodgement. This reliance on low velocities implies that lodgement will only occur beneath slowly moving ice, for example, at the glacier margins, the interior of icecaps, or perhaps at the very end of a glaciation just before a possible stage of stagnation. The existence of two models with clearly contrasting predictions offers great scope for testing by field geomorphologists.

The secondary processes indicated in Table 17.4 are intimately associated with the deposition of till. *Deformation till* has been deformed by glacier movement after its primary deposition. The most common situation is beneath glaciers where lodgement occurs. Such a layer is massive, relatively poorly consolidated and, when saturated, easily collapses beneath the weight of a person! This is the 'mud' surrounding so many glaciers. In front of Skallafjellsjökull, in Iceland, Sharp (1982b) found the layer was consistently present and up to 70 cm in thickness.

*Sediment flow* and *gravitational slumping* are two processes associated with supraglacial moraine. Sediment flows occur when fine-grained, basal debris is exposed on the surface of a glacier. They are sometimes called flow tills (Hartshorn, 1958; Boulton, 1968). The till becomes saturated and flows down the local ice slope. Lawson (1982) studied such processes on the Matanuska Glacier in Alaska and recognized four main types of flow, depending on the proportion of water they contained and, as a result, the amount of sorting present (Figure 17.21). At one extreme a plug of till flows over a thin shear zone lubricated by water (type 1). At the other extreme the water content is sufficiently high for lique-

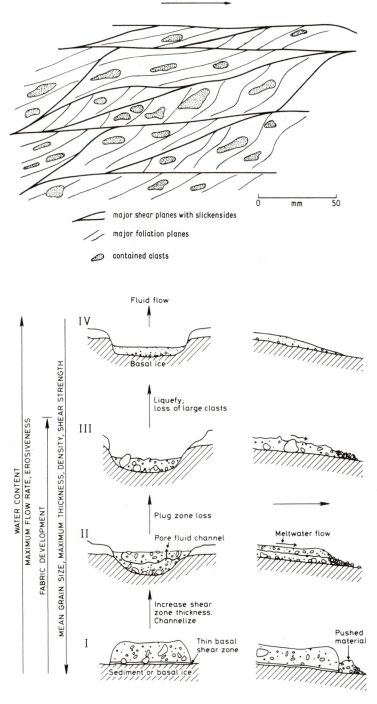

direction of ice movement

/ major shear planes with slickensides

/ major foliation planes

contained clasts

0        mm        50

**Figure 17.20** Schematic diagram showing the structures in lodgement till.
*Source:* Boulton, 1970, figure 8, p. 241.

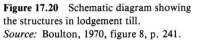

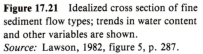

Fluid flow

IV

Basal ice

Liquefy;
loss of large clasts

III

Plug zone loss

Meltwater flow

II

Pore fluid channel

Increase shear
zone thickness.
Channelize

I          Thin basal
shear zone

Pushed
material

Sediment or basal ice

WATER CONTENT

MAXIMUM FLOW RATE, EROSIVENESS

FABRIC DEVELOPMENT

MEAN GRAIN SIZE, MAXIMUM THICKNESS, DENSITY, SHEAR STRENGTH

**Figure 17.21** Idealized cross section of fine sediment flow types; trends in water content and other variables are shown.
*Source:* Lawson, 1982, figure 5, p. 287.

faction and for grain sorting to occur. Such flows most commonly originate at the base of ice slopes which are retreating due to ice melting. The scarp slope may be arcuate in shape. The four types of flow have distinct morphological and sedimentary characteristics (Figure 17.22). Most are from tens to hundreds of metres in length. It is difficult to appreciate the scale and importance of sediment flows. Where basal sediment is on the glacier surface, all slopes of less than 1° may have been affected by flow. They occur in the marginal zones of certain glaciers, especially those with strong upward compressive flow associated with a cold margin or surging. On the Matanuska Glacier they account for 95 per cent of the till.

Where supraglacial debris on the glacier surface has been derived from the valley sides and has not touched bottom, the secondary process of sedimentation is gravitational slumping. Such debris does not contain enough fine material to support flow. Instead it is stable on relatively steep ice slopes and melting causes part periodically to slump down the ice slope. On glaciers in South Georgia Birnie (1978) found that debris which originated from the valley side was characterized by a hummocky ice topography with slopes of 35°. Coarse angular debris accumulated in depressions on the ice surface. These angles contrasted with maximum slope angles of only 15°–20°

associated with basal debris in immediately adjacent positions.

A further secondary process is settling through a water column. This is the situation common around the Antarctic where basal debris melts out from the bottom of an ice shelf or icebergs and falls to the sea floor. Anderson *et al.* (1980) distinguished such a marine deposit from normal basal tills in the same sea area on several criteria (Table 17.5). Two distinctive characteristics are the horizontal pebble fabric of the dropstones and the distinctive marine microfauna.

## 17.8 Landforms of glacial deposition

The processes that release debris from a glacier often construct distinctive landforms. The morphology and the sedimentary characteristics depend on the blend of processes involved at any particular location, the dynamics of the glacier and position in relation to the glacier. These is a huge and often bewildering range of literature on glacial sedimentary landforms and here it is proposed to highlight only the key forms. Figure 17.23 shows a simple classification based primarily on whether the glacier is actively flowing or stagnant, and secondarily on whether the landforms are subglacial or at the glacier surface margin.

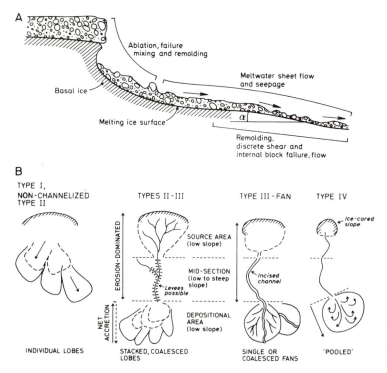

**Figure 17.22** The geometry of fine-grained sediment flow types. A. Most typical situation for sediment flow initiation; types II and III are primarily the result of failure and flow of downslope edge of sediment accumulated at base of laterally retreating ice-cored slope; type I flows result from failure in a thin zone at the base of a sediment pile with low-water content. B. Idealized characteristics of sediment flow system and modes of deposition; in general, source and depositional areas are of low-angled slope (1° to 7°), whereas the mid section may be steep and locally irregular in slope; stacking and coalescing of deposits is common; different flow types may originate and be deposited from the same source area; preservation of the system and sediments is limited to the depositional area; meltwater availability determines the extent of preservation.
*Source:* Lawson, 1982, figures 3b and 13, pp. 282, 296.

**Table 17.5** Contrasts between normal basal till and basal debris which has settled on the sea floor from floating ice

| Normal basal till | Sea floor deposition from floating ice |
| --- | --- |
| massive | crudely stratified |
| poorly sorted | sorted mud fraction |
| texture homogeneous downcore | texture inhomogeneous downcore |
| mineralogically homogeneous downcore | mineralogically inhomogeneous downcore |
| unfossiliferous | distinctive microfossil assemblages |
| overcompacted | normal compaction |
| variable pebble fabric | horizontal pebble fabric |

*Source:* Anderson *et al.*, 1980.

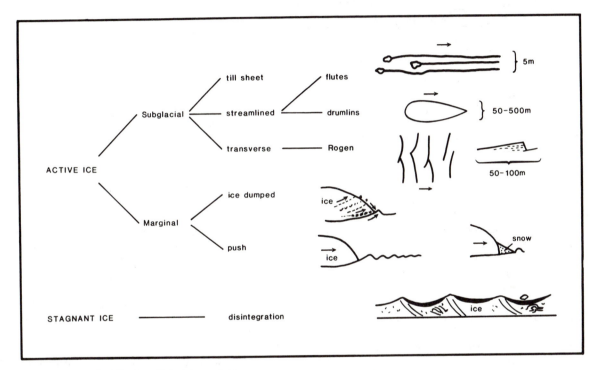

**Figure 17.23** A classification of glacial moraines.

Subglacial landforms formed beneath an actively flowing glacier are the result of lodgement and deformation at the base of a warm-based glacier. Bottom melting provides a supply of debris which is lodged on the glacier bed. The most widespread yet least obvious landform is the till sheet. Such sheets, 0·5 m to several metres thick, may mantle a landscape with a spread of till with uniform sedimentary characteristics. The uniformity may persist over thousands of square kilometres, as for example in Iowa and Illinois (Kemmis, 1981). Evidence of a basal origin is often revealed by the way local bed material is incorporated into the base of the till sheet. The

origin of such till sheets is far from clear. Kemmis (1981) explained the uniformity of the sediment by thorough mixing of the sediment at the glacier base owing to repeated phases of melting and freezing during transport. Presumably the uniform thicknesses often revealed by the sheets means that depositional processes operated uniformly over large areas. In the case of thick till sheets this may be the result of constant bed and glacier conditions over long periods of time. On the other hand, thinner sheets could represent the rather sudden deposition of the glacier bedload, for example, at the end of a glaciation.

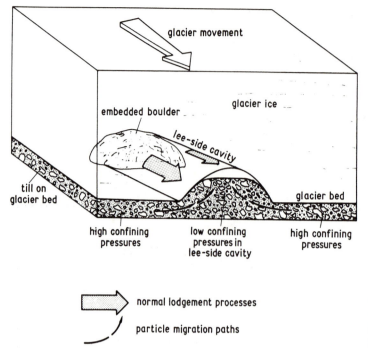

glacier movement

glacier ice

embedded boulder

lee-side cavity

till on glacier bed

glacier bed

high confining pressures

low confining pressures in lee-side cavity

high confining pressures

→ normal lodgement processes

particle migration paths

**Figure 17.24** An explanation of flute formation.
*Source:* Sugden and John, 1976, figure 12.2, p. 239.

A number of characteristic landforms are the result of selective deposition due to spatially varying bed or glacier conditions. Such selective deposition commonly forms mounds and these may conveniently be sub-divided into those which are streamlined parallel to ice-flow and those which are transverse to iceflow. Glacial flutes are small-scale examples of the former group and are well exposed around retreating glacier margins. Generally they form ridges a few tens or hundred metres long and up to 2 m high. They build up in the lee of boulders which become lodged on the bed and thereby create a cavity or low-pressure zone of ice in their lee. A variety of processes, such as deformation of the under-lying till in relation to pressure gradients and lodgement build up the ridge (Figure 17.24). The importance of boulders in the formation of flutes may be illustrated by the way flutes are diverted round boulders which are themselves creating new flutes (Figure 17.23). Flutes seem especially common at the edge of glaciers where boulders lodge, and where the relatively thin ice is rigid and unable to suppress the pressure differentials set up in the lee of the boulder.

A common larger form related to overriding ice is the drumlin, a streamlined mound of glacial debris, gen-erally 5–50 m high and 10–3000 m long. Chorley (1959) has drawn attention to their characteristic form which, like an aerofoil, is blunt on the upstream side and tapered

on the lee side, offering a minimum of resistance to over-riding ice (Figure 13.25). Although such classic forms are by no means uncommon, irregular forms abound and in some cases small drumlins build up on the backs of larger ones. They are often composed wholly of till, but there are many instances of others with gravel or bedrock cores. There are two main problems to solve; first, the cause of the initial deposition, and, second, the cause of the streamlining. In view of the fact that drumlins commonly occur in fields, the original location of depo-sition within a field is likely to reflect local factors, such as those induced by changes in bed roughness. It can be inferred that such local factors are sufficient to arrest the bedload and cause deposition. At one extreme a rock bump is likely to encourage lodgement in its upstream (stoss) side, while deposition may also be expected in the zone of low ice pressures in the lee. Thus stoss and lee moraines (Hillefors, 1973), and the classic crag and tail, may be seen as members of the same group of landforms as the rock-cored drumlins illustrated in Figure 17.25. At another extreme changes may take place within the basal till which increases its stiffness or resistance to movement by overrunning ice. Possibilities are the build-up of a 'traffic-jam' of boulders behind one large lodged boulder, an increase in the stiffness of the till due to the escape of subglacial meltwater into permeable bedrock or gravel (Menzies, 1979) and a rapid consolidation of

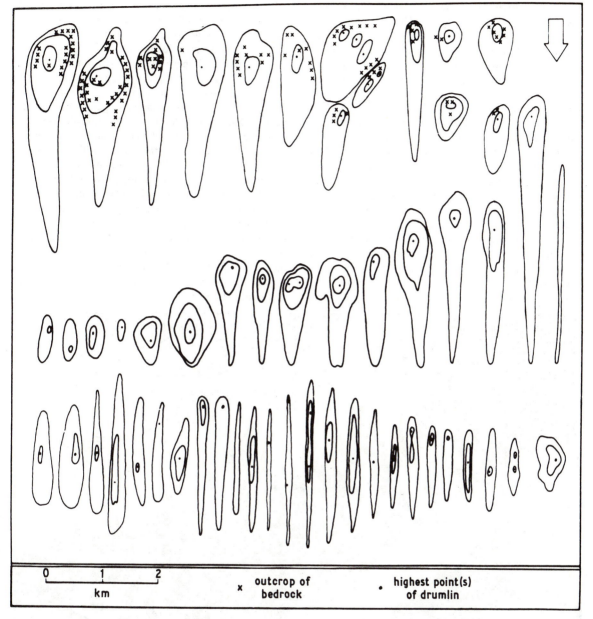

**Figure 17.25** Types of drumlin in central Finland.
*Source:* Glückert, 1973.

the till in response to mechanical changes related to changing stress levels (Smalley and Unwin, 1968). Once any of the above or other factors cause initial lodgement, the bump that results increases bed roughness and so accentuates the process of lodgement. Further, the streamlined form can be explained in terms of overriding ice removing or preventing small-scale irregularities from

building up during a process of progressive lodgement. Such irregularities tend to locally increase ice pressures and these, in turn, are sufficient to remove the irregularity (Plate 23).

There is a whole family of moraine ridges which have been constructed transverse to ice flow in a subglacial position. They are commonly termed cross-valley moraines or Rogen moraines after a lake in Sweden where they abound (Figure 17.26). The moraines consist of asymmetric ridges with the gentle slope facing up-glacier. The moraines frequently consist of discrete ridge sections, some of which are arcuate and slightly concave upglacier. Individual ridges are commonly 5–20 m high and 100–3000 m long. The ridgetops have often been fluted or drumlinized in the direction of iceflow. The basal till often bears evidence of shearing and thrusting and, indeed, may contain slabs of bedrock which have been removed from the underlying bedrock and stacked one above the other (Moran, 1971). One likely explanation is that the moraines result from a zone of thrusting beneath the ice. The basal ice and its bedload is arrested in one place and becomes stagnant. As a result the upstream ice shears over the obstruction along a thrust plane. This, in turn, leads to further lodgement on the upstream side of the obstruction as a result of the increased bed roughness. If the pattern of events is repeated, then a succession of transverse forms can be created. If this explanation is approximately correct,

then one can predict that transverse moraines will be associated with locally intense compressive flow associated, for example, with a change from a warm-based sliding basal regime to a cold-based non-sliding regime, a change from an impermeable to permeable bed which reduces the basal water pressures, flow across a topographic depression, or a submarginal position near a glacier snout. Perhaps the basal ice simply becomes too heavily charged with debris to deform any longer.

Active glaciers which end on land build up moraines at the ice margin. The size of the moraine will depend on the period that the margin lies in the same location and also the amount of debris being transported to the margin by the glacier. Thus a fast-flowing glacier bringing a large amount of debris to a stable margin will build a prominent moraine ridge, while a slow-flowing debris-free glacier may produce little more than the occasional boulder. The former type of glacier is common in high mountains of the world, while the latter type is exemplified by the cold-based glaciers of Victoria Land, Antarctica (Plate 24).

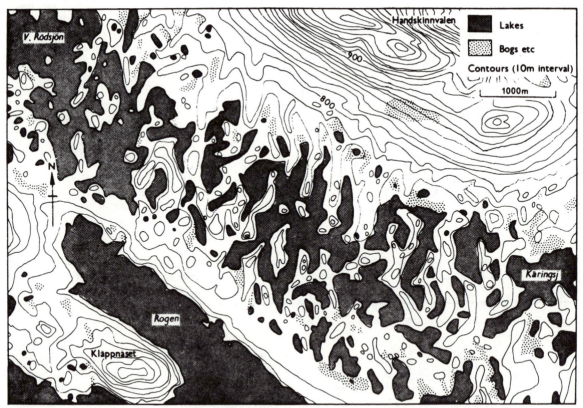

**Figure 17.26** Map of part of the classic Rogen moraine area around Lake Rogen, Sweden; the direction of ice movement was from south-east to north-west.
*Source:* Sugden and John, 1976, figure 12.7, p. 244.

The normal 'ice-dumped' marginal moraine is a complex landform reflecting a combination of many different processes. Much rock debris may melt from the glacier surface and slump or flow down to the ice margin; the ridge may consist of basally derived material raised to the ice surface by compressive flow or material which has not touched the bottom. Part of the moraine is likely to consist of lodgement or deformation till formed beneath the ice margin. It is important to note that such marginal moraines also include the lateral margins of a glacier in the ablation zone. Thus they typify most terminal and lateral moraines.

A second very common type of ice-marginal moraine results from deformation of sediments by the active forward motion of the glacier snout. Glaciers tend to advance in winter when ablation rates at the snout fall to a minimum. This advance may physically push ice-marginal sediments into a small ridge up to 2 m high. Sometimes it may push a frontal snowbank which itself deforms sediment (Birnie, 1977). If a glacier is in overall retreat, such annual advances are marked by a succession of small ridges, whose spacing can be directly related to annual variations in net ablation. Often the ice-pushed sediments are mixed with normal ice-dumped debris. At a larger scale push moraines may involve a large area of sediments in front of a glacier. Some of the best examples are associated with surging glaciers where longitudinal stresses can build up as the wave of surging ice ploughs into the stagnant marginal area. Such stresses can cause shearing, overthrusting and arcs of folded marginal sediments, glacial and proglacial (Sharp, 1982a).

Disintegration moraine describes a situation where debris-bearing ice is stagnant and melts *in situ*. Under such circumstances moraines form in relation to the processes of melting at both the base and the upper surface. Basal meltout till can be expected at the base. At the surface the processes will depend on the nature of the debris. If it is basal till, then sediment flow processes will dominate, while if it is material that has avoided contact with the bottom, slumping will predominate. While the ice is still present, debris-bearing zones will form high points with the debris shielding the underlying ice from melting. Instead the debris will flow or slump into depressions. When the ice has disappeared, there will be an inversion of relief with the former accumulations of debris in ice depressions now forming the hummocks (Figure 17.27). In certain situations these are exceedingly complex patterns of landforms and sediments.

Disintegration moraine is favoured by stagnant ice conditions combined with an abundance of surface debris. Such conditions characterize surging glaciers where stagnant ice exists in the lower part of the glacier between surges and is heavily charged with debris as a result of upward compressive flow (Clapperton, 1975). It is no surprise to notice that the best descriptions of disintegration moraine, for example in Svalbard (and Figure 17.27), come from such surging glaciers. But disintegration moraine is also important in other situations, for example, on glaciers with frozen margins causing strong compressive flow near their snouts, or on those mountain glaciers which are bounded by stagnant lateral moraines.

## 17.9 The glacier meltwater subsystem

Meltwater is an integral subsystem within most glaciers and fulfils several critical roles. It is the main ablation product of most glaciers; it is intimately involved in glacier sliding and it is responsible for flushing out debris from the ice–rock interface and carrying it beyond the confines of a glacier.

Glacier meltwater is derived from melting on the surface and within a glacier. Surface sources are more important than internal sources by an order of magnitude and are usually highly seasonal in character reflecting the effect of summer ablation on the glacier surface. In effect the bulk of the annual precipitation of a glacier may be discharged in only a few summer months. Meltwater of internal and basal origin is small in comparison but may be produced throughout the year. Internal sources involve ice melting in conduits and water produced by internal deformation of ice at the pressure melting point. Basal water is derived from geothermal heat melting the base of ice at the pressure melting point. On average this heat is enough to melt a layer of ice about 6 mm in thickness. Although not strictly meltwater, a varying proportion of the water flowing through a glacier in summer may be derived from rainfall or groundwater flow in a glacier catchment.

Meltwater soon finds its way into channels. On glaciers consisting of cold ice the meltwater is unable to penetrate without freezing and tends to be confined to the surface and the sides. When the ice is warm, the water penetrates and flows in tunnels within the glacier. Surface streams tend to flow in distinct channels less than about 2 m in width. The channels may exploit structural weaknesses in the ice and display straight sections or they may meander freely. The channel sides excavated in the ice are exceedingly smooth and water velocities are higher than is normal for the gradient. Such a relationship is readily experienced by anyone unfortunate enough to fall into even a small stream, only to find themselves swept rapidly downglacier! On glaciers at the pressure melting point this pursuit is not recommended since surface

A

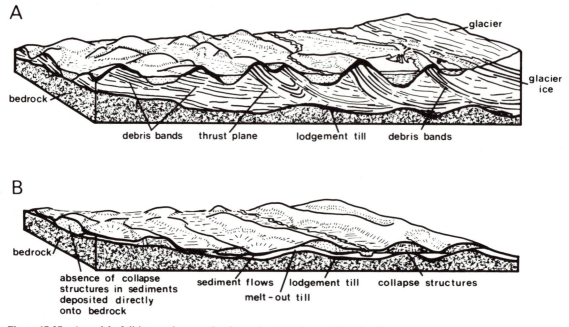

B

**Figure 17.27** A model of disintegration moraine formation A. before and B. after ice melt.
*Source:* Boulton, 1972b, figure 3c and d, p. 367.

channels quickly find their way into the glacier via a vertical cylindrical shaft – a *moulin*. Within warm ice the meltwater makes its way into discrete drainage tunnels, usually leaving the glacier via one or several streams flowing at the ice–rock interface. Such discrete systems and their lag times are well illustrated by dye-tracing studies, whereby dye is inserted into a moulin and later picked up at the glacier margin. Stenborg (1969) mentioned an average lag of 40 min for water to flow 1·7 km in Mikkaglaciären, in Sweden. Once at the glacier bed, surface meltwater mixes with water originating from the base. This latter water is chemically enriched, and Collins (1979) has shown how this may be used to assess the proportion and transit times of water from basal and surface sources.

There is considerable evidence that water levels within a glacier are often linked to form a piezometric surface. Regular fluctuations of water pressures in subglacial locations have been correlated with changing water levels in boreholes and meltwater discharge. For example, Vivian (1970) noted that in summer there was a regular 3–6 bar diurnal variation in water pressure beneath Glacier d'Agentière in the French Alps, with high values coinciding with high melt conditions in the late afternoon.

Conduits within a glacier which are full of water are believed to have a circular cross section. The dimensions

of the tunnel will assume some sort of equilibrium between two processes. On the one hand, ice pressures will tend to close the tunnel through ice deformation, a process which would close a tunnel over a matter of months. On the other hand, water flowing in the tunnel will melt the walls through heat generated by friction and from any warmth carried into the glacier in terms of positive water temperatures. This process can widen a tunnel significantly over a matter of hours.

Although the balance of these two sets of processes explains how a tunnel is maintained, once formed, it does not explain how a tunnel is initiated. Shreve (1972) suggested that the water first flows through the ice via grain intersections. Bigger passages tend to get bigger at the expense of smaller ones, largely because more heat relative to wall area is carried by water in the larger passages than in the smaller ones. Assuming that there is equilibrium between ice and water pressures in the ice, the direction of flow of the water is controlled primarily by the surface slope of the glacier, and secondarily by the slope of the underlying topography. The pressure gradient driving the water is of the potential $\phi$, which is given by the equation:

$$\phi = \phi_0 + P_W + \rho_W g z$$

where $\phi_0$ is an arbitrary constant; $P_W$ is the water pressure in the conduit; $\rho_W$ is the density of water; $g$ the acceleration due to gravity; $z$ is the elevation of the point considered. Under an ice sheet or icecap the subglacial water flow is approximately radial in the direction of iceflow, but there are deviations in sympathy with the underlying topography as the water avoids hills and bumps and flows in valleys or through saddles. In a valley glacier the flow runs down glacier in mid-valley. However, in the ablation zone where the transverse glacier surface profile is convex, subglacial streams may, if the valley floor is sufficiently wide and gentle, flow at either side of the glacier. The steeper the slope of the glacier sides and the gentler the rock floor, the stronger this tendency is.

There is still considerable uncertainty about conditions at the bottom of a glacier. However, it seems likely that a series of channels, some incised into the ice, some into bedrock, discharge the bulk of surface meltwater flow. In between these channels the ice is separated from the bed by a film of water associated with bottom melting and sliding. On the basis of the chemical deposits associated with this flow, Walder and Hallet (1979) argue that the latter is a largely independent system. Movement of a glacier over an irregular bed means that subglacial channels may be pinched out while water may be stored in cavities which become separated from the main system. It is possible that considerable volumes of meltwater can be in storage at the bottom of a glacier at any one time. Under certain conditions the storage reservoir may be a permanent feature. This applies in the case of subglacial lakes. Radio echo-sounding has allowed the discovery of lakes occupying rock basins beneath parts of the Antarctic ice sheet. One such lake measures 180 × 45 km in size (Robin *et al.*, 1977).

The discharge of glacial meltwater streams fluctuates greatly from time to time. On a short time scale there are fluctuations lasting a matter of seconds or minutes. These are thought to be related to the tapping and drainage of pockets of water in the glacier. The best-known fluctuation is diurnal, one which is responsible for marooning numerous fieldworkers at the end of a day. Discharge is low in the early morning and rises in the late afternoon or evening (Figure 17.28). The diurnal fluctuation is suppressed in winter but increases towards late summer when daily ablation rates are highest. Elliston (1973) made the important point that, however characteristic the fluctuation, it is superimposed on a baseflow accounting for the bulk of the discharge. The peaks are related to the height and gradient of the piezometric surface in a glacier. The daily ablation has the effect of topping up the meltwater reservoir and thus increases the rate of outflow from the snout.

Another marked fluctuation is seasonal. When compared to ice-free river basins, glaciers have the effect

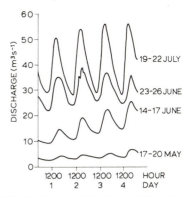

**Figure 17.28**　Discharges of the Matter–Vispa River, Switzerland during four four-day periods in May–July 1959 showing how baseflow and diurnal fluctuations increase during the summer melting season.
*Source:* Elliston, 1973, figure 1, p. 80.

of storing winter precipitation and discharging it in the summer. For example, the meltwater discharge beneath Glacier d'Argentière increases from 0.1 to 1.5 m³/s in winter to 10–11 m³/s in summer. A special category of meltwater fluctuation occurring on an annual or longer-term basis concerns the periodic draining of glacier-dammed lakes. Typically such lakes drain suddenly within 24 hours as the outlets through the glacier are initiated and quickly widened. In Iceland such an outburst is known as a *jökulhlaup*. In 1934 the drainage of Grímsvötn lake was described vividly by Thorarinsson (1953, p. 268): 'forty to fifty thousand cubic metres of muddy, grey water plunged forth every second from under the glacier border bringing with it icebergs as big as three storeyed houses. Almost the whole of the sandur or outwash plain, some 1,000 km² in area was flooded.'

The sediment loads of meltwater streams are legendary and easily believed by anyone looking at the greyish-brown colour of a meltwater stream. Suspended sediment concentrations of 3800 mg/l have been measured in front of Norwegian glaciers. Moreover, the bedload is probably even more impressive and may account for up to 90 per cent of the total load. An impressive sound near bigger Greenland meltwater streams is the noise and vibration associated with boulders trundling down the stream. Such high debris loads are favoured by the peaked runoff regime and the ready availability of suitable sediment beneath a glacier.

Glacial meltwater has important erosional effects beneath a glacier. At the smallest scale the water film beneath warm-based ice allows both physical and chemical rock removal. Most particles finer than 0.2 mm are flushed out in the water film, thus allowing the efficient removal of rock flour which might otherwise clog up the tools of abrasion. Also the water allows chemical modification of the rock surface. Hallet (1976) noted how $CaCO_3$ is derived during pressure melting on the upstream side of bumps and precipitated by refreezing. Meltwater which is not refrozen carries this solute load out of the glacier system.

A number of erosional forms relate to direct erosion by subglacial meltwater. *Sichelwannen* or crescent-shaped scour marks and potholes 15–20 m across occur widely in Scandinavia and have been attributed to scour by separated flow at the high velocities common in subglacial streams (Allen, 1971a). Meltwater channels reflect the action of subglacial streams as a whole. Such channels occur at a small scale and may be a few centimetres deep and a few metres long, wending their way up and between hummocks abraded by the action of ice (Figure 17.29). At a larger scale subglacial meltwater channels abound in formerly glaciated areas and are witness to subglacial meltwater flow across the underlying topography. In upland areas they may be barely discernible to 100 m in depth with rock-cut sides and with lengths from tens to thousands of metres. Many channels have convex long profiles and, as predicted by Shreve (1972), they tend to follow low points in the underlying topography and some of the best examples are sharply incised into saddles and

cols (Figure 17.30). Similar channels are found incised into the floors of glaciated valleys both beneath present-day glaciers like Glacier d'Argentière and in formerly glaciated valleys. Shreve (1972) has shown how meltwater flowing at the base of a glacier will tend to increase its velocity as it flows over a convexity in the underlying relief. This favours erosion and accounts for the way channels are best displayed on interfluves and incised into rock steps.

Some of the sediments carried by meltwater channels are deposited beneath glaciers in two main forms – *eskers* and *sheets*. Eskers are the channel deposits of former subglacial rivers (Figure 17.31). There are often slightly sinuous ridges with an irregular long profile. They vary from a few metres to tens of metres in height and from less than 1 km to hundreds of kilometres in length. The sediments contain several elements. Frequently there is a core of poorly sorted matrix-supported gravel with a great range of particle size, including clay and sand. Saunderson (1977) utilized engineering studies of the transmission of solids in pipes to explain the characteristic material as a sliding bed facies where all available particle sizes are in motion together. Finer particles have a buoyant effect on larger particles. Deposition of large particles is impossible because the constriction in the

**Figure 17.29** Presently exposed area lying in front of the Blackfoot Glacier, Montana, showing the locations of inferred previous subglacial channels, cavities and depressions identified on morphological grounds; the white areas are bedrock which were in contact with the ice and were either altered by $CaCO_3$ or were abraded.
*Source:* Walder and Hallet, 1979, figure 7, facing p. 342.

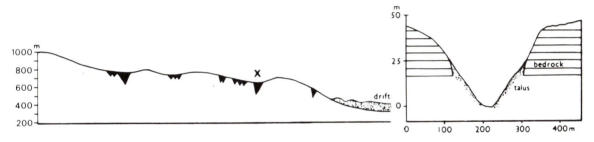

**Figure 17.30** The sites of subglacial meltwater channels crossing a spur on the northern flank of the Cairngorm Mountains, Scotland; the cross section of channel X is shown in detail; the deepest channels are sited in cols.
*Source:* Sugden and John, 1976, figure 15.3b and c, p. 304.

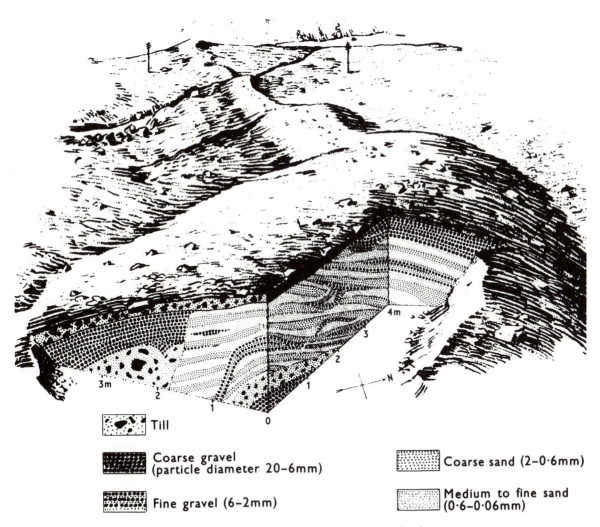

**Figure 17.31** Cross-profile and longitudinal section through an esker in Sylälvsdalen, Sweden.
*Source:* Mannerfelt, 1945, figure 93, p. 194.

conduit that would result increases the water velocity until it is capable of moving the particles again. Deposition of all the material in transit takes place if flow in the pipe is suddenly obstructed. A second element is bedding of sands and gravels with arched bedding, a feature predicted to occur in pipes with a declining discharge. A final element is glacial till, usually on top of the esker, which reflects the subglacial origin of the landform. In view of the high sediment loads of meltwater streams and the extreme variability of flow it is possible that eskers build up rather rapidly; indeed, Allen (1971b) has calculated deposition rates of 2 m in 26 h for parts of an esker near Uppsala. It is tempting to suggest that whereas a matrix-supported core suggests a sudden blocking of a meltwater route, deposition of bedded gravels may take place towards the end of a melt season as the discharge falls. In such a case the tunnel can be envisaged as full of water and enlarging its conduit until the maximum discharge is attained at the peak of the ablation season. Progressive sedimentation then occurs on the falling limb of the discharge graph. Individual deposition phases may occur diurnally again on the falling discharge limb.

Subglacial meltwater deposition of sheets of sediments are not commonly discussed. Yet in eastern Scotland there are whole valleys or embayments lying athwart the overall direction of ice movement which appear to consist of poorly sorted, coarsely bedded gravels interspersed with well-sorted beds of sand and gravel. On top of such deposits one can find clear evidence of subglacial landforms. It is tempting to suggest that such deposits reflect subglacial meltwater deposition, perhaps in meltwater tunnels which are migrating under the ice.

Because for most populated regions there is evidence of past glaciation rather than active glaciers and also because glacial landscapes are only revealed following deglaciation, glacial landscapes are considered in Chapter 19 rather than in this chapter. For the most part glacial landscapes are a record of past climatic conditions.

## References

Addison, K. (1981) 'The contribution of discontinuous rock-mass failure to glacier erosion', *Annals of Glaciology*, vol. 2, 3–10.

Allen, J. R. L. (1971a) 'Transverse erosional marks of mud and rock: their physical basis and geological significance', *Sedimentary Geology*, vol. 5, 1–385.

Allen, J. R. L. (1971b) 'A theoretical and experimental study of climbing – ripple cross-lamination, with a field application to the Uppsala esker', *Geografiska Annaler*, vol. 53A, 157–87.

Anderson, J. B., Kurtz, D. D., Domack, E. W. and Balshaw, K. M. (1980) 'Glacial and glacial marine sediments of the Antarctic continental shelf', *Journal of Geology*, vol. 88, 399–414.

Birnie, R. V. (1977) 'A snow-bank push mechanism for the formation of some "annual" moraine ridges', *Journal of Glaciology*, vol. 18, 77–85.

Birnie, R. V. (1978) 'Rock debris transport and deposition by valley glaciers in South Georgia', unpublished Ph D dissertation, University of Aberdeen.

Boulton, G. S. (1968) 'Flow tills and related deposits on some Vestspitsbergen glaciers', *Journal of Glaciology*, vol. 7, 391–412.

Boulton, G. S. (1970) 'The deposition of subglacial and melt-out tills at the margins of certain Svalbard glaciers', *Journal of Glaciology*, vol. 9, 231–45.

Boulton, G. S. (1972a) 'The role of thermal regime in glacial sedimentation', *Institute of British Geographers Special Publication* 4, 1–19.

Boulton, G. S. (1972b) 'Modern arctic glaciers as depositional models for former ice sheets', *Journal of the Geological Society of London*, vol. 128, 361–93.

Boulton, G. S. (1974) 'Processes and patterns of glacial erosion', in D. R. Coates (ed.), *Glacial Geomorphology*, Binghamton, NY, State University of New York, 41–87.

Boulton, G. S. (1979) 'Processes of glacier erosion on different substrata', *Journal of Glaciology*, vol. 23, 15–38.

Boulton, G. S. and Jones, A. S. (1979) 'Stability of temperate ice caps and ice sheets resting on beds of deformable sediment', *Journal of Glaciology*, vol. 24, 29–43.

Brochu, M. (1954) 'Lacs d'érosion différentielle glaciaire sur le Bouclier Canadien', *Revue de Géomorphologie dynamique*, vol. 6, 274–9.

Budd, W. F. (1975) 'A first simple model for periodically self-surging glaciers', *Journal of Glaciology*, vol. 14, 3–21.

Budd, W. F., Jenssen, D. and Radok, U. (1970) 'The extent of basal melting in Antarctica', *Polarforschung*, vol. 6, 293–306.

Chorley, R. J. (1959) 'The shape of drumlins', *Journal of Glaciology*, vol. 3, 339–44.

Clapperton, C. M. (1975) 'The debris content of surging glaciers in Svalbard and Iceland', *Journal of Glaciology*, vol. 14, 395–406.

Collins, D. N. (1979) 'Quantitative determination of the subglacial hydrology of two alpine glaciers', *Journal of Glaciology*, vol. 23, 347–62.

Drewry, D. J. (1982) 'Antarctica unveiled', *New Scientist* vol. 95 (22 July), 246–51.

Drewry, D. J., Jordan, S. R. and Jankowski, E. (1982)

'Measured properties of the Antarctic ice sheet: surface configuration, ice thickness, volume and bedrock characteristics', *Annals of Glaciology*, vol. 3, 83–91.

Elliston, G. R. (1973) 'Water movement through Gornergletscher', *Symposium on the Hydrology of Glaciers, 9–13 September 1969, International Association of Scientific Hydrology*, vol. 45, 79–84.

Glen, J. W. (1955) 'The creep of polycrystalline ice', *Proceedings of the Royal Society of London*, series A, vol. 228, 519–38.

Glückert, G. (1973) 'Two large drumlin fields in central Finland', *Fennia*, vol. 120, 1–37.

Gordon, J. (1981) 'Ice-scoured topography and its relationships to bedrock structure and ice movement in parts of northern Scotland and west Greenland', *Geografiska Annaler*, vol. 63A, 55–65.

Haeberli, W. (1981) 'Ice motion on deformable solids', *Journal of Glaciology*, vol. 27, 365–6.

Hagen, J. O., Wold, B., Liestøl, O., Østrem, G. and Sollid, J. L. (1983) 'Subglacial processes at Bondhusbreen, Norway', *Annals of Glaciology*, vol. 4, 91–8.

Haldorsen, S. (1981) 'Grain-size distribution of subglacial till and its relation to glacial crushing and ablation', *Boreas*, vol. 10, 91–105.

Hallet, B. (1967) 'Deposits formed by subglacial precipitation of $CaCO_3$', *Bulletin of the Geological Society of America*, vol. 87, 1003–15.

Hallet, B. (1979) 'A theoretical model of glacial abrasion', *Journal of Glaciology*, vol. 23, 39–50.

Hallet, B. (1981) 'Glacial abrasion and sliding: their dependence on the debris concentration in basal ice', *Annals of Glaciology*, vol. 2, 23–8.

Hambrey, M. J. and Müller, F. (1978) 'Structures and ice deformation in the White Glacier, Axel Heiberg Island, Northwest Territories, Canada', *Journal of Glaciology*, vol. 20, 41–66.

Hartshorn, J. H. (1958) 'Flowtill in southeastern Massachusetts', *Bulletin of the Geological Society of America*, vol. 69, 477–82.

Hillefors, Å. (1973) 'The stratigraphy and genesis of stoss and lee-side moraines', *Lund Studies in Geography*, series A, vol. 56, 139–54.

Hooke, R. L. and Hudleston, P. J. (1978) 'Origin of foliation in glaciers', *Journal of Glaciology*, vol. 20, 285–99.

Hutter, K. (1982) 'Glacier flow – recent developments', *Mitt. der Versuchsanstalt für Wasserbau, Hydrologie und Glaziologie*, vol. 57, 27–33.

Iken, A., Flotron, A., Haeberli, W. and Röthlisberger, H. (1979) 'The uplift of Unteraargletscher at the beginning of the melt season – a consequence of water storage at the bed?', *Journal of Glaciology*, vol. 23, 430–42.

Jahns, R. H. (1943) 'Sheet structure in granite: its origin and use as a measure of glacial erosion in New England', *Journal of Geology*, vol. 51, 71–98.

Jouzel, J. and Souchez, R. A. (1982) 'Melting – refreezing at the glacier sole and the isotopic composition of the ice', *Journal of Glaciology*, vol. 28, 35–42

Kamb, B. and LaChapelle, E. R. (1964) 'Direct observations of the mechanism of glacier sliding over bedrock', *Journal of Glaciology*, vol. 5, 159–72.

Kemmis, T. J. (1981) 'Importance of the regelation process to certain properties of basal tills deposited by the Laurentide ice sheet in Iowa and Illinois, USA', *Annals of Glaciology*, vol. 2, 147–52.

Lawson, D. E. (1981a) 'Sedimentological characteristics and classification of depositional processes and deposits in the glacial environment', *CRREL Report 81-27*.

Lawson, D. E. (1981b) 'Distinguishing characteristics of diamictons at the margin of the Matanuska Glacier, Alaska', *Annals of Glaciology*, vol. 2, 78–84.

Lawson, D. E. (1982) 'Mobilization, movement and deposition of active subaerial sediment flows, Matanuska Glacier, Alaska', *Journal of Geology*, vol. 90, 279–300.

Lliboutry, L. (1965) *Traité de Glaciologie*, Paris, Masson, Vol. 2.

Mannerfelt, C. M. (1945) 'Några glacialmorfologiska Formelement', *Geografiska Annaler*, vol. 27, 1–239.

Menzies, J. (1979) 'The mechanics of drumlin formation with particular reference to the change in pore-water content of the till', *Journal of Glaciology*, vol. 22, 373–84.

Moran, S. (1971) 'Glaciotectonic structures in drift', in R. P. Goldthwait (ed.), *Till, A Symposium*, Ohio State University Press, 127–48.

Nye, J. F. (1952) 'The mechanics of glacier flow', *Journal of Glaciology*, vol. 2, 82–93.

Paterson, W. S. B. (1981) *The Physics of Glaciers*, 2nd edn, Oxford, Pergamon.

Rastas, J. and Seppälä, M. (1981) 'Rock jointing and abrasion forms on roches moutonnées, SW Finland', *Annals of Glaciology*, vol. 2, 159–63.

Robin, G. de Q. (1976) 'Is the basal ice of a temperate glacier at the pressure melting point?', *Journal of Glaciology*, vol. 16, 183–96.

Robin, G. de Q., Drewry, D. J. and Meldrum, D. T. (1977) 'International studies of ice sheet and bedrock', *Philosophical Transactions of the Royal Society of London*, series B, vol. 279, 185–96.

Röthlisberger, H. and Iken, A. (1981) 'Plucking as an effect of water-pressure variations at the glacier bed', *Annals of Glaciology*, vol. 2, 57–62.

Rudberg, S. (1973) 'Glacial erosion forms of medium size – a discussion based on four Swedish case studies', *Zeitschrift für Geomorphologie*, vol. 17, 33–48.

Saunderson, H. C. (1977) 'The sliding bed facies in esker sands and gravels: criterion for full pipe (tunnel) flow?', *Sedimentology*, vol. 24, 623–38.

Sharp, M. J. (1982a) 'Modification of clasts in lodgement tills by glacial erosion', *Journal of Glaciology*, vol. 28, 475–81.

Sharp, M. J. (1982b) 'A comparison of the landforms and sedimentary sequences produced by surging and non-surging glaciers in Iceland', unpublished Ph D dissertation, University of Aberdeen.

Shaw, J. (1977) 'Tills deposited in arid polar environments', *Canadian Journal of Earth Sciences*, vol. 14, 1239–45.

Shreve, R. L. (1972) 'Movement of water in glaciers', *Journal of Glaciology*, vol. 11, 205–14.

Smalley, I. J. and Unwin, D. J. (1968) 'The formation and shape of drumlins and their distribution and orientation in drumlin fields', *Journal of Glaciology*, vol. 7, 377–90.

Smith, H. T. U. (1948) 'Giant glacial grooves in northwest Canada', *American Journal of Science*, vol. 246, 503–14.

Smith, J. M. (1982) 'A geomorphological study of glacial striations in Coire Lagan, Skye', undergraduate dissertation, University of Aberdeen.

Souchez, R., Lemmens, M., Lorrain, R. and Tison, J-L. (1978) 'Pressure-melting within a glacier indicated by the chemistry of re-gelation ice', *Nature*, vol. 273, 454–6.

Stenborg, T. (1969) 'Studies of the internal drainage of glaciers', *Geografiska Annaler*, vol. 51A, 13–41.

Sugden, D. E. (1977) 'Reconstruction of the morphology, dynamics and thermal characteristics of the Laurentide ice sheet at its maximum', *Arctic and Alpine Research*, vol. 9, 21–47.

Sugden, D. E. and John, B. S. (1976) *Glaciers and Landscape: A Geomorphological Approach*, London, Arnold.

Thorarinsson, S. (1953) 'Some new aspects of the Grímsvötn problem', *Journal of Glaciology*, vol. 2, 267–74.

Vivian, R. (1970) 'Hydrologie et érosion sous-glaciaires', *Revue Géographie Alpine*, vol. 58, 241–64.

Walder, J. and Hallet, B. (1979) 'Geometry of former subglacial water channels and cavities', *Journal of Glaciology*, vol. 23, 335–46.

Weertman, J. (1957) 'On the sliding of glaciers', *Journal of Glaciology*, vol. 3, 33–8.

Zumberge, J. H. (1952) 'The lakes of Minnesota, their origin and classification', *Minnesota Geological Survey Bulletin*, vol. 35.

# Part Four  *Climatic geomorphology*

In Part III the variety of geomorphic processes was considered and it became apparent that climate must significantly influence not only the type of process that may act under given conditions, but also its intensity and rate of operation. In Part IV the cumulative effect of climate and of climatic change on landforms is considered specifically, with reference to variables 4 and 5 in Table 1.1. An examination is made of the extent to which different climatic regimes are potentially capable of exerting direct and indirect influences on geomorphic processes, and thereby of generating different 'morphogenetic' landform assemblages (Chapter 18). Of the eight morphogenetic regions identified, four are treated in detail in Chapter 18: arid and semi-arid in Section 18.4, tropical wet–dry in Section 18.3 and humid tropical in Section 18.2. Chapter 19 discusses the important morphogenetic class of glacial landscapes which, for the most part, are the result of past climatic conditions in that they are only fully revealed following deglaciation. Periglacial landforms are examined in Section 18.5.1 and 20.2.2. The humid mid-latitude landform assemblage is composed of well-studied features which have formed much of the subject-matter for Chapters 10–13. In contrast, the dry continental landforms are the most indistinct and poorly studied of the eight classes, being intermediate between widely contrasting climatic regimes. Chapter 20 is dominantly concerned with 'polygenetic' landforms which exhibit the morphological effects of multiple climates, particularly those involving recent changes in precipitation and the lowering of temperature under periglacial conditions.

# Eighteen *Morphogenetic landforms*

The initial question to which those concerned with climatic geomorphology must address themselves is the extent to which distinctive climatic types are capable of producing, over a period of time, distinctive regional assemblages of landforms which, to some appreciable extent, transcend regional differences in lithology, structure and relief (Stoddart, 1969).

The morphometric feature which has appeared to be particularly diagnostic of climatic influences is drainage density. These influences may either act directly (e.g. through precipitation intensity) or indirectly (e.g. through vegetational effects). For example, differences in drainage density between sandstone areas of Exmoor, in south-west England, the Appalachian Plateau of Pennsylvania and northern Alabama (3, 5 and 8·5 mile/sq. mile, respectively –1·9, 3, 5·3 km/km²) (see Figure 8.1), and between the granite areas of Dartmoor, south-west England, and the Unaka Mountains of Tennessee (3 and 11 mile/sq. mile, respectively –1·9, 6·8 km/km²) can be statistically 'explained' by contemporary differences in average amounts, intensities and effectiveness of precipitation. In addition, a study by Melton (1957) of a large sample of drainage basins from the American south-west showed that observed variations in drainage density (349 mile/sq. mile on shale to 4 mile/sq. mile on dacite and rhyolite ₊217 to 1·6 km/km²) could be most effectively explained by reference to:

Log precipitation – effectiveness index of Thornthwaite, explaining 89 per cent. (This index involves monthly average precipitation amounts, and although being directly climatic, it is commonly used as a measure of vegetation amount – an indirect influence.)

Percentage bare area, explaining 81 per cent (an indirect influence).

Log infiltration capacity, explaining 63 per cent (an indirect influence).

Log runoff intensity, explaining 43 per cent (an influence involving a trade-off between direct precipitation intensity and indirect surface properties controlling infiltration) (Figure 18.1).

A multiple correlation showed that all four variables together explain some 92 per cent of the spatial variance in drainage density, which implies that although rock type is important (especially differences between shale and schist on the one hand, and granite, volcanics, clastics and limestones on the other), contemporary climate and surface properties appear to be dominant in controlling differences of drainage density. These two studies immediately involve us in the following speculations:

(1) that distinctions between direct and indirect climatic processes are difficult to make;

(2) that the high degree of correlations between drainage densities and contemporary and surface properties raises many problems regarding possible rates of landform adjustment to climatic change (see Chapter 20) or to what extent contemporary climatic and surface conditions can be viewed as surrogates for past ones.

Another important question relates to what attributes of climate and what magnitudes and frequencies of climatically generated processes are mainly responsible for generating characteristic landforms. This is a complex question because two different average climates may exercise little significant difference in respect of important landscape processes such that, for example, landforms developed in semi-arid regions may differ little from dry continental ones. It is also important to remember that whereas it is customary to distinguish between different climatic regimes on the basis of average climatic characteristics, it may be the *extreme*

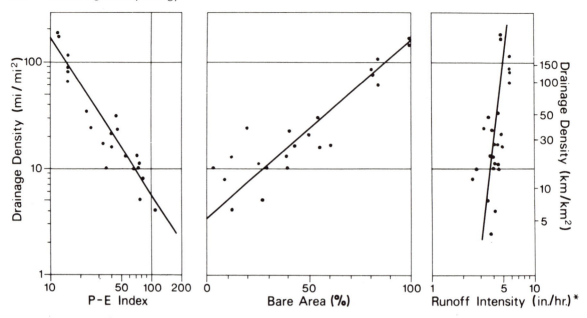

**Figure 18.1** Regressions of drainage density as a function of precipitation effectiveness index, percentage of bare area and runoff intensity (5-year, 1-h rainfall minus assumed infiltration capacity of 5 in/h).
*Source:* Melton, 1957, figures 1, 11 and 2, pp. 73, 75.

conditions in any regime which are responsible for the production of characteristic landforms. In addition, not only do different levels of magnitude and frequency of process have different geomorphic effects in different environments, but within a single environment different attributes of morphometry (e.g. hydraulic geometry, slope forms and divide configuration) may themselves be formed by processes of different magnitude and frequency. Nevertheless, the identification of regions, where climate may determine the dominant geomorphic processes and thereby significantly influence landform production, has been the goal of many investigators.

## 18.1 Morphogenetic regions

*Morphogenetic regions* are large areal units within which distinctive associations of geomorphic processes (e.g. weathering, frost action, mass movements, fluvial action and wind action) are assumed to operate, tending towards a state of *morphoclimatic equilibrium* wherein regional landforms reflect regional climates. Clearly, the realism of such an assumption in practice is very much open to question, particularly as we have shown in Chapters 7 and 8 that structural and lithological variations are responsible for considerable variety of landforms, and in Section 3.3 that these factors plus relief and vegetational variations may result in regional rates of

erosion which may vary by between ten and a hundred times. It is, therefore, proposed to review those morphogenetic regions which have been identified (see especially Büdel, 1948, 1982; Peltier, 1950; Troll, 1958; Tanner, 1961; Birot, 1968; Stoddart, 1969; Wilson, 1969; Tricart and Cailleux, 1972).

It is significant that the average number of potential morphogenetic regions which have been suggested (namely 9: Büdel 5, Wilson 6, Peltier 9; Butzer 13; and Tricart and Cailleux 13) is smaller than the mean number of current major climatic regions (namely 13: Flohn 8; Köppen 11; Strahler 14; Trewartha 16; and Thornthwaite 18). This suggests that greater climatic differences are necessary to generate distinctive landforms than to produce effects which are appreciable by plants, animals and man. It is also apparent that any simple morphogenetic classification might reasonably be constructed on a threefold basis of temperature, precipitation and seasonality. Figure 18.2 shows a rough, schematic attempt to delineate eight world morphogenetic regions. These regions may conveniently be divided into two groups:

First-order morphogenetic regions. Glacial, arid and humid tropical. These have non-seasonal processes, generally low average erosion rates, highly infrequent and episodic erosional activity (e.g. glacial surges, desert rainstorms and slope mass failures,

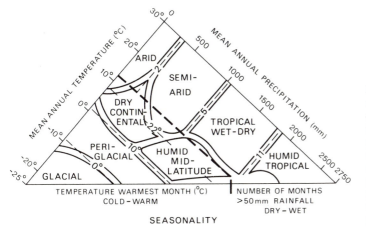

**Figure 18.2** The possible present morphogenetic regions of the world roughly classified according to mean annual temperature (degC), mean annual precipitation (mm), mean number of wet months (>50 mm) and mean temperature of the warmest month (degC).
*Sources:* Peltier, 1950, figure 7, p. 222; Tanner, 1961, figure 3, p. 255; especially Wilson, 1969, figure 3, p. 310; see also Büdel, 1948; Troll, 1958.

respectively), and a tendency for the location of their cores to persist latitudinally (at 90°, 25° and 0°, respectively) during climatic changes, even if the climatic type is completely obliterated on occasions.

Second-order morphogenetic regions. Tropical wet–dry, semi-arid, dry continental, humid mid-latitude and periglacial. These have processes which are more distinctly seasonal in their operation; in places high average erosion rates; erosional activity which, although it may be episodic, shows some consistency over a period of years; and a tendency for considerable changes of their size and location accompanying global climatic changes. Second-order regions can be divided into two groups:

(1) warmer climates (tropical wet–dry and semi-arid) where geomorphic processes differ most significantly in terms of the length of the wet season;

(2) cooler climates (dry continental, humid mid-latitude and periglacial) whose geomorphic processes differ mainly in respect of summer temperatures, as well as with some regard to precipitation amounts.

It may be helpful here to summarize briefly the proposed features of each of the above regions, before proceeding with a fuller discussion. The current distribution of these eight morphogenetic regional types is shown in Figure 18.3, and their description owes much to the work of Büdel (1948, 1982) Peltier (1950), Tanner (1961), Wilson (1969, 1973), Butzer (1976) and Tricart and Cailleux (1972).

### GLACIAL

Other names given to whole or part of this region – subglacial; Köppen region EF.

Geomorphic processes – frost weathering (max.); mechanical weathering (mod.); chemical weathering

(min.); mass wasting (min.); fluvial processes (min., except for seasonal meltwater); glacial scour (max.); wind action (max.).

Morphological features – alpine topography; abrasion surfaces; kames; eskers; till forms; fluvio-glacial features.

### ARID

Other names – desert; true desert; tropical and sub-tropical desert; Köppen region BWh.

Geomorphic processes – frost weathering (min., except at high latitudes); mechanical weathering (max., especially salt and thermal); chemical weathering (min.); mass wasting (min.); fluvial processes (min. at present; commonly evidence of previous action, especially gullying and sheetwash); glacial scour (nil); wind action (max., perhaps overestimated by past workers).

Morphological features – dunes; playas; deflation basins; angular, debris-covered slopes; fossil fluvial forms (e.g. fans and arroyos).

### HUMID TROPICAL

Other names – selva; rainforest; inner tropical zone; Köppen regions Af and Am.

Geomorphic processes – frost weathering (nil); mechanical weathering (min.); chemical weathering (max., giving thick, fine-grained debris even on steep slopes covered with thick vegetation); mass wasting (max., but highly episodic and irregular on steep slopes, especially in the form of earthflows and slips); fluvial processes (mod.–min. slopewash and rainbeat; min. stream corrasion due to lack of coarse debris; max. chemical and suspended-load transport); glacial scour (nil); wind action (nil).

Morphological features – low-gradient rivers; wide, flat, or gently undulating floodplain floors up to several kilometres wide; steep slopes (frequently

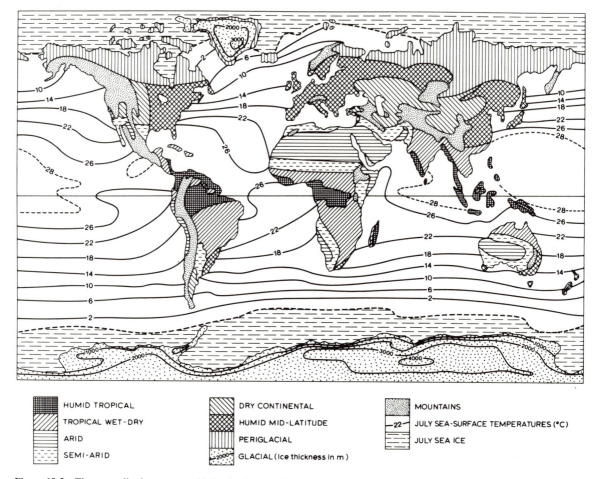

**Figure 18.3**   The generalized present world distribution of climates corresponding to the morphogenetic regions shown in Figure 18.2; cf. Pleistocene glaciation map in Figure 20.7.

*Sources:* Tricart and Cailleux, 1972, map III, copyright © Longman, London; Stoddart, 1969, figure 8, p. 173.

>40°) rising abruptly from valley floors, stabilized by dense vegetation but susceptible to infrequent landslides; knife-edged ridges maintained by parallel retreat of steep slopes.

TROPICAL WET–DRY

Other names – savanna; tropical sheetwash zone; moist and dry savanna; tropical savanna; Köppen region Aw.

Geomorphic processes – frost weathering (nil); mechanical weathering (min.–mod.); chemical weathering (seasonal max. – accelerated by standing water, giving deep weathering and two weathering zones, at the surface and at the weathering front; surface crusts of Al and Fe generated during dry seasons); mass wasting (mod.–max.); fluvial processes (mod.–max. – highly seasonal sheetfloods, rill wash and channel flow); glacial scour (nil); wind action (min.–mod.).

Morphological features – steep irregular slopes of coarse debris retreating parallel by basal chemical weathering and sheetflood removal; wide planation surfaces (<3·5°) bounded by steep slopes and inselbergs; irregular bedrock surfaces broken by inselbergs and torlike features probably produced by stripping along the weathering front; badlands on shales and other impermeable rocks.

The morphological features of this region may not be greatly different from those of the semi-arid region *in appearance*, but may differ in terms of their generating processes. This introduces an important matter of principle in morphogenetic classification.

## SEMI-ARID

Other names – peripheral or marginal hot deserts; desert margins; thorn savanna; semi-desert; semi-arid steppes; Mediterranean or summer-dry subtropical zone (lower latitude parts only); Köppen regions BS, BW and Cs (part).

Geomorphic processes – frost weathering (min., except at higher latitudes); mechanical weathering (min.–mod., especially thermal and salt); chemical weathering (min.–mod.); mass wasting (mod., but infrequent); fluvial processes (max. but episodic in the form of sheetwash, gullying and ephemeral stream action giving high overall erosion rates); glacial scour (nil); wind action (mod.–max.).

Morphological features – pediments (1°–4°) backed by cliffs and angular talus slopes (25°–35°) and inselbergs; integrated ephemeral stream systems; arroyos; badlands; alluvial fans; local dunes.

## DRY CONTINENTAL

Other names – steppe zone; mid-latitude grasslands; semi-arid steppes; degraded steppes; extratropical deserts (part); savanna (part); Köppen regions BSk and BWk.

Geomorphic processes – frost weathering (min.–mod., but highly seasonal with bedrock affected at times); mechanical weathering (min.–mod.); chemical weathering (min.–mod.); mass wasting (mod., but seasonal activity important); fluvial processes (mod.–max., especially in the form of spring and summer sheetwash and gullying); glacial scour (nil); wind action (mod., fossil loess in places).

Morphological features – pediments flanked by steep, scree-covered slopes; badlands; alluvial fans; arroyos; like tropical wet–dry landforms, these may not appear to differ greatly from the semi-arid ones, but processes may do so (especially those associated with frost and spring thaw).

## HUMID MID-LATITUDE

Other names – oceanic and continental zones of mature soils; temperate marine and continental regions; moist subtropical; humid temperate; maritime; mid-latitude forests; Mediterranean zone and summer-dry subtropical (higher latitude parts only); moderate (part); forest on Pleistocene permafrost (part); Köppen regions Cf, Da, Db, Cs (part) and Dc (part).

Geomorphic processes – frost weathering (min.–max., especially in higher latitude and continental locations where this region is transitional with the periglacial region); mechanical weathering (min.–mod.); chemical weathering (mod.); mass wasting (mod.–max., creep dominant); fluvial processes (mod.); glacial scour (nil, except on mountains and as fossil evidence); wind action (min.).

Morphological features – smooth, soil-covered slopes; ridges and valleys; streams with wide range of grain sizes.

## PERIGLACIAL

Other names – frost debris zone; tundra; subpolar zone; high arctic barrens; forest with permafrost; humid microthermal; boreal, mature soil zone with permafrost; Köppen regions ET, Dd and De (part).

Geomorphic processes – frost weathering (max., permafrost conditions or subsoil freezing during significant parts of the year; many months under snow cover; spring thaw phenomena); mechanical weathering (max., especially nivation); chemical weathering (min.); mass wasting (max., talus creep and solifluction everywhere during the spring and in northern parts during the summer); fluvial processes (mod., slopewash and valley-cutting concentrated in limited thaw season); glacial scour (min., except on mountains and recent fossil evidence); wind action (mod.–max.).

Morphological features – scree slopes ($\simeq$ 25°, sometimes gullied), solifluction slopes (15°–20°) and cryoplanation surfaces (<5°) with residuals rising above them; solifluction lakes and terraces; outwash plains; patterned ground; loess and dunes.

Besides these eight possibly distinct morphogenetic regions, there exist certain *azonal mountain landforms* (see Figure 18.3). In lower latitudes (e.g. Andes, East Africa, South-west Asia and East Indies), except at the highest elevations the landforms appear dominantly fluvial with steep slopes, sharp ridges, frequent landslides (especially in earthquake zones) and rapid stream incision. Even in the driest regions these mountainous areas exhibit organized fluvial erosional landforms (e.g. Ahaggar, Ethiopian highlands and north Chilean Andes). At high latitudes even low mountains exhibit the residual influences of Pleistocene mountain glaciation and high mountains possess subdued but still-active valley glacial processes.

Before proceeding to a more detailed examination of the landforms of the major morphogenetic types, it is instructive to return to the matter of the conflicting influences of climate on the one hand, and of structure and lithology on the other. This subject is best referred to the context of spatial scale influences, which were introduced in Chapter 1, when we try to determine over what scales of space may similar morphogenetic influences be appreciable. An answer to this question can only be given in a very general sense such that on a large scale the effects of geological variety may be outweighed by climate, whereas on smaller scales the reverse is true. At scales of $10^6$–$10^7$ km$^2$ landscape variety is probably best explained by climatic differences; at $10^4$–$10^5$ km$^2$ structural and lithological differences are dominant; and at less than $10^2$ km$^2$ landform differences are increasingly the result of spatial variations in different erosional mechanisms (i.e. stream action; slope mass movements; creep on divides; etc.).

## 18.2 Humid tropical landforms

The humid tropics (see Figure 18.2) are characterized by an absence of low temperatures, a small annual temperature range ($< 10°C$; $18°F$) which is less than the diurnal range and by the small number of dry months – usually not more than one to two months with less than 50 mm precipitation (Colombo, Sri Lanka, has only three months with less than 100 mm) (Fosberg *et al.*, 1961). Rainfall intensity is high especially on tropical mountains; for example, in El Salvador intensities of 5 mm/min are common. The humid tropics are defined both on the basis of their climate and characteristic rainforest vegetation, which makes their delimitation rather difficult, especially where they border on the humid subtropics (e.g. Vietnam, the Caribbean, south-east Brazil, etc.) and where human destruction of vegetation has generated geomorphic processes characteristic of the adjacent wet–dry tropics (e.g. the marginal areas of Central, East and West Africa, and in areas like the Philippines, where extensive hardwood logging has taken place).

The rapidity of nutrient cycling and the intensity of chemical weathering in the humid tropics have been stressed in Section 9.1, but it is now necessary to draw attention to the geomorphic implications of such action (Douglas, 1969). The depth of chemical weathering is usually great in the humid tropics, with depths in excess of 100 m common in Brazilian crystalline rocks, 5–6 m on steep slopes in New Guinea and Central America, and even up to 40 cm of soil on slopes as steep as 60°. Bare rock outcrops are noticeably absent, showing that in the short term decomposition is more rapid than transport. The reasons for this are (Tricart, 1972):

(1) The generally high rate of infiltration (over 90 per cent of rainfall in places), due to the considerable vegetation cover (even on 70° slopes) and high humidity, leading to more or less complete saturation over long periods.
(2) The high ground surface and soil temperatures.

Van't Hof's Law postulates that a temperature increase of $10°C$ ($18°F$) increases the rate of chemical reactions by 2·5 times. At a depth of about 1 m soil temperatures are constant at about 2° to 4°C above the free air average, due to the insulating effects of the layered vegetation canopies and to the generation of heat by the rotting of biomass.

(3) The intensity of biochemical action due to the rate and amount of nutrient cycling. The humid tropics provide the optimum conditions for photosynthesis and the production of biomass. The decomposition of humus promotes high soil $CO_2$ levels (on average five times that of temperate soils) and makes the soil act on the bedrock as a giant acidic sponge. In the Ivory Coast the average groundwater pH values for granite are 4·2 at the surface and 5·3 at a depth of 25 m; equivalent figures for shales are 5·3–5·7 and for basic igneous rocks 6·0–6·4.

(4) Possibly a long period of stable ecosystems during which soils and vegetation in core climatic areas have been adjusted to little-changing climate. This, however, is speculative.

One of the major effects of this accentuated chemical action is to produce excessive leaching, including that of quartz. This leaching commonly results in some 2 m of sandy regolith immediately under the humus cover in rainforests and the restriction of nutrients to the upper weathered layers such that tree roots rarely extend deeper than about 2 m, causing large trees to be potentially unstable. The acid character of the soil water, associated with humus decay, allows iron oxides to be very mobile and iron pans are often formed in the debris layers. Another feature of tropical chemical weathering is the leaching of silica from the silicate minerals (e.g. feldspars, rather than from mineral quartz), resulting in there being relatively less weathered clay in the humid tropics than in the wet–dry tropics and in there being considerable denudation by solution action. Although the average concentration of dissolved material in tropical areas is not in excess of that in extratropical rivers (Table 18.1), the higher discharges per unit area in the tropics imply considerably greater tropical solution denudation rates. For example, the Malayan archipelago (2 per cent of land surface of the world) provides 12 per cent of the world's runoff, the Amazon basin 20 per cent and the Congo 4·5 per cent. In Malaysia the total dissolved stream load may be as much as one-third of the suspended load and up to 22 per cent of the $SiO_2$ weathered from Hawaiian basalts is removed in solution (the remaining 78 per cent as hydrous aluminium silicates), compared with only 1–3 per cent for the rocks of the Mississippi basin. Surface and near-surface solution of rocks in the humid tropics results in a variety of solutional furrows (lapiès) in rocks including granite and basalt, as well as limestone.

**Table 18.1** Dissolved material in rivers (p.p.m.)

|  | Silica | Total dissolved solids |
|---|---|---|
| Tropical rivers | 8–40 (mean 17) | 34–198 (70) |
| Extratropical rivers | 3–30 (11) | 100–900 (436) |

*Source:* Douglas, 1969, table 2, p. 8.

Despite the special character of humid tropical weathering, it would be wrong to give the impression that the associated valley-side slopes present features which clearly distinguish them from other humid slopes. In Johore, West Malaysia, slopes lying at less than 100 m elevation exhibit characteristics dependent on rock type with deeply weathered basic rocks suffering little surface runoff and presenting convex slopes, whereas arenaceous sandstones and granites appear more resistant to erosion, support surface runoff and possess significant concave elements (Swan, 1972). Having said this, it is clear that there is a widespread tendency for humid tropical landforms in areas of high relief to exhibit flat-floored valleys upon which steep forested valley sides abut (exceeding 45° and even 70° in some headward parts), leading up to sharp interfluves. Such tendencies can be attributed in large part to the general vegetational inhibition of overland flow and creep, together with the importance of infrequent slope failures by landslides, earthflows, soil avalanches and mudflows. The effect of vegetation canopies is to decrease the rainfall directly reaching the forest floor to 5–50 per cent of that experienced in the upper canopy, and the thick humus layer ($A_1$ horizon) promotes infiltration and inhibits surface runoff and splash erosion. On the island of Dominica, West Indies, no true Hortonian overland flow is observed even during intense rainfalls (up to 25·9 mm/30 min) and saturation overland flow is limited to channel margins and valleyheads. Considerable throughflow has been observed at depths of 45–60 cm in the andesitic soil with wet antecedent conditions, together with some soil piping (Walsh, 1980). In Selangor, West Malaysia, saturation overland flow in restricted headwater areas (5 per cent basin area) occurs when rainfall intensities exceed 16 mm/h, but there is considerable throughflow at intensities exceeding about half this. On some tropical slopes gully erosion is important, for example, on slopes as short as 10 m on Viti Levu, Fiji (mean annual rainfall 2540 mm), as well as where surface splash and drip is concentrated by the channelling effect of vegetation cover. Near-surface rooting inhibits soil creep in humid tropical forests even where there is very deep weathering, but where the vegetation is cleared it becomes very important. The upper 60 cm of a 10° cleared sandstone and shale slope in West Malaysia had a creep rate of 12·4 cm² cm⁻¹ year⁻¹, comparing with 4·4 cm² cm⁻¹ year⁻¹ for the northern territory of Australia and 0·5 for the Pennine Hills in England.

For tropical slopes whose steepness is maintained by a dense vegetation cover, by far the most important processes are infrequent catastrophic slope failures. Certain conditions in the humid tropics are particularly favourable to such processes as follows (Plate 25):

(1) A dense vegetational cover with a relatively shallow root system. This allows a slope to remain steep until some trigger action causes a threshold of stability to be transgressed. Such triggers may be external (e.g. heavy rains or earthquakes), or internal, such as slope loading by the accumulated weight of growing vegetation itself. It has been estimated that natural tree falls in Panama forests affect 62–125 per cent of a given area every 100 years (i.e. some places more than once) and that, assuming a vegetation recovery time of twenty to fifty years, some 12–62 per cent of any given area can show tree-fall disturbance at any one time (Garwood *et al.*, 1979). Vegetational clearance by tree falls, lighting fires, volcanic destruction and human activity can expose steep slopes to catastrophic failure.

(2) Steep slopes. Shallow landslides are both a cause and an effect of steep slopes. In the basaltic Koolau Range, Oahu (Hawaii), reaching elevations of 300–450 m (900–1300 ft), lower talus slopes (15°–30°) grade upward into slopes in excess of 40° and into headwalls of 70°–80° approaching 100 m high. Hill crests are knife-edged, with gradients increasing to 60°–85° within some 16 m of the divides. Active sliding occurs along these headwalls and in valleyheads following prolonged rain, but about 80 per cent of all landslides have been observed to occur on slopes with angles 42°–48° (Wentworth, 1943). These slides and soil avalanches are only 1–2 ft thick (rarely more than 1 m) and 100–500 ft (30–150 m) in downslope length and 25–125 ft (7–38 m) wide, but parts of the range have 25–50 per cent exposed bedrock due to their action. In eight years some 200 slides of about 0·4 ha (1 acre) each were observed in an area of 39 sq. km (15 sq. mile) of the Koolau Range and these were estimated to represent a possible removal of 0·3 m (1 ft) over the whole area every 400 years. Similarly, the steep upper slopes (relief >60 m) in Johore, West Malaysia, appear to be retreating in a parallel manner under the influence

of shallow slides (Swan, 1972).

(3) Heavy rain. High-intensity tropical precipitation (sometimes associated with hurricanes) is very commonly responsible for sudden slope failures due to increase of unit mass, increase of porewater pressure and moistening of clays. However, very high moisture contents are required to cause slope failure, especially where the regolith contains much kaolin. At Santos, Brazil, 250 mm of rain in 10 h caused the stripping of a 20 m (66 ft) debris layer from a slope. In Nigeria a boulder 8 m wide and 5 m high slid 80 m down a 10°–30° granite slope during heavy rain. In June 1966 120 mm of rain fell over Hong Kong in 1 h causing 702 landslides (558 slips and 144 smaller washouts), 40 per cent of which occurred on slopes steeper than 27°. The highly episodic character of storm-generated failures is highlighted by the huge fossil mudflow (*diamicton*) which occurred 20,000–30,000 years BP in the volcanic and Miocene sediments of the Kaugel Valley, Papua New Guinea, removing an average of 65 cm from 345 km² to produce a deposit of 3·5 m average thickness over 64 km² (Pain, 1975).

(4) Earthquakes. Some 18 per cent of tropical rainforests lie in zones of high seismic activity, including 38 per cent of the Indo-Malayan rainforests, 14 per cent of the American and 1 per cent of the African. In 1976 two shallow earthquakes (magnitudes 6·7 and 7) occurred along the southern Panama Pacific coast causing slides over 12 per cent (54 km²) of the affected region of 450 km². Figure 18.4 shows the extent of the resulting landslides in an area of steep slopes. It has been estimated that earthquakes of at least magnitude 6 occur once every 250 years in each area of 10³ km² in Central

America and that they disturb at least 2 per cent of the forest area per century (Garwood *et al.*, 1979). In 1935 two shallow earthquakes (magnitudes 7·9 and 7) denuded 8 per cent (130 km²) of an afforested region of 1662 km² in Papua New Guinea. Earthquakes of at least magnitude 6 appear to occur here once every 56–111 years in each area of 1000 km², disturbing 8–16 per cent of the forest cover per century (compared with 3 per cent by other causes). Table 18.2 compares earthquake landslides with other failures in the two regions. In November 1970 an earthquake (magnitude 7) at Madang, Papua New Guinea, caused intense debris avalanches to occur over 60 km² of steep slopes in an affected area of 240 km². Debris, of average thickness 11·5 cm over 240 km² (i.e. 27·6 × 10⁶ m³), was deposited in the valley bottoms of which 50 per cent had been washed into the sea within six months. Assuming that effective landslips cannot occur much more frequently than once per 1000 years at any given site, an average earthquake denudation rate might be 57·5 cm/1000 years, which is some 60–70 per cent of the estimated total denudation rate for the region (Pain and Bowler, 1973).

(5) Debris character. Despite their common occurrence, landslides in the humid tropics appear to be favoured by certain debris characteristics. The most unstable slopes occur on rocks in which weathering has only progressed far enough to produce montmorillonite and illite, as distinct from the more mature kaolinite which does not shrink and crack so easily and needs more moisture to reach its liquid limit. Landslides also appear to be most common where relatively thin layers of weathered material lie on unweathered bedrock. It has been

**Table 18.2** Landslides in Papua New·Guinea and Panama

| Type of disturbance | Location | Rate of occurrence (% of area affected per 100 years) | Recovery time of canopy (years) | % total area disturbed during recovery time |
|---|---|---|---|---|
| Tree falls | Panama | 62–125 | 20 | 12–24 |
| | | | 30 | 19–38 |
| | | | 50 | 31–62 |
| Earthquake, landslides | Papua New Guinea | 8–16 | 200 | 16–33 |
| | | | 300 | 24–49 |
| | | | 500 | 41–89 |
| | Panama | 2 | 200–500 | 4–10 |
| 'Erosional' landslides | Papua New Guinea | 3 | 200–300 | 6–15 |

*Source:* Garwood *et al.*, 1979, table 1, p. 998.

**Figure 18.4**  A part of southern Panama showing (in black) landslides resulting from an earthquake; information from an aerial photograph taken three days later.
*Source:* Garwood *et al.*, 1979, figure 2, p. 998.

observed in Papua New Guinea that earthquake slope failures mainly occur where poorly developed soils 40–50 cm thick lie directly on bedrock. Various authors (e.g. White, 1949) have proposed a cycle of active humid tropical slope failure, roughly along the following lines:

(a) bedrock exposed after landslide at slope of greater than 35°;

(b) weathering and invasion of mosses, ferns and trees;

(c) evolution of soil (perhaps as much as 10 m thick);

(d) surface failure due to rainstorms, earthquakes, etc.; or alternatively

(e) subsurface earthflows from beneath the root system;

(f) deterioration of remaining vegetation and invasion of grasses;

(g) surface fluvial erosion and exposure of bedrock.

The length of time for such a hypothetical cycle is unknown and estimates as great as 400,000 years have been made. At the other end of the timescale large landslides in Papua New Guinea are still evident from air photographs after fifty years, showing that vegetational cover has not been restored during this period.

Despite the vegetational inhibition of surface runoff in the humid tropics, drainage densities are generally greater there than in humid temperate regions (Figure 18.5). There has been little work on the initiation of first-order channels in the humid tropics but many must result from gullies produced by occasional high-intensity storms, some develop from slides and some from subsurface solution or washouts. Valley sides on Oahu, Hawaii, are intersected by small, closely spaced, parallel valleys and on Viti Levu, Fiji, significant first-order channels develop on slopes 30 m long with 15–25 m relief. On permeable volcanic rocks in Hawaii competition takes place between channels for groundwater and the first channel to incise down to the water table rapidly develops into a major valley. The special problems presented by vertical stream incision in the humid tropics will be treated on p. 477. Although humid tropical drainage densities tend to be higher on average than those of other humid areas, they exhibit a considerable range, mainly due to lithology and climatic differences. Rocks producing permeable waste (e.g. granites and sandstones) generally exhibit convex slopes and lower drainage densities, whereas where the soil has a higher clay content (e.g. schists and shales) the slopes are more concave and the drainage density greater. The general effect of higher mean annual precipitation on drainage density is well shown in Figure 18.6A for Dominica, West Indies, and in West Malaysia the wetter east coast has generally higher drainage densities (greater than 23 km/km²) than the drier west (less than 8 km/km²). The incidence of infrequent higher-intensity storms is generally recognized to increase drainage densities, as is shown in Figure 18.6B for the Central Highlands of Sri Lanka, but in some areas (such as West Malaysia) this effect is not clearly marked. Figure 18.6 suggests an interesting threshold in the precipitation/drainage density relationship at a mean annual precipitation value of some 2500 mm.

River activity in the humid tropics contrasts to some degree with that in other climatic regions. The transportation of solid load is especially varied both in time and space. Most load is transported during short periods of high discharge with, on average, 2 per cent of the highest discharges removing 50 per cent of the total solid load. It has been estimated that West Malaysian rivers transport 50 per cent of the annual suspended load in an average of twenty-four days, comparing with a figure of seven days in northern Queensland. Ei Creek, Papua New Guinea, performs 99 per cent of its annual solid load transport in less than five days (Turvey, 1975). This episodic character of fluvial erosion, combined with similarly episodic slope failure, may mean that whereas denudation rates may be low when observed over short periods, long-term averages (including high-intensity events) may be much greater. There is also a strong contrast between the action of mountain rivers and lowland ones, the former transporting more and coarser

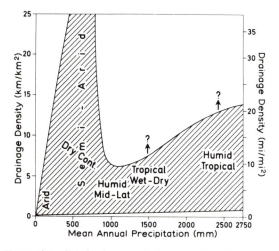

**Figure 18.5** Relation between drainage density and mean annual precipitation.
*Source:* Gregory in E. Derbyshire (ed.), *Geomorphology and Climate*, 1976, figure 10.3, p. 297; copyright © John Wiley and Sons, by permission.

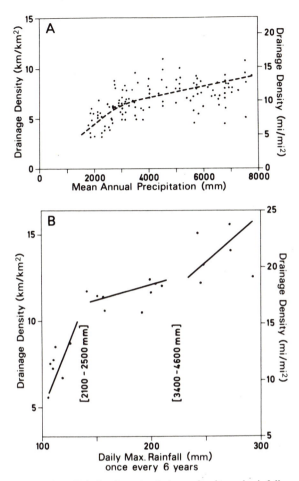

**Figure 18.6** Relation between drainage density and rainfall in two tropical areas: A. for two volcanic centres of southern Dominica; drainage lines derived from contour crenellations and each dot represents the drainage density of a 0·65 km² grid square; B. for the Central Highlands of Sri Lanka; related mean annual precipitation is given in brackets. *Sources:* Walsh, 1980, figure 6, p. 199; Madduma Bandara, 1971.

material consisting of higher proportions of quartz and less-weathered clay minerals (e.g. montmorillonite), whereas the load of the latter is often characterized by suspended kaolinite (Figure 18.7 and Table 18.3). Humid tropical rivers are generally lacking in coarse debris due to the efficacy of chemical weathering and this may seriously inhibit their ability to incise. It is, therefore, an observed feature of tropical rivers that the effects of bedrock in producing rapids and sustaining local base-levels is more marked than in temperate regions and that tropical river profiles are commonly composed of reaches of low gradient separated by rapids. However,

deep-river incision has produced striking valleys in the humid tropics, and Sparks (1972) has suggested that this may be due in part to deep weathering occurring beneath valley floors. It is natural that the foregoing features of stream activity in the humid tropics exercise an influence over both hydraulic geometry and the form of flood-plains. Plots of samples of velocity $m$, width $b$ and depth $f$ exponents for at-a-station and downstream records in humid temperate, humid tropical and semi-arid regions (see Section 12.5.1) (Figure 18.8) suggest (Park, 1977):

(a) At-a-station – little difference between rates of change of velocity, width and depth with discharge in humid temperate and humid tropical regions. Depth changes more readily than width due to high bank resistance produced by vegetation and high clay content. By contrast, semi-arid channels increase in width more readily with increasing discharge.

(b) Downstream – width changes less strikingly with increased discharge downstream in humid tropical rivers than in humid temperate ones because of the lack of coarse debris in tropical rivers. Bank coherence leads to a greater rate of increase of velocity downstream in humid tropical rivers and probably to more frequent overbank flows.

Valley floors in the humid tropics are a response to frequent flooding, and entire floodplains are commonly affected by high-velocity flood flows. Relief is provided by small-scale depositional features, such as levees, and there is commonly an abrupt transition with lower valley-side slopes at about 10°. In this zone slump and wash debris from the slopes is frequently removed, even where river terraces flank the floodplain. In eastern Jamaica, where rainfall of 1000 mm in three days is not unknown, a low terrace (1·75–3 m above the rivers) made of coarse material is inundated by hurricane floods every ten to fifteen years by rapid flow which covers it to a depth of 4·5 m (Gupta, 1975).

## 18.3 Tropical wet–dry landforms

The landforms of the seasonally wet–dry tropics make a transition between the more distinct morphometries of semi-arid areas and those of the humid tropics; the pediments and inselbergs of the former being to some extent integrated with the valley-side and floodplain features of the latter. There is thus considerable variety among 'savanna' landforms (Thomas, 1974), even though they have been characterized as possessing widely spaced, saucer-shaped valleys rising gently to steeper upper slopes outlining inselbergs, flared interfluves and flattened hilltops. The major geomorphic distinction

**Table 18.3**  Transport by two Amazon tributaries

| Amazon tributary | Location | Basin area (10³ km²) | Concentration of suspended solids (p.p.m.) dry season | wet season | Mean discharge (10¹² m³/year) | Dissolved salts erosion rate (10⁶ g km⁻² year⁻¹) | Solids erosion rate (10⁶ g km⁻² year⁻¹) |
|---|---|---|---|---|---|---|---|
| Xingu | Near Amazon mouth | 540 | 1 | 3 | 0·243 | 2·8 | 0·9 |
| Marañon | Andes headwaters | 407 | 95 | 464 | 0·343 | 92·8 | 251·5 |

*Source:* Gibbs, 1967.

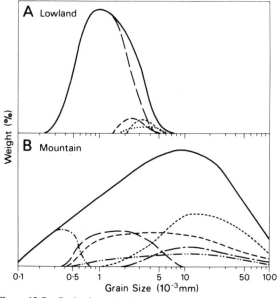

**Figure 18.7**  Grain size and mineral composition frequency (schematic) of suspended solids in A. a lowland humid tropical, and B. a mountain Andean tributary of the Amazon River.
*Source:* Gibbs, 1967, figure 14, p. 1224.

both the extent and character of tropical wet–dry landforms is that the conditions which produce and sustain them exhibit considerable spatial transition and temporal oscillation. For example, Tricart (1972) has suggested that savanna surficial processes have clearly extended into the humid tropical belt recently due to rainfall decreases being especially destructive to vegetation in areas of sandy soil (e.g. Assinie, Ivory Coast, and Pará, Brazil), due to accentuated river flooding of lowlands (e.g. the Llanos of the Orinoco, where the seasonal river rise may be 16–18 m) and due to human destruction of tropical rainforest (e.g. Batéké Plateau, Brazzaville).

Weathering in the wet–dry tropics is primarily chemical and the average depth of deeply weathered profiles decreases from an average of some 25 m in the wetter savannas to about 6 m in the drier areas. This weathering occurs more particularly in the zone of seasonal water table fluctuation where the water level rises dramatically at the beginning of the wet season and falls more slowly later, allowing a complex capillary migration of dissolved constituents in the continuous flow zone. Local weathering depths are influenced by the following factors (see especially Thomas, 1974):

(1) Rock type. Maximum local weathering depths are commonly 100 + m for some sandstones and 30 + m for igneous and metamorphic rocks, although quartz diorite in Colombia has been weathered *in situ* down to 90 m and parts of the Zambian Copper Belt basement complex to more than 1000 m, but this latter instance is perhaps due to complex hydrothermal metamorphic processes. Granites often present deeply weathered profiles consisting of corestones separated by sandy kaolin debris. In general corestones are best developed in coarse porphyritic granite and granodiorite, and less well in gneiss. Figure 18.9 gives a typical weathered granite profile from Hong Kong.

(2) Rock porosity. Certain porosities favour wide and

between the tropical wet–dry zone and those adjacent to it results mainly from its processes, there being on the one hand too sparse a vegetation cover to prevent active surface runoff by rainsplash and overland flow, and on the other considerable deep chemical weathering, especially in seasonally flooded lowland areas or localities of closely jointed rock. A further problem in identifying

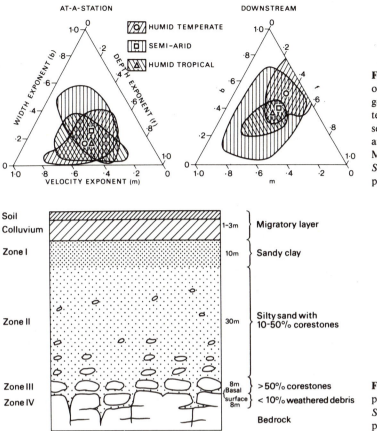

**Figure 18.8** Triaxial graphs showing ranges of at-a-station and downstream hydraulic geometry exponents in respect of humid temperate (Britain and the United States), semi-arid (Great Plains and New Mexico) and humid tropical (Puerto Rico and West Malaysia) regions.
*Source:* Park, 1977, figures 2 and 3, pp. 138–9.

**Figure 18.9** Characteristic deep-weathered profile on jointed granite in the tropics.
*Source:* Ruxton and Berry, 1961, figure 2, p. 18.

rapid water table fluctuations. Gneiss in Dahomey has been observed to be capped by a 10-m weathered zone, with smaller pores permanently filled below 2 m and no clear water table. This allows for a very rapid rise of the zone of saturation in the July rainy season and for a relatively rapid fall of 1·5 m/month immediately afterwards. In some instances increase of porespace by weathering allows the water table to fluctuate more readily, whereas in others the water table tends to stabilize more at higher porosities.

(3) *Rock jointing.* Jointing clearly facilitates weathering and, for example, the form of the subweathered surface in granite near Jos in northern Nigeria has been shown to reflect joint trends and spacing. Below about 100 m, however, confining pressure tends to close joints and inhibits both groundwater circulation and weathering.

(4) *Rainfall.* Although confused by other factors, there is a general tendency for deeper maximum depths of weathering to occur in areas of higher rainfall. In Australia locations of 250 mm mean annual precipitation commonly possess maximum weathered thicknesses of

about 30 + m, whereas in northern Nigeria (750 mm precipitation in a five-month wet season) maximum thicknesses of 50 m are common.

(5) *Topographic location.* This controls to a considerable extent the rate of surface inflow of water. In the wetter areas the deepest weathering is often beneath small seasonal channels with low gradients or under interfluves of low to moderate elevation unprotected by duricrusts, rather than beneath the larger stream courses where fine sediment may seal the surface. In drier savanna areas, by contrast, the deeper weathering is commonly under the larger stream channels.

(6) *Time and climatic change* (see Chapter 20).

The above factors produce a complex spatial pattern of deep weathering in the tropical wet–dry regions which may or may not reflect the surface topography (Figure 18.10). For example, the basal weathered surface of diorite near Medallin, Colombia, exhibits only 43–78 per cent of the relief possessed by the overlying topographic surface. This basal weathered surface is sometimes

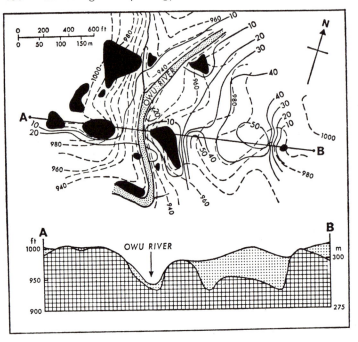

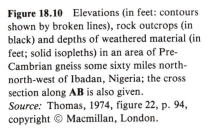

**Figure 18.10** Elevations (in feet: contours shown by broken lines), rock outcrops (in black) and depths of weathered material (in feet; solid isopleths) in an area of Pre-Cambrian gneiss some sixty miles north-north-west of Ibadan, Nigeria; the cross section along **AB** is also given.
*Source:* Thomas, 1974, figure 22, p. 94, copyright © Macmillan, London.

termed the 'weathering front', and may be locally very irregular exhibiting domes and basins.

An important geomorphic effect of markedly seasonal precipitation regimes and water table fluctuations in tropical wet–dry and semi-arid regions (as distinct from the core areas of the adjacent arid and humid tropical regions) is the production of duricrusts, which are indurated layers produced at or near the surface by the local concentration of mobile ions of aluminium, iron, silica and calcite (Goudie, 1973; Thomas, 1974; McFarlane, 1976). The chemistry of this concentration, much of which may have occurred under much earlier different conditions, is acknowledged to be of considerable complexity and a source of conjecture, but it seems likely that two major mechanisms of primary concentration may be identified:

(a) As a residuum, due to concentration in the zone

of leaching during selective downwearing of the surface.

(b) As a precipitate, due to vertical or horizontal transfer of ions by percolating water or water table fluctuations with reference to an already formed erosional surface.

In addition, secondary duricrusts occur on valley sides and at lower levels resulting from the erosion, transportation, deposition and renewed induration, of primary duricrust material. The composition of duricrusts varies in the proportions of Al, Fe, Si and Ca involved and, in particular, many combinations of the first three are found in the field, as the profile developed from granite at Samaru, Nigeria, indicates (Table 18.4). Nevertheless, it is helpful to identify four major types of duricrusts according to chemical composition.

**Table 18.4** Duricrust composition at Samaru, Nigeria

| Designation | Depth (cm)(ins. in brackets) | % composition | | | |
|---|---|---|---|---|---|
| | | $SiO_2$ | $Al_2O_3$ | $Fe_2O_3$ | Kaolin |
| Hard laterite | 0–81 ( 0–32) | 35·6 | 18·8 | 33·9 | 15·0 |
| Soft laterite | 81–163 (32–64) | 55·2 | 21·4 | 13·7 | — |
| Granite saprolite | 163–305 (64–120) | 67·3 | 18·4 | 6·3 | 40·0 |

*Source:* Thomas, 1974, table 8, p. 62, after Sivarojasingham *et al.*, 1962.

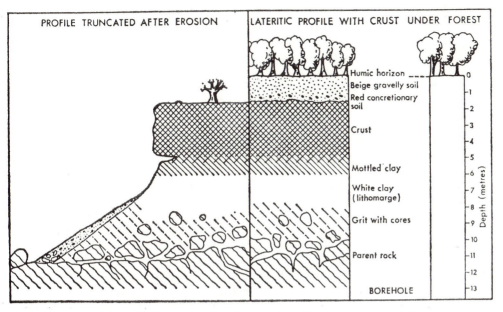

**Figure 18.11** Lateritic profile developed under a forest cover and its equivalent truncated by erosion in the more arid Sudanese region.
*Source:* Thomas, 1974, figure 12, p. 54, after Millot, 1970.

(1) Bauxite. This product of weathering has already been mentioned in Chapter 9 and seems to have developed from rocks rich in Al and poor in Fe (e.g. Phyllites and basic igneous rocks), occurring at well-drained sites of intense leaching, where rainfalls in excess of 1200–1500 mm/year enable the differential removal of the more mobile Fe ions. Bauxite, however, is usually a poorly indurated deposit which has little significance in producing or sustaining distinctive landforms.

(2) Laterite. This iron concentrate is the most widespread and geomorphically significant of duricrusts and – in distinction to semi-arid iron incrustations – does not appear to form under mean annual precipitations of less than 1200 mm on acid rocks or 1000 mm on basic rocks. On the other hand, it appears to form under tropical forests and may become exposed and hardened by the stripping off of soil and vegetation. Laterite is very variable in composition, thickness and resistance even over short distances, but Figure 18.11 shows an ideal profile consisting of the following (see Thomas, 1974, pp. 53–4):

surface soil : 0–2 m;
laterite : 1–10 m of indurated iron concentrate (30–80 per cent iron oxides). This layer sometimes exceeds 15 m, but seldom exceeds 2 m thick over wide areas;

mottled clay : 1–10 m, often containing abundant quartz grains;
pallid zone : 0–25 m of kaolin and quartz sand. This zone is very variable in composition and thickness, in places being as much as 60 m thick;
partly weathered
bedrock : up to 60 m thick.

Laterite appears to form in the surficial decomposition zone of iron-rich rocks where the seasonal rainfall regime causes considerable water table fluctuation, where the rainfall is heavy enough to cause leaching and deep weathering; in the presence of organic acids; and in locations where water can move laterally. Vertical weathering involving concentration by residual iron accumulation (it is estimated that the weathering of 20 m of granite in Buganda, East Africa, yields 1 m of laterite) and upward iron migration from the pallid zone cannot account for the observed thicknesses of laterite, and lateral migration of iron is locally important on slopes and valley bottoms. Laterite is by far the most geomorphically significant of the duricrusts and has been especially well studied in East Africa and Australia. It seldom extends over wide areas, but occurs locally in the following situations (McFarlane, 1976):

(a) Tabular capping on hill tops and undulating elevated surfaces. These form characteristic mesas, for example, in Uganda, eastern Zaire and French Guinea. Subsequent subsidence and cambering may produce tilts of a few degrees. Laterite is most commonly associated with surfaces of low relief, but it appears to be formed on slopes as steep as 20° with a local relief of 60 m in south-western Australia. It is natural that fragmented laterite-capped surfaces should have formed the basis for the identification and attempted correlation of supposed erosion surfaces, particularly in East Africa and around Bouré, Upper Volta. Laterite-capped mesa surfaces occur primarily at 1650 m (5400 ft) and 1420 m (4660 ft) in western Uganda, at 1430–1460 m (4700–4800 ft) to the west of Lake Victoria, at several levels between 1280 m and 1340 m (4200 ft and 4400 ft) north of Lake Victoria and at about 1420 m and 1525 m (4650 ft and 5000 ft) in central Uganda. Attempts to correlate these surfaces locally and regionally by elevation have variously involved postulates of undulating single surfaces, different surfaces, differential warping and faulting of single surfaces, and the irregular formation of protecting laterite on the higher parts of single undulating surfaces.

(b) Exposed ledges and valley-side benches; for example, around Rwampara, Uganda. In areas of higher rainfall (e.g. Buganda, Uganda and the Sula Mountains, Sierra Leone) lateritic formations on slopes become very prone to slumping, so complicating their stratigraphic locations.

(c) Footslope and pediment recemented accumulations of lateritic debris. These pediment accumulations are particularly notable around Lake Kyoga, Uganda.

(3) Silcrete. This is a silica-rich accumulation (95 per cent $SiO_2$), often overlying a kaolinized zone, formed within the weathered zone. It is up to 3 m thick, extremely resistant and having columnar structure. It appears to form under drier conditions (250 mm mean annual precipitation), where drainage is poor, or in areas of frequent sheetwash and where the groundwater has a high pH. Like laterite, it appears to be both accumulative and residual, and may even form in association with laterite or as a secondary product of it. Silcrete sheets are widespread in central and southern Australia (e.g. southern Flinders Range and Arcoona Plateau, South Australia) and in Southern and Central Africa, where they form important topographic caprocks.

(4) Calcrete (caliche). Although mainly a product of semi-arid and arid conditions, calcrete may be included in the above list. It is composed of thin layers of 80 per cent rich calcium carbonate formed by the growth and coalescence of small carbonate nodules in the soil. It appears to be formed in areas of carbonate-rich groundwater where sheetfloods dry out, capillary rise occurs and the water table oscillates close to the surface. Calcrete produces minor topographic capping in such regions as the south-western United States and southern and western Australia.

Rivers in the tropical wet–dry region may be instructively compared with those of the adjacent morphogenetic regions. A study of drainage densities in Guyana obtained from 1:50,000 photogrammetric maps of the rainforest west of the Essequibo River and of the savanna grassland in the south of the country contains a faint suggestion that they may be less in the savanna region but above 3000 mm/year there is certainly no further increase (Figure 18.12; but see also Figure 18.6). However, the variance is such that regional variations in drainage density may be due to other causes, or may be absent particularly because of the prevalence of wide, shallow valleys in the savanna region not easily identifiable from maps. Studies of the Rufigi River, the largest in Tanzania, draining a large area within which the mean annual precipitation varies spatially from 200 mm to more than 1800 mm show (Temple and Sundborg, 1972):

(i) A great variation of mean daily discharge. At Stiegler's Gorge the range was 69–6637 m³/s during the period 1954–70, with a daily value of 5000 m³/s, having a return period of five years. During the January to April flood period rising limb differences of more than 3000 m³/s and falling limb differences of 2000 m³/s were observed over consecutive 24-hour averages.

(ii) Sediment transport showed similar wide variations (Figure 18.13). Although rising limb transport for a given discharge was significantly greater than falling limb transport for the same discharge, the variances indicated that other factors (e.g. time of year and vegetational cover) were also very important controls over sediment availability.

The large size of this drainage basin (156,000 km²) ensures a lower than expected average denudation rate of 0·06 mm/year, but mean annual sediment yields for small savanna catchments in Tanzania tend to be significantly greater than those for basins in eastern Wyoming of equivalent size greater than about 0·5 sq. mile (1·3 km²) (Figure 18.14).

The integrity of wet–dry tropical landforms as a geomorphic class is brought into question by their range of morphometry and by the variety of erosional environments in which they are found. Difficulties are compounded by the exaggerated influence of rock type on

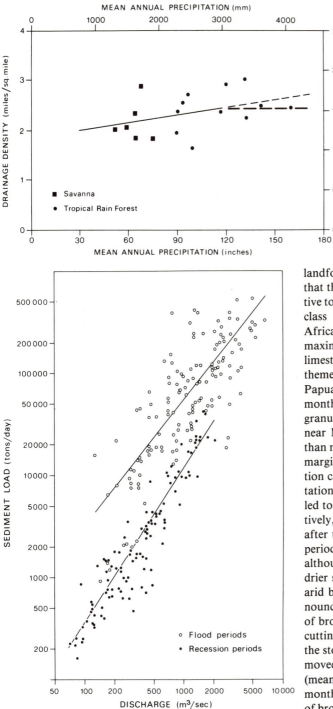

**Figure 18.12** Relationship between drainage density and mean annual precipitation in respect of savanna and tropical rainforest environments in Guyana; the regression has a coefficient of correlation of 0·42, and it is very questionable whether drainage density varies with precipitation, especially above 3000 mm.
*Source:* Daniel, 1981, figure 2, p. 152, copyright © Elsevier, Amsterdam.

**Figure 18.13** Sediment rating curves for flood and recession periods of the Rufigi River, Tanzania, at Stiegler's Gorge. *Source:* Temple and Sundborg, 1972, figures 17 and 18, p. 364.

landforms, especially in drier regions, and by the fact that the tropical wet–dry zone has been especially sensitive to past climatic oscillations. Slopes attributed to this class range from pediment-like features in the drier African savanna, through convexo-concave slopes of maximum angle 15°–21° on moderately dipping marls, limestones and mudstones covered with grass and themeda–eucalyptus savanna near Port Moresby in Papua New Guinea (1150 mm in more than five wet months) (Figure 18.15A), to steep V-shaped valleys in granulite cleared of forest with soils less than 25 cm thick near Morogoro, Tanzania (approx. 1500 mm in more than nine wet months) (Figure 18.15B). In drier savanna margins the coarse weathered debris gives high infiltration capacities such that, for example, artificial precipitation on a weathered gneissic slope at Ibadan, Nigeria, led to 284·5 mm/h, 198 mm/h and 82·5 mm/h, respectively, during three successive half-hour periods; and after the equivalent of 132 mm of rain during the first period no overland flow was observed by Thomas (1974), although there was some seepage at the slope foot. In the drier savanna regions, which would be classed as semiarid by the criteria of Figure 18.1 except for their pronounced wet season, the terrain is commonly composed of broad, shallow basins and valleys with stream downcutting possibly inhibited both by the sporadic nature of the storm rainfall and by the abundance of infrequently moved debris. In the region near Dodoma, Tanzania (mean annual rainfall 567 mm during a four-to-five-month wet season), much of the granitic terrain consists of broad gentle valleys dominated by seasonal rainsplash and sheetwash erosion, with gullying only becoming important on the steeper hillslopes close to the interfluves. In somewhat wetter areas the break of slope is less marked or absent and the landforms consist of broad

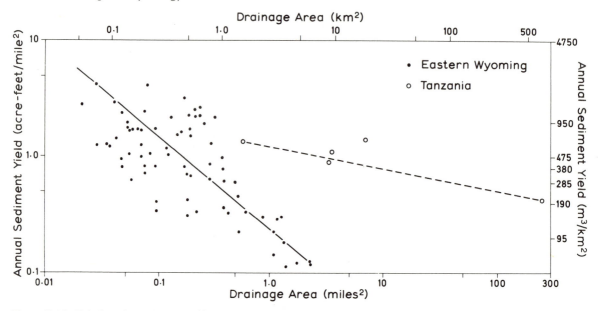

**Figure 18.14** Relation of mean annual sediment yield to drainage area for five basins in Tanzania and seventy-three basins in eastern Wyoming.
*Source:* Rapp *et al.*, 1972b, figure 48, p. 313; Wyoming data from Schumm and Hadley.

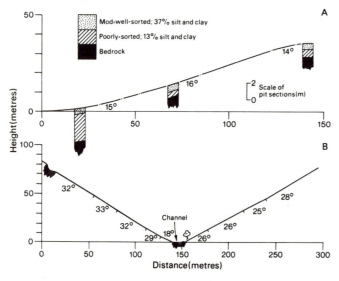

**Figure 18.15** Surveyed valley-side profiles in the tropical wet–dry region: A. on the marls, limestones and mudstones of the Port Moresby Group, Papua New Guinea; B. on granulite in the Morogoro River catchment, Tanzania.
*Sources:* Mabbutt and Scott, 1966, figure 4, p. 74; Rapp *et al.*, 1972a, figure 6, p. 132.

open valleys, sometimes with deep regolith, made up of wide alluviated valley floors and gentle bedrock slopes leading to steep interfluve slopes (up to 25°) in the dissected divide areas where vertical jointing (e.g. in crystallines and sandstones) can produce ramps and scarps. In parts of Malawi (mean annual precipitation 1000 mm in four wet months) a series of broad open valleys (up to 1600 m wide) have maximum slopes of only 2°–8°.

Process studies on low savanna slopes on sandstone and granite 150 km south-east of Darwin, northern territories of Australia (mean annual precipitation 1235 mm in five wet months) have yielded the process rates shown in Table 18.5. These figures suggest that in this savanna region there are no great differences between creep and slopewash action on granite and sandstone, and that slopewash is five times more effective as an erosive agent

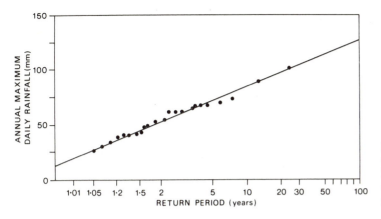

**Figure 18.16** Return periods in years for annual maximum daily rainfalls at Mgeta Mission, Tanzania (elevation 1020 m). *Source:* Temple and Rapp, 1972, figure 3, p. 166.

than creep. Comparing these figures with equivalent measurements in temperate New South Wales (in brackets), there is a suggestion that in the savanna region creep on granite may be greater and the erosion of sandstone by slopewash less. In general, however, there is a surprising lack of difference between the efficacy of savanna and temperate creep and slopewash processes on low-angle slopes. In interfluve valley areas the processes are somewhat different. At Mgeta Mission, Tanzania (mean annual precipitation 950 mm in seven wet months), a twenty-five-year rainstorm (Figure 18.16) of 100·7 mm in 24 h produced slides, mudflows, rilling and sheet erosion but not much gullying on steeper slopes (maximum angle > 36°) in granulites and meta-igneous rocks. Similarly, a storm of 472 mm in 72 h at Rio de Janeiro (mean annual precipitation 1093 mm in six wet months) produced several hundred slides. In the Ivory Coast average rains produce little overland flow but storms of 90–100 mm during the wet season may give rise to 30 per cent unconcentrated surface runoff. It is clear that in such areas one is verging on humid tropical geomorphic slope conditions.

**Table 18.5** Creep and slopewash in northern Australia

| Rock | Slope angle | Creep rate (cm³ cm⁻¹ year⁻¹) | Slopewash erosion rate (mm/1000 year) |
|------|-------------|------------------------------|----------------------------------------|
| Granite | 1°–3° | 7·33 (1·90)* | 54 ± 40 ( 54 ± 73) |
| Sandstone | 1°–11°35′ | 4·39 (3·25) | 56 ± 30 (103 ± 90) |

*Figures for temperate New South Wales.
*Source:* Williams, 1973.

## 18.4 Arid and semi-arid landforms

It has already become apparent that the simplistic attempt in Figure 18.2 to subdivide both climates and climatic landform assemblages into neat groups on the basis of four average climatic variables has been less than satisfactory. The difficulties become especially acute when we are considering climate and landforms within the drier tropical wet–dry, semi-arid and arid range, as well as the warmer dry continental regions (see Figure 18.2). This is due to:

(1) the lack of any satisfactory correlation between mean annual rainfall amounts and the length of the wet season;

(2) the especial susceptibility of these environments to recent and erosionally drastic climatic changes;

(3) the marked variability, both in time and space, of precipitation – annual, monthly and in terms of individual storms. Although, for example, much of the Utah plateau province experiences less than a dozen storms a year capable of producing any erosion at all, the figures in Table 18.6 for extreme monthly ranges for a period of only forty-three years show the possibility of rainy months occurring at any time of year.

**Table 18.6** Precipitation: Springdale, south-west Utah (43-year period) (elevation 1234 m) Extreme monthly ranges (mm) (mean annual: 381 mm)

| | |
|---|---|
| January 0–124 | July 1–91 |
| February 0–127 | August 0–108 |
| March 1–108 | September 0–110 |
| April 4–138 | October 0–135 |
| May 0–54 | November 0–82 |
| June 0–26 | December 0–118 |

*Source:* Gregory, 1950, p. 31.

The south-western United States, as a whole, has an average of only two months (July and August) which receive more than 50 mm precipitation, but for the most part there are usually none (Table 18.6).

The precipitation table introduces another complexity in terms of attempts to define the environments within which landforms in the dry tropical wet–dry to arid ranges develop. This involves the considerable elevations capable of sustaining generally low precipitation, together with hot summer conditions. Much of the south-western United States experiences such conditions, which also exhibit certain nival features in winter which are unknown in most other dry areas at lower elevations and at lower latitudes. Taking all the above problems into account, what follows is an attempt to generalize the landform types associated with a wide range of subhumid conditions, excluding the aeolian processes and landforms which appeared in Chapter 16.

Hot arid and semi-arid climatic regions, plus adjacent dry savanna (i.e. dry tropical wet–dry) types and dry continental areas with hot summers and cold winters cover more than 33 per cent of the earth's surface (see Figure 18.3; note projection is not equal area), including some 9 per cent having dry continental climates in the south-western United States, Central Asia and southern South America. Despite the popular conception, tracts of sand form only a small part of dry areas, 15 per cent of the Sahara and only 2 per cent of the American southwest. The remainder of the regional climatic types listed above are currently made up of:

(a) Mountain fronts, often intensely dissected by ephemeral channel systems, and exhibiting badlands where uncapped shale formations are exposed. Some slopes are bare, some boulder-controlled.

(b) Pediments, low-angle (4° or less) surfaces of bare or thinly mantled bedrock sloping away from the mountain fronts or low residuals.

(c) Residual hills (inselbergs, mesas, buttes, etc.), commonly steep-sided, situated on or near the pediments, especially in the vicinity of low divides.

(d) Debris-covered surfaces of low slopes (less than 2°) made up of residual weathered material on the pediment or of transported material further out from the mountain front. Many of these surfaces are mantled with coarse debris and in the Sahara are termed the *hamada* and the *reg* (see pp. 493–4).

(e) Playas are the flat, fine-grained deposits of ephemeral lakes in closed desert basins.

(f) Mountain and upland surfaces which are either dissected by valleys cut by ephemeral streams (possibly during past pluvial episodes) or present low relief controlled by horizontal strata or lateritic covers.

However, many landforms that are assumed to be typical of more arid areas are not, but they are best exposed in deserts. For example, alluvial fans can be found wherever a sediment-source area delivers large quantities of sediment to an adjacent lowland (see Chapter 14.1). Inselbergs and pediments have been described in more humid regions (e.g. tropical wet–dry and even humid temperate), and sand dunes can form wherever there is a sand supply, wind and sparse vegetation, for example, along coasts. Similarly, despite the wide variety of landforms found in the present tropical wet–dry regions, an accident of geology and climatic zonation has conjoined large areas of granite and crystalline rocks with contemporary savanna conditions to exhibit the supposedly classic tropical wet–dry morphogenetic inselberg terrain; as is found in such areas as Uganda, Madras, South Australia, southern Nigeria, Mozambique, eastern and southern South Africa and south-eastern Brazil. It is here that the central dilemma of climatic geomorphology finds its sharpest expression, namely to what extent are the landform characteristics attributable to distinctive climatic regimes and to what extent to structural, lithological, or past climatic causes?

Desert rocks are frequently dark, being stained by varnish, which is primarily a weathering phenomenon that involves solution of manganese and iron from the rock and its precipitation on the surface, possibly associated with bacterial action. The moisture may be provided by dew. As has been suggested in Chapter 9, by no means all of the rock weathering in deserts is due to physical processes alone. Recently it has been found, at least on rocks collected in the deserts of North America and Australia, that under these conditions of extreme heat and dryness bare rock surfaces are colonized by lichens. In addition, algae grow in crevices in desert rocks and endolithically beneath the rock surface. Even in the dry valleys of Antarctica, desert lichens grow between the crystals of porous rocks. The rocks must be sufficiently porous to permit colonization, as well as translucent, in order that photosynthesis may occur. A typical growth layer shows an upper black zone 1 mm thick, then a white zone 2–4 mm thick and below this a green zone. All these zones are formed by fungi and unicellular green algae, which form a symbiotic lichen association. The presence of these micro-organisms assists rock weathering by dissolving the cementing materials at the lichen level. The rock then exfoliates at the lichen level, and a new lichen zone is formed at depth. Even in extreme hot and cold desert regions, processes of physical and chemical weathering are significantly enhanced by biological activity.

### 18.4.1 Pediments

Perhaps the landform that has created the greatest interest in arid-land erosional processes is the pediment. Early travellers in the desert south-western United States were surprised to find their wagon-wheels grinding over relatively flat bedrock surfaces at the foot of nearby mountains (Figure 18.17). The obvious question is how did such features form?

Pediments were first named by G. K. Gilbert, in 1877, in respect of the flights of gravel-covered shale surfaces north of the Henry Mountains in Utah (e.g. soft-rock pediments), but similar low-angle features backed by mountain fronts had previously been noted in very different granite terrains to the south-west in Arizona (e.g. hard-rock pediments) (Plate 26). Pediment-type features thus occur in a variety of geological situations, but certain conditions seem necessary for the production of clearly defined pediments:

(1) Conditions which allow a backing slope to develop from which material is constantly and uniformly being removed – e.g. a repose slope of granite boulders augmented by weathering beneath and dissipated by weathering break-up into individual, easily transported constituent crystals at the surface; or by a resistant caprock overlying a slowly retreating shale slope.

(2) Erosional events which will regularly and uniformly strip the steep slope of transportable material, allowing it to retreat parallel.

(3) The ability of the slope base (piedmont junction) to migrate.

(4) Processes available to allow the slope base to migrate in a controlled manner without excessive stream dissection. These processes are commonly associated with a lack of trapping vegetation, occasional intense sheetflood runoff or basal stream sapping (dominant in valleys and near the exit point of rivers draining from the highlands). This active migration

permits the maintenance of a well-defined piedmont angle.

Parallel slope development and the consequent production of simple skirting soft-rock pediments can take place in shales and other fine-grained rocks of low resistance when they are capped by rock which is resistant to surface stream action (e.g. sandstone or permeable gravel) or where the shale tends to weather seasonally into a fragmental layer over the whole slope (e.g. due to the swelling and cracking of sodium-rich clay minerals or to frost action) and to be completely removed by surface runoff during the rainy season. Where shales which are susceptible to stream incision are exposed by the stripping off of a more-resistant cover (e.g. the Wingate Sandstone from the Chinle Shales which form the Painted Desert) or near the base of a retreating cliff (see Figure 11.35B and C), then *badlands* are formed in areas of sparse vegetation. Badlands are so called because their very high drainage densities and steep slopes make them very difficult to traverse. Geomorphologists have shown considerable interest in badlands because they provide a model of arid erosion, and because erosion is rapid. During a period of a few years information can be collected on the manner and rate of hillslope retreat (see Chapter 11), as well as on alluvial fan and pediment formation. The experimental studies discussed in Chapter 13 are in reality a study of badland processes in the laboratory.

Studies of pediment formation in Badlands National Monument, South Dakota, provide information on miniature pediment formation by slope retreat. During a nine-year period erosion of the siltstone slopes and pediments extended the pediment. In addition, there was regrading of the upper part of the pediment and burial of the pediment as the alluvial cover advanced upslope (Figure 18.18). This process must be occurring at many locations at a larger scale, and the badland study provides an insight into one type of pediment development.

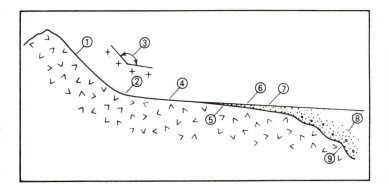

**Figure 18.17** Terminology of pediments and related features: 1, backing hillslope; 2, piedmont junction; 3, piedmont angle; 4, pediment; 5, mantled pediment (suballuvial bench); 6, alluvial mantle (feather-edge); 7, alluvial plain (bajada); 8, alluvial fill; 9, suballuvial floor. *Source:* J. A. Mabbutt, *Desert Landforms*, 1977, figure 16, p. 82, copyright © MIT Press, Cambridge, Mass.

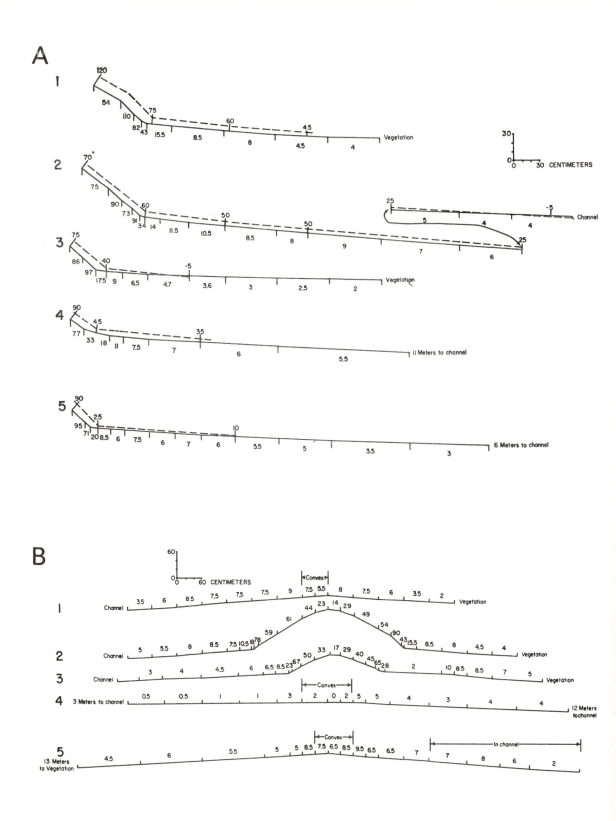

Using the ergodic hypothesis, it was possible also to study longer-term badland slope retreat and pediment formation. Figure 18.18B shows a series of profiles that demonstrate the eventual reduction of a miniature mountain range and coalescence of pediments to form a gently sloping pediplain.

A somewhat different type of pediment formation occurs along the Book Cliffs in western Colorado and eastern Utah, and around the flanks of the Henry Mountains, Utah. The pediments form on readily erodible Mancos Shale, and they are capped by as much as 10 m of poorly sorted bouldery alluvium derived from the overlying Mesa Verde sandstone. The pediments are formed by normal fluvial erosion, stream capture and source area rejuvenation. These are 'soft-rock' pediments that form as follows (Figure 18.19): a stream draining from the highlands delivers coarse sediment to the piedmont. Because of its coarse sediment load, it has a steeper gradient than the piedmont streams that are eroding only shale (Figure 18.19A). One of the piedmont streams captures the upland stream leaving its former course as a pediment remnant (Figure 18.19B). The rejuvenation of the upland stream produces a flood of gravel that buries the irregular fluvial topography and forms a relatively smooth gravel surface. Streams in unmantled areas continue to erode and eventually another capture occurs. In this way with one baselevel lowering, or indeed continuous baselevel lowering, multiple pediment surfaces form (Figure 18.19C). The dissection of the pediment remnants provides excellent exposures of the alluvium–bedrock contact, and it is possible to identify old drainage divides that separated shale drainage basins now buried beneath the alluvial cover. In this area the bedrock surface had as much as 30 m of relief before burial. Hence the relatively smooth alluvial surface is misleading. It is a depositional surface that obscures the irregularity of the pediment, in contrast to the smooth surfaces of pediments resulting from parallel scarp retreat. The Book Cliffs pediment closely resembles pediments elsewhere, and they conform to the definition of pediments as a bedrock suface mantled with gravel; however, they provide an excellent example of convergence (see Chapter 1) because they are a piedmont landform that is basically unlike the pediments formed by parallel scarp retreat (see Figure 18.18) or lateral stream planation.

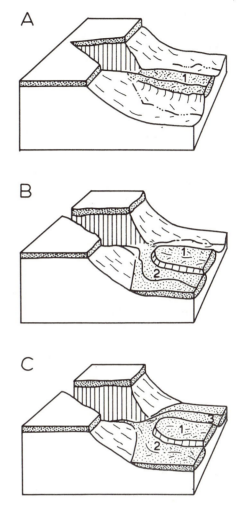

**Figure 18.19** Pediment formation on shales from an initial gravel-covered surface, 1; by successive captures, 2 and 3; giving a flight of pediments.
*Source:* Carter, 1981.

Pediments are characteristic of a wide variety of climatic situations but appear to be especially prominent in the dry savanna (i.e. drier tropical wet–dry) and semi-arid locations. However, it is just in such regions that past climatic changes have been most active and aggressive, and it is therefore impossible to link pediment formation with a narrow range of present climates.

**Figure 18.18** Slope profiles in Badlands National Monument, Dakota A. profiles in May 1961; numbers below profiles are slopes in percentages, those above profiles are erosion depths in millimetres at stake positions; broken lines represent profile in July 1953. B. Profiles illustrating headward extension and coalescence of miniature pediments: the profiles were measured at the same time in different locations; numbers above profiles give slope in percentages.
*Source:* Schumm, 1962, figures 1 and 2, pp. 721, 723.

Pedimentlike features, perhaps residual, have been identified in presently humid temperate regions. An example of a hard-rock pediment has been described on a granite at Balos in east-central Sudan (mean annual rainfall 650 mm in less than five wet months) (Figure 18.20) (Ruxton, 1958). Pediments in such an environment exhibit gradients of 1°–9°, grade downslope into basal

plains covered with less than 2 m of clay which appears to have been washed laterally out of the weathered debris covering the pediment and hillslope. At the head of the pediment the piedmont angle is not as sharp as in truly arid areas, and it is the site of localized chemical weathering associated with the seeping out of throughflow from the debris-covered hillslope (10°–40°) (see Figure

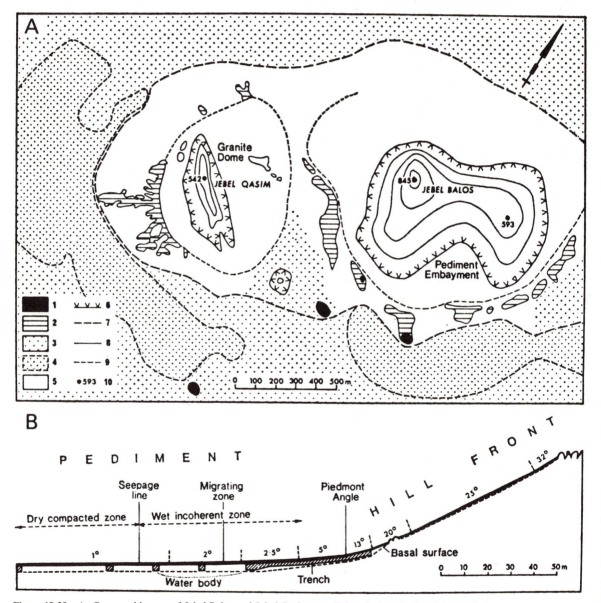

**Figure 18.20**   A. Geomorphic map of Jebel Balos and Jebel Qasim near Balos, Sudan: 1, closed depressions; 2, eluviated clay; 3, long grass/thin bush; 4, tall grass/thick bush; 5, short grass/some trees; 6, piedmont angle; 7, sand/clay boundary; 8, form lines; 9, seepage line; 10, heights: in feet above sea level. B. Slope profile on the south-west side of Jebel Qasim.
*Source:* Thomas, 1974, figure 25, p. 105, after Ruxton, 1958, figures 2 and 3a, pp. 355, 356.

18.20B). Little overland flow appears to occur on these slopes even during the intense storms of the short wet season, but on the pediments and basal plains rainsplash and sheetfold erosion are dominant, especially during the early part of the wet season before the vegetation cover has been augmented. It should be noted that pediments can be modified by bedrock weathering beneath the alluvial mantle and subsequent erosion of this material during a period of higher runoff or during a baselevel fluctuation. The result will be an irregular bedrock surface. Hence the originally smooth surface can become irregular as a result of differential weathering and stream incision.

The most striking feature of many of the drier regions of the lower and lower-middle latitudes are extensive plains, best developed on granites, gneisses, schists and phyllites. Such surfaces have low gradients (generally less than 3°, except in association with some gullies); cut across different rock types; are variously composed of bedrock, weathered bedrock and thin alluvial covers; and occur over wide areas (e.g. in Southern Africa, Guyana, the Rio Branco region of Brazil and the Indian Deccan). The most continuously extensive of such surfaces has been proposed to extend from the southern Sahara southwards to Upper Guinea and Asande (total extent 5000 km EW and 500 km NS). Such surfaces have been attributed to:

(a) Pediplanation by scarp retreat and the coalescence of pediments under more arid conditions, as suggested by King (1967). An example given of such a surface is that of the 'African' planation surface forming the Highveld of South Africa (900–1675 m; 3000–5500 ft) and possibly formed during the Early Tertiary.

(b) Büdel's (1970) double surface mechanism. An upper wash surface of erosion cuts across planed bedrock and deeply weathered profiles (Figure 18.21A) above which rise 'outlying inselbergs' formed from the irregular lateral retreat of the debris-covered mountain slopes by basal sapping (Figure 18.21B) and 'shield inselbergs'. The latter are more resistant parts of a lower basal weathering surface (weathering front) which are exhumed by progressive erosion of the upper wash surface. On occasions large areas of the basal weathering surface may be exposed by the stripping off of the *in situ* weathered debris (due to uplift or climatic change) to form an irregular 'etchplain'. Such a surface has been proposed to exist around Adrar des Iforas in the southern Sahara (Thomas, 1974).

### 18.4.2 Inselbergs

Rising above flanking pediments and pediplains are *inselbergs* (German, 'island mountains'), the isolated remnants of once more extensive upland areas (see Twidale, 1964; Büdel, 1965; Thomas, 1974).

Inselbergs occur on a wide range of resistant outcrops but are most common on granites, gneisses, syenites and

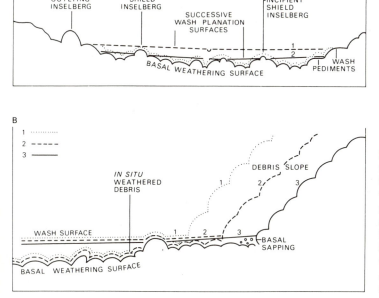

**Figure 18.21** Formation and extension of tropical planation surfaces: A. double levelling plain showing the formation of Rim washpediments and shield inselbergs as a result of a lowering of the upper wash surface; B. retreat of the debris slope by basal sapping processes, with simultaneous lowering of upper wash surface and basal weathering surface.
*Source:* Thomas, 1974, figure 40, p. 232, after Büdel, 1970, figure 11, p. 51.

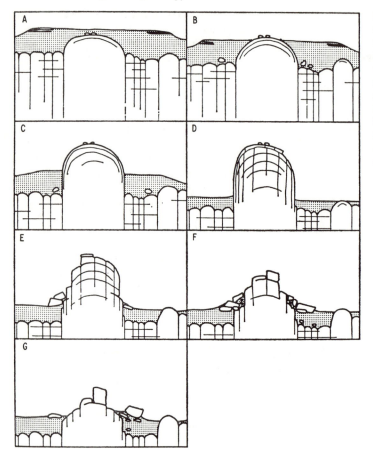

**Figure 18.22**   A schematic diagram to show one mode of development and decay for domed inselbergs. In diagrams A.–D. both the landsurface and the basal surface of weathering are differentially lowered around the developing domed summit which is stripped of regolith at an early stage. Stages D.–G. show the disintegration of the dome into a kopje as a result of tensional jointing and peripheral collapse.
*Source:* Thomas, 1974, figure 36, p. 201 and 1965, figure 2, p. 72.

diabase rocks (Plate 27). In terms of their geometry inselbergs can be divided into two classes:

(1) Domed inselbergs (bornhardts). These are broadly rounded forms, the shape of which is largely controlled by massive jointing and sheeting. These exhibit a variety of basal conditions, some having steep junctions whereas others do not, and some rise from bedrock pediments whereas others appear to be surrounded by deeper *in situ* weathering (Figure 18.22).

(2) Tors. These consist of piles of smaller blocks. They frequently occur together with domed inselbergs and their mode of formation may be similar, their differences possibly being due to the differences in size and resistance of the jointed corestones (Figure 18.23).

Inselbergs possibly develop by two mechanisms which frequently merge into each other – as subaerial residuals, or as exhumation features. The first mechanism may produce inselbergs as residuals on more massively jointed or preferentially located rock, remaining either during downwearing accompanying accelerated differential *in situ* weathering (as has been proposed for parts of south-east Brazil and the Ivory Coast) or by scarp retreat (as King has proposed for parts of southern Africa). The occurrence of structurally controlled inselbergs rising from little-weathered bedrock pediments supports such a mechanism. It has been suggested that, for example, the 'sugarloaves' of Rio de Janeiro may be the exposed cores of otherwise fractured rock masses (Twidale, 1982). Figure 18.24 shows how, during folding, a rock mass, although under general compression, develops tension fractures that permit weathering to weaken the rocks surrounding the massive cores. When the weathered rock is eroded, the inselbergs appear as steep-sided residuals on an erosional plain. On the other hand, exhumation mechanisms involve the association of differential deep weathering and surface degradation either concurrently or consecutively (Figures 18.22 and 18.23). The reality of such mechanisms is supported by the observed depths of local weathering in the wet-dry

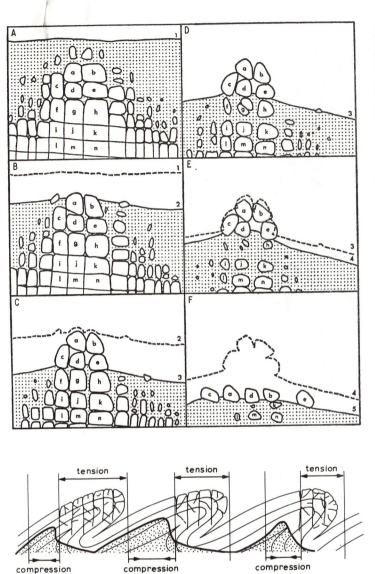

**Figure 18.23** Schematic representation of the development and decay of granite tors A. Differentially weathered granite (joint blocks are indicated a–n; B. emergence of joint blocks **a**, **b** as a result of ground surface lowering from 1 to 2; C. tor comprised of blocks **a–e** exhumed from regolith following further surface lowering from 2 to 3; D. foundations of tor become rotted as blocks **f–n** remain subject to continued groundwater weathering; E. tor subsides as ground surface is lowered from 3 to 4; F. tor becomes reduced to a group of residual corestones as further lowering of surface and continued groundwater weathering take place (3 to 4 to 5).
*Source:* Thomas, 1974, figure 34, p. 188, after Linton, 1955 and Thomas, 1965.

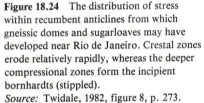

**Figure 18.24** The distribution of stress within recumbent anticlines from which gneissic domes and sugarloaves may have developed near Rio de Janeiro. Crestal zones erode relatively rapidly, whereas the deeper compressional zones form the incipient bornhardts (stippled).
*Source:* Twidale, 1982, figure 8, p. 273.

tropics, the fact that domed forms appear to be favoured in some circumstances by subsurface weathering, the existence of weathering or laterite remains on some inselberg summits, and the occurrence of weathered zones surrounding many inselbergs (Figure 18.25). This instance highlights the problems of attempting to relate specific landform assemblages unequivocally to present climates. The inselberg region shown in Figure 18.25 at present has a tropical wet–dry climate (mean annual rainfall approx. 1300 mm in more than eight wet months), yet only 480 km north the mean rainfall is 689 mm in four wet months. Thus there is here a relatively narrow zone of considerable climatic gradient across which climates have oscillated violently between moist savanna (i.e. virtually humid tropical) and semi-arid even during the last few thousand years.

### 18.4.3 Desert surfaces

As has been suggested, apart from minor areas of sand (which may be locally important), desert surfaces are mainly composed of bedrock or fragmental material. This latter material may be either the residual deposits of weathering or selective evacuation, or those resulting from debris transportation and deposition.

*Hamada* surfaces, characterized by cobbles and

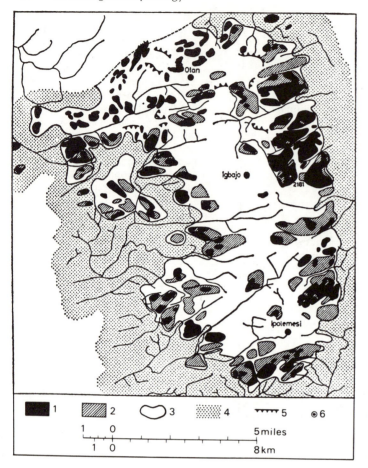

**Figure 18.25**  Distribution of domed inselbergs on the granite massif of the Igbajo Plateau, southern Nigeria: 1, bare, domed inselbergs; 2, soil-covered, domed hills; 3, weathered, dissected granite; 4, surfaces on basement gneiss; 5, major breaks of slope; 6, settlements.
*Source:* Thomas, 1974, figure 30, p. 169, copyright © Macmillan, London.

boulders, are established by the nature of the underlying rock, or rock derived from a nearby escarpment and they are residual. Hence the hamada is a mantle protecting the underlying sediments or rock. The *reg*, on the other hand, is usually deposited by streamflow or sheetfloods and the coarser component is concentrated to form a *desert pavement*. The concentration process can be by deflation, as the finer sediments are blown away. This sorting process leaves the coarser fraction of the sediment on the surface with unsorted sediments beneath the armour, and is effected by rainbeat and surface runoff that carries away the finer sediments.

As one follows the pediment or alluvial fan into a closed desert basin eventually an ephemeral lake will be reached. These *playa* lakes fill with water during periods of runoff from the adjacent highlands, but the water quickly evaporates leaving a deposit of fine sediments and evaporites. Playa-like surfaces may provide excellent natural landing-fields for aircraft when they are dry, but as they are alternately wetted and dried they can behave

like mudpies and they form mud cracks of huge dimensions. For example, the cracks are often 5 m deep. Desiccation polygons can be 300 m across, and in 1962 50-m diameter polygons formed in Rogers Playa, California (Neal *et al.*, 1967, 1968). It is interesting that polygonal troughs up to 20 km across have been observed on Mars (Baker, 1981).

## 18.5  Cold region landforms

Ice is the distinctive feature of the world's cold morphogenetic regions both at high latitudes and high altitudes (see Chapter 17). Here ice may be dominant in the process of rock weathering, in disturbance of the soil associated with both freezing and thawing and in modifying the earth's topography by building up a cover of snow or glacier ice. In combination these processes create a distinctive suite of landforms. The degree of overall landscape modification that results will vary according to the intensity of the processes, their combination and the length of time they have operated. In many areas of the world the modification is modest and little more than the regolith bears signs of the cold morphogenetic system. In other areas the impact of the system is overwhelming and the whole landscape is distinctive. The boundaries of the 'cold morphogenetic region' are difficult to define and at one extreme can include every area subjected to a frost, or at the other areas permanently below freezing. Usually a compromise between these two extremes is found most helpful, as shown in Figures 18.2 and 18.3.

The cold morphogenetic system can be subdivided according to one clear threshold. If the snow that falls in winter can be melted during the following summer, then 'periglacial' processes operate and modify the surface soil. If the snow that collects in winter is too abundant to be melted in the following summer, then *glaciers* build up. The threshold is shown schematically in Figure 18.26, where one can appreciate the range of environments that may be occupied by glaciers or exposed to periglacial conditions. Glaciers are favoured by a combination of high snowfall and cold summers while periglacial environments are favoured by warmer summers, low snowfall, or ideally a combination of both.

### 18.5.1 The periglacial system

There are two main ways in which ice in periglacial areas can modify the landscape. The first is through landforms associated with permafrost, or ground which is continuously below 0°C in both winter and summer. The second is through landforms associated with freezing and thawing on an annual or shorter-term basis.

Permafrost is continuous in those polar areas where the mean annual temperature is −6°C to −8°C or colder. It thickens poleward and may exceed depths of 600 m. Any understanding of permafrost is helped by appreciation of the temperature regime (Figure 18.27). The upper ground layers are subjected to an annual fluctuation in temperature which is greatest at the surface and falls off exponentially with depth until it is usually negligible at a depth of 6 m. The surface layer which melts each year is the *active* layer and the processes associated with it are considered later. Under the active layer is the permafrost proper, where the temperature rises with depth in response to geothermal heat flow.

Permafrost is of two fundamental kinds, depending on whether or not there is excess ice in the ground. Excess ice forms when ice builds up in the ground and physically separates soil particles. The ice is described as *segregated ice*. Common values of excess ice in unconsolidated sediments lie between 15 and 50 per cent by volume. This means that any melting of the permafrost would lead to subsidence by a commensurate amount. In other situations the ice may build up in pores in the spaces between soil particles and is known as *pore ice*. In these situations thawing would lead to no settlement of the ground. The most critical factor determining whether or not excess ice

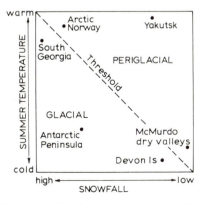

**Figure 18.26** Snowfall and summer temperature threshold which separates the periglacial and glacial morphogenetic zones.
*Source:* Sugden, 1982, figure 4.3, p. 66.

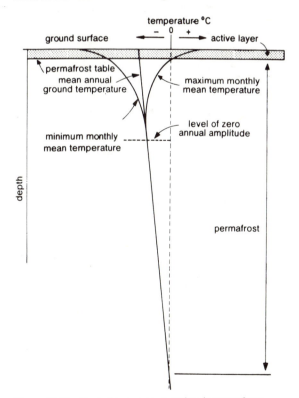

**Figure 18.27** Typical temperature regime in permafrost.
*Source:* Sugden, 1982, figure 5.4, p. 92, after Brown, 1970.

builds up in the ground is the grain size of the soil. If it is
fine-grained, then it is frost-susceptible and tends to
contain segregated ice (Mackay, 1972).

A number of landforms are associated with the growth
of segregated ice in the ground and are perhaps best
appreciated by perusal of the fine photographs in Washburn (1979). Mounds are distinctive and come in many
sizes and forms with a bewildering multilanguage
nomenclature. The mounds are due to the growth of ice
lenses, or hydrolaccoliths, which lift the ground above
them as they grow (Figure 18.28). Some are only a few
metres across, while the better-known type, the pingo,
may be as much as 300 m in diameter and 60 m in height.
The mounds have limited lives. Small ones may last a few
years, large ones several thousand years. Eventually the
uplifted soil is eroded away to expose the icecore, which
then melts.

Much more common in terms of areal coverage are
ice-wedge polygons, which bequeath a mosaic surface
pattern to extensive low-lying areas of the Arctic. The
polygons are commonly 15–40 m across and have three
to seven sides, as befits a random pattern. Their sides are
uplifted and separated by a narrow ditch. In summer
their centres are marshy or even covered by water. The
rims are underlain by ice wedges and the mode of formation is shown in Figure 18.29. In winter low temperatures
below approximately –19°C cause the ground to
contract. As a result it cracks in a more or less random
fashion (assuming uniform sediments and horizontal
temperature gradients). In the next melt season the crack
fills with water and a vein of ice forms in the permafrost
below the active layer. In future years the ground cracks
along the same line of weakness and allows the vein to
widen into a wedge, eventually forcing sediment upwards

to form ice-wedge rims. Ice-wedge polygons are part of a
family of contraction-crack polygons. In arid areas the
cracks may not fill with water and ice, but sand or other
debris.

When the ground contains excess ice, thawing leads to
subsidence which results in thermokarst. Perhaps the
most common form is the thaw lake (Figure 18.30).
These are circular or oval in shape, generally less than
1–2 km across and 3–4 m deep. Thaw lakes originate
when an ice lens is exposed to the air and melts, often as a
result of some chance factor, such as slumping of surface
materials or stream migration. The lake quickly assumes
an approximately circular form as it melts and undercuts
the adjacent ice-rich permafrost. There is a cycle whereby
the vegetation growth around the lake eventually protects the underlying ice from melting and the lake begins
to be infilled and revegetates once more. Eventually the
dried-out lake may be marked by the site of a pingo, as
shown in Figure 18.28.

The ground ice environment is an interesting example
of a landscape in dynamic equilibrium. There is a
complex cycle involving build-up of ice in the ground, its
subsequent exposure and melting, followed by renewed
build-up. The morphological features associated with
each stage are readily identifiable but together they
operate to produce a landscape in constant oscillation.

An understanding of the landforms associated with
thawing and freezing within the active layer is helped by
appreciation of the annual cycle. In winter the ground is
frozen and insulated by snow. In early summer the
surface snow and upper ground layers melt. Drainage is
usually impeded by the underlying permafrost and the
regolith is saturated. In autumn the ground freezes once
more from the surface and to a lesser extent from below.

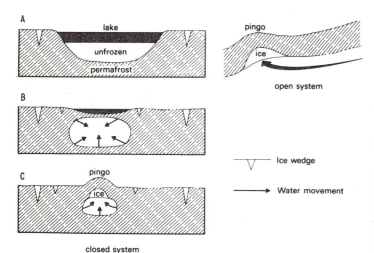

**Figure 18.28**   The origin of pingos. Closed-
system pingos result when a lake is drained
and infilled and the unfrozen ground (talik)
beneath the lake is progressively frozen from
all sides. The resultant volume increase is
taken up by the growth of an ice lens which
lifts the ground surface (Mackay, 1982).
Open-system pingos occur when the ice lens
is nourished by extraneous water. These
processes probably account for most ice-
cored mounds.
*Source:* Sugden, 1982, figure 5.7, p. 96.

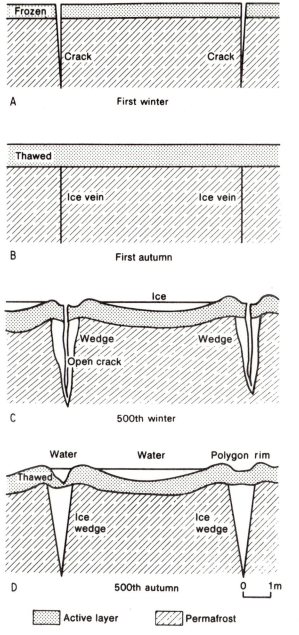

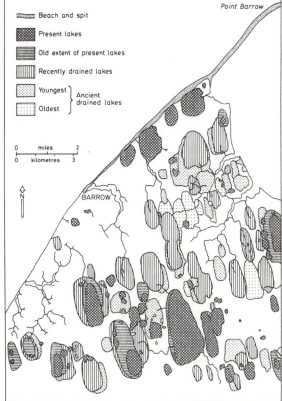

**Figure 18.30**   Oriented lakes and former thaw lakes on the Arctic Coastal Plain in the Barrow area, Alaska.
*Source:* Washburn, 1979, figure 12.3, p. 272, after Carson and Hussey, 1962.

**Figure 18.29**   The origin of ice-wedge polygons.
*Source:* Sugden, 1982, figure 5. 10, p. 98, after Lachenbruch, 1962.

This annual cycle produces a number of imperfectly understood processes of heaving, sorting and flow.

A common active-layer landform is the vegetated hummock with a core of mineral soil. They vary from a few centimetres to several metres in height, with the most common dimensions perhaps 30 cm across, 20 cm high and with a spacing of around 1 m. Their origin is far from clear, but some sort of differential frost heaving or the expulsion of unfrozen material to the surface during the autumn freeze-up seem important possibilities.

Another distinctive active-layer landform is sorted patterned ground. Again the mechanisms are far from understood but it seems that the process is for coarse stones to move to the surface and outwards from freezing centres. If the centres are scattered widely, then stone circles form. If they impinge on one another, then polygons of coarse stones form around the centres; generally they are 1–10 m in diameter, with the larger sizes associated with larger constituent stones. If the sorting takes place on a slope, then the polygons are elongated downslope until at gradients above 4°–11° they form stone stripes (Figure 18.31).

Other characteristic landforms are associated with the flow of regolith during the summer thaw. In permafrost

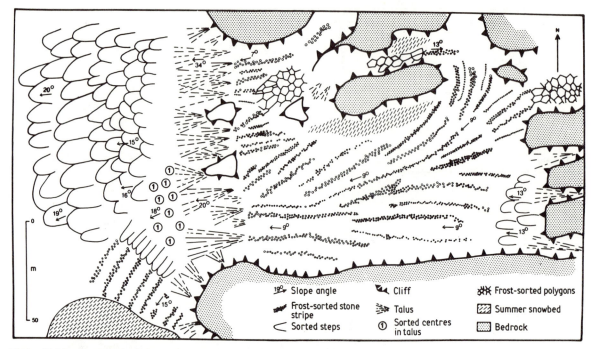

**Figure 18.31** An example of the complexity of sorted patterned ground and its relationship to local topography, Byers, Peninsula, Livingston Island, South Shetland Islands, Antarctica.
*Source:* Sugden, 1982, figure 5.19, p. 106, mapped by Gordon Thom.

areas this process is known as gelifluction (an icebound version of solifluction). It describes the slow flow down-slope of saturated regolith, as well as downslope displacement due to repeated freezing and thawing of particles on a slope. Generally gelifluction occurs at rates of 1–5 cm per year and can occur on slopes as gentle as 1°. The most common features associated with gelifluction are sheets, uniform monotonous expanses of regolith with angles of 1°–3° often continuous over several square kilometres, as in Arctic Canada. Sometimes they are bounded on the downslope side by a small vegetated riser some 10–50 cm high, but commonly they merge into the valley floor (French, 1976). On steeper slopes gelifluction produces steps in the regolith. Measuring tens of metres across, these are commonly bounded on their downslope sides by large stones or vegetation (Figure 18.32).

It is clear from the above that a number of different processes contribute to the landforms of the periglacial morphogenetic zone. Some processes are favoured by certain climatic and ground conditions, while others require different climates and ground conditions. It follows that the particular blend of processes and landforms in any one place depends on these other factors. Karte (1979) has attempted to distinguish the main

climatic and topographic factors involved and expressed these in the form of a block diagram (Figure 18.33) and a map of the northern hemisphere periglacial morphogenetic region (Figure 18.34). In the maritime environment, as in south-west Alaska, the periglacial morphogenetic zone extends beyond the limit of continuous permafrost and is represented by features like earth hummocks, gelifluction and sorted patterned ground. This boundary coincides well with the −1°C to −2°C mean annual isotherm. With increasing thermal continentality (i.e. when the mean annual air temperature range exceeds 25°C), the boundary of the periglacial zone runs through the boreal forest close to the boundary of continuous permafrost. Here dominant periglacial features are those associated with segregated ground ice – pingos, ice-wedge polygons and thermokarst. With increasing continentality, the mean annual air temperature along the southern boundary falls from −2°C to −8°C. Towards the pole these regional types give way, first, to tundra, and then to the frost debris periglacial zone. The latter zone is arid and can support little ice in the ground. This has the effect of suppressing regolith movements and encouraging the existence of block fields and other rock weathering forms. An extreme version of such an environment occurs in the dry valleys of

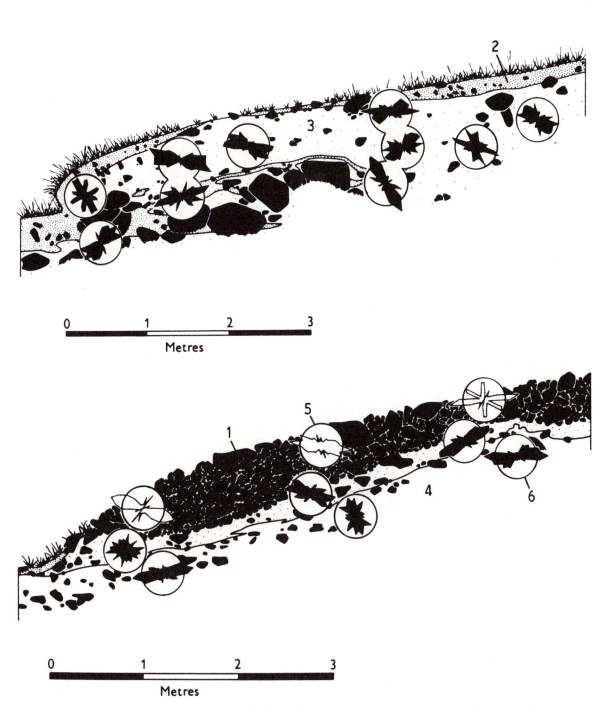

**Figure 18.32** Details of sections through steps bounded by turf (above) and stones (below) in Niwot Ridge, Colorado Front Range: 1, cobbles and boulders (to scale); 2, Ah horizon; dark brown sandy loam; 3, moving soil; sandy loam becoming finer near lobe fronts; weakly platy, finely vesicular structure; 4, Cox horizon; gravelly, sandy loam; 5, orientation diagram of stone long axes; circle radius = 10 per cent frequency; 6, orientation diagram of elongate sand grain axes; circle radius = 10 per cent frequency.

*Source:* Washburn, 1979, figure 6.22, p. 218, after Benedict, 1970, 1976.

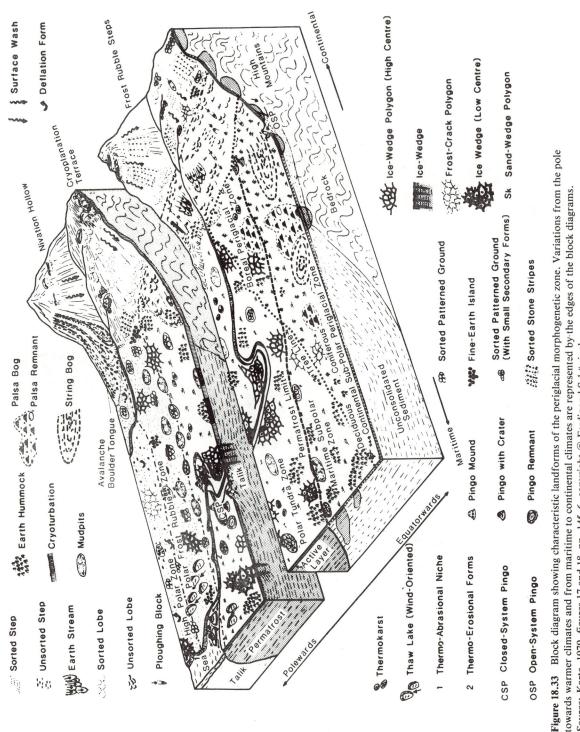

**Figure 18.33** Block diagram showing characteristic landforms of the periglacial morphogenetic zone. Variations from the pole towards warmer climates and from maritime to continental climates are represented by the edges of the block diagrams.
*Source:* Karte, 1979, figures 17 and 18, pp. 145–6, copyright © Ferdinand Schöningh.

Surface Wash

Dellation Form

Frost Rubble Steps

Nivation Hollow

Cryoplanation Terrace

Avalanche Boulder Tongue

Palsa Bog

Palsa Remnant

String Bog

High Mountains

Continental

Boreal Periglacial Zone

Bedrock

Coniferous Periglacial Zone

Deciduous Sub-Polar Zone

Continental

Unconsolidated Sediment

Maritime

Maritime Subpolar Zone

Permafrost Limit

Tree Line

Polar Tundra Zone

Active Layer

High Polar Frost Zone

High Polar Zone

Rubble Zone

Talik

Sea

Talik

Permafrost

Equatorwards

Polewards

Sorted Step

Unsorted Step

Earth Hummock

Earth Stream

Cryoturbation

Sorted Lobe

Mudpits

Unsorted Lobe

Ploughing Block

Thermokarst

Thaw Lake (Wind-Oriented)

1 Thermo-Abrasional Niche

2 Thermo-Erosional Forms

CSP Closed-System Pingo

OSP Open-System Pingo

Pingo Mound

Pingo with Crater

Pingo Remnant

Sorted Patterned Ground

Fine-Earth Island

Sorted Patterned Ground (With Small Secondary Forms)

Sorted Stone Stripes

Ice-Wedge Polygon (High Centre)

Ice-Wedge

Frost-Crack Polygon

Ice Wedge (Low Centre)

Sk Sand-Wedge Polygon

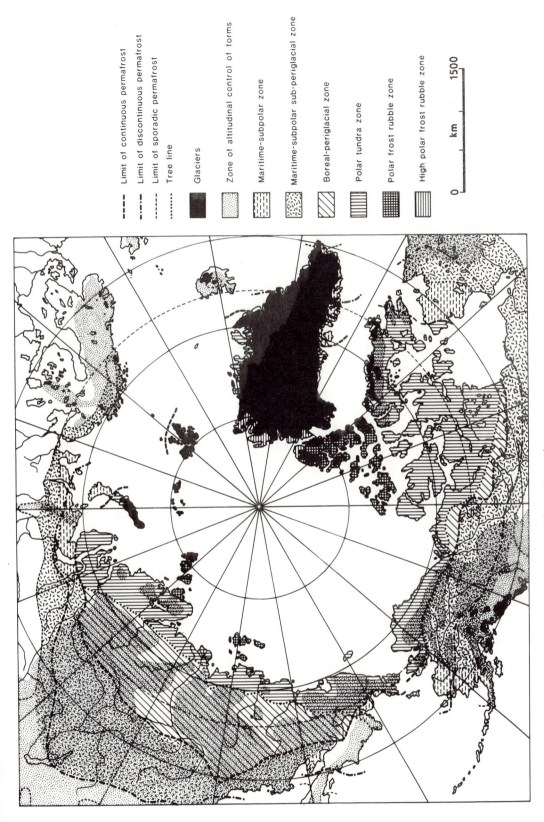

**Figure 18.34** Regional variation in the northern periglacial morphogenetic zone.
*Source:* Karte, 1979, figure 20, p. 149, copyright © Ferdinand Schöningh.

Limit of continuous permafrost

Limit of discontinuous permafrost

Limit of sporadic permafrost

Tree line

Glaciers

Zone of altitudinal control of forms

Maritime-subpolar zone

Maritime-subpolar sub-periglacial zone

Boreal-periglacial zone

Polar tundra zone

Polar frost rubble zone

High polar frost rubble zone

km 1500

McMurdo, Victoria Land in Antarctica (Tedrow and Ugolini, 1966).

Karte (1979) has also discussed how the periglacial zone varies with altitude in the world's mountains. Figure 18.35 is an idealized transect from the pole to the equator. In subpolar and middle latitudes there are three divisions. The lower zone displays gelifluction features covered by vegetation. The middle zone contains more sorting and vegetation tends to cover gelifluction forms incompletely. In the upper zone sorted forms and gelifluction predominate in the virtual absence of vegetation. With the transition into the high mountains of the subtropics with dry summers, the lower periglacial zone becomes discontinuous and finally ceases to exist. In the high mountains of the arid zone there is insufficient moisture for any significant periglacial activity, although temperatures are low enough. In the increasing humid equatorial zone there is a distinctive periglacial zone where diurnal freezing and thawing creates small-scale gelifluction and sorted forms. The climatic thresholds along the lower boundary of the mountain periglacial zone vary considerably. In subpolar latitudes the mean annual air temperature is $-2°C$, in mid-latitudes $-1$ to $+5°C$, in the semi-humid tropics $+3°$ to $+6°C$ and in the humid tropics $0°C$.

### 18.5.2 The glacial system

The glacier morphogenetic system covers approximately 10 per cent of the world's land area. All but a tiny portion of the area is accounted for by the Antarctic and Greenland ice sheets and the remaining 3–4 per cent is confined to high-latitude coastal areas or high mountains.

The factors most suitable for glacier growth are high snowfall, low summer temperatures and a favourable topographic site for ice to accumulate. Glacier distribution illustrates this. At the scale of an ice sheet it is obviously difficult to generalize since there are only two main examples – in Antarctica and Greenland. But one can point out that they both lie on continents in high latitudes and that most of their nourishment is obtained from air masses crossing subpolar seas. Going back to earlier ice ages in the world's history, it seems that one prerequisite for an Ice Age is the presence of a continent surrounded by open sea in a polar latitude – as exists in Antarctica today.

Smaller glaciers in high latitudes clearly reflect the distribution of snowfall. It can be seen in Figure 18.34 how islands in the vicinity of the North Atlantic and Eurasian Arctic Ocean are glacier-covered and this is the main source of moisture in the Arctic. Away from this axis glaciers become progressively less common in spite of lower summer temperatures. Thus islands in the north-western Canadian Arctic archipelago may be ice-free.

Outside high latitudes glaciers are confined to high mountains which offer low summer temperatures and often high snowfall. The latitudinal tendency of the snowline is shown in Figure 18.36, where the combination of temperature and precipitation causes the snowline

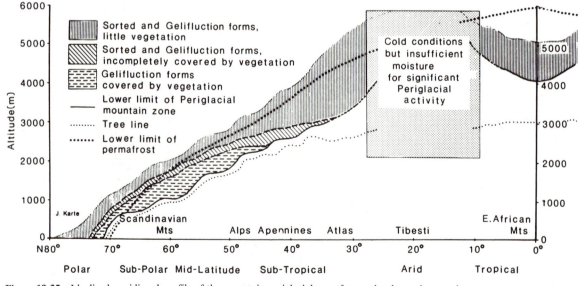

**Figure 18.35** Idealized meridional profile of the mountain periglacial zone from subpolar regions to the equator.
*Source:* Karte, 1979, figure 23, p. 169, copyright © Ferdinand Schöningh.

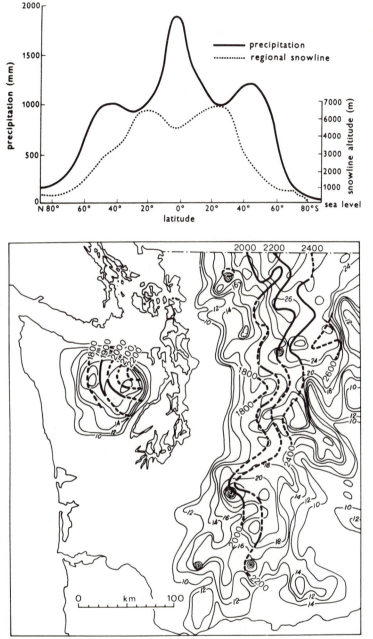

**Figure 18.36** Generalized curves showing the latitudinal variations in the regional snowline, together with precipitation averages.
*Source:* Sugden and John, 1976, figure 5.5, p. 88, after Paschinger and Sellers.

**Figure 18.37** Map of western Washington State showing gross summit topography (100-m contours) and the increasing elevation of the glaciation threshold (full lines, heights in metres) with distance from the sea.
*Source:* Porter, 1977, figure 1, p. 103.

to rise equatorwards but with irregularities related to snowfall amounts. Within any one mountain range conditions suitable for glaciation rise with distance from the sea in sympathy with falling precipitation (Figure 18.37).

The forms displayed by the glaciers depend on their type. Ice sheets and icecaps have a distinctive form and the morphology of one, the Antarctic ice sheet, is shown in Figure 17.2. The basic form is a convex-upwards dome, which on a continental scale may reach an altitude in excess of 4000 m, as in Antarctica. The slopes of the dome reflect the flow properties of ice and may be modelled closely. Figure 18.38 shows some actual profiles of continental-sized domes and some theoretical

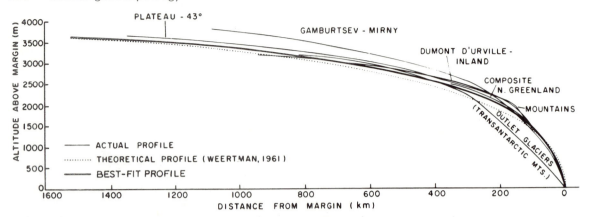

**Figure 18.38** Surface profiles of the ice domes in Greenland and east Antarctica, compared with theoretical curves. The unusually high ice profile over the Gamburtsev Mountains is ignored for the purposes of best-fit profile.
*Source:* Sugden, 1977, figure 1, p. 24. Reproduced with permission of the Regents of the University of Colorado from *Arctic and Alpine Research*, vol. 9.

generalizations. The best-fit profile has the form:

$$y = a + bx + cx^2$$

where $a = 0{\cdot}607$; $b = 0{\cdot}005$; and $c = -0{\cdot}2 \times 10^{-5}$.

On smaller icecaps the dome has a steeper parabolic profile and as a generalization, using simplifying assumptions such as that of a horizontal bed and perfect plastic behaviour of ice, can be expressed by the formula $h = \sqrt{(2h_0 s)}$, where $h$ = ice altitude in metres; $h_0 = 11$ m; and $s$ = horizontal distance from the margin in metres (Nye, 1952).

The peripheral zone of an ice sheet or icecap is often marked by a radiating pattern of outlet glaciers which may extend some way beyond the dome margin. Within the ice dome they may occupy a depression, while beyond the ice margin they commonly occupy a glacier trough and, indeed, often may end as a calving fjord glacier. The outlet glaciers which cut across the Transantarctic Mountains are some of the world's most spectacular landforms. One of them, Beardmore Glacier is 200 km long and 23 km wide and flows at a velocity of 1 m per day. Outlet glaciers are also common round smaller icecaps, for example, in Iceland and Norway. Usually such glaciers leave the ice dome by means of an ice fall and then flow in a rock trough.

Ice shelves are in effect ice sheets afloat. They are confined to cold environments with mean summer temperatures below about 0°C and to topographic embayments where rock promontories and/or islands provide anchoring-points. The largest examples occur in the Ross and Weddell Sea embayments in Antarctica. The Ross Ice Shelf extends 900 km inland and is approximately 800 km across. In Antarctica 30 per cent of the coastline is occupied by ice shelves. In the Arctic small ice shelves

exist along the northern coast of Ellesmere Island and Greenland. The main characteristics of ice shelves are well described by Swithinbank and Zumberge (1965). The seaward margin forms a sheer cliff rising some 30 m above the sea. The surface of the shelf is virtually flat. Unretarded by friction at the base, ice shelves flow fast at velocities of 1–3 km per year and calve to produce large tabular icebergs.

Mountain glaciers exhibit a bewilderingly varied morphology and reflect the varied mountain scenery which they inhabit. The key component is the 'alpine' glacier which is everywhere overlooked by rock walls. Beginning in an icefield, or in an armchair-shaped hollow – a *cirque* – the glacier may extend downvalley 10–30 km, although a few approach 100 km in length. Typically such glaciers receive avalanches from overlooking mountain slopes and have a high altitudinal range. Both these factors encourage relatively high rates of flow when compared to glacier size. In more marginal situations the glacier may be confined to the shaded cirque basin itself. As a generalization the longest valley glaciers will mark mountains with heavy snowfalls, for example, in south-east Alaska, while shorter valley glaciers and cirque glaciers will characterize more continental mountains, for example, in the Brooks Range in north Alaska or in Central Asia.

A characteristic feature of mountain slopes overlooking glaciers is their precipitous nature. Above the equilibrium line a glacier is able to evacuate any debris that falls on it. This, together with erosion by the glacier itself, favours undercutting and rock failure on a large scale. Steep fretted ridges, or *arêtes*, and residual mountain peaks, or *horns*, abound in the glaciated mountains of the world. The rock slopes overlooking major valley glaciers may themselves be the sites of an infinitely varied suite of small glaciers or ice patches. They may cling to hollows on steep valley sides, form sheets of ice on rock slopes, nestle in depressions, fringe a coastline or occupy a summit only a few square metres in size. As a rule such small glaciers are most common in mountain areas, such as the Antarctic Peninsula, where there is high snowfall and a high percentage of cloud cover. They become progressively less common towards continental interiors.

# References

Baker, V. R. (1981) 'The geomorphology of Mars', *Progress in Physical Geography*, vol. 5, 473–513.

Benedict, J. B. (1970) 'Downslope soil movement in a Colorado Alpine region: rates, processes and climatic significance', *Arctic and Alpine Research*, vol. 2, 165–226.

Benedict, J. B. (1976) 'Frost creep and gelifluction: a review', *Quaternary Research*, vol. 6, 55–76.

Birot, P. (1968) *The Cycle of Erosion in Different Climates*, London, Batsford.

Bremer, H. (1980) 'Landform development in the humid tropics: German geomorphological research', *Zeitschrift für Geomorphologie*, Supplement 36, 162–75.

Brown, R. J. E. (1970) *Permafrost in Canada: Its Influence on Northern Development*, University of Toronto Press.

Büdel, J. (1948) 'Das System der klimatischen Geomorphologie', *Verhandlungen Deutscher Geographentag*, vol. 27, 65–100.

Büdel, J. (1963) 'Klima-genetische Geomorphologie', *Geographische Rundschau*, vol. 15, 269–85.

Büdel, J. (1965) 'Die relieftypen der Flächenspülzone: Süd-Indiens am Ostabfall Dekans gegen Madras', *Colloquium Geographicum*. (Bonn), vol. 8.

Büdel, J. (1970) Pedimente, Rümpflächen und Rückland-Steilhänge', *Zeitschrift für Geomorphologie*, vol. 14, 1–57.

Büdel, J. (1980) 'Climatic and climamorphic geomorphology', *Zeitschrift für Geomorphologie*, Supplement 36, 1–8.

Büdel, J. (1982) *Climatic Geomorphology*, Princeton, NJ, Princeton University Press.

Burke, K. and Durotoye, B. (1970) 'Boulder slide on an inselberg in the Nigerian rain forest', *Journal of Geology*, vol. 78, 733–5.

Butzer, K. W. (1976) *Geomorphology from the Earth*, New York, Harper & Row.

Carson, C. E. and Hussey, K. M. (1962) 'The oriented lakes of Arctic Alaska', *Journal of Geology*, vol. 70, 417–39.

Carter, T. E. (1981) '*Pediment development along the Book Cliffs, Utah*', unpublished MS dissertation, Colorado State University.

Chorley, R. J. (1957) 'Climate and morphometry', *Journal of Geology*, vol. 65, 628–38.

Chorley, R. J. and Morgan, M. A. (1962) 'Comparison of morphometric features, Unaka Mountains, Tennessee and North Carolina, and Dartmoor, England', *Bulletin of the Geological Society of America*, vol. 73, 17–34.

Cooke, R. U. and Warren, A. (1973) *Geomorphology in Deserts*, London, Batsford.

Cotton, C. A. (1942) *Climatic Accidents in Landscape Making*, Christchurch, NZ, Whitcombe & Tombs.

Cotton, C. A. (1961) 'The theory of savanna planation', *Geography*, vol. 46, 89–101.

Cotton, C. A. (1962) 'Plains and inselbergs of the humid tropics', *Transactions of the Royal Society of New Zealand (Geol.)*, vol. 1, 269–77.

Daniel, J. R. K. (1981) 'Drainage density as an index of climatic geomorphology', *Journal of Hydrology*, vol. 50, 147–54.

Derbyshire, E. (ed.) (1973) *Climatic Geomorphology*, London, Macmillan.

Doornkamp, J. C. (1968) 'The role of inselbergs in the geomorphology of southern Uganda', *Transactions of the Institute of British Geographers*, no. 44, 151–62.

Douglas, I. (1968) 'Erosion in the Sungei Gombak catchment, Selangor, Malaysia', *Journal of Tropical Geography*, vol. 26, 1–16.

Douglas, I. (1969) 'The efficiency of humid tropical denudation systems', *Transactions of the Institute of British Geographers*, no. 46, 1–16.

Douglas, I. (1980) 'Climatic geomorphology: present-dry processes and landform evolution', *Zeitschrift für Geomorphologie*, Supplement 36, 27–47.

Engle, C. E. and Sharp, R. P. (1958) 'Chemical data on desert varnish', *Bulletin of the Geological Society of America*, vol. 69, 487–518.

Eyles, R. J. and Ho, R. (1970) 'Soil creep on a humid tropical slope', *Journal of Tropical Geography*, vol. 31, 40–2.

Fosberg, F. R., Garnier, B. J. and Küchler, A. W. (1961) 'Delimitation of the humid tropics', *Geographical Review*, vol. 51, 333–47.

French, H. M. (1976) *The Periglacial Environment*, London, Longman.

Friedmann, E. I. (1982) 'Endolithic microorganisms in the Antarctic cold desert', *Science*, vol. 215, 1045–53.

Garwood, N. C., Janos, D. P. and Brokaw, N. (1979) 'Earthquake-caused landslides: a major disturbance to tropical forests', *Science*, vol. 205, 997–9.

Gibbs, R. J. (1967) 'The geochemistry of the Amazon River system', *Bulletin of the Geological Society of America*, vol. 78, 1203–32.

Goudie, A. (1973) *Duricrusts in Tropical and Subtropical Landscapes*, Oxford University Press.

Gregory, H. E. (1950) 'The Zion Park region', *US Geological Survey Professional Paper* 220.

Gregory K. J. (1976) 'Drainage networks and climate', in E. Derbyshire, (ed.). *Geomorphology and Climate*, Chichester, Wiley, 289–315.

Gupta, A. (1975) 'Stream characteristics in eastern Jamaica; an environment of seasonal flow and large floods', *American Journal of Science*, vol. 275, 825–47.

Harvey, A. M. (ed.) (1968) 'Geomorphology in a tropical environment', *British Geomorphological Research Group, Occasional Paper 5*.

Holzner, L. and Weaver, G. D. (1965) 'Geographic evaluation of climatic and climato-genetic geomorphology', *Annals of the Association of American Geographers*, vol. 55, 592–602.

Hunt, C. B., Averitt, P. and Miller, R. L. (1953) 'Geology and geography of the Henry Mountains region, Utah', *US Geological Survey Professional Paper 228*.

Karte, J. (1979) *Räumliche Abgrenzung und regionale Differenzierung des Periglaziärs*, Paderborn, Ferdinand Schöningh.

King, L. C. (1948) 'A theory of bornhardts', *Geographical Journal*, vol. 112, 83–7.

King, L. C. (1966) 'The origin of bornhardts', *Zeitschrift für Geomorphologie*, vol. 10, 97–8.

King, L. C. (1967) *The Morphology of the Earth*, 2nd edn, Edinburgh, Oliver & Boyd.

Krynine, P. D. (1936) 'Geomorphology and sedimentation in the humid tropics', *American Journal of Science*, vol. 32, 297–306.

Lachenbruch, A. (1962) 'Mechanics of thermal contraction cracks and ice wedge polygons in permafrost', *Geological Society of America Special Paper 70*.

Linton, D. L. (1955) 'The problem of tors', *Geographical Journal*, vol. 121, 470–87.

Mabbutt, J. A. (1977) *Desert Landforms*, Cambridge, Mass, MIT Press.

Mabbutt, J. A. and Scott, R. M. (1966) 'Periodicity of morphogenesis and soil formation in a savanna landscape near Port Moresby, Papua', *Zeitschrift für Geomorphologie*, vol. 10, 69–89.

Mackay, J. R. (1972) 'The world of underground ice', *Annals of the Association of American Geographers*, vol. 62, 1–23.

McFarlane, M. J. (1976) *Laterite and Landscape*, London, Academic Press.

Madduma Bandara, C. M. (1971) 'The morphometry of dissection in the central highlands of Ceylon', unpublished PhD dissertation, Cambridge University.

Melton, M. A. (1957) 'An analysis of the relations among elements of climate, surface properties, and geomorphology', *Columbia University, Department of Geology, Office of Naval Research Technical Report 11*.

Miller, A. A. (1961) 'Climate and the geomorphic cycle', *Geography*, vol. 46, 185–97.

Millot, G. (1970) *Geology of Clays*, Paris, Masson.

Morgan, R. P.C. (1973) 'Soil–slope relationships in the lowlands of Selangor and Negri Sembilan, West Malaysia', *Zeitschrift für Geomorphologie*, vol. 17, 139–55.

Morgan, R. P. C. (1976) 'The role of climate in the denudation system: a case study from West Malaysia', in E. Derbyshire (ed.), *Geomorphology and Climate*, Chichester, Wiley, 317–43.

Neal, J. T. and Motts, W. S. (1967) 'Recent geomorphic changes in playas of western United States', *Journal of Geology*, vol. 75, 511–25.

Neal, J. T., Langer, A. M. and Kerr, P. F. (1968) 'Giant desiccation polygons of Great Basin playas', *Bulletin of the Geological Society of America*, vol. 79, 69–90.

Nye, J. F. (1952) 'A method of calculating the thickness of ice sheets', *Nature*, vol. 169, 529–30.

Ollier, C. D. (1960) 'The inselbergs of Uganda', *Zeitschrift für Geomorphologie*, vol. 4, 43–52.

Pain, C. F. (1975) 'The Kaugel Diamicton – a late Quaternary mudflow deposit in the Kaugel Valley, Papua New Guinea', *Zeitschrift für Geomorphologie*, vol. 19, 430–42.

Pain, C. F. and Bowler, J. M. (1973) 'Denudation following the November 1970 earthquake at Madang, Papua New Guinea', *Zeitschrift für Geomorphologie*, Supplement 18, 92–104.

Park, C. C. (1977) 'World-wide variations in hydraulic geometry exponents of stream channels: an analysis and some observations', *Journal of Hydrology*, vol. 33, 133–46.

Passarge, S. (1926) 'Morphologie der Klimazonen oder Morphologie der Landschaftsgürtel?', *Petermanns Geographische Mitteilungen*, vol. 72, 173–5.

Peltier, L. C. (1950) 'The geographic cycle in periglacial regions as it is related to climatic geomorphology', *Annals of the Association of American Geographers*, vol. 40, 214–36.

Peltier, L. C. (1962) 'Area sampling for terrain analysis', *Professional Geographer*, vol. 14, 24–8.

Porter, S. C. (1977) 'Present and past glaciation threshold in the Cascade Range, Washington, USA: topographic and climatic controls, and paleoclimatic implications', *Journal of Glaciology*, vol. 18, 101–16.

Rapp, A., Axelsson, V., Berry, L. and Murray-Rust, D. H. (1972a) 'Soil erosion and sediment transport in the Morogoro River Catchment, Tanzania', *Geografiska Annaler*, vol. 54A, 125–55.

Rapp, A., Murray-Rust, D. H., Christiansson, C. and Berry, L. (1972b) 'Soil erosion and sedimentation in four catchments near Dodoma, Tanzania', *Geografiska Annaler*, vol. 54A, 255–318.

Ruxton, B. P. (1958) 'Weathering and sub-surface erosion in granite at the piedmont angle, Balos, Sudan', *Geological Magazine*, vol. 95, 353–77.

Ruxton, B. P. (1960) 'The basal rock surface on weathered granitic rocks', *Proceedings of the Geologists Association*, vol. 70, 285–90.

Ruxton, B. P. and Berry, L. (1961) 'Weathering profiles and geomorphic position on granite in two tropical regions', *Revue de Géomorphologie Dynamique*, vol. 12, 16–31.

Schumm, S. A. (1956) 'The role of creep and rainwash in the retreat of badland slopes', *American Journal of Science*, vol. 254, 693–706.

Schumm, S. A. (1962) 'Erosion on miniature pediments in Badlands National Monument, South Dakota', *Bulletin of the Geological Society of America*, vol. 73, 719–24.

Schumm, S. A. (1964) 'Seasonal variations of erosion rates and processes on hillslopes in western Colorado', *Zeitschrift für Geomorphologie*, Supplement 5, 215–38.

Schumm, S. A. and Chorley, R. J. (1966) 'Talus weathering and scarp recession in the Colorado plateaus', *Zeitschrift für Geomorphologie*, vol. 10, 11–36.

Simonett, D. S. (1967) 'Landslide distribution and earthquakes in the Bewani and Torricelli Mountains, New Guinea', in J. N. Jennings and J. A. Mabbutt (eds), *Landform Studies from Australia and New Guinea*, Canberra, Australian National University Press, 64–84.

Sivarojasingham, S., Alexander, L. T., Cady, J. G. and Cline, M. G. (1962) 'Laterite', *Advances in Agronomy*, vol. 14, 1–60.

So, C. L. (1971) 'Mass movements associated with the rainstorm of June 1966 in Hong Kong', *Transactions of the Institute of British Geographers*, no. 53, 55–65.

Sparks, B. W. (1972) *Geomorphology*, 2nd edn, London, Longman.

Staley, J. T., Palmer, F. and Adams, J. B. (1982) 'Microcolonial fungi: common inhabitants on desert rocks?', *Science*, vol. 215, 1093–5.

Stoddart, D. R. (1969) 'Climatic geomorphology: review and re-assessment', *Progress in Geography*, vol. 1, 159–222.

Sugden, D. E. (1977) 'Reconstruction of the morphology, dynamics and thermal characteristics of the Laurentide ice sheet at its maximum', *Arctic and Alpine Research*, vol. 9, 21–47.

Sugden, D. E. (1982) *Arctic and Antarctic*, Oxford, Blackwell.

Sugden, D. E. and John, B. S. (1976) *Glaciers and Landscape*, London, Arnold.

Swan, S. B. St C. (1972) 'Land surface evolution and related problems with reference to a humid tropical region; Johore, Malaya', *Zeitschrift für Geomorphologie*, vol. 16, 160–81.

Swithinbank, C. W. M. and Zumberge, J. H. (1965) 'The ice shelves', in T. Hatherton (ed.), *Antarctica*, London, Methuen, 194–220.

Tanner, W. F. (1961) 'An alternate approach to morphogenetic climates', *Southeastern Geologist*, vol. 2, 251–7.

Tedrow, J. C. F. and Ugolini, F. C. (1966) 'Antarctic soils', in J. C. F. Tedrow (ed.), *Antarctic Soils and Soil Forming Processes*, American Geophysical Union, Antarctic Research Series 8, 161–77.

Temple, P. H. and Rapp, A. (1972) 'Landslides in the Mgeta area, western Uluguru Mountains, Tanzania', *Geografiska Annaler*, vol. 54A, 157–93.

Temple, P. H. and Sundborg, A. (1972) 'The Rufiji River, Tanzania: hydrology and sediment transport', *Geografiska Annaler*, vol. 54A, 345–68.

Thomas, M. F. (1965) 'Some aspects of the geomorphology of domes and tors in Nigeria', *Zeitschrift für Geomorphologie*, vol. 9, 63–81.

Thomas, M. F. (1974) *Tropical Geomorphology*, London, Macmillan.

Thorbecke, F. (1927) 'Die Formenschatz im periodisch trockenen Tropenklima mit überweigender Regenzeit', *Düsseldorfer Geographische Vorträge und Erörterungen*, vol. 3, 10–18.

Tricart, J. (1972) *Landforms of the Humid Tropics, Forests and Savannas*, London, Longman.

Tricart, J. and Cailleux, A. (1972) *Introduction to Climatic Geomorphology*, London, Longman.

Troll, C. (1958) 'Climatic seasons and climatic classification', *Oriental Geographer*, vol. 2, 141–65.

Turvey, N. D. (1975) 'Water quality in a tropical rain forested catchment', *Journal of Hydrology*, vol. 27, 111–25.

Twidale, C. R. (1964) 'A contribution to the general theory of domed inselbergs', *Transactions of the Institute of British Geographers*, no. 34, 91–113.

Twidale, C. R. (1982) 'The evolution of bornhardts', *American Scientist*, vol. 70, 268–76.

Walsh, R. P. D. (1980) 'Runoff processes and models in the humid tropics', *Zeitschrift für Geomorphologie*, Supplement 36, 176–202.

Washburn, A. L. (1979) *Geocryology: A Survey of Periglacial Processes and Environments*, London, Arnold.

Wentworth, C. K. (1928) 'Principles of stream erosion in Hawaii', *Journal of Geology*, vol. 36, 385–410.

Wentworth, C. K. (1943) 'Soil avalanches on Oahu, Hawaii', *Bulletin of the Geological Society of America*, vol. 54, 53–64.

White, S. E. (1949) 'Processes of erosion on steep slopes of Oahu, Hawaii', *American Journal of Science*, vol. 247, 168–86.

Williams, M. A. J. (1969) 'Production of rainsplash erosion in the seasonally wet tropics', *Nature*, vol. 222, 763–5.

Williams, M. A. J. (1973) 'The efficacy of creep and slopewash in tropical and temperate Australia', *Australian Geographical Studies*, vol. 11, 62–78.

Wilson, L. (1969) 'Relationships between geomorphic processes and modern climates as a method in paleoclimatology', *Revue de Géographie physique et de Géologie dynamique*, vol. 11, 303–14.

Wilson, L. (1973) 'Variations in mean annual sediment yield as a function of mean annual precipitation', *American Journal of Science*, vol. 273, 335–49.

Wright, L. W. (1973) 'Landforms of the Yavuna granite area, Viti Levu, Fiji', *Journal of Tropical Geography*, vol. 37, 74–80.

Zonneveld, J. I. S. (1975) 'Some problems of tropical geomorphology', *Zeitschrift für Geomorphologie*, vol. 19, 377–92.

# Nineteen   *Geomorphological effects of former glacier expansion*

## 19.1 Introduction

Areas formerly covered by glaciers represent an extreme example of climatic change. With climatic cooling, a threshold is crossed whereby snow accumulates year by year in formerly ice-free areas and builds up glaciers (see Figure 18.26). As a consequence the glacier sedimentary system (see Chapter 17) superimposes its landforming processes on a non-glacial landscape. The geomorpho-

logy of such landscapes requires appreciation of the interplay between glacial processes and the pre-existing topography. As much as 30 per cent of the earth's surface has been covered by glaciers in the last 3 million years or so. The largest areas are in the northern hemisphere and include large areas of North America and Asia (Figure 19.1), but glacier expansion also occurred in Antarctica and the surrounding areas (Figure 19.2). In warmer lati-

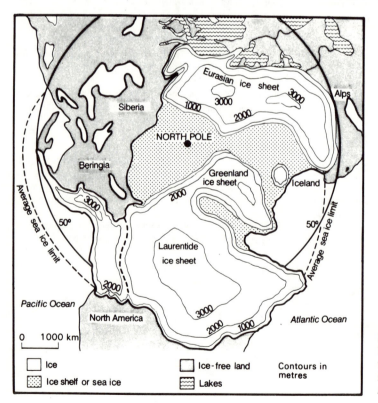

**Figure 19.1** The geography of the high latitudes in the northern hemisphere during the Pleistocene maximum; altitudes are given in metres.
*Source:* Sugden, 1982, figure 7.3, p. 169.

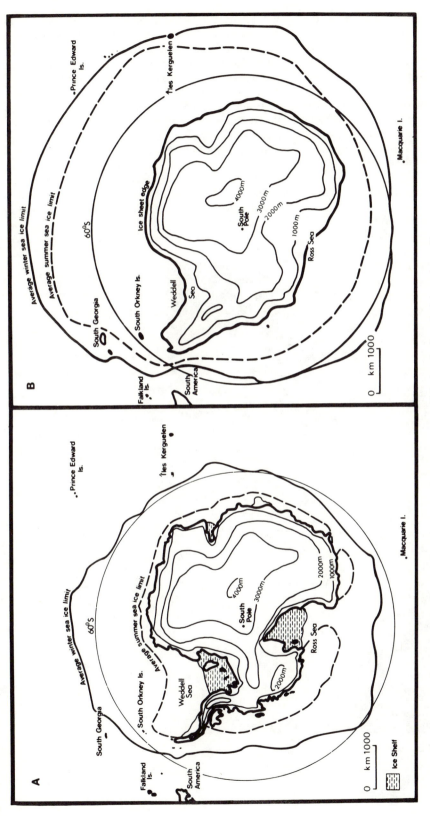

**Figure 19.2** Antarctic glacier and glaciomarine conditions at the present time A. and during the last glacial maximum B.
*Source:* Sugden, 1982, figure 7.2, p. 168.

tudes sizeable ice caps built up in high mountains, the largest in the Andes of Chile and Argentina and in the Himalayas and Pamirs. Until recently it was thought that there were four glacial advances, but new work on loess deposits and deep-sea cores has suggested that there were some seventeen in the last 2–3 million years. Little is known about the extent and behaviour of any other than the last glaciation, and so Figures 19.1 and 19.2 are idealized illustrations of ice maximum conditions which may have persisted on several occasions. During other maxima, as indeed during the last, the ice may have been slightly less extensive in places. Figure 19.3 shows a more detailed reconstruction of the Laurentide ice sheet during the last glaciation.

It is necessary to place this chapter in perspective by noting that this last (Pleistocene or Cenozoic) ice age is but one of several other ice ages in the earth's history. In other parts of the world there are landscapes which bear a glacial imprint of these older glaciations. One of the finest examples is in Saharan Africa, where landforms of an Ordovician glaciation are exposed over hundreds of square kilometres. Striations, grooves, fluted till surfaces and meltwater features are perfectly preserved in the landscape of one of the hottest climates of the world, and tell of a time when this part of the world was sited over the South Pole (Beuf, *et al.*, 1971, summarized in Fairbridge, 1970). There was also an extensive Carboniferous glaciation in Southern Africa. It is even more sobering to realize that it is not just our planet which has experienced landscape modification due to the expansion of glaciers. Several spectacular landforms on Mars are thought to reflect more widespread glaciation there

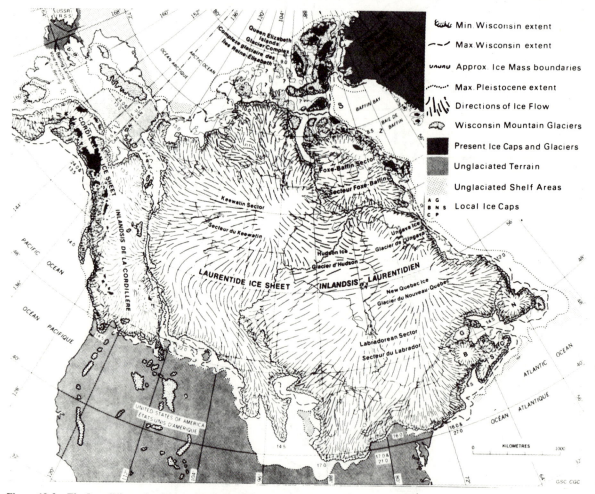

**Figure 19.3**  The Late Wisconsin glacier complex of North America.
*Source:* Prest, 1983, figure 1.

(Lucchitta, 1981). Returning to earth, this chapter discusses the *direct* effects of the action of Pleistocene glacier fluctuations on the landscape, as well as the *indirect* effects resulting from glacioisostasy and glacioeustasy.

## 19.2  Direct erosional effects

### 19.2.1  Alpine glaciation

Former alpine landscapes reflect the greater expansion of Alpine glaciers during the glacial age. The distribution is different to that of present-day Alpine glaciers only in that it encompasses lower altitudes, more mountain ranges and, in mid-latitudes, low-lying foothills. The landscape type is characterized by local glaciers which exploit the pre-existing morphometry of the mountains (Figure 19.4). Glacial erosion is concentrated in depressions and valleys, while subaerial processes of rock weathering attack the valley sides overlooking the former glaciers.

Glacial troughs are perhaps the most common erosional form and often are incised into a broader valley depression. Frequently they have a dendritic plan of master channels and tributaries directly inherited from the pre-existing river system. The cross-section of a trough is flat-floored and steep-sided. The cross-section is sometimes called U-shaped, but this depends on one's handwriting! – and in practice the shape is more akin to a parabola of the form $y = ax^b$, with each half of the trough treated separately, where $y$ = vertical distance; $x$ = horizontal distance from the origin in mid-valley; $a$ = the coefficient; $b$ = the exponent (Graf, 1970). In the Beartooth Mountains of Montana and Wyoming the depth–top width ratios of glacial troughs vary between 0·2 and 0·4. The cross-section of a glacial trough is analogous to that of a river channel, with the need for a larger channel a function of the slower flow of ice. A parabolic profile is favoured by iceflow beneath a warm-based glacier. As shown in Figure 19.5, basal velocity and ice thickness are greatest in the middle, as also will be subglacial meltwater activity. This favours both abrasion and the ability to transport eroded debris and will tend to deepen the bottom of the channel more rapidly than the sides. The steepness of the valley sides probably reflects structural and lithological factors. It has long been observed that troughs are deep and narrow where the bedrock is resistant, and wider and shallower where the bedrock is less resistant (Matthes, 1930). The widening of the bottom of a pre-existing meandering river valley produces a number of characteristic features, such as truncated spurs and hanging tributary valleys (Figure 19.4D).

The long profile of an Alpine glacial trough is often described as overdeepened in comparison to the profile of a river valley. Many have a 'down-at-heel' profile with a steep gradient near or at their heads and a gentler slope, sometimes even a reverse slope, towards their mouths. The 'overdeepening' is explained by the relative position of the equilibrium line along the trough which determines the location of the zone of greatest ice discharge. Here ice velocity or thickness will tend to be greatest, thus favouring higher than normal rates of erosion.

Glacial troughs are good examples of equilibrium landforms related to the amount of ice being discharged and, indeed, in 1905 Penck drew attention to a 'law of adjusted cross sections'. He pointed out that although hanging valley floors were obviously discordant, the surfaces of the glaciers which cut them were not.

The slopes overlooking valley glaciers have characteristic forms as a result of glacier undercutting and the evacuation of debris by the glacier. The slopes are straight and rectilinear. In such situations the mountains may form upstanding horns with three or four distinct faces, depending on the geometry of the surrounding glaciers. Ridges between glaciers may be crenulated and are known as *arêtes*. Linton (1963) suggested that horns are stable equilibrium forms. If, as he argued, the stress release of the rock parallel to each face is an important mechanism of rock fracture, then the process will become decreasingly effective as the horn gets smaller. *Arêtes* may be relatively stable forms also, partly for the same reason; but also because the lower they become, the less snow they trap for the nourishment of adjacent glaciers.

In mountain areas which were only marginally glaciated typical alpine features as described above may not occur. Instead the glacial imprint may be confined to discrete cirque basins. In such a situation snow accumulation has only been able to maintain a glacier in the most topographically suitable location. In mid-latitudes of the northern hemisphere the most suitable locations are topographic hollows facing north-east (Derbyshire and Evans, 1976). This provides the most shade from summer insolation and also is suitable for the collection of snow on the lee side of massifs exposed to prevailing westerly winds. In the southern hemisphere the favoured orientation is south-east for the same reasons. The more marginal the conditions for former cirque formation, the more tightly constrained are cirque basins. Thus in areas like the Tatra Mountains of Eastern Europe, cirques are closely linked to favourable pre-existing river valley heads. In the Falkland Islands strong winds allowed the accumulation of snow in only the most favourable lee-side location, those facing east and backed by broad plateaus (Roberts, 1984). In such situations the glacial imprint on the overall landscape is limited to a scatter of

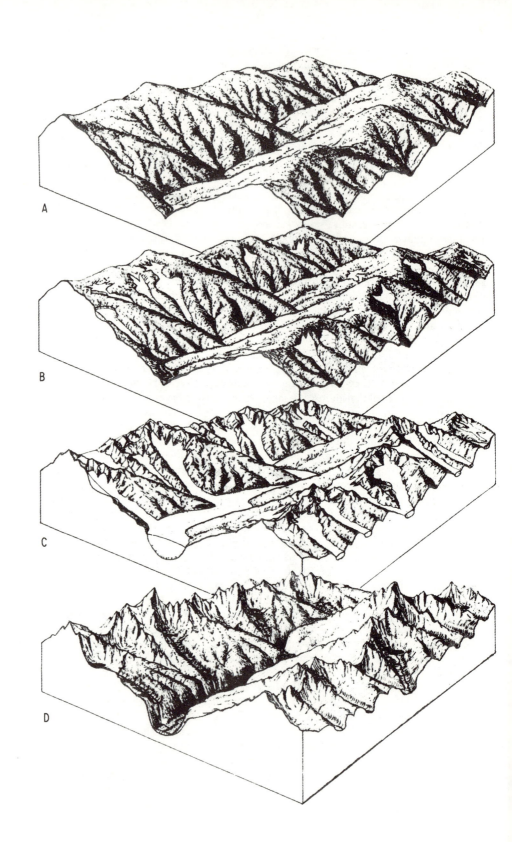

**Figure 19.4**   The effects of alpine glaciation: A. mountains before glaciation; B. the growth of glaciers in topographic hollows; C. network of valley glaciers; D. area after deglaciation showing glaciated troughs, cirque basins, truncated spurs, hanging valleys, *arêtes* and horns.
*Source:* R. F. Flint, 1971, *Glacial and Quaternary Geology*, figure 6-3, p. 140, copyright © John Wiley and Sons, by permission.

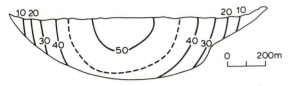

**Figure 19.5**   Distribution of longitudinal velocity in metres per year as measured in the Athabasca Glacier, Canada.
*Source:* Raymond, 1971, figure 10, p. 69.

armchair-shaped hollows which sharply truncate the lines of the pre-existing relief. As conditions become more conducive to glaciation, cirque orientation and the range of their topographic situations increases. Cirques may face in all directions, although there will still be a dominance in the shaded sector. Also they will create basins in less-favoured sites, for example, on the flank of a glacial trough or horn (Plate 28).

The morphology of cirques is a reflection of geological and topographic factors, as well as the type and duration of glaciation. Jointing and rock structure are particularly important factors which dominate the cliff form and the depth of the cirque basin. Figure 19.6 shows some ideal-

ized basin profiles based on a study of jointing in a selection of Scottish cirques and serves as an example. Further, it is likely that cirques will maintain their characteristic form over long periods of glaciation. Gordon (1977) carried out a morphometric study of cirques in Scotland. Making the ergodic assumption that larger cirques represented a more advanced stage of evolution, he predicted that cirques will tend to become more enclosed and deeper with time (see Figure 19.7).

### 19.2.2   Ice sheet glaciation

The great northern ice sheets that expanded on many occasions in northern and mid-latitudes eroded the landscape in characteristic ways. Table 19.1 shows a landscape classification with the chief variables responsible for each.

A   PLAN

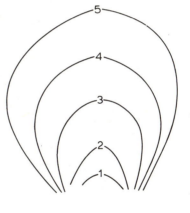

B   PROFILE

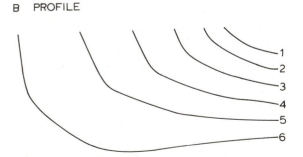

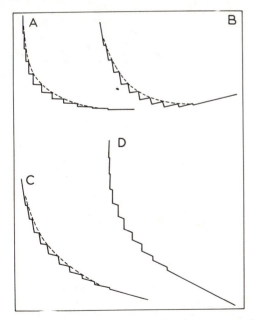

**Figure 19.6**   Idealized downvalley profiles based on a study of jointing in Scottish cirques.
*Source:* Haynes, 1968, figure 5, p. 225.

**Figure 19.7**   Suggested model of cirque evolution based on a morphometric study of Scottish cirques.
*Source:* Gordon, 1977, figure 7, p. 192.

*Landscapes of areal scouring* everywhere bear signs of glacial erosion. Joints, faults and dykes are the master features which are generally eroded to form depressions. In between are upstanding bosses of rock which are often shaped like roches moutonnées with blocky, craggy downstream sides and smoothly convex upstream slopes (Plate 29). *Landscapes of selective linear erosion* describe situations where glacial erosion has been confined to the excavation of troughs usually along pre-existing river valleys, and where intervening uplands may support old regolith and pre-glacial tors. *Landscapes with little or no sign of glacial erosion* include those known to have been covered by ice but which bear no sign of the event. Instead gentle, regolith-covered slopes, river valleys and tors are characteristic. Finally, there are *composite landscapes* where any of the above may also bear signs of

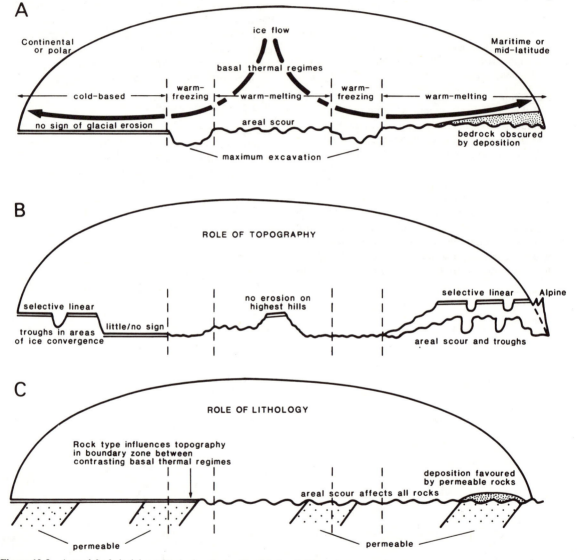

**Figure 19.8** A model of glacial erosion by ice sheets. The left-hand side represents continental or polar conditions and the right-hand side maritime or mid-latitude conditions: A. the main basal regimes and the idealized erosional effects; B. the schematic effect of topography showing how uplands experience minimal erosion or selective linear erosion; lower uplands in maritime sectors may, however, be scoured; massifs rising above the ice sheet are sculpted by alpine glaciers; C. the schematic effect of the permeability of the bedrock, showing how it can modify patterns of erosion and deposition, especially at the boundary between basal thermal zones.

**Table 19.1**  Classification of landscapes eroded by ice sheets and the role of the main variables involved.

| Landscape type | Ice characteristics | Most common topography | Geology |
|---|---|---|---|
| Areal scouring | Warm-based | Low-lying, flattish | Crystalline, jointed rocks; impermeable |
| Selective linear erosion | Warm-based in troughs, cold-based over plateaus | Upland plateau | Not relevant |
| Little or no sign of glacial erosion | Cold-based/no basal debris | Upland/ peninsula | Permeable? |
| Composite ice sheet/alpine glaciation | Fluctuating glacier types | Mid-latitude upland | Hard rocks |

local alpine glaciation. Usually these are landscapes with cirques incised into uplands which have themselves been covered by an ice sheet.

It seems that the ice sheet landscape types can be related to the variables shown in Table 19.1, but in an as yet imperfectly understood way. The key factor is the state of the base of the ice. Where the ice is warm-based, it may slide and achieve erosion. Areal scouring occurs where the basal ice is at the pressure melting point (except perhaps on the tops of small hummocks, as described in Chapter 17). At the scale of an ice sheet basal melting occurs beneath the centre and near the periphery in warmer latitudes or seas. Probably during ice sheet build-up warmer temperatures extend over most of the bed. The broad pattern is modified at a smaller scale by topography and, in particular, by its role in influencing the thickness of overlying ice and in favouring ice divergence or convergence. Generally an upland which is only thinly covered by ice, and which causes ice to diverge round it will tend to be protected beneath cold-based ice. Conversely, depressions channel more ice, which favours higher basal temperatures and erosion. Geology plays an as yet unclear role. Landscapes of areal scouring are well developed on shield rocks and this may relate to fracture characteristics. On the other hand, permeable rocks will reduce the water present at the ice–rock interface and suppress glacial erosion. It seems likely that geology can play a local role on a scale similar to that of topography.

Some of these relationships are combined into a model in Figure 19.8, which attempts to provide a basis for predicting the impact of ice sheets on landscape. At the scale of an ice sheet as a whole there is a tendency for areal scouring to occur beneath the warm-based portion of the ice sheet, namely beneath the centre and near the warm margin. In between is a zone of cold-based ice which experiences 'freezing-on' and may represent a zone of maximum debris entrainment, and thereby of erosion. Towards the dry continental or polar periphery the ice is cold-based and may achieve no erosion. Topography and geology have local effects which are most obvious in a zone of thermal transition. Thus a hill or zone of permeable rock in the midst of a zone of warm-melting may undergo erosion because the local effects are submerged by large-scale ice sheet conditions favouring erosion. In a marginal thermal zone, however, where the ice is only just at the pressure melting point, then the hill and permeable rock may influence the pattern and stand out as areas of minimal erosion.

The distribution of landscape types lends some support to these generalizations. Areal scouring occurs beneath the middle of the former Laurentide ice sheet and in warmer maritime margins like Labrador and West Greenland (Figure 19.9). Areas of little or no erosion are common in 'dry' northern Arctic Canada and Greenland. Landscapes of selective linear erosion are associated with uplands in intermediate areas – e.g. Baffin Island and east and north Greenland. In the south in Labrador, and in west Greenland, lower uplands are scoured by ice. Together these uplifted continental margins provide some of the world's classic fjord landscapes, the fjords representing normal glacial troughs only with their floors excavated below sea-level, sometimes to considerable depths (Figure 19.10) (Plate 30).

A contrast in ice sheet scenery exists on the small scale, for example, in an area no larger than Scotland. Whereas the western Highlands consists of a classic ice-scoured landscape, including some high uplands and fjords, landscapes of selective linear erosion occur in eastern uplands like the Cairngorm Mountains, where the dramatic pre-glacial tors highlight the protection that has occurred on the plateau tops (Sugden, 1968). In the eastern lowlands of Buchan there are places where there is no evidence of glacial erosion and, indeed, clear pre-glacial gravels and deeply weathered rocks remain intact despite submergence by an ice sheet (Hall, 1983) (see Plate 8).

In mid-latitude areas the fluctuations between full ice sheet cover, no glaciers and local alpine glaciation lead to further complications. It used to be thought that cirques were created at the beginning and end of a glacial hemicycle: cirques → valley glaciers → ice sheet → valley glaciers → cirques. Probably there are few grounds for such a general model and it is likely that the complex fluctuations over the last 2–3 million years have offered many occasions when cirque glaciers could thrive. The

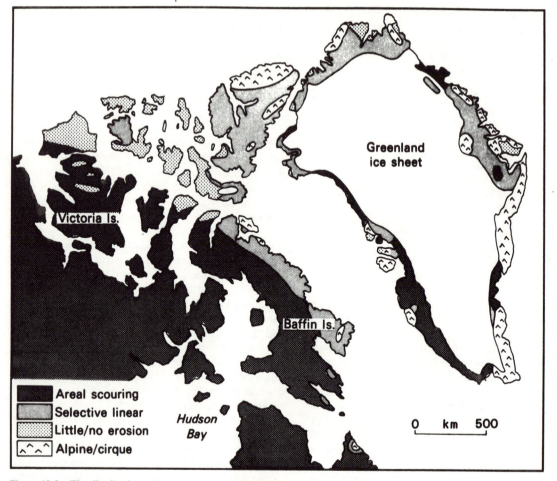

**Figure 19.9**   The distribution of landscapes of glacial erosion in part of North America and Greenland.
*Source:* Sugden, 1982, figure 4.16, p. 84.

landscape that results depends very largely on the effects of intervening ice sheet phases. In areas of selective linear erosion cirques are likely to be well preserved, since they occur in mountains which by virtue of their bulk and altitude would be covered by thin and diverging ice favouring a cold-based temperature regime. A fine example of some forty well-preserved cirques exists in the Cairngorm Mountains of Scotland and it can be demonstrated by the lack of cirque moraines that many were excavated before the last ice sheet glaciation. In the west of Scotland where ice sheet conditions favoured erosion even over uplands, the cirque forms have been heavily modified by overrunning ice.

There has been considerable debate about the amount of glacial erosion achieved in mid-latitudes. One view is that here ice sheets have been agents of massive landscape modification and White (1972), for example, suggested

that over 1000 m of rock was removed from over the site of Hudson Bay. There is formidable evidence to suggest, however, that the depth of material removed is of the order of tens to a hundred metres. Pre-glacial river patterns, landforms and Tertiary-dated deposits can still be recognized in areas of areal scouring and imply only limited lowering (Sugden, 1976). One particularly interesting piece of evidence occurs at the southern edge of the Scandinavian shield, where tors and huge corestones due to deep weathering can be shown to be pre-Cretaceous in age. Not only can they be traced beneath adjacent Cretaceous deposits, but some of the corestones actually have Cretaceous oysters in growing positions on their flanks! The implication for glacial erosion is that this marginal part of the shield has undergone little Pleistocene glacial erosion (Rapp, 1981, personal communication). Another example comes from

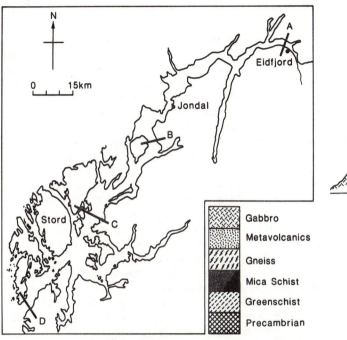

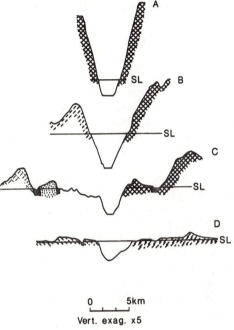

**Figure 19.10**   Map and cross-sections of Hardanger Fjord, western Norway.
*Source:* Holtedahl, 1975, figures 5 and 6, pp. 14, 15.

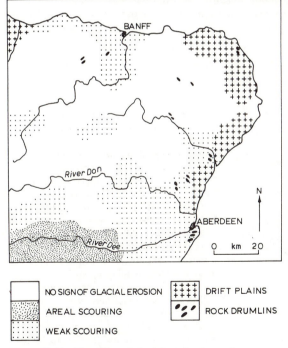

**Figure 19.11**   The effect of glacial erosion in north-eastern Scotland.
*Source:* Hall, 1983, figure 15.2.1.

recent work in north-east Scotland (Figure 19.11). In places there are clear landscapes of areal scouring, particularly in the vicinity of the Dee and other valleys, which channelled ice flow. There is a gradation from this fresh, ice-scoured bedrock, through lightly modified hills with some streamlined forms and exposed bedrock, to a landscape underlain entirely by deeply rotted rock, overlain in places by gravels of pre-glacial age. Common weathering depths in this area range from 17 m to 50 m. Bearing in mind reconstructions of the pre-glacial relief, Hall (1983) has shown that glacial erosion by ice sheets in this part of Scotland has done little more than remove the pre-existing weathered regolith and, indeed, represents a lowering of less than 50 m. Similar ideas have been argued in North America (Feininger, 1971) and it may be that the bulk of the products of ice sheet erosion consist of this pre-glacial regolith. This conclusion has important implications for understanding glacial deposits, especially fine-grained tills which are so widespread in mid-latitudes. It may explain the long-observed contrast between these fine-grained mid-latitude tills and the coarser-grained tills associated with modern ice sheets which are eroding fresh bedrock.

## 19.3 Direct depositional effects

### 19.3.1 Glacial deposition

Deposits laid down by formerly more expanded alpine glaciers and ice sheets are widespread and, in places, dominate the land surface characteristics. The degree to which they dominate depends on many factors, such as the shape of the underlying topography, the geology, debris-bearing capacity of the glacier, duration of the glaciation and the processes of deposition that actually operated. Nevertheless, it is possible to recognize certain common associations. It is helpful for discussion to distinguish former *ice margin landscapes*, where the dominant landforms relate to ice margin processes over a relatively long period of time. Within the ice margin landscape we can distinguish two further zones: *stagnant ice landscapes*, and *active ice retreat landscapes*; the former are characterized by irregular hummocky moraine, and the latter by a distinctive blend of subglacial landforms crossed by minor marginal ridges.

Figure 19.12 is a model of the features associated with valley glacier retreat. Perhaps the most important feature is the *marginal (lateral or terminal) moraine* with varying proportions of debris accumulated from the valley walls and basal debris. The terminal moraine has a regular slope and characteristically is arcuate across a valley floor at the glacier's extremity. Its size is likely to reflect both the duration of the glacier still-stand and the amount of debris being transported to the margin by the glacier. The moraine may be discontinuous, particularly where removed by meltwater activity and may give way to an irregular, hummocky landscape where debris was abundant on the glacier surface. In such areas typical stagnation features, such as kettle holes, appear. Within the outer moraine limit glacier retreat may expose streamlined landforms of subglacial origin, such as flutes, crossed by annual moraines associated with a winter advance. The more rapid the retreat, the more likely these are to escape modification by marginal processes. Often glacier retreat may be marked by still-stands or readvances of several years in which case a succession of moraines interspersed with areas of outwash or fluted till are typical (Figure 19.13).

Depositional landscapes associated with ice sheets are less well understood and are complicated, in particular, by the differential coverage of various glaciations and their varying and uncertain ages. Figure 19.14 is an example of attempts to relate former ice extent to the main deposits in Britain. It shows the direction of ice-flow, distribution of main drumlin fields and fluvio-glacial deposits, moraine ridges, proglacial lakes and the relationships with former sea levels. In addition to these forms, till sheets mantle much of the low-lying landscape

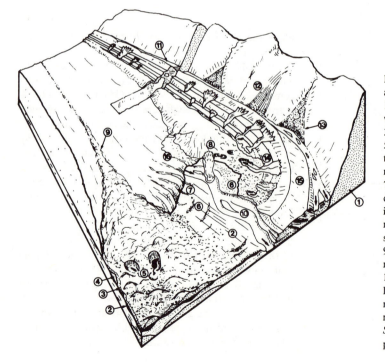

**Figure 19.12** Landforms and sediments associated with the retreat of a valley glacier: 1, bedrock; 2, lodgement till with fluted and drumlinized surface; 3, icecores; 4, supraglacial morainic till; 5, icecored and kettled supraglacial morainic till; 6, bouldery veneer of supraglacial morainic till with deposited crevasse fills; 7, dump moraine ridges showing internal deformed scree-bedding having formed as ice-contact screes; 8, supraglacial lateral moraine and till flows into meltwater streams resulting in till complexes; 9, supraglacial medial moraine; 10, proximal meltwater streams; 11, kame terrace; 12, truncated scree; 13, fan; 14, gullied lateral terrace; 15, lateral-terminal dump moraine; 16, fines washed from supraglacial moraine till into crevasses and moulins. *Source:* Boulton and Eyles, 1979, figure 3, p. 15.

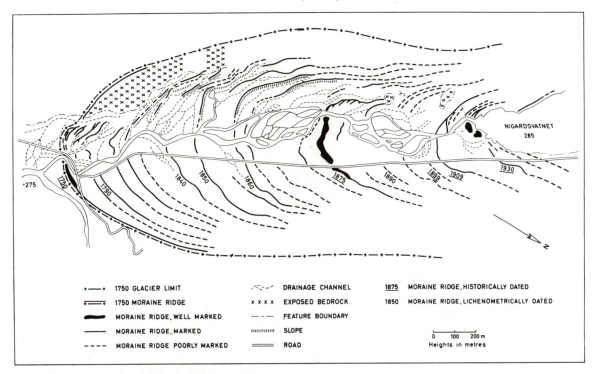

**Figure 19.13**  Moraines reflecting the glacial retreat in the Nigardsdalen Valley, south Norway.
*Source:* Andersen and Sollid, 1971, figure 33, p. 30.

covered by each glaciation and may overlie former inter-glacial sites, which in East Anglia are of at least two distinct ages.

Ice-marginal ice sheet landscapes are well studied in North America where a sometimes exceedingly complex pattern of moraines and tills occurs (Figure 19.15A.) Here along the southern margin of the Laurentide ice sheet total till thicknesses have been observed to exceed 400 m and individual till sheets may exceed 40 m (Figure 19.15B). In places the till that culminates in a broad undulating moraine ridge is up to several kilometres across and up to 200 m in height. Complications arise when a till sheet and culminating ridge have themselves been covered by another more extensive till of younger age, as has been demonstrated in Ohio. One striking characteristic of the till is its regular lithological makeup often over hundreds of square kilometres of terrain, which indicates a more or less continuous supply of debris from the same source areas and the mixing of the debris during transport, presumably in the zone of sliding. Further west in the high plains of Saskatchewan and Alberta ice-marginal moraine landscapes associated with the former Laurentide ice sheet have greater evidence of thrusting of pre-existing sediments which out-

line the ice margin (Moran *et al.*, 1981).

Ice stagnation landscapes can occur on a massive scale. Again good examples occur on the high plains of the Prairies in Canada, where hummocks and hollows with a relief of 30–45 m occur in a random pattern (Prest, 1983). More commonly such hummocky moraine may have a relief of 2–10 m and may extend over hundreds of square kilometres. The debris in the mounds demonstrates that much of it is derived from the basal ice load and the inference is that it has been uplifted in the ice by compressive flow and later melted away *in situ*. Meltwater activity associated with the stagnation can contribute irregular fluvioglacial mounds to the overall complex.

Landscapes of active ice retreat are perhaps the most widespread within the limits of former ice sheets. Huge areas of Arctic Canada and northern Scandinavia are marked by subglacial landforms, streamlined parallel to and transverse to former ice movement, as well as a series of marginal moraines of varying significance representing annual or periodic retreat of the ice border. Figure 19.16 shows a typical association of such landforms in Finland and the Kola Peninsula area of the Soviet Union. In this case streamlined drumlins, rock striations and

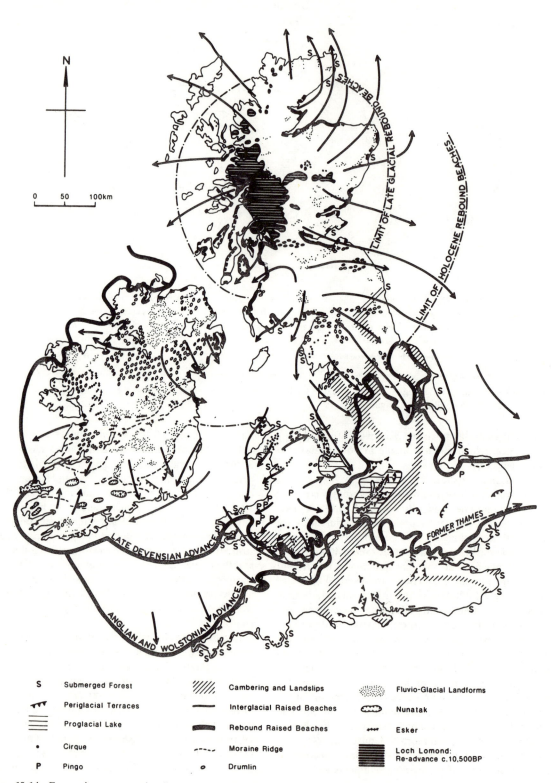

**Figure 19.14** Former ice extent, major directions of ice movement, depositional features and raised beaches in the British Isles. *Source:* Bowen, 1977, pp. 690, 691, 693.

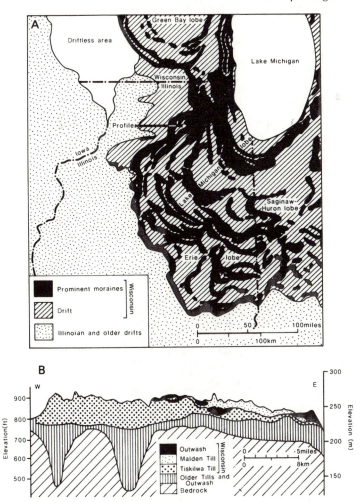

**Figure 19.15** Glacial deposits at the southern end of Lake Michigan: A. moraines associated with fluctuations of the Lake Michigan lobe of the Laurentide ice sheet; B. profile across some till sheets in northern Illinois; the location is shown in A. *Sources:* Frye and Willman, 1973, figure 1, p. 138; Wickham and Johnson, 1981, figure 3, p. 178.

eskers represent the direction of iceflow within lobes which remained active during the process of deglaciation. The positions of such lobes during still-stands in retreat are clearly picked out by arcuate lines of marginal deposits. At a scale which cannot be shown on the map is a transverse pattern of small moraine ridges. Some of these are subglacial thrust ridges, while others represent the annual retreat of the ice margin. A virtually identical series of landforms reflect the deglaciation of the Labrador–Ungava Peninsula in sub-Arctic Canada, where the lineation parallel to iceflow and transverse elements in the landforms is clearly displayed (Figure 19.17). In most such areas the landforms and drift are thin and the underlying ice-moulded bedrock is readily apparent and frequently exposed at the surface.

A further distinctive landscape of deposition is associated with the retreat of glaciers with their snouts ending in lakes or in the sea. In such a situation the moraine of a land ice margin is replaced by another distinct suite of landforms and sediments. Some of the main characteristics are summarized in Figure 19.18, which shows an active calving glacier in relatively shallow water. In this case a moraine bank is formed as the result of the melting out of debris from beneath, within and on top of the glacier. In addition, debris is dumped from icebergs as they calve from the glacier. Meltwater streams issuing from the base of the glacier build up coarse fans of debris at their exits from the glacier. If the meltwater is sufficiently loaded with suspended sediment, it may form an underflow and transport fine sand and mud well away from the glacier snout. Winter advances of the ice front can cause minor push ridges on the morainic bank, while the exits of the meltwater channels may change from year to year building up an overlapping sequence of deltas.

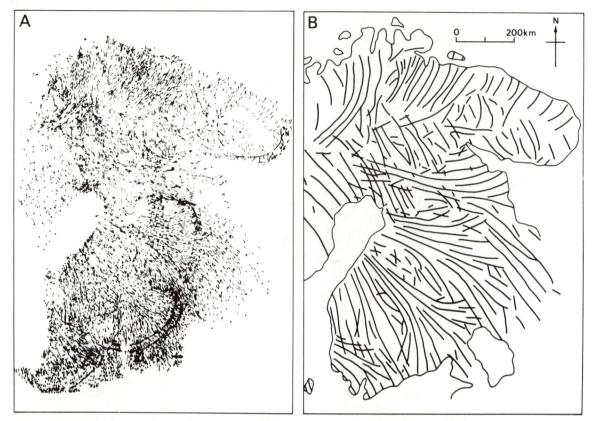

**Figure 19.16**  The glaciation of Finland and the north-western Soviet Union: A. streamlined drumlins and flutes (single lines); striae (arrows) and fluvioglacial deposits (black); B. flowlines of the active lobes during overall retreat. *Source:* Punkari, 1982.

Slumping and sediment gravity flows are common on the front of the slope, when angles are close to 20° or above. Iceberg-dropped debris (mud and dropstones) will occur away from the ice front.

Such subaqueous deposition forms are common in glaciers retreating in fjord country like Alaska, Chile and Norway, and are often exposed above sea level by isostatic recovery. They are also common around the flanks of Hudson Bay and the Baltic Sea where ice retreat was intimately associated with sea calving and again where isostatic recovery has subsequently led to exposure. They are also common in former ice-marginal lakes. Such lakes were particularly common in North America as the ice retreated eastward from the high central plains. Indeed, between one-quarter and one-third of the land vacated by the Laurentide ice sheet was covered by proglacial lakes during ice retreat.

Floating ice shelves are capable of building distinctive moraines. Such moraines are characteristically hori-

zontal and may be deposited along a coastline where the ice shelf runs aground. Figure 19.19 shows an idealized example from George VI Sound, Antarctica, where the ice sweeps across the Sound to ground on Alexander Island. There is abundant surface ablation, and this allows the upward movement of debris at the edge and bottom freezing. Characteristically the deposit consists of far-travelled rocks mixed with local rocks derived at the point of grounding. Roughly horizontal moraines of this type have recently been discovered in Arctic Canada and eastern Greenland. It seems likely that, as with other glaciers, such moraines can only build up in the surface ablation areas of ice shelves.

### 19.3.2 Models of ice retreat

The previous section has described actual landscape features that can be found in formerly glaciated areas. The nature of these landscapes depends on the mode of

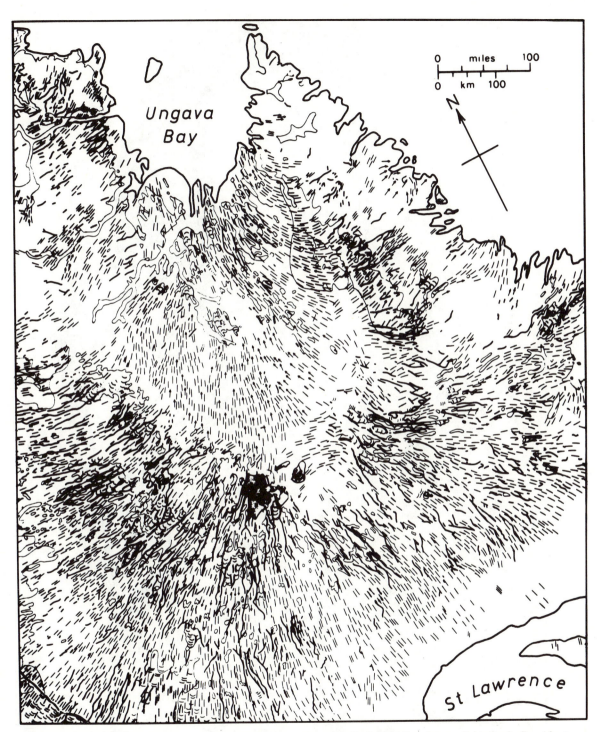

**Figure 19.17** Part of the glacial map of Canada in Labrador–Ungava showing glacial lineation parallel to the iceflow (short, straight lines), glacial striations (arrows), morainic lineation transverse to the iceflow (arcuate lines) and eskers (longer, less-regular lines).

*Source:* © Geological Society of Canada, *Glacial Map of Canada*, 1253A, 1:5,000,000, 1968, reproduced by permission.

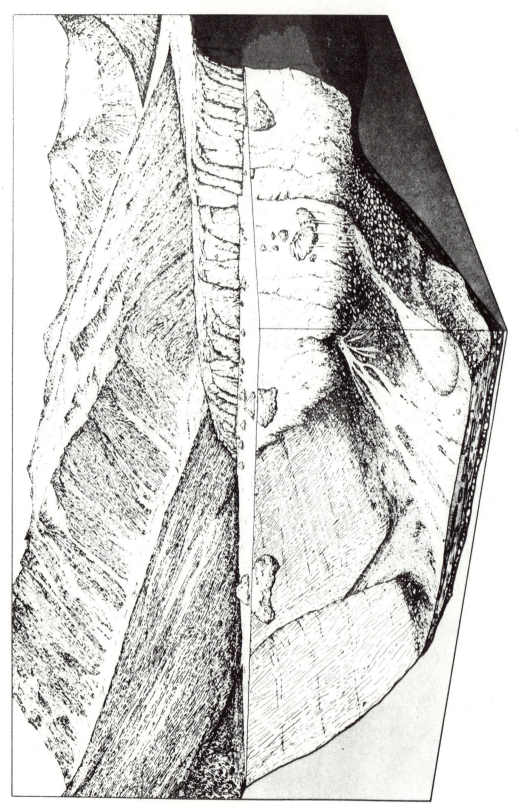

**Figure 19.18** Depositional processes at the snout of a glacier calving into shallow water.
*Source:* Powell, 1981, figure 2, p. 131, drawn by R. W. Tope.

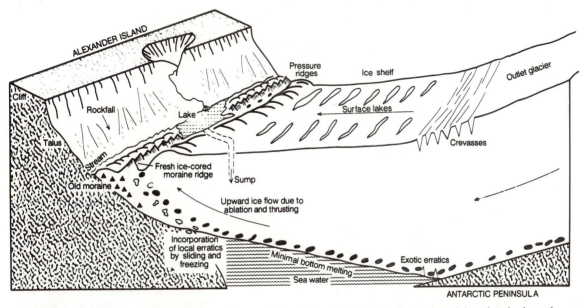

**Figure 19.19**  Model of an ice shelf which crosses George IV Sound from the Antarctic Peninsula and deposits a horizontal moraine on the shore of Alexander Island.
*Source:* Sugden and Clapperton, 1981, figure 8, p. 140.

deglaciation and this, in turn, depends on several factors. The main variables are the type and behaviour of the glacier involved, the nature and duration of the climatic amelioration, the underlying topography and its lithology. These all represent areas of current research, and this section can only point to the sort of relationships involved.

The type of glacier is critical in determining how it will retreat. Perhaps the most fundamental contrast of all is between a land-based and marine-based ice sheet. The former is securely sitting on land, while the latter is grounded on rock, which may as in the case of west Antarctica be as much as 2000 m below sea level. Weertman (1974) and Hughes (1975) have drawn attention to the potential instability of a marine-based ice sheet. In particular, thinning for climatic or other reasons or a rise in sea level may assist in flotation and begin a progressive disintegration due to rapid calving into a 'calving bay'. Presumably in such a case of deglaciation marine and icefront deposits will be dominant. Predictions of ice sheet collapse within only a few hundred years are borne out by just such a collapse of ice in Hudson Bay between 8000 and 7000 years BP. Indeed, it is interesting that the mid-latitude ice sheets centred over sea basins – Hudson Bay, Baltic, Barents Sea – have all disappeared. Only the land-based Greenland ice sheet survives. Apart from the speed of collapse, the other main contrast is that a landbound ice sheet will respond

only to climatic fluctuations, whereas a marine-based ice sheet may also be controlled by sea level.

A second fundamental contrast is between an alpine glacier and an icecap. The critical difference is the altitudinal range of the accumulation zone which tends to be high for an alpine glacier and low for an icecap. Generally speaking, glaciers with large parts of their accumulation area close to the equilibrium line are more susceptible to change than those with a more even altitude–area distribution. Thus a modest rise in the equilibrium line could wipe out an icecap leading to widespread stagnation and yet cause only a steady retreat of an alpine glacier (Figure 19.20).

The internal characteristics of a glacier are likely to affect the landscape they expose during retreat. The contrast between a margin which is warm-based and one that is cold-based is critical. A warm-based margin is likely to lead to the deposition of uniform till sheets and is likely to fluctuate freely leading to the build-up of marginal moraines. A cold-based margin is likely to involve thrusting of basal material, and because of its effect on compressive upward flow transporting dirty ice to the surface, widespread ice stagnation. This contrast is precisely that observed around the Laurentide ice sheet, with till sheets representative of the warm-based southern margin contrasting with the thrusting and ice stagnation features characteristic of the cold-based margin in the high western plains.

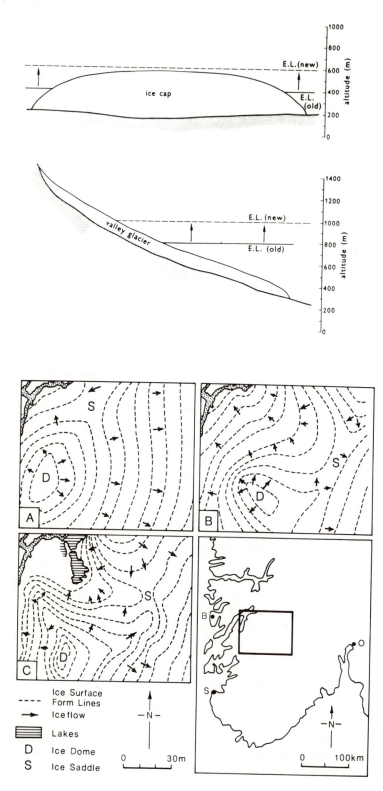

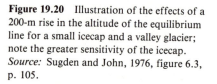

**Figure 19.20** Illustration of the effects of a 200-m rise in the altitude of the equilibrium line for a small icecap and a valley glacier; note the greater sensitivity of the icecap.
*Source:* Sugden and John, 1976, figure 6.3, p. 105.

**Figure 19.21** Ice surface form lines, directions of iceflow and locations of lakes during three phases of deglaciation (A.–C.) in the Hardangervidda, south-west Norway; the migration of the ice dome D and saddle S are most striking.
*Source:* Vorren, 1979, figure 16, p. 30.

Another factor which can lead to intense ice compression with the subsequent stagnation of debris-bearing ice is surging (Sharp, 1982). Perhaps some complex stagnant ice landscapes may turn out to be related to surges during retreat.

The climate during deglaciation and particularly the time required for a glacier to respond to any change will affect the landforms on the ground. One aspect is the time needed for ice centres to migrate during deglaciation. Given sufficient time, a large overall ice dome may break down and be succeeded by a succession of smaller ice domes over high ground. An example of adjusting ice divides on a small scale is shown in Figure 19.21 associated with the deglaciation of part of the Norwegian ice sheet over Hardangervidda. Larger-scale examples are the migration of Laurentide ice centres on to the Labrador–Ungava Peninsula during deglaciation or on to Baffin Island where the Barnes icecap is an actual Pleistocene ice remnant.

The glacier response to climatic change will vary according to the glacier's local climatic environment. A good example is the Younger Dryas cold period of

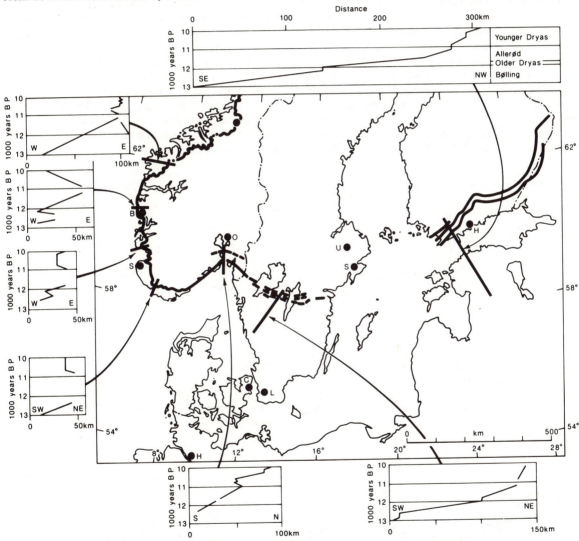

**Figure 19.22** End-moraines of Younger Dryas age (11,000–10,000 years BP) around the Scandinavian ice sheet. The time–distance diagrams show the response of the ice edge to the Younger Dryas climatic deterioration and reveal a sharper readvance of the margin in the maritime west than in the more continental east.
*Source:* Reprinted with permission from Mangerud, 1980, figures 1 and 2, pp. 24–5; copyright © Pergamon Press.

around 10,500 years BP which interrupted the overall retreat of the Scandinavian ice sheet. The cold phase lasted some 500 years. In western Norway the greater precipitation and quicker response time of a more active ice margin caused a sharp ice advance (Figure 19.22). In the east in Finland the overall retreat of several thousand years was interrupted by a minor still-stand of the ice. In intermediate Oslo and Västergötland there was a modest advance of the ice front. In this case different ice-marginal landscapes result from the same climatic event.

The underlying topography is bound to influence the landscape of glacial deposits. When the glacier retreats across a regular plain, then one might expect subglacial and ice-marginal features to be clearly displayed. However, irregularities are introduced by hilly topography. During deglaciation an ice sheet will retreat at its margins and at the same time it will become lower and thinner. Areas of higher relief will protrude through the ice surface and isolate patches of stagnant ice in deep valleys and other depressions (Figure 19.23). Once isolated, such stagnant ice will be charged by fluvioglacial deposits

derived from both the ice surface and surrounding hills and finally melt away *in situ* to leave a complex topography of hummocky moraine and fluvioglacial deposits. Such a model of deglaciation seems common in the uplands of Scandinavia and Scotland.

The lithology of the glacier bed may influence glacier morphology and thereby deglaciation. Boulton and Jones (1979) have suggested that glacier gradients on soft deforming sediments are lower than normal. The hypothesis may explain the low-gradient lobes postulated on geomorphological evidence in areas such as the southern Baltic. What is clear is that a low-gradient lobe is highly susceptible to climatic warming, and one would expect faster than average rates of retreat and perhaps widespread stagnation as well.

Ice retreat is far removed from the simple idea of regular glacier withdrawal up a valley. Whereas such a model is applicable to many alpine glaciers, it is clear that it fails to explain landscapes deglaciated by ice sheets. In the latter case many other as yet poorly understood variables are relevant to an understanding of the complexities displayed in the real world.

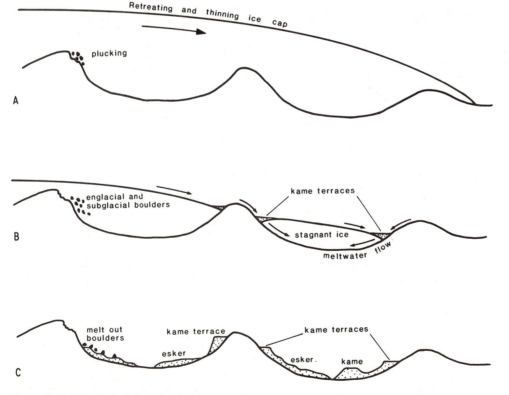

**Figure 19.23** The deglaciation of an ice cap over dissected uplands leading to stagnation of ice in valleys and the formation of kame terraces.

## 19.4 Indirect effects: glacioisostasy and glacioeustasy

The build-up and decay of ice sheets during the Pleistocene has had implications on geomorphology over the whole globe through the effects of glacioisostasy (see Section 3.4.2.) and glacioeustasy (see Section 15.5.1). Glacioisostasy describes the deformation of the earth's crust beneath the load of an ice sheet, while glacio-eustasy describes the influence on sea level of the water periodically stored in and discharged from the ice sheets. As will be seen, there are complex worldwide effects of immense importance to coastal lands.

When an ice sheet builds up on land, it depresses the earth's crust beneath its weight. Bearing in mind the average densities of ice and rock then the equilibrium depression is approximately 0·267 times the thickness of the ice. Thus an ice sheet 3000 m thick will depress the rock beneath its centre by as much as 800 m. If one ignores irregularities caused by the underlying topography, then depression will tend to be greatest beneath the centre of the ice sheet and decrease towards the margins. When the ice melts, the land rebounds, with most recovery taking place near the former ice sheet centre. Generally the time taken for isostatic recovery to achieve equilibrium is of the order of 15,000–20,000 years with recovery beginning rapidly and tailing off exponentially. A relaxation time of this sort of magnitude can be inferred from the fact that isostatic recovery is still occurring around Hudson Bay and the Baltic. The isostatic depression and recovery described so far is hydrostatic in that it assumes depression is directly equivalent to the weight of overlying ice. For such depression to occur, there must be a compensatory movement of material within the earth and this causes a bulge around the ice sheet, as suggested by Daly (1940). In the case of the Laurentide ice sheet the forebulge was of the order of 80 m and occurred at some 1000 km from the ice margin. Following deglaciation, the forebulge sank as the central ice sheet area rose. The basic results of isostatic recovery following deglaciation can be inferred from this simple model. There will be raised beaches representing former sea or lake levels within the formerly ice-covered area which are highest near the centre and submerged beaches in the area of forebulge. The evolution of one such shoreline is illustrated schematically in Figure 19.24A.

The above isostatic model is oversimplified, in one important respect, in that the earth's crust has a certain rigidity and the load of an ice sheet is partly borne beyond its margin. Figure 19.24B shows a transect across an ice margin, representing an elastic, as opposed to a hydrostatic, crust. For the diagram, the ice sheet is assumed to

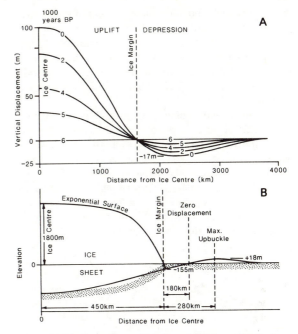

**Figure 19.24** Isostatic models of crustal deformation by ice load assuming hydrostatic and elastic behaviour. A. The isostatic modification of a 6000 years BP shoreline extending from an ice sheet centre (left) to beyond the ice margin (right). Deformation (in metres) is shown at 6000, 5000, 4000, 2000 and 0 years BP assuming complete ice removal and the hydrostatic behaviour of the crust. B. The elastic deformation of the crust at an ice sheet edge showing proglacial depression and the forebulge. The shaded area shows depression assuming hydrostatic behaviour of the crust alone, and is included for comparison. See text for explanation of heights and distances.
*Sources:* Walcott, 1972, figure 4, p. 7; Walcott, 1970, figure 10, p. 723.

have a radius of 450 km and a central elevation of 1·8 km. There are two important resulting implications. First, the earth's surface is depressed beyond the ice margin. At the margin the amount of depression can be calculated from $H/11·5$, where $H$ is the elevation of the centre of the ice sheet. In this case with an elevation of 1·8 km the depression is 155 m. The amount of depression decreases away from the ice front until in this example it is zero at a distance of 180 km and positive beyond. Second, there is a forebulge due to this elastic effect. Its elevation is $H/100$, while the distance from the ice margin is related to the flexural parameter of the lithosphere. In this example it is 280 km from the ice margin (Walcott, 1970). The elastic response of the earth's crust is rapid and the pattern of depression will respond immediately to changes in the ice load. What is important from a geomorphological point of view is that low-elevation shorelines outside the ice margin can exist during a glaciation, and that such apparently 'interglacial' beaches can be uplifted following deglaciation. Examples of such a situation are known from certain peninsulas in eastern Baffin Island which have escaped glaciation (Nelson, 1980).

The waxing and waning of ice sheets has an effect on the volume of water in the ocean and its displacement. For simplicity, it has often been tempting to regard the oceans as a tank where water level is lowered some 100–130 m during a glaciation and topped up to its present level in interglacial episodes. This view tended to regard sea level as a datum which moved up and down uniformly depending on the amount of ice stored in ice sheets on the land. However, things are not so simple and

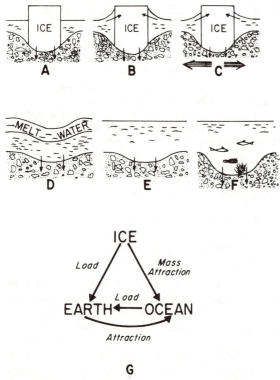

**G**

**Figure 19.25** Interactions among ice loads, water loads and the deformable crust: A. the weight of the ice deforms the crust; B. the ice mass attracts the water; C. the transfer of matter within the earth distorts the crust; D. the weight of meltwater depresses the crust differentially; E. more water flows into this depression increasing the water load; F. additional ocean floor deformation due to E; G. the flow diagram expressing A.–E. from which a numerical model was derived which was used to produce Figures 19.26 and 19.27. *Source:* Clark *et al.,* 1978, figure 1, p. 266.

Figure 19.25 shows the main factors which influence sea level. Changing ice and meltwater loads, the gravitational attraction of sea-water mass to an ice sheet and the movement of material within the earth's surface play critical roles in influencing the form of the ocean geoid (the constant gravitational potential). These various processes operate with different response times and it is necessary to appreciate the combined mass transfer of ice, water and subsurface material in sympathy with each glacial cycle in order to predict sea level at any one time in any one place.

The volume of water in the ocean has increased since the wasting of the ice sheets, and following the maximum of the last glaciation about 18,000 years BP. The rise was complete about 5000 years BP when overall glacier wastage ceased. The way this water has been distributed throughout the world's oceans is shown in Figure 19.26, which models the amounts and location of glacier melting at 1000-year intervals. The sea-level change is plotted in relation to the sea level of 16,000 years ago (which was taken in this model to be the same as the ice maximum condition). It can be seen that the increase in sea-level rise is far from uniform and varies from zero in parts of the northern hemisphere in the vicinity of ice sheets to a high of 90 m in the northern Pacific. This model does not take into consideration the effect of any melting in Antarctica and is accepted as being only a first step. Nevertheless, it gives an important indication of the main trends of post-glacial sea level.

The results become even more meaningful if they are related to zones with specific sea-level responses (Figure 19.27). Zone 1 is a zone of continuous coastal uplift predominantly due to isostatic recovery. Uplift has caused raised shorelines to altitudes over 250 m in Hudson Bay, and large areas of Arctic Canada have been below the sea since deglaciation. The pattern of isostatic uplift of former raised beaches is also brought out in Figure 19.14, which shows the centre of uplift to be over Scotland, the centre of the former ice sheet (Plate 31). Zone II is the zone of the collapsing forebulge. This is the zone where the sea has encroached steadily on the land since deglaciation. Submerged terrestrial remains are common on coasts of such regions (Walcott, 1972). In reality between zones I and II is a transitional zone where there is first emergence and then submergence, the relative importance of each depending on location with regard to the ice load. Zones III, IV and V are places where relative sea-level movements reflect the differential effects of glacial and isostatic response. Zone III is where an emerged beach formed some few thousand years ago. Zone IV has seen continual submergence since deglaciation, while zone V has seen submergence followed by minor emer-

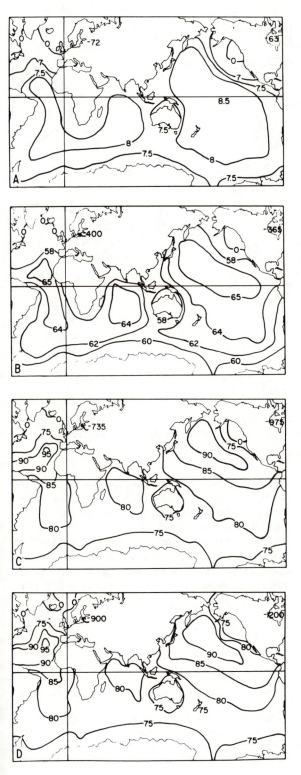

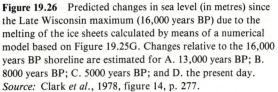

**Figure 19.26** Predicted changes in sea level (in metres) since the Late Wisconsin maximum (16,000 years BP) due to the melting of the ice sheets calculated by means of a numerical model based on Figure 19.25G. Changes relative to the 16,000 years BP shoreline are estimated for A. 13,000 years BP; B. 8000 years BP; C. 5000 years BP; and D. the present day. *Source:* Clark *et al.*, 1978, figure 14, p. 277.

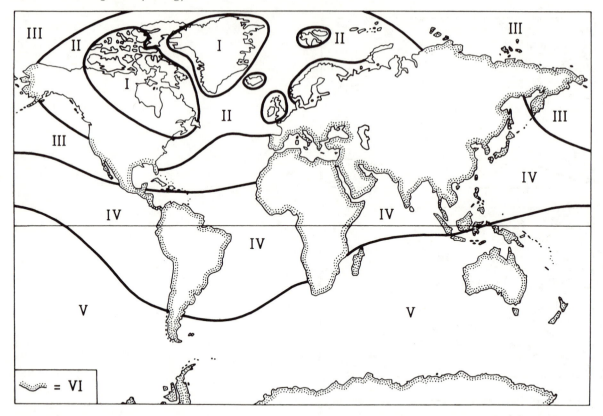

**Figure 19.27**    The six sea-level zones predicted by the numerical model suggested in Figure 19.25 in respect of sea-level changes during the last 16,000 years. Emerged coasts are predicted for zones I, III, V and VI: submerged coasts for zones II and IV. *Source:* Clark *et al.*, 1978, figure 15, p. 278.

gence after meltwater ceased to flow into the oceans about 5000 years BP. Finally zone VI encompasses all main continental shelves other than those in zone II. This is a different type of zone in that it is not related to the actual position of the ice load. Rather, it is a result of isostatic uplift of the coastal margins as a result of the meltwater load in the oceans. It is marked by recent modest uplift of the continental margins.

It can be appreciated from the above that the build-up and decay of ice sheets in the Pleistocene has had an indirect and important influence on most coasts of the world. As we have seen, the understanding of coral reefs, barrier beaches, deltas, the lower courses of rivers and high-latitude coasts demands appreciation of relative sea-level movements. When one considers that one-third of the land area has also been directly modified by the action of glaciers, then it is appreciated just how significant ice has been in modifying the geomorphology of the world.

## References

Andersen, J. L. and Sollid, J. L. (1971) 'Glacial chronology and glacial geomorphology of the marginal zones of the glaciers, Midtdalsbreen and Nigardsbreen, South Norway', *Norsk Geogr. Tidsskr.*, vol. 25, 1–38.

Beuf, S., Biju-Daval, B., de Charpal, O., Rognon, P., Gariel, O. and Bennacef, A. (1971) *Les Grès du Paléozoique*

*Inférieur au Sahara*, Paris, L'Institut français du pétrole Paris.

Boulton, G. S. and Eyles, N. (1979) 'Sedimentation by valley glaciers: a model and genetic classification', in Ch. Schlüchter (ed.), *Moraines and Varves*, Rotterdam, Balkema, 11–23.

Boulton, G. S. and Jones, A. S. (1979) 'Stability of temperate ice caps and ice sheets resting on beds of deformable sediment', *Journal of Glaciology*, vol. 24, 29–43.

Bowen, D. Q. (1977) 'Hot and cold climates in prehistoric Britain', *Geographical Magazine*, vol. 49, 685–98.

Clark, J. A., Farrell, W. E. and Peltier, W. R. (1978) 'Global changes in postglacial sea level: a numerical calculation', *Quaternary Research*, vol. 9, 265–87.

Daly, R. A. (1940) *Strength and Structure of the Earth*, New York, Prentice-Hall.

Derbyshire, E. and Evans, I. S. (1976) 'The climatic factor in cirque variation', in E. Derbyshire (ed.), *Geomorphology and Climate*, New York, Wiley, 447–94.

Fairbridge, R. W. (1970) 'South pole reaches the Sahara', *Science*, vol. 168, 878–81.

Feininger, T. (1971) 'Chemical weathering and glacial erosion of crystalline rocks and the origin of till', *US Geological Survey Professional Paper* 750-C, 65–81.

Flint, R. F. (1971) *Glacial and Quaternary Geology*, New York, Wiley.

Frye, J. C. and Willman, H. B. (1973) 'Wisconsinian climatic history interpreted from Lake Michigan lobe deposits and soils', *Memoirs of the Geological Society of America*, vol. 136, 135–52.

Gordon, J. E. (1977) 'Morphometry of cirques in the Kintail-Affric-Cannich area of northwest Scotland', *Geografiska Annaler*, vol. 59A, 177–94.

Graf, W. L. (1970) 'The geomorphology of the glacial valley cross section', *Arctic and Alpine Research*, vol. 2, 303–12.

Hall, A. M. (1983) 'Weathering and landform evolution in north-east Scotland', unpublished PhD dissertation, University of St Andrews.

Haynes, V. M. (1968) 'The influence of glacial erosion and rock structure on corries in Scotland', *Geografiska Annaler*, vol. 50A, 221–34.

Holtedahl, H. (1975) 'The geology of the Hardangerfjord, west Norway', *Norges geologiske undersøkelse*, vol. 323, 1–87.

Hughes, T. (1975) 'The west Antarctic ice sheet: instability, disintegration and initiation of Ice Ages', *Reviews of Geophysics and Space Physics*, vol. 13, 502–26.

Linton, D. L. (1963) 'The forms of glacial erosion', *Transactions of the Institute of British Geographers*, no. 33, 1–28.

Lucchitta, B. K. (1981) 'Mars and earth: comparison of cold-climate features', *Icarus*, vol. 45, 264–303.

Mangerud, J. (1980) 'Ice-front variations of different parts of the Scandinavian ice sheet, 13,000–10,000 years BP', in J. J. Lowe, J. M. Gray, and J. E. Robinson (eds), *The Lateglacial of Northwest Europe*, Oxford, Pergamon, 23–30.

Matthes, F. E. (1930) 'Geologic history of the Yosemite Valley', *US Geological Survey Professional Paper* 160.

Moran, S. R., Clayton, L., Hooke, R. Le B., Fenton, M. M. and Andriashek, L. D. (1981) 'Glacier-bed landforms of the prairie region of North America', *Journal of Glaciology*, vol. 25, 457–76.

Nelson, A. R. (1980) 'Chronology of Quaternary landforms, Qivitoo Peninsula, Northern Cumberland Peninsula, Baffin Island, N.W.T., Canada', *Arctic and Alpine Research*, vol. 12, 265–86.

Penck, A. (1905) 'Glacial features in the surface of the Alps', *Journal of Geology*, vol. 13, 1–17.

Powell, R. D. (1981) 'A model for sedimentation by tidewater glaciers', *Annals of Glaciology*, vol. 2, 124–34.

Prest, V. K. (1983) 'Canada's heritage of glacial features', *Geological Survey of Canada, Miscellaneous Report* 28.

Punkari, M. (1982) 'Glacial geomorphology and dynamics in the eastern parts of the Baltic Shield interpreted using LANDSAT imagery', *Photogrammetric Journal of Finland*, vol. 9, 77–93.

Raymond, C. F. (1971) 'Flow in a transverse section of Athabasca glacier, Alberta, Canada', *Journal of Glaciology*, vol. 10, 55–84.

Roberts, D. E. (1984) 'The Quaternary history of the Falkland Islands', unpublished PhD dissertation, University of Aberdeen.

Sharp, M. J. (1982) 'A comparison of the landforms and sedimentary sequences produced by surging and non-surging glaciers in Iceland', unpublished PhD dissertation, University of Aberdeen.

Sugden, D. E. (1968) 'The selectivity of glacial erosion in the Cairngorm Mountains, Scotland', *Transactions of the Institute of British Geographers*, no. 45, 79–92.

Sugden, D. E. (1976) 'The case against deep erosion of shields by ice sheets', *Geology*, vol. 4, 580–2.

Sugden, D. E. (1982) *Arctic and Antarctic: A Modern Geographical Synthesis*, Oxford, Blackwell.

Sugden, D. E. and Clapperton, C. M. (1981) 'An ice-shelf moraine, George VI Sound, Antarctica', *Annals of Glaciology*, vol. 2, 135–41.

Sugden, D. E. and John, B. S. (1976) *Glaciers and Landscape*, London, Arnold.

Vorren, T. O. (1979) 'Weichselian ice movements, sediments and stratigraphy in Hardangervidda, Norway', *Norges geologiska undersøkelse*, vol. 350, 1–117.

Walcott, R. I. (1970) 'Isostatic response to loading of the crust in Canada', *Canadian Journal of Earth Sciences*, vol. 7, 716–34.

Walcott, R. I. (1972) 'Past sea levels, eustasy and deformation of the earth', *Quaternary Research*, vol. 3, 1–14.

Weertman, J. (1974) 'Stability of the junction of an ice sheet and an ice shelf', *Journal of Glaciology*, vol. 13, 3–11.

White, W. A. (1972) 'Deep erosion by continental ice sheets', *Bulletin of the Geological Society of America*, vol. 83, 1037–56.

Wickham, S. S. and Johnson, W. H. (1981) 'The Tiskilwa till, a regional view of its origin and depositional processes', *Annals of Glaciology*, vol. 2, 176–82.

# Twenty  *Climatic change and polygenetic landforms*

## 20.1  Climatic change

It is true to say that there is no matter of prime significance to the geomorphologist on which he is so profoundly ignorant as that of climatic change. There are a number of very obvious reasons for this:

(1) As far as we can tell, climatic change has been continuous at all relevant scales of space and time throughout the whole span of earth history.

(2) The mechanisms of climatic change – extraterrestrial, global, regional and local – operate in a very complex manner such that, even where the major general cause has been extraterrestrial, the effects have commonly been globally unequal, conflicting or out of phase. Thus it is difficult to seek a general model for world climatic change.

(3) Climatic change can occur in respect of many aspects and combinations of means, extremes and seasonal frequencies of temperature, precipitation, humidity, evapotranspiration, and the like. Only until the last few hundred years – and in a limited number of areas only – do we possess anything but the coarsest and most speculative paleoclimatic information.

(4) We know very little about the length of time during which given landscapes have to be exposed to given climatic conditions until various attributes of these landscapes begin to reflect the operation of these climatic processes. For example, it has been suggested that some existing landscapes can only be interpreted in terms of conditions assumed to have existed during the Early Tertiary or even the Cretaceous.

(5) Even when we are in possession of quite detailed climatic information of very recent date, it is not at all clear which aspects of climate have the most significance as generators of distinctive temporal rates and spatial patterns of erosion. This difficulty is largely created by the complex role of vegetation in controlling erosion which was treated in Chapter 3. Although earlier chapters have suggested that rates of erosion in mountains and high latitudes are primarily controlled by amounts of winter snowfall and/or the frequency of cycles of freezing and thawing and that high fluvial erosion rates generally result from a combination of frequent high-intensity rainstorms and a deficient vegetation cover, these generalizations have severe limitations.

The recognition of the importance of climatic change raises certain fundamental questions for the geomorphologist. The first is: *what periods of time are required for landform assemblages to become adjusted to new climatic regimes?* Landforms related to current processes occur most obviously in two types of location.

(1) In limited core climatic regions which have been relatively unaffected by the past shifting of climatic belts (e.g. arid regions in continental locations at about 25° latitude, and humid tropical regions at about 0° latitude).

(2) In regions where the most active contemporary processes have been operating on the least-resistant surface materials (e.g. on shale and clay locations where postglacial moister semi-arid processes have been dominant). As has been shown in Chapter 3, there is no general agreement regarding the absolute efficacy of present or past erosion rates or regarding the relation of present basin erosion rates to sediment production by contemporary weathering. However, it seems reasonable to assume that the greatest contemporary erosion rates occur where the seasonal character of rainfall inhibits a stabilizing vegetational cover and yet allows intensive spring and summer runoff. It has been suggested by

Wilson (1973) that the highest erosion rates (approx. 800 tonne/km$^2$/year) occur on the Mediterranean climate borders of the semi-arid regions (rainfalls of 1270–1650 mm; 50–65 in) and in tropical wet–dry regions (1780–1905 mm; 70–75 in). High erosion rates (approx. 450 tonne/km$^2$/year) may also occur in dry continental regions (250–500 mm; 10–20 in). Compared with this possible adjustment of certain morphometric features (especially drainage density) in some 10$^4$ years, it is likely that the higher-latitude humid mid-latitude landforms may require 10$^6$ or more years to adjust. The extent and rapidity of climatic changes in the higher mid-latitudes suggest that in most periglacial regions the distinctive processes have only had time to embroider minor landforms on pre-existing terrains and that most landscapes are composed of a palimpsest of forms variously inherited from a succession of previous climates.

A second question is: *to what extent do landforms consist of such superimposed palimpsests of landform assemblages inherited from a succession of different climates so as to produce polygenetic features to be studied in terms of climatogenetic geomorphology?* Many terrains have been proposed as exhibiting polygenetic climatic features, and it is clear that climatogenetic geomorphology can often bear a striking resemblance to the denudation chronology studies discussed in Chapter 2 in that it is based on the assumption that landform assemblages represent a variously lagging cavalcade of superimposed forms associated with a sequence of Tertiary and Quaternary climatic changes (Büdel, 1948). The following are examples of a range of such polygenetic landforms:

(a) Humid tropical processes are at present operating on previous tropical wet–dry surfaces in the Ivory Coast and Zaire and on talus slopes in Guinea.

(b) Humid tropical processes are similarly superimposed on previous semi-arid alluvial fans on the edge of Lake Maracaibo in Venezuela.

(c) Tropical wet–dry conditions at 2500–3500 m in northern Peru occur on previous humid tropical landforms still exhibiting steep slopes, knife-edged ridges and the remains of large fossil landslides.

(d) Extensive arid and semi-arid regions show fossil landform evidence of previous moister conditions.

(e) Certain arid and semi-arid regions (e.g. southern Sahara and Somaliland) exhibit uneven bedrock surfaces possibly resulting from the stripping of weathered debris from above weathering fronts developed in previously tropical wet–dry conditions.

(f) Many humid mid-latitude regions (e.g. central and eastern England and the north-eastern United States) have suffered fluvial dissection of earlier glacial landforms.

(g) Certain chalk dry valleys in the humid mid-latitude regions (e.g. southern England and northern France) were cut by surface runoff during periglacial conditions.

(h) Mid-latitude areas close to the previous Pleistocene ice margin (e.g. the Driftless Area of Wisconsin, Dartmoor in south-west England and the uplands of Missouri) have fluvial landforms significantly modified by periglacial processes, giving rounded hilltops and filled valleys, the latter most recently trenched due to the contemporary humid mid-latitude conditions.

(i) Much of the contemporary periglacial zone has dominantly glacial landforms due to the recent and intense character of its inundation by ice.

(j) The climatically marginal current humid mid-latitude regions have been identified as being particularly rich in polygenetic climatic landforms. In particular, Büdel (1980) has proposed the following four-stage climatogenetic history for the Hercynian Massifs of Central Germany:

(i) Late Cretaceous–Early Pliocene etchplains formed by deep weathering under tropical wet–dry conditions, in places interrupted by inselbergs but standing generally at elevations of about 300 m.

(ii) Pliocene–Early Pleistocene terraces, being remnants of broad shallow valleys formed by climates varying from tropical wet–dry to periglacial.

(iii) Pleistocene periglacial cutting producing asymmetrical valleys incised at rates of about 0·6 m/1000 year and flanked by river terraces correlated with the main glacial phases elsewhere.

(iv) Recent Holocene valley-cutting.

Paleoclimatic information is available to the geomorphologist from a variety of sources. Biannual lake sediment varves and annual tree rings provide detailed indications of ice-free conditions and temperature/moisture combinations, respectively. Pollen and macrofossils give an often-biased record of plant life assemblages. Faunal remains, particularly from rivers, lakes and seas, reflect water temperatures and other ecological conditions. $0^{18}/0^{16}$ ratios from faunal carbonates give accurate water temperature during growth and similar ratios from ice-cores indicate air temperatures of condensation during snowfall and ice formation. In addition, there is a great body of more vague geomorphic and geologic climatic indicators; till and moraines, ancient lake levels, fossil dunes, past flood levels, fossil periglacial features, previous hydraulic geometries and textures of dissection,

**Table 20.1** Isotope methods of dating

| Isotope | Dating range (Years BP) | Datable material (Examples) | Halflife (Years) |
|---|---|---|---|
| Radiocarbon ($C^{14}$) | 0–50,000 | Peats, muds, wood, shells, etc. | 5730 $\pm$ 40 |
| Thorium ($Th^{230}$) | 0–400,000 | Deep-sea cores, corals, molluscs, etc. | 75,000 |
| Uranium ($U^{234}$) | 50,000–1,000,000 | Corals, marine molluscs, etc. | 250,000 |
| Potassium/argon ($K^{40}$) | >20,000 | Igneous rocks | $1 \cdot 3 \times 10^9$ |

*Source:* Goudie, 1977, table 1.2, p. 10.

and so on, which present problems of both interpretation and dating. Archaeological dating of floods, floor levels and cultivation layers in the soil merges into more strictly historical records of harvest amounts, floods and storms. Most accurate, of course, are the detailed climatic records which go back only a mere 200 years or so in the best-documented localities. As will be seen later, the geomorphic interpretation of even these detailed records is not without problems. A selection of precise dating methods is given in Table 20.1.

### 20.1.1 Tertiary climatic changes

At the end of the Cretaceous (65 million years BP) the climate of the earth was much more equable and less spatially varied than today, and meridional temperature gradients were less steep. Sea level was more than 100 m higher than at present, sea surface temperatures may have been some 10°C greater and there was no significant continental glaciation. Europe and Africa were more than 10° latitude further south than today, the Indian Deccan lay south of the equator, Australia was only recently separated from Antarctica, and North and South America were separated and situated further east than today (see Section 5.2). The significant climatic changes which occurred during the Tertiary were due to a combination of causes relating to solar, earth planetary motion, plate tectonic and regional tectonic mechanisms.

During the Paleogene (Paleocene, Eocene and Oligocene: 65–22·5 million years BP) (Figure 20.1) temperature fluctuations occurred, with warmer periods punctuated by abrupt and successively greater drops in temperature leading to a general decline of world temperatures (Figure 20.2). These declines were greatest where they were reinforced by poleward plate movements and Western Europe, much of which was subtropical or tropical wet–dry at the end of the Cretaceous, cooled more appreciably than the westward-moving North America (Figure 20.2B). Along with these temperature changes, there was a general increase of high-latitude humidity. By the Early Oligocene (35 million years BP) world precipitation was at a maximum. By the Middle Oligocene (30 million years BP) there was an end to high-latitude 'subtropical' climates, glacial tendencies had begun in some regions and the variety of our present climatic belts was becoming apparent (Figure 20.3A). By the Late Oligocene (25 million years BP) world temperatures were more constant than hitherto and there was a decrease of precipitation in higher latitudes; Antarctica had been finally isolated allowing a circumpolar current to develop and the Antarctic icecap was growing.

During the Neogene (Miocene, Pliocene, Pleistocene and Holocene) there was continued cooling, but with significant and complex oscillations; for example, warming in the Early Miocene (20 million years BP) and Late Miocene (7 million years BP), and cooling in the Middle Miocene (14 million years BP) and Late Pliocene (2·5 million years BP) (see Figure 20.1). Global generalizations become difficult and, for example, during the Early Miocene there was warming in the western United States, cooling in Japan, little change in Europe and the establishment of the Sahara Desert. The Middle Miocene appears to have been a time of high humidity and increased river loads. During the first half of the Miocene sea level rose, only to fall again in the Late Miocene. By the Early Pliocene (6 million years BP) the Mediterranean Sea had been isolated by the fall in sea level or by tectonic movements, leading to regional desiccation; and by the Middle Pliocene sea-level glaciation had occurred in the Arctic and glaciers had been initiated in the southern Andes and the Sierra Nevada (Frakes, 1979).

### 20.1.2 Pleistocene climatic changes

The idea of four similar and synchronous worldwide glacial stages has been severely modified of recent years.

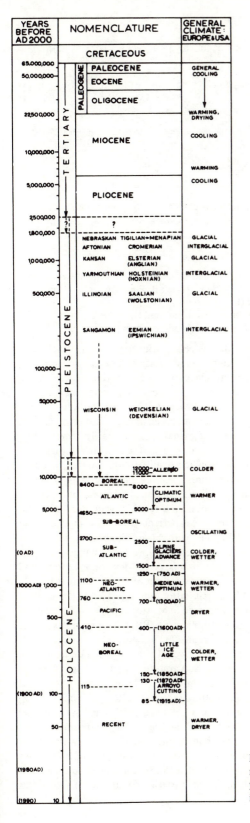

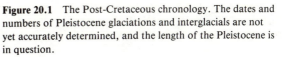

**Figure 20.1** The Post-Cretaceous chronology. The dates and numbers of Pleistocene glaciations and interglacials are not yet accurately determined, and the length of the Pleistocene is in question.

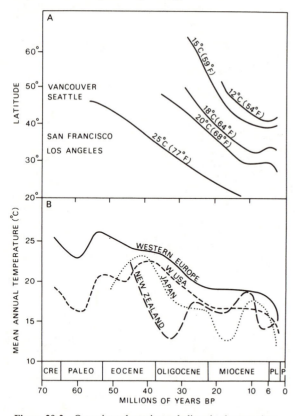

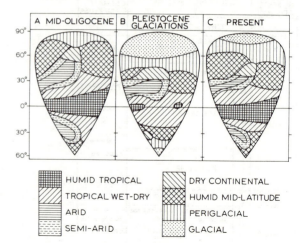

**Figure 20.3** Speculative models of world climatic zonation during the A. Mid-Oligocene; B. Pleistocene glacial maxima; C. Present (Late Holocene).
*Source:* A. and B. freely inferred from Butzer, 1976, figure 16-2a and c, p. 327; C. modified after Trewartha.

**Figure 20.2** Oceanic and continental climatic changes since the Cretaceous: A. general southward movement of February isotherms for northern Pacific surface waters during the Tertiary as the world climate became colder; B. average mean annual temperatures since the Cretaceous for four regions of the world.
*Sources:* Goudie, 1977, figure 1.5, p. 21, after Kurtén, 1972; Frakes, 1979, figures 7-1 and 7-6, pp. 191, 203.

Individual glacial stages are now seen to have involved significant climatic oscillations, especially the last and best-known Wisconsin (Weichselian in Europe and Devensian in eastern England) (see Figure 20.1). Variously generated temperature curves for different localities do not always exhibit clear synchroneities (Figure 20.4), and the dating and correlation of glaciations is very indeterminate (Bowen, 1978). Distinctions between glacial and interglacial stages have blurred to some extent and, for example, in north-western Europe three later major glacial stages are now recognized (Elsterian, Saalian and Weichselian) preceded by some 700,000 years of earlier Pleistocene severe climatic oscillations containing other glacial stages. Some authorities believe that each of the major North American glaciations was progressively less intense, whereas others

believe that the successive European interglacials may have been progressively cooler. There is also no doubt that certain interglacial times had higher general world temperatures than at present. The relations between conditions in the lower and higher latitudes during the Pleistocene are more clear than hitherto. Glacial stages are now recognized to be primarily generated by extraterrestrial (solar cyclical) and general planetary (volcanic) causes which resulted in a lowering of ocean surface temperatures, less evaporation, less tropical precipitation and the expansion of lower-latitude deserts. Moist tropical mountains (e.g. Hawaii, New Guinea, Peru) appear to have experienced glacial expansions and contractions in phase with those of higher latitudes under the influence of general temperature changes, but the drier tropical mountains (e.g. Mexico, Ethiopia) and some dry polar areas were out of phase, being primarily influenced by precipitation amounts. Recent maximum ice extensions occurred in Mexico at 35,000–32,000 years BP and 12,000–10,000 years BP. Changes in the size of tropical lakes (e.g. in Queensland, Mauritania, Chad, Bolivia and Ethiopia) were largely (but not entirely) controlled by precipitation amounts, with high levels before the last glacial maximum, low levels during the glacial maximum (21,000–12,500 years BP) and most rising rapidly after about 12,500 years BP. During the period 12,000–7000 years BP the East African lakes were much more extensive than now. Lake Chad, in west-central Africa, rose to some 150 m above its present level about 10,000 years BP (Figure 20.5), and this should be compared with the greater Pleistocene extent of Great Salt

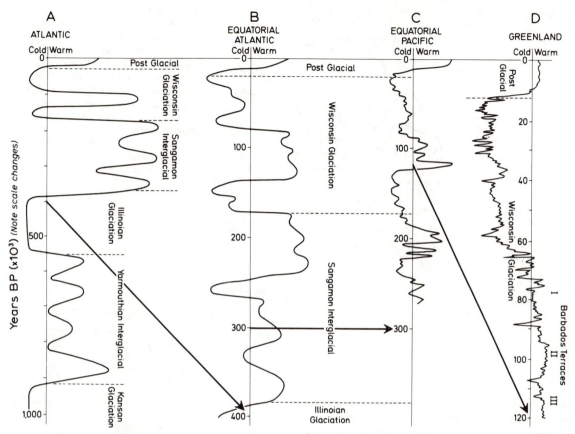

**Figure 20.4** Temperature changes since 1,000,000 years BP: A. generalized curve from Atlantic Ocean cores based on frequency of *Globorotalia menardii* Foraminifera; B. generalized curve from equatorial Atlantic Ocean cores based on ratios of cold to warm Foraminifera during the past 400,000 years; C. curve from a core in the equatorial Pacific Ocean based on oxygen-isotope variations during the past 270,000 years; D. curve from an icecore at Camp Century, Greenland, based on oxygen-isotope variations during the past 120,000 years; note the rapidity of climatic changes and the warming in the middle of the Wisconsin Glaciation (c. 100,000 years BP) associated with high marine terraces in Barbados; also degrees of cold and warm are approximate and illustrative only; and timescales differ between A.–D.
*Sources:* Goudie, 1977, figures 2.2 and 1.4, pp. 29 and 14; Street, 1981, figure 2, p. 161; Barry, 1981, figure 2, p. 4, after Dansgaard *et al.*, 1971.

Lake, Utah (Lake Bonneville)(see Figure 5.3), although it is clear that the chronologies of their levels are quite different (Figure 20.6).

It is natural that most Pleistocene interest centres on conditions during the last glacial stage and Figure 20.7 provides a rough estimate of world conditions during the Wisconsin maximum (August 18,000 years BP), which can be compared with those at present (see Figure 18.2; also Figure 20.3B and C). At the Wisconsin maximum global temperatures were 3° to 6°C lower than at present; ocean surface temperatures 2·3°C lower; sea level at least 85 m lower; large ice sheets exceeded 4000 m thick in Antarctica and 3000 m in the northern hemisphere; sea ice was much more extensive than at present and there

was increased upwelling of cold water in the Atlantic and Pacific; there was a steeper thermal gradient along the polar front, which was displaced equatorward; the subtropical high-pressure cells over the oceans were more or less in their present positions; continents were substantially cooler and drier than at present with grasslands, steppes and deserts spreading at the expense of forests; the tropics were particularly drier than today with humid tropical forests present only as shrunken remnants in a greatly extended area of tropical wet–dry savanna lands. The latter condition contrasts strikingly with the extension of humid tropical conditions during the interglacials and, especially, during Tertiary times, and it is now clear why the distinctions between humid tropical

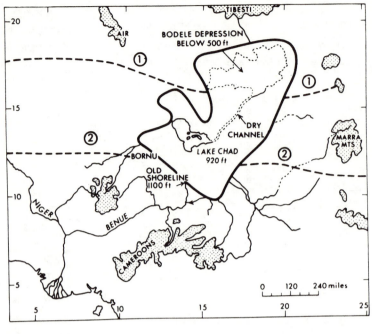

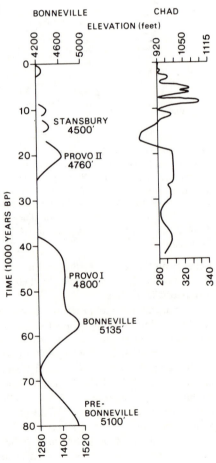

**Figure 20.5** Post-Pleistocene (i.e. 20,000–5000 years BP v. the Present) changes of lake size and the southern limit of moving sand dunes in the Chad Basin, Central Africa. Present Lake Chad (920 ft: 280 m) is a shrunken remnant of the former MegaChad (1100 ft: 335 m; thick line) which overflowed south-west into the Benue. Broken line 1 is the present southern limit of moving sand dunes; line 2 is that of ancient moving dunes, now vegetation-covered; uplands are dotted.
*Source:* Goudie, 1977, figure 3.6, p. 78, after Grove.

and tropical wet–dry landforms are so difficult to make (see Chapter 18). In North America the tundra zone extended thousands of kilometres further south than today, the regional snowline dropped as much as 1500 m in places, but the south-western desert appears to have been more restricted than today and the Appalachian spruce forests were only displaced southward by a few tens of kilometres. In Europe the mean annual temperature was as much as 10° to 12°C lower than at present, Australia was more arid than today, especially in the tropical north, and the southern Mediterranean and Red Sea areas remained moister for a longer period than most other dry tropical regions. At the end of the last glaciation there was a return of cold conditions following the Allerød Interstadial (12,000–11,000 years BP) in Northern Europe, when Iceland and Scotland were covered by ice, central England and Wales by periglacial spruce forest, and southern England by birch and pine.

**Figure 20.6** Changing elevations of A. Lake Bonneville, Utah, and B. Lake Chad, Central Africa, during the last 80,000 and 40,000 years, respectively; absolute elevations are given in feet (above) and in metres (below).
*Sources:* Crittenden, 1963, figure 8, p. 20, based on Servant from Street, 1981, figure 10, p. 176.

**Figure 20.7** Climatic conditions deduced (and guessed!) for the northern summer in 18,000 BP. Part of this map was derived from the CLIMAP project; see text for description; the extent of periglacial conditions is especially speculative.

*Source:* Partly from McIntyre, 1976, figure 1, p. 1132.

### 20.1.3 Holocene climatic changes

The period following the last glacial stage in the mid-latitudes has been one of general amelioration of temperatures whose land conditions have been analysed primarily by means of fossil pollen linked to C$^{14}$ dating, with past climatic conditions being inferred in increasing detail by analogy with those accompanying present pollen distributions. In Europe and North America this rapid amelioration led to a Climatic Optimum (8000–5000 years BP) (Figure 20.8) followed by cooler and wetter conditions which pollen analysis has allowed us to estimate in some considerable seasonal detail (Figure 20.9). In the tropics there was a time of maximum precipitation about 9000–7000 years BP, during which rainfall in the southern Sahara and most of tropical Africa was 25–85 per cent greater than that at present. Central African lakes stood at high levels until as late as 4500 years BP, by contrast with those of south-western North America which have been low for most of the Holocene (see Figure 20.6), but since that time there has been a general desiccation in the tropics.

### 20.1.4 Historical climatic changes

Of course, we have increasing knowledge of climatic changes the more recent the time period, and it is also clear that these more proximate records should reveal many of the complexities of climatic change hidden by the coarser less-direct earlier records. However, more sophisticated information requires more sophisticated means of analysis and it is somewhat of a paradox that more precise climatic data can be analysed in ways which produce ambiguous views of the nature of the continuity of climatic change which have considerable geomorphic significance (but see Hooke and Kain, 1982). Historical climatic changes are those which are obtained either directly or indirectly as the result of human action, and although some such information is available from as long ago as Ancient Egypt, it is convenient to identify the period since 1500 years BP (AD 500) as that most clearly exhibiting historical climatic changes. It is also convenient to identify three types of geomorphically significant climatic change; changes of mean annual conditions, changes of population distributions of climatic variables (i.e. magnitude/frequency changes), and changes of the seasonal occurrence of climatic variables.

A long record of *mean annual* climatic tendencies is provided by tree rings, which are somewhat analogous to the biannual varves in proglacial lake deposits. The width of tree rings in the south-western United States has been shown to be affected in a complex manner by the inter-

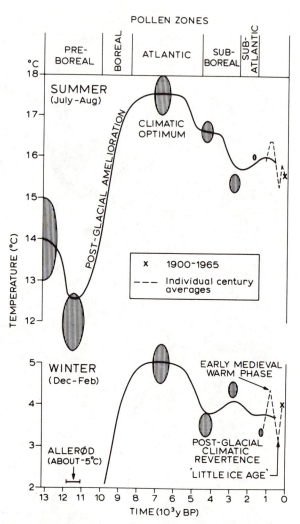

**Figure 20.8** Trends of air temperatures for the lowlands of central England for summer and winter months. Supposed 1000-year averages are shown by full lines in respect of the past 13,000 years, and 100-year averages by broken lines for the past 1000 years; shaded ovals indicate the approximate ranges within which the temperature estimates lie and error margins of the radiocarbon dates.
*Source:* Barry and Chorley, 1982, figure 8.3, p. 327, after Lamb, 1966.

action of temperature, moisture and altitude, with moist and warm conditions tending to give wide rings, cold and dry narrow rings. At high elevations rates of growth seem to be directly related to temperature, whereas at lower levels wide rings may be produced by a combination of high moisture and low temperature (Table 20.2). Tree-ring widths also decline exponentially with age and they are standardized for comparison by producing a ring-

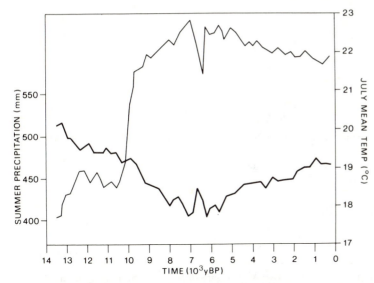

**Figure 20.9** Reconstructed values of summer precipitation (in millimetres) (i.e. during the growing season) (full line) and of July mean temperature (degC) since about 13,000 BP for Kirchner Marsh, mid-western United States, using pollen data.
*Source:* Webb and Bryson, *Quaternary Research*, 1972.

width index. By linking the ring widths of living trees with those of archaeological wood used in building, it has been possible to construct a record going back some 1500 years in the south-western United States (Figure 20.10). From such records, periods of temperature and precipitation anomaly can be inferred (Figure 20.11). Two other long records of climatic inferences are given in Figure 20.12, one from measured flood levels and the other from weather measurements, general observations, harvest records, and the like. The interpretation of the Nile Roda Gauge records presents two main problems: first, to what extent the step about AD 1550 is due to a change in datum level (if any), and, second, to what extent the increased subsequent trend is the result of accelerated river bed aggradation. Figure 20.12B shows both the Little Ice Age and, to a lesser degree, the Medieval Warm Phase (Little Climatic Optimum) which were so marked in north-western Europe.

**Table 20.2** Width of tree rings in the United States related to moisture, temperature and altitude

| Moisture | Altitude | Temperature | |
| --- | --- | --- | --- |
| | | *Warm* | *Cool* |
| Moist | High | Wide | Narrow |
| | Low | Wide | Wide |
| Dry | High | Wide | Narrow |
| | Low | Narrow | Narrow |

*Source:* LaMarche, 1974.

Where annual climatic or hydrologic data are involved, various techniques can be employed to generalize their trends, but a problem arises in that the use of each technique may serve to accentuate some different aspect of the time series – i.e. long-term change, continous variation, cyclical oscillation, abrupt transitions, random fluctuations, and so on. Figure 20.13 fits generalized continuous curves to hydrological data, whereas Figure 20.14 shows how annual variations can be successively generalized and damped by the use of 5-, 10-, and 20-year running mean values and by a single fitted linear regression. Some 'neo-catastrophist' geomorphologists (especially Dury) are increasingly impressed by the possibility of sudden climatic changes which are better described by means of step functions based on cumulative variations from a general mean value (Figure 20.15). Such analysis assumes that the time series is not drawn from a single climatic population, but from different subpopulations separated from adjacent populations by discrete thresholds.

Climatic change can occur not only as the result of changes in the mean value of subpopulations of rainfall and temperature events through time, but also by changes in the statistical distribution of the *frequency* of occurrences. Thus it is possible for geomorphically significant climatic change to occur purely by changes in the frequency distribution of events, while their mean value remains constant. We have seen in Chapter 3 how frequency distributions of geomorphic events, such as rainfall and stream discharge, are characteristically right-skewed, and Figure 20.16 shows two important features of such distributions: first, as the mean decreases the skewness commonly increases (A), and,

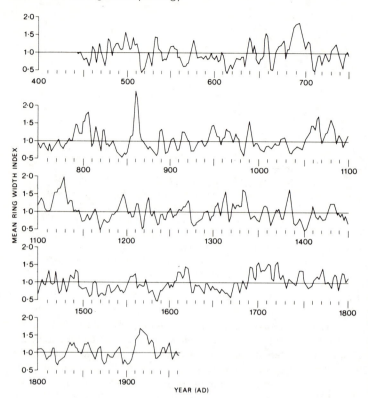

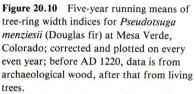

**Figure 20.10** Five-year running means of tree-ring width indices for *Pseudotsuga menziesii* (Douglas fir) at Mesa Verde, Colorado; corrected and plotted on every even year; before AD 1220, data is from archaeological wood, after that from living trees.
*Source:* Fritts, 1976, figure 8.8, p. 406, copyright © Academic Press, London.

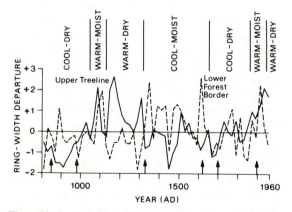

**Figure 20.11** Estimates of the twenty-year average tree-ring growth of *Pinus longaeva* (Bristlecone pine) on the lower forest border (broken line) and the upper treeline (full line) in the White Mountains, California: data expressed in standard normal values of ring-width departure, and climatic variations inferred from ring-width departures; arrows show dates of glacial moraines in the mountains.
*Source:* LaMarche, 1974, figure 6, p. 1047.

second, that when two distributions have the same mean but differing standard deviations and skewness, the more-skewed distribution may have a greater frequency of *both* high and low magnitudes of events (B). The characteristics of such distributions of magnitudes can be expressed by Gumbel plots (see Chapter 3), and Figure 20.17 shows how both the means and skewness of annual precipitation in Sydney, Australia, differs drastically in the successive periods 1935–48 and 1949–63 (see Figure 20.15C). The problem of fitting infrequent high-intensity events into a climatic series is dramatically illustrated in Figure 20.18, where the flood of 24–29 June 1954 caused the Pecos River near Comstock, Texas, to rise 17.5 m above the previous recorded high level and to inundate a 1300-year-old archaeological site, suggesting a return period of some 2000 years.

Such speculations lead us to return to a consideration of which *magnitudes* of applied stress are most effective as agents of geomorphic work (see Chapter 3) and, therefore, how shifts in climatic frequency distributions might affect geomorphic processes and forms. The idea that total work is greatest in the middle range of the magnitude of applied stress is true only for a low threshold

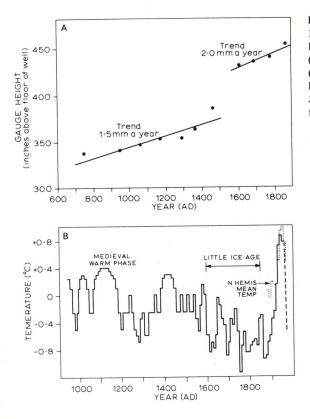

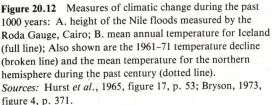

**Figure 20.12** Measures of climatic change during the past 1000 years: A. height of the Nile floods measured by the Roda Gauge, Cairo; B. mean annual temperature for Iceland (full line); Also shown are the 1961–71 temperature decline (broken line) and the mean temperature for the northern hemisphere during the past century (dotted line).
*Sources:* Hurst *et al.*, 1965, figure 17, p. 53; Bryson, 1973, figure 4, p. 371.

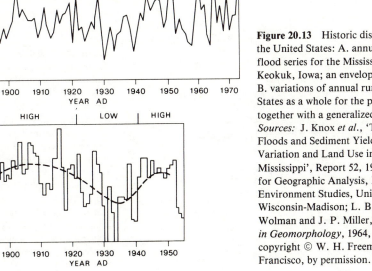

**Figure 20.13** Historic discharge records for the United States: A. annual maximum flood series for the Mississippi River at Keokuk, Iowa; an envelope curve is given; B. variations of annual runoff in the United States as a whole for the period 1895–1955, together with a generalized trend curve.
*Sources:* J. Knox *et al.*, 'The Response of Floods and Sediment Yields to Climatic Variation and Land Use in the Upper Mississippi', Report 52, 1975, The Center for Geographic Analysis, Institute for Environment Studies, University of Wisconsin-Madison; L. B. Leopold, M. G. Wolman and J. P. Miller, *Fluvial Processes in Geomorphology*, 1964, figure 3-13, p. 62, copyright © W. H. Freeman & Co., San Francisco, by permission.

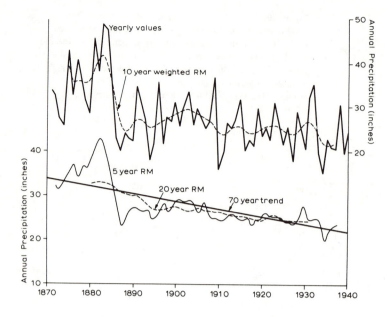

**Figure 20.14** Annual precipitation data from Omaha, Nebraska (1871–1940) showing the relative smoothing effects of three types of running means (**RM**), together with a generalized seventy-year trend. *Source:* Chorley and Kennedy, 1971, figure 7.11, p. 267, after Foster.

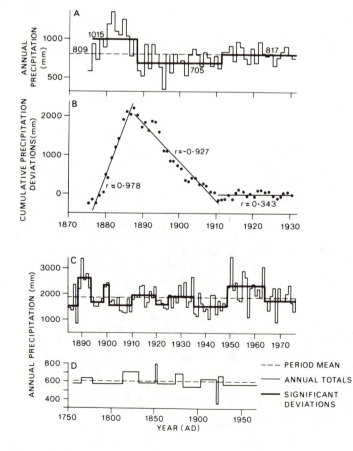

**Figure 20.15** Climatic step functions suggesting rapid fluctuations of climate: A. and B. the precipitation at Madison, Wisconsin (1875–1930); histogram A. shows annual totals (thin line) and blocks (full line) whose cumulative precipitation deviations B. differ from adjoining means at a 5 per cent confidence level or better; the period annual mean (809 mm) and the block period means (1015 mm, 705 mm and 817 mm) are shown; C. annual precipitation figures for Sydney, Australia, (1885–1976) with fitted step functions; D. step functions fitted to the annual precipitation of London for about a 200-year period. The period mean is shown. *Sources:* Dury, 1980, figures 1 and 2, pp. 63, 65; Dury, 1981, figure 3, p. 50.

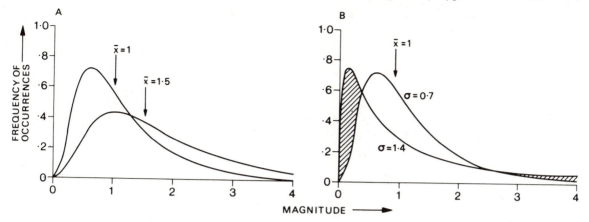

**Figure 20.16** Characteristics of right-skewed frequency distributions: A. two distributions with different means $\bar{x}$ and skewness; B. two distributions with the same mean $\bar{x} = 1$ but different standard deviations $\sigma$ and skewness; excess values of the more skewed distribution are shaded.

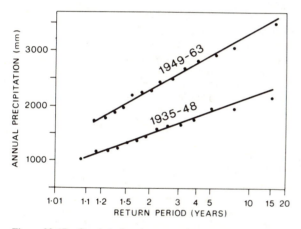

**Figure 20.17** Gumbel plots (e.g. magnitude/frequency) for two successive subpopulations of annual precipitations for Sydney, Australia (see Figure 20.15C).
*Source:* Dury, 1980, figure 13, p. 74.

of movement rates ($t_1$) (Figure 20.19A and B) and especially in respect of humid, low-relief terrains with deep, vegetated soils (Figure 20.19A). More arid conditions, especially applying over small areas with low infiltration capacities, exhibit highly skewed frequency distributions, and when these are associated with high thresholds of movement rates ($t_2$) the maximum total work is produced by the infrequent high magnitudes of applied stress (Figure 20.19D).

Significant changes of climate do not only occur by changes in the distribution of annual events, but also by monthly (i.e. *seasonal*) and even daily changes. It has been suggested that a deficiency of individual (daily) low-intensity rainstorms in the spring growing season prior to the end of May may inhibit vegetation growth later in the

year and allow accelerated erosion by the summer rainstorms in the south-western United States. Figure 20.20 shows for Santa Fé how such nuances which are masked by annual precipitation amounts (A) become well marked when daily rainstorm values are plotted (B and C). Such considerations are important in attempts to understand the causes of accelerated arroyo-cutting in the American south-west which occurred at the end of the last century.

## 20.2 The geomorphic effects of climatic change

As we have seen, climatic change always involves a complex interplay of precipitation and temperature variations. However, in examining the geomorphic effects of such changes, it is convenient to separate them into those in which precipitation is the dominant variable and those where temperature changes predominate. Precipitation changes have especial geomorphic significance when they occur around the humid–semi-arid margin, as has been stressed in Chapter 18 in respect of the oscillations characteristic of the savanna regions. Under such conditions small variations of precipitation amount and regime can, mainly through their vegetational effects, exercise great changes over hydrology and erosion rates. Temperature changes have the greatest geomorphic significance when they transgress the freezing temperature, causing processes to oscillate between fluvial, periglacial and glacial. The following two sections thus treat the dominant effects of precipitation and temperature changes, respectively. As we have seen, the glacial effects are so profound that they have already been treated in Chapter 19.

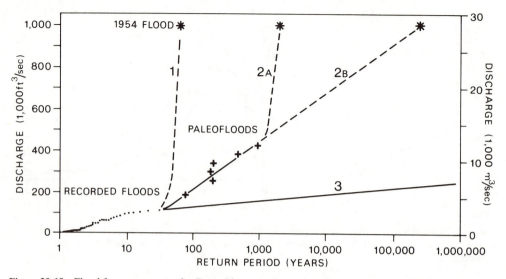

**Figure 20.18** Flood frequency curves for Pecos River near Comstock, Texas. Curve 1 is the flood frequency curve including the 1954 flood with the other fifty-three years of record; based on this record, the 1954 flood would have a recurrence interval of fifty-five years. Curve 3 is the extrapolated pre-1954 flood frequency curve; based on this curve the 1954 flood would have a recurrence interval of millions of years. Curve 2A is the paleoflood frequency curve estimated from the alluvial stratigraphy. The alluvial chronology refines the flood frequency between 100 and 1,000 years, but individual paleoflood frequency estimates (crosses) may be in error by as much as 150 × 1000 ft³/s (5000 m³/s) and several hundred years. Curve 2B includes the 1954 flood with the generalized paleoflood record.
*Source:* Patton and Baker, 1980, figure 3, p. 196.

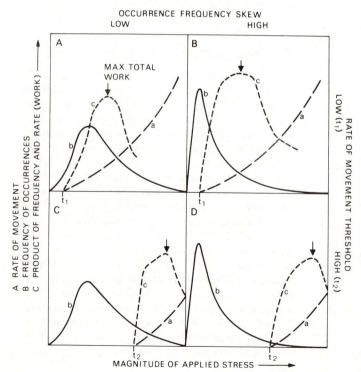

**Figure 20.19** Illustrations of how the maximum total work (arrow) **C** can change with variations in the skewness of the frequency of occurrences **B** and shifts in the rate of movement curve **A** as the threshold of movement rate $t_1$–$t_2$ changes: see text, for description.
*Source:* Baker, 1977, figures 1 and 3, pp. 1057, 1059.

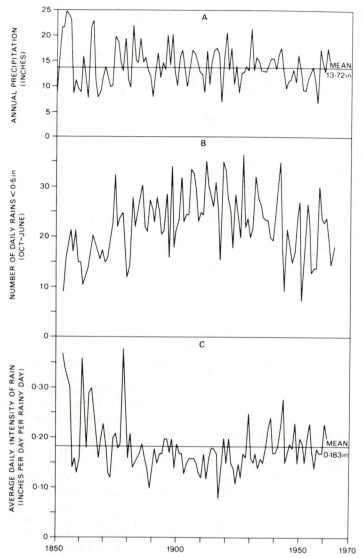

**Figure 20.20** Precipitation data for Santa Fé, New Mexico, showing how the mean annual precipitation A. masks significant variations both in the number of daily rains B. especially during the early growing season, and in daily rainfall intensity C. For example, the period approximately 1850–90 was probably one of deficient spring rains and of higher than average rainstorm intensities, perhaps resulting in high erosion rates.
*Source:* Leopold *et al.*, 1966, figures 171, 173 and 174, pp. 241, 242, 243.

## 20.2.1 Precipitation changes dominant

The major climate changes of the past undoubtedly had a significant effect upon geomorphic processes and their rates of operation. As described earlier (Section 12.7.1), the North Platte and South Platte rivers have adjusted dramatically to short-term hydrologic variations. However, the morphologic evidence for climate change and the effect of climate change on landforms is to a large extent viewed through hydrologic and vegetational filters. Therefore, before the effect of climate change on landforms can be discussed, there must be some consideration of the effect of climate on hydrology. This can be viewed as an extension of modern hydrologic relations to the past – i.e. paleohydrology (Gregory, 1983).

In the United States vegetational bulk or weight increases approximately as the cube of mean annual precipitation (Langbein and Schumm, 1958), and a significant reduction in erosion and sediment yield can be anticipated as precipitation increases or as vegetation increases in a given climatic situation. The asymptotic approach to the *y*-axis in Figure 11.13 is indicative of the rapid increase of erosion at small percentages of ground cover. However, there must exist a maximum rate of erosion for a given soil or rock type that will be reached before ground cover is zero. The curve should plateau at

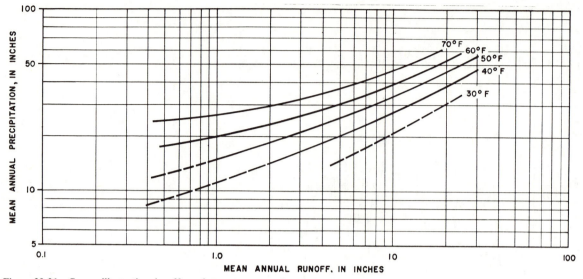

**Figure 20.21** Curves illustrating the effect of temperature on the relation between mean annual runoff and mean annual precipitation.
*Source:* Langbein *et al.*, 1949.

that value, as the broken line suggests. If this is true, then Figure 11.13 leads us to the conclusion that with less than about 8·0 per cent cover vegetation is probably relatively ineffective in controlling erosion. On the other hand, above about 70 per cent cover additional vegetation will not significantly reduce erosion.

Runoff is that part of precipitation that appears in surface streams, and Figure 20.21 illustrates the general relation between climate and runoff. The position of each of the curves depends on mean annual temperature, which was adjusted to remove the effects of seasonal rainfall on runoff. For example, a given amount of precipitation yields less runoff during the warmer months of summer than during the cool winter months. The family of curves shows that, as expected, when annual precipitation increases, annual runoff increases and that as temperature increases with constant precipitation, runoff decreases. These relations are straightforward and need no elaboration.

Sediment yield is the total amount of sediment moved from a drainage basin (Figure 3.21). The relations between annual sediment yield and temperature and precipitation for drainage basins averaging about 1500 square miles (3885 km²) in the conterminous United States are shown in Figure 20.22. At 50°F sediment yield is a maximum between 5 in and 25 in of precipitation with a peak at about 12 in (300 mm) of precipitation. The variation in sediment yield with precipitation can be explained by the interaction of precipitation and vegetation on runoff and erosion. As precipitation increases

above zero sediment yields increase at a rapid rate because more runoff becomes available to move sediment. Opposing this tendency is that of the vegetation, which becomes more abundant as precipitation increases. Above 12 in of precipitation, sediment-yield rates decrease under the influence of the more effective grass and forest cover, but where monsoonal climates prevail, sediment yield rates may increase again above 50 in (1270 mm) of precipitation under the influence of highly seasonal climates (Figure 3.19).

A more recent analysis by Dendy and Bolton (1976) of data from 800 drainage basins scattered throughout the United States reveals that a peak of sediment yield exists at between 1 in and 3 in of runoff. At a temperature of 50°F this corresponds to about 18–20 in (455–510 mm) of precipitation. This more comprehensive set of data suggests that the peak of sediment yield occurs at slightly higher precipitation conditions.

One of the most important factors that influence sediment yield is, of course, land use. For example, Wilson (1972) shows that the sediment yield data for areas of the humid eastern United States, where strip mining or urbanization has had an important effect on the watershed, are similar to those of arid regions (Figure 3.21). Wilson's (1973) analysis of sediment yield data available in 1969 reveals that the Langbein–Schumm relationship tends to be valid for continental types of climates, but where there is marked seasonal variability of precipitation, sediment yields can be much greater than indicated by mean annual precipitation alone. Hence sediment

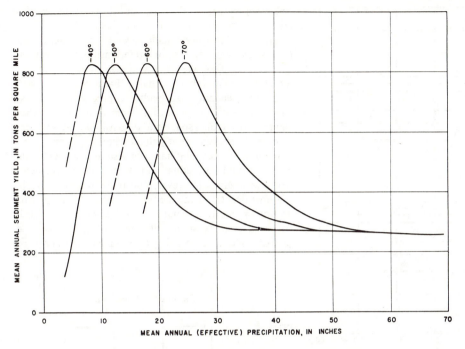

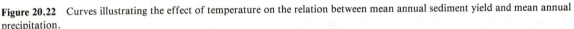

**Figure 20.22**   Curves illustrating the effect of temperature on the relation between mean annual sediment yield and mean annual precipitation.
*Source:* Schumm, 1965, figure 2, p. 785, in H. E. Wright and D. G. Frey (eds), *The Quaternary of the United States*, copyright © 1965 by Princeton University Press. Reproduced by permission of Princeton University Press.

yields in Mediterranean climates and monsoonal climates probably exceed those of Figure 20.22. An attempt by Jansen and Painter (1974) to estimate and develop empirical equations to predict sediment yield from the world's climate zones for drainage basins exceeding 5000 km² shows that erosion rates are the lowest in the continental and tropical forest climates, intermediate in arid and semiarid climates, and very high in Mediterranean climates as suggested by Wilson. The Langbein–Schumm relation, despite its weaknesses, is nevertheless a reasonable approximation of the effect of climate on erosion and sediment yield in the United States. Remember, however, that the variables relief and lithology can greatly influence erosion rates.

Figures 20.21 and 20.22 also provide a basis for estimating the effects of Quaternary climate change. This is relatively straightforward as vegetation has not evolved significantly during late Tertiary and Quaternary time. As annual temperature increases, the peak of sediment yield occurs at higher amounts of annual precipitation (Figure 20.22); that is, because of higher rates of evaporation and transpiration, less of the precipitation at higher temperatures is available to support vegetation, runoff is less and so the peak rate of sediment yield shifts to the right. These curves may be used to estimate

changes in sediment yield as temperature or precipitation or both changed during the past. However, only the direction and magnitude of the changes are meaningful because variations of geology and geomorphology also effect the sediment yields.

In addition to the amount of sediment moved, its concentration in the water by which it is moved is important. As might be expected, for a given annual precipitation, sediment concentration increases with annual temperature, and for a given annual temperature, sediment concentration decreases with an increase in annual precipitation (Figure 20.23). This means that more sediment is moved by a unit of water in arid regions, but because runoff is infrequent, less total sediment is transported per year.

If Quaternary climates are known, they permit the use of Figures 20.21 and 20.22 to determine changes in the hydrology of non-glacial continental regions. For example, a change from present climatic conditions to a cooler, wetter climate causes an increase in runoff. This increase is greater for areas with initially drier climates (semi-arid regions). For example, if temperature is 10°F less and precipitation is 10 in (254 mm) greater, a twenty-fold increase in runoff will occur in a region with a present annual precipitation of 10 in and a present annual

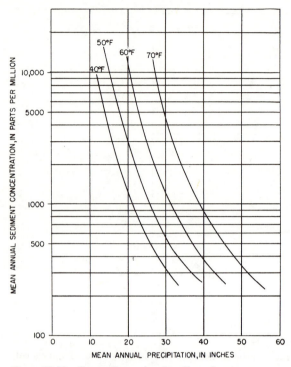

**Figure 20.23**   Curves illustrating the effect of temperature on the relation between mean annual sediment concentration and mean annual precipitation.
*Source:* Schumm, 1965, figure 3, p.786, in H. E. Wright and D. G. Frey (eds), *The Quaternary of the United States*, copyright © 1965 by Princeton University Press. Reproduced by permission of Princeton University Press.

temperature of 50°F; however, only a twofold increase in runoff will occur in a region of 40 in (1016 mm) of precipitation with an annual temperature of 50°F.

Sediment yields generally decrease with increased precipitation and decreased temperature because this type of climate change improves the vegetal cover (Figure 20.22). An exception occurs, however, when the change is from an arid (60°F and 10 in) to a subhumid climate (50°F and 20 in). In this case the increase in sediment yield may be explained by the assumption that the increase in runoff is more effective in promoting erosion than the improved vegetal cover is in retarding erosion.

In summary, the curves of Figures 20.21 and 20.22, which are based on modern runoff and sediment yield data, suggest the changes in runoff and sediment yield that can be expected with a change in climate. One major limitation to the use of the curves is that they are applicable only to nonglacial or to pluvial regions. For these regions, a decrease in temperature and an increase in precipitation cause an increase in runoff and a decrease in the sediment concentration. Sediment yield, however,

can either increase or decrease depending on the temperature and precipitation before the change.

Attempts have been made recently to estimate past river discharge by the use of a relationship that has been found to exist between the meander characteristics of modern rivers and their discharge (Figure 12.19), as well as between channel size (width, depth, area) and discharge. However, because of the numerous variables that can control channel shape and dimensions only an order of magnitude estimate of paleodischarge is possible.

In all the preceding it has been assumed that a climate change would be gradual and the hydrologic response would be gradual. However, Knox (1972) suggests that in the north-central United States climate change can be abrupt or step-functional (Figure 20.24), and this will trigger a response that, although short, may be the opposite of that suggested by the sediment yield curves of

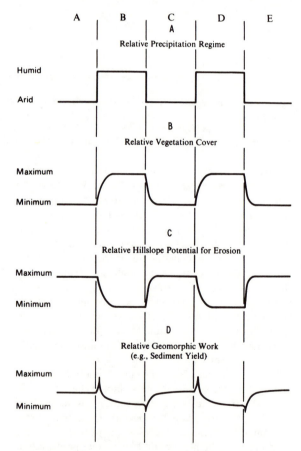

**Figure 20.24**   Vegetational and geomorphic responses to abrupt changes in climatic regimes. The curves would seem to be most applicable to climatic regions that have annual precipitation between 10 in and 60 in (250–1520 mm).
*Source:* Knox, 1972, figure 6, p. 408.

Figure 20.22. He concludes that a sharp and sustained increase of precipitation, as a result of a changed circulation pattern, will cause higher slope and channel erosion until the vegetational cover can improve (Figure 20.24B). The result is a short period of very high sediment yields followed by a decrease (Figure 20.24D). It is undoubtedly true that a change of climate of any type will trigger a period of landscape instability that will produce higher erosion rates and sediment yields as the landscape and stream channels respond. However, such fluctuations may be difficult to distinguish from the effects of very large floods of rare occurrence.

Another aspect of hydrology that can be changed significantly by climate alteration is the nature of flood events. The very lack of vegetation that permits maximum erosion is also responsible for larger floods (Figure 20.25). In dry areas a given amount of precipitation will run off more quickly (lag time is less) and it will produce a higher flood peak. These flashfloods are capable of moving large quantities of sediment, and it is likely that a climate change from relatively wet to dry will cause channel enlargement to accommodate the larger floods, as well as channel-straightening or a gradient increase to accommodate larger sediment loads.

Normally hydrologic records are too short to provide information on extreme events and, therefore, a geomorphic approach to this problem is necessary because in dealing with long timespans the possibility of rare, extreme geomorphic events, such as catastrophic floods,

cannot be ignored. For example, the chance of a 1000-year flood occurring in any single year is slight, only 0.1 per cent. But in a 1000-year period the chance of the 1000-year flood occurring is 63 per cent, and in 5000 years the chance is near certainty, over 99 per cent (Costa, 1982). An even more severe flood (5000-year flood) has an 18 per cent chance of occurring in 1000 years. Using conventional hydrologic analysis, it is difficult or impossible to identify the magnitude of a 1000-year flood as well as the recurrence intervals of large catastrophic floods. However, geomorphic techniques offer an alternative solution to these questions.

In order to design for long-term security from flooding (e.g., radioactive waste-disposal sites) the concept of the probably maximum flood (PMF) is used. The PMF is derived from estimates of probable maximum precipitation (PMP). The PMFs are calculated from data based on the historical period of rainfall and runoff records, and thus change as the history of large rainstorms change. On numerous occasions PMFs have been exceeded and recurrence intervals for PMFs vary over at least four orders of magnitude. However, the geomorphic record of landforms and deposits in a valley can give reasonable estimates of the magnitude and frequency of extreme events. Results of hydrologic reconstructions using sediments deposited by known large catastrophic floods are within 20–30 per cent of accepted peak discharge values. Climate and land use do not seem to be controlling factors in determining catastrophic

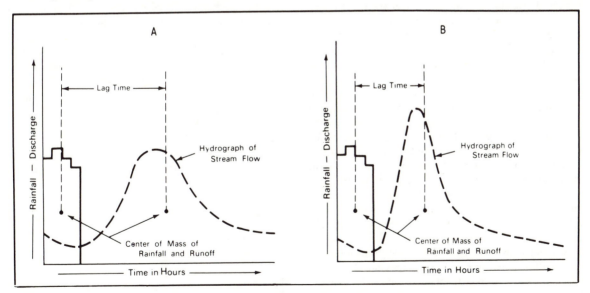

**Figure 20.25** Hypothetical hydrographs for runoff in A. humid and B. arid climate regime. The precipitation and drainage basin areas are assumed to be equivalent.
*Source:* Knox, 1972, figure 2, p. 403.

flood peaks in small basins in the United States; drainage area is the most important variable. Maximum floods for basins up to 500 square miles (1300 km²) in the United States follow the power function:

$$Q_{max} = 11,000 \text{ (drainage area)}^{0.61} \text{ (cfs and sq. mile}$$

units).

Not all basins will have appropriate sedimentological and/or stratigraphical evidence of extreme floods. This could be because evidence has not been preserved, or because no such floods have occurred. However, investigations in other basins in the region may well provide paleodischarge information which can be extrapolated to the site under consideration (Costa, 1978).

It is now possible to suggest how the variations in hydrology, resulting primarily from precipitation changes, may influence fluvial landforms and sedimentary deposits.

*Interfluves.* The density of vegetation, which is an indication of the effectiveness of vegetal cover as a protection against raindrop impact and runoff, increases with annual precipitation. The rate of erosion of hillslopes should, therefore, decrease with increased precipitation. Exceptions occur, however, and in periglacial regions during a glacial advance intense frost action and solifluction increased slope erosion and even altered the form of the slope profile by the dominance of mass movement over runoff (Peltier, 1950; Cotton, 1958). In arid and semi-arid regions the increase in precipitation also increased erosion by causing landslides and slumping along escarpments (Strahler, 1940; Watson and Wright, 1963).

The effects of Quaternary climatic changes on slopes were as variable as the processes operative and the materials involved. During pluvials, increased vegetal cover probably decreased slope erosion in semi-arid and humid regions; however, in initially arid regions the increase in runoff probably more than compensated for the increase in vegetal cover, and initially arid slopes would have been subjected to more intense erosion. Although this last statement is conjectural, the peaks of the sediment yield curves (Figure 20.22) demonstrate that more sediment is exported from semi-arid than from arid regions, and some of this increase must be derived from the slopes.

A recent study by Toy (1977) provides some information on the complex interaction of slopes and climate. In order to study slopes through a range of climatic influences it was necessary to find geologically similar situations. He located outcrop areas of marine shales with dips less than 3°. Data on slope morphology were collected on slopes with a southern exposure in order to eliminate the effect of aspect. The sample sites extended east–west along the 37th parallel for about 1600 miles from Kentucky to Nevada, which provided a considerable precipitation contrast, and north–south along the 105th meridian for about 900 miles from New Mexico to Montana, which provided a range of temperature. His conclusions are that 'hillslopes in arid regions tend to be shorter, steeper and have a smaller radius of curvature of the crest' (Figure 20.26). In this case a major control is exerted by drainage density, which is lower in the humid regions (see Figure 18.6).

Studies of the effect of microclimatic variations on soil moisture, vegetational differences and on slope morphology further confirm the climatic influence.

Although hillslopes may have a characteristic mean or maximum angle, it is true that in semi-arid regions slope aspect or the direction faced by the slope may be an important factor determining its steepness. Earlier we discussed the effect of aspect on mass movement frequency with the wetter or north-facing slopes being for the most part more susceptible to mass movement (see Figure 10.18).

In a semi-arid climate where stresses on vegetation are great, it is apparent in many locations that the north-facing slope is more densely vegetated than the south-facing slope. The great erosion of the less well protected south-facing slope delivers large quantities of sediment to the channel forcing it to the south, thereby undercutting and steepening the north-facing slope. A study in Wyoming shows that the north-facing slopes are 4·5° steeper, but this is only in the situation where channel gradients are relatively gentle. For example, where channel gradients exceeded 6°, the valleys are symmetrical. Hence a steep, presumably incising channel, will not permit the asymmetry to develop. This is another example of the need for a multivariate approach to the landscape.

The slope differences detected between climatic regions by Toy (1977) (Figure 20.26) suggest that there is a transition not only of climate, but of erosional process from slopes that are significantly influenced by surficial

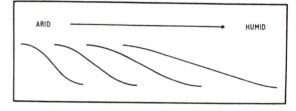

**Figure 20.26** Conceptualized hillslope profiles.
*Source:* Toy, 1977, figure 4, p. 22.

creep in the humid regions to rainwash-dominated slopes in the arid regions (Chapter 11). In addition, the increase of porewater pressure and seepage forces will trigger mass movement and slope decline from the friction angle in dry areas to a semi-friction angle in humid regions (Chapter 10).

*Drainage network.* As vegetational protection is removed by a decrease of precipitation or as the intensity of precipitation events increases, the drainage network should expand to accommodate the increased runoff. Gregory (1976) (see Figure 18.6) shows how drainage density varies with mean annual precipitation and with precipitation intensity, and it is apparent that drainage density varies with precipitation as does sediment yield (see Figure 20.22).

If the relation obtained by Gregory is correct, and it is supported by the work of Peltier (1959), then a major increase in precipitation will either increase or decrease drainage density. The increase in precipitation and run-off accompanying a climatic change from arid to semi-arid should increase drainage density. Similarly, in a humid region a further major increase in precipitation and runoff should increase the drainage density. However, in semi-arid regions a major increase in precipitation could cause a change from shrubs to grass or from grass to trees, which would inhibit further channel development and might even cause obliteration of existing small channels thereby decreasing drainage density.

In summary, Figures 18.6 and 20.22 indicate that both drainage density and sediment yields are greatest in semi-arid regions. The concurrence of maximum drainage densities and maximum sediment yields at the same climate suggests that the high sediment yields are a reflection of increased channel development and a more efficiently drained system. Hence a shift to a semi-arid climate from either an arid or a humid one should allow extension of the drainage network with increased channel and hillslope erosion.

High-intensity storms will also, at least temporarily, alter the drainage network (Figure 20.27), and this suggests that a change in storm intensity without a change of annual precipitation could alter drainage density significantly, and it further suggests that the uppermost parts of a drainage network could be the result of mass movement.

*Rivers.* If annual runoff increases significantly as a result of climatic change, a larger river channel must exist to carry it. The relations among mean annual discharge and channel width and depth of modern rivers (Chapter 12) indicate that width will increase as the 0·5 power of mean annual discharge direction, and depth will increase as the 0·4 power. The relations indicate that as discharge increases both channel width and depth will increase, however, an accompanying change of sediment load will have a major effect. The influence of climate on channel morphology can be anticipated from equations (12·21) through (12·24).

Climatic fluctuations have produced major changes of river morphology, as the discussion of river terraces in Chapter 14 has shown. In Table 20.3 an effort has been made to summarize the nature of river change during the Quaternary period as a result of climate change, glaciation and climatically induced baselevel influences. For example, it is certain that for rivers that drained from continental and valley glaciers (Table 20.3, locations 1 and 2), fluvial deposition was associated with glaciation (Zeuner, 1959, pp. 45–6). Meltwater from the glaciers transported huge loads of sediment and formed valley trains (Peltier, 1962; Krigström, 1962) and outwash plains. Following cessation of glacial erosion and the flushing of glacial debris from the valleys, the reduced sediment loads caused river incision to form terraces.

In periglacial regions (Table 20.3, location 3), Zeuner (1959, p. 49) concluded that as the climate became colder, and as the forests disappeared, periglacial processes or erosion, which produced much slope erosion, caused deposition of sediment in the valleys.

Near coasts (Table 20.3, location 4) late interglacial and early glacial lowering of sea level in response to climate changes caused river erosion, and conversely the rise of sea level accompanying melting of the ice sheets caused deposition (Fisk and McFarlan, 1955). For a closed interior basin (Table 20.3, location 5), river activity was the reverse of that for the coastal situation because during late interglaciation and early glaciation the formation of lakes due to increased runoff raised the baselevel of the streams entering the lakes.

The description of river activity in non-glaciated regions beyond the effects of changing sea level (locations 6–8) is based on the estimated hydrologic changes that accompany climate changes (see Figures 20.1 and 20.22). Initial climate or the climate before the change is of critical importance here, for it determines the type of river to be considered. For example, in the humid region of location 6, the rivers were perennial, and they remained perennial following a shift to a wetter climate. The increased runoff, with a minor decrease in sediment yield, produced incision, or at least channel enlargement.

The situation is different for an initially semi-arid region (Table 20.3, location 7), for the rivers were

## A

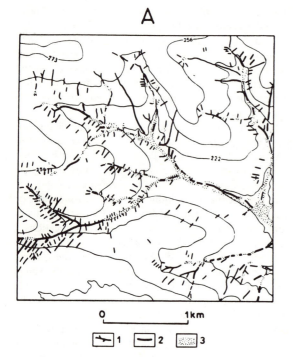

0         1km

⊟1    ⊟2    ⊟3

## B

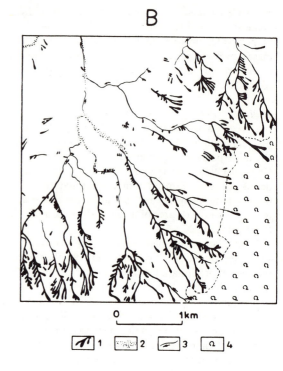

0         1km

⊟1    ⊟2    ⊟3    ⊟4

## C

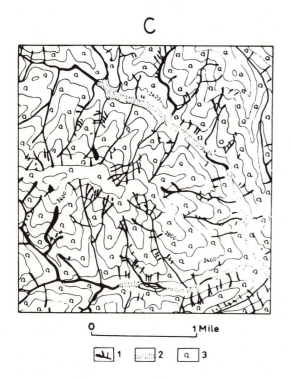

0         1 Mile

⊟1    ⊟2    ⊟3

## D

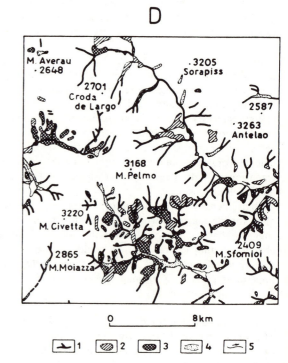

M. Averau
· 2648

2701
Croda
de Largo

· 3205
Sorapiss

2587

· 3263
Antelao

3168
M. Pelmo

3220
M. Civetta

2865
M. Moiazza

2409
M. Sfornioi

0         8km

⊟1    ⊟2    ⊟3    ⊟4    ⊟5

**Table 20.3** Summary of changes in channel behaviour (erosion–deposition) during climate change

| Location or type of basin | Climate change to | Dominant effect | River activity |
|---|---|---|---|
| 1 Basin partly occupied by ice sheet | Cooler, wetter or cooler, drier | Increased sediment load | Deposition |
| 2 Basin partly occupied by valley glacier | Cooler, wetter or cooler, drier | Increased sediment load | Deposition |
| 3 Periglacial | Cooler, wetter or cooler, drier | Increased sediment load | Deposition |
| 4 Coastal | Cooler | Fall of sea level | Erosion |
| 5 Interior basin | Cooler, wetter | Rise of lake levels | Deposition |
| 6 Unglaciated continental interior humid, perennial streams | Cooler, wetter | Increased water discharge Minor increase of sediment yield (Figures 20.21 and 20.22) | Erosion |
| 7 Unglaciated continental interior semi-arid, ephemeral streams | Cooler, wetter | Significant decrease of sediment yield (Figure 20.22) | Erosion |
| 8 Unglaciated continental interior arid, ephemeral streams | Cooler, wetter | Significant increase of water discharge (Figure 20.21) | |
| (a) tributaries | | Increase of drainage density (Figure 18.5) | Erosion |
| (b) main rivers | | Significant increase of sediment load (Figure 20.22) | Deposition |

initially either intermittent or characterized by long periods of flow. The smaller tributaries were ephemeral, as are many in humid regions, but many more drainage channels were present. With a shift to a wetter climate and greater runoff, the major rivers and many of the tributaries became perennial. Vegetation density increased and, if a change of vegetation type from grass to trees occurred, many small channels and the headward portions of larger channels were obliterated. As runoff increased sediment yield decreased, as did sediment concentration (Figure 20.22). The result was undoubtedly the enlargement and incision of the main channels. With a return to semi-arid conditions, sediment yield increased as both hillslope and channel erosion increased and runoff decreased. Deposition in the main channels probably resulted, as the tributaries again reached their maximum extent and number.

It has not always been recognized that the changes in tributary channels might not conform to changes along the larger rivers. This may not be important in humid regions, but it is of major importance in arid regions. Increased precipitation during the Pleistocene epoch converted ephemeral rivers to perennial ones. Increased runoff eroded the tributaries (Table 20.3, location 8a) and extended the drainage network. Sediment was flushed out of the tributary valleys into the main channels during local storms, and because the loss of water into the alluvium of the main channels was appreciable (Cornish, 1961), aggradation of the main channels probably resulted (Table 20.3, location 8b). The situation was probably similar to modern conditions along the Rio Grande, where trenching of the Rio Puerco and other tributaries during the past century has caused deposition in the Rio Grande itself (Rittenhouse, 1944).

**Figure 20.27** Details from maps expressing geomorphological changes after downpours A.–C. or continuous rains D. A. Effects of downpour on 23 June 1956 in Piaski Szlacheckie, loess plateau, eastern Poland: 1, erosional rills; 2, gullies and new channels; 3, debris accumulation: B. Landslides near Mgeta, Tanzania, after the rainstorm of 23 February 1970: 1, slides and mudflows; 2, channel sedimentation and debris fans; 3, streams; 4, woodland and forest. C. Effects of the 1949 flood in the Little River basin, Virginia: 1, chutes and channels created after flood; 2, valley bottoms with erosion and debris accumulation (ornament added); 3, forest; altitudes in feet. D. Effects of the November 1966 flood in the Italian Alps: 1, forms of linear erosion; 2, rejuvenated older landslides; 3, newly formed landslides; 4, fluvial accumulation; 5, rivers.
*Source:* Starkel, in E. Derbyshire (ed.), *Geomorphology and Climate*, 1976, figure 77, p. 217, copyright © John Wiley and Sons, by permission.

Obviously, only the simplest cases are presented in Table 20.3, for example, a major river may drain a glaciated basin, cross through a semi-arid or humid region, and then enter the sea. The terrace sequence and sedimentary deposits found along this river course should be complicated and difficult to correlate. It may be concluded that the correlation of river terraces by height or number of terraces over long distances under such circumstances is unwise (Hadley, 1960).

The Nile River basin provides a good example of climatic variability within a major drainage basin. This exotic river derives its discharge from Central Africa, and any basinwide increase of precipitation will increase runoff from the headwaters, and the increased vegetal cover will decrease the sediment load of the Upper Nile. However, it will also greatly increase the sediment produced by tributaries in arid Sudan and Egypt. This influx of sediment from the tributaries should cause deposition in the Lower Nile Valley and, in fact, Butzer and Hansen (1968) have concluded that the Lower Nile aggraded during wetter periods.

The great variability of climate in large drainage basins

prevents any simplistic application of the relations in Table 20.3, and this is also true for smaller drainage basins with sufficient relief to produce climatic zones within the watershed. For example, climate may range from arid or semi-arid to humid or periglacial at the highest elevations, and Richmond (1962) has described the vertical shifting of such climate zones in the La Sal Mountains of Utah (Figure 20.28).

When rivers are confined to valleys, any evidence related to the dimensions of the ancient channel is destroyed by the channel adjustment. However, when the rivers flow across an alluvial plain or in a wide valley and the position of the river shifts with time, the possibility of preservation of the paleochannel is improved. For example, in many valleys of the Gulf Coast of the United States large meanders are preserved on the Pleistocene Deweyville terrace. These reflect a major decrease in discharge to the present. In the Prosna River valley in Poland (Figure 20.29) a series of meander scars of decreasing size provide clear evidence of the decrease of discharge since the Würm glaciation. In fact, the highest terraces preserve traces of braided stream patterns sug-

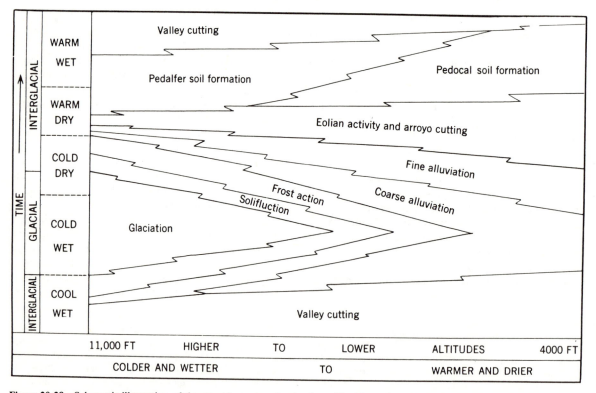

**Figure 20.28** Schematic illustration of the range in process domination with altitude, time and inferred climate during a typical glacial–interglacial cycle in the La Sal Mountains, Utah.
*Source:* Richmond, 1962, figure 7, p. 18.

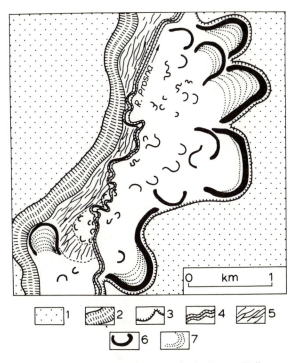

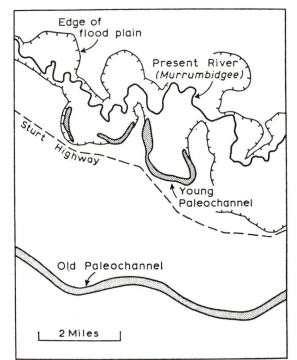

**Figure 20.29** Paleochannel patterns in the Prosna Valley near Wieruszów, Poland: 1, Würm terraces with traces of braided paleochannels; 2, gentle slope between terraces (1) and main valley floor; 3, steep slope to main valley floor; 4, present river channel incised 3·5–4 m into the main valley floor; 5, traces of braided paleochannels; 6, paleomeanders filled with organic deposits; 7, paleopoint bars.
*Source:* Korzarski and Rotnicki, 1977, figure 6.

**Figure 20.30** Diagram made from an aerial photograph of a portion of the Riverine Plain near Darlington Point, New South Wales. The sinuous Murrumbidgee River, which is about 200 ft wide, flows to the left across the top of the figure. It is confined to an irregular floodplain on which a large oxbow lake (youngest paleochannel) is preserved. The oldest paleochannel crosses the lower part of the figure (see Plate 32).
*Source:* Schumm, 1969, figure 11.II.4, p. 531.

gesting a major change of sediment load as well a discharge.

An excellent example of river metamorphosis as a result of climate change is provided by the Riverine Plain of New South Wales, Australia. Here the sinuous Murrumbidgee River (Figure 20.30) drains from the highlands of south-eastern New South Wales towards the west. It crosses the Riverine Plain, an alluvial plain that slopes at about 1·5 ft per mile to join the Murray River at the New South Wales–Victoria border. The channel is about 200 ft (60 m) wide, and it is confined to an irregular floodplain on which are preserved large oxbow lakes which are evidence of a past time of high discharge (young paleochannel, Figure 20.30). The trace of an older low-sinuosity stream channel (old paleochannel) crosses the lower quarter of the figure. In the upper half of its drainage basin the Murrumbidgee River is confined within a valley, and no evidence pertaining to its past condition exists because the old channels have been

destroyed. On the alluvial plain, however, the position of the channel has shifted and the three different types of channels are visible. The morphology of these channels reflects the hydrologic regimen of the time when each channel was functioning. Pedologic and geomorphic evidence from the Riverine Plain indicates that the oldest of the paleochannels was functioning during a climate drier than that of the present and that the youngest paleochannel was functioning during a climate more humid than that of the present (Plate 32).

The channel changes occurred not only because the water discharge increased or decreased, but also because climatic changes significantly altered the type of sediment load moved from the headwaters across the alluvial plain. At present erosion is not a problem in the Murrumbidgee drainage basin. A good cover of vegetation protects the source area, and the river transports small quantities of sand, silt and clay. An increase in precipitation will further improve the vegetation, and although runoff will increase, the sediment yield will decrease. On the other hand, a decrease in precipitation will decrease the amount of runoff, but will also greatly increase the yield of sediment from the drainage basin (see Figures 20.21 and 20.22).

As fascinating as the paleochannels of the Riverine Plain may be, they are of concern here only as an illustration of the types of river changes that can occur naturally. If, for simplicity, it is considered that the paleochannels represent changes from a channel which initially was like that of the modern Murrumbidgee River, we can discuss the channel changes which will occur if the present climate of south-eastern Australia becomes wetter (young paleochannel) or drier (old paleochannel). An increase in precipitation in the Murrumbidgee River headwaters will result in increased annual discharge and increased mean annual flood, but only small change in sediment concentration and in the type of sediment load moved through the channel. In response to this hydrologic change, which is similar in effect to that described by equation (12·23), the shape of the channel will not change, but it will become wider, deeper, and meander wavelength and amplitude will increase. The initial response will undoubtedly be severe channel erosion until the new dimensions are established. The result will be a relatively stable channel, but one larger than the modern river. On the Riverine Plain it is significant that although deepening of the channel did occur, the increased discharge did not cause major incision. Gradient remained essentially constant.

On the other hand, a decrease in precipitation in the headwaters of the Murrumbidgee River will not only cause a decrease in mean annual discharge, but through reduction of vegetation density it will increase peak discharges and greatly increase the amount of sand moved through the channels. The result will be a complete transformation of the river system. To transport the increased sand load with less water, the channel will become wider and shallower. The greater slope required to transport this load will not be developed by deposition and steepening of the Riverine Plain, which would require movement of immense quantities of sediment. Rather, the channel pattern will be changed. The gradient of the channel will be doubled without significant deposition by a reduction of the sinuosity from about 2 to about 1. These changes are in complete accord with equation (12·30). In fact, the events that occurred on the surface of the Riverine Plain were the reverse of the above sequence with the wide, straight, steep channel changing under the influence of altered hydrologic conditions to a relatively deep and sinuous channel. The documented changes are in accord with those suggested by equation (12·29).

The Riverine Plain paleochannels provide an example of adjustment of gradient by a major change of channel pattern. This supports the suggestion made earlier that, if unconfined, rivers may decrease their gradient by meandering rather than by major degradation. Therefore, modern bank erosion problems may be only the initial stage of river metamorphosis if a reduction of river gradient is required by man-induced changes of river regimen.

In contrast to the changing channel patterns of the Prosna, Mississippi and Murrumbidgee Rivers major incision does occur especially in semi-arid and arid areas where sediment is stored in valleys. The great arroyos of the south-western United States that began to incise in the late nineteenth century, although perhaps caused by land use change, are the modern equivalents of Holocene incision caused by climate change. Considerable discussion of the type of climate change required has led to little agreement probably because, as Table 20.3 shows, upstream and downstream reaches may be out of phase.

### 20.2.2 Temperature changes dominant

During the ice ages cold conditions drastically increased their hold on the earth. The geomorphological effects of the growth of glaciers have been discussed in Chapter 19. In this section it is proposed to examine the geomorphological effect of the expansion of the periglacial morphogenetic system into mid-latitudes and, in mountains, into lower altitudes. In all 20 per cent of the earth's surface may have been so affected. The most important areas are in Eurasia and North America, but upland areas in the southern continents and southerly cool temperate islands like the Falkland Islands were also involved. In all situations the temperature change was significant because it caused water to be frozen for a greater part of the year than previously. This, in turn, encouraged the operation of processes currently typical of the periglacial morphogenetic zone, whose effects were superimposed on the pre-existing landscape. Although the number of such cold phases in the Pleistocene is still unknown, it seems likely that there were at least seventeen in approximately the last 3 million years.

When seaching for evidence of periglacial conditions in mid-latitudes, it is tempting to look for direct analogies with current periglacial areas in high latitudes. However, the latitudinal contrast between the two areas leads to problems. French (1976) has emphasized some of the more important contrasts. Whereas the temperature regime in high latitudes is marked by a strong seasonal amplitude with diurnal fluctuations suppressed, in mid-latitudes the opposite is likely to have been the case. Processes reacting to frequent oscillations above and below the freezing point are likely to be more common in mid-latitudes than in high latitudes. French, therefore, suggested that frost action and gelifluction were likely to be relatively more important in mid-latitudes. However, this does not take into account the intensity of the fluctuations, and Washburn (1979) is unconvinced by the argument. A second possible contrast is that permafrost is likely to have been less extensive and shallower in mid-latitudes than in present high latitudes. This is because the depth of the permafrost is strongly influenced by the long polar night when air temperatures stay below 0° for months on end. In mid-latitudes such long periods of low temperature are less likely to have occurred in the shorter more variable winter. French suggested that permafrost would have been little in evidence in maritime environments and only discontinuous or locally important elsewhere.

Another contrast concerns wind. It is likely that wind action was relatively more important than in polar peri-glacial areas today. The growth of mid-latitude ice sheets is likely to have displaced the westerlies southward and, as a result of the increased intensity of climatic gradients, made them stronger. In addition, strong downslope kata-batic winds, such as occur in Antarctica today, would have affected the ice-marginal areas. Furthermore, glacier action in these marginal areas would have provided ample fine, loose material ideal for aeolian entrain-ment. A final contrast concerns the activity of rivers. Whereas flow in high latitudes is highly seasonal, mid-latitude flow is likely to have persisted for a longer summer period. It is difficult to equate this difference with intensity of fluvial activity. The floods in the Arctic are high-magnitude events but they often occur when much of the ground is frozen and protected. In mid-latitudes the flood may have been more prolonged, but the ground less frozen and more susceptible to erosion. What does seem clear is that many rivers in former mid-latitude periglacial zones flowed in broad braided channels.

Evidence of permafrost landforms is the best evidence of periglacial conditions extending into mid-latitudes. Casts of ice- or sand-wedge polygons are the most

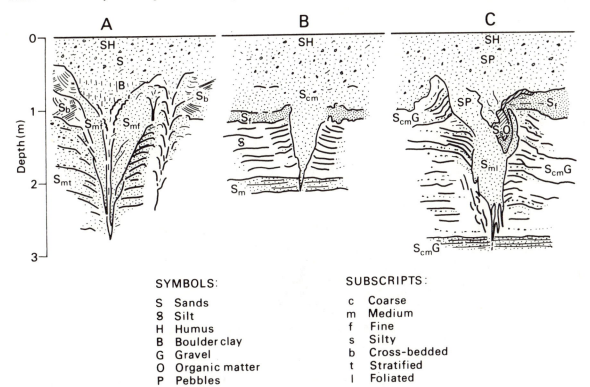

SYMBOLS:

| | |
|---|---|
| S | Sands |
| ẟ | Silt |
| H | Humus |
| B | Boulder clay |
| G | Gravel |
| O | Organic matter |
| P | Pebbles |

SUBSCRIPTS:

| | |
|---|---|
| c | Coarse |
| m | Medium |
| f | Fine |
| s | Silty |
| b | Cross-bedded |
| t | Stratified |
| l | Foliated |

**Figure 20.31** Examples of different types of wedge casts: A. ice wedge which has been infilled and whose sides have slumped; B. sand wedge which has experienced no slumping; C. composite wedge.
*Source:* French, 1976, figure 12.7, p. 241, after Gozdik, 1973.

conclusive evidence (Figure 20.31) and often these may be related to polygonal patterns picked out in crop patterns in fields. The best examples in Europe lie outside the limits of the last glaciation (Figure 20.32), although smaller examples occur in lowland Scotland and Scandinavia outside more recent glacial limits. Ice-wedge polygons are well displayed in central Europe, especially in Germany and Poland. Pingo scars can be recognized by narrow ramparts surrounding a depression. The rampart represents the material swept off the pingo by mass wasting. Good examples of such features have been recognized in the Ardennes in Belgium (Figure 20.33). The importance of such features is that they are evidence of permafrost. Further, since cold temperatures are required to cause ground contraction, the wedges can be used to infer mean annual temperatures of the order of $-6°$ to $-8°C$.

Active-layer periglacial landforms are widely recognized but sometimes less easy to interpret. Frost-churned soils frequently produce characteristic involutions which can be seen in sections (Figure 20.34). Sometimes they can be related to patterned ground visible on the surface.

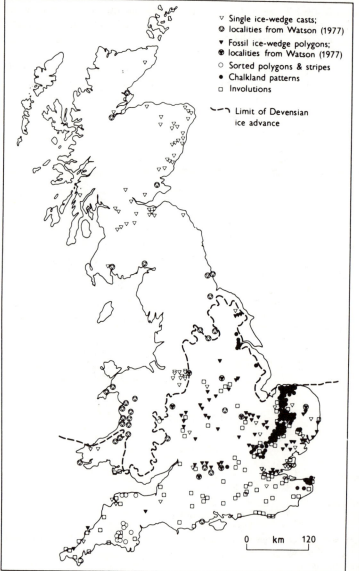

∇  Single ice-wedge casts;
⊚  localities from Watson (1977)
▼  Fossil ice-wedge polygons;
⊛  localities from Watson (1977)
○  Sorted polygons & stripes
•  Chalkland patterns
□  Involutions

⌐  Limit of Devensian
   ice advance

0    km    120

**Figure 20.32**  Distribution of patterned ground of the last glacial age (Devensian) in Britain.
*Source:* Washburn, 1979, figure 14.1, p. 281, after West *et al.*

Identification of such involutions has been common in both Europe and North America. Gelifluction deposits are widely recognized. A good example is the 'head' of southern England. It forms a mantle of regolith, particularly on lower slopes where it may attain thicknesses of up to 33 m. Characteristically the deposit is poorly sorted and of local origin, with angular stones, many of them oriented downslope. In places it may contain varying quantities of sediment related to water activity on the slope. Head is widespread south of the limit of the last glacial maximum, but is more limited to the north. This implies that the periglacial phase was coincident with the glaciation and must have been short-lived after deglaciation commenced. Gelifluction deposits equivalent to

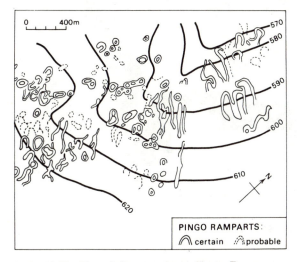

**Figure 20.33** Plan of pingo remains on Hautes Fagnes Moor, Ardennes, near Malmedy, eastern Belgium; contours are shown in metres.
*Source:* Pissart, 1953.

the British head are well known in continental Europe, especially Poland where they are well studied (Dylik, 1969). In places gelifluction deposits are related to relict block streams. Although several good examples are known in Britain, some of the world's best examples occur in the Falkland Islands in the South Atlantic. Here 0·3–2·0-m-sized boulders of quartzite and sandstone originate at rock outcrops on ridge crests and are arranged into broad stripes running down the gentle valley sides. In certain places the boulders accumulate in the valley floors and one such accumulation, called 'Princes Street', is 6 km long and about 800 m wide. The Falkland Island examples are related to periglacial conditions during the last glacial period, when glaciers occupied only a few small cirques (Roberts, 1984) (Plate 33).

Gelifluction is particularly well developed in the vicinity of snow patches which keep the regolith moist throughout the summer. Small valleys up to several hundred metres in length scallop the flanks of larger valleys in many parts of the Chalk uplands of southern Britain and also the Lódz Plateau of Poland. They may represent excavation selectively associated with seasonal snow accumulation.

It is increasingly clear that wind-blown deposits are associated with the periglacial environment in both North America (Figure 20.35) and Europe (Figure 20.36). Aeolian sands consist of particles usually 0·06–1·0 mm in diameter which may form a mantle of cover-sand. In places it has accumulated as dunes, and a magnificent series of dunes covers much of central North America, especially Nebraska. Aeolian silt, generally known as loess, is the most widespread aeolian deposit. It is a buff-coloured, well-sorted deposit, with particle sizes between 0·01 mm and 0·05 mm in diameter. It forms a swathe south of and parallel to the glacier limits in North America and Europe (Figure 20.37). In places it attains a

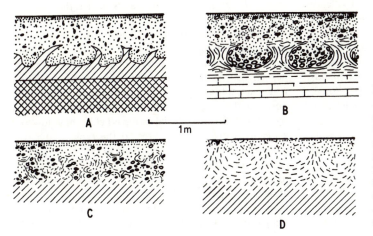

**Figure 20.34** Various types of involution: A. injection of tongues of silt and clay into overlying sand and gravel; B. frost pockets, frost kettles, or pillar involutions, with ascending tongues of finer sediments; C. irregular or amorphous involutions; D. festoons, with tongues of frost-shattered disintegrating rock rising into finer sediments.
*Source:* West, 1977, figure 5.13a–d, p. 92, after Sekyra, 1961.

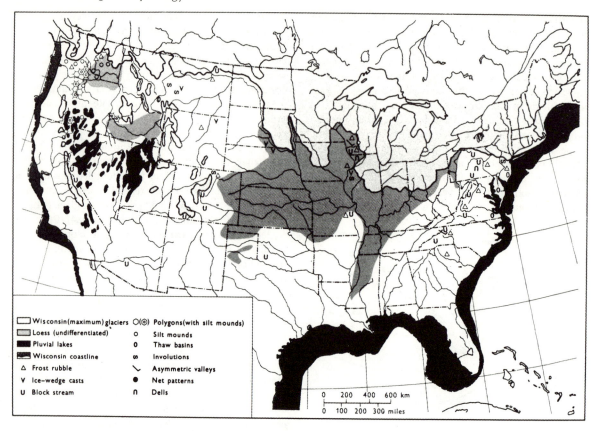

**Figure 20.35** Pleistocene periglacial features in the United States.
*Source:* Washburn, 1979, figure 14.19, p. 305, after Brunnschweiter, 1964, figure 1, p. 224.

thickness of 30–35 m. Probably most of the loess was derived from the braided courses of meltwater and periglacial rivers.

In mountainous areas Pleistocene periglacial features occur at altitudes lower than those active today. Evidence of former enhanced frost action is demonstrated by blockfields, or 'felsenmeer', in present temperate environments. Such accumulations of angular boulders occur, for example, in many temperate European uplands (Pissart, 1953) and in the central interior uplands and Appalachian plateaus in the United States. Sometimes distinctive scree accumulations exist. Relict protalus ramparts, ridges of boulders accumulating at the foot of a snow patch, are known at low altitudes in Scotland (Sissons, 1976), the Alps and the Rockies of North America. Sometimes former low-level limestone screes have been cemented and are currently being eroded by stream gullies. Good examples occur in the Pindus Mountains of Greece, for example, on Mt Parnassus. Relict gelifluction features are also common on the slopes of cool temperate mountains and date from a colder period in the past. Sorted steps are particularly common and, for example, may be seen in the world's major mountain ranges, areas formerly glaciated by ice sheets such as Scotland, and islands such as the Falkland Islands.

Whereas relict periglacial landforms can often be recognized, discussion as to their significance in landform evolution has generated enough heat to melt much permafrost. At one extreme the argument has tended to regard accentuated mass wasting and frost action as so powerful that the whole landscape has been transformed by such activity. Periglacial action is held to have lowered watersheds dramatically, created the typical concavo-convex slopes of southern England, to have created asymmetry in many previously existing river valleys and to have cut new valleys. At the other extreme some have argued that periglacial activity has operated so spasmodically that it has done little more than modify regolith patterns. As in so many discussions of this type the

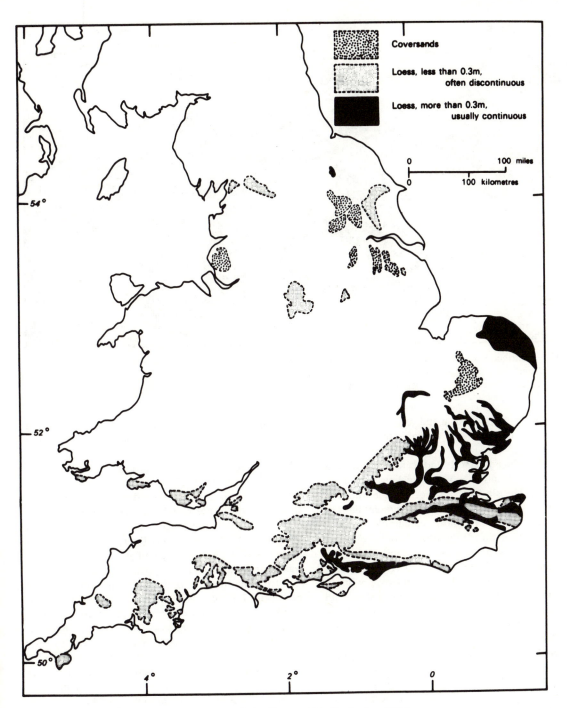

**Figure 20.36** Distribution of loess and coversands (aeolian sands) in England and Wales.
*Source:* Catt, 1977, figure 16.1, p. 222.

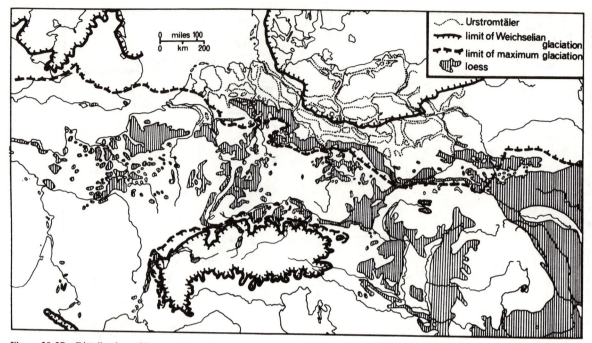

**Figure 20.37**    Distribution of loess and of major meltwater rivers (Urstromtäler) associated with the limits of the Pleistocene maximum and the Weichselian glaciations in Europe.
*Source:* West, 1977, figure 5.3, p. 75, after Grahmann and Woldstedt.

answer probably lies in between and varies according to location and the length of time available for periglacial activity.

It seems that the Falkland Islands can be regarded as an example of a location subjected to long periods of periglaciation. The islands lie just north of the Antarctic Convergence, which in effect marks the present limit of sea-level glaciation. Unlike most comparable locations which were glaciated during Pleistocene cold phases, the islands were not glacier-covered (except for a few cirques) (Sugden and Clapperton 1976). The surprising lack of glaciation probably reflects limited snowfall in the lee of the South American Andes, but the islands did experience cold periglacial conditions. Nivation hollows and valleys abound as do protalus ramparts, thick geli-fluction deposits (sheets and steps) and crenulated ridges with bedrock exposed. And yet the overall topography is still clearly fluvial with open river valleys which continue below sea level. Indications that the topography is old come from the dating of a tree deposit beneath 2 m of gelifluction material as 38 million years old (Birnie and Roberts, 1984). The apparent conclusion to be reached is that here, in a particularly fine and long-lived periglacial environment, periglacial landscape modification has been relatively subdued and limited to the enhanced

stripping of upper valley slopes and the accumulation of regolith in low-lying areas. Over a long timescale fluvial activity has been dominant.

The degree of landscape modification will depend on the duration of periglacial conditions, the climate and lithology. Areas that are immediately outside the ice sheet limits of the northern hemisphere are likely to have experienced both the longest and severest periglacial conditions. Within the glacial limits periglaciation has played only a limited role, and was restricted to the waxing and waning phases of glacials. The climatic variation from maritime to continental climate also affected periglacial efficacy. Williams (1968, 1969) concluded that permafrost was widespread in eastern England but restricted to uplands in the west. One result was the apparent increase in gelifluction activity from west to east. For example, gelifluction moved boulders 5–10 km in Cambridgeshire on slopes averaging less than 1°, while on Dartmoor in the west boulders moved a maximum of 1 km on slopes with angles as high as 8–11°.

A similar contrast has been recognized within Europe, with the coldest conditions occurring in southern Poland and eastern Germany and warmer conditions to the west and east (Poser, 1948). There is an interesting contrast between the mid-latitude periglacial areas of Europe and

North America. Evidence of permafrost and paleo-ecology suggests that in Europe periglacial conditions were severe and open tundra extensive, while in North America the zone was narrow and less severe. Indeed, trees may have been growing close to the ice margin. Probably such a contrast reflected the anticyclonic circulation over the ice sheets. The western position of the European ice sheet attracted winter air masses from the continental interior, while the centrally located Laurentide ice sheet drew in maritime air masses into the eastern United States.

Lithology plays an important role in the formation and preservation of periglacial landscapes. The Chalk in southern England is particularly suitable on both counts, and it is no surprise to find that it is in these areas that periglacialists wax lyrically. The chalk is highly frost-susceptible and, furthermore, its underground drainage in subsequent times has allowed relict periglacial features to survive. The unconsolidated drift terrains of the earlier glaciations also provided easily modified landscapes for further periglaciation. Thus East Anglia and central Poland are examples of drift landscapes well modified by periglacial processes. It seems fitting that the latter country where both lithology and climatic conditions favoured intense periglacial activity has produced the first journal specializing in periglacial geomorphology, namely *Biuletyn Peryglacjalny*.

## 20.3 Conclusion

Landform assemblages consist of a complex of interlocking systems having different threshold conditions, response times, relaxation paths and time trajectories. Such an assemblage may be viewed as forming a spectrum with the following end-members (Trudgill, 1976):

(1) Labile landforms, which are operated on by relatively powerful local processes and/or possess weak surface resistance to change. These are very prone to change, are commonly small scale and whose forms always remain close to equilibrium in respect of changing climatic processes.

(2) Sluggish landforms, which are locally subjected to relatively weak processes and/or possess high surface resistance. These respond only very slowly to change and only reflect the operation of highly significant climatic changes after a long timelag (i.e. they possess a long relaxation time). Figure 20.38 shows how landform systems may respond differently to climatic changes with true equilibrium between forms and process being only achieved for the whole assemblage of systems after a long period $t_e$ of virtual climatic stability, where a false impression of equilibrium may occur by the association of suitable lagged and non-lagged systems $t_f$ and how for the rest of the time the total landform assemblage is composed of

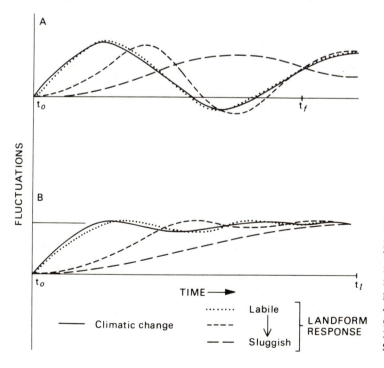

**Figure 20.38** Labile to sluggish landform responses over time to A. cyclic and B. pseudo step-functional changes of climate. This shows the differing lags between the responses of landforms of differing resistance to erosion to different temporal patterns of climatic change.
*Source:* Trudgill, in E. Derbyshire (ed.), *Geomorphology and Climate*, 1976, figure 3.20, p. 92, copyright © John Wiley and Sons, by permission.

differentially adjusted systems. The effect of climatic changes on such a complex of systems is thus to produce a hierarchy of systems, each differentially related to the existing climatic environment and, hence, each carrying an historical signal of different strength. The relaxation time of some landform systems is short (e.g. those involving hydraulic geometry and beach characteristics); for others it is longer (e.g. those involving drainage density); but for others it is very long (e.g. changes in the relief of granite terrains). Thus one local landscape may be made up of systems of widely differing behaviour, so that it must be viewed as a *palimpsest of systems* whose operation is interlocked, but whose history is superimposed. Just because some of the landform systems or their variables exhibit a certain equilibrium with respect to contemporary climatic processes, it does not necessarily imply that this is true for all others.

# References

Baker, V. R. (1977) 'Stream channel response to floods; with examples from central Texas', *Bulletin of the Geological Society of America*, vol. 88, 1057–71.

Barry, R. G. (1981) *Climatic Change*, London, Macmillan Education.

Barry, R. G. and Chorley, R. J. (1982) *Atmosphere, Weather and Climate*, 4th edn, London, Methuen.

Birks, H. J. B. and Birks, H. H. (1980) *Quaternary Palaeoecology*, London, Arnold.

Birnie, J. F. and Roberts, D. E. (1984) 'Evidence of a Tertiary forest in the Falkland Islands (Islas Malvinas)', *Palaeogeography, Palaeoclimatology and Palaeoecology*.

Bowen, D. Q. (1978) *Quaternary Geology*, Oxford, Pergamon.

Bowles, F. A. (1975) 'Paleoclimatic significance of quartz/illite variations in cores from the eastern equatorial North Atlantic', *Quaternary Research*, vol. 5, 225–35.

Brunnschweiter, D. (1964) 'Der pleistozane Periglazialbereich in Nordamerika', *Zeitschrift für Geomorphologie*, vol. 8, 223–31.

Bryson, R. A. (1973) 'Drought in Sahelia: who or what is to blame?', *Ecologist*, vol. 3, 366–71.

Büdel, J. (1948) 'Das System der klimatischen Geomorphologie', *Verdhandlungen Deutscher Geographentag*, vol. 27, 65–100.

Büdel, J. (1980) 'Climatic and climamorphic geomorphology', *Zeitschrift für Geomorphologie*, Supplement 36, 1–8.

Butzer, K. W. (1976) *Geomorphology from the Earth*, New York, Harper & Row.

Butzer, K. W. and Hansen, C. L. (1968) *Desert and Rivers in Nubia*, Madison, Wis., University of Wisconsin Press.

Catt, J. A. (1977) 'Loess and coversands', in F. W. Shotton (ed.), *British Quaternary Studies: Recent Advances*, Oxford University Press, 222–9.

Chorley, R. J. and Kennedy, B. A. (1971) *Physical Geography: A Systems Approach*, London, Prentice-Hall.

Cornish, J. H. (1961) 'Flow losses in dry sandy channels', *Journal of Geophysical Research*, vol. 66, 1845–53.

Costa, J. E. (1978) 'Holocene stratigraphy in flood frequency analysis', *Water Resources Research*, vol. 14, 626–32.

Cotton, C. A. (1958) 'Alternating Pleistocene morphogenetic systems', *Geological Magazine*, vol. 95, 125–36.

Crittenden, M. D. (1963) 'New data on the isostatic deformation of Lake Bonneville', *US Geological Survey Professional Paper* 454-E.

Dansgaard, W., *et al.* (1975) 'Climatic changes, Norsemen and modern man', *Nature*, vol. 225, 24–8.

Dendy, F. E. and Bolton, G. C. (1976) 'Sediment-yield runoff drainage area relationships in the United States', *Journal of Soil Water Conservation*, vol. 31, 264–6.

Derbyshire, E. (ed.) (1976) *Geomorphology and Climate*, New York, Wiley.

Dury, G. H. (1980) 'Step functional changes in precipitation at Sidney', *Australian Geographical Studies*, vol. 18, 62–78.

Dury, G. H. (1981) 'Climate and settlement in late-Medieval England', in C. D. Smith and M. Parry (eds), *Consequences of Climatic Change*, University of Nottingham, Department of Geography, 39–53.

Dylik, J. (1969) 'Slope development under periglacial conditions in the Lódz region', *Biuletyn Peryglacjalny*, vol. 18, 381–410.

Fisk, H. N. and McFarlan, E. Jr (1955) 'Late Quaternary deltaic deposits of the Mississippi River', *Geological Society of America Special Paper* 62, 279–302.

Frakes, L. A. (1979) *Climates throughout Geologic Time*, Amsterdam, Elsevier.

French, H. M. (1976) *The Periglacial Environment*, London, Longman.

Frenzel, B. (1979) 'Europe without forests', *Geographical Magazine*, vol. 51 (August), 756–61.

Fritts, H. C. (1976) *Tree Rings and Climate*, London, Academic Press.

Gates, W. L., (1976) 'Modelling the ice-age climate', *Science*, vol. 191, 1138–44.

Goudie, A. S. (1972) 'The concept of post-glacial progressive desiccation', *Oxford University, School of Geography Research Paper* 4.

Goudie, A. S. (1977) *Environmental Change*, Oxford University Press.

Gozdzik, J. (1973) 'Origin and stratigraphic position of periglacial structures in Middle Poland', *Acta Geographica Lódziensia*, vol. 31, 104–17.

Gregory, K. J. (1976) 'Drainage networks and climate', in

E. Derbyshire (ed.), *Geomorphology and Climate*, New York, Wiley, 289–310.

Gregory, K. J. (ed.) (1983) *Background to Palaeohydrology*, Chichester, Wiley.

Hadley, R. F. (1960) 'Recent sedimentation and erosional history of Fivemile Creek, Fremont County, Wyoming', *US Geological Survey Professional Paper* 352-A.

Hooke, J. M. and Kain, R. J. P. (1982) *Historical Change in the Physical Environment*, London, Butterworth.

Hurst, H. E., Black, R. P. and Simaika, Y. M. (1965) *Long-Term Storage: An Experimental Study*, London, Constable.

Jansen, J. M. L. and Painter, R. B. (1974) 'Predicting sediment yield from climate and topography', *Journal of Hydrology*, vol. 21, 371–80.

Kennedy, B. A. (1976) 'Valley-side slopes and climate', in E. Derbyshire (ed.), *Geomorphology and Climate*, New York, Wiley, 171–99.

Knox, J. C. (1972) 'Valley alluviation in southwestern Wisconsin', *Annals of the Association of American Geographers*, vol. 62, 401–10.

Knox, J. C., Bartlein, P. J., Herschboeck, K. K. and Muckenhirn, R. J (1975) 'The response of floods and sediment yields to climatic variation and land use in the Upper Mississippi Valley', *University of Wisconsin-Madison, Institute of Environmental Studies Report* 52.

Kozarski, S. and Rotnicki, K. (1977) 'Valley floors and changes of river channel patterns in the north Polish Plain during the Late Würm and Holocene', *Questiones Geographicae*, vol. 4, 51–93.

Krigström, A, (1962) 'Geomorphological studies of sandur plains and their braided rivers in Iceland', *Geografiska Annaler*, vol. 44, 328–46.

Kurtén, B. (1972) *The Ice Age*, Stockholm, International Book Production.

LaMarche, V. C. (1974) 'Paleoclimatic inferences from long tree-ring records', *Science*, vol. 183, 1043–8.

Lamb, H. H. (1966) *The Changing Climate*, London, Methuen.

Langbein, W. B. *et al.* (1949) 'Annual runoff in the United States', *US Geological Survey Circular* 52.

Langbein, W. B. and Schumm, S. A. (1958) 'Yield of sediment in relation to mean annual precipitation', *Transactions of the American Geophysical Union*, vol. 39, 1076–84.

Leopold, L. B. (1951) 'Rainfall frequency: an aspect of climatic variation', *Transactions of the American Geophysical Union*, vol. 32, 347–57.

Leopold, L. B., Emmett, W. W. and Myrick, R. M. (1966) 'Channel and hillslope processes in a semiarid area New Mexico', *US Geological Survey Professional Paper* 352-G, 193–253.

Leopold, L. B., Wolman, M. G. and Miller, J. P. (1964) *Fluvial Processes in Geomorphology*, San Francisco, Calif., Freeman.

Manley, G. (1964) 'The evolution of the climatic environment', in J. W. Watson and J. B. Sissons (eds), *The British Isles: A Systematic Geography*, London, Nelson, 152–76.

McIntyre, A., *et al.* (1976) 'The surface of the ice-age earth (CLIMAP Project)', *Science*, vol. 191, 1131–7.

Nicholson, S. E. and Flohn, H. (1980) 'African environmental and climatic changes in the Late Pleistocene and Holocene', *Climatic Change*, vol. 2, 313–48.

Ninkovitch, D. and Shackleton, N. J. (1975) 'Distribution, stratigraphic position and age of ash layer "L" in the Panama Basin region', *Earth and Planetary Science Letters*, vol. 27, 20–34.

Patton, P. C. and Baker, V. R. (1980) 'Geomorphic response of central Texas stream channels to catastrophic rainfall and runoff', in D. O. Doehring (ed.), *Geomorphology in Arid Regions*, London, Allen & Unwin, 189–217.

Peltier, L. C. (1950) 'The geographic cycle in periglacial regions as it is related to climatic geomorphology', *Annals of the Association of American Geographers*, vol. 40, 214–36.

Peltier, L. C. (1959) 'Area sampling for terrain analysis' (abstract), *Bulletin of the Geological Society of America*, vol. 70, 1809.

Peltier, L. C. (1962) 'Area sampling for terrain analysis', *Professional Geographer*, vol. 14, 24–8.

Pissart, A. (1953) 'Less coulées pierreuses du Plateau des Hautes Fagnes', *Annales Société Géologique de Belgique*, vol. 76, 203–19.

Poser, H. (1948) 'Boden und Klimaverhältnisse in Mittel und Westeuropa der Würmeiszeit', *Erdkunde*, vol. 2, 53–68.

Richmond, G. M. (1962) 'Quaternary stratigraphy of the La Sal Mountains, Utah', *US Geological Survey Professional Paper* 324.

Rittenhouse, G. (1944) 'Sources of modern sands in the middle Rio Grande Valley, New Mexico', *Journal of Geology*, vol. 52, 145–83.

Roberts, D. E. (1984) 'The Quaternary History of the Falkland Islands', unpublished Ph D thesis, University of Aberdeen.

Rodda, J. B. (1970) 'Rainfall excesses in the United Kingdom', *Transactions of the Institute of British Geographers*, no. 49, 49–60.

Schumm, S. A. (1965) 'Quaternary paleohydrology', in H. E. Wright and D. G. Frey (eds), *The Quaternary of the United States*, Princeton, NJ, Princeton University Press, 783–94.

Schumm, S. A. (1968) 'River adjustment to altered hydrologic regimen – Murrumbidgee River and paleochannels, Australia', *US Geological Survey Professional Paper* 598.

Schumm, S. A. (1969) 'Geomorphic implications of climatic changes', in R. J. Chorley (ed.), *Water, Earth and Man*, London, Methuen, chapter 11.II, 525–34.

Sekyra, J. (1961) 'Periglacial phenomena', *Instytut Geologiczny (Warszawa), Prace*, vol. 34, 99–107.

Sissons, J. B. (1976) *Scotland*, London, Methuen.

Starkel, L. (1976) 'The role of extreme (catastrophic) meteorological events in contemporary evolution of slopes', in E. Derbyshire (ed.), *Geomorphology and Climate*, New York, Wiley, 203–41.

Strahler, A. N. (1940) 'Landslides of the Vermilion and Echo Cliffs, Northern Arizona', *Journal of Geomorphology*, vol. 3, 285–301.

Street, F. A. (1981) 'Tropical palaeoenvironments', *Progress in Physical Geography*, vol. 5, 157–85.

Street, F. A. and Grove, A. T. (1979) 'Global maps of lake-level fluctuations since 30,000 years BP', *Quaternary Research*, vol. 12, 83–118.

Sugden, D. E. and Clapperton, C. M. (1976) 'The maximum extent of glaciers in part of West Falkland', *Journal of Glaciology*, vol. 17, 73–7.

Toy, T. J. (1977) 'Hillslope form and climate', *Bulletin of the Geological Society of America*, vol. 88, 16–22.

Trudgill, S. T. (1976) 'Rock weathering and climate: quantitative and experimental aspects', in E. Derbyshire (ed.), *Geomorphology and Climate*, London, Wiley, 59–99.

Washburn, A. L. (1979) *Geocryology*, London, Arnold.

Watson, R. A. and Wright, H. E., Jr (1963) 'Landslides on the east flank of the Chuska Mountains, northwestern New Mexico', *American Journal of Science*, vol. 261, 525–48.

West, R. G. (1977) *Pleistocene Geology and Biology*, 2nd edn, London, Longman.

Williams, R. B. G. (1968) 'Some estimates of periglacial erosion in southern and eastern England', *Biuletyn Periglacjalny*, vol. 17, 311–35.

Williams, R. B. G. (1969) 'Permafrost and temperature conditions in England during the last glacial period', in T. L. Péwé (ed.), *The Periglacial Environment*, Montreal, McGill-Queens University Press, 399–410.

Wilson, L. (1972) 'Seasonal sediment yield patterns of United States rivers', *Water Resources Research*, vol. 8, 1470–9.

Wilson, L. (1973) 'Variations in mean annual sediment yield as a function of mean annual precipitation', *American Journal of Science*, vol. 273, 335–49.

Wollin, G., Ericson, D. B. and Ryan, W. B. F. (1971) 'Variations in magnetic intensity and climatic changes', *Nature*, vol. 232, 549–51.

Zeuner, F. E. (1959) *The Pleistocene Period*, London, Hutchinson.

# **Appendix** *Applied geomorphology*

# Appendix *Applied geomorphology*

## A1.1 Nature of applied geomorphology

Applied geomorphology may be defined as the application of geomorphic understanding to the analysis and solution of problems concerning land occupancy, resource exploitation, environmental management and planning (Jones, 1980). It has a surprisingly long and varied history (Coates, 1972, 1973) and has recently attracted the increasing interest of geomorphologists (Fels, 1965; Jennings, 1966; Brown, 1970; Cooke and Doornkamp, 1974; Hails, 1977; Gregory and Walling, 1979; Goudie, 1981; Craig and Craft, 1982; Verstappen, 1982). Applications of geomorphology can be divided broadly into two classes:

(1) Man as a geomorphic agent, in terms of his inadvertent and planned effects on geomorphic processes and forms.
(2) Geomorphology as an aid to resource evaluation, engineering construction and planning.

The inadvertent effects of man's geomorphic activities include such matters as the modification of terrain by engineering works, quarrying, mining, excavation and dumping; subsidence due to water extraction, oil-drilling, mining and drainage; increased erosion and arroyo-trenching due to changes in the vegetation cover; desertification due to overintensive agriculture (Biswas and Biswas, 1980); increased weathering rates due to pollution; accentuated mass movements and slope failures due to engineering activities; effects on hydraulic geometry due to changes of river load by dams or accelerated erosion; interference with coastal debris movement (see Section 15.7); erosion and deposition by coastal engineering works; permafrost modification due to high-latitude construction; and so on. Planned effects include all manner of engineering works, such as coastal pro-

tection; the fixing of sand dunes; the use of vegetation to control erosion; flood control; attempts to minimize the effects of natural catastrophes (Scheidegger, 1975), such as volcanic eruptions, seismic activities, tsunamis and hurricanes; the control of the engineering properties (e.g. shearing resistance, drainage, frost heaving, etc.) of surface materials; and river channellization and stabilization.

The resource evaluation and planning aspects of applied geomorphology are concerned with such matters as the role of geomorphology in resource inventories and environmental management; soil and land evaluation (Dent and Young, 1981); the production of maps for hydrological, erosional and stability control; geomorphic mapping, the mapping of land systems and terrain evaluation (Mitchell, 1973) concerned with the acquiring, classification and retrieval of terrain and other information for the use of civil and military engineers, earth scientists and planners; the preparation of analogue maps to assist in the interpretation of territory too expensive or too inaccessible (e.g. politically or extraterrestrially) to analyse in detail; the dumping of waste products, including nuclear waste (Schumm and Chorley, 1983); and urban geomorphology (Coates, 1976; Cooke *et al*, 1983).

## A1.2 Objectives of applied geomorphology

It is clear from the above brief survey that the scope of applied geomorphology is as vast as that of the discipline as a whole, and we have chosen to limit this Appendix mainly to certain applied aspects of fluvial geomorphology. Above all, consideration of the practical applications of geomorphology brings us back to the subject with which we began this book – the objectives of geomorphology. If geomorphology has as its role the identification and description of landforms, the explanation of their origin and the prediction of future change, then applications of geomorphology should fall within these categories of description, postdiction and prediction.

Identification can be the simple quantitative description of river or drainage basin characteristics with the purpose of relating the description to hydrologic, climatic, or other variables. For example, Figure 10.15 shows how hydrologic data are related to landform characteristics, and these relations can be used to predict the hydrologic character of small ungauged drainage basins. Furthermore, the simple description of a river channel may lead to a quantitative evaluation of its stability (see Figure 11.25). For example, a braided river has a large stream power and sediment load, and it is subject to avulsion and rapid lateral shift, whereas a meandering stream will be characterized by considerable stability if gradient is low and banks resistant, or by rapid meander shift and cutoffs if the gradient is steep and banks are easily eroded.

The geomorphologist's perspective over long time-spans and large areas give him an awareness of the causes of landscape form and of the complexity and potential for landscape change in time and space. Two examples will suffice to explain the practical advantages of this approach.

When a river changes course abruptly by avulsion (for example, by a meander cutoff), a boundary formed by the original river, remains fixed, Hence there are many portions of states and countries that lie on the opposite sides of rivers that are presumed to be their boundaries (e.g. Mississippi and Louisiana; the United States and Mexico). However, when a river shifts position relatively slowly (for example, by meander shift), the boundary shifts with the river. Obviously, the opportunities for litigation are numerous, and geomorphologists have an important role to play in such legal problems. They can develop a history of the channel and present evidence for change or the lack of it. The Snake River in Jackson Hole, Wyoming, is a gravel-bed braided stream with a gradient of about 4 m per kilometre. In the late nineteenth century land surveys were made in the area, with boundaries marked as following the bank of the meandering river. Depending on the accuracy of the original work, these surveys can provide a valuable base for evaluating subsequent channel changes. A few years ago it was recognized that the present banks of the Snake River did not correspond to the earlier surveys. This should not have been surprising, but the lack of agreement between the present position of the river banks and

the earlier surveys led to claims that these earlier surveys had been either fraudulent or grossly in error. The assumption was being made by these critics that river channels are fixed in space and time, and it was necessary to demonstrate at law that the Snake River had moved in its valley. If this could not have been done a landowner could have lost access to the river. The changing position of the channel was proved by assembling all available maps and aerial photos which then demonstrated that the river had moved, and soil, sediment and dendrochronological data were also used as supporting evidence. The judge was convinced that the original surveys were accurate and agreed consequently that the property boundaries should be deemed to shift as the river shifted.

Another example is the recent litigation concerning the cause of bank erosion along the Ohio River. Twenty-two landowners claimed that the erosion of their property was caused by the raising of water levels behind navigation locks and dams. In order to maintain navigation on the Ohio River during low water a series of dams with locks maintain a minimum nagivation depth of 9 ft. The pool level behind the dam, therefore, never falls to the old low-water levels. It was alleged that the maintenance of the pools at a constant level causes bank erosion by wave action. Preliminary studies showed that, indeed, erosion was occurring on the litigants' lands, and their claims seemed valid. However, when the river as a whole was considered, rather than twenty-two limited portions of the bank, it became clear that the river has eroding, stable and healing banks, and the type and extent of erosion that is occurring is expectable. In fact, much of the erosion was due to the landowners' activities behind the bankline which added water to the banks and caused slumping well above the pool level. In this case the ability to consider the entire river rather than a few specific locations permitted the development of a strong argument that the bank erosion was natural or that in some cases it was induced by the landowners themselves. The landowners lost the case because the judge found that the geomorphic arguments were convincing. However, in 1981 an additional ninety-two landowners decided to renew the litigation, showing the continued need for work by applied geomorphologists!

### A1.3 Geomorphic hazards

Over both long and short periods of time a major practical contribution of geomorphology lies in the identification of stable landforms, or sites that are free from the threat of geomorphic hazards (Scheidegger, 1975). A geomorphic hazard can be defined as any change, natural or man-induced, that may affect the geomorphic stability of a landform to the adversity of living things. This definition is very broad, and it includes those geomorphic changes that occur so slowly that they are a hazard only in the sense that over thousands of years their cumulative effect may threaten site stability. The identification of geomorphic hazards is in reality based on the prediction of landform change. The change may be of little or of great significance to man. For example, river changes can be considered hazardous if they threaten highways, bridges and other riparian structures. In addition, the disposal of toxic and radioactive wastes increasingly requires identification of landforms having very long-term stability. In this case consideration must be given to the potential for climate, baselevel and tectonic change. Schumm and Chorley (1983) attempted to consider geomorphic hazards in relation to the disposal of uranium mill tailings and dangerous wastes, and Shen and Schumm (1981) used the same approach for evaluation of the shorter-term effect of geomorphic hazards on bridges and highways.

Table 1.2. (see Chapter 1, p. 14) listed the variables that control landform morphology and stability, showing that the normal erosional evolution of a landscape through time (variable 1) will depend on tectonics (variable 2), geology (variable 3), climate (variable 4) and hydrology (variable 7). Information on modern rates of erosion and on the understanding of the manner in which the landscape evolves and responds to change should provide a means of predicting future change. However, as the timespan of concern increases such cannot be assumed to be the case. Relief will change as a result of tectonic uplift, depression, by faulting, or by baselevel change thereby changing the rate at which the erosion processes operate. Climate may change, influencing the erosion rate, but perhaps also changing the dominant process itself (e.g. surface runoff to mass movement; fluvial to glacial, etc). Of the variables listed in Table 1.2, the following appear to be most significant for landform stability:

(1) time – natural landform (evolution rate and processes);
(2) relief – change of relief as a result of tectonics (uplift or subsidence) and baselevel change (rise or fall of sea level, lake level, or abrupt change of stream profile);
(3) climate change – change of vegetation cover and type and hydrologic regimen.

In addition, changes of land use or deforestation by man, fire, or disease can also have significant influences on the fluvial system. The effect of these changes on a site can be expressed by changes of runoff (discharge), sediment yield and baselevel. Depending on the direction of change, the result can also be beneficial or detrimental to sites in all three zones of the fluvial system (Figure 1.2).

In Table A1.1 the effect of climate, tectonic and land-management changes upon the variables of runoff, sediment yield and baselevel are shown by plus, minus, or zero; thereby indicating an increase, decrease, or no change. The effect of climatic change on the water discharge from a drainage basin is based on the relations given in Figure 20.21, which shows that a decrease of temperature and/or an increase of precipitation will increase runoff. The changes of sediment yield are more complex and are based on a relationship between precipitation and sediment yield (Figure 20.22) that shows maximum sediment yield occurring in semi-arid climates. It is also accepted that glacial and periglacial activity increase sediment production as does an increase in the seasonality of precipitation. Baselevel will also be affected as more or less water is delivered to lakes and as large quantities of water are stored in continental glaciers thereby significantly lowering sea level. Tectonic activity that either raises or lowers a portion of the landscape will affect a site differently depending on the location of the change of relief (Table A1.1B). If downstream it is a baselevel change, but if it is upstream it will affect relief and sediment yield. Table A1.1C attempts to summarize a variety of man-induced changes on the three variables. All the changes are well known and their consequences can be predicted with reasonable confidence. Channelization is the straightening of sinuous channels with the result that gradients are steepened and the effect downstream is similar to baselevel lowering.

Table A1.2 presents a list of twenty-eight geomorphic hazards, and the effect of four variables on each hazard for each zone of the fluvial system. The hazards are grouped according to the landforms affected (drainage network, slopes, channels, plains) and the results of the hazard (erosion, deposition, pattern change, metamorphosis). Using Table A1.1, it is possible to determine if under a given set of circumstances, discharge, sediment load and baselevel change will be affected. If so, the effect of a change on the geomorphic hazards can be determined, and the hazards affected by a change of one of the variables are indicated by X. Time is included as a variable because landforms are altered naturally through time.

Table A1.2 also indicates, by X, the hazards that will affect surface, subgrade (i.e. shallow-burial) and deep-burial waste-disposal sites, during three periods of time. There is no attempt to quantify the relations, but if the extent of the change is assumed or known, then an

**Table A1.1**  Effect of climate, tectonic and land management changes on discharge, sediment load and baselevel at a site.

(a)  *Climate changes*

| Variables | Arid to semi-arid | Semi-arid to subhumid | Subhumid to humid | Non-glacial to glacial | Uniform to seasonal |
|---|---|---|---|---|---|
| Discharge | + | + | + | + | + |
| Sediment load | + | − | − | + | + |
| Baselevel |
| lake | + | + | + | + | 0 |
| ocean | 0 | 0 | 0 | − | 0 |

(b)  *Tectonic changes*

| Variables | Upstream uplift | Upstream downwarp | Downstream uplift | Downstream downwarp |
|---|---|---|---|---|
| Discharge | 0 | 0 | 0 | 0 |
| Sediment load | + | − | 0 | 0 |
| Baselevel | 0 | 0 | + | − |

(c)  *Land management changes*

| Variables | Increased land use | Dam construction upstream | Dam construction downstream | Channellization upstream | Channellization downstream |
|---|---|---|---|---|---|
| Discharge | + | − | 0 | + | 0 |
| Sediment load | + | − | 0 | + | 0 |
| Baselevel | 0 | 0 | + | 0 | − |

estimate of the magnitude of the hazard can be made. The bases for the tables are the simplest geomorphic-hydrologic relations, and the discussion that follows is based on a single change of one variable. Note that for 1000 and 10,000 years surface and subgrade sites are considered, whereas for 100,000 years only subgrade and deep-burial sites are considered. It is assumed that in most cases surface sites will be affected after 10,000 years.

Depending on the position of the surface site, the potential for failure occurs for virtually all hazards. However, if the site is not located in an alluvial valley or on a delta or alluvial fan, then all channel, piedmont and coastal plain hazards can be eliminated. Much, clearly, depends on the location of the site. For both 1000-year and 10,000-year time periods subgrade disposal will be affected by only eight of the twenty-eight hazards. After 10,000 years the possible effects of hillslope retreat become significant and this hazard must be considered for subgrade sites, and at 100,000 years all slope hazards

are potentially important. Deep-burial of wastes will not be affected by any geomorphic hazards for 100,000 years unless if by deep burial is meant the placement of material in the wall of a canyon. Depending on depths of burial, the hazards related to rejuvenation and channel incision may become significant. In the worst case major climatic change will affect all of the hazards, and rapid uplift or baselevel change will also cause rapid channel incision thereby rendering subgrade and perhaps deep-burial ineffective. For example, if river incision can keep pace with rapid uplift (1 m/1000 year), then channels could theoretically incise 100 m in 100,000 years and 1000 m in 1,000,000 years. Depending on the nature of bedrock, the 100-m incision seems probable, but the 1000-m incision will be significantly affected by variation of uplift rates and the complex response of the fluvial system. In summary, the integrity of any site depends on its location within the fluvial system. Sites can be selected that will be safe regardless of climate and baselevel change. Some surface sites can be safe for thousands of

**Table A1.2** Variables affecting geomorphic hazards and associated site risk

| Geomorphic hazards | Time | Discharge + | – | Sediment load + | – | Base-level up | down | 1000 years surface | sub-grade | 10,000 years surface | sub-grade | 100,000 years sub-grade | deep burial |
|---|---|---|---|---|---|---|---|---|---|---|---|---|---|
| **1 Drainage networks** | | | | | | | | | | | | | |
| **(a) Erosion** | | | | | | | | | | | | | |
| (i) rejuvenation | | X | | | X | | X | X | X | X | X | X | X |
| (ii) extension | | X | | | X | | X | X | X | X | X | X | |
| **(b) Deposition** | | | | | | | | | | | | | |
| valley filling | | | | X | | X | | X | | X | | | |
| **(c) Pattern change** | | | | | | | | | | | | | |
| capture | X | | | | X | X | X | X | X | X | X | X | X |
| **2 Slopes** | | | | | | | | | | | | | |
| **(a) Erosion** | | | | | | | | | | | | | |
| (i) denudation-retreat | X | X | | | X | | X | X | | X | | X | |
| (ii) dissection | | X | | | X | | X | X | | X | | X | |
| (iii) mass failure | X | X | | | X | | X | X | | X | | X | |
| **3 Channels** | | | | | | | | | | | | | |
| **(a) Erosion** | | | | | | | | | | | | | |
| (i) degradation (incision) | | X | | | X | | X | X | X | X | X | X | X |
| (ii) nickpoint formation and migration | X | X | | | X | | X | X | X | X | X | X | |
| (iii) bank erosion | X | X | | X | X | X | X | X | | X | | | |
| **(b) Deposition** | | | | | | | | | | | | | |
| (i) aggradation | | | X | X | | X | | X | | X | | | |
| (ii) back and downfilling | | | X | X | | X | | X | | X | | | |
| (iii) berming | | | | X | | | | X | | X | | | |
| **(c) Pattern change** | | | | | | | | | | | | | |
| (i) meander growth and shift | X | | | | | | | X | | X | | | |
| (ii) island and bar formation and shift | X | | | X | | | | X | | X | | | |
| (iii) cutoffs | X | X | | X | | X | X | X | | X | | | |
| (iv) avulsion | | X | | X | | X | | X | X | X | X | X | |
| **(d) Metamorphosis** | | | | | | | | | | | | | |
| (i) straight to meandering | | X | | X | | | X | X | | X | | | |
| (ii) straight to braided | | | X | X | | X | X | X | | X | | | |
| (iii) braided to meandering | | X | | | X | | X | X | | X | | | |
| (iv) braided to straight | | | X | | X | | X | X | | X | | | |
| (v) meandering to straight | | X | | X | | X | X | X | | X | | | |
| (vi) meandering to braided | | | X | X | | X | | X | | X | | | |
| **4 Piedmont and coastal plains** | | | | | | | | | | | | | |
| **(a) Erosion** | | | | | | | | | | | | | |
| dissection | | X | | | X | | X | X | X | X | X | X | X |
| **(b) Deposition** | | | | | | | | | | | | | |
| (i) aggradation | | | X | X | | X | | X | | X | | | |
| (ii) progradation | X | | | | | | X | X | | X | | | |
| **(c) Pattern change** | | | | | | | | | | | | | |
| (i) development of pattern | | X | | | X | | X | X | X | X | X | X | |
| (ii) avulsion | X | X | | X | | X | | X | | X | | | |

years. Carefully selected subgrade sites will be safe for tens of thousands of years, and deep-burial sites will not be affected by geomorphic hazards for millions of years. There are exceptions to all of the above, but a careful geomorphic assessment of the sites will greatly reduce the risk of failure.

## A1.4 Fluvial hazard evaluation

A great variety of landform changes can occur naturally through time or as a result of changed runoff, sediment loads, or baselevel change. A geomorphic evaluation of each site is required in order to determine which hazards are present and the probability that they may, in fact, affect site stability. Clearly, the geologic, climatic and geomorphic history of the area is of concern and must be considered for long-term site evaluation. Here we will discuss only short-term evaluation of site stability.

There are basically two ways to identify hazards and to assess their potential effect. The first is an *historical approach* that utilizes existing information to recognize change. The second is more specific and relies on the recognition of the hazard from existing *site-specific conditions*. Much more emphasis has been placed on the identification of river hazards and mass movement hazards than any others: indeed, examination of Table A1.2 reveals that, of the twenty-eight hazards listed, twenty-seven of them relate directly to river behaviour. Only mass failure (2a3) need not, except if precipitated by stream action. Therefore, the following discussion will emphasize river hazards and mass wasting.

In many areas there is historical information available that will permit identification of a potential hazard and estimation of the relative seriousness of that hazard. Aerial photographs, old maps and surveying notes, and bridge design files and river survey data as well as gauging-station records and interviews of long-time residents, can provide useful historical information. These resources can be used in the following ways:

(1) If a sequence of aerial photographs can be obtained that was taken during a period of several decades, and particularly if the photographs were taken during low water, a comparison of the photographs may show changes in channel morphology during that period. If maps or aerial photographs of good quality are available, it is frequently possible to measure the amount of bank and hillslope erosion that has occurred in the past. Recent mass movement features can be identified. Measurement from a known point (road junction, building etc.) can provide information on bank and hillslope erosion, and rapid hillslope erosion, change of channel width through

time also will provide a basis for calculating the rate of bank erosion, although it may not be possible to determine whether both banks are eroding or only one bank is eroding.

(2) Old maps and fieldnotes of surveys undertaken by the General Land Office (in the United States) and other agencies can provide information on river width and position and perhaps on the position of nickpoints at the time of the survey. Comparison of the old maps with more recent Survey topographic maps may provide information on the rate of channel change or the lack of it. Series of cross sections have been surveyed along some rivers by agencies such as the US. Army Corps of Engineers (e.g. Mississippi, Missouri and Ohio rivers) and the US Bureau of Land Management (e.g. Rio Grande). If repeat surveys have been made, they will provide a means of evaluating channel stability. If not, new surveys will provide a basis for comparison.

(3) The history of bridge failures or bridge problems may provide clues as to the nature of river change or the lack of it through time. At most bridge sites the height of the crown of the road above the channel is given on the plans of the bridge. A simple measurement from the road crown to the bed will provide a check on this distance and perhaps information on degradation or scour at that site.

(4) Gauging-station records for a given discharge will show if there is a change in the water surface elevation through time. For example, Borsuk and Chalov (1973) reported on the downcutting of the Lena River in the Soviet Union. They were able to establish that for a given discharge, there had been a decrease in the gauge height between 1946 and 1966 (Figure A1.1). This progressive lowering of the water surface at a given discharge is clear evidence that the Lena River was degrading during that period.

(5) Long-time residents of the area may remember channel changes that have occurred during their lives. Photographs from family collections and archives can frequently provide information on landform characteristics in the nineteenth century.

(6) When no historical information is available, it may be possible to determine rate of channel shift by a study of vegetation distribution on the flood plain (see Figure 2.16). Hickin (1974), Hickin and Nanson (1975) and Everitt (1968) have made use of dendrochronology to date the migration of the Beatton and Little Missouri rivers. Figure 2.16 shows the present channel of the Little Missouri river with isochrones showing the age of the valley floor. The downvalley shift of two meanders is clearly displayed. The smaller upstream

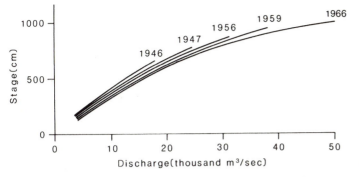

meander was cut off about twenty years before the study. Both meanders are confined by the bedrock walls of the valley.

(7) A final method of determining channel or slope change is to measure erosion by repeated surveys at the site of interest. This, of course, is an expensive and time-consuming operation, but if there is sufficient time it may be worthwhile to document site behaviour over such short timespans.

In summary, a sequence of maps and photographs and a study of vegetation distribution are of great value in determining landform change. These techniques can provide information on lateral and longitudinal change.

Bank erosion and nickpoint migration can also be identified in this manner. However, slight or slow changes due to degradation and aggradation can only be identified from data that reveal the altitude of the bed. For this purpose, cross sections and gauging-station data are needed. The same techniques can be used in some areas to study the slower processes of drainage network rejuvenation and extension, valley-filling and the potential for stream capture, as well as hillslope dissection and mass failure.

Table A1.3 provides a basis of the evaluation of slope and scarp stability from aerial photographs, and in the field where evidence of past or present slope movement is a primary factor to be used to indicate potential

**Table A1.3** Site evaluation of slope and scarp stability

| | *Observed feature* | *Indicated type and/or cause of failure* | *Severity of failure* | *Information needed for analysis* |
|---|---|---|---|---|
| 1 | Changes in historic levels, elevations of structure or monuments | Creep | Variable – slight if movement is slow | Movement rates and amounts |
| 2 | Increased required maintenance of roads, pipelines, buildings, etc. | Any type or cause; site-specific | Depends on magnitude of slide; if movement is rapid, imminent progressive failure is indicated | Cause of movement; location of shear surface |
| 3 | Tension cracks parallel to slope | Rotational or block failure; potential infinite slope type of sliding | Failure imminent; severity depends on magnitude | General slope conditions and subsurface profile |
| 4 | Tilted vertical features | Creep | Low if movement is slow | General material properties |
| 5 | Curved tree trunks | Creep | Same as 4 | Same as 4 |
| 6 | Misalignment of roads, fences, or other linear features | Indicates relative slope movement; nature of slide is site-specific | Depends on magnitude | Cause of movement; general slope conditions |

**Table A1.3** — *continued*

| | Observed feature | Indicated type and/or cause of failure | Severity of failure | Information needed for analysis |
|---|---|---|---|---|
| 7 | Open jointed rocks | Potential rockfalls due to water, ice, or weathering | Site-specific | Susceptibility of rock to weathering; joint patterns, spacing and inclination |
| 8 | Presence of escarpments | Rotational or block failure | Depends on magnitude of sliding mass and age of escarpment; old revegetated escarpment indicates present stability | General slope geometry and subsurface conditions |
| 9 | Scalloped topography (due to series of escarpments) | Rotational failure; probable toe erosion causing successive failures | Failure will progress headward; severity depends on height of slope | General subsurface conditions; toe erosion and drainage conditions |
| 10 | Abrupt cross-slope changes in vegetation | Old rotational failures; changes in drainage patterns | Generally low; failure can be induced by changes in environment of construction | Depends on potential changes or construction |
| 11 | Lobate, hummocky topography on slope | Creep failure | Low; increases with rate of movement | General material properties |
| 12 | Isolated longitudinal areas bare of vegetation | Avalanche, rockslide, mudflow | Low; failure of adjacent areas may be possible | General subsurface profiles and material properties |
| 13 | Single age, young vegetation on slope | Large past mass movement such as mudflow | Low if slide is currently stable; future movement may be induced by construction | Material types, magnitude and origin of flow |
| 14 | Steep slopes of fresh bare soil or sparse vegetation | Progressive failure of steep residual soils | In proportion to height of slope | General subsurface conditions |
| 15 | Fresh unweathered rocks on cliff faces | Spalling due to weathering | Generally low; depends on magnitude of potential failure | Same as 7 |
| 16 | Seepage on face | High phreatic surface | Low to high, depending on slope conditions | Phreatic surface location; soil profile and permeabilities |
| 17 | Swampy areas on slope or near toe | High phreatic surface and poor drainage | Low to high, depending on slope conditions | Phreatic surface location; soil profile and permeabilities |
| 18 | Abnormal termination of drainage features on surface | Piping erosion | May affect large area | Soil types; source of water |

*Source:* Nelson and Martin, 1981, pp. 129–32.

instability. The features listed in Table A1.3 vary in the information provided concerning slope stability. For example, some are direct evidence of instability (features 1–7), others indicate that failure may have occurred in the past (features 8–15) and others suggest future problems (features 16–18). Needless to say, if a river shifts against the base of a slope, this basal erosion will probably decrease the stability of the slope, as will removal of vegetation, irrigation above or on a slope, or loading of the slope by stockpiling of waste materials or by construction. Toe erosion may indicate future potential instability problems even in currently stable slopes, and is the single greatest factor causing slope failure (Thomson and Morgenstern 1977). Toe erosion at a particular site is relatively easy to identify by visual inspection of the site. The potential for changes in the river location to initiate erosion at a site must be considered. An example would be the initiation of erosion due to upstream meander cutoff. Although slowly progressing creep may or may not be indicative of an incipient sudden massive failure, creep may accelerate in the future to a point where relatively rapid movement or full-scale failure does occur.

Although the general evaluation of a site is summarized in Table A1.3, the characterization of a slope failure according to this table is very general. It must be recognized that each situation will be unique and must be evaluated individually. Also the general nature of the solution will vary from site to site depending on site-specific considerations and ease of implementation. Table A1.3 must be recognized to be only a summary and should not be implemented indiscriminately. The features noted therein indicate potential slope instability, and evidence for past slope movements. These features are only indicators that may forecast a probability of failure. The importance of each feature is site-specific depending on site conditions. An outline of a procedure for site evaluation follows:

(1) Collect and review the immediately available pertinent data; these data may include the following items:

(a) locational, topographic, geologic and soil maps and reports; also, special land use and river survey maps may be available for the site and the surrounding area;

(b) aerial photographs and satellite images;

(c) hydrologic data, especially river stage and flow records; data on sediment, precipitation, and wave and tide frequency and magnitude may be useful, although the wave and tidal data are needed only for certain locations;

(d) meteorological data, including rainfall, snowfall and wind data; temperature variations and solar data may also be useful;

(e) any specific reports which consider variables, hazards and land use, as related to the site.

A preliminary data analysis is essential for planning a site visit. Examine the aerial photographs and maps to determine the types of landforms that occur at and near the site.

The need for a general reconnaissance of the site is obvious, and it is essential for evaluation of future stability at the site. The reconnaissance of the reach can be performed by the use of aerial photographs and by aerial reconnaissance. Obviously, if hazards are recognized either upstream or downstream, these reaches of the river should be visited in order to verify the nature of the hazard.

(2) The nature of the landforms must be determined. The site of the construction of the proposed works should be chosen in a place where potential hazards are minimal. Once the landforms associated with the site have been identified, the potential for change during three time periods can be assessed by using Table A1.2. The possibility of climate change (runoff or sediment yield) or tectonics and baselevel change (Table A1.1) will provide an indication of those hazards that will be involved.

The information presented on Table A1.2 provides a basis for geomorphic evaluation of a site. The potential for occurrence of each of the hazards must be considered, and evidence for each geomorphic hazard must be sought. It is important to consider not only the geomorphology of the site, but also the character of the region in which the site is located. For example, it is possible that increased precipitation at the site might render it more stable because of increased vegetative cover. The same rainfall increase upstream could increase runoff, but decrease sediment loads to the extent that rivers near the site might begin to incise thereby affecting the stability of the site. Also a river reach many miles below the site might be incising. This incision would move upstream, eventually affecting site stability.

Even in depositional situations, such as alluvial fans, there is no guarantee that uninterrupted deposition will continue. The history of depositional landforms is one of repeated interruption of deposition by erosion just as erosion is frequently interrupted by deposition. The geomorphic threshold conditions at which these changes occur can be identified by comparative studies of stable and unstable landforms. Of course, although climatic and tectonic effects are of primary importance, the landforms at a site also reflect the geologic, climatic, hydrologic, and vegetational character of the area.

## A1.5 Case studies

Although some of the preceding discussion has considered long-term site stability, in most cases prediction for the normal life-expectancy of a structure (i.e. fifty years) is adequate. Two cases of bridge failure are presented, first , because although the failures were subsequently explained by geomorphic evaluation, they could have easily been predicted. The third case involves pattern change and sediment transport.

### Case study 1

Meander shift (hazard 3c1: Table A1.2) is one of the major problems at bridge crossings. Needless to say, this hazard should be one of the easiest to recognize if maps and aerial photographs for a period of years are available. Utilizing the procedure set forth earlier, the US Highway 177 crossing of the Cimarron River near Perkins, Oklahoma, is a good example of this hazard.

*Background*

In 1953 a new bridge was constructed downstream from an old bridge, which in 1949 was judged to be in poor condition with erosion concentrated on the south bank about 1500 ft above the bridge abutment. The following is a summary of site conditions before and after construction of the new bridge:

1950 – five pile diversion structures had been constructed upstream from the new bridge site (Figure A1.2) in order to stabilize the south bank.

1957 – continued erosion of south bank immediately upstream of south abutment occurred during a period

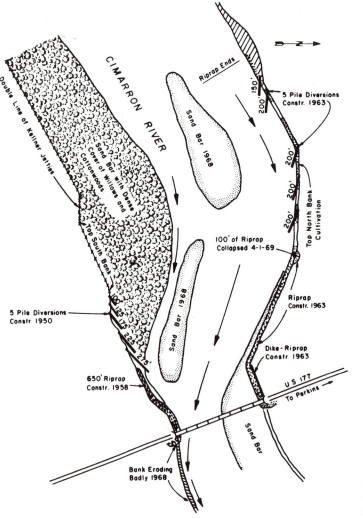

**Figure A1.2** Map of the Cimarron River at US 177 road crossing.
*Source:* Keeley, 1971, figure 27, p. 74.

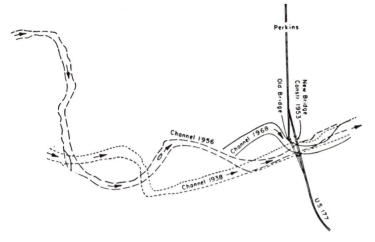

**Figure A1.3** Meander shift of the Cimarron River at Perkins, Oklahoma, as shown by aerial photographs taken in 1938, 1956 and 1968.
*Source:* Keeley, 1971, figure 28.

of large floods. Following the floods, 650 ft of riprap (i.e. coarse, protective rock debris) was emplaced on the south bank between the piles and bridge abutment.

1959 – the second-highest flood on record occurred in which all five pile diversions were damaged. Some bank erosion occurred on north-west bank 1500 ft upstream of north abutment.

1959–62 – this was a period of high discharge. The point of bank attack shifted from the south bank to north bank, with up to 325 ft of erosion on the north bank. Five pile diversion structures were constructed on the north bank, and riprap was extended upstream from the bridge.

1965 – further scour of north bank.

1969 – 100 ft of riprap were lost by scour.

This continuing problem of channel stability at this crossing could have been anticipated if the following evaluation of the stability of the channel had been made before or after construction.

*Procedure*

All available maps and photographs were obtained. In 1976 the *US Geological Survey Topographic Map Index* for Oklahoma shows that the only maps available for the area prior to bridge construction were the Perkins (1907) and Agra (1906) 15 min quadrangles. The Perkins $7\frac{1}{2}$ min quadrangle was published in 1978. In 1953 only 1938 photography was available. Subsequently aerial photographs were taken in 1956 and 1968.

The aerial photographs show that the channel was straight and braided at the site at the time of bridge construction, but there was a large meander about 1 mile upstream (Figure A1.3). Obvious hazards associated with this pattern are meander shift, bar and island shift, chute cutoffs, and a general instability of the thalweg

and channel in the wide alluvial valley.

Available hydrologic and climatologic data suggest a flashy type of flow regime that is likely to produce channel change. Analysis of the stage–discharge relationship for the Perkins gauging-station shows that the stage at 100 cfs decreased 4·0 ft between 1939 and 1959 (Figure A1.4), which is evidence of degradation. Between 1959 and 1970 the water level at a discharge of 100 cfs (2·8 m³/s) was constant, but it decreased between 1970 and 1975. The obvious conclusion is that degradation is a major hazard at the site.

However, in 1953 a cursory inspection of the channel upstream and a comparison with the 1939 aerial photographs would have revealed that the major hazard was meander shift. With this readily available information and maps and photographs, observations should have been made on the extent and location of bank erosion, channel-shifting, presence of gravel pits, and so on. In 1953 the floodplain appeared to be little affected by man, but inspection of the channel would have revealed that the large bend visible on the 1938 aerial photographs had moved downstream about one-half mile. The change of gauge height at a discharge of 100 cfs for the period of record, 1939–75 (Figure A1.4) probably reflects the shift of the meander towards the bridge. As the bend approached the bridge, the cross section changed from that of a relatively flat bed (i.e. a crossing) to a deeper channel characteristic of a bend.

Little study is required to conclude that the Cimmaron River is a relatively unstable channel at this site and that a major problem is meander shift. In 1968 the apex of the meander had almost reached the bridge, and the shift of the maximum erosion from the south to the north side of the bridge between 1959 and 1962 was the result of the movement of the bend into the area of the bridge. There

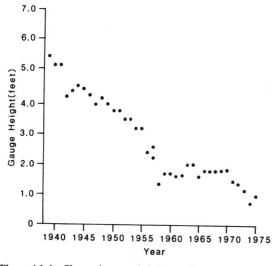

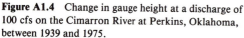

**Figure A1.4** Change in gauge height at a discharge of 100 cfs on the Cimarron River at Perkins, Oklahoma, between 1939 and 1975.

has been no damaging scour at the bridge piers in spite of the evidence of degradation supplied by the gauging-station record. Therefore, this is a simple example of meander shift with the degradation being related to the development of a thalweg at the site. Examination of the 1938 aerial photographs, together with a field survey of the channel, would have revealed the potential problem of meander shift, which posed the major hazard. However, it was the resulting shifting pattern of bank erosion and scour which attracted the most attention.

From the point of view of the engineer, the site selected in 1953 was a reasonable one, as the channel was straight and it was near bedrock on the south side of the channel.

Only if the upstream changes in the channel position had been recognized and the hazard identified could the engineer have anticipated the problems that later developed at this site. This emphasizes the need to consider and evaluate several miles of adjacent river in addition to the specific bridge site.

**Case study 2**

Channel incision (hazard 3a1: Table A1.2) and nickpoint migration (3a2) are obviously also major problems at bridge crossings. They are usually less easily identified than meander shift, but a clearly hazardous situation existed, or was permitted to develop, below the Interstate Highway 10 Bridge over the Salt River at Phoenix, Arizona (Figure A1.5). Man-induced changes had significantly altered the Salt River in the Phoenix reach, thereby causing changes of flow alignment, constriction of the channel and degradation (Arizona Department of Transportation, 1979), a major impact being extensive gravel-mining within the channel.

*Background*

The Interstate 10 bridge over the Salt River was constructed in 1962. The Salt River Bridge was designed to accommodate a fifty-year flood with a peak discharge of 175,000 cfs. Discharges had been relatively low or non-existent for a number of years, but a large flood (67,000 cfs) occurred in January 1966, and a 22,000 cfs flood in April 1973. Following these, the river was essentially dry until in March 1978 there was a 115,000 cfs flood, followed by a 120,000 cfs flood in December 1978, a 80,000 cfs flood in January 1979 and a 48,000 cfs flood in March 1979.

During the latter flood, scour undermined the footing

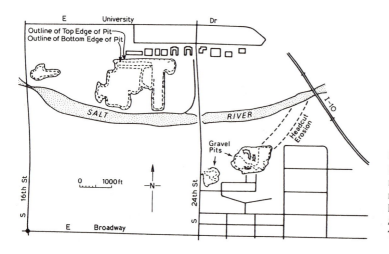

**Figure A1.5** Map showing the location of I-10 Bridge at Phoenix, Arizona, with reference to the Salt River which flows from right to left; gravel pits and the path of headcut erosion are shown.
*Source:* Arizona Department of Transportation, 1979.

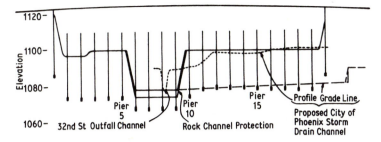

**Figure A1.6** Cross section at I-10 Bridge, Phoenix, Arizona, showing channel cross section and location of low-water channel and bridge piers and footings.
*Source:* Arizona Department of Transportation, 1979.

of the no. 11 pier (Figure A1.6), causing subsidence and tilting of one of the spans. The spread footing (i.e. pier foundation) was 20 ft below the channel in 1962. However, a low-water channel was dredged artificially to the north of this pier, between piers 5 and 10 (Figure A1.6), where the spread footings were 10 ft deeper than those of piers 11 to 19. When the bridge was designed, it was assumed that the thalweg would remain fixed in position above the deepest piers.

### Procedure

Available aerial photographs and maps of the Phoenix reach were obtained. The Phoenix and Tempe *US Geological Survey* topographic maps showed the river as it was in 1951, the 1:250,000 *US Army Map Service Phoenix* topographic map was available in 1954 and aerial photographs available for 1960 and 1961. Hydrologic data were available upstream and downstream but not in the reach of concern. Nevertheless, the upstream gauge provided information of the occurrence of major floods.

The river pattern is braided and the sediment in the channel ranges in size from sand to boulders. The river makes a gentle turn to the right at the site, and upstream left-bank erosion with bend shift could have been a problem (Figure A1.5). The USGS 1952 Phoenix topographic maps show extensive gravel-mining in the channel downstream from the bridge site. Hydrologic data reveal that the river only flows during floods. It is, therefore, highly susceptible to change and the details of the channel (bar and thalweg patterns) change rapidly during large floods. The character of the site cannot be reconstructed as it was in 1962, but the erodible bank and bed material, the flashy nature of the discharge, and extensive downstream gravel-mining all indicate a tendency for change, and especially degradation.

A 30-ft-deep gravel pit was dredged on the south side of the river about 2000 ft downstream and 750 ft south of the low-water channel (Figure A1.5). Floodwater flowed into the gravel pit and nickpoint erosion resulting from the lowering of baselevel caused development of a new channel, which was centred on pier 11 (Figure A1.6). Scour and undermining of pier 11 followed, with serious damage to that span of the bridge.

The nature of the channel pattern and its bed and bank materials clearly indicate that the river is relatively unstable. Because of the intensified constriction of the channel upstream and gravel-mining both upstream and downstream both due to urban expansion, the low-water channel changed position and degraded during floods.

These two case studies indicate the necessity for a geomorphic approach to practical problems that considers the potential for change at a site that is based on observations made *elsewhere*, i.e. both upstream and downstream from the bridge crossings.

### Case study 3

The identification of threshold conditions of relative geomorphic stability and instability is of great practical value to civil and agricultural engineers. For example, if a channel is very unstable, the appropriate question is: how near to a condition of relative stability is the channel? The answer is not only of academic and practical interest, but considering the huge sums spent on erosion control and channel stabilization structures, it is of great concern to administrators.

For example, if a river lies near a pattern threshold (see Figures 12.27 and 12.29), then it may be shifted from one pattern to another (e.g. meandering to braided and *vice versa*). The most commonly desirable condition is the meandering one because a meandering stream provides good fish habitats and its lower gradient is associated with less bank erosion. Experimental work shows that for a given discharge, the greater the stream power, the more likely is the stream to be braided (see Figure 12.27), and it shows also that there is a gradient or discharge threshold above which rivers are braided (Figure A1.7). Therefore, if one can identify the natural range of patterns along a river, then within that range the most appropriate channel pattern and sinuosity can probably be identified (see Chapter 12). If so, the engineer can work with the river to produce its most efficient or most stable channel. Obviously, a river can be forced into a straight configuration or it can be made more sinuous, but

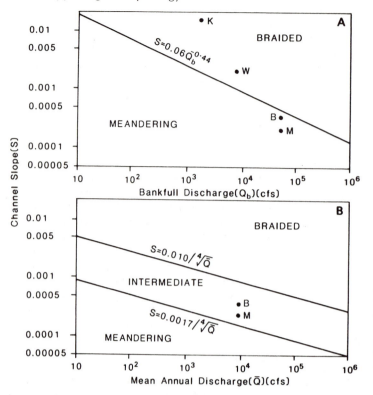

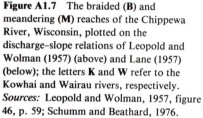

**Figure A1.7** The braided (**B**) and meandering (**M**) reaches of the Chippewa River, Wisconsin, plotted on the discharge–slope relations of Leopold and Wolman (1957) (above) and Lane (1957) (below); the letters **K** and **W** refer to the Kowhai and Wairau rivers, respectively. *Sources:* Leopold and Wolman, 1957, figure 46, p. 59; Schumm and Beathard, 1976.

there is a limit to the changes that can be induced beyond which the channel cannot function without a radical morphologic adjustment, as suggested by Figure 12.27 and the northern Mississippi channels (see below). Identification of rivers that are near the pattern threshold is useful, for example, because a braided river near the threshold might be converted to a more stable single-thalweg stream. On the other hand, a meandering stream near the threshold should be identified in order that steps can be taken to prevent a tendency to braiding, perhaps due to changes in land use.

An example of the way that this could be done is provided by the Chippewa River of Wisconsin (Schumm and Beathard, 1976), a major tributary to the Mississippi River. The Chippewa River rises in northern Wisconsin and flows 320 km to the Mississippi River, entering it 120 km below St Paul. It is the second largest river in Wisconsin, with a drainage basin area of 24,600 km².

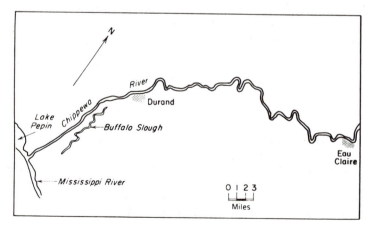

**Figure A1.8** Map of the lower Chippewa River between Eau Claire, Wisconsin, and the Mississippi River. *Source:* S. A. Schumm, *The Fluvial System*, 1977, figure 5-26, p. 148, copyright © John Wiley and Sons, by permission.

**Table A1.4** Chippewa River morphology

| Location | Channel pattern | Channel width (m) | Sinuosity | Valley slope tan (ft/mile) | Channel slope tan (ft/mile) |
|---|---|---|---|---|---|
| Below Durand | Braided | 333 | 1·06 | 0·00035 (1·83) | 0·00033 (1·76) |
| Above Durand | Meandering | 194 | 1·49 | 0·0004 (2·14) | 0·00028 (1·46) |
| Buffalo Slough | Meandering | 212 | 1·28 | 0·00035 (1·83) | 0·00027 (1·43) |

The Chippewa River is braided from its confluence with the Mississippi to the town of Durand 26·5 km up the valley (Figure A1.8 and Table A1.4). The main channel is characteristically broad and shallow, and it contains shifting sand bars. The bankfull width of this reach is 333 m and the sinuosity is only 1·06. However, in the 68 km reach from Durand to Eau Claire, the Chippewa River has a meandering configuration with a bankfull width of 194 m and a sinuosity of 1·49. The valley slope and channel gradient are different for each reach of the river, with the braided section having a gentler valley slope than the meandering reach upstream: 0·00035 as opposed to 0·00040. This is contrary to what might be expected from Figure A1.7. However, the situation is reversed for channel slope in that the braided reach has a channel gradient of 0·00033, whereas the meandering reach had a gradient of 0·00028.

There is no evidence to suggest that the Chippewa River is either progressively degrading or aggrading its channel at present. In fact, the river below Durand has remained braided during historic time. It has maintained its channel position and its pattern, but a significant narrowing as the result of the attachment of islands and the filling of chute channels has occurred downstream of Durand, which resulted in a recent decrease in channel width of over 40 per cent. The only other change in the lower river is a noticeable growth of its delta into the Mississippi Valley. The delta deposits show increased vegetational cover, as well as progradation into the Mississippi River valley.

The relations described by Leopold and Wolman (1957) and Lane (1957) provide a means of evaluating the relative stability of the modern channel patterns of the Chippewa River. The bankfull discharge was plotted against channel slope on Figure A1.7A for both the braided and the meandering reaches of the Chippewa. The value used for the bankfull discharge was 53,082 cfs, which is the flood discharge having a return period of 2·33 years. The braided reach plots higher than the meandering reach, but both are well within the meandering zone, as defined by Leopold and Wolman (1957). This suggests that the braided reach is anomalous in that, according to this relation, the lower Chippewa would be expected to display a meandering pattern rather than a braided one. Even when the twenty-five year flood of 98,416 cfs was used, the braided reach still lay within the meandering region of Figure A1.7A. When the Chippewa data were plotted on Lane's (1957) graph (Figure A1.7B), the same relation was found to exist. The Chippewa fell in the intermediate region, but within the range of scatter about the regression line for meandering streams. Again the braided reach was seen to be anomalous because it should plot much closer to, or above, the braided stream regression line. The position of the braided reach as plotted on both figures indicates that, theoretically, this reach should be meandering.

An explanation for this apparent anomaly can be based on the geomorphic history of Chippewa River. There is an as yet unmentioned significant morphologic feature on the Chippewa River floodplain. This is Buffalo Slough, which occupies the south-eastern edge of the floodplain (Figure A1.8). It is a sinuous remnant of the Chippewa River that was abandoned, and is evidence of a major channel change in the Chippewa River valley (Table A1.4). Flow through Buffalo Slough has decreased during historic time and was completely eliminated in 1876, when the upstream end of Buffalo Slough was permanently blocked. The abandonment of the former Buffalo Slough channel by the Chippewa River is the result of an avulsion, but one that took many years to complete. The channel shifted from Buffalo Slough to a straighter, steeper course along the north-western edge of the floodplain. This more efficient route gradually captured more and more of the total discharge. The new Chippewa channel produced a divided flow or braided configuration due to a higher flow velocity and the resulting bank erosion. As the new channel grew the old course deteriorated, and eventually its discharge was so reduced that the Mississippi was able to dam the old channel mouth effectively with natural levees and plug its outlet.

The sinuosity of Buffalo Slough is approximately 1·28. Although sinuosity is defined as the ratio of stream length to valley length, it is also expressed by the ratio of valley slope to channel slope. A sinuosity of 1·28 is, therefore, the ratio of the present valley slope (i.e. 0·00035) to a channel slope of about 0·00027. This calculated channel slope value is very similar to that of the meandering reach upstream of Durand, 0·00028, therefore, the original meandering pattern of the Buffalo Slough channel was appropriate. Although delta construction was responsible for the lower valley slope of the lower Chippewa River valley, the Buffalo Slough channel had a gradient that was not appreciably different from that in the upstream reach. Channel sinuosity decreased from 1·49 to 1·28 in a downstream direction thereby maintaining a channel gradient of about 0·00027. This sinuous channel could not have trans-

ported the large amounts of sediment that the present braided channel carries to the Mississippi River, or it too would have followed a straight, braided course. Therefore, the present sediment load carried by the Chippewa River is greater than that conveyed by the Buffalo Slough channel, but this is due almost entirely to the formation of the new straight, steep, braided channel, which is 121 m wider than the old sinuous Buffalo Slough channel. It appears that the lower Chippewa has not been able to adjust as yet to its new position and steeper gradient, and the resulting bed and bank erosion has supplied large amounts of sediment to the Mississippi.

The normal configuration of the lower Chippewa is sinuous, and if it could be induced to resume such a pattern, the high sediment delivery from the channel widening and bank erosion could be controlled. The upstream sediment production must also be controlled, especially where the upper Chippewa River is cutting into Pleistocene outwash terraces. If the contribution of sediment from these sources were reduced, the lower Chippewa could be stabilized by resuming its original sinuous course.

## References

Arizona Department of Transportation (1979) 'Salt River bed study report', unpublished report.

Biswas, M. R. and Biswas, A. K. (eds) (1980) *Desertification*, Oxford, Pergamon.

Borsuk, O. A. and Chalov, R. S. (1973) 'Downcutting of the Lena River channel', *Soviet Hydrology*, no. 5, 461–4.

Brown, E. H. (1970) 'Man shapes the earth', *Geographical Journal*, vol. 136, 74–85.

Coates, D. R. (ed.) (1972) *Environmental Geomorphology and Landscape Conservation*, Stroudsburg, Pa, Dowden, Hutchinson & Ross, Vol. 1.

Coates, D. R. (ed.) (1973) *Environmental Geomorphology and Landscape Conservation*, Stroudsburg, Pa, Dowden, Hutchinson & Ross, Vol. 3.

Coates, D. R.(ed.) (1976) 'Urban geomorphology', *Geological Society of America Special Paper* 174.

Cooke, R. U. and Doornkamp, J. C. (1974) *Geomorphology in Environmental Management*, Oxford University Press.

Cooke, R. U., Doornkamp, J. C., Brunsden, D. and Jones, D. K. C. (1983) *Urban Geomorphology in Drylands*, Oxford University Press.

Craig, R. G. and Craft, J. L. (eds) (1982) *Applied Geomorphology*, London, Allen & Unwin.

Dent, D. and Young, A. (1981) *Soil Survey and Land Evaluation*, London, Allen & Unwin.

Everitt, B. L. (1968) 'Use of cottonwood in the investigation of the recent history of a flood plain', *American Journal of Science*, vol. 206, 417–39.

Fels, E. (1965) 'Nochmals: Antropogen Geomorphologie', *Petermanns Geographische Mitteilungen*, vol. 109, 9–15.

Goudie, A. (1981) *The Human Impact: Man's role in environmental change*, Oxford, Blackwell.

Gregory, K. J. and Walling, D. E. (eds) (1979) *Man and Environmental Processes*, Folkestone, Dawson.

Hails, J. R. (ed.) (1977) *Applied Geomorphology*, Amsterdam, Elsevier.

Hickin, E. J. (1974) 'Development of meanders in natural river channels', *American Journal of Science*, vol. 274, 414–42.

Hickin, E. J. and Nanson, G. C. (1975) 'The character of channel migration on the Beatton River, north-east British Columbia', *Bulletin of the Geological Society of America*, vol. 86, 487–94.

Jennings, J. N. (1966) 'Man as a geological agent', *Australian Journal of Science*, vol. 28, 150–6.

Jones, D. K. C. (1980) 'British applied geomorphology: an appraisal', *Zeitschrift für Geomorphologie,* Supplement 36. 48–73.

Keeley, J. W. (1971) *Bank Protection and River Control in Oklahoma*, Report, Federal Highway Administration, Oklahoma Division.

Lane, E. W. (1957) 'A study of the shape of channels formed by natural streams flowing in erodible material', *US Army Corps of Engineers Missouri River Division, Sediment Series* 9.

Leopold, L. B. and Wolman, M. G. (1957) 'River channel patterns: braided, meandering and straight, *US Geological Survey Professional Paper* 282-B, 39–85.

Mitchell, C. (1973) *Terrain Evaluation*, London, Longman.

Nelson, J. D. and Martin, J. P. (1981) *Slope Stability as Related to Stream Hazards to Bridges*, Federal Highway Administration Report FHWA/RD-80/160, appendix A, 87–143.

Scheidegger, A. E. (1975) *Physical Aspects of Natural Catastrophes*, Amsterdam, Elsevier.

Schumm, S. A. (1977) *The Fluvial System*, New York, Wiley.

Schumm, S. A. and Beathard, R. M. (1976) 'Geomorphic thresholds: an approach to river management', in *Rivers 76, 3rd Symposium of the Waterways, Harbors and Coastal Engineers Division of the American Society of Civil Engineers*, vol. 1, 707–24.

Schumm, S. A. and Chorley, R. J. (1983) *Geomorphic Controls on the Management of Nuclear Waste*, US Nuclear Regulatory Commission Report NUREG/CR-3276.

Shen, H. W. and Schumm, S. A. (1981) *Methods for Assessment of Stream-Related Hazards to Highways and Bridges*, Federal Highway Administration Report FHWA/RD-801/160, 1–86.

Thomson, S. and Morgenstern, N.R. (1977) 'Factors affecting distribution of landslides along rivers in southern Alberta', *Canadian Geotechnical Journal*, vol. 14, 508–23.

Verstappen, H. Th. (1982) *Applied Geomorphology*, Amsterdam, Elsevier.

# Plates

**Plate 30** Coronation Fjord, Baffin Island, produced by selective linear erosion by a glacier. (See Chapter 19.2.2.)

**Plate 31** Rebound raised beaches at the north-west end of the Black Isle, Cromarty Firth, north-east Scotland. View looking north-east with Cromarty Bay in the foreground. (See Figure 19.14 and Chapter 19.4.) (Cambridge University Collection: copyright reserved.)

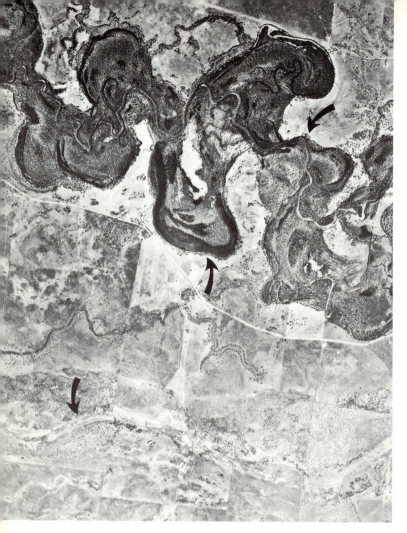

Plate 32 Vertical aerial photograph of the Murrumbidgee River, New South Wales, Australia (upper arrow). The middle arrow shows an oxbow lake (Yarrada Lagoon) formed from larger ancestral meanders which developed during an earlier, wetter climate. The lower arrow shows the older, straighter paleochannel formed during a dryer phase. (See Figure 20.30.) (Courtesy New South Wales Department of Lands.)

Plate 33 Sorted fossil gelifluction lobes in the Cairngorm Mountains, Scotland. (See Chapter 20.2.2.)

# Index

# Index

ablation, 434
abrasion, by ice, 446; by wind, 413, 415
accumulation, of ice, 434
acid, acetic, 219–20; humic, 219; organic, 204; oxalic, 220
acid, bacterial, 220; citric, 220; fulvic, 219
acid lavas, 137
active ice-retreat landscapes, 518–19
adjustment of landforms, to climatic change, 534–5
aeolian bedforms, wavelengths of, 419
aeolian environments, 410
albite, weathering of, 212
alcove, 192
alluvial fan, dry, 341–7; morphometry, 341, 345–7; structure, 342–4, 346; wet, 347–8
alpine glacial erosion, 511–13
alpine structures; 168
amphibolite, 96
anabranching channels, 295
anastomosing channels, 295
anatexis, 123, 126
andesite line, 136
angle of internal friction, peak, 250–1, 271–2; residual, 250–1, 271–2; semi-frictional, 272
angle of internal sliding friction, 249–50
angle of repose, 235, 238, 250
anomalies, gravity, 105; magnetic, 109
Antarctic, present conditions, 509
antecedence, 21
arête, 511–12
arid morphogenetic region, 469
arkose, 92
asthenosphere, 111
atoll, 404–6; theories of formation, 406
Atterberg limits, 248, 251
attrition of river debris, 287
avulsion, 302, 347, 366–8, 574
azonal mountain landforms, 471

backshore, 371–2
backwash, 379, 382
bacteria, activity in soil, 182
badlands, 487, 489
bajada, 486–7
bar shift, 302
barrier islands (offshore bars), 387–9
barrier reef, 404–5
basal creep, of ice, 438
basal temperature, of ice, 435–6
basalt, alkali, 124
basaltic plateau, 139–40
baseflow, 327
baselevel changes, 20
baselevel lowering, effect on stream networks, 334–5
basin morphometry, 317–24
basin unit, 316
batholith, 129
bauxite, 480
beach cusp, 382
beach face, 371–2
beach profile, changes through time, 38; cycle of change, 38, 379, 381
beach sediment movement, 376–83
beach slope, 371–2, 379–80
bedding, characteristics, 90; cross, 90–1; graded, 90; mechanisms, 90
bedload, of rivers, 51–2; measurement, 52
bedload transport in rivers, 286–9
Benioff zone, 105
berm, 371–2
Bernoulli equation, 280–1
bifurcation ratio, 318, 329
biochemical weathering, in deserts, 486
biomes, 203
block fields, 498
block glide, 235–6
block streams, 563
bornhardts, 492

braided channels, 295; deposits, 349
braiding, conditions for, 585–8
breaker zone, 371–2; debris transport in, 381–2
breakers, types of, 375
bumps, iceflow over, 445, 447; under ice, 459
butte, 150
bysmalith, 130–1

Calabrian shoreline, 25–6
calc-alkaline rocks, 124
calcrete (caliche), 482
caldera, 144–6
cambering, 236, 238
capacity, river, 290
carbonate rocks, 93–4
cascade, hillslope hydrological, 44; solar energy, 44
catastrophe theory, 440–1
cation exchange, 210
cauldron subsidence, 130, 135
chalk, 188
channel classification, 303
channel incision, practical example, 584–5
channel patterns, 295, 297–8, 303–5
channel roughness, 279–81, 283, 292–3
channel shifting, rate of, 351, 353
channel stability, 302–3
chemical weathering, products of, 214, 218
Chézy equation, 280–1, 412
chute, 351
chute cutoff, 302
circularity ratio, 319
cirque, 511–13, 515
clay, structure, 207; swelling, 207
clay minerals, 83–4; production by weathering, 210, 212–14; properties of, 209
clay slopes, engineering properties, 250–2
cliff, critical height, 274

cliff evolution, 32–3
climate, effect on baselevel, 575–6; effect on discharge, 575–6; effect on sediment load, 575–6
climatic accidents, 20
climatic change, geomorphic response to, 552–3; historical, 540–1, 543–7; Holocene, 540; Pleistocene, 536, 538–40; problems, 534; Tertiary, 536
climatic optimum, 540
coastal classification, by D. W. Johnson, 403; by F. P. Shepard, 403; by H. Valentin, 403; by J. C. Davies, 403; by V. R. Dolan, 403–4
coastal erosion, 388–9
coastal management, 407–8
coastal sediment budgets, 383
coastal types, classification of British, 397; classification of United States, 403
coefficient of sliding friction, 249
cohesion, 250
cold-based ice sheet, 514
Colorado Plateau, 152–5
competence, 169; river, 287–8
complex response, 11, 39–40
compressive flow, 437–8, 441
cone sheet, 130
conglomerate, 91–2
continental drift, 112
continental platform, 102
continental rise, 101
continental shelf, 101, 401
continental slope, 100
continuity equation, 280
convection cell, 116
coral reef, ecology, 404; structure, 404
coral reefs, geomorphology of, 405–6
correlation structures, rivers, 290–1
correlation structures, stream-slope, 338
coversands, 563, 565
craton, 118
creep, rock, 239; soil, 238–40

crevasse-splay, 352–3
crevasses, in glaciers, 441–2
critical tractive force, 287
crust, elastic behaviour, 529; hydrostatic behaviour, 529
crustal movement rates, Eastern Europe, 69
crustal spreading, 114–15
crustal strength, 68
crustal structure, 110–11
cryptovolcanic features, 148
crystal growth stresses, 206
cuesta, relief of, 159
cumulo lava domes, 140–1
cuspate foreland, 385–6
cycle, biological, 203; calcium, 204
cycle of erosion, 4, 17–18, 27, maturity, 19–20; modified, 39; old age, 20; youth, 19
cycling, nutrient, 204–5

Darcy's law, 48–9
Darcy-Weisbach coefficient, 281
dating, by isotopes, 537
debris flow, 236, 241; rainfall induced, 243–5
debris in glaciers, 442
debris slides, 241
debris topple, 235–7
decomposition, geometry of, 221
deglaciation, 527; icecap, 528
delta, bird's foot, 367–8; high-constructive, 362–3; high-destructive, 362–3; lobate, 363; structure, 362–3, 367; submarine slope of, 368
delta development, Mississippi delta, 366–8
delta formation, mechanisms of, 361–2
denudation chronology, 27; Appalachians, 22–5; South-East England, 25–6
denudation rates, climatic averages, 61; continental averages, 60, 63; drainage-basin averages, 60

depositional landforms, of ice, 452–7
desert pavement, 415
desertification, 425
diagenesis, mechanisms of, 91
diapir, 123, 125–6
diastrophism, 98
diatremes, 148
dip, 169
discharge, bankfull, 289, 293; glacial meltwater, 459; mean annual, 285
discharge records, historical, 543, 545
discharges, 278, 289, 293
dissolved load, in tropical rivers, 472–3; of rivers, 51–3, 55, 59
doline, 187
domed strata, 152
double surface mechanism, of Büdel, 491
drag velocity, 412
drainage basin landforms, hierarchy of, 10
drainage basin variables, status through timespans, 14
drainage density, 257, 264, 319–21, 325; controls over, 467–8; effect on hydrology, 325–7; effects of climatic change, 555–6; influence of precipitation, 476–7, 483
drainage network, 317–18, 325; development, 37, 328–35; effects of climatic change, 555–6
drainage patterns, 318
drift plain, 517
drumlin, 453–5, 518–19, 522; rock, 517
dry continental morphogenetic region, 471
DuBoys equation, 412
dune, aklé, 416, 418–19; barchan, 416, 418–19, 421, 424–5; barchanoid, 416, 418–19, 420–1, 424; blowout, 416, 423, 425–6; complex, 422, 425; compound, 422, 424; dome, 416, 418, 421; draa, 415, 419, 421–3; longitudinal, 416, 418, 421; parabolic, 416, 423–6; reversing, 417–18, 422; seif, 416, 421–3; star, 417–18, 422, 424; thourd, 422–3; transverse, 416, 418–21, 423, 425; vegetated, 426
dune morphometry, 419
dunes, active, world distribution, 411; classification of, 426; classification of coastal, 425–6; coastal, 425–6; periglacial,

563; sand, 415–26; snow, 429; types of, 415–18
duricrusts, 480–2
dyke (dike), 130, 132
dyke swarms, 135

earth heatflow, 107–8
earth magnetism, 107
earthfall, 235
earthflow, 236, 241
earthquake waves, 108, 110
earthquakes, 104, 106; cause of mass movements, 238, 240;
    effects on erosion, 473–5
eclogite, 96
effective diameter, 283
elongation ratio, 319
endogenetic processes, 43
energy, kinetic of streamflow, 281
entrainment of debris in glaciers, 443–4
entropy, 4, 18–19
environments of stream particle trapping, 286–8
epeirogenic movements, 98
episodic erosion, 39
equifinality, 273
equilibrium, 9–10, 17; types of, 11, 38
equilibrium line, of glacier, 434, 441–2, 526
equilibrium, dynamic, 40; dynamic metastable, 40;
    morphoclimatic, 468
erg, 410
ergodic approach, to drainage network development, 328–30;
    to slope development, 330
ergodic method, 32–3
ergodic reasoning, and slope development, 269–70
erosion, by subglacial erosion, 459–61
erosion and isostatic adjustment, 71–2
eskers, 460–2
eugeosyncline, 118–19
eustasy, glacial, 530–2

eustatic changes, 98; mechanisms, 398; Pleistocene, 398–400;
    Tertiary, 398
evolution, 17
exfoliation, 275
exogenetic processes, 43
extending flow, 437–8, 441

facies, metamorphic 96–7; sedimentary, 94
fan, alluvial, 341–8
fan, deep sea, 368–9
fault, 168; normal, 169–70; strike-slip (wrench), 169–70; tear,
    162; thrust (reverse), 169; transcurrent, 172–3
fault scarp, erosion of, 269–70
faulting, posthumous, 172
fault-line scarp, resequent, 173
faults, physiographic expression, 174–5
feedback, 9–10
feedback effects, of shorelines, 384
felsenmeer, 564
fenster, 165
fjord, 399, 401, 515, 517
flood, mean annual, 289, 293; probable maximum, 553–4
floodplain, age, 33, 35; formation, 351, 353; meandering
    deposits of, 350–2; size, 311
flow, homopycnal, 361; hyperpycnal, 361; hypopycnal, 361
flutes, glacial, 453–4, 522
fluvial system, 6–7; variables, 5–6
flysch sedimentation, 118
fold, allochthonous, 165; autochthonous, 165; drainage
    development on, 166–7; plunging, 161–2; recumbent, 165;
    simple, 160–1
folded mountains, 102
folding, Appalachians, 163; complex, 165, 168; disharmonic,
    162; Juta, 164; parallel, 162; similar, 162
fractionation, 123
free face, of slope, 256
freezing-on, 515

frequency, of occurrence, 543–4, 548
frequency distribution, of events, 543–4, 547
fringing reef, 404–5
Froude number, 281–2

gelifluction, 241, 498, 563–4, 566
geomorphic system, 5
geomorphology, conceptual bases, 4; definition, 3; functional
    approach, 3; historical approach, 3
glacial deposition, 448–52, 518–22
glacial erosion, 445–8; in mid-latitudes, 516–17
glacial grooves, 445, 447
glacial morphogenetic region, 469, 495
glacial striations, 522–3
glacial traction, Boulton's model, 444; Hallet's model, 444–5
glacial trough, 511
glaciation, British Isles, 520; Finland, 522; Labrador-Ungava,
    523; pre-Pleistocene, 510
glacier, calving, 521, 524; cold-based, 525; valley, 431–2;
    warm-based, 525
glacier flow, of internal deformation, 437–8
glacier fluctuations, topographic effects, 515–16
glacier retreat, alpine, 525–6; icecap, 525; land-based, 525;
    marine-based, 525
glacier velocity, 440
glaciers, 440; as landforms, 503–5; distribution, 502
Glen's law, 437
gneiss, 95, 196
gneissic terrain, 196
grade, metamorphic, 95
graded condition, 19
granite gneiss, weathering of, 213
granite pluton, erosion of, 193–4
granulite, 95
gravitational sliding, 115
gravitational slumping, till, 450
gravity tectonics, 168, 232–3

Gumbel plot, 544, 547–8

Haldenhang, 29–30
hanging valley, 511–12
hardware models, delta growth, 364–6; drainage network
    evolution, 330–4
harmada, 486, 493–4
hazard, definition, 574–5
hazards, historical approach, 578; site and specific condition,
    578; timespans, 576–7; variables affecting, 575–8
heat-pump effect, of ice movement, 439–443
hillslope, classification, 256; components of, 257; frequency
    distribution, 256, 258
hillslope denudation, recurrence intervals, 59
hillslope hydrology, 48, 259, 265; dynamic partial-area model,
    48; Horton model, 48
historical information, use in hazard evaluation, 578–9
homoclinal structure, drainage development, 158
horizontal strata, 150, 152
horn, 511–12
humid mid-latitude morphogenetic region, 471
humid tropical landforms, 472–7
humid tropical morphogenetic region, 469
hummock, periglacial, 497
hydration, 210
hydraulic geometry, effects of climatic change, 555;
    equations, 292; humid tropics, 477, 479
hydraulic radius, 279, 281
hydraulic variables, 291
hydrograph, storm, 47
hydrology, 289–90; influence of climatic change, 553
hydrolysis, 210
hypsometric curve, of the earth, 99
hypsometric integral, 319, 323; basin, 330–1

ice, cold, 435; warm, 435
ice-margin landscapes, 518–19
ice retreat, 522, 525, 527–8
ice sheet, 431–2; Antarctic, 433
ice-sheet glaciation, 513–17
ice shelf, 423, 431, 522, 525
ice wedge, 561–2
igneous movements, 98
igneous rocks, classification, 81–2; history of formation, 81; texture, 81
ignimbrite, 142
infiltration, effect of surface cover, 47
infiltration capacity, 467
infiltration rate, 260
inselberg, 486, 491–3; domed, 492–3
interception, effect of vegetation, 47
interior of the earth, composition, 105
inversion of relief, 192
involutions, periglacial, 562–3
island arc, 102
isostasy, glacial, 529–30
isostatic equilibrium, 111–12
isostatic movements, 98; fold mountains, 68
isostatic rebound, 68, 103–4, 529

joints, 168; effect on cirques, 513; in folds, 170; influence on drainage patterns, 180
jokulhlaup, 459

kame terrace, 518
karst, cone, 189–90; cycle, 191; hydrology, 184, 186; ideal conditions, 181; rate of denudation, 188–9; tower, 189–90; tropical, 189
kettle hole, 518
King model, 29–31

klippe, 165
Knickpunkte, 28

labile landforms, 567
laccolith, 130–1, 133–4, 194
lake, ice-marginal, 522; oriented, 497
Lake Bonneville, 103–4
lake levels, during Pleistocene, 538–40, 542
land management, effect on baselevel, 575–6; effect on discharge, 575–6; effect on sediment load, 575–6
landscape, recovery, 8–9; sensitivity, 8
landscapes, areal scouring, 514–17; selective linear erosion, 514–16
landscapes eroded by ice sheets, classification of, 515
landslide terrain properties, 233–4
landslides, 241; aspect of, 246; rapid, 242–3; significant historical, 230–1
lapiés (karren, grikes), 183
lateral spreading, 236
laterite, 480–2
law of adjusted cross sections, 511
length of overland flow, 318
levée, 352
limestone caves, 185–6, 191
liquidity index, 248
lithic arenites, 92–3
lithosphere, 111
litigation, resulting from river action, 574
Little Ice Age, 540, 543, 545
load-discharge changes of river morphology, 306–9
lodgement, deposition by, 449–50
loess, 427–8, 563–6; China, 428; United States, 428; world distribution, 411
long profiles of rivers, 289, 296
longitudinal profile, of glacial valley, 511
longshore current, 375–6
longshore movement of debris, 379, 381–3

longshore trough, 371–2
lopolith, 130

magnetic reversals, 107, 109
magnitude and frequency, of events, 8, 57, 543–4, 547–8
man, as geomorphic agent, 573
mangrove swamps, 407
Manning equation, 279–81
mantle-bedrock contract, geometry of, 226–7
marble, 95
marginal sea basins, 102
marine cliffs, 390–5; clay, 394; climatic controls of, 391; geological classification of, 390–1; hog's back, 392; rate of retreat, 392–4, 395
marine erosion, maximum depth of, 396
mass movements, causes of, 247; classification of, 234; location of, 244–7
mean annual flood, 327
meander, incised, 312–13; ingrown, 312–13; intrenched, 312
meander development, 295, 297–302
meander shift, 302; practical example, 582–4
meander wavelength, 293
meandering, changes resulting from, 33, 35; conditions for, 585–8
measurement, 33–5
Medieval Warm Phase, 540, 543, 545
mélange, 119
meltwater, glacial, 457–60, 521
mesa, 150
mesosphere, 111
metamorphic processes, 94
metamorphic rock terrains, 195
metamorphic rocks, 96
metamorphism, contact, 194; regional, 94–5
meteorite impacts, 148
mineral solubility, 210–11
minerals, composition and structure, 79–80
miogeosyncline, 118–19

model, ice sheet erosion, 514–15; mathematical, 36–7
models, analogue, 36; hardware, 35–6; mathematical of hillslopes, 266–8; white, black and grey box, 43
Mohorovičić discontinuity, 105, 110
molasse sedimentation, 118
monocline, 156
moraine, cross-valley, 455–6; disintegration, 453, 457–8; dump, 518; glacial retreat, 519; ice shelf, 522; lateral, 518; marginal, 453, 457, 518–19; push, 453, 457; Rogen, 453, 456; terminal, 518
morphogenetic regions, 468–72; first order, 468–9; second order, 469
morphometry, Horton's laws of, 324
moulin, 518
mountain climates, erosion associated with, 558
movements, horizontal, 65
movements, verticial, 65
mudflow, 232, 241; submarine, 368
mudstone, 93
multiple working hypotheses, 4

nappe (decke), 119, 165
neck, 130–1
neck cutoff, 302
neo-catastrophism, 543
neo-tectonism, 298
net balance, of glacier, 434
net balance gradient, of glacier, 434
Nile floods, 543, 545
nivation hollow, 566
nuclear-waste disposal, 553, 573, 575–6
null point, 378
nutrient cycle, 44

obstacles, to ice flow, 438
Occam's Razor, 27

ocean basin, 101
ocean trench, 101
oceanic ridge, 100, 109, 116
oceanic sedimentation, 116
offshore zone, 371–2
open channel flow, regions of, 281, 283
ophiolites, 119
organic acid, solution by, 218
organic activity, in soil, 219
orogenic movements, 64–8, 98
orogeny, 118–20
orthoclase, weathering of, 212
orthoquartzite, 92
overland flow, 259–61, 265; erosion by, 263, 266
oxidation, 211

P–E index, 467–8
paleochannels, 558–60
paleohydrology, 558–60
palimpsest, of systems, 568
paradigm, 4
paraliageosyncline, 116
partial area, 265–6
patterned ground, 497–500, 561–3
pediplanation, 30–1, 491
pediment, 486–91; conditions of formation, 487; on granitic
    rocks, 487, 490–1; on shale, 488–9
Penck model, 28–30
peneplain, Mio-Pliocene, 25–6
peneplains, Appalachians, 24
periglacial landforms, 500
periglacial morphogenetic region, 471, 501
periglacial valley forms, 564
periglacial zone, mountain, 502
periglaciation, 560–7
permafrost, 495
permeability, 49–50

phacolith, 130–2
phyllite, 95
physiographic regions, 14; of the United States, 151
Piedmontfläche, 28–9
Piedmonttreppen, 28
pingo, 496, 562–3
piping, 265
plate movements, 31, 114–15
plate tectonics, 113
playa, 486, 494
Pleistocene climates, low latitude, 538
Pleistocene maximum, Antarctic, 509; in high latitudes, 508
plucking, by ice, 446
plume, 115, 124–5
plunge, 169
point bar, 351–2, 354
pollen analysis, 535, 540, 543
polycyclic landforms, 14, 20
polygenetic landscapes, due to climatic change, 535
polygon, ice-wedge, 496–7, 561–2; sand-wedge, 561
polygons, dessication, 494
pool, 295, 297–8
porewater pressure, 247, 248–9
post-Cretaceous chronology, 537
pot hole (jama), 186
potential energy, of water, 281
precipitation changes, aspects of, 547, 549
Primärrumpf, 28–9
process-response system, 317
profile of equilibrium, 10
protalus rampart, 566
provenance, of sedimentary rocks, 82–4
pyroclastic material, 140–1

quartzite, 95
quartzite terrain, 195
quick clay, 252

radioactive decay, 106–7
rainfall erosion facility, 330, 332, 334
rainsplash, 260–3
raised beach, 401–2, 529
rate of erosion, effect of altitude, 54; effect of relief, 54–5; United States, 53–4
recovery, of landforms, 335–6
recurrence interval (see return period), 8, 57
reg, 486
regelation slip, of ice, 438–9
rejuvenation, 20
relaxation, path, 567; time, 7–8
relief, 257, 319, 324
resistance to weathering, related to rock properties, 216–17
resource evaluation, 573
response time, 567–8
return period, 57
Reynolds number, 281–2
riffle, 295, 297–8
rift, 171–2
rill erosion, 264
ring dyke, 130, 132, 135
rip current, 375–6
ripples, wind, 415, 419, 421
river activity, climatic change, 555, 557; humid tropics, 476–8; wet-dry tropics, 482–4
river bars, 299
river bedforms, 283–4
river capture, 21, 333
river flow, round bends, 300–1
river terrace, 355–9; alluvial fill, 355–6; glacial, 357; interglacial, 357; overslag, 358; paired poly-cyclic, 355–6; periglacial, 357; stratigraphy of, 356, 358–9; thalassostatic, 355, 357; unpaired non-cyclic, 355–6, 359–60
rivers, as boundaries, 574
roche moutonnée, 445, 447–9
rock slide, 241
rock strength, 196–8; angle of slope, 198; classification of, 197

rock unit, 94
rockfall, 235–7
ruggedness number, 319
running mean, 543, 546
runoff, function of precipitation, 550

sackung, 236, 238
safety factor, 247
salt, anticline, 147; crystallization, 206; marsh, 406–7; weathering, 224
saltation, 411–12
sand, sheet, 416; stringer, 416; wedge, 561
sand transport, by wind, 410–13, 427; effect of wind velocity, 413–14; with height, 413
sandstone, classification, 92; Colorado Plateau, 179; provenance, 93; topographic resistance of, 177–8; weathering rates, 225
sandur, 349
sastrugi, 429
scarp, retreat, 273–6; stability, 579–80
schist, 95; terrain, 196
Schmidt-Hammer test, 197
sea floor spreading, 116, 125–6; rates, 65–6
sea level, glacial fluctuations, 530–2; recent changes, 398; short-term changes, 371–3; zones, 530, 532
sea temperatures, 538
seamount, 136
sediment, form, 88; maturity of, 218; roundness, 89; size, 83, 84–8; sphericity, 88–9; statistical description, 85–7
sediment flow, in till, 450–2
sediment load, and river discharge, 57–9, 285–6
sediment yield, 325, 327, 336–7; basin sources, 56; effect of vegetation, 56; function of climate, 550–1; function of precipitation, 550–2; function of temperature, 551–2; related to climate, 62, 64
sedimentary fabric, 89
sedimentary packing, 89

segregated ice, 495–6
semi-arid morphogenetic regions, 471
sensitivity, of landforms, 335
serpentine, 96
shear stress, 247
sheet, of fluvio-glacial sediments, 460, 462
shield volcano (basaltic dome), 139
shore platform, types of, 395–6; width, 396, 398
silcrete, 482
sill, 130–1
silt-clay percentage, 294
sinuosity, of rivers, 295, 299–300
site evaluation, procedure, 581
slate, 95
slate terrain, 195–6
slope, 255; characteristic, 255–6; constant, 256; stream bed, 280–1, 294; stream energy gradient, 280–1; stream surface, 280–1; wash, 256
slope erosion, 261–2
slope failure, cycle of in humid tropics, 476; rapid, 231
slope processes, humid tropics, 473–6; wet-dry tropics, 483–5
slope segment, mechanisms of production, 273
slopes, effect of climatic change on, 554–5; maximum valley-side, 256, 258, 271, 322; shale badlands, 33–4, 266–7 269
sluggish landforms, 567
slump, 236, 252–3
snow drifting, 428–9
snowline, regional, 503
soil creep, 258–9
soil permeability, 227
soil slides, 241
soil slips, precipitation induced, 241, 243
soil water, 48–9; velocity, 50; zones, 49
solar energy, 43–7; balance, 43
solubility level, 211
solution, and hillslope hydrology, 211; calcareous materials, 181–3
solution velocity, 211
south-east England, structure, 156–7

spacial hierarchy of landforms, 15
spatial taxonomy, 15
spit, 384–7; cycle of, 385; marine, 33–4; mechanisms of formation, 386–7
stable channels, 290–2
stagnant ice, 528; landscapes, 518–19
step, periglacial, 498–9
step function, 543, 546
stock, 130, 135
Strandflat, 390
strato volcano, 142–3
stream, frequency, 319, 321, 330–1; magnitude, 318; order, 318; orientation, 318–19; power, 280, 299
stream debris transport, regimes of, 288
stream–slope relationships, 258, 337–8
strike, 169
subaqueous glacial deposits, 522, 524–5
subduction zone, 115, 118; evolution, 127
sublimation, 429, 449
submarine canyon, 368–9, 401
superimpositions, 21
surf zone, 371–2; debris transport in, 382
surface heating, 45–6; bare surfaces, 45–6; vegetated surfaces, 46
surging of glaciers, 440
suspended sediment load, of rivers, 51–2
suspended sediment transport, 283–6
suture zone, 121
swallow hole (sink hole, ponor), 186
swash, 378–9, 382
swash zone, 371–2; debris transport in, 382
Swedish break, 250
syneresis, 91
system, cascading, 43; closed, 19; process-response, 317

talus, 235
talus creep, 241
tectonics, effect on sediment load, 575–6

temperature changes, since 1,000,000 years BP, 539
tholeiite, 124
threshold, 11, 39–40, 567; fluid, 412; glacial, 503; impact, 412; periglacial, 495
threshold condition, practical example, 585–8
throughflow, 259, 265
thrust, 168
tidal range, 373
till, deformation, 450; lodgement, 451, 458, 518; meltout, 449–50
till sheets, mid-West, 521
time, cyclic, 13; graded, 13; steady, 13
time unit, 94
time-rock unit, 94
tor, 194, 492–3
tors, origin of, 227
tractive force, 280–1
transcurrent (transform, transverse) fault, 115
transport limited environments, 220
tree-ring analysis, 535, 541, 543–4
tropical wet-dry morphogenetic region, 470
turbidity current, 368, 401

U-shaped valley, 511
unstable channels, effects of changes, 305–6
uplift and erosion, rates, 71
upper convexity, of slope, 256
Urstromtäler, 566

valley meanders, 310
valley morphology, 309–14
valley sediments, classification of, 353
varves, 535
velocity of streamflow, 279, 281–2, 288
ventifacts, 414
vertical movements, Great Britain, 70; Japan, 65–6; young fold mountains, 66–8

viscosity, dynamic, 282; kinematic, 281
volcanic arc, 136
volcanic ash, 141
volcanic chain, 136
volcanic cluster, 137
volcanic cone, erosion of, 192–3
volcanic eruptions, classification, 138
volcanic extrusive rocks, Tertiary and Recent, 128
volcanic lines, 137
volcanoes, 126; active, 136

warm-based ice sheet, 514
water pressure, under ice, 458
water temperatures, 535
wave energy, 379, 381
wave mechanics, 373–6
wave pressure, 390
waves, deep-water, 374; shallow-water, 374
weathered mantle, structure of, 205; thickness of, 224–5
weathered products, 205
weathered profiles, mineralogy of, 215
weathering, arid regions, 486; biological, 204–5; definition, 205; humid tropics, 472–3; marine, 390; pre-glacial, 517; rates of, 221–4; thermal, 206; wet-dry tropics, 478–80
weathering front, 226
weathering limited environments, 220
weathering potential index, 208
wetted perimeter, 279, 281
whale-back forms, 445
width-depth ratio, 293–4, 300
wind regimes, of dome formation, 423
Wisconsin glaciation, North America, 510
world climate, mid-Oligocene, 538; Pleistocene maxima, 538, 542

zones of spreading, 115–18

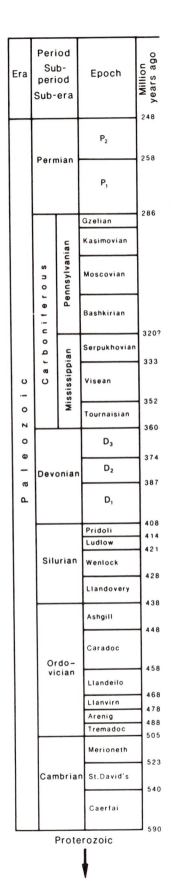
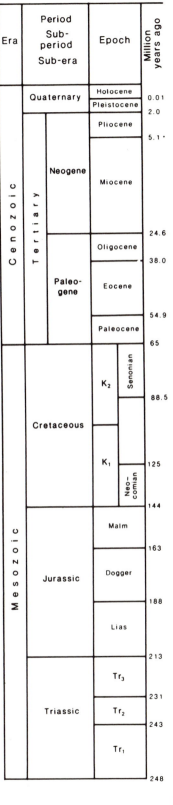

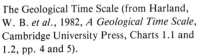
The Geological Time Scale (from Harland, W. B. *et al.*, 1982, *A Geological Time Scale*, Cambridge University Press, Charts 1.1 and 1.2, pp. 4 and 5).